X-RAY LASERS 1994
FOURTH INTERNATIONAL COLLOQUIUM

AIP CONFERENCE PROCEEDINGS 332

X-RAY LASERS 1994

FOURTH INTERNATIONAL COLLOQUIUM
WILLIAMSBURG, VA MAY 1994

EDITORS: **DAVID C. EDER**
DENNIS L. MATTHEWS
LAWRENCE LIVERMORE
NATIONAL LABORATORIES

American Institute of Physics   New York

Authorization to photocopy items for internal or personal use, beyond the free copying permitted under the 1978 U.S. Copyright Law (see statement below), is granted by the American Institute of Physics for users registered with the Copyright Clearance Center (CCC) Transactional Reporting Service, provided that the base fee of $2.00 per copy is paid directly to CCC, 27 Congress St., Salem, MA 01970. For those organizations that have been granted a photocopy license by CCC, a separate system of payment has been arranged. The fee code for users of the Transactional Reporting Service is: 0094-243X/ 87 $2.00.

© 1994 American Institute of Physics.

Individual readers of this volume and nonprofit libraries, acting for them, are permitted to make fair use of the material in it, such as copying an article for use in teaching or research. Permission is granted to quote from this volume in scientific work with the customary acknowledgment of the source. To reprint a figure, table, or other excerpt requires the consent of one of the original authors and notification to AIP. Republication or systematic or multiple reproduction of any material in this volume is permitted only under license from AIP. Address inquiries to Series Editor, AIP Conference Proceedings, AIP, 500 Sunnyside Boulevard, Woodbury, NY 11797-2999.

L.C. Catalog Card No. 95-76067
ISBN 1-56396-375-2
DOE CONF-940592

Printed in the United States of America.

Contents

Preface xiii

Improvement in Efficiency and Coherence of Collisional Excitation X-Ray Lasers. 1
    H. Daido, Y. Kato, R. Kodama, K. Murai, G. Yuan, M. Takagi, H. Takabe, S. Nakai, C. H. Nam, I. W. Choi, D. Neely, and A. G. MacPhee

Status of Collision-Pumped X-Ray Lasers 9
    C. L. S. Lewis

Ray Tracing Calculations of the Output from Germanium Slab Lasers 17
    J. A. Plowes, P. B. Holden, G. J. Pert, S. B. Healy, A. E. Kingston, and E. Roberts

Multiple Pulse Traveling Wave Excitation of Neon-Like Germanium 21
    J. C. Moreno, J. Nilsen, and L. B. DaSilva

High Gain-Production Efficiency and Large Brightness X-UV Laser at Palaiseau 25
    P. Jaeglé, A. Carillon, P. Dhez, P. Goedtkindt, G. Jamelot, A. Klisnick, B. Rus, Ph. Zeitoun, S. Jacquemot, D. Mazataud, A. Mens, and J. P. Chauvineau

Soft X-Ray Amplification in a Plasma Waveguide. 35
    Y. Kato, R. Kodama, H. Daido, K. Murai, G. Yuan, S. Ninomiya, D. Neely, A. G. MacPhee, C. L. S. Lewis, I. W. Choi, C. H. Nam, and T. Kawachi

Recent Progress in Table-Top EUV Lasers at MIT 41
    P. L. Hagelstein, J. Goodberlet, M. Muendel, T. Savas, M. Fleury, S. Basu, and S. Kaushik

A Survey of the Theory of Recombination Lasers. 49
    G. J. Pert

Progress Towards Large Gain-Length Products on the Li-Like Recombination Scheme 55
    P. Zeitoun, G. Jamelot, A. Carillon, P. Goedtkindt, H. Guennou, P. Jaeglé, A. Klisnick, C. Möller, B. Rus, and A. Sureau

Studies of ~ ps Laser Driven Plasmas in Line Focus Geometry 60
    G. J. Tallents, Y. Al-Hadithi, L. Dwivedi, A. Behjat, A. Demir, M. Holden, J. Krishnan, J. Zhang, M. H. Key, D. Neely, P. N. Norreys, C. L. S. Lewis, and A. G. MacPhee

Recent Progress on Shorter Wavelength Recombination X-Ray Laser Research at SIOFM 68
    Z. Xu, Z. Zhang, P. Fan, R. Li, X. Wang, P. Lu, S. Han, L. Zhang, B. Shen, W. Zhang, X. Feng, A. Qian, and H. Xiang

A Computational Investigation of Radiative Cooling in Freely Expanding Recombination Lasers 76
    S. B. Healy and G. J. Pert

Experiments of High Gain C VI X-Ray Lasing in Rapidly Recombining Plasmas 80
    J. Zhang, M. H. Key, P. N. Norreys, G. J. Tallents, A. Behjat, C. Danson,
    A. Demir, L. Dwivedi, M. Holden, P. B. Holden, C. L. S. Lewis,
    A. G. MacPhee, D. Neely, G. J. Pert, S. A. Ramsden, S. J. Rose,
    Y. F. Shao, O. Thomas, F. Walsh, and Y. L. You

Theory of Recombination X-Ray Lasers Based on Optical-Field Ionization 87
    D. C. Eder, P. Amendt, L. B. DaSilva, T. D. Donnelly, R. W. Falcone,
    R. A. London, M. D. Rosen, and S. C. Wilks

Optical-Field-Induced Ionization X-Ray Laser Using Preformed Li^+ Plasma 94
    K. Midorikawa, Y. Nagata, M. Obara, H. Tashiro, and K. Toyoda

X-Ray Spectroscopic Investigation of Optical-Field Ionized Plasmas 101
    S. Borgström, E. Fill, J. Larsson, T. Starczewski, S. Svanberg,
    and C.-G. Wahlström

Plasmas for Short-Wavelength Lasers Driven by Ultra-Short, High-Intensity Laser Pulses 106
    T. D. Donnelly, T. E. Glover, M. Hofer, E. A. Lipman, R. W. Lee,
    L. B. DaSilva, D. C. Eder, S. Mrowka, and R. W. Falcone

Application of a Plasma Waveguide to X-Ray Lasers 113
    H. M. Milchberg, C. G. Durfee III, and J. Lynch

Short-Pulse Laser-Produced Plasma from C_{60} Molecules 121
    C. Wülker, W. Theobald, D. Ouw, F. P. Schäfer, and B. N. Chichkov

Ultrashort Pulse Laser Produced Plasmas 126
    J. Davis, R. W. Clark, and J. L. Giuliani

Controlled Power Compression in Materials for X-Ray Amplification 134
    A. B. Borisov, A. McPherson, K. Boyer, and C. K. Rhodes

Simulations of X-UV Gain in a Plasma Pumped by a Pulse-Train Laser 137
    A. Klisnick, J. Virmont, N. Grandjouan, H. Guennou, and A. Sureau

Ionization State of an Aluminium X-Ray Laser Plasma by $K\alpha$ Spectroscopy 142
    A. Klisnick, C. Chenais-Popovics, C. A. Back, P. Zeitoun,
    P. Renaudin, O. Rancu, J. C. Gauthier, and P. Jaeglé

A Self-Consistent Model for Line Trapping Effect in X-Ray Lasers 147
    D. E. Benredjem, A. Sureau, C. Möller, and H. Guennou

Experimental Study of Neon Like Zinc J=0−1 Soft X-Ray Lasing at 21.2 nm 152
    B. Rus, A. Carillon, P. Dhez, B. Gauthé, P. Goedtkindt, P. Jaeglé,
    G. Jamelot, A. Klisnick, M. Nantel, A. Sureau, and P. Zeitoun

Dielectronic Spectra for Ne-Like Ions from F-Like Low-Lying States 157
    M. Cornille and S. Jacquemot

Hydrodynamic Simulation of a Laser-Produced Magnesium Recombination Plasma and Soft X-Ray Lasing 161
    H. Kitazawa, S. Morinaga, H. Takakuwa, K. Shimizu, and S. Karashima

**Effect of Plasma Nonuniformity in Recombination Pumping Soft
X-Ray Lasers.** 166
    A. Sasaki, H. Yoneda, K. Ueda, and H. Takuma

Fluorine Recombination X-Ray Laser Pumped by 10 ps KrF Laser Pulse 171
    T. Tomie, E. Miura, I. Okuda, and Y. Owadano

**Observation of Motional Doppler Decoupling Effect in a 10ps KrF
Laser Produced Plasma** 176
    E. Miura, T. Tomie, I. Okuda, and Y. Owadano

Compact Soft X-Ray Laser Pumped by a Pulse-Train Laser 181
    T. Hara, K. Ando, and Y. Aoyagi

**Near Field Beam Characteristics of the Beam from the Amplifier
of an Injector/Amplifier Germanium XXIII XUV Laser System** 186
    C. G. Smith, M. H. Key, G. F. Cairns, C. L. S. Lewis, D. Neely,
    and A. G. MacPhee

**Coupling Between Remote Plasmas in an 'Injector-Amplifier'
XUV Laser System** 191
    G. F. Cairns, M. J. Lamb, C. L. S. Lewis, A. G. MacPhee, D. Neely,
    C. Pichler, M. Holden, J. Krishnan, G. J. Tallents, C. G. Smith, J. Zhang,
    M. H. Key, P. N. Norreys, P. B. Holden, G. J. Pert, S. A. Ramsden,
    R. E. Burge, M. T. Browne, and G. E. Slark

Enhanced Output of Soft X-Ray Lasers Using Double Slab Targets 196
    J. C. Moreno, J. Nilsen, and E. Chandler

**Analytic Models for Beam Propagation and Far-Field Patterns
in Slab and Bow-Tie X-Ray Lasers** 200
    E. A. Chandler

Light Scattering Measurement in a Ge X-Ray Laser Plasma 205
    C. H. Nam, I. W. Choi, Y. Oshikane, K. Murai, G. Yuan, R. Kodama,
    H. Daido, Y. Kato, and S. Nakai

Refractive Compensation in Slab Target Geometries 210
    D. Neely, Y. Kato, R. Kodama, H. Daido, K. Murai, G. Yuan, C. L. S. Lewis,
    A. G. MacPhee, M. J. Lamb, P. B. Holden, G. J. Pert, and A. Djaoui

**Na-Like Autoionizing Levels: Plasma Diagnostics and Prospects
for Photopumped Soft X-Ray Lasers.** 215
    A. L. Osterheld, J. Dunn, B. K. F. Young, S. B. Libby, A. Szoke,
    R. S. Walling, W. H. Goldstein, R. E. Stewart, A. Ya. Faenov, I. Yu. Skobelev,
    and S. Ya. Khakhalin

**Recombination X-Ray Lasers Driven by Ultra-Short Pulses
in Hydrogen-Like Ions: Theoretical Analysis** 220
    Y. Li and J. Zhang

Soft X-Ray Laser of He-Like Aluminum Ions 225
    B. Shen, Z. Xu, and S. Han

A Numerical Method for Obtaining the Fine Structure of X-Ray Spectra 229
    L. Zhang, S. Han, C. Jiang, Z. Xu, Z. Zhang, and L. Sun

Possible Approaches to the Recombination X-Ray Lasers
with Large GL Value 235
    S. Han, B. Shen, Z. Xu, Z. Zhang, H. Teng, W. Zhang, and L. Zhang

Z-Dependent Spatial Behavior of $4d$-$3p$ and $4f$-$3d$ Transitions of Li-Like
Ions in Laser-Produced Plasmas and the Electron Density Measurement 241
    R. Li, B. Shen, Z. Xu, P. Fan, Z. Zhang, X. Wang, P. Lu, and S. Han

Simulation of X-Ray Laser for Li-Like Silicon 245
    B. Shen, X. Wang, Z. Xu, and H. Teng

Investigation on the Homogeneity of Line Plasmas Created
with the Aid of a Cylindrical Lens Array 248
    A. Glinz and J. E. Balmer

Observation of Strong Emission from Ne IX and NeX Transitions
in a Laser-Driven Plasma 253
    J. K. Crane, T. Ditmire, H. Nguyen, and M. D. Perry

Towards a 38 Å X-Ray Laser 258
    L. Bonnet, S. Jacquemot, and A. Decoster

Inner-Shell Photo-Ionized X-Ray Laser Schemes for Low-Z Elements 262
    S. J. Moon, D. C. Eder, and G. L. Ströbel

The Monochromaticity and Intensity Scaling of the Neon-Like
Yttrium Laser 267
    P. B. Holden, M. Nantel, B. Rus, and A. Sureau

Hyperfine Splittings, Prepulse Technique, and Other New Results
for Collisional Excitation Neon-Like X-Ray Lasers 271
    J. Nilsen, J. C. Moreno, J. A. Koch, J. H. Scofield, B. J. MacGowan,
    and L. B. DaSilva

Theory of Ne-Like Collisional X-Ray Lasers 279
    S. Jacquemot

Enhancement of the $J=0-1$ (19.6nm) Transition Relative to the $J=2-1$
(23.6nm) One Using a Prepulse with the Ne-Like Germanium XUV
Laser System 289
    G. F. Cairns, M. J. Lamb, C. L. S. Lewis, A. G. MacPhee, D. Neely,
    P. N. Norreys, M. H. Key, C. Smith, S. B. Healy, P. B. Holden, G. J. Pert,
    and J. A. Ploues

Overview of Research on Ne-Like Ge Soft-X-Ray Laser in China 293
    S. Wang, G. Zhou, G. Zhang, J. Sheng, and S. Chunyu

A Linearly Polarized Soft X-Ray Laser 301
    B. Rus, G. F. Cairns, P. Dhez, P. Jaeglé, M. H. Key, C. L. S. Lewis,
    D. Neely, A. G. MacPhee, S. A. Ramsden, C. G. Smith, and A. Sureau

EBIT X-Ray Spectroscopy Studies for Applications to Photo-Pumped
X-Ray Lasers 307
    S. R. Elliott, P. Beiersdorfer, and J. Nilsen

High-Order Harmonic Generation 312
    C.-G. Wahlström

Prospects for High Power Linac Coherent Light Source (LCLS)
Development in the 1000Å-1Å Wavelength Range 320
 R. Tatchyn, K. Bane, R. Boyce, G. Loew, R. Miller, H.-D. Nuhn,
 D. Palmer, J. Paterson, T. Raubenheimer, J. Seeman, H. Winick,
 D. Yeremian, C. Pellegrini, J. Rosenzweig, G. Travish, D. Prosnitz,
 E. T. Scharlemann, S. Caspi, W. Fawley, K. Halbach, K.-J. Kim,
 R. Schlueter, M. Xie, R. Bonifacio, L. De Salvo, and P. Pierini

Numerical Simulation of X-Ray Zone Plates with High Aspect Ratio 330
 Y. V. Kopylov, A. V. Popov, and A. V. Vinogradov

Self-Induced Spatially-Hole-Burned Distributed Feedback Resonator
for X-Ray Lasers 345
 F. Ráksi

Development of X-Ray Laser Architectural Components.................... 350
 A. S. Wan, L. B. DaSilva, J. C. Moreno, R. C. Cauble, E. A. Chandler,
 H. E. Dalhed, S. B. Libby, R. W. Mayle, J. Nilsen, R. P. Ratowsky,
 H. A. Scott, and B. Van Wonterghem

Soft-X-Ray Amplification in a Capillary Discharge Plasma................ 359
 J. J. Rocca, F. G. Tomasel, V. A. Shlyaptsev, O. D. Cortázar,
 J. L. A. Chilla, and G. Giudice

Soft X-Ray Lasing in a Capillary Discharge 367
 T.-N. Lee, H.-J. Shin, and D.-E. Kim

Modeling of a Capillary Discharge Soft X-Ray Amplifier 375
 V. N. Shlyaptsev, J. J. Rocca, and A. L. Osterheld

Lasing at Short Wavelength in a Capillary Discharge
and in a Dense Z-Pinch... 380
 H.-J. Kunze, S. Glenzer, C. Steden, H. T. Wieschebrink,
 K. N. Koshelev, and D. Uskov

Nonstationary Argon Plasma, Containing Ne-Like and Na-Like Ions.
"Fast Compression" and Population Inversion. 388
 L. N. Ivanov and L. V. Knight

Non-Maxwellian Plasma Electrons in the Inversion Population
of Ne-Like Ions.. 393
 E. P. Ivanova, L. V. Knight, and B. G. Peterson

X-Ray Laser Research Using Z Pinches 399
 J. P. Apruzese

Heavy Ion Beam Pumping of Charge Transfer Lasers..................... 404
 A. Ulrich, R. Gernhäuser, W. Krötz, M. Salvermoser, J. Wieser,
 and D. E. Murnick

X-Ray Lasers Active Media Formation by Ponderomotive Force
of Coherent Optical Pumping.. 410
 V. V. Korobkin and M. Yu. Romanovsky

Reduction of Driver Energy for X-Ray Lasers 423
 M. H. Key and C. G. Smith

On Developing a Table Top Soft X-Ray Laser............................ 432
 A. Morozov, K. Krushelnick, L. Polonsky, C. H. Skinner, S. Suckewer,
 C. M. Falco, and J. M. Slaughter

CCD Imaging from 20eV to 8keV 441
    A. G. MacPhee and C. L. S. Lewis

X-Ray Holography with High Resolution 446
    J. Chen, P. Zhu, T. Xiao, and Z. Xu

Space-Resolved Electron Density Measurements Using the Stark-Broadened Line Wings of Hydrogenic Ions in Line-Shaped Laser Plasmas 451
    L. Zhang, S. Han, Z. Xu, Z. Zhang, P. Fan, and L. Sun

An Experimental Study on Line-Shaped Laser-Plasma 457
    S. Han, L. Zhang, Z. Xu, Z. Zhang, P. Fan, and L. Sun

Observation of Resonant Photo-Excited Fluorescence and Application to Plasma Diagnostics 463
    H. Yashiro, T. Tomie, Y. Matsumoto, I. Matsushima, and Y. Owadano

Higher Order Structure Analysis of Nano-Materials by Spectral Reflectance of Laser-Plasma Soft X-Ray 468
    H. Azuma, A. Takeichi, and S. Noda

Bright Picosecond X-Rays from Intense Sub-Picosecond Laser-Plasma Interactions 473
    A. Maksimchuk, J. Workman, X. Liu, U. Ellenberger, S. Coe, C.-Y. Chien, and D. Umstadter

X-Ray+Optical Nonlinear Mixing in Plasma 478
    P. L. Shkolnikov and A. E. Kaplan

Gain Measurements and Spatial Coherence in Neon-Like X-Ray Lasers 483
    J. Krishnan, C. Cairns, L. Dwivedi, M. Holden, M. H. Key, C. L. S. Lewis, A. G. MacPhee, D. Neely, P. N. Norreys, G. J. Pert, S. A. Ramsden, C. G. Smith, G. J. Tallents, and J. Zhang

Comparison of the Spectral Response of a Thinned, Backside Illuminated CCD with a CsI Coated MCP System and Kodak 101 Film 488
    Y. Li, J. R. Crespo López-Urrutia, G. D. Tsakiris, R. Sigel, R. Volk, and L. Pina

Study of Multilayer Structures as Soft X-Ray Mirrors 493
    D. Kim, H. W. Lee, D. Cha, J. J. Lee, and J. H. Je

Space and Time Resolved Investigation of Recombination X-Ray Lasers 498
    X. Wang, Z. Xu, Z. Zhang, P. Fan, B. Shen, S. Han, L. Zhang, P. Lu, R. Li, X. Feng, and A. Qian

Stark Line Broadening of the $n=4$ to 3 Transitions in High-Z Helium-Like Ions 505
    P. A. Loboda, V. A. Lykov, and V. V. Popova

Measurements of Line Overlap for Resonant Spoiling of X-Ray Lasing Transitions 512
    P. Beiersdorfer, S. R. Elliott, B. J. MacGowan, and J. Nilsen

Generation of Tunable XUV Radiation by High-Order Frequency Mixing..... 517
    C. Momma, H. Eichmann, and B. Wellegehausen
X-Ray Nonlinear Optics with High-Order Harmonics 522
    P. L. Shkolnikov and A. E. Kaplan
Nonresonant Photopumping Using Heavy Ion Beam Produced Soft X-Rays ... 525
    W. Krötz, A. Ulrich, M. Salvermoser, J. Wieser, and D. E. Murnick
Neck Type Instabilities in Axial Discharges and Population Inversion 529
    K. N. Koshelev and H.-J. Kunze
**Evaluation of Imaging Properties of Soft-X-Ray Multilayer Mirrors
and their Application to Highly Dispersive Spectral Imaging 533**
    N. N. Kolachevsky, M. M. Mitropolsky, E. N. Ragozin,
    N. N. Salashchenko, V. A. Slemzin, and I. A. Zhitnik
**Characterization of a Plasma Produced Using a High Power Laser
with A Gas Puff Target for X-Ray Laser Experiments...................... 538**
    H. Fiedorowicz, A. Bartnik, K. Gac, P. Parys, M. Szczurek, and J. Tyl
Three-Dimensional X-Ray Imaging with Off-Axis X-Ray Zone Plate 543
    H.-R. Lee, E. Anderson, L. B. DaSilva, and J. E. Trebes
X-Ray Laser Interferometry Experiments at LLNL....................... 549
    P. Celliers, F. Weber, L. B. DaSilva, T. W. Barbee, Jr., S. Mrowka,
    and J. E. Trebes
X-Ray Lasers for Imaging and Plasma Diagnostics 553
    L. B. DaSilva, T. W. Barbee, Jr., R. C. Cauble, P. Celliers, J. Harder,
    S. B. Libby, H. R. Lee, R. A. London, D. L. Matthews, S. Mrowka,
    J. C. Moreno, D. Ress, J. E. Trebes, A. S. Wan, and F. Weber
Demonstration of Ultra High Resolution Soft X-Ray Tomography............ 559
    W. S. Haddad, I. McNulty, J. E. Trebes, E. H. Anderson, L. Yang,
    and J. M. Brase
**Micron-Scale Resolution Radiography of Laser-Accelerated
and Laser-Exploded Foils Using an Yttrium X-Ray Laser 562**
    R. C. Cauble, L. B. DaSilva, T. W. Barbee, Jr., P. Celliers, J. C. Moreno,
    S. Mrowka, T. S. Perry, and A. S. Wan
Some Potential Applications of X-Ray Lasers in Atomic Physics 566
    B. Crasemann
Collisional Redistribution Effects on X-Ray Laser Saturation Behavior 574
    J. A. Koch, B. J. MacGowan, L. B. DaSilva, D. L. Matthews,
    J. H. Underwood, P. J. Batson, R. W. Lee, R. A. London, and S. Mrowka
**Application of X-Ray Lasers to Current and Future Experiments
in Atomic and Molecular Physics.. 579**
    C. D. Caldwell
Applications of Subpicosecond Soft X-Ray Sources 587
    R. Haight and P. F. Seidler
Author Index.. 609

Preface

The 4th International Colloquium on X-Ray Lasers was held in Williamsburg, VA, May 15–20, 1994. There were 140 participants from the U.S.A., China, France, Germany, Korea, Japan, Poland, Russia, Sweden, Switzerland, and the United Kingdom. The papers in this proceedings review the progress in the development of x-ray lasers and their applications. The colloquium marked the 10th anniversary of the initial demonstration of x-ray lasing and this collection of papers shows the outstanding progress that has been made during these 10 years.

An exciting result, discussed in this proceedings, is the first demonstration of an electron-capillary discharge XUV laser. Work done at Colorado State University using a discharge achieved lasing in Ne-like Ar at 46.88 nm with an impressive gain-length product (GL) of 7. A less dramatic (GL~4) but still convincing demonstration of a CVI Balmer Alpha laser (operating at 18.2 nm), again pumped by a fast discharge capillary, was reported by researchers from the Pohang University of Science and Technology. The capillary discharge x-ray laser system immediately improves the efficiency of x-ray lasers up to $\sim 10^{-5}$ at the wall plug. Using the more conventional laser approach, a demonstration of a GL=6, recombination-pumped CVI Balmer alpha laser (18.2 nm) by using a 2 psec KrF laser to pump a carbon fiber at Rutherford-Appleton Laboratory is also discussed.

The Ne-like missing or weakly observed J=0−1 transition mystery, dating back to the original demonstration of Ne-like Se x-ray lasers, appears to be finally solved. Using solid targets irradiated by optical pulse trains of 100 psec pulses separated by 300 psec, researchers at Rutherford-Appleton Laboratory and at Osaka consistently observed the J=0−1 transition to strongly lase early in time when the plasma density is highest. The key ingredient missing from prior work and which fosters the J=0−1 line appears to be the preparation of high density but non-refractive plasma by using a prepulse. Researchers at LLNL have observed hyperfine splitting of the J=0−1 line and they explain why low-Z ions with odd Z have not been observed to lase. Multiple optical pumping pulses were also used by researchers from Osaka to demonstrate Nd, Sm, Gd, and Td Ni-like lasers with gains as high as 3.0 cm^{-1} and wavelengths as short as 6.7 nm. Even more important, these lasers were made using solid targets and with laser energies of 150 J/pulse, again with 100 psec pulses separated by 300 psec. These results indicate it is possible to produce short wavelength lasers with relatively small pump laser facilities provided the use of prepulses is exploited.

The optical field ionization x-ray laser work done at RIKEN laboratory in Japan was also widely discussed with corroborating results presented by a few others attending the conference. This scheme has shown some promising indications (GL products of ~2 to 3, but L is only 2 mm) that one can obtain sizable gain on the H-like Li Lyman alpha transition which occurs at 13 nm. A problem with this scheme is ionization-induced refraction, but development of plasma optical fibers discussed by researchers from University of Maryland are a possible solution.

The first demonstration of a linearly polarized x-ray laser (Ne-like Ge operating at 23.2 and 23.6 nm) was revealed at this conference by researchers from Palaiseau working in collaboration with the Rutherford-Appleton Laboratory. The development of circular or even elliptically polarized x-ray lasers should also be possible, given high quality, 1/4 waveplate, multilayer optics. In addition, both the Rutherford-Appleton Consortium and a Chinese Group reported achieving a very high degree of spatial coherence for the Ne-like Ge 23.2 nm x-ray laser.

An important application of x-ray lasers, that was discussed at this conference, is the development of plasma probing and imaging systems. Researchers at LLNL demonstrated the first x-ray laser interferometer. In addition, measurements of large-plasma electron density gradients using Moiré diffractometry were shown, as well as evidence of hydrodynamic instabilities on a very fine spatial scale by using the Ne-like yttrium laser (15.5 nm) as an illumination source for a laser heated Al plasma.

The future of x-ray lasers depends on their applications and this proceedings contains an outstanding set of papers on current and future applications in biological imaging, high and low photon energy atomic physics, and semiconductor research. A common theme is that significant research can be done now, even at the modest wavelengths of 10 to 30 nm, by taking advantage of the narrow linewidths, high brightness, polarizability, coherence and line by line wavelength tuning. All these researchers stated the need for a powerful, near diffraction-limited x-ray source that is compact and user-friendly. We feel that this goal can be obtained in the near future and hope that this proceedings will help in this quest.

The Program Chairmen wish to express their sincere appreciation for the invaluable contribution of the Conference Administrator, Kathleen Tarlow, for her organization and management of this conference. In particular, we would like to thank Kathleen Tarlow and Patrece Talley for their expert coordination efforts and administrative support in the successful operation of this colloquium.

David C. Eder
Dennis L. Matthews
Editors

Improvement in Efficiency and Coherence of Collisional Excitation X-Ray Lasers

H. Daido[1], Y. Kato[1], R. Kodama[1], K. Murai[1], G. Yuan[1],
M. Takagi[1], H. Takabe[1], S. Nakai[1], C. H. Nam[2], I. W. Choi[2],
D. Neely[3], and A. MacPhee[3]

1) Institute of Laser Engineering, Osaka University,
2-6 Yamada-oka, Suita, Osaka 565 Japan
2) Korea Advanced Institute of Science and Technology
373-1, Gusang-Dong, Yusang-Gu, Taejon, Korea
3) Department of Pure and Applied Physics,
Queens University of Belfast,
Belfast BT-7, Northern Ireland

Abstract. We have obtained improved performance in terms of intensity and beam divergence (6 mrad) at J=0-1 line of a germanium soft x-ray laser with a curved slab target which compensates beam deflection due to plasma density gradient. At the first stage of this experiment, the pump laser pulse width was 1 ns and the average irradiance on target was 1.7×10^{13} W/cm^2. The lasing performance of the J=0-1 line suggested improvement in efficiency with shorter pulse pumping. In order to optimize the plasma density and temperature profiles of the expanding germanium plasma, we have employed double pulse pumping of 100 ps pulse width with 300 ps separation having a total energy of 300 J. This has resulted in a further improvement in the x-ray laser intensity and the beam divergence (2 mrad). We also describe briefly the experimental result of Ni-like collisional excitation lasers having wavelengths of 6-8 nm.

INTRODUCTION

Successful experimental results on the saturated soft x-ray amplification have been achieved at several laboratories with the electron collisional excitation scheme(1-3). These results were obtained using large scale Nd:glass lasers making repetitive operation difficult. Improvement in pumping efficiency is necessary for realization of a modest size repetitive x-ray laser pumping system such as diode laser pumped zigzag slab lasers (4).

NEON-LIKE GERMANIUM LASER

We have obtained improved performance in terms of intensity and beam divergence at J=0-1 line of a germanium soft x-ray laser with a curved slab target which compensates a beam deflection from a gain medium by refraction due to the plasma density gradient (5) using Gekko XII laser. Schematic diagram of the line focusing system and diagnostic system are shown in Fig. 1. Figure 2(a) and (b) shows the measured time histories of 19.6 nm and 23.6 nm lines obtained with a flat slab target (dotted lines) and a curved slab target (solid lines). The pump laser pulse width was 1 ns and the average irradiance on target was 1.7×10^{13} W/cm^2. However the time at the peak intensity of J=0-1, 19.6 nm line was 500 ps prior to the pump peak, while that of J=2-1, 23.2/23.4 nm lines were placed at the pump peak. This suggested an improvement in pumping efficiency for the 19.6 nm transition with shorter pulse pumping. In order to reduce the pump laser energy, we propose the spatially and temporally optimized plasma as a gain medium controlled by a refraction wave guiding (5) in conjunction with a pre-pulse or a double pulse pumping technique which were originally used for an improvement of a lasing condition with control of a density scale length of an active medium (6-8).

Multi-Pulse Pumping with a Curved Slab Target

We tested a double pulse pumping technique coupled with the curved slab target (9) in order to optimize the density and temperature profiles of the expanding germanium plasma. In addition to the transverse control of the gain medium, control of the axial configuration of the gain medium has been done with a curved target. An individual duration of the double pulse was 100 ps separated by 300 ps at a wavelength of 1.05 µm. The pulse to pulse intensity ratio was almost 1 : 1, which gave the highest 19.6 nm laser intensity obtained by the experiment. The total pumping energy on target was 200-300 J. The beam was focused to a line focus of 6 cm long and approximately 100 µm wide; this gave rise to an intensity of $[2-3] \times 10^{13}$ W/cm^2 per beam. The diameter of the vacuum target chamber was 1.7 m and the pressure in it was 10^{-5} Torr. An on-axis spectrometer was translated in the horizontal plane perpendicular to the soft x-ray laser beam axis with an x-ray streak camera with a 650 µm wide slit-like cathode which could measure the time dependent angular intensity distribution of the x-ray laser beam at an individual lasing line (10). The time resolution of the streak camera was less than 70 ps determined by the lasing line width on the cathode given by the resolution of the spectrometer. The line focusing conditions were monitored for each shot by recording x-ray slit camera images; these cameras were fitted with a 40 um thick vacuum-tight beryllium filter and a 10 µm thick aluminum filter to cut the visible and ultra-violet light and to avoid the

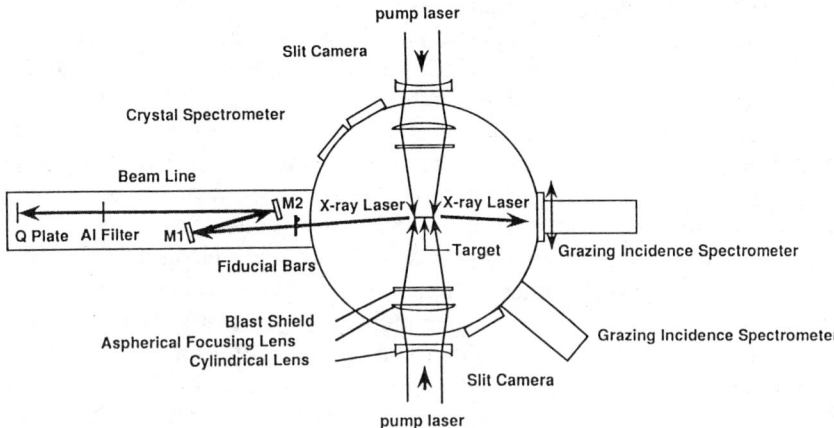

FIGURE 1. Schematic diagram of the line focusing system and diagnostic system.

(a) 19.6nm(J=0-1)

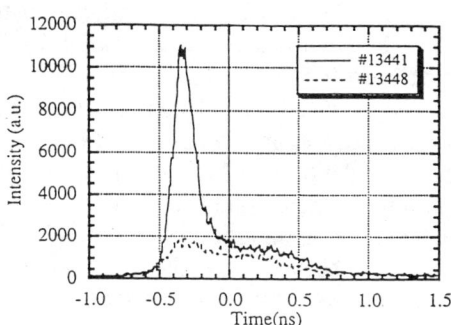

(b) 23.6nm(J=2-1)

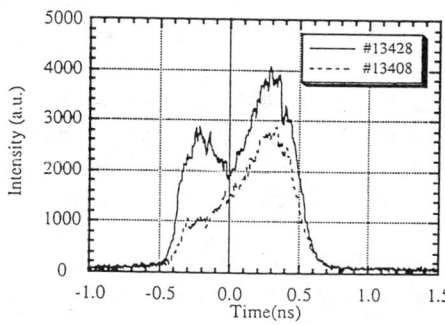

FIGURE 2. Measured time histories of the 19.6 nm and 23.6 nm lines obtained with a flat target (dotted lines) and a curved slab target (solid lines).

over-exposure by an x-ray emission on the film. The images therefore corresponded to photon energies greater than 1.5 keV. This resulted in a significant improvement in the performance of the germanium curved slab target at 19.6 nm line giving 12 times higher peak intensity with a smaller beam divergence of 2 mrad compared with those given by the 1.1 kJ, 1 ns pumping. Time integrated soft x-ray laser spectra are shown in Fig. 3. The upper spectrum was taken with 1 ns pumping. The lower one was taken with 100 ps double pulse pumping. Significant enhancement of the 19.6 nm is clearly visible.

We measured the lasing line intensity at 19.6 nm wavelength as well as beam divergence with a streak camera coupled with a spectrometer from a 4 cm long curved slab target as a function of bent angle, giving a clear peak at 17 mrad. The measured beam divergence gave the minimum angle of 2 mrad at the bent angle of 17 mrad. The value corresponds to the coherence length of 15 µm at the end of the gain medium. In contrast to this clear result, the result from the 1 ns laser pumped targets showed that the intensity increased monotonously as the bent angle increased up to 25 mrad (9).

The intensity of the 19.6 nm line as a function of the target length showed exponential growth up to 3 cm due to the short gain duration compared with the traverse time along the gain medium. We have calculated the temporal and spatial profiles of the electron density and the temperature using one-dimensional hydrodynamic simulation code ILESTA coupled with a non local thermodynamic equilibrium atomic physics package (11). Figure 4 shows the spatial profiles at the time of the first and the second pumping pulse peaks. The profiles given by 1 ns pumping are also shown at the time of 19.6 nm lasing line for comparison.

In the double pulse pumping scheme, the first pulse creates a preformed plasma with a steep density gradient and it explodes. At the suitable density gradient, the second pulse is efficiently absorbed over the absorption scale length and it ionizes the plasma to produce conditions suitable for creating high gain as shown in Fig. 4.

NI-LIKE ION LASERS

Desired electron density and temperature for the Ni-like laser medium of Lanthanide series are almost the same as those of germanium laser (12). We tested the Ni-like lasers using the multi pulse pumping technique with a curved slab target resulted in intense Ni-like ion soft x-ray lasers whose wavelengths were below 8 nm. The 2.8 cm long line focused energy on target was 300 J at 1.05 µm. The individual pulse width was 100 ps with a separation of 400 ps giving average irradiance on target was 6×10^{13} W/cm^2. In our experiment, the

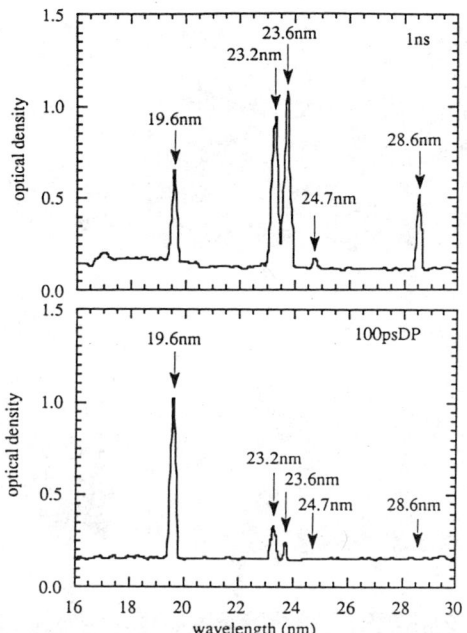

FIGURE 3. Time integrated spectra of the germanium x-ray laser with a curved slab target. The upper spectrum was taken with 1 ns pumping. The lower one was taken with 100 ps double pulse pumping.

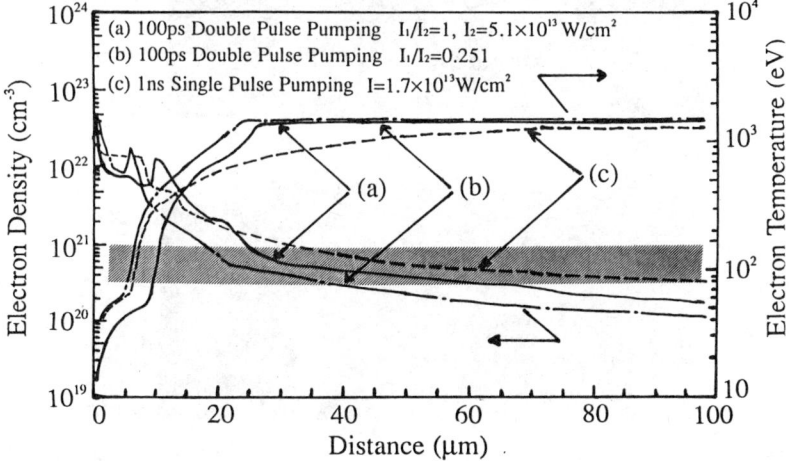

FIGURE 4. Density and temperature profiles are improved significantly using double pulse pumping of appropriate pump-intensity ratio. The notation (a), (b) and (c) are the case of 100 ps double pulse pumping whose intensity ratio are 1:1 and 1:4. The notation (c) is that of 1 ns single pulse pumping.

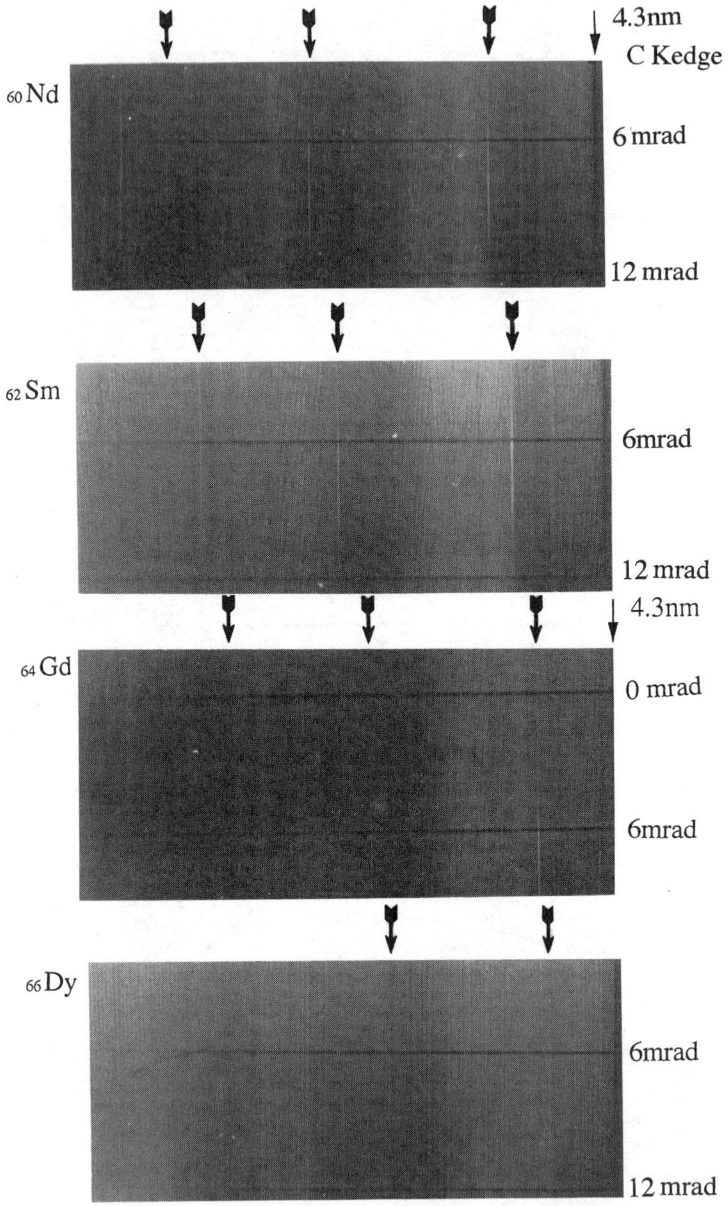

6   Excitation X-Ray Lasers

FIGURE 5. Soft x-ray spectra of the various species of the Lanthanoid series. The laser wavelengths of Nd, Sm, Gd, and Dy are 7.95, 7.36, 6.90, and 6.43 nm, respectively. Strong J=0-1 line of the Ni-like x-ray lasers with narrow divergence are visible. Note that the arrows indicate the first, second and third order diffractions from the right to the left in each spectrum.

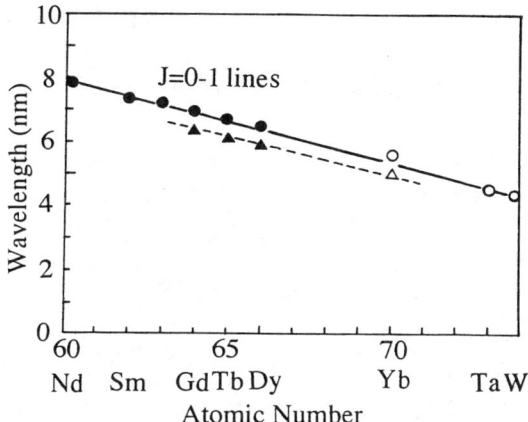

FIGURE 6. The lasing wavelengths of Ni-like soft x-ray lasers as a function of atomic number. The solid and the open symbols are the data taken by the authors and the Livermore group.

longer wavelength line of J=0-1 transitions was much stronger than shorter one. Figure 5 shows the intense lines of Ni-like Nd, Sm, Gd and Dy J=0-1 lines indicated by arrows whose wavelengths are 7.95 nm, 7.36 nm, 6.90 nm and 6.43 nm, respectively. The right most line shows the first order spectrum and the second and the third one are clearly visible except for Dy. The divergence of the lasers are around 3 mrad measured with the angular fiducial bars. For Nd, the gain coefficient was around 3 cm^{-1} and the gain length product was 7.5. Figure 6 shows the lasing wavelengths as a function of atomic number. The solid circles and the solid triangles denote the present data. The open circles and triangles denote the data taken by Livermore group (1). The shorter wavelength laser may need shorter wavelength and higher intensity pumping lasers. The longer wavelength lasers given by the lighter atomic species are very interesting to investigate because lighter species need less and less pumping energy leading to the table top soft x-ray laser with reasonably strong power.

SUMMARY

In summary, we have described double pulse pumping (100 ps pulse width with 300 ps pulse to pulse separation) of a germanium curved slab target, giving significant improvement of lasing performance, especially for the J=0-1, 19.6 nm line. The pump laser energy was 200-300 J. The x-ray beam intensity with 2 mrad divergence was 12 times larger than that of 1 ns, 1.1 kJ pumping. The x-ray pulse width was 70 ps FWHM. We also described Ni-like x-ray laser whose wavelengths of 6-8 nm using double pulse pumping and a curved target.

ACKNOWLEDGMENTS

The authors are indebted to Y. Oshikane, J. Goto, The Gekko X II glass laser operation group, and the target fabrication group for their technical support.

REFERENCES

1. MacGawan, B. J., DaSilva, L. B., Fields, D. J., Keane, C. J., Koch, J. A., London, R. A., Matthews, D. L., Maxon, S., Mrowka, S., Osterheld, A. L., Scofield, J. H., Shimkaveg, G., Trebes, J. E., and Walling, R. S., Phys. Fluids **B4**, 2326-2337 (1992) and references therein.
2. Carillon, A., Chen, H. Z., Dhez, P., Dwivedi, L., Jacoby, J., Jaegle, P., Jamelot, G., Zhang, Jie, Key, M. H., Kidd, A., Klisnick, A., Kodama, R., Krishnan, J., Lewis, C. L. S., Neely, D., Norreys, P., O'Neill, D., Pert, G. J., Ramsden, S. A., Roucourt, J. P., Tallents, G. J. and Uhomoibhi, J., Phys. Rev. Lett. **68**, 2917-2920 (1992).
3. Da Silva, L. B., MacGowan, B. J., Mrowka, S., Koch, J. A., London, R. A., Matthews, D. L. and Underwood, J. H., Opt. Lett. 18, 1174- 1176 (1993).
4. Eder, D. C., Amendt, P., Bolton, P. R., Da Silva, L. B., Hackel, L. A., London, R. A., MacGowan, B. J., Matthews, D. L., Rosen, M. D., Stewart, R. E., and Wilks, S. C., Proc. Photo-Opt. Instrum. Eng. **2012**,165- 171 (1993).
5. Lunney, J. G., Appl. Phys. Lett. **48**, 891-893 (1986).
6. Boehly, T., Russotto, M., Craxton, R. S., Epstein, R., Yaakobi, B., Da Silva, L. B., Nilsen, J., Chandler, E. A., Fields, D. J., MacGowan, B. J., Matthews, D. L.., Scofield, J. H., and Shimkaveg, G., Phys. Rev. **A42**, 6962-6965 (1990).
7. Nilsen, J., MacGowan, B. J., Da Silva, L. B., and Moreno, J. C., UCRL-JC -112704 Preprint Jan. 12, 1993 (submitted for publication in Phys. Rev. Lett.).
8. Da Silva, L. B., Cauble, B., Frieders, G., Koch, J. A., MacGowan, B. J., Matthews, D. L., Mrowka, S., Ress, D., Trebes, J. E., and Weiland, T. L., Proc. Soc. Photo-Opt. Instrum. Eng. **2012**, 158- 164 (1993).
9. Kato, Y., Daido, H., Kodama, R., Murai, K., Yuan, G., Schulz, M. S., Yamanaka, M., Takagi, M., Kanabe, T., Nakai, S., Neely, D., MacPhee, A., Lewis, C. L. S., Slark, G., Niibe, M., Tsukamoto, M., Fukuda, Y., Tsunemi, H., Nomoto, S., Kodama, I., Honda, T., Shinohara, K., Iwasaki, H., and Yoshinobu, T., Proc. Soc. Photo-Opt. Instrum. Eng. **2012**, 12-21 (1993).
10. Kodama, R., Neely, D., Dwivedi, L., Key, M. H., Krishnan, J., Lewis, C. L. S., O'Neill, D. M., Norreys, P., Pert, G. J., Ramsden, S. A., Tallents, G. J., Uhomoibhi, J., and Zhang, J., Opt. Commun. **90**, 95-98 (1992).
11. Takabe, H., "Radiation transport and atomic modeling for laser produced plasmas," ILE Research Report ILE9008P (Institute of Laser Engineering , Osaka University, Osaka, Japan, 1990).
12. MacGowan, B. J., Maxon, S., Hagelstein, P. L., Kean, C. J., London, R. A., Matthews, D. L., Rosen, M. D., Scofield, J. H., and Whelan, D. A., Phys. Rev. Lett. **59**, 2157-2160 (1987).

Status of Collision-Pumped X-ray Lasers

Ciaran L.S. Lewis

School of Mathematics and Physics
The Queen's University of Belfast
Belfast BT7 1NN, N. Ireland

Abstract. Collision-pumped soft X-ray lasers, based on transitions found in neon-like and nickel-like ions, continue to provide the highest gain-length products achievable with terawatt class optical lasers providing the necessary primary pump power. Recent efforts to improve the production efficiency of such systems and to characterise and manipulate their optical properties are described.

RECENT DEVELOPMENTS

Several significant advances in the field of collision-pumped soft X-ray lasers (XRL's) have been made in the last two years, since the 3rd International Colloquium on X-ray lasers (1), and have been based on access to some of the largest lasers in the world including those in China, France, Japan, the UK and the USA. Nearly all the elements from Ti (Z=22) to Ag (Z=47) have now lased on neon-like transitions and many elements from Nb (Z=41) to Au (Z=79) have lased on nickel-like transitions giving coverage over a wide spectral range (3.5-35 nm) on a large number of discreet spectral lines. Recent work by many groups has concentrated on four main issues viz. improving the efficiency of these lasers, characterising and improving their optical properties, demonstrating applications for these radiation sources and developing the theory associated with all these aspects.

Efficiency aspects have been studied in many ways including the use of half-cavity geometries, curved slab targets, low density aerogel targets, travelling wave pumping, prepulses and picket-fence pulse trains. Beam properties studied include polarisation, line-widths, hyperfine splitting and spatial coherence. Applications have concentrated on the use of XRL's as sources for microscopy and holography but a new and important development is the use of XRL's as plasma probe beams for imaging and interferometry experiments. Theory is improving on all fronts but with new emphasis on wave propagation in amplifiers and coupling between amplifiers.

Almost all the topics mentioned above, and many more, are reported in detail in these proceedings (2). In the material below, I will concentrate on some issues not appearing elsewhere.

Probing plasmas with an XRL

The high brightness associated with soft X-ray laser beams makes them suitable as probe beams to diagnose density structures in hot, dense transient plasmas, as may be produced, for example, by direct or indirect drive in inertial confinement fusion experiments. The two important parameters making soft XRL's suitable for such an application is their short wavelength enabling propagation through plasma regions with density up to the high critical density limit and their unparalleled brightness which allows the probing or backlighting soft XRL beam to "outshine" any probed plasma self emission or any thermal source of alternative probe radiation. The critical density limit can be expressed in terms of the laser wavelength in Angstrom units as

$$n_{ec} \approx 10^{25} / [\lambda/100]^2 \text{ cm}^{-3} \quad or \quad \rho_c \approx 33 / [\lambda/100]^2 \text{ g cm}^{-3} \quad (1)$$

for nearly fully ionised plasmas of low atomic number species. Propagation distances are limited by absorption and refraction effects in practice. The "equivalent" black-body radiation temperature T_{BB} of a thermal source emitting the same spectral intensity (W/m²/ster/Hz) as an X-ray laser with relative bandwidth B can be expressed as

$$T_{BB}(\text{GeV}) \approx [\lambda^3(100 \text{ Å}) \text{ E(mJ)}] / [d^2(100\mu\text{m}) \theta^2(\text{mrad}) \text{ B(\%)} \tau(100 \text{ psec})] \quad (2)$$

where the parameters characterising the laser pulse are expressed in the units shown in brackets. For systems operating near saturation T_{BB} can be several GeV. This equation can also be used to estimate the performance of a thermal X-ray source coupled with optics of bandwidth B and acceptance angle θ assuming the mean photon energy is much less than the source Planckian temperature for the spectral region (typically kT < 200eV for a 1kJ / 1nsec heating a 200 μm spot on a high Z target).

Some preliminary experiments have been carried out at Rutherford-Appleton Laboratory which serve to demonstrate the potential of bright soft X-ray laser beams to probe target structures irradiated with high intensity laser beams (3). A schematic of the experimental geometry used is shown in Figure 1. An XRL was used to backlight microballoon and foil targets which were irradiated at $\approx 6 \times 10^{14}$

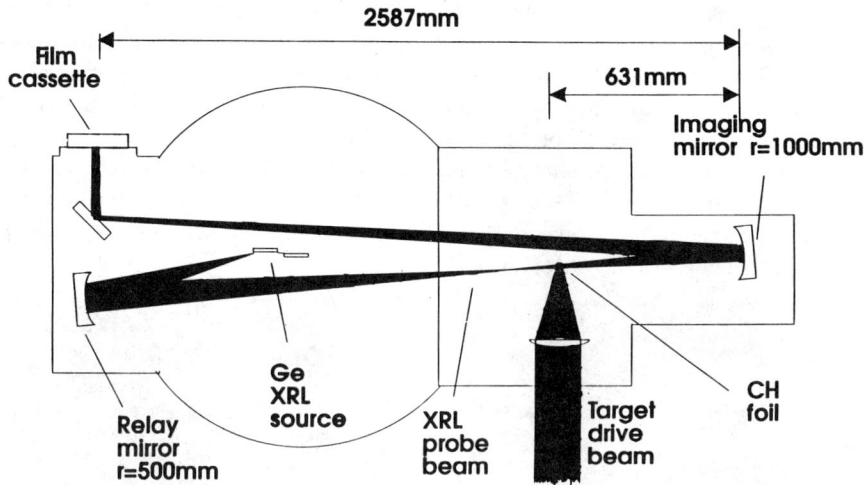

FIGURE 1. Experimental arrangement at RAL to demonstrate imaged radiography of a dense plasma using an XRL as a bright backlighting source. The Ge XRL beam from a double target injector is relayed to a plane ≈9 cm in front of the object plasma and the object plane is imaged onto film (or CCD). The object (a lollipop foil or microballoon shell) is irradiated at ≈5×10^{14} Wcm^{-2} by one of the VULCAN beams normally used in line focus mode to heat one of two possible additional XRL amplifiers.

Wcm^{-2} by a single beam (250J/1µm/1nsec). The ≈300 psec duration XRL pulse was generated from a double, refraction-compensating Ge stripe target where each component was 22 mm long and irradiated at ≈1.6×10^{13} Wcm^{-2}. X-ray optics with a bandpass matched to the J=2-1 lines at ≈23.4 nm and a reflectance of ≈15% were used to relay the XRL output to the target plane with a spot size of ≈1mm diameter. The target plane was then imaged onto film with a magnification of ≈ 4.2X and with a demonstrated intrinsic spatial resolution at the target plane of ≈2.5 µm. All the optical laser beams used had known relative timings so that the probe time could be controlled. Sample images recorded are shown in Figure 2. In one case the target was a CH lollipop foil, 150 µm diameter and 7 µm thick, supported on a 6-7 µm diameter carbon fibre and backlit edge-on about 1.5 nsec after the peak of the foil drive. In the second case, the target was a CH microballoon of 150 µm diameter and 6 µm wall thickness. The target was shot with the extreme rays of the F/2.5 optics tangential to the balloon surface and the XRL was timed to arrive ≈600 psec after the peak of the target drive pulse. Quantitative interpretation is difficult in these demonstration shots as the line of sight opacity is too high (»1) over most of the 2-D image and the ionisation state of the material probed is not known. Nevertheless at the probe time for the foil shot we deduce that material with density ≳2 mgcm^{-3} is spread out over 200-300 µm around the initial target position. For the balloon shot the conditions are intermediate to the limiting cases of 1-D accelerations of a planar foil and a uniformly irradiated spherical shell. Simulations using the hydrocode MEDUSA for these extremes are also shown in Figure 2. The image recorded shows a cone of material, blown off the irradiated surface, centred on the beam axis and also absorbing material beyond the opposite, non-irradiated surface. The simulations indicate that there has been time for the irradiated wall to accelerate and collide with the opposite wall causing the shocked spallation of some material from the outer surface of the shell, in general agreement with the observation.

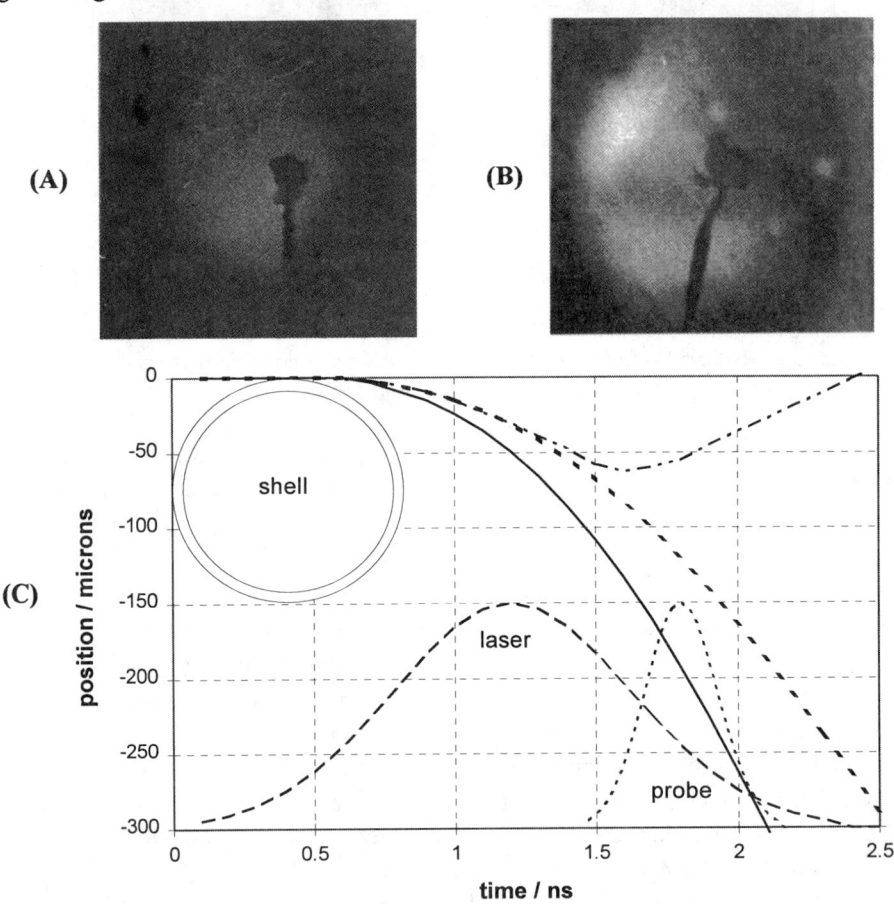

FIGURE 2. Radiographs of a CH foil (A) and a CH microballoon (B) taken with a Ge XRL. Both targets are 150 μm in diameter and the drive laser beam is incident from the left : (C) Shell wall trajectories calculated by MED101 for image (B) are shown in the planar (solid) and spherical (dash-dots) limits with the dashed trajectory an estimate of the actual situation. Relative timings of the drive laser and the probing XRL are shown.

These proof-of-principle images suggest that well-designed probing experiments can take advantage of the XRL brightness to make quantitative measurements. Such an experiment, for example, could be based on in-line probing where all the beams are arranged to come from the same side with the XRL in the centre of a cluster of target drive beams. Frequently spectrometers are filtered with up to 6 µm of cold Al foil in Ge XRL experiments, so it is possible to accelerate Al foils up to this thickness and make in-line 2-D opacity images as a function of time throughout the foil flight. Planar targets driven at 10-20 Mbar pressures, relevant to ICF experiments, experience shock heating to 15-30 eV and the in-flight density due to the combined effects of preheat induced decompression and shock compression as well as the spatial uniformity of the mass distribution can be studied. At these temperatures opacity changes in Al below the L-edge are most sensitive to density changes as may occur during shock compression. Since we can probe material with high average opacity we can study very low levels of mass redistribution, which may occur due to inhomogeneous irradiation or due to Rayleigh-Taylor instability after shock break-out and acceleration of the whole target. A major advantage of this in-line geometry is that spatial blurring due to material motion during the XRL pulse is much less a problem compared to the transverse probing geometry depicted above.

Multipulse Plasma Pumping

In recent years several groups have studied the use of temporally modulated pump power profiles to optimise the plasma conditions for generation of ASE outputs with enhanced efficiency. Two main approaches have been taken. In one case where long pulse (i.e. ≈1 nsec) drive is used, a prepulse is generated to form an initial plasma with which the second and main pulse interacts. The net effect is that a larger plasma volume is heated to the temperature required to produce the ideal ionisation balance and that, furthermore, within a zone where ionisation and density are best matched to pump a particular transition then the density gradient tends to be relaxed with the bonus that the ASE can amplify over longer path lengths before refraction effects dominate. Modelling indicates that for slab targets a delay between prepulse and main pulse of order 1 - 10 nsec is useful for prepulse levels of 1 - 10 % (4). Experiments at LLNL using 0.3-0.5% prepulse levels with 0.53 µm light on slab targets have led to lasing on many low and even Z (22-34) targets (2,5). Surprisingly, experiments at RAL and at LULI have shown that prepulse levels as low as 0.02% with 1.06 µm light (or at the 10^9 Wcm^{-2} level) are effective for optimising the conditions for efficient production of J=1-0 transitions in Ne-like Zn and Ge slab targets (2,6). Time-resolved spectra of such shots show that the J=1-0 lines are comparatively short-lived (50-100 psec) and much brighter (>10X) than the J=2-1 lines for targets of ≈ 2 cm length. This has advantages for applications demanding short pulse, monochromatic radiation such as imaging microscopy and probing of fast transient events.

In the second approach picket-fence trains of short pulses ($\approx$100 psec) of comparable intensity and separated by 300-400 psec have been used to irradiate both foil (Y) and slab (Ge and Se) targets at LLNL and ILE (7,8) showing short duration and strong J=1-0, Ne-like XRL output on all pump pulses except the first. This is promising for future closed cavity development. Perhaps the most significant development in this area has been the application of multipulse trains in ILE to prepare and heat Ni-like plasmas using slab targets of elements with Z=60-66 (8). With 1.06 µm laser light (150 J/100 psec per pulse) gain-length products of $\approx$7 have been achieved at several wavelengths in the region of $\approx$7 nm. This represents $\approx$10X enhancement of the efficiency in producing gain at these wavelengths compared to long ($\approx$0.5 nsec) single pulse irradiation of foil targets at 0.53 µm.

Spatial Coherence

There have been several reports on spatial coherence measurements on the soft X-ray beams produced in several laboratories where typically 1 mm^2 areas of the beam have been sampled with a uniformly redundant slit array (LLNL and RAL) or a 2-D array of pinholes with several repeat period patterns (ILE) to produce interference patterns which can then be interpreted in terms of the spatial coherence function of the part of the beam sampled. Recent work at RAL has extended these ideas to simultaneously sample several parts of a beam on the same shot leading to the possibility of studying how the local, time-averaged spatial coherence is related to the paths amplifying rays have taken in inhomogeneous and refracting plasma media (9). As illustrated in Figure 3, the basic element in the diagnostic is a pair of Young's slits where each slit is also a narrow diffraction grating, enabling, in principle, measurements on each of the lasing transitions. Arrays of these elements, separated by 2.5 mm, were positioned 58 cm from the end of a single 14 mm Ge amplifier used in the standard injector-amplifier configuration at RAL (10). Figure 3 also shows that the fringe visibility increases for parts of the axially injected beam exiting the amplifier at the largest angles relative to the amplifier surface. As seen, this is fairly reproducible over four separate shots and using the 88% criterion indicates a transverse coherence length (at the diagnostic plane) of $\approx$100 µm in the less intense and most strongly refracted part of the beam. This suggests that the "effective" source size within the amplifier is not constant but is being reduced by refraction effects. This idea is supported by the observations at ILE where a half-cavity employing a refraction compensating curved Ge slab target and irradiated with either a single long pulse or multiple short pulses generated sub-milliradian divergence beams (2,11). Similarly, at the

Shenguang facility in China a half-cavity arrangement employing four zig-zag, refraction compensating slab target elements has produced saturated output on the J=2-1 lines of Ge with a 1.5 mrad divergence and a near diffraction limited output of Fresnel number N≈1.7 (2,12). Future developments in this area should aim to make amplifiers more homogeneous.

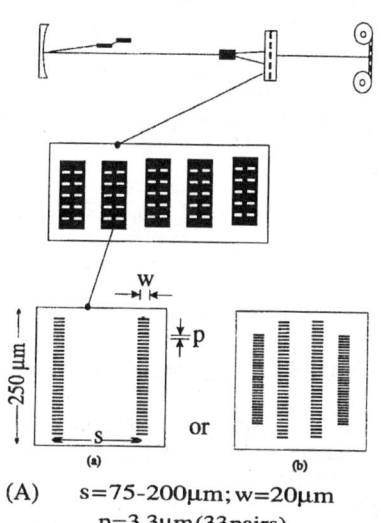
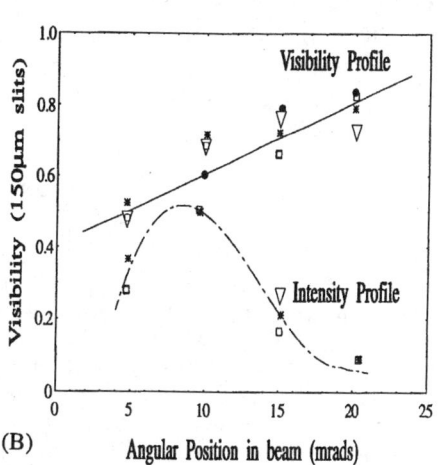

FIGURE 3. (A) Young's slits structures cut in gold and supported on silicon nitride membranes. Double slit pairs (b) have different grating pitch to separate fringe patterns. : (B) Fringe visibility across the XRL beam profile in the horizontal plane using data from array of slits with 150 μm spacing. Each different symbol relates to a particular shot.

CONCLUSIONS

Collision pumped XRL's continue to provide the most intense beams but still with efficiencies less than $\approx 10^{-6}$ and thus requiring terawatt class lasers as pumps. Much ingenuity is being applied to improve this position and with some success. Indeed it would seem imperative that breakthroughs are made soon to open up the field to a wider community with access to smaller laboratory-scale pumps. There are indications that this is happening and that the success of the collision-pumped systems over the last decade has spurred general interest in the subject area with complimentary research investigating high harmonic generation and their use for injection seeding plasma amplifiers, capillary discharge pumps, optical field ionisation schemes and many others. Future workers may well look back to the initial era of brute force techniques and marvel at how Xrayasaurus Lux evolved into the efficient and reliable table-top devices of the day.

ACKNOWLEDGEMENTS

It is a pleasure to acknowledge the help given by many colleagues and friends in providing material used or referred to in this paper. In particular, I thank RE Burge, P Jaeglé, Y Kato, MJ Lamb, J Nilson, S Wang, and the many members of the UK X-ray laser consortium who work hard converting VULCAN's 1 eV photon beams to 60 eV photon beams and who have contributed many more detailed papers in these proceedings.

REFERENCES

1. *Proc.3rd Int. Coll. X-ray Lasers* held at Schliersee, Germany, May 1992 : ed. Fill E.E., Inst. of Physics Conf. Series No. **125**, pp.1-457, 1992.
2. *Proc.4th Int. Coll. X-ray Lasers* held at Williamsburg VA, USA, May 1994 : eds. Matthews D. and Eder D. Publ. AIP, 1994.
3. MacPhee A.G., Cairns G.F., Key M.H., Lewis C.L.S., Neely D., Norreys P. and Smith C. *Rutherford-Appleton Lab. Central Laser Facility Annual Rept.* (RAL-94-042), p.15, 1994.
4. Pert, G. J., Private communication.
5. Nilson J., MacGowan B.J., DaSilva L.B. and Moreno J.C., *Phys. Rev. A* **48**, 4682-8, (1993).
6. Cairns G.F., Lewis C.L.S., MacPhee A.G. et al., *Rutherford-Appleton Laboratory Central Laser Facility Annual Rept.* (RAL-94-042), pp. 16-18, 1994
7. Nilson J. et al. and DaSilva L.B. et al in reference 2 above.
8. Daido H. et al. and Kato Y. et al. in reference 2 above.
9. Burge R.E., Browne M.T. and Slark G.E. *Rutherford-Appleton Laboratory Central Laser Facility Annual Rept.* (RAL-94-042), pp. 19-20, 1994.
10. Cairns G., Lewis C.L.S., MacPhee A., Neely D. et al. *J. Appl. Phys. B.* **58**, pp. 51-56, (1994).
11. Kato Y., Daido H. Kodama R. et al., *Proc. S.P.I.E.*, **2012**, pp. 12-21, 1993.
12. Wang S., Gu Y., Mao C., Zhou Z. et al. *Chinese J. of Lasers*, **B2 (No. 6)**, pp. 481-4, (1993).

Ray Tracing Calculations Of The Output From Germanium Slab Lasers

JA Plowes[1], PB Holden[1], GJ Pert[1] and SB Healy[1]
AE Kingston[2] and E Roberts[2]
[1]Department of Physics, University of York, York, Y01 5DD, UK.
[2]Department of Mathematics, Queens University, Belfast, BT7 1NN, UK.

Abstract

A 3D Raytracing code is used, as a post processor, to simulate experimental observables, such as divergences, deflected angle and output intensity, from a $1\frac{1}{2}$D fluid code. The latter self consistently treats the plasma expansion with the atomic physics of the Ne-like ion.

The results presented relate to two seperate experiments. Firstly, an experiment carried out at R.A.L, where Ge slab targets, of varying lengths, were irradiated at driving laser intensites in the range $0.8 \rightarrow 2.3 \times 10^{13}$ Wcm^{-2}. Results presented here are for the 236Å line and good agreement is found with experiment.

Also presented, are simulations which relate to an experiment carried out at Osaka University, where a 4cm Ge slab target, with a curvature from 0 to 20 mrad, along the lasing axis, was irradiated. General agreement with experiment is obtained. Tightening in the output beam, with increasing curvature, can clearly be seen.

Introduction

High electron density gradients generated in collisional excitation lasers lead to refraction of the x-ray laser beam. This requires the use of a raytracing technique[1,2,3,4], or wave technique[5,6], to generate experimental observables, for example, divergences and intensities, from an analytical expression for the plasma properties, or values obtained from a hydro/atomic code.

The model used here is an extension, to three dimensions, of the two dimensional raytracing code decribed by Holden et al[4]. The model calculates the refraction of ray bundles across a plasma mesh, generated by the $1\frac{1}{2}$ hydro/atomics code EHYBRID. These bundles are then amplified as they pass through regions of gain. When rays exit the plasma they are added to the output profiles to build up the final far and near field profiles. The model includes curvature of the target and the effects of saturation.

Initially modelled were a series of experiments conducted at R.A.L[7,8,9,10], where a 100μm wide Ge slab target of lengths 0.9, 1.4 and 2.2cm respectively were irradiated at intensities between $0.8 \rightarrow 2.3 \times 10^{13}$ Wcm^{-2}. Characteristics of the J=2$\rightarrow$1 236Å line were principally measured. Typical experimental measurements were a horizontal divergence $\sim$10-12mrads, vertical divergence $\sim$20-25mrads, a deflected angle of $\sim$10-12mrads, and a maximum gain-length $\sim$8. Modelling was carried out with a driving laser pulse that rose linearly for 200ps,

18 Ray Tracing Calculations

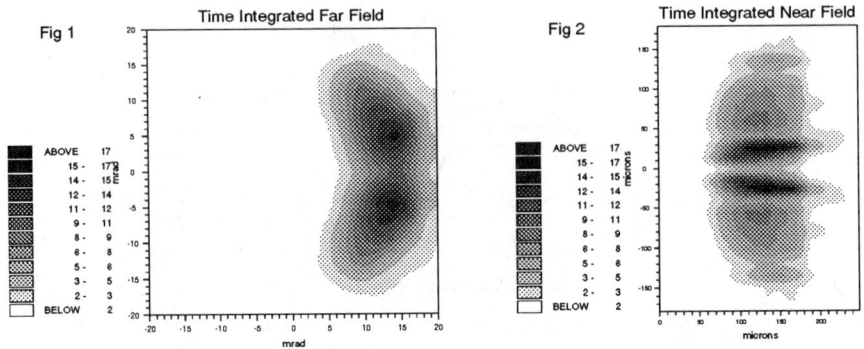

Figure 1: Far field, Figure 2: Near field, for 2.2cm Ge slab target

was flat topped for 300ps, and then fell linearly in 200ps, in line with the experimental configuration. A selection of results will be presented for comparision.

Refraction of the x-ray laser beam by the high electron density gradients limits the maximum length of target that can be used. Refraction compensation by bending the target in a toroidal surface was suggested by Lunney[11]. In recent experiments[12], carried out at Osaka, a 4cm Ge slab target, 200μm wide, was irradiated at 1.7×10^{13}Wcm^{-2}. The target was curved by 3 point mechanical pressure along its length up to curvatures of 5mrad/cm, to compensate for refraction in the radial direction.

Experimentally, this gave enhancement on all lasing lines, including a marked increase in the intensity ratio of the J=0→1 196Å to J=2 →1 236Å line, at a similar curvature. At larger curvatures the 196Å line intensity exceeded the 236Å line intensity by a factor $\sim$ 3.

The model was altered to account for a homogeneous curvature along the target length, by altering the ray angles accordingly at each cell exit. The theoretical outputs, from the raytracing code for the above conditions, were calculated for a variety of curvatures for both the 196Å and 236 Å line. Due to space limitations only a selection of 236Å results are presented. A full account will appear at a later date.

Results

1. R.A.L Slab Target Simulations

Calculations were carried out at lengths of 0.9, 1.4 and 2.2cm respectively at intensities of 1.0, 1.5, 2.0 and 2.5×10^{13}Wcm^{-2} with the conditions stated. Fig 1, shows a time integrated far field for the 236Å line generated from a 2.2cm target irradiated with a driving laser intensity of 2.0×10^{13} Wcm^{-2}. The general shape of the pattern is typical for flat, single pulse irradiation, slab targets. The output has moved off the laser axis, defined by (0,0); the straight through direction, due to the density gradients in the radial and transverse directions. The calculated horizontal and vertical divergences are $\sim$9 and $\sim$20mrad respectively with a deflected angle of $\sim$13mrad. This shows good agreement with experiment.

Fig 2, shows the corresponding near field pattern. The active region of the plasma can clearly be seen to be $\sim$60μm wide and situated $\sim$100μm from the target surface.

A measurement of the gain coefficent was calculated by fitting the output intensities at the three plasma lengths to a Linford formula[13]. Fig 3, shows calculated gain coefficients from the 3D raytrace at varying intensities compared to an experimental curve[9]. General

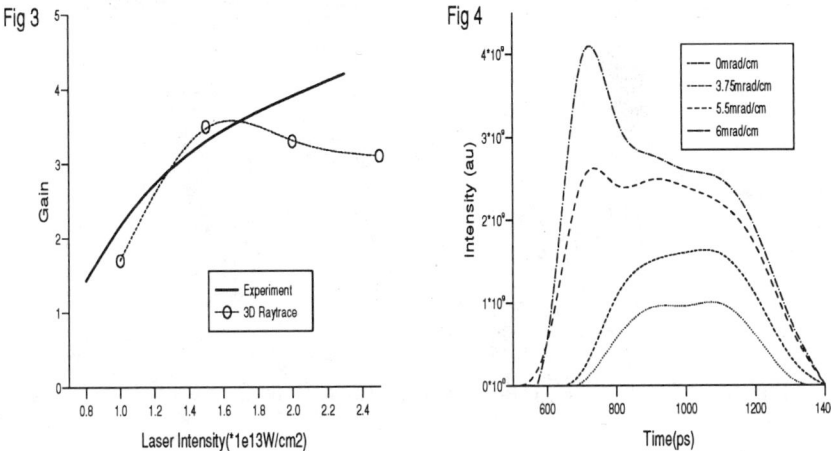

Figure 3: Raytraced gains vs experimental curve, Figure 4 : Intensity vs curvature for 4cm Ge slab target

agreement of the magnitude of the gain coefficients, between theory and experiment, is found.

2. Osaka Simulation Results

Fig 4, shows time resolved intensities for the 236Å line at various curvatures. The intensity for highly curved targets peaks at ~700ps, compared to ~1100ps for flat targets. This is due to the existence of higher gain at earlier times that cannot normally be accessed due to the density gradients.

Due to space limitations, far and near fields of the 236Å line calculated for a flat target are not presented here, since they are of the same general shape as Fig 1 and 2. Fig 5 and 6, show the far and near fields respectively when the target was curved to 6mrad/cm. A marked change is seen in the patterns. Due to the refraction compensation of the curvature certain ray bundles have a much straighter path through the plasma and can experience a larger gain length product than would be possible for a flat target. This not only leads to a generally higher output signal than for the flat case but, since ray bundles which pass through higher gain regions amplify far more than ray bundles which travel through nearby lower gain regions, a small proportion of the total number of rays dominate the output signal. This gain area dominance, on the output, can clearly be seen in the near field, Fig 6. The output is a very tightly confined area on the end face of the plasma, in contrast to a flat target where a far larger area is observed. To maximise the gain-length product, a ray must follow a straighter path through the plasma. This is clearly seen is Fig 5, as the lobes are now much closer to the (0,0) direction. It should be noted that the curvature only compensates for refraction in the radial direction and that the lobes are brought closer together, in the far field, in the transverse direction, due to the greater gain-length product of these straighter rays.

As the curvature is increased the beam tightens in the far field and the bright peak in the near field is pushed into areas of higher gain. This is an aspect of the gain area

Ray Tracing Calculations

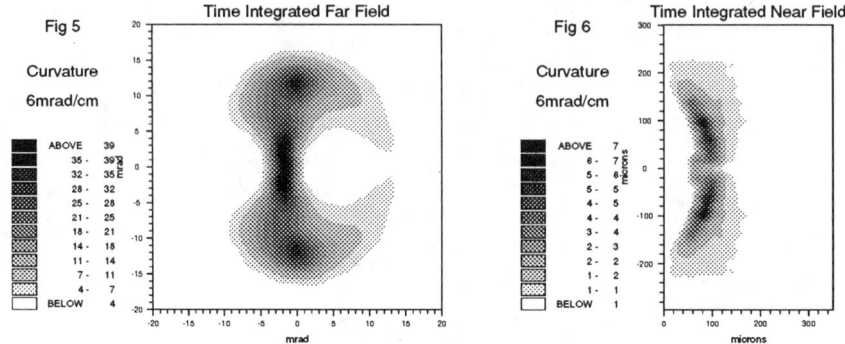

Figure 5: Far field, Figure 6 : Near field for 4cm Ge slab target

dominance on the output profile; ray bundles which pass on straighter paths through areas of higher gain acquire a greater gain length and so dominate the signal.

Experimentally the horizontal divergence, of a flat target, was ~ 8mrad, which decreased to ~4mrad when the curvature was increased to 5mrad/cm. This is approximately the width of the central region of Fig 5, and is in good agreement with experiment given that we have assumed homogeneous curvature in the model.

The temporal width of the pulse is in slight disagreement with experimentally observed results. The general shape of the pulse, double peaked time resolved intensity structure, observed with curvature, and peak intensity ratio between flat and curved targets all agree well with experiment.

Conclusion

It has been shown that the general agreement of results from the 3D raytracing code and experimentally observed results is good. For curved target modelling the observed effects are explained in terms of gain area dominance on the signal, leading to large amplification of specific ray paths. This has been shown to lead to a tighter output beam and an increased output intensity.

References

1. RA London, Phys.Fluids v31, 1988, p184
2. O Zahavi et al, J.Opt.Soc.Am.B v10, 1993, p271
3. A Dulieu, X-ray lasers 1992, IOP.conf.series.125, p243
4. PB Holden et al, J.Phys.B v27, 1994, p341
5. MD Feit and JA Fleck, J.Opt.Soc.Am.B v7, 1990, p2048
6. JW Greene, Phys.Rev.E v48, 1993, p3130
7. DM o'Neill et al, Opt.Comm v75, 1990, p406
8. CLS Lewis et al, X-ray lasers 1990, IOP.conf.series.116, p37
9. D.Neely et al, X-ray laser 1992, IOP.conf.series.125, p37
10. D.Neely et al, Opt.Comm v87, 1992, p231
11. JG Lunney, Appl.Phys.Lett v48, 1986, p891
12. R Kodama et al, to be published
13. GJ Linford et al, Appl.Opt v13, 1974, p379

Multiple Pulse Traveling Wave Excitation of Neon-like Germanium

J. C. Moreno, J. Nilsen, and L. B. Da Silva

*Lawrence Livermore National Laboratory
P. O. Box 808, Livermore, CA 94550*

Abstract. Traveling wave excitation has been shown to significantly increase the output intensity of the neon-like germanium x-ray laser. The driving laser pulse consisted of three 100 ps Gaussian laser pulses separated by 400 ps. Traveling wave excitation was employed by tilting the wave front of the driving laser by 45 degrees to match the propagation speed of the x-ray laser photons along the length of the target. We show results of experiments with the traveling wave, with no traveling wave, and against the traveling wave and comparisons to a numerical model. Gain was inferred from line intensity measurements at two lengths.

INTRODUCTION

Tilting the wave front of the driving laser to match the propagation velocity of the x-ray laser can be an effective method of increasing the effective gain length to extract the maximum energy out of the x-ray laser. This technique for exciting the plasma, usually called traveling wave excitation, is particularly useful and important when the duration of the driving laser divided by the length of gain medium is ≤ 33 ps/cm (the propagation time of x-ray laser photons). It is expected that a table-top x-ray laser will most likely require a short duration, high intensity driving laser to produce the plasma gain medium. A short duration x-ray laser has many potential applications, such as probing a high density, laser-produced plasma or imaging a biological sample.

The traveling wave technique has been employed in various short pulse lasers to shorten the pulse duration and increase the amplification (1-5). Traveling wave excitation of an x-ray laser has been demonstrated for the Ne-like yttrium exploding foil laser at 155 Å (6). In these experiments we demonstrate the application of the traveling wave technique to the Ne-like Ge x-ray laser using slab targets. We observe a factor of ten increase in x-ray laser line intensity compared to no traveling wave and a factor of seventy increase compared to having the traveling wave go the opposite direction. Transit time effects are include in a numerical model that agrees well with the experiments.

EXPERIMENT AND DISCUSSION

Experiments were performed using the Nova laser at Lawrence Livermore National Laboratory. The laser pulse consisted of three 100 ps Gaussian pulses separated by 400 ps peak to peak. Each pulse had an intensity on target of 1.1×10^{14} W/cm^2 at a laser wavelength of 0.53 μm. A 3 cm long line focus of width 120 μm was used. Targets consisted of a 1 μm thick layer of Ge deposited on a 125 μm thick foil of Ni. Because of a 16% gap in the laser beam, the effective length of the target was 2.52 cm. We also used half length targets for gain measurements. Our diagnostics consisted of a 1-meter grazing incidence spectrograph (McPigs) with a microchannel plate detector viewing one end of the x-ray laser axis and a flat field spectrograph (SFFS) using a varied line space grating and a streak camera detector viewing the opposite end of the x-ray laser.

The traveling wave was implemented by inserting a diffraction grating along with additional mirrors in the path of the driving laser (Nova), before the main amplifiers and KDP frequency doubling crystal. First order diffraction from the grating produces the tilted wave front due to a varying path length. The angle ϕ of the tilted wave front is given by $\tan(\phi) = \lambda_L/D \cos(\vartheta_o)$, where λ_L is the laser wavelength (1.06 μm), D is the groove spacing and ϑ_o is the diffraction angle. The grating groove spacing and incident angle ϑ_i, was selected in this case so that the wave front was tilted to ~ 45 degrees to match the transit time of photons along the laser medium.

Germanium spectra taken from multiple pulse irradiated targets show a very strong J=0-1 line unlike earlier experiments with a single pulse (7). In Fig. 1 we show a comparison between a multiple pulse spectrum and a single longer (600 ps) pulse spectrum. For the 600 ps Gaussian pulse experiment we observe the J=2-1 lines at 232 Å and 236 Å but do not clearly see the J=0-1 line. Other recent experiments of Ne-like Ge with a slab target have also exhibited stronger lasing in the J=2-1 lines (8,9). Lasnex hydrocode simulations indicate that the multiple pulses are acting similar to the prepulse in that a larger more uniform density plasma is created compared to the standard single pulse (10). Streaked spectra of the plasmas show that the x-ray laser lines do not show up during the first pulse, but only appear during

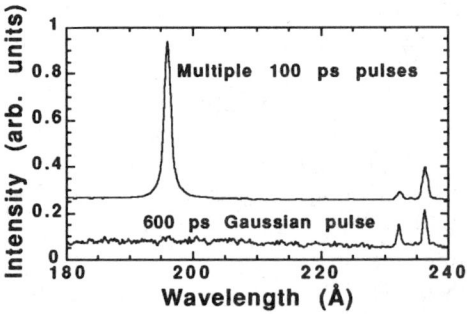

FIGURE 1. Comparison of Ne-like Ge x-ray laser spectrum for two pump laser conditions.

the second and third pulse of the driving laser. The first pulse ionizes the plasma but the conditions are unsuitable for gain. The second and third pulse create a somewhat hotter plasma with more gradual density gradients that are better suited for good amplification. The J=0-1 line appears earlier in time than the J=2-1 lines and has a somewhat shorter time duration.

We compared laser irradiation of Ge slab targets for three different wave front angles: first with the traveling wave going toward the detector (ϕ= 45 degrees); second with no traveling wave (ϕ = 0 degrees); and third with the traveling wave going away from the detector (ϕ = -45 degrees). A simple numerical model was developed to simulate the effect of the transit time of the x-ray laser photons on the integrated output line intensity. In Fig. 2 we compare the measured integrated intensities of the J=0-1 line at 196 Å, taken with the McPigs spectrograph, to this numerical model. The points show the measured data while the lines are from the numerical model. We measured a gain of 4.1 cm^{-1} for the 196 Å line on data taken at two different lengths for the case with the traveling wave going toward the detector since there is no transit time effect for this case. For cases with no traveling wave or the traveling wave going in the opposite direction, there is a fall-off in the increase of intensity as the length of the plasma column is increased.

Our simulation assumed the time history of the emissivity and gain was Gaussian with a full-width at half-maximum (FWHM) of 85 ps for the J=0-1 line at 196 Å. This emissivity time duration gave the best fit to our data for the x-ray laser going against the traveling wave. Experimental data in Fig. 2, for the case with no traveling wave, falls slightly below the expected values. This may be due to inhomogeneous

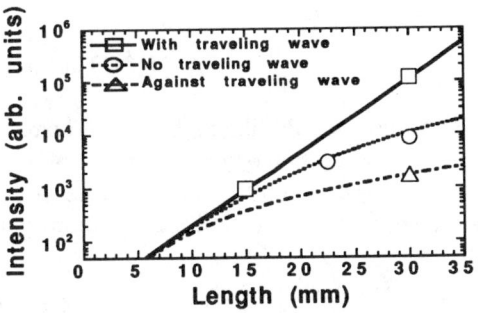

FIGURE 2. Intensity of the J=0-1 line plotted vs length. The curves are from the numerical model while the points are experimental data.

plasma effects and imperfect matching of the traveling wave pump to the propagation speed of the x-ray laser photons as well as approximations in the model. It should be noted that our model includes the 16% gap in the middle of the target since the propagation time across the gap is non-negligible.

The J=2-1 line at 236 Å was also measured for the three cases and exhibited a similar change in intensity due to the traveling wave. Again we adjusted the FWHM of the emissivity and gain time history in order to agree with the experimental data for the traveling wave going against the direction of the x-ray laser photon. In this case we have a somewhat longer duration (FWHM = 115 ps) in our model consistent with the longer measured duration of the 236 Å laser compared to the 196 Å laser. The time duration of the J=2-1 laser line at 236 Å was measured for the three traveling wave configurations (at a plasma length of 2.52 cm) and compared to the numerical model using the measured gain of 4.3

cm^{-1}. Our measurements give a laser duration (FWHM) of 55 and 77 ps for the cases with and against the traveling wave as compared to model predictions of 36 and 64 ps, respectively. Agreement with our model is reasonable given the approximations in our model and the uncertainties in the plasma conditions.

CONCLUSION

We have compared x-ray laser experiments with, without and against the traveling wave. An order of magnitude enhancement of the Ne-like Ge x-ray laser intensity is observed when the traveling wave technique is used to compensate for the short gain duration of these plasmas. Gains of 4.1 cm^{-1} for the 196 Å line and 4.3 cm^{-1} for the 236 Å line were inferred from integrated line intensity measurements at two lengths. These measurements for the various traveling wave configurations and plasma lengths agree well with a numerical model including transit time effects. This technique can be important for applications of x-ray lasers where a short pulse laser is required.

ACKNOWLEDGMENTS

The authors would like to thank S. Alvarez, T. Demiris, H. Louis, J. Ticehurst and the Nova operations crew for providing support for these experiments. This work was performed under the auspices of the U. S. Department of Energy by Lawrence Livermore National Laboratory under contract No. W-7405-ENG-48.

REFERENCES

1. Wyatt, R., and Marinero, E. E., *Appl. Phys.* **25**, 297-301 (1981).
2. Bor, Zs., Szatmari, S., and Muller, A., *Appl. Phys. B* **32**, 101-104 (1983).
3. Polland, H. J., Elsaesser, T., Seilmeier, A., and Kaiser, W., *Appl. Phys. B* **32**, 53-57 (1983).
4. Barty, C. P. J., et al, *Phys. Rev. Lett.* **61**, 2201-2203 (1988).
5. Sher, M. H., Macklin, J. J., Young, J. F., and Harris, S. E., *Opt. Lett.* **12**, 891-893 (1987).
6. Da Silva, L. B., et al, "Imaging with x-ray lasers," Proceedings of the SPIE, Ultrashort Wavelength Lasers II, **2012**, 158-164 (1993).
7. Nilsen, J., Moreno, J. C., MacGowan, B. J., and Koch, J. A., *Appl. Phys. B* **57**, 309-311 (1993).
8. Lee, T. N., McLean, E. A., and Elton, R. C., *Phys. Rev. Lett.* **39**, 1185 (1987).
9. Neely, D., et al, *Opt. Commun.* **87**, 231 (1992).
10. Nilsen, J., MacGowan, B. J., Da Silva, L. B., and Moreno, J. C., *Phys. Rev. A* **48**, 4682-4685 (1993).

High Gain-production Efficiency and Large Brightness X-UV Laser at Palaiseau

P. Jaeglé, A. Carillon, P. Dhez, P. Goedtkindt, G.Jamelot, A. Klisnick, B. Rus, Ph. Zeitoun.

Laboratoire de Spectroscopie Atomique et Ionique, Bât. 350, Université Paris-Sud, 91405 Orsay, France

S. Jacquemot, D. Mazataud, A Mens

C.E.A., Centre d'Etudes de Limeil-Valenton, 94190 Villeneuve-St-Georges, France

J.P. Chauvineau

Institut d'Optique Théorique et Appliquée, 91405 Orsay ,France

Abstract: A large gain has been measured for the J=0-1 line of neonlike Zn at $\lambda = 21.2$ nm. The time evolutions and the localisation of emission zones of the J=0-1 and J=2-1 lines are compared. It is shown that a train of very small prepulses before the main pulse has an important role in the J=0-1 emission. A half-cavity has been successfully used to attain a nearly saturated intensity with a 2 cm long plasma. The X-UV pulse energy is of 400 µJ, the laser power of 5 MW. The driving laser is the 0.4 KJ, 600 ps laser of LULI

INTRODUCTION

It is well established that collisional pumping of neon-like ions provides large gain coefficients in the X-UV range [1]. So the first nearly saturated emission of an X-UV laser has been achieved at the 23.6 nm line of neonlike germanium, in using a mutilayer mirror, double-pass system [2]. However, in previous experiments, to bear large gain factor by collisional pumping required a pumping energy larger than 1 KJ. Therefore an important issue of the X-ray laser research is to substantially reduce the pumping energy so as to the size and the cost of future installations were economically reasonable. Here we will show that a stride is made in this direction with the recent development of a neonlike zinc laser at Palaiseau.

The Nd-glass laser of LULI at Palaiseau (France) provides 0.45 KJ energy, in 0.6 ns, at $\lambda = 1.06$ µm. The energy is delivered by six 90-mm-diameter beams. Focusing optics centre all the energy in the same 150 µm x 2.4 cm line focus. The length of the single-side illuminated target.is 2 cm. The available heating flux (~1.4 10^{13} W/cm^2) is not sufficient to yield large neonlike populations of elements like selenium or others having larger atomic numbers. Nevertheless a preliminary study on germanium showed a gain coefficient of 2,5 cm^{-1} at the J=2 to J=1 neonlike lines ($\lambda = 23.2$, 23.6 nm). Then one could expect from plasma double-passing a gain factor, gl, of 7 or so, i.e. a value remaining below saturation threshold.

Therefore we decided to go down slightly into Z scale in investigating neonlike zinc (Z=30), for which a preliminary study by the N.R.L. [3] had shown a gain of 2 cm^{-1} for the two J=2 to J=1 transitions and 2.3 cm^{-1} for the J=0 to J=1 transition.

Hereafter we present the main results of our experimental study of X-UV lasing in zinc. This study has been performed for the three lasing lines at 21.2 nm (J_{1-0}), 26.2 nm (J_{2-1}) and 26.7 nm (J_{2-1}). The first section below presents gain coefficient measurement, beam refraction, beam divergence and localisation of emission region. Some data are reported with more details in a joined paper in the same volume [4]. In the last section we present the main characteristics of the nearly saturated, double-pass zinc laser at 21,2 nm

MEASUREMENT OF GAIN COEFFICIENTS

The details on the experimental setup have been given elsewhere [5]. Let us recall that the targets are 1.25 mm thick, 2cm long Zn slabs Length variation of the plasma column is controlled by beam aperturing. Two main diagnostics are used for soft X-ray emission analysis and gain measurement. A Rowland-circle grazing incidence spectrometer is used to photographically record a large spectral range around the lasing lines. On the other hand, time-resolved line intensities are measured with a Wadsworth-geometry spectrometer coupled to a streak camera [6]. This spectrometer has been designed for the study of parallel X-ray laser emission with a fixed direction. His typical acceptance angle is 1.6 mrad. A collecting optics consisting of two confocal elliptical mirrors under grazing incidence, counter-balances the refractive deviation of the X-ray laser beam through the plasma [7].

The time-integrated spectrum of neonlike zinc, obtained with the Rowland-circle spectrometer between 18 nm and 28 nm, is displayed in figure 1. One sees immediately that the J_{0-1} line at λ =21.2 nm is far the most intense line of the spectrum. Its intensity ratio with respect to the two other lasing lines at 26.2 nm and 26.7 nm is near 25. When we observed this large intensity for the first time (see ref. 5) it differed strongly from all previous experimental results with neonlike ions, in which the intense lines were the J_{2-1} ones though, as it is well known, numerical modeling did predict features like shown in figure 1 long ago [8]. However J. Nielsen et al [9] recently reported that using a prepulse technique increases dramatically the J_{0-1} intensity in neonlike nickel.

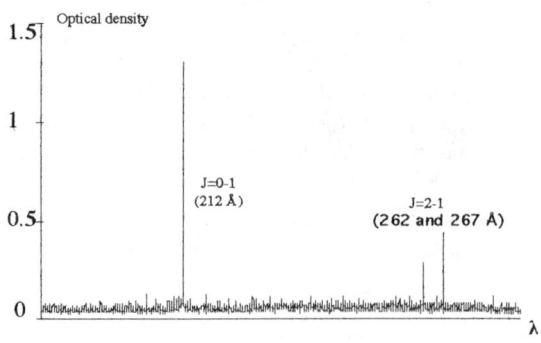

FIGURE 1: Photographic recording of neonlike zinc lasing lines.

1) Time-dependent gain measurement

Gain coefficients deduced from the non-linear variation of line intensity against plasma length corroborate this new behaviour. The values corresponding to the instant of maximum intensity, obtained from streak-camera records, are presented for increasing plasma length in figure 2. Discussion of measurement accuracy and fitting calculation can be found in ref.5. One sees that the values of gain coefficient are 2,3 cm^{-1} and 2.6 cm^{-1}, at 26.2 and 26.7 nm respectively, but it reaches 4,9 cm^{-1} at 21.2 nm. Thus, for the first time, the gain coefficient measured for a J_{0-1} line is considerably larger than for the J_{2-1} ones.

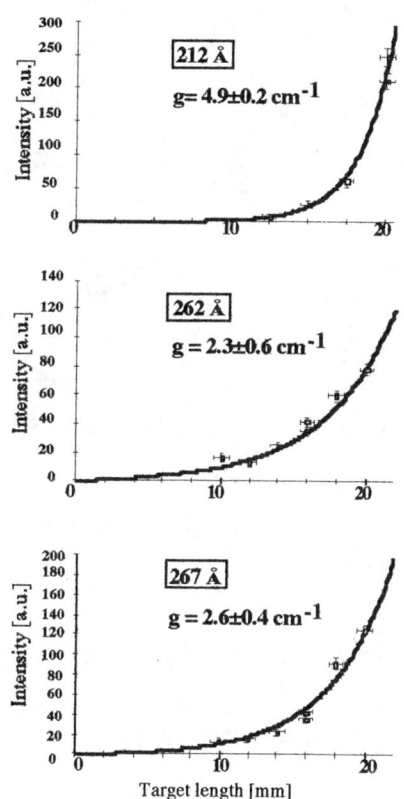

FIGURE 2: Gain coefficient for the 3 lasing lines at the instant of the intensity maximum

The temporal intensity variations also show large differences among the three lines. While the J_{2-1} emissions last 300-350 ps (FWHM), the duration of the J_{0-1} is ~ 100 ps. The 21.2 nm time maximum precedes the peaking of the J_{2-1} lines. This appears in the upper part of figure 3 where the intensities of the 21.2 nm and 26.7 nm lines are plotted as a function of time.

Divergence and refraction angle exhibit differences among J_{0-1} and J_{2-1} lines quite as significant as time-evolutions. For our 2 cm plasma the refraction angle is found at ~ 7 mrad for the J_{0-1} line and at ~ 15 mrad for the J_{2-1}. The beam divergences are respectively ~2.5 and ~10 mrad.

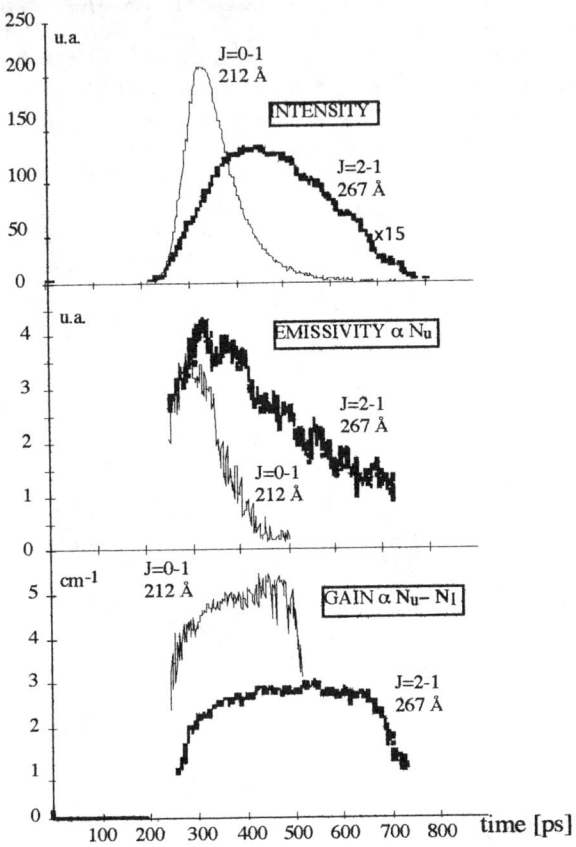

FIGURE 3: Temporal dependence of intensity, emissivity factor (proportional to upper level population) and gain coefficient (proportional to population inversion)

2) Discussion;

Emitting zones

The analysis of the lasing-line temporal dependence shows that the 21.2 nm and 26.7 nm lines must be emitted from two different regions of the plasma. As a matter of fact these lines have the same 3s (J=1) lower level. Calculating the time evolution of the gain coefficicents:

$$g(t) = \left(N_u(t) - \frac{g_u}{g_l} N_l(t)\right) \frac{A_{ul} \lambda_0^2}{8\pi} \frac{1}{\Delta v(t)} \qquad (1)$$

as of the emissivity factors:

$$j(t) = N_u(t) A_{ul} h\nu_0 \frac{\Omega(t)}{4\pi} \frac{1}{\Delta v(t)} \qquad (2)$$

from measured intensities I(t) such as:

$$I(t) = \frac{j(t)}{g(t)} e^{(g(t).L-1)} \qquad (3)$$

leads to the curves shown in figure 3 (middle and bottom part). N_u, N_l, g_u, and g_l are respectively the populations of the upper and lower lasing levels and their statistical weights, A_{ul}, λ_0 and ν_0 respectively the Einstein coefficient of the spontaneous emission, the wavelength and the frequency of lasing transition, Ω is the solid angle of observation which corresponds to the divergence angle of the X-ray beam. These curves show that the gains of both lasing transition maintain basically the same values in time (4.8-5.2 cm^{-1} for 21.2 nm line and 2.5-3 cm^{-1} for 26.7 nm line) while their emissivity factors undergo significant evolutions which differ strongly from each other.

Now the reason for "quenching" the J_{0-1} line must be a diminishing population, N_u, of the upper level J=0. Furthermore, regarding that $g(t) \propto (N_u(t)-g_u/g_l N_l(t))/\Delta v(t)$ and considering that the factor $1/\Delta v$ cannot significantly change in course of the time of interest, the only way to maintain the gain is simultaneously reducing the populations of both levels J=0 and J=1. Stated otherwise, the temporal evolution of the lower level J=1 must roughly follow the J_{0-1} emissivity factor curve from Fig. 3. On the other hand, the duration of the emissivity of the J_{2-1} line is much longer and its evolution much slower. Arguing in same way as in the case of the J_{0-1} line, the main factors determining such behaviour is that the populations of both upper J=2 and lower J=1 levels evolve in a similar fashion as the J_{2-1} emissivity factor curve, thus maintaining a roughly constant value of population inversion term.

Then from comparing the characters of temporal evolutions of one and the same J=1 level relative to both transitions we infer that these evolutions are utterly inconsistent which each other if the lines are emitted from the same plasma element. The necessary conclusion is that the J_{0-1} and J_{2-1} lasing lines don't originate from the same plasma region.

We recently confirmed this conclusion by direct measurements. We observed that the emitting zone of the J_{0-1} peaks at 30 μm from target surface and extends to 50 μm whereas the J_{2-1} peaks near 100 μm and extends to more than 200 μm from target [4]. On the other hand the results of numerical modeling [11] are in good agreement with these observations as this can be seen in Table 1.

TABLE 1: Main characteristics of zinc lasing emission

Zn^{20+}	J = 0-1 line		J= 2-1 lines	
	Measured	Calculated	Measured	Calculated
target distance: r	< 50 µm	< 50 µm	< 200 µm	< 120 µm
Δr	50 µm	≈ 40 µm		≈ 100 µm
N_e		$8\ 10^{20}$ cm^{-3}		$5\ 10^{19}$ cm^{-3}
T_e		300 ev		500 ev
duration: T	100 ps	110 ps (local)	300 ps	370 ps (local)
refraction: θ	7.2 mr	7.5 mr	15 mr (flat)	6 mr (flat)
divergence (hor.): α	2.mr		≈ 10 mr	
gain coefficient: g	4.9 cm^{-1}	4.9 cm^{-1}	2.6 cm^{-1}	2.6 cm^{-1}

Prepulses

Numerical modeling accounts very well for gains as measured at Palaiseau for the three lasing lines of neonlike zinc, especially as regards the large gain coefficient of the J_{0-1} line at 21.2 nm. The pulse of the driving laser used in the calculation is a single pulse of 600 ps width which properly reproduces the main characteristics of the pulses of the LULI laser. On the other hand the experimental arrangement used in our work did not include any device designed for prepulse emission. However Nielsen et al.[9] reported that the large enhancement of the J_{0-1} lasing line at 23.1 nm in neonlike nickel needed a prepulse technique. The prepulse energy was of 6 J in 600 ps and it preceded the main pulse by 7 ns. Similar observations were reported on copper and germanium. Very small prepulse effect has also been observed [10].

A further detailed examination of target illumination in our experiment has shown that the rejection of the mode-locked oscillator pulse train was realized with a contrast about 10^{-3} at the wavelength of 1.06 µ. Oscillator pulses were separated by 10 ns. Then a train of pulses reached the target with 30 - 70 mJ energy/pulse, the last one attaining the target 10 ns before the main pulse. Now a crucial observation is that the increase of the rejection contrast up to 10^{-6}, with the help of an additional Pockel's cell, results in the quasi-vanishing of the J_{0-1} line of zinc. Thus we are led to conclude that prepulses of energy less than 100 mJ, coming to the target 10 ns or more before the main pulse, are necessary to the lasing action at 21.2 nm. We observed that, at the same time, the presence of the prepulse train broadened the Kev pinhole-image of the plasma column by a factor 1.3.

Standard numerical models are not designed for accurate predictions about interactions relative to such very small pulses. Therefore we cannot be sure that the physics implied here is the same as in experiments where the prepulse energy is of several joules. In addition, discussing the prepulse effect should include the fact that the detailed numerical modeling of our experiment [11] does not require any prepulse to account for the experimental results. This suggests to look for prepulse effects making the real plasma more similar to its theoretical model which, of course, does not involve some possible accidental defects. One can think that the J_{0-1} emitting zone narrowness (~50 µm) causes some brittleness of the amplifying plasma column. Then a slowly expanding preplasma heated by the main pulse might produce a lasing plasma of more stable features.

DOUBLE-PASS X-RAY LASER AT 21.2 nm

The half-cavity

A sketch of the double-pass system carried out to arrive at saturation at 21.2 nm is shown in figure 4. The figure shows the detection device used for measuring the absolute number of photons, i.e. a thined CCD camera previously calibrated with synchrotron radiation. Time-resolved measurements were performed with the same streak-camera as previously, the 45° plane mirror being then removed. This part of the system is not represented on the figure. The Mo/Si cavity mirror is aligned to return the beam refracted at ~ 7 mrad from axis to the amplifying region.

It is important to point out that comparing the time-dependant signals [4] for single- and double pass shows the cavity mirror to be efficient during the whole duration of the X-ray laser emission. There is no indication of mirror destruction during the pulse .

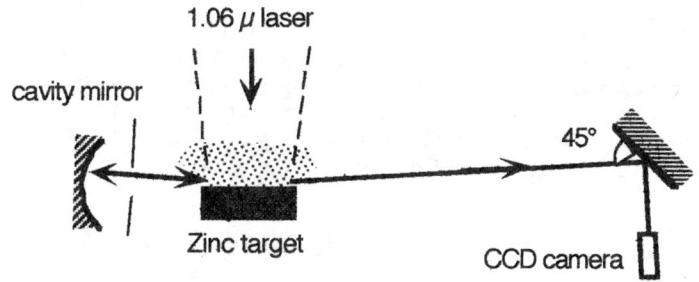

FIGURE 4: Diagram of the double-pass system.

The effective gain factor, gl_{eff}, after double-pass is approximately given by:

$$gl_{eff} = 2gl + LnR_{eff} \qquad (4)$$

where l is the plasma column length and R_{eff} an effective reflexion coefficient characteristics of the half-cavity. As regards the reflectivity band of the spherical mirror, it is shown in figure 5 as measured at the N.I.S.T. (Gaithersbourg, Md) with synchrotron radiation. At 21.2 nm the reflectivity is about 30% .

However the geometry of the cavity, which is fixed by the mirror curvature radius, ρ, and the mirror-to-plasma distance, d, generally limits the single-pass beam fraction which can be re-injected in the plasma. For the present experiment ρ is of 13 cm. On the other hand, seeing the 100 ps duration of the single-pass XRL pulse, we fixed d at 0.9 cm for most of the shots. This was the smallest value for which adjustment and protection of the mirror remained manageable.

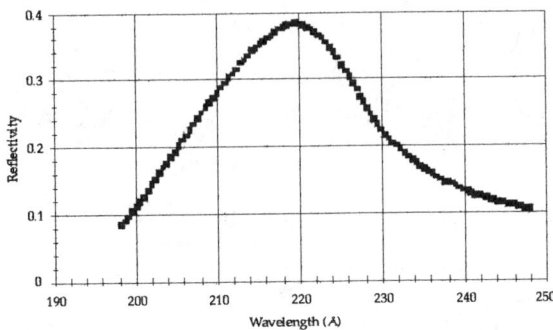

FIGURE 5: Reflectivity band of the spherical mirror [12]

Now, r being the transverse size of the amplifying region and α the divergence of the beam, the geometrical coupling between mirror and plasma is optimum when:

$$d \sim \rho - \frac{r}{\alpha} \qquad (5)$$

In the present case, $r/\alpha \sim 1 - 2$ cm according as vertical or horizontal divergence is considered. Thus the best coupling would require a mirror-to-plasma distance of 11 - 12 cm which is much too large for being conformable to the time allowed for beam returning to the plasma. From calculation, the reflected energy loss due to this difference is of 60 -70%, which reduces R_{eff} at about 10%. Thus the maximum gain factor, gl_{eff}, that the half-cavity can supply in our experimental conditions is ~ 17.

Results

The gain coefficient of the plasma being known from single-pass measurements, the effective gain factor produced in the half-cavity can be calculated from (4). As long as the intensity dependence versus the plasma length remains exponential, the R_{eff} value can be obtained by measuring the ratio I_2/I_1 between double-pass and single-pass intensities for a fixed plasma length l. R_{eff} is given by:

$$R_{eff} = \frac{I_2}{I_1} \exp(-gl) \qquad (6)$$

The value of l must obviously be chosen so that the X ray laser is NOT saturated. In fact gl must lie in the interval:

$$6 < gl < 10$$

where 6 is the lower validity limit of approximations involved in (6). Thus for the R_{eff} estimation we used a plasma length of 1.5 cm, for which gl ~ 7.

In order to minimize the timing change of the return beam through the plasma we kept constant the distance from the exit plasma end to the mirror, i.e. the quantity l+d. Therefore changing the plasma length from 2 cm to 1.5 cm was attended by changing d from 0.9 cm to 1.4 cm. The resulting change in the geometrical mirror-plasma coupling is obtained from the calculation model. One finds:

$$R_{eff}(d=0.9 \text{ cm}) \sim 1.6 \times R_{eff}(d=1.4 \text{ cm}) \qquad (7)$$

The typical value of I_2/I_1 observed for a plasma length of 1.5 cm and for $d = 1.4$ cm is 30. Then, for $d=0.9$ cm, (6) and) (7) give $R_{eff} = 4\%$ and it results from (4) that double-passing the 2-cm-long plasma will produce the gain factor:

$$gl_{eff} = 16.4$$

Now, from (6)and (7), increasing l from 1.5 to 2 cm should cause the increase of I_2/I_1 from 30 to 900. But the I_2/I_1 value which we measured was about 60. This indicates that the intensity increase has left off being exponential. The intensity is 900/60 =15 times smaller than expected from exponential increase. Thus we observe a gain decrease which results from intensity saturation.The results of the measurements are represented in figure 6 where the X-UV laser intensitiy is plotted versus the effective gain factor for single- as well as double-pass shots. The deviation from exponential increase at large gl_{eff} is clearly seen.

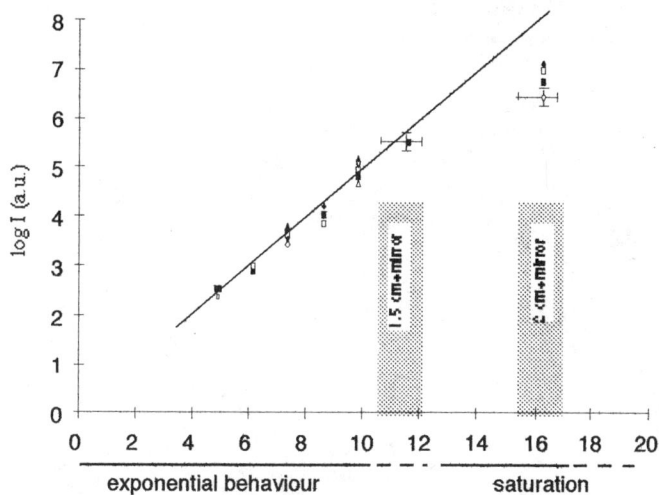

FIGURE 6: 21.2 nm emission intensity as a function of effective gain factor gl_{eff}.

In using the CCD camera we completed this study by the absolute measurement of the photon number output. We found that the X-UV laser supplies about $4 \cdot 10^{13}$ photons per pulse. The resulting energy is 400 μJ. The pulse length being of 80 ps in double-pass, the beam power is as high as 5 MW and the brightness,is of $3 \cdot 10^{15}$ W/cm^2/ster.

In conclusion the J=0-1 line of neonlike zinc appears a one of the most efficient transitions for an X-UV laser in the region of 20 nm. Further works are necessary to understand the effect of a tiny prepulse train to allow zinc plasma lasing. The pumping energy required for arriving at saturated emission is smaller than for lasers previously investigated. This is to the advantage of the compacity of the zinc laser.

References

[1] D.L. Matthews et al., Phys. Rev. Lett., **54** (1985) 110; J. Opt. Soc. Am., B **4** (1987) 575; see also X-ray Lasers by R.C. Elton, Acacemic Press, Boston, MA (1990).

[2] A. Carillon et al., Phys. Rev. Lett., **68**, (1992) 2917; P. Jaeglé et al., in X-ray Lasers 1990, G.J. Tallents ed., Institute of Physics Conference Series N° 116 (IOP Publishing Ltd Techno House, Redcliffe Way, Bristol, England), p.1.

[3] E.A McLean, J.A. Stamper, C.A. Brown, C.K. Manka, H.R. Griem, J.C. Rife, B.H. Ripin, in X-ray Lasers 1990, G.J. Tallents ed., Institute of Physics Conference Series N° 116 (IOP Publishing Ltd Techno House, Redcliffe Way, Bristol, England), p. 339.

[4] B. Rus et al, in this volume.

[5] B. Rus, A. Carillon, B. Gauthé, P. Goedtkindt, P. Jaeglé, G. Jamelot, A. Klisnick, A. Sureau, P. Zeitoun, J. Opt. Soc. Am. B, **11**.(1994) 564.

[6] A. Carillon et al., in Physics, Astrophysics and Synchrotron Radiation, R. Benattar et al. ed. Proc. Soc. Photo-Opt. Instrum. Eng. **1140**, (1989) 271.

[7] A. Carillon in LULI Annual Report 1992, (LULI, Ecole Polytechnique, 91128 Palaiseau Cedex, Fr.) p. 160.

[8] see for instance M.D. Rosen, et al. Phys Rev. Lett., **54** (1985) 106.

[9] J. Nielsen, J.C. Moreno, B.J. MacGowan, J.A. Koch, Appl. Phys. B **57** (1993) 309.

[10] G. Cairns et al., in this volume.

[11] S. Jacquemot, in this volume.

[12] Courtesy of T. Lucatorto, C. Tarrio and R.N. Watts, N.I.S.T., Gaithersburg (Md)

Soft X-Ray Amplification in a Plasma Waveguide

Y.Kato[1], R.Kodama[1], H.Daido[1], K.Murai[1], G.Yuan[1],
S.Ninomiya[1], D.Neely[2], A.MacPhee[2], C.L.S.Lewis[2], I.W.Choi[3],
C.H.Nam[3] and T.Kawachi[4]

1) Institute of Laser Engineering, Osaka University, Suita,Osaka 565, Japan
2) Department of Physics, Queen's University of Belfast, Belfast BT7 INN, UK
3) Department of Physics, Korean Advanced Institute for Science and Technology, Taejon, Korea
4) Faculty of Engineering, Kyoto University, Kyoto 606, Japan

Abstract. Narrow divergence soft x-ray laser has been generated by double-pass amplification in a curved slab target with a neon-like germanium laser. Considering the electron density profile with the curved slab target, the plasma acts as a one-dimensional waveguide to the x-ray laser beam. Beam parameters such as radius and wavefront curvature of the Gaussian beam are derived from the analysis on beam propagation in the waveguide. Possibility for generating a single-mode x-ray laser beam by multiple-pulse pumping of the curved target is discussed.

INTRODUCTION

X-ray laser beam propagation in amplifying plasmas is strongly affected by refraction due to density gradient, resulting in large beam divergence of collisional excitation x-ray lasers with planer exploding foil (1) and slab (2) targets. Improvement in the divergence has been achieved with the double slab target (3,4) which compensates refraction in a single slab targets. Prior to the double target, bending a planer target was proposed by Lunny (5) for refraction compensation. This configuration has not been tested in spite of its early proposal, until our recent experiment where it has been shown that the curved slab target is very effective in increasing the x-ray laser intensity and decreasing the beam divergence (6,7).

Considering the exponential profile of the electron density, the plasma in the curved target acts as a one-dimensional waveguide to the x-ray laser beam. Characteristics of the x-ray beam propagation in a medium having spatially dependent refractive index and gain profiles may be analyzed using the formula developed by Casperson (8) and applied to x-ray lasers by Fill (9). Application of these formula to the curved slab target provides us with the Gaussian beam parameters such as the beam stability, beam radius, wavefront curvature and effective gain. The analysis shows that the x-ray laser radiation is stably propagated in the waveguide resulting in a beam with small diameter and narrow divergence along the horizontal direction.

Although the plasma waveguide in the curved target is effective for generating a narrow divergence beam, it does not always lead to generating a spatially coherent beam since the waveguide is in general multi-mode. The high power coherent x-ray laser beam could be generated when a low power coherent beam is injected into an amplifying waveguide which does not disturb the coherence during propagation.

In this paper we briefly describe the experimental results obtained with the curved target, present the analysis on beam propagation in a plasma waveguide, and report on initial experiment on multiple-pulse pumping of the curved target which may lead to generation of coherent x-ray laser beam.

GERMANIUM X-RAY LASER WITH CURVED TARGETS

Germanium stripe was coated on a fused quartz substrate which was bent up to 5 mrad/cm with the target length varied up to 4 cm. A single laser beam of 1.053 μm wavelength and 1ns width was focus to a line of 100 μm average width and 6 cm length with the average peak intensity on the target of 1.7×10^{13} W/cm^2. A grazing incidence flat-field XUV spectrometer was placed on the x-ray laser axis with the spectral dispersion along the vertical direction. An x-ray streak camera with its horizontally-oriented slit cathode positioned on one of the x-ray laser lines monitored the time dependence of the angular intensity distribution. Either the J=0-1, 19.6 nm line or the J=2-1, 23.6 nm line was chosen in these experiments.

With the flat slab targets, the peak intensity of the 0-1 line was approximately 10% of the 2-1 line. With increase in the bending angle, increase in the intensity was more pronounced for the 0-1 line than the 2-1 line, reaching approximately to half of the 2-1 line at 5 mrad/cm. The peak of the 0-1 line was in the rising part (400 ps before the peak) of the pumping pulse with the pulse width at half maximum intensity of ~200ps. On the other hand, the 2-1 line had a longer duration (~800 ps half width) with the peak at 250 ps after the peak of the pumping pulse. The horizontal beam divergence decreased from 8 mrad at 0 mrad/cm to 4 mrad at 5 mrad/cm for both of the lines. These results show that the refraction compensation is more effective for the 0-1 line which favors the higher density, lower temperature region for the optimum gain in comparison to the 2-1 line and thereby is affected more strongly by the density gradient.

Double-pass amplification in a curved plasma was tested using either a flat, concave on convex multilayer soft x-ray mirror. The narrow divergence of slightly less than 1mrad has been achieved with the flat mirror for both the 0-1 and 2-1 lines. The beam direction was close to tangential to the target end. Figure 1 shows the angular intensity distributions of the 0-1 line at its temporal peak taken at

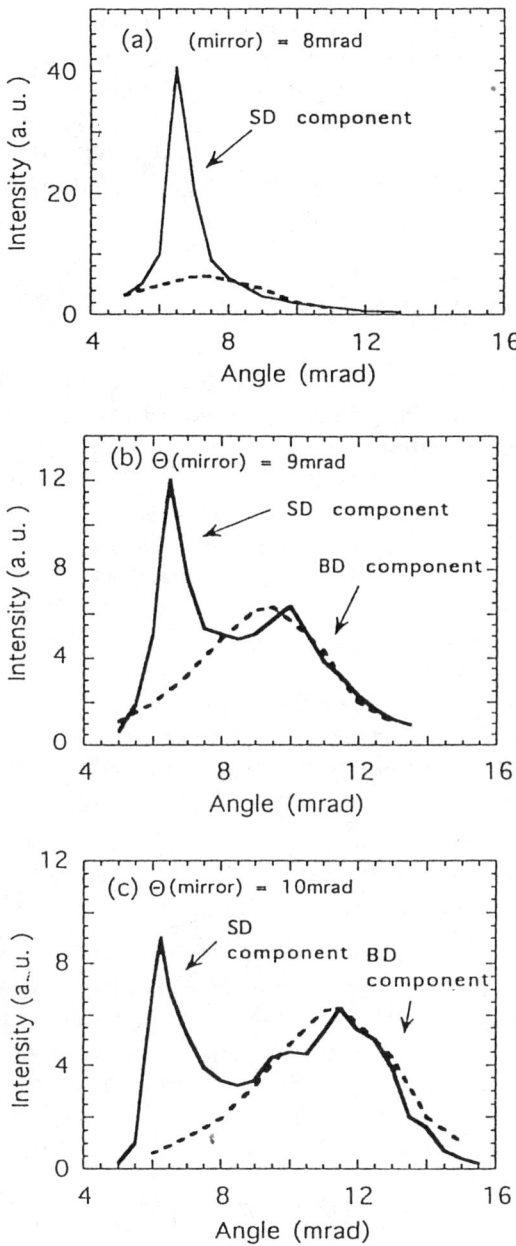

FIGURE 1 Angular distributions of the J=0-1 Ge laser line generated by double-pass amplification in curved slab targets obtained at the different mirror setting angles.

different mirror setting angles. In addition to small divergence (SD) component, a broader divergence (BD) component appears when the mirror is tilted from the optimum angle (8mrad). The BD component corresponds to the specular reflection of the double-passed signal by the plasma. On the other hand the direction and divergence of the SD component is not affected by the mirror angle although its intensity decreases off the optimum mirror angle, suggesting that this component corresponds to the guided mode in the plasma.

BEAM PROPAGATION IN THE PLASMA WAVEGUIDE

Propagation of a Gaussian beam through a homogeneous media with quadratic refractive index and gain profiles is described by simple formulae, derived from the Maxwell equations, for the complex parameters characterizing the Gaussian beam (8, 9). The quadratic profiles were determined from the 2D simulations applicable to our experimental condition (10). At the bending angle of 5 mrad/cm, the effective refractive index profile along the horizontal direction has the maximum at $\sim$120 μm from the target surface, whereas the refractive index has the minimum along the vertical direction. Since the gain maximum position is $\sim$50 μm closer to the target surface and does not coincide with the center of the waveguide, the effective gain for the guided mode becomes less than the maximum value. Although beam propagation is slightly affected by the presence of gain, the gain guiding is far weaker than the refractive wave guiding.

Figure 2 shows an example of the calculation, showing variations of the beam radii w_x and w_y along propagation in the plasma waveguide. The input beam has a beam radius of 160 μm and a diverging wavefront curvature of 8 cm simulating the double pass amplification with the curved target. It shows that the horizontal radius w_x decreases rapidly reaching to the matching value of $\sim$10 μm with the damping length of 30 cm, whereas the vertical radius w_y remains at a large value of $\sim$120 μm along propagation. At the propagation distance of 4 cm, which is still in a transient stage, the beam divergence is 2 mrad and 8 mrad along the x and y directions, respectively, in close agreement with the experiment. The beam is astigmatic due to the anisotropy of the refractive index profile. The astigmatism could be corrected with external optics.

POSSIBILITY FOR COHERENT BEAM GENERATION

For high power, coherent x-ray laser beam generation, a single mode beam has to be amplified in a plasma which preserves the coherence of the input beam. One of the possible arrangements is to use a curved slab target pumped with multiple laser pulses in order to generate and amplify a seed pulse, as shown in Fig.

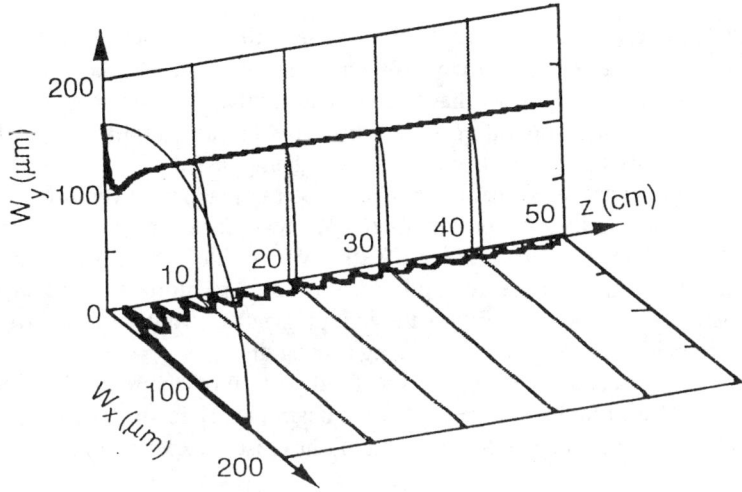

FIGURE 2 Variation of the beam radius along the horizontal (w_x) and vertical (w_y) directions with propagation in a waveguide formed with a curved target.

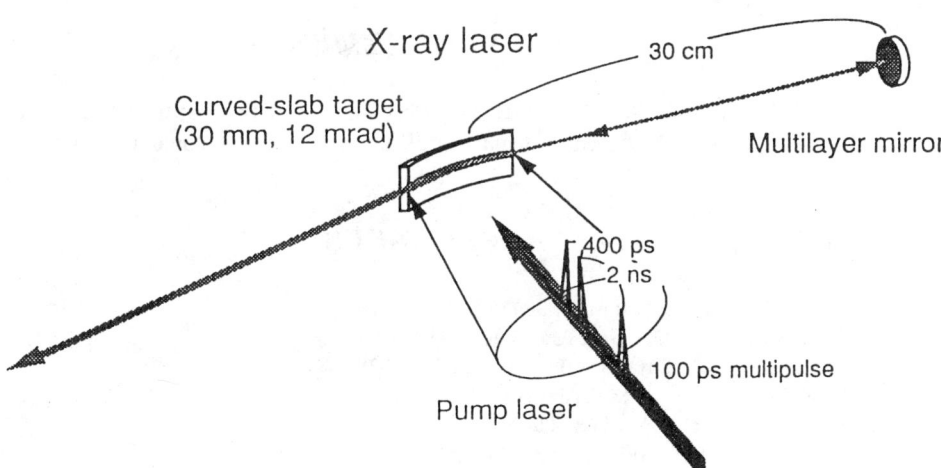

FIGURE 3 Double pass amplification with a curved target pumped with multiple pulses for generating coherent x-ray laser.

3. A spatially coherent beam could be injected into the amplifier when the seed pulse is propagated in a free space over ~60 cm. With the multiple, short pulse irradiation of the curved target, we have found that the Ge laser has a gain at least up to the pumping pulse separation of 2 ns. Double-pass amplification with a mirror placed at various distances from the target is being tested.

The important issue is the optical quality of the plasma waveguide. Since the quadratic profile is only an approximation to the real index profile, the beam has to be propagated only at the center of the waveguide where the quadratic approximation is valid. The waveguide is not uniform along the propagation direction when the line focus profile of the pumping pulse is not uniform over the length, causing spatial mode conversion along propagation.

Our initial measurement of the near field pattern of the x-ray laser beam under multiple pulse pumping indicated that the beam pattern is strongly influenced by the spatial variation of the line focus width in our present experiment.

CONCLUSION

We have demonstrated that higher intensity and narrower divergence x-ray laser can be generated with a curved slab target. The experimental results are explained, at least qualitatively, as the x-ray laser beam propagation in a plasma waveguide. Possibility for generating a high power, coherent x-ray laser using a curved slab target has been discussed.

ACKNOWLEDGEMENTS

We are indebted to Prof. G. Pert for sending manuscripts prior to publication, and to Dr. A.Wan for making 2D simulation results available to us.

REFERENCES

1. London, R. A., Phys. Fluids **31**, 184 (1988)
2. Murai, K., *et al.*, J. Opt. Soc. Am. B, to be published.
3. Carillon, A., *et al.*, Phys. Rev. Lett. **68**, 2917 (1992).
4. Wang, S., *et al.*, J. Opt. Soc. Am. B **9**, 360 (1992).
5. Lunny, J. G., Appl. Phys. Lett. **46**, 891 (1986).
6. Kato, Y., *et al.*, Proc. SPIE **2012**, 12 (1993).
7. Kodama, R., *et al.*, submitted for publication.
8. Casperson, L.W., Appl. Opt. **12**, 2434 (1973).
9. Fill, E. E., Opt. Commun. **67**, 441 (1988).
10. Wan, A., private comunication.

RECENT PROGRESS IN TABLE-TOP EUV LASERS AT MIT

P. L. Hagelstein, J. Goodberlet, M. Muendel, T. Savas, M. Fleury, S. Basu, and S. Kaushik

Massachusetts Institute of Technology
Research Laboratory of Electronics
Cambridge, Massachusetts 02139

1. Introduction

The goal of the research that we have pursued during the past several years is to develop relatively small-scale EUV lasers that would be useful for scientific applications in industry and academia.

Short wavelength lasers offer novel capabilities and research opportunities in the EUV. As lasers, they promise high brightness and good coherence, both longitudinal and transverse. As EUV radiation sources, they offer the potential of directed high power monochromatic EUV light. Based on highly stripped ions in laboratory plasmas, they provide an opportunity to develop an improved understanding of atomic physics, plasma physics and laser physics. In the university environment, EUV laser research provides a multidisciplinary topic, rich in applied physics, that challenges students in preparation for professional technical careers.

In the near-term, applications of EUV lasers might include: EUV interferometry as a diagnostic tool for dense plasmas, gases and thin films; EUV microscopy, both conventional and holographic; the development of novel holographic gratings; nonlinear optics in the EUV; EUV laser photoelectron spectroscopy for surface analysis; and the development of an EUV laser precision test facility for EUV optics. In the long-term, one might envision a future in which EUV lasers are used much as optical and UV lasers are currently used. Alignment with an EUV laser for precision vacuum applications could be considerably more accurate than with optical lasers. Micromachining might one day be done with an EUV laser.

Many of the important contributions to the science of short wavelength lasers has come from laboratories with access to multi-kilojoule pump lasers. It has become evident recently that significant research can be carried out on a much smaller scale. In this article, we review experiments carried out at a level of ~1 Joule per pulse. Recent work with femtosecond lasers has shown that quite interesting results can be obtained with even less total pump energy. It is clear that the coming decade will produce very significant developments in the field. Most of this progress can be expected to take place in Japan and European laboratories, as reductions in US funding in the area have forced most US laboratories to scale back or terminate their efforts.

2. Progress in Ni-like Nb

Electron collisional excitation schemes have proven to be successful in x-ray laser experiments at many laboratories. The nickel-like ions can give gain at a shorter wavelength than the neon-like ions at fixed pump intensity, although the highest gains observed in the neon-like ions are greater. We opted to pursue mid-Z Ni-like ions in our experiments at MIT, with the hope of seeing gain near 200 Å using a ~1 Joule per pulse pump source.[1-5]

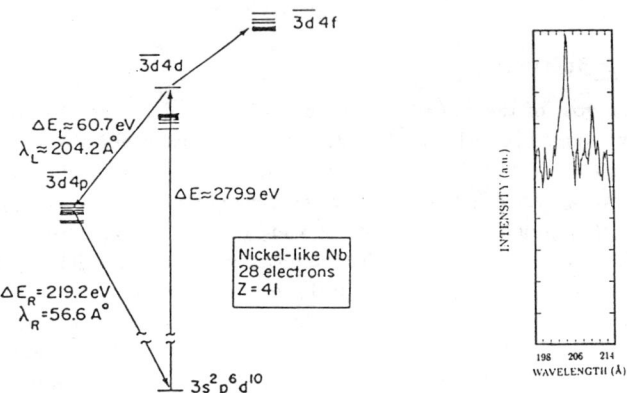

Figure 1: Laser scheme in Ni-like Nb.

The operating conditions for low-Z Ni-like ions can be determined from an optimization of the gain-length product[6-8]

$$\alpha L = \sigma_{SE} N^*$$

at fixed total power. To do this, a model is required to relate temperature to pump intensity. Matching the absorbed pump laser intensity with the flux-limited thermal heat conduction at critical density into a cold slab yields

$$I_{abs} = f N_c v_{th} kT$$

This model produces a dependence $T_e \sim I^{2/3}$, which ultimately leads to the optimization:

$$\frac{\partial}{\partial T_e}(\alpha L) = 0 \quad \longrightarrow \quad kT_e = \frac{1}{2}\Delta E$$

where ΔE is the $3d - 4d$ excitation energy; under the assumption that the fractional population of Ni-like ions is slowly varying in temperature near the optimum. The optimum comes about since the temperature must be sufficiently large to produce net excitation, but overheating the plasma is wasteful of pump energy. The optimum electron temperature for Ni-like Nb is $T_e = 140 eV$.

The Cu-like 4-4 lines are evident in our spectra, and we have observed Ni-like 3-4 emission, but we have so far no independent determination of the electron temperature.[9-11] The equivalent pump intensity estimated for these experiments is $I = (2\ \text{J})/(60\ \text{ps})\ (0.9\ \text{mm})\ (35\ \mu) \sim 10^{13}$ Watts/cm^2 (our linewidth on target is better than 20 μ, but energy is transported laterally by electron conduction). This number is higher by about a factor of two from the steady-state planar estimate used above, which is not unreasonable considering that the pump pulse duration is short compared to the observed emission time (1 ns) of the Ni-like Nb 204.2 Å 4d-4p J=0 line. Our picture is that the coronal temperature is established rapidly by the laser pulse, and that it subsequently cools relatively slowly, primarily due to hydrodynamic expansion.[11]

The optimum electron density can also be determined by maximizing the total gain, given a model for the population kinetics. We have described previously a simple analytical model, and we find[7]

$$\frac{\partial}{\partial N_e}(\alpha L) = 0 \quad \longrightarrow \quad N_e = \frac{A_u}{<\overline{\sigma v}>_u} f(\zeta)$$

where $f(\zeta)$ depends on ratios of radiative lifetimes and electron collisional rate coefficients, and is of order unity. This model also assumes that the fractional population of Ni-like ions is slowly varying in N_e. The optimum electron density for Ni-like Nb is $N_e = 8 \times 10^{18}$ cm^{-3} ($f(\zeta) = 2.5$).

We use multiple pulses, separated by 7.75 ns, to establish a shallow density gradient. No amplification is seen following the first pulse; there is evidence for some amplification (about 2.7 cm^{-1}) on the second pulse, as shown below in Figure 2. Data taken at longer length and higher energy seems to be consistent with $\alpha L \sim 3$. A retrospective examination of the data indicates that there is likely more gain after the third pulse, but this has not been quantified.

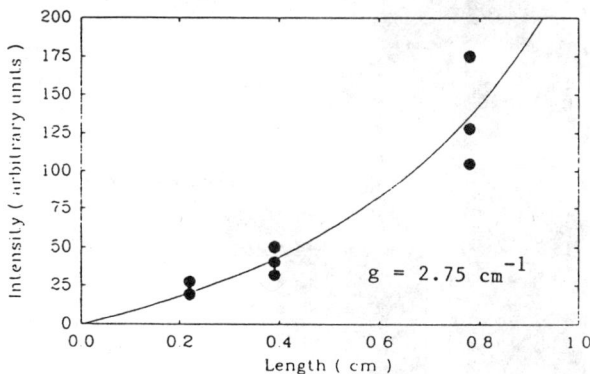

Figure 2: Intensity versus length experiment in Ni-like Nb at 204.2 Å (from Ref. 9).

The predicted gain under optimum conditions for this experiment is more than 10 cm^{-1}, and to date the observed gain is much lower than this value. To account for this we have considered some possible effects, including: (1) T_e being too low; (2) insufficient Ni-like population at time of peak gain; (3) uniformity of path length; (4) refraction; and (5) radiation trapping.

On theoretical grounds, the ionization time for the Cu-like sequence is fast, and we see Cu-like emission soon after the onset of the laser pulse. Delayed pulse experiments were attempted, in which a second beam irradiated the plasma with a several hundred psec delay from the second pulse; no significant increase in the 204.2 Å signal was detected.[11] There has been no evidence in support of the degradation of gain due to radiation trapping in Ne-like and Ni-like experiments. The third pulse results point to possible refraction problems, and this should be followed up.

We currently believe that a major source of the difficulties are associated with insufficient temperature. To address this, a pulse compression system is under development that will allow higher intensities by an order of magnitude (3 Joules/pulse at 10-15 ps). If it is so that the coronal temperature is set initially by the local intensity at the critical surface, then we should obtain a much higher starting temperature.

3. Progress in H-like B

Recombination schemes were recognized early on as having the potential for low pump energy requirements. Experiments in H-like and Li-like ions have indicated that gain can be produced at quite modest pump energies, but there has so far been difficulties with achieving a high total gain length product.

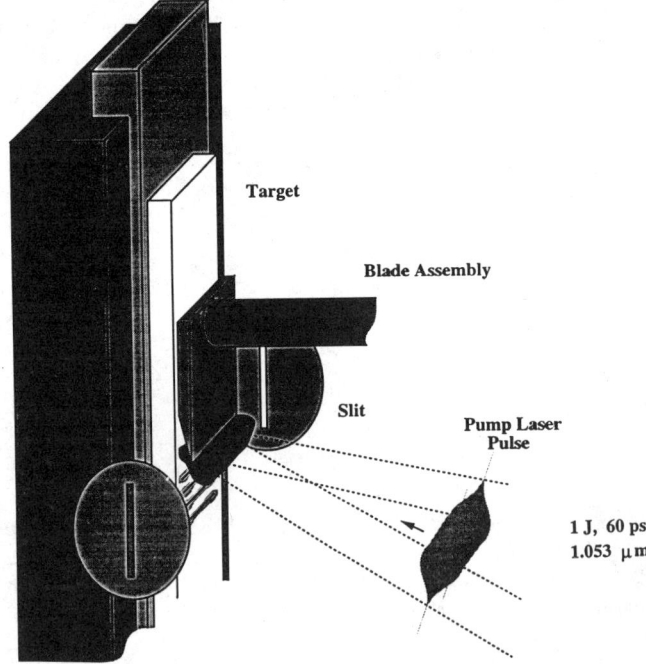

Figure 3: Laser target arrangement for H-like boron experiments.

Our initial work focused on recombination in H-like C, seeking gain at 182 Å. This effort was stimulated to some degree by the low-power results reported by the Princeton group; with encouragement from Professor Suckewer, we wanted to see whether gain was present under our pump conditions. Although we observed emission at 182 Å, it became clear that we did not have enough pump power to obtain significant gain. It was decided to try again at lower Z (boron). Some exponentiation on the analogous $3d - 2p$ transition in H-like B at 262 Å was observed.[12]

In these experiments, an incident 60 ps 700 mJ laser pulse irradiates a line less than 20 μ wide and 0.2 - 0.8 cm long. The plasma expands out approximately cylindrically, and cools predominantly due to expansion cooling. EUV emission is observed at 262 Å for more than 1 ns, and there is evidence for weak gain. A blade placed about 400 μ from the surface increases the local cooling rate (through conduction to the blade, which is a heat sink); this is shown in Figure 3. In this case, the gain is stronger and more reproducible, as shown in Figure 4. The best of these experiments produced signals consistent with a total gain length product $\alpha L \sim 4$.

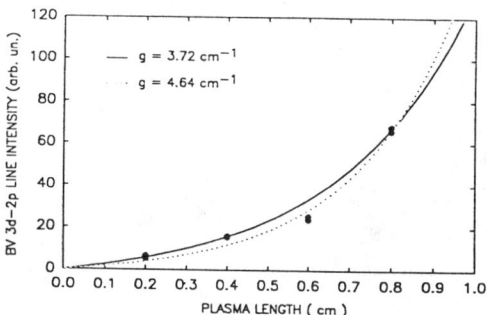 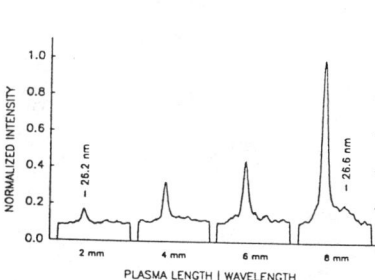

Figure 4: Intensity versus length for H-like B $3d - 2p$ emission at 262 Å (from Ref. 12).

4. Efficient End-pumped Lasers

The majority of x-ray laser experiments to date have been carried out so far using a side-pumped plasma. There are a number of reasons for this, some historical, some practical; ultimately, it is recognized that such an approach works. But it might be asked whether this is the optimum approach. For example, it may be that an end-pumped arrangement might be able to achieve a considerably higher conversion efficiency of pump radiation to EUV laser output.

Side-pumped lasers have so far achieved IR to EUV efficiencies on the order of 10^{-5}. Such a low efficiency comes about due to the concatenation of significant losses in absorption, hydrodynamic expansion, kinetics losses, and incomplete extraction. In the case of the collisional and recombination lasers that we have so far explored, there is a severe mismatch between the operating electron density and the critical density (assuming a 1 μ pump). We have considered improving the overall efficiency through end-pumping in a low density pre-expanded plasma. Under conditions of efficient EUV energy extraction, there is no fundamental reason why the conversion efficiency could not exceed 0.01.

We previously described pumping geometries that would permit re-use of the pump radiation.[8] Some preliminary experiments have been carried out on hollowed-out slab targets with a glancing-angle pump, as shown in Figure 5.

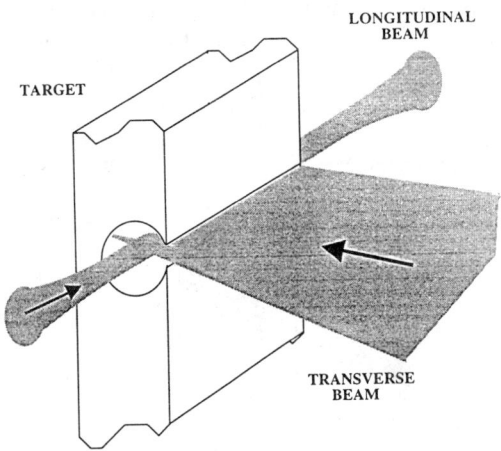

Figure 5: Glancing-angle pump arrangement for a hollowed-out slab target (J. Goodberlet, unpublished).

A new design for a gas target that would employ a cylindrical tube to implement a zig-zag pumping arrangement was studied. The basic idea was to bring the pump beam into a 2-5 cm tube focusing to a line focus in the gas on-axis at shallow angle. A cylindrical tube will collect the emergent beam and re-focus it on-axis further down the tube. Such an arrangement could produce a plasma in gas that could be more than 30 cm long, although rather stringent constraints on the precision of the cylindrical figure is required to do so.

Stimulated by the presentation at this conference by Professor Milchberg on channel confinement, it became clear that the combination of an axicon and tube would allow for the efficient production of a very long confining plasma channel. A cylindrical tube that is properly aligned with an axicon would re-focus the axicon beam again and again down the length of the tube. Such a configuration is illustrated in Figure 6. This type of system would have the potential for a very high IR to EUV conversion efficiency.

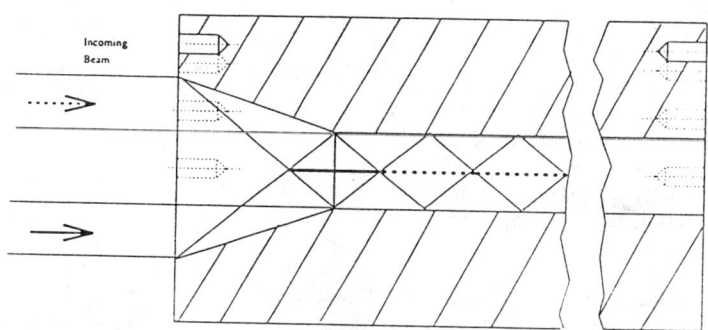

Figure 6: Axicon plus tube arrangement for channel-confining end-pumped gas laser (M. Fleury, unpublished).

5. Discussion

There are considerable advantages working with a small-scale EUV laser system, as a high shot rate affords the ability to perform optimizations that would be expensive on a large system. Additionally, if we succeed in obtaining higher gain and are able to generate a well-collimated beam, our system is well-posed for applications studies. The drawback to such a system has so far been that we have been relatively under-powered compared to the requirements of the collisional and recombination schemes tested so far. Scaling further down in Z in the Ni-like and H-like sequences is problematic; at lower Z in the Ni-like sequence, the optimum density goes quickly lower, and the gain lower still; in the H-like sequence the next stop is Be, which we prefer not to work with for safety reasons.

We are presently upgrading our system to increase the intensity available on target, and hopefully increase the number of gain lengths. Other recombination lasers in the Li-like and Na-like sequence may be better matched to our system. Accessible collisional lasers in the Ne-like sequence will be at longer wavelength; the Nd-like scheme yet remains untested. Perhaps the end-pumped systems will prove to be interesting. The issues are the subject of our current and future investigations at MIT.

Acknowledgements

This work was supported by grants from the MIT Lincoln Laboratory, DOE Basic Energy Sciences, the Lawrence Livermore National Laboratory (O-Group and R-Program), the Bose Foundation, Newport Corporation, and AT&T. PLH acknowledges support from the MIT EE&CS Department.

References

1. P. L. Hagelstein, "Short Wavelength Lasers: Something New, Something Old," *Proceedings of the OSA Conference on Short Wavelength Coherent Radiation: Generation and Applications*, Falmouth, MA, September 1988, 28-35.

2. M. Muendel and P. Hagelstein, "High Repetition Rate, Tabletop X-ray Lasers," *Proc. of OE-Lase 90: Optics, Electro-optics and Laser Applications in Science and Engineering*, January (1990).

3. P. L. Hagelstein, S. Basu, J. P. Braud, S. Kaushik, M. Muendel, D. Tauber, and A. Salas, "Status Report on the MIT EUV Laser Project," *Proc. of the Second International Conference on X-ray Lasers*," York, September, (1990); *Inst. Phys. Conf. Series No 116*, 255-262, (1990).

4. D. Tauber, S. Basu and P. L. Hagelstein, "An Alignment System for Solid Surface EUV Laser Targets," *Appl. Optics*, **30** 4337, (1991).

5. S. Basu, M. Muendel, J. Goodberlet, S. Kaushik and P.L. Hagelstein, "A Search for Gain at 191 Å," *Proc. SPIE 1991 Int. Symposium*, San Diego, CA (1991).

6. P. L. Hagelstein, "Development of the MIT tabletop soft X-ray laser," *Proc. SPIE 1991 Int. Symposium*, San Diego, CA (1991).

7. P.L. Hagelstein, "Short Wavelength Laser Studies at MIT: An Update," *Proc. SPIE Solid State Lasers III* **1627** 340 (1992).

8. P. L. Hagelstein, "XRL Research at MIT: Recent Progress," *3rd International Conference on XRLs*, Schliersee, Germany, May 1992.

9. S. Basu, P. L. Hagelstein, J. Goodberlet, M. Muendel and S. Kaushik, "Amplification in Ni-Like Nb at 204.2 Å Pumped by a Table-Top Laser," *Appl. Phys. B* **57** 303-307 (1993).

10. P. Hagelstein, M. Muendel, J. Goodberlet, S. Basu, and S. Kaushik, "Gain Observations in 204.2 Å in Ni-like Nb" *Proc. SPIE 1993 Int. Symposium*, San Diego, CA (1993).

11. M. Muendel, "Short Wavelength Laser Gain Studies in Plasmas Produced by a Small Nd:Glass Slab Laser," *MIT PhD Thesis, Physics Department* (1994).

12. J. Goodberlet, S. Basu, M. H. Muendel, S. Kaushik, T. Savas, M. Fleury, and P. L. Hagelstein, "Observation of Gain in a Recombining H-like Boron Plasma," submitted to *J. Opt. Soc. Am. B* (1994).

A Survey of the Theory of Recombination Lasers

G J Pert

*Department of Physics University of York, Heslington,
York YO1 5DD, England*

Abstract A review of theoretical studies of recombination lasers will be given identifying both the success and inadequacies of current attempts. The experimental difficulties associated with obtaining accurate data makes validation of the modelling uncertain in many cases. Nonetheless, at present good agreement is only possible with a limited set of the experimental configurations which have been studied.

The development of recombination lasers using short pulse laser pumping will be discussed, with particular reference to the problems of retaining the essential low electron temperature.

INTRODUCTION

Recombination lasers have been widely seen as an attractive route for high efficiency, short wavelength soft X-ray lasers. The underlying principles of their operation are conceptually simple and well understood. A relatively low atomic number atom is strongly ionised to a fully stripped or closed shell ion. As the ion recombines population inversions form in the resultant cascade from high-lying states to the ground state. However in detail the behaviour is significantly more complex, and in consequence the gain is generated as a compromise between effects which promote and destroy the inversion. In practice recombination lasers have to-date been disappointing, and the full realisation of their promise not achieved. Despite this some encouraging results have been obtained, and a practical device constructed.

Recombination lasers have essentially been considered by two very different approaches depending whether the ionisation processes is direct, or due to thermal processes in a hot plasma. At present nearly all experimental studies have investigated the latter approach, however there is now considerable interest in the former. We will discuss these two methods separately.

© 1994 American Institute of Physics

THERMAL IONISATION

Most current designs rely on thermal ionisation to generate the fully stripped ion. Since recombination must take place with a non-equilibrium ionisation system, ionisation must be followed by rapid cooling. A number of different methods have been proposed to accomplish this, with varying degrees of success. However before examining these in detail it is convenient to make some general comments regarding lasers of this type.

Recombination lasers are, in principle, attractive devices because of their favourable scaling with wavelength, and intrinsic efficiency[1,2]. This arises from the fact that they operate on transitions involving a change in the principle quantum number, n, and in essentially hydrogenic systems. Their design is conceptually simple, and in principle allows one to avoid many of the problems associated with collisional systems

Unfortunately there are many problems of detailed design, which are reflected in the several radically different schemes. These principally arise from the central compromise made in the concept, namely that a cold plasma be created hot. The consequent demand for rapid cooling inhibits the control of many important parameters in the design. Thus the outset of gain is controlled by two conditions[3]. Firstly that the upper state be collisionally pumped, whilst the lower state has a radiative decay - typically establishing an upper bound to the electron density of $2 \times 10^{14} Z^7$ cm^{-3} for an H_α transition in a hydrogenic ion of charge Z. Secondly that radiative decay of the lower laser state is not inhibited by self-absorption. In the usual case that the decay is to the ion ground state this places a limit on the population density of the lasant ion. Good design attempts to ensure that both conditions are simultaneously satisfied, in order to maximise the gain.

The experimental measurement of the performance of these devices is difficult due to a number of factors. The most important of these is the widespread inability to generate large gain-length products, and demonstrate saturation[4]. The reasons for this remain unexplained, but it has been observed with many dissimilar systems. As a result gain measurements have usually been made at Gl values of about 4, and may not be always reliable due to reproduc- ibility, alignment and high background (spontaneous) emission. Many of these problems arise from the experimental difficulty induced by systems required to satisfy the basic conflict of design.

a) Expansion Cooling

This simplest method of cooling is adiabatic resulting from the free expansion of the hot plasma. The expansion may be explosive or steady. The plasma is generally produced by the irradiation of a suitable target by an appropriate high power laser focused onto a line. In the <u>explosive</u> mode the laser pulse is short compared to the expansion time, and no heating occurs during the recombination phase of the laser. In the <u>steady mode</u> the laser irradiates the target continuously showing the expansion phase, the cooling occurs in fluid elements which are thermally isolated from the zone of heating, within the downstream rarefaction.

Explosion Mode

The explosive mode[3] has been more thoroughly analysed for both hydrogenic and lithium-like ions. Two types of targets have been proposed, fibres and foils. Although experimentally more difficult to use, fibres have been more successful, as the size, and therefore mass of the plasma is more carefully controlled by the transverse dimension of the target.

In this mode the target rapidly expands from a nearly fully stripped mode, rapidly cooling with relatively little recombination during the expansion. Radiative re-absorption of the decay transition is inhibited by the reduction in optical depth resulting from Doppler shifts induced by the differential expansion velocity across the plasma. The gain switches once the population inversion is established, limited by the constraints imposed by the electron density or the optical depth. Since the gain is proportional to density, the peak gain rapidly follows gain onset, falling thereafter with a duration determined by the expansion rate. As a consequence of the weak population inversion at peak gain, saturation intensities are relatively low, and spontaneous emission is not cut-off at saturation.

Some obvious criteria for this mode are that the expansion rate should be as rapid as possible but allowing ionisation to take place. This is favoured by short, but not ultra-short, pulses in the range of a few psec. Under these latter conditions high gains of up to about 50 cm^{-1} are predicted[6], which considerably ameliorate the experimental difficulties. Experiments using hydrogenic[7,8] and lithium-like[9] systems have been carried out using thin carbon fibres as the basis of the target, and with pulse lengths in the range of 2-100 psec. Good agreement between experiment and simulation, has been obtained, provided a complete description of the physics is attempted[10,11]. The development of this system based on fibres has been limited by the inability to generate targets of length greater than 1 cm. The use of short pulse drive may overcome this hurdle by enabling the generation of high gain lengths with short plasma length. However the use of short pulses may also introduce difficulties due to refraction in the larger density gradients and transit time effects.

Experiments have also been carried out using foil targets and narrow line foci, with varying success. In general the expansion is limited by a comparatively wide plasma line induced either by the focal spot or lateral transport. Simulations show that the expansion rate is more planar with a reduction in the temperature decrease rate and increased line trapping. Reasonable agreement between experiment and simulation was found for polymer foils at RAL, but the gain was too low for accurate analysis, or application (~1 cm^{-1})[12]. At Osaka agreement was obtained for the sodium H$_\alpha$ line at low (Gl), but not for the fluorine line[13]. In consequence the theoretical interpretation is uncertain.

Steady Mode

There is extensive experimental investigation of the steady mode of recombination laser in which the laser pulse is relatively long compared to the characteristic flow time. The pioneering work in this area is due to the French group at Orsay[5], who have studied a number of lithium-like ions, and has been followed elsewhere. The data is limited to low (Gl) ≤ 4, and attempts to scale to longer lengths have been unsuccessful for reasons which are not fully understood[4]. The theoretical position is also confused: extensive simulations have been carried out principally at Orsay[14], but also elsewhere, with two major problems. Firstly gains of the right order of magnitude can be obtained only if line trapping is neglected. Secondly the gain on the 3d-5f line in Al XI is observed to be more intense than the 3d-4f line, simulation predicts the latter line to be much the stronger. The overall theoretical position regarding these systems must be considered unsatisfactory, and may well be due to lack of a detailed description of a complex system, and this undoubtedly limits our ability to resolve the length scaling problem. However the possibility must be recognised that some of the data may be in error due to the low (Gℓ) values attained.

Two methods of alleviating the trapping problem have been investigated. Jaeglé and co-workers have used a higher lying lower laser state, in Li-like S, 4f-5g to allow a decay transition to a state (3d) which itself has a strong decay[15]. However the laser transition is of longer wavelength, substantially reducing the efficiency of the scheme. Alternatively it has been suggested that the inclusion of a light impurity reduces the density of the lasant, whilst maintaining the electron density and expansion rate, and may allow earlier onset of gain compensating for the loss of lasant ions[2]. Simulations show a modest increase of gain can be achieved if hydrogen is used as impurity.

b) Radiative Cooling

Radiative cooling involves the loss of energy from the plasma by direct radiative emission. It has been envisaged in two different contexts: directly to cool a stationary plasma[16], and as an adjunct to improve the heat loss in an expanding system[17,18]. In the context of recombination lasers radiative loss suffers from a major deficiency. In most plasmas radiation is produced in an equilibrium situation in which collisional pumping sustains the radiative rate, the heat loss is therefore indirectly from the free electrons and lowers their temperature. In a recombining plasman the free-electrons are already cool, and cannot sustain collisional pumping, the radiation is produced during the recombination cascade, the energy is supplied from the potential energy of the ionic levels, and does not give rise to cooling of the free-electrons: in contrast the three body recombination gives rise to electron re-heating. In fact strong cooling may occur when the plasma is hot and dense, ie during the ionisation phase. This requires additional heat input, and is consequently disadvantaged. Simulations have been carried out in both configurations. Keane and Suckewer[16] showed that small gains of not exceeding 1 cm^{-1} could be generated in a stationary carbon plasma with aluminium coolant. Healy has confirmed this result[19]. Radiative cooling as an adjunct to expansion via added impurity ions has also been examined in detail by a number of workers with disappointing results[18,19]. The essential limitation is the one referred to above. If a heavy ion impurity is introduced with complex structure, radiative cooling may take place, but the increased overall mass reduces the expansion rate, and the heat loss at ionisation decreases the efficiency, making the overall scheme of little practical value.

Experimental results of sufficient detail to validate code predictions are lacking, and the theoretical support for schemes based on this approach is weak.

c) Direct Cooling

There are a number of other approaches to cooling in recombination lasers which have been studied experimentally. These include direct cooling on surfaces inserted in the plasma, and diffusion processes in magnetically confined plasmas[20,21]. These systems are extremely complex involving complicated flow regimes with heat transfer and population dynamics. In consequence they have not been fully modelled, and it would be inappropriate to comment on them here.

DIRECT IONISATION

Conventional recombination lasers are limited by two handicaps: the high initial electron temperature required for ionisation, and the trapping of the lower state decay transitions. Direct ionisation avoids both difficulties[22]. The first by generating the plasma by tunnel ionisation in gas using an ultra-short pulse high intensity laser, which, in theory, produces a very cold electron distribution if the beam is plane polarised. Secondly by using a transition to the ground state on a self-terminating line the trapping problem is avoided. There are two essential requirements for this approach, namely that the ion ground state (lower laser level) be completely stripped, and that the electron temperature be maintained low. The experimental and theoretical evidence indicates that the first condition can be satisfied. Electron heating remains a serious problem particularly at the densities required for significant gain. Heating can arise from a number of sources not yet well understood: inverse bremsstrahlung, stimulated Raman scattering, ponderomotive force arising from focal spot structure, elliptic polarisation and axial field components[23]. In addition the experimental configuration poses difficulties due to the length of focus required for gain and dispersion between pump and X-ray beam. Assuming none of these problems are more serious than anticipated Eder et al have shown that Li-like Ne is a promising candidate for laser action in this mode[23]. Recently experiments in H-like Li have given indications of gain at long wavelength generated in this mode[24].

CONCLUSIONS

Recombination lasers have generally proved disappointing in their performance in comparison to collisional devices. Gains have been experimentally measured, and theoretically calculated indicating that their basic theory is understood. However problems remain in describing steady systems. The principle difficulty remains the failure to scale to long lengths and achieve saturated behaviour with high output powers. Theoretical understanding at present is unable to fully account for these results.

Recent developments using short pulse pumping are encouraging as this is the natural mode for this approach. However uncertainties remain with both such schemes, and further experiments and understanding of fundamental theory is necessary before a clear indication of their potential is obtained.

ACKNOWLEDGMENTS

It is a pleasure to acknowledge many valuable discussions over many years with graduate students and fellows, and with colleagues in the UK X-ray laser consortium, which have contributed to the views expressed in this paper. This work is supported by SERC.

REFERENCES

1. Key, M.H., *Nature* (London) **316**, 314 (1985)
2. Pert, G.J., *Laser Plasma Interactions Vol 3*, Edinburgh: Scottish Universities Summer School Press, 1985, p315
3. Pert, G.J., *J Phys B* **9**, 3301 (1976); **12**, 2067 (1979); JOS A Pt B **4**, 602 (1987)
4. Jaeglé, P. et al, *X-ray Lasers*, Bristol, IoP Publishing, Conference Series 116, 1991, p43

Jamelot, G. *et al*, *X-ray Lasers*, Bristol, IoP Publishing, Conference Series 125, 1992, p89
5. Jamelot, G. *et al*, *J Phys B* **18**, 4647 (1985)
6. Zhang, J. and Key, M.H., *Appl Phys B* **58**, 13 (1994)
7. Chenais Popovics, C. *et al*, *Phys Rev Lett* **59**, 2161 (1987)
8. Tallents, G.J. *et al*, this conference
9. Carillon, A. *et al*, *J Phys B* **23**, 147 (1990)
10. Eder, D.C., *Phys Fluids B* **2**, 3086 (1990)
11. Borovskiy, A.V. *et al*, *J Phys B* **25**, 4991 (1992)
12. O'Neill, D.M. *et al Central Laser Facility Annual Report*, Rutherford Appleton Laboratory Report RAL-91-025, 1991, p7
13. Azuma, H. *et al*, *Optics Letters* **15**, 1011 (1990)
14. Klisnick, A. *et al*, *X-ray Lasers*, Bristol, IoP Publishing, Conference Series 116, 1991, p17
15. Zeitsun, P. *et al*, *X-ray Lasers*, Bristol, IoP Publishing, Conference Series 125, 1993, p129
    Jamelot, G. *et al*, *Ultrashort Wavelength Lasers II*, SPIE Proceedings, Vol 202, 1994, p2
16. Keane, C.J. and Suckewer, S., *JOSA Pt B* **8**, 201 (1991)
17. Nam, C.H. *et al*, *JOSA Pt B* **3**, 1199 (1986)
18. Epstein, R., *Phys Fluids B* **1**, 214 (1989)
    Thornhill, J.W. *et al*, *J Appl Phys* **68**, 33 (1990)
19. Healy, S.B. and Pert, G.J., this conference
20. Suckewer, S. *et al*, *Phys Rev Lett* **57**, 1004 (1986)
21. Suckewer, S., this conference
22. Corkum, P.B. *et al*, *Phys Rev Lett* **62**, 1259 (1989)
23. Eder, D.C. *et al*, *Phys Rev A* **45**, 6761 (1992)
24. Nagata, Y. *et al*, *Phys Rev Lett* **71**, 3774 (1993)

Progress Towards Large Gain-length Products on the Li-like Recombination Scheme

P. Zeitoun, G. Jamelot, A. Carillon, P. Goedtkindt, H. Guennou,
P. Jaeglé, A. Klisnick, C. Möller, B. Rus[#], A. Sureau

*Laboratoire de Spectroscopie Atomique et Ionique, URA 775 CNRS,
Université Paris-Sud, Bâtiment 350, 91405 Orsay Cedex, France*
[#]*on leave from Department of Gas Lasers, Institute of Physics, 18040 Prague 8, Czech Republic*

Abstract. Investigating possibilities of attaining large gain-length products on the recombination scheme using lithium-like ions, we have examined two approaches aimed at overcoming the problem of plasma non-uniformity susceptible to destroy gain by a number of processes. In the first approach we studied amplification on the transitions 5f-3d and 4f-3d in Li-like Al^{10+} plasma column produced by smoothing optics using lens arrays. Employing this device resulted in the gain holding up significantly longer than when no smoothing optics was used. Secondly, we have investigated numerically and experimentally the 5g-4f transition in Li-like S^{13+}, as the gain should be barely affected by the plasma nonuniformities. Encouraging results were obtained and their various aspects are discussed.

INTRODUCTION

The recombination scheme in Li-like ions is attractive for producing X-ray lasers owing to the relatively low pumping energy needed, but until now it has exhibited only small gain-length values, well short of saturation, in contrast to the collisional scheme. The high sensitivity of the gain value to the plasma parameters and the poor homogeneity of laser-produced plasma columns seem to be the most probable cause of the lethargy of amplification in recombining plasmas.

Population inversions on the transitions 4f-3d and 5f-3d depend upon the fast radiative de-excitation from the lower lasing level 3d to the near-ground state 2p. However, the inverse process of excitation 2p→3d, either radiative or collisional, may become very probable in a number of experimental situations and destroy population inversions. Previous simulations on 5f-3d and 4f-3d lines in Li-like aluminium (Al^{10+}) have shown that the plasma is strongly absorbing at the beginning of the interaction and becomes amplifying later. The time and duration of the absorbing period depend on the pumping flux. As a consequence, inhomogeneity of the plasma column due to the non-uniformity of pumping laser deposition may result in successive amplifying and absorbing plasma zones along the X-ray laser path, reducing dramatically the X-ray laser intensity.

The aim of the present work was to overcome the problems originating from pump laser non-uniformity. We performed firstly a theoretical study of 4f-3d and 5g-4f transitions of Li-like sulphur. Simulations were run using a 1.5D hydrocode (FILM) post-processed by detailed atomic calculations (POPEXC). The pump laser pulse duration was 600 ps, the wavelength 1.06 μm, and the focal line width 100 μm. Effect of pump laser non-uniformity was investigated by comparing spatial and temporal evolution of gain for three different laser intensities between 2.5 and 5×10^{12} Wcm^{-2}. Figure 1 shows the spatial distribution of gain of

© 1994 American Institute of Physics

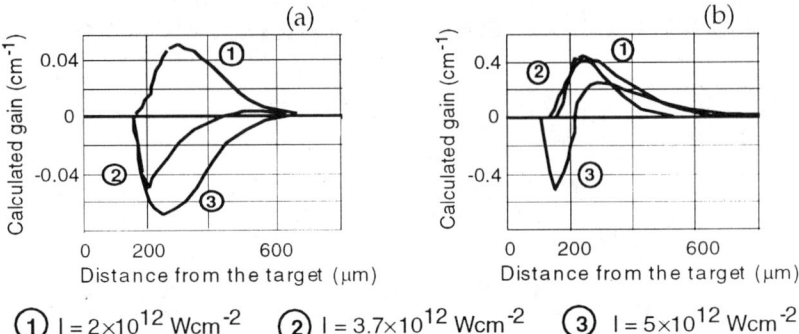

FIGURE 1. Spatial distribution of 4f-3d (a) and 5g-4f (b) gains in Li-like sulphur calculated by the 1.5D hydrocode FILM, post processed by atomic calculation code POPEXC.

transitions 4f-3d (resp. 5g-4f) at times t=1.8 ns (resp. 2 ns) after the peak of the driving laser pulse. These instants were chosen as 4f-3d (resp. 5g-4f) calculated gain reaches a maximum for the lowest intensity (2.5×10^{12} Wcm^{-2}).

We see in Figure 1a that plasmas produced by the two highest intensities are absorbing for the 4f-3d transition, whereas the lowest intensity gives rise to amplification. This suggests that hot parts of the plasma column may strongly reduce or even destroy the 4f-3d amplification. On the contrary, the 5g-4f transition exhibits positive gain for all the three intensities, at distances larger than 200 µm from the target, as shown by Figure 1b. Simulations performed at later times show positive gain for both lines and all intensities. Two important features are worthy of mention: (i) pump laser non-uniformity reduces the measured gain at earlier times, and (ii) 5g-4f amplification is much less affected by "hot" parts of the plasma column, even at earlier times.

Considering this analysis, two types of experiments were performed. Firstly, we have measured the 5f-3d and 4f-3d gain of an aluminium plasma column in which the homogeneity was improved by using a smoothing focusing device (a cylindrical lens array plus a spherical lens). Secondly, we have studied the 5g-4f transition in a sulphur plasma produced by a classical, non-smoothing device (2 crossed cylindrical lenses).

5F-3D AND 4F-3D GAIN IN A SMOOTHED PLASMA COLUMN

A massive slab aluminium target was used for the line focus smoothing experiments. Two laser beams were focused onto a 150 µm × 3 cm line focus by smoothing optics improving the energy deposition uniformity along the line focus. These consisted of a cylindrical lens array and a spherical convex lens[1]. The intensity contrast C=(I_{max} - I_{min})/(I_{max} + I_{min}) was thus improved from 40% to 20%. The 1.06 µm driving laser pulse was 600 ps in duration. The on-axis X-UV intensity was recorded using a Wadsworth spectrometer ($\Delta\lambda \approx 0.5$ Å) coupled to a streak camera. A vertical slit of 200 µm width concealed the most dense plasma zone from the spectrometer.

The gain history of the 5f-3d, 4f-3d, 5d-3p, 4d-3p lines was studied for 5 to 30 mm long plasma columns for a driving intensity of ~1.6×10^{12} Wcm^{-2}. All the

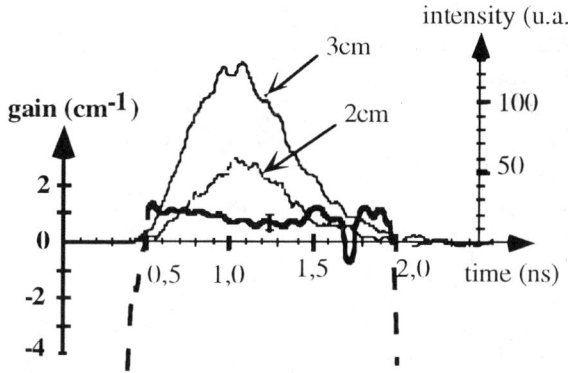

FIGURE 2 Temporal evolution of Al^{10+} 5f-3d intensity (plain) for 2 and 3 cm long plasma columns. The gain evolution (bold) was deduced by solving Equation 1. Employing the smoothing optics improves the plasma homogeneity and leads to a quasi-constant gain holding up throughout the line duration.

lines exhibited qualitatively similar temporal evolutions. Time variation of the gain was deduced from time resolved intensity measurements by solving at each instant the equation:

$$\frac{I_L}{I_\ell} = \left(\frac{e^{g_0(t).L}}{e^{g_0(t).\ell}} \right)^{3/2} \times \left(\frac{\ell.e^{g_0(t).\ell}}{L.e^{g_0(t).L}} \right)^{1/2} \quad (1)$$

where g_0 is the gain value at time t, L and ℓ the lengths of the two considered plasma columns. Figure 2 exhibits typical results for the 5f-3d transition in the case of a smoothed pump laser. The gain value (bold line) deduced from intensities (plain lines) is almost constant throughout the duration of the line emission. The duration of gain is then ~1.5 ns. A temporal plateau of gain was not observed in previous experiments which were performed without smoothing devices[2]. In the latter, 4f-3d or 5f-3d gains were found to grow slowly from negative values up to their maximum near the end of the line emission (Figure 3).

This difference in the behaviour is probably due to the presence of hot spots in the earlier experiments, which were suppressed with the use of smoothed beams. To verify this hypothesis, we used a third laser beam to add an overintensity on a 5mm-long zone situated at the spectrometer-side end of the plasma column, in

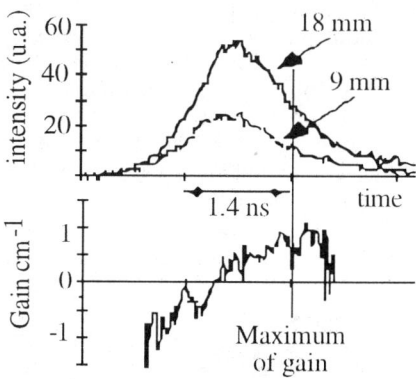

FIGURE 3. Intensity of the 5f-3d line for 9 and 18 mm long plasmas and temporal variation of the gain when no smoothing optics is used (from Ref 2); the experimental parameters were otherwise identical to those used in this work. The lack of gain in earlier times is due to the presence of hot plasma zones along the line focus.

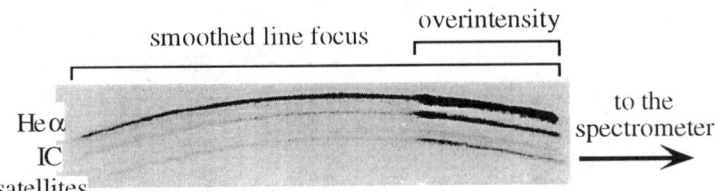

FIGURE 4. Spectrally resolved 1D image of the plasma column, irradiated by two smoothed beams over 3 cm and an extra beam over 5 mm at the right end of the column.

order to examine the absorption effects. The intensity contrast in this case is C~135%, compared to C~20% along the smoothed line focus. The plasma was monitored by a pinhole camera and by a crystal spectrometer working in the imaging mode. A spectrally resolved 1D image of the homogeneous plasma plus overintensity is shown in Figure 4. Time variations of 3p-5d line intensity and gain are plotted in Figure 5. The 5f-3d gain is negative during the first 500 ps of the emission, reaching its maximum ~100 ps later. Figures 3 and 5, both obtained when non-uniform line focii are used, show similar time variations of gain. This is in good agreement with the simulations mentioned earlier, suggesting the desynchronisation of gain for different local pump laser intensities.

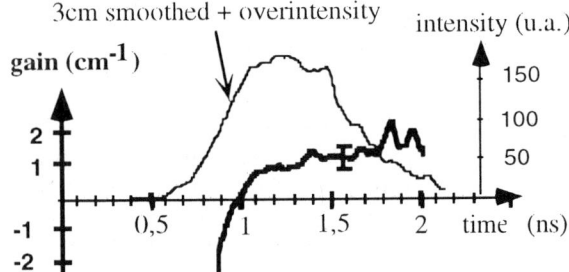

FIGURE 5. Time variation of the 5f-3d line intensity and gain. Effect of the controlled hot zone placed in front of the plasma is to turn the plasma into an absorbing medium during the earliest 500 ps.

5G-4F GAIN FOR A NON-UNIFORM LINE FOCUS

Two different experiments of 5g-4f measurements in S^{13+} are described here, differing in target composition. In both experiments, the pump laser was **not** smoothed and consequently the line focus was non-uniform. The mean value of the laser flux on the target was $~7.5\times 10^{12}$ Wcm^{-2}. In the first experiment, the 20 mm-long target consisted of a 7500 Å-thick sulphur layer deposited on an aluminium alloy slab substrate. Figure 6 shows the exponential growth of the 5g-4f line intensity with the plasma column length, yielding a gain value of 2.5 ± 0.7 cm^{-1}, and a g×L of ~5. We point out that this gain-length value is one of the largest ever measured using the Li-like recombination scheme.

Figure 7 shows intensity (plain line) and resulting gain (bold line) time variations of the 5g-4f transition. The plateau behaviour of gain history deduced from Equation 1 is consistently reproducible. For the X-ray laser lines terminating at the 3d level, such a plateau behaviour was observed exclusively in plasmas produced using a smoothing device. In the present case, of a non smoothed line focus, the constant gain during the whole 5g-4f emission gives clear experimental evidence of its low sensitivity to pump laser non-uniformity. This result is in agreement with the simulations of Figure 1.

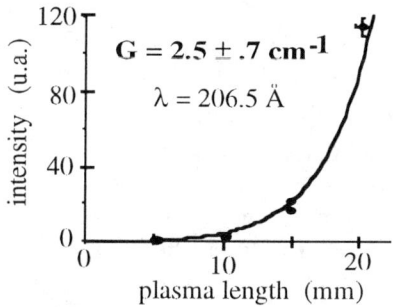

FIGURE 6. Recorded intensity of the 5g-4f line in Li-like sulphur. The Linford function fitted on the experimental points corresponds to a gain of 2.5±0.5 cm^{-1}.

A second experiment was performed in order to increase the gain×length on 5g-4f line. For an intensity of ~7.5×10^{12} Wcm^{-2}, the maximum plasma length accessible at LULI is 40 mm. All driving laser parameters remained unchanged. For technical reasons this experiment was performed using a massive sulphur (99.999% pure) 40 mm-long target. The measured gain of 5g-4f was only 1 ± 0.2 cm^{-1}, yielding a g×L~4, i.e. significantly lower than that obtained with a 7500 Å-thick layer (g~2.5 cm^{-1}). As the only modification of the experimental set-up was the nature of the target, we believe that a possible explanation of such a difference is the high sulphur-aluminium reactivity, which results into a layer of Al_2S_3 on the target surface. As the substrate was an alloy in the first experiment, the plasma in that case was composed of sulphur and aluminium, with traces of copper and magnesium, in contrast to the second experiment which used a pure sulphur plasma. The enhanced gain (by a factor 2.5) in the mixed plasma might be due to enhancement of radiative cooling by copper ions[3]. However, the effects of substrate-originating impurities on the gain need further investigation.

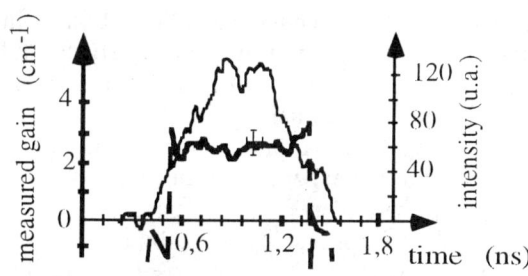

FIGURE 7. Time evolution of the 5g-4f line intensity (plain) and gain (bold) in Li-like sulphur, for a 2 cm-long plasma.

Acknowledgments : This work was partially supported by the DRET under contract n° 91-109. Authors acknowledge B. Gauthé and the Laboratoire des Cibles de l'Institut de Physique Nucléaire d'Orsay (Université Paris-Sud), who provided us with the thin-layer targets of sulphur. We also gratefully acknowledge many sparkling discussions with Phil Holden during preparation of this paper.

REFERENCES

[1] Carillon, A. , *Rapport LULI 1993*, (1994)
[2] Jamelot, G. , Carillon, A., Klisnick, A., and Jaeglé, P., *Appl. Phys. B*, **50**, 239-246 (1990)
[3] Moreno, J.C., Goldsmith, S., and Griem, H.R., *Phys. Rev.A*, **40**, 8, 4654 (1989)

Studies of ~ ps laser driven plasmas in line focus geometry

G J Tallents[*], Y Al-Hadithi[*], L Dwivedi[*], A Behjat[*], A Demir[*], M Holden[*], J Krishnan[*], J Zhang[+], M H Key[+,@], D Neely[@], P A Norreys[@], C L S Lewis[#] and A G MacPhee[#]

[*]Department of Physics, University of Essex, Colchester CO4 3 SQ, UK.
[+]Clarendon Laboratory, University of Oxford, Oxford OX1 3PU, UK.
[@]Central Laser Facility, Rutherford Appleton Laboratory, Chilton OX11 0QX, UK.
[#]Department of Pure and Applied Physics, Queen's University Belfast, Belfast BT7 1NN, UK.

Abstract. Measurements of X-ray emission along linear plasmas produced in short pulse (2 -12 ps) experiments using the Rutherford Appleton Laboratory glass (1.06 μm) and KrF (0.268 μm) lasers are interpreted to provide information about the uniformity and lateral and axial energy transport of X-ray laser gain media. For fibre targets, the difficulties of achieving uniform irradiation and accurate plasma length measurements are illustrated and discussed. For slab targets, it is shown that the ratio of the distance between the critical density surface and the ablation surface to the laser focal width controls lateral transport in a similar manner as for spot focus experiments.

Introduction

X-ray lasing is usually pumped by optical lasers focussed to a line. Long thin plasmas are formed by laser ablation of solid fibre, foil or slab targets and population inversions are produced by adiabatic expansion and cooling of the plasmas (recombination pumping) or by electron collisional excitation. X-ray amplification occurs along the plasma lengths. Although many X-ray laser observations have been made, there has been less experimental investigation into the plasmas *per se* produced in X-ray laser experiments. In this paper, we report on investigations of the uniformity of plasmas producing recombination X-ray lasing and present results from a study of the extent of lateral transport from line focus irradiation of slab targets by ~ ps lasers.

The plasmas produced by short pulse lasers are of topical importance in X-ray laser studies. Recently obtained record gain length products ~ 6 for the CVI n = 3 -2 transition pumped by recombination in adiabatically cooling plasmas created by 2 ps driver laser pulses (1) indicate that the newly developed ~ ps driver lasers should be able to produce saturation with recombination lasers. The limit on plasma width with short pulse lasers has become important for collisionally pumped X-ray lasers because of the need to reduce the required optical laser pump energy (2).

Line focus uniformity and length

In a recent experiment (1), 2 ps, 20 TW laser pulses at 1.053 μm wavelength were generated using the Nd:glass laser VULCAN at the Rutherford Appleton Laboratory by chirped pulse amplification and used to drive recombination X-ray lasing on the CV1 n = 3 - 2 (Balmer α) transition. An off-axis parabolic mirror reflected the beam to a spot focus which was imaged by an off-axis spherical mirror to produce a line focus 7 mm long and 20 μm wide. The targets were 7 μm carbon fibres supported at one end. The fibre targets were positioned with better than ± 2 μm spatial accuracy and ± 1 mrad angular accuracy using a modified split-field microscope system (3).

In this experiment, a flat field spectrometer (4) viewing down the fibre target axis recorded carbon Balmer α emission 70 times more intense from a 5 mm long plasma than from a 1 mm long plasma which implies a gain length product of 6.5. Length scans of the Balmer α emission (figure 1) can be fitted by a Linford formula with a gain coefficient of 12 cm^{-1}.

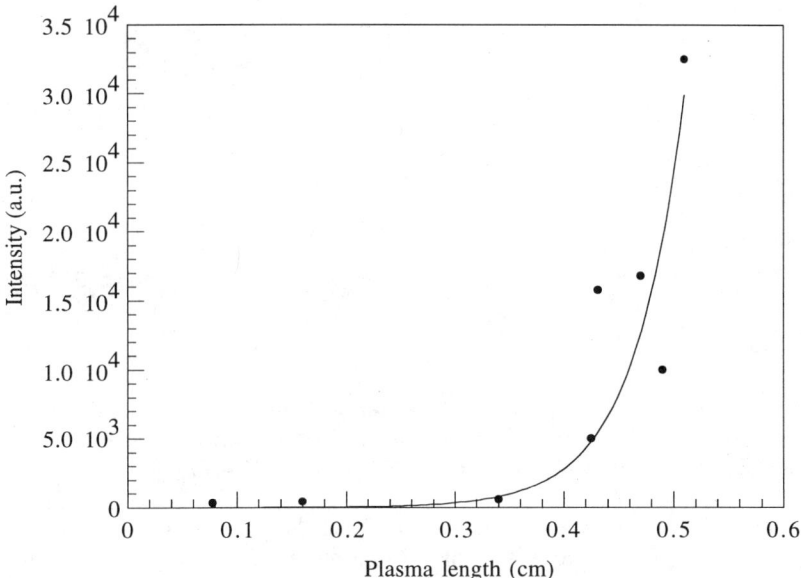

Figure 1. Measured intensity of CVI n = 3 -2 intensity as a function of plasma length (data points). The curve is a Linford equation fit to the experimental data points with a gain coefficient of 12 cm^{-1}.

Figure 1 illustrates that for accurate determinations of X-ray laser gain, measurements of the plasma length in the lasing direction need to be accurate. A pinhole camera filtered with 200 nm of aluminum and 50 nm of SiNi recorded on X-ray film the spatial variation of spectrally integrated CV and CVI resonance line emission along the target lengths (figure 2) and was used to determine the irradiated plasma lengths. A large diameter pinhole ($\approx$ 400 µm) was employed because of the small emission from the line plasmas. The optical densities on the film (Kodak 104-02) were converted to incident flux onto the film (5). The length was taken to be the full width half maximum (fwhm) of the pinhole camera recorded emission variation.

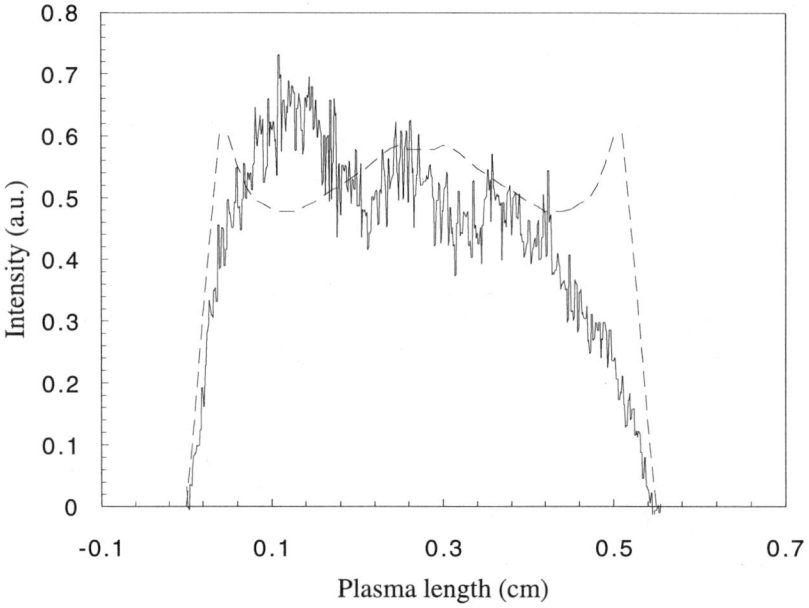

Figure 2. An example of the variation of X-ray emission along the plasma length from carbon fibre targets. The solid curve represents the experimentally observed intensity variation and the broken curve shows a calculated variation (see text).

To calibrate the length measurements, a mylar slab target of 4.2 mm length was irradiated and an emission length of 4.2 mm recorded by the pinhole camera. However, the pinhole camera was positioned at 45° to the fibre axis so a (cos45°) foreshortening of the observed plasma length combined with the effect of the spatial resolution needs to be allowed for in longer or shorter length measurements. For example (6), an actual plasma length of 0.77 mm would record as 1.1 mm (fwhm), while a 5.1 mm length plasma would record as 5.0 mm (fwhm).

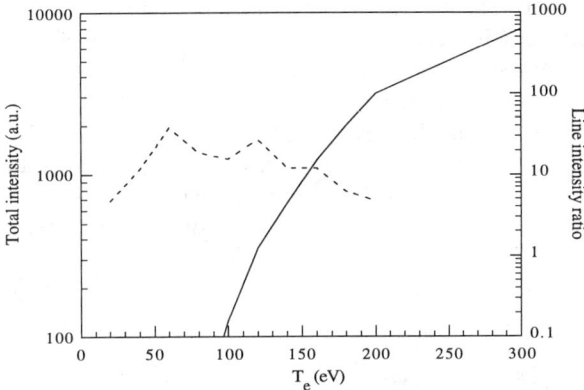

Figure 3. The intensity ratio of CVI 1s - 3p and CV $1s^2$ - 1s3p spectral lines (solid curve) and the total CVI and CV resonance line emission (broken curve) as a function of electron temperature calculated using the RATION code (8).

The spatially resolved X-ray emission recorded by the pinhole camera is not expected to be uniform along the fibre length because of the variation of focussed irradiance of the laser. Assuming a flat beam footprint, the variation of driving laser irradiance $I(x)$ along distance x of the line focus is given by

$$I(x) = I(0) [1 - (2x/L)^2]^{1/2} \qquad (1)$$

where L is the length of the line focus and x is the distance from the centre of the line focus. We can expect this variation of irradiance to produce a variation of peak electron temperature $T_e(x)$ according to the scaling of Fauquignon and Floux (7), viz.

$$T_e(x) = T_e(0) \, [I(x)/I(0)]^{2/3} \qquad (2)$$

where the electron temperature $T_e(0)$ at the centre of the line focus can be determined from the ratio of time and space averaged spectral line intensities recorded from a grazing incidence flat field spectrometer viewing the plasma at an angle of 30° to the fibre axis. The RATION code (8) was used to determine the emitted intensity ratio of CVI 1s - 3p and CV $1s^2$ - 1s3p spectral lines as a function of electron temperature (figure 3). Experimentally, this ratio was observed to be 1.25 which implies (figure 3) an electron temperature of 120 eV. The pinhole camera filtering ensured that the CVI and CV resonance lines exposed the film. The RATION code was used to calculate the total emission from these lines as function of electron temperature and so using equations (1) and (2), the variation of emission along the fibre axis was calculated. Finally, the instrument spatial resolution of 400 μm was convoluted into the calculated X-ray emission intensities and these intensites normalised in amplitude to the experimental values (figure 2). A reasonable agreement of the calculated and measured X-ray emission along the fibre length has been obtained. In particular, the three peak structure of X-ray emission is often seen in the experimental X-ray emission (figure 4). The increase in emission of the CV spectral lines with decreasing laser irradiance and, consequently, electron temperature towards the fibre ends produces the peaks at the fibre ends, while the increase in emission of the CVI line with electron temperature produces the central peak.

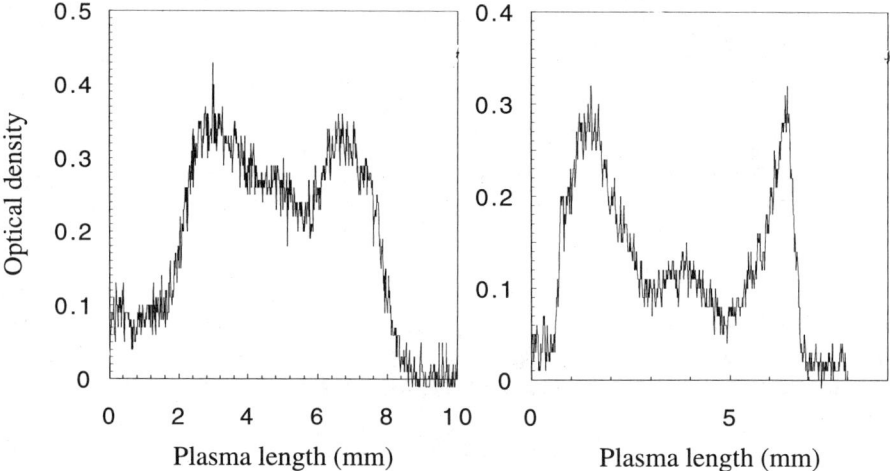

Figure 4. Densitometer traces of the emission along the plasma length from carbon fibre targets recorded by the pinhole camera.

Lateral transport

In X-ray laser research, a narrow line of plasma, not broadened by transport of the laser energy is often required. Lateral transport from a line plasma is of particular concern in recombination X-ray laser experiments where, for example, it is required to produce a narrow line plasma from a slab target so that rapid cooling and subsequent recombination-produced population inversions are created. Scaling studies (9) show that shorter pumping laser pulse lengths increase X-ray laser gain.

The limit on plasma width for collisional X-ray lasers has become important because of the need to reduce the required optical laser pump power by restricting the width of laser irradiation in order to obtain laser action with higher atomic number elements lasing close to the carbon K edge at 4.5 nm. For X-ray lasers to become useful for biological and other imaging, lasing close to the carbon K edge needs to be achieved with small to medium sized pumping laser facilities and this will require narrow line plasmas not broadened by energy transport.

The Rutherford Appleton Laboratory KrF pumped Raman laser with a pulse length of 12 ps, contrast ratio of 10^{10}, low beam divergence (3 X diffraction limited) and 0.3 TW power was focussed using an f/3 lens and random phase plate to produce line foci of 10 µm X 660 µm or 10 µm X 330 µm (2). Energy transport was studied using targets with a tracer layer of 0.25 µm thick sodium flouride buried below overlay plastic. The laser was focussed onto the overcoat and the penetration of the heat front was measured by recording the depth of overcoat for which the tracer layer can be observed to emit characteristic X-ray spectral lines or simply more emisssion due to its higher Z. The spatial distribution of the X-ray emission was measured using a pinhole camera (figure 5).

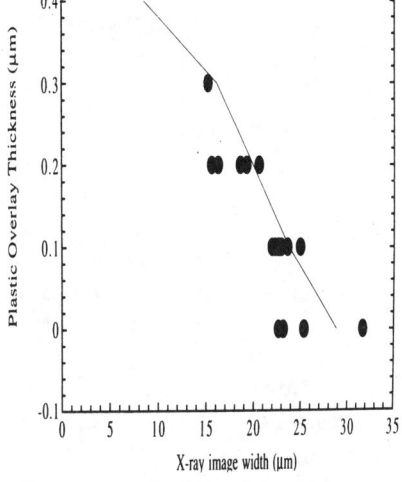

Figure 5. The experimentally measured width of the X-ray image for different plastic overlay thicknesses.

It has been proposed[1] that lateral transport is controlled by the ratio of the laser focus width (W) to the distance between the critical surface and the ablation surface (which we will label Δr). A larger ratio means the plasma is more one-dimensional in its expansion and transport dynamics and hence the lateral transport is less. The critical surface to ablation surface distance is difficult to measure as it is only approximately a micron in dimension (for the present experiments with 12 ps laser pulses) and requires probing or observation with X-ray radiation which can penetrate above the critical density. For the following analysis, we estimate Δr by multiplying the ion acoustic speed v_s by the laser pulse duration t, i.e.

$$\Delta r = v_s t \qquad (3)$$

The ion acoustic speed is given by

$$v_s = (kT_i/M_A m_h)^{1/2} \qquad (4)$$

where M_A is the mass number of the plasma material, m_h is the proton mass and T_i is the plasma ion temperature. The ion temperature T_i can be estimated from the work of Fauquignon and Floux (7) who first calculated the temperatures to be expected in laser-interactions when energy is deposited primarily at the critical density. Equating the absorbed irradiance fraction producing plasma expansion with the power per unit area outflow from the critical density region (α P v_s, where P is the plasma pressure), it can be shown following reference (7) that the ion temperatures T_i in ev are given by

$$T_i = 2.9 \times 10^{-7} \, I_a^{2/3} \, \lambda^{4/3} \, M_A^{1/3}$$

(5)

where I_a is the absorbed laser irradiance in Wcm^{-2} and the laser wavelength λ is in μm. Using this expression for T_i gives that

$$\Delta r = 0.11/(M_A)^{1/3} \, (I_a/10^{13})^{1/3} \, \lambda^{2/3} \, t$$
(μm)   (Wcm^{-2})   (μm) (ps)

(6)

A plot of W/Δr as a function of the X-ray emission width (determined from pinhole camera results, see figure 5) to laser focal width $W_{emission}$/W for the present experiment is shown in figure 6. Other data taken for spot focus geometry and published elsewhere is also shown for comparison. In spot focus geometry the plasma widths are the diameters across the laser or X-ray emission circular profiles. Figure 6 shows that lateral transport as evidenced by the value of $W_{emission}$/W increases with decreasing values of W/Δr. We can also see that the lateral transport for the line focus geometry is similar to that for the spot focus experiments.

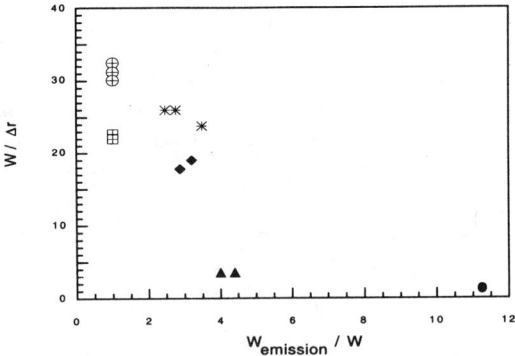

Figure 6. Focused laser line focus width W divided by the critical density to ablation surface distance Δr (see text) as a function of the width of X-ray emission $W_{emission}$ divided by the laser line focus width W. The results obtained in the present experiment with line foci of length 330 μm (♦) and 660 μm (∗) are shown, together with data for spot focus experiments as follows: • 40 ps pulselength, 0.268 μm wavelength, ref(10); ▲ 12 ps, 0.268 μm, ref (11); ⊞ 20 ps, 0.53 μm, ref(12); ⊕ 20 ps 0.35 μm, ref. (12).

Conclusion

Measurements of X-ray emission along the linear plasmas produced in short pulse (2 -12 ps) experiments using the Rutherford Appleton Laboratory glass (1.06 μm) and KrF (0.268 μm) lasers have been interpreted to provide information about the uniformity and lateral and axial energy transport of X-ray laser gain media. For fibre targets, the difficulties of achieving uniform irradiation and accurate plasma length measurements have been illustrated and discussed. For slab targets, it has been shown that the ratio of the distance between the critical density surface and the ablation surface to the laser focal width controls lateral transport in a similar manner as for spot focus experiments. The results are relevant to the topically important ~ ps laser-pumped X-ray laser experiments now producing significant gain length products by recombination (1) and potentially efficient short pulse collisional X-ray lasers (9).

Acknowledgements

It is a pleasure to acknowledge the target manufacturing and laser staff at the Rutherford Appleton Laboratory. The work was supported by research grants from the United Kingdom Science and Engineering Research Council.

References

1. Zhang, J., Norreys, P.A., Key, M.H., Tallents, G.J. et al, *these proceedings*.
2. Al-Hadithi, Y., Tallents, G.J., Zhang, J., Key, M.H., Norreys, P.A. and Kodama, R., Phys. Plasmas, **1**, 1279-86 (1994).
3. Grande, M., Key, M.H., Kiehn, G., Lewis, C.L.S., Pert, G.J., Ramsden, S.A., Regan, C., Rose, S.J., Smith, R., Tomie, T. and Willi, O. Optics Commun., **74**, 309 (1990).
4. Kita, T., Harada, N., Nakano, N. and Kuroda, H. Appl. Optics **22**, 512 (1983).
5. Krishnan, J., Neely, D., Danson, C., Dwivedi, L., Lewis, C.L.S. and Tallents, G.J., Rutherford Appleton Laboratory Annual Report **RAL-92-020** (1992).
6. Dwivedi, L., PhD thesis *unpublished* (1994).
7. Fauquignon, C. and Floux, F., Phys. Fluids **13**, 386 (1970).
8. Lee, R.W., Whitten, B.L. and Strout, R.E. J. Quant. Spectrosc. Radiat. Trans. **32**, 91 (1984).
9. Key, M.H., Tragin, N. and Rose, S.J., in *X-ray Lasers 1990*, edited by G. J. Tallents (Institute of Physics, Bristol, 1991), p. 163.
10. Tallents, G.J., Key, M.H., Norreys, P., Jacoby, J., Kodama, R., Baldis, H., Dunn, J. and Brown, D., Optics Commun. **89**, 410 (1992).
11. Tallents, G.J., Key, M.H., Norreys, P., Tragin, N., Kodama, R., Baldis, H., Dunn, J and Brown, D., Proc SPIE **1413**, 70 (1991).
12. Tallents, G.J., Key, M.H., Norreys, P., Brown, D., Dunn, J and Baldis, H., Phys. Rev. A**40**, 2857 (1989).

Recent Progress on Shorter Wavelength Recombination X-Ray Laser Research at SIOFM

Zhi-zhan Xu, Zheng-quan Zhang, Pin-zhong Fan, Ru-xin Li, Xiao-fang Wang, Pei-xiang Lu, Shen-sheng Han, Linqing Zhang, Bai-fei Shen, Wen-qi Zhang, Xian-ping Feng, Ai-di Qian, and Hui-zhu Xiang

Shanghai Institute of Optics and Fine Mechanics, Academia Sinica, P.O.Box 800-211, Shanghai 201800, P.R.China

Abstract. We will report recent progress on shorter wavelength recombination X-ray laser research at SIOFM, including the observation of lasing at 46.8Å of the lithium-like Ti^{19+} 4f-3d transition, which was conducted at LF12 laser facility of SIOFM with Ti slab targets driven by an about 110 ps-duration FWHM quasi-Gaussian pulse ($\lambda=1.05\mu m$). Comparison of long-pulse (~900ps) and short pulse(~110) driving Li-like Si and Ca ions and H-like F ion recombination X-ray lasers will also be presented.

INTRODUCTION

Recently, there has been considerable progress in the development of recombination X-ray lasers (1-11). The transitions, about which most of the investigations have been concerned, are 3-2 transitions in hydrogenic ions and 4f-3d and 5f-3d transitions in lithium-like ions and 5-4 and 6-4 transitions in Na-like ions. For recombination pumping scheme, the population inversion is achieved by fast radiative decay of lower level and recombination from the ground state of ions whose ionization stage are one more higher than the lasing ions. Accordingly, high density and fast cooling are desired. When scaling to shorter wavelength with higher atomic number element, the condition for lasing becomes more and more restricted. Fortunately, our previous study of Ca^{17+} 4f-3d lasing transition using 900ps (FWHM) 1.05μm pulse (11), indicated that additional cooling mechanism besides free expansion makes the plasma cool rapidly. The additional cooling may originate from heat conduction and radiation losses which becomes important in high Z plasma.

At Shanghai Institute of Optics and Fine Mechanics(SIOFM), we have successfully carried out the recombination soft X-ray laser gain experiments of the Li-like Al, Si, K and Ca ions (9-11). In this paper we present our results of shorter wavelength recombination X-ray lasers. To begin with, we present briefly the experimental conditions and diagnostics in the experiment. And then we will display the results on X-ray gain experiments with Ti slab targets driven by an about 110 ps-duration FWHM quasi-Gaussian pulse(1.05μm). The lasing transition is determined to be 4f-3d transition of Li-like Ti ions and to our knowledge the lasing wavelength (46.8Å) is the shortest X-ray laser wavelength in recombination pumping Li-like ions scheme. A previous paper (12) also claimed ASE behaviors of 4f-3d transition in Ti^{19+} ions accompanied by a lasing line at 326.5Å by Ne-like Ti ions, but the result is rather preliminary. We will present the spatial distribution of the gain coefficients of 4f-3d (46.8Å) transition of Li-like Ti ions and the temporal history of the lasing line emission. At the forth part, we will make a comparison of long-pulse (~900ps) and short pulse(~110) driving Li-like Si and Ca ions and H-like F ion recombination X-ray lasers. At the end the conclusion will be made.

EXPERIMENTAL ARRANGEMENT

The experiment was carried out at the LF12 laser facility of SIOFM. The experimental arrangement was essentially the same as before (9-11). In the experiment, only the north beam of this two-beam facility was used to produce line-shaped plasma as lasing medium. The energy of the 1.05μm laser beam was changed from 10 to 90J and the duration (FWHM) of the quasi-gaussian pulse varied in the range of 90 to 140ps. The beam was line-focused by a six-element cylindrical-lens array and an F/1.7 aspherical lens to form a 12.5mm × 120μm uniform focus on target surface, and the corresponding laser intensities on the target surface were about 6.5×10^{12} — $4.3 \times 10^{13} W/cm^2$. Two kinds of targets were adopted in the experiment (i.e. 5mm thick slabs with polished surfaces and strip-shaped target of 300 nm thick and 100μm wide, coated on the base of glass). Different lengths of targets up to 10 mm were adopted.

A flat field grazing incidence grating spectrograph(FFGIGS) (13) with a grazing incidence pre-optics was used to measure the on-axis spectrum of line-shaped plasmas. 1-D spatially resolved spectra were recorded on X-ray film or by a soft X-ray streak camera. All of the measured spectral intensity is relative one. In an independent experiment, the 1-D spatial resolution and spatial magnification of the FFGIGS were measured by using a stepped target. The resolution and spatial magnification were evaluated to be better than 35μm and 2.1, respectively, in the whole recorded spectral region. The SIOFM-5FW X-ray film (14) was used to record the time-integrated spectrum. We have calibrated this kind of film in the region of 50-80Å, the response characteristics of the film down to 46.8Å and up to 88.8Å were obtained by extrapolating. When the streak camera (with time resolution of 50ps) was used, a novel adjustment technique was taken to obtain space- and time-resolved spectrum (10). Because the spatial magnification of the FFGIGS was 2.1, the 100μm width of the scanning slit corresponds about 48μm width at the source.

At the another end of the plasma column, namely the opposite side of the FFGIGS, was positioned a 1m Rowland circle grazing incidence grating (2400l/mm) spectrograph for obtaining the axial spectra with higher resolution than FFGIGS. A pinhole transmission grating spectrograph was positioned 45 with the normal of the target(in the same horizontal plane as the axis of the driving laser), the diameter and line spacing of the pinhole grating being 25μm and 1/2000 mm, respectively, for monitoring the uniformity of the plasma along its axis (using the film recording) and the time history of the emission from the plasma (using the soft X-ray streak camera recording). Two flat-crystal X-ray spectrographs, one with a pinhole, also used to monitor the X-ray emission from the plasma column.

Normal methods were used to obtained relative intensity of spectral line. The evaluation of gain coefficients was performed by numerically matching the data to the Linford formula, and the error of gain were mainly caused by fluctuations in output parameters of the drive laser system from shot to shot and relative intensity measurement.

LI-LIKE TI ION X-RAY LASER

The energy of the driving laser beam was chosen to be about 71J for Li-like Ti ion x-ray laser experiment. The duration (FWHM) of the quasi-gaussian pulse was about 110ps and the corresponding laser intensity on the Ti target surface was about $4.3 \times 10^{13} \text{W/cm}^2$. We have measured the time-integrated but space-resolved gain coefficients.

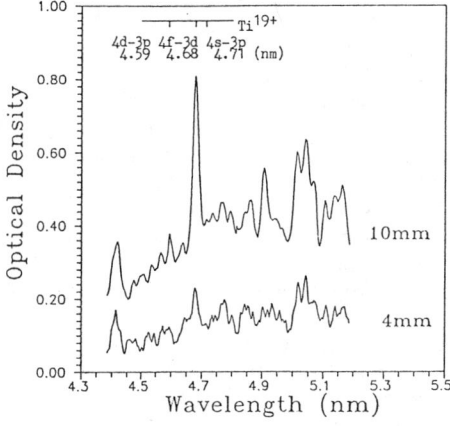
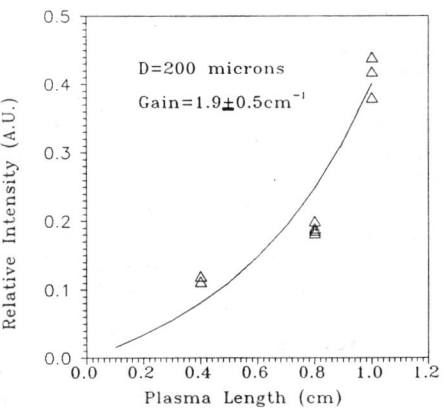

FIGURE 1. Spatially resolved on-axis spectra of Ti for 4 and 10mm plasma lengths at 200μm from target surface.

FIGURE 2. Curves of time-integrated axial intensities of Ti^{19+} 4f-3d emission as a function of plasma length at 200μm from target surface.

Time-integrated on-axis spectra are shown in Fig.1. The spectra correspond to two different lengths of line-shaped plasma, 4 and 10mm, at a position 200μm from the target surface. The spectral lines of 4-3 transitions of Li-like Ti ion are labeled in the figure and the wavelength measured agree well with the calculated wavelength using the Cowan Program (15). The nonlinear increase of the Li-like Ti ion 4f-3d line intensity with plasma length is clear. Curves of growth obtained from the spatially resolved, time-integrated axial emission at 200μm from the target surface are shown in Fig.2. The distribution of gain coefficients with regarding to the distance from target surface are plotted in Fig.3. Starting from the target surface, absorption region emerges first because of too high density and insufficient cooling, the amplification onset at about 125μm and reaches its peak value at 200μm, the gain coefficients gradually decreases with increasing distance from the target surface and amplification vanishes at about 300μm from the target

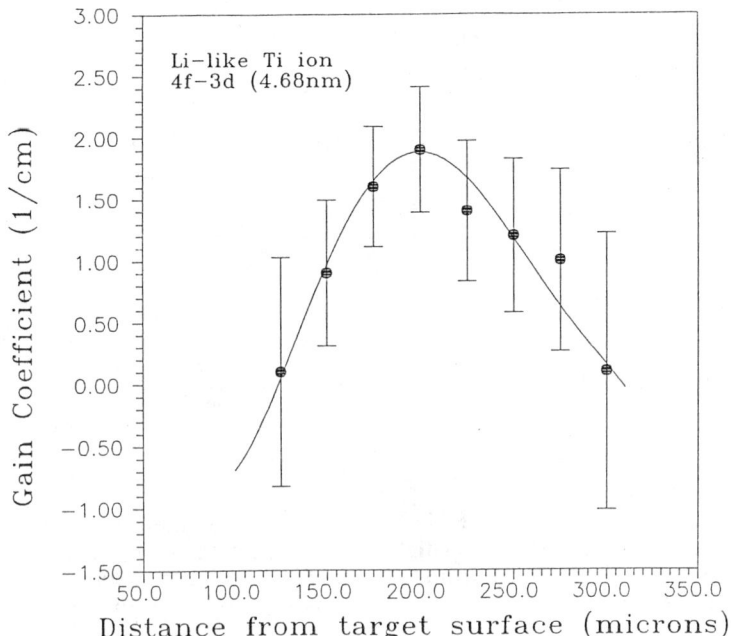

FIGURE 3. Gain coefficients of Li-like Ti 4f-3d transition as a function of distance from target surface.

surface. The highest gain was found to be 1.9±0.5 cm^{-1} at 200 μm from the target surface and the half-maximum width of the amplification region is deduced to be about 110μm. There may be two main reasons responsible for the relatively low peak gain coefficient. One is that the drive laser intensity on target is not optimal. In this experiment, we have not investigated the gain dependence on the laser intensity on target surface. With the intensity of 4.3×10^{13} W/cm^2 in this experiment, the spectral identification showed that many line emissions originated from those ions whose ionization stage are lower than Li-like ion in the axial spectrum, indicating a too low drive laser intensity which seems to be not enough to reach the optimal initial conditions of the plasma as a gain medium. Another reason is that the electron density of gain region in this experiment maybe too low. The optimal density is scaled as Z^7 for the recombination hydrogenic ion laser, there is a similar isoelectronic scaling of electron density for the recombination Li-like laser as $N_e \sim (Z-2)^7$. The optimal electron density for Li-like Al 4f-3d laser was determined to be of the order of 10^{19} cm^{-3}, so the density for optimal gain of Li-like Ti laser must be of the order of 10^{21} cm^{-3}. In this experiment, at the position of 200μm from target surface, the density is obviously not so high as 10^{21} cm^{-3}. So increasing the cooling rate to ensure the electron temperature at high density region to be low enough for lasing, is strongly desired. However, in order to understand the lasing mechanism involved, the detailed parameters of the lasing plasma need to be deduced by more diagnostics.

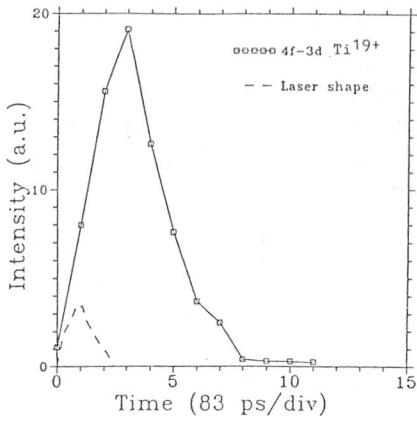
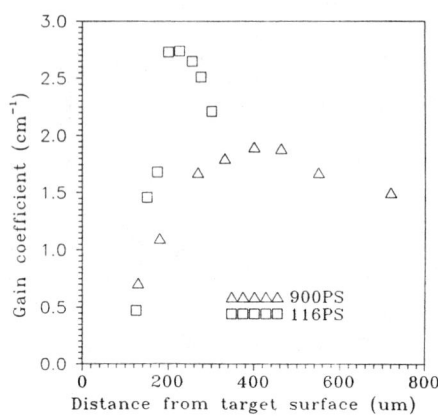

FIGURE 4. Temporal history of Ti^{19+} 4f-3d transition at the spatial region of 48μm around 170μm from the surface of a 6mm target.

FIGURE 5. Spatial dependence of the time-integrated gain coefficients of the Si^{11+} 5f-3d transition under two kinds of driving condition.

The temporal behavior of the lasing line emission obtained by streak camera is shown in Fig.4. The covering region of the slit of streak camera corresponds the region of 48μm around 170μm away from target surface. As can be seen, the delay of the Ti^{19+} 4f-3d emission relative to the drive laser pulse is about 170ps, and the duration of this transition is 290ps(FWHM). The time resolved gain coefficients measurements will be done in the future experiment.

It may be mentioned that, in order to increase the expanding cooling rate, we have tried to observe gain with strip target, whose parameters were mentioned in the second part of the paper, driven at the same laser intensity as slab targets. However, the on-axis intensity of the emission of 4f-3d transition was very weak because of the nonuniformity of the strip.

COMPARISON OF LONG-PULSE AND SHORT-PULSE DRIVING X-RAY LASERS

1. Li-Like Si Ion X-ray Laser

In the Li-like Si ion X-ray laser experiment, the energy of the driving laser beam and the duration (FWHM) of the quasi-gaussian pulse were about 11J and 116ps, respectively, and the corresponding laser intensity on the Si target surface was about $6.5 \times 10^{12} W/cm^2$. Only slab targets of Si were used in the experiment. We have not only measured the time-integrated space- resolved gain coefficients but also the time- and space- resolved ones.

The results of time-integrated measurements can be seen in Fig.5 which displays the distribution of gain coefficients of 5f-3d lasing line with regarding to the distance from target surface, also shown are the data of long-pulse (900ps) driving case (9). For the short-pulse driving case, the highest gain were found to be 2.7±0.9 cm^{-1} and 2.8±1.1 cm^{-1} for 5f-3d and 5d-3p lasing lines, respectively, locating at about 200 μm from the target surface. And the half-maximum width of the amplification region is deduced to be about 150μm. In comparison with the long-pulse driving, the position where the highest amplification locates is nearer to the target surface and the spatial region of amplification is narrower. Although in this experiment we have demonstrated slightly higher gain coefficients than that for long pulse driving, but the present results are still not ideal. The electron density at peak gain position was determined to be about 3×10^{19}cm^{-3} by measuring the Stark broadening. Obviously, the driving conditions must be optimized in the future work.

The results of time-resolved gain coefficients of 5f-3d lasing line were deduced from comparison of two plasma length. The highest gain coefficients of 1.9 cm^{-1} and 2.5 cm^{-1} were found for the 150μm and 250μm positions, respectively. The delay of gain peak to the peak of heating pulse was measured to be about 700ps.

2. H-Like F and Li-like Ca ions X-ray laser

In the experiment, the energy of the driving laser beam and the duration (FWHM) of the quasi-gaussian pulse were about 43J and 110ps, respectively, and the corresponding laser intensity on target surface was about 2.6×10^{13}W/cm^{2}. We used CaF2 slab target in the experiment. Only the time-integrated space-resolved gain coefficients have been measured.

The highest gain coefficient for Li-like Ca^{17+} 4f-3d transition was found to be 2.7±0.8 cm^{-1}, locating at about 100 μm from the target surface, and the half-maximum width of the amplification region is deduced to be about 100μm. Comparing with the results obtained by 900ps driving pulse, It is found that the position with the highest amplification is nearer to the target surface and the spatial region of amplification is narrower. On the other hand, the highest gain coefficient was found to be 3.5±0.7 cm^{-1} for H-like fluorine ion Hα transition, locating at about 160 μm from the target surface. And the half-maximum width of the amplification region is deduced to be about 300μm. Similar to the case of Li-like Calcium ion X-ray laser, the position of highest amplification is nearer to the target surface and the spatial region of amplification is narrower than the results obtained by 900ps driving pulse, however the gain of Ca^{17+} transition become lower while the gain of F^{8+} Hα transition become higher. In the experiment driven by a 900ps pulse, the highest gain coefficient for Li-like Ca^{17+} 4f-3d transition in CaF2 plasma was found to be 4.3±0.9 cm^{-1} when the laser intensity on target surface was 4×10^{13}W/cm^{2}, accompanied by a relatively low gain (1.4±0.3 cm^{-1}) of F^{8+} Hα (11). For the present experiment, the laser intensity may be too low for the Li-like Ca^{17+} 4f-3d transition to realize high gain, but the plasma condition is better for F^{8+} Hα lasing and Ca ions maybe play an important role in improving cooling rate by radiation and increasing the electron density of gain region.

CONCLUSION

The driving conditions and results are summarized in Table.1. In conclusion, we have demonstrated the soft X-ray amplification of Li-like Ti^{+19} ion 4f-3d transitions at 46.8Å, using a 110ps(FWHM)-duration laser pulse. We have also obtained the spatial distribution of the gain and the temporal history of the lasing line emission. It is worth noticing that the lasing wavelength is very near the "water window", so a cheap amplifier around the "water window" region using recombination pumping Li-like scheme is very promising. Although as mentioned above, the optimal electron density for Li-like Ti laser maybe of the order of 10^{21} cm^{-3} and may accompany a serious refraction effect, the experimental results reveals a gain region of relatively wide in space and long in time, which offers the possibility for multi-pass amplification, and then saturated amplification could be realized in gain region of moderate gain coefficient and less refraction.

TABLE 1. Summary of the driving conditions and experimental results.

Ion	Transition	Wavelength (A)	Target	Driving laser			Gain (cm^{-1})	GL
				λ (μm)	τ (ps)	$I(W/cm^2)$		
Ti^{19+}	4f-3d	46.8	slab	1.05	110	4.3×10^{13}	1.9 ± 0.5	1.9
Si^{11+}	5f-3d	88.8	slab	1.05	116	6.5×10^{12}	2.7 ± 0.9	2.7
	5d-3p	87.3					2.8 ± 1.1	2.8
Ca^{17+}	4f-3d	57.7	CaF_2 slab	1.05	110	2.6×10^{13}	2.7 ± 0.8	2.7
F^{8+}	3-2	81.0					3.5 ± 0.7	3.5

We have made comparisons of long-pulse (~900ps) and short pulse(~110) driving Li-like Si and Ca ions and H-like F ion recombination X-ray lasers. The preliminary experiments shows that slightly higher gain of Si^{11+} 5f-3d and 5d-3p transitions and F^{8+} Hα transition. However, as the gain region for short laser driving pulse is much narrower, a more uniform line-shaped plasma should be provided. It was found that nonuniformities still occur during laser heating even with uniform irradiation provided by a cylindrical lens array (16). Further work should be done to account for the temporal intensity profile of the laser beam to realize more uniform line-focused laser irradiation.

ACKNOWLEDGMENTS

The authors gratefully acknowledge the LF12 Laser Facility Operation Group for their assistance and cooperation in the experiment. This work was supported by the National Nature Science Foundation and the National High Technology Program.

REFERENCES

1. P. Jaegle, G. Jamelot, A Carillon, A. Klisnick, A. Sureau, and H. Guennou,

J.Opt. Soc. Am. **B4**, 563-574(1987).
2. G. Jamelot, P. Jaegle, A. Carillon, F. Gadi, B. Gauthe, H. Guennou, A. Klisnick, C. Moller, and A. Sureau, IEEE. Trans. on Plasma Sci. **16**, 497-504(1988).
3. C. L. S. Lewis, R. Corbett, D. O'Neil, C, Regan, S. Saadat, C. Chenais-Popovics, T. Tomie, J. Edwards, G. P. Kiehn, R. Smith, O. Willi, A. Carillon, H. Guennou, P. Jaegle, G. Jamelot, A. Klisnick, A. Sureau, M. Grande, C. Hooker, M. H. Key, S. J. Rose, I. N. Ross, P. T. Rumsby, G. J. Pert, and S. A. Ramsden, Plasm Phys. Controlled Fusion **30**, 35-44(1988).
4. P. R. Herman, T. Tachi, K. Shihoyama, H. Shiraga, and Y. Kato, IEEE. Trans. on Plasma Sci.**16**, 520-528(1988).
5. D. Kim, C. H. Skinner, A. Wouters, E. Valeo, D.Voorhees, and S. Suckewer, J. Opt. Soc. Am. **B6**, 115-125(1989).
6. J.C. Moreno, H. R. Griem, S. Goldsmith, and J. Knauer, Phys. Rev. **A39** 6033-6036(1989).
7. T. Hara, K. Ando, N. Kusakabe, H. Yashiro, and Y. Aoyagi, Japanese J. Appl. Phys. **28**, L1010-1021(1990).
8. C. J. Keane, N. M. Ceglio, B. J. MacGowan, D. L. Matthews, D. G. Nilson, and D. A. Whelan, J. Phys. **B22**, 3343-3362(1989).
9. Z. Z. Xu, Z. Q. Zhang, P. Z. Fan, S. S. Chen, L. H. Lin, P. X. Lu, X. P. Feng, X. F. Wang, J. Z. Zhou, and A. D. Qian, Appl. Phys. **B50**, 147-154(1990).
10. Z. Z. Xu, P. Z. Fan, Z. Q. Zhang, S. S. Chen, L. H. Lin, P. X. Lu, L. Sun, X. F. Wang, J. J. Yu, and A. D. Qian, *X-ray Lasers 1990*, ed. by G. J. Tallents (IOP, Hilger, Bristol, England, 1991), pp.151-158.
11. Z. Z. Xu, P. Z. Fan, L. H. Lin, Y. L. Li, X. F. Wang, P. X. Lu, R. X. Li, S. S. Han, L. Sun, A. D. Qian, B. F. Shen, Z. M. Jing, Z. Q. Zhang, and J. Z. Zhou, Appl. Phys. Lett., **63**, 1023-1025(1993).
12. T. Boehly, D. McCoy, M. Russotto, J. Wang, and B. Yaakobi, *Laser Interaction* and Related *Plasma Phenomena*, Vol.9, ed. by H.Hora and G.H.Miley (Plenum Press, New York, 1989), pp.185-196.
13. P. Z. Fan, Z. Q. Zhang, J. Z. Zhou, J. R. Jin, Z. Z. Xu, and X. Guo, Appl. Opt., **31**, 6720-6723(1992).
14. P. X. Lu, P. Z. Fan, Z. Z. Xu, R. X. Li, X. F. Wang, Y. L. Li, Z. Q. Zhang, and S. S. Chen, Rev. Sci. Instrum., **64**, 2879-2882(1993).
15. P.Yuan, Y.C.Wang, P.Z.Fan, Chinese J. At. Mol. Phys., **9**, 2174-2182(1992).
16. Z.Z.Xu, Z.Q.Zhang, P.Z.Fan, X.F.Wang, R.X.Li, P.X.Lu, L.Q.Zhang, A.D.Qian, C.H.Jian, S.S.Han and X.P.Feng, Appl. Phys. **B57**, 319-323(1993).

A Computational Investigation of Radiative Cooling in Freely Expanding Recombination Lasers.

SB Healy and GJ Pert
Department of Physics, University of York, York, Y01 5DD, UK.

Abstract

Simulations of freely expanding carbon and aluminium hydrogenic recombination lasers, doped with heavy ion impurities for increased radiative cooling, are presented. The results indicate a number of difficulties associated with these systems. Radiative losses during the pumping pulse reduce the overall efficiency of these systems. Expansion cooling rates are slower due to the increased inertia of the mixtures. In addition, the impurity is shown to reheat plasma during the expansion cooling. This is due to increased three-body and collisional de-excitation reheating, whilst increased radiative losses from the impurity are due to cascading electrons, which do not modify the free electron temperature.

Introduction

The general concepts of recombination lasers are well understood. In the simplest form, the hydrogenic recombination laser, a lasant material is irradiated with a high powered pumping laser forming a plasma of fully stripped ions and free electrons. If the plasma cools sufficiently rapidly, the free electrons preferentially populate the high lying Rydberg states due to three-body recombination. A population inversion is formed between the $n = 3$ and $n = 2$ levels as the upper is populated by a collisional cascade, whilst the lower level is depopulated by rapid radiative decay. The cooling rate is of extreme importance in these systems as the three-body recombination rate, R_3, scales with the electron temperature T_e as $R_3 \propto T_e^{-9/2}$. Hence, in general faster cooling leads to a larger population inversion and bigger gain.

Two distinct approaches to cool the plasma have been studied extensively. In the first, the plasma adiabatically expands and the free electrons cool as their thermal energy is converted into the kinetic energy of the expanding ions(1,2). In the second approach, the plasma is magnetically confined and the plasma cools through radiative losses which can be increased with the inclusion of cooling blades(3,4). The aim of this work is to investigate the possibility of increasing the cooling rate in freely expanding plasmas by doping them with heavy ion impurities in order to enhance the radiative losses. Although these systems have not been investigated experimentally, the observation of anomalously high gain on the 182Å line of hydrogenic from a plastic foil coated with selenium by Seely et al (5) was attributed to the selenium radiatively cooling the carbon. Subsequent modelling by Nam et al (6) appeared to confirm this possibility, but this was not reproduced in further work by Epstein (7) and Thornhill et al(8). In this work a number of possible heavy impurities have been investigated for both hydrogenic carbon and aluminium recombination lasers.

The mechanism of radiative cooling is usually visualised in the following way. If an ion is collisionally excited by an electron across an energy separation, ΔE, followed by a radiative decay, there is a net reduction in the free electron thermal energy, E_e, and electron temperature. This is given by

$$\Delta E_e = -\Delta E = -3/2 Z^* \Delta T_e \qquad (1)$$

where Z^* is the mean ionisation per ion. More generally, the plasma may be ionising or recombining and the change in free electron thermal energy over timestep Δt due to electron transitions is given by,

$$\Delta E_e = -(\Delta E_{ion} + W_{rad}\Delta t) \qquad (2)$$

where ΔE_{ion} is the change in potential energy stored in ionisation/ excitation and W_{rad} is the radiated power including bound-bound, free-bound and free-free transitions. This form includes the reduction in free electron thermal energy as the plasma is being ionised and correctly treats three-body recombination reheating. It also emphasises that radiative losses do not always cool the free electrons. If radiative losses are due to bound electrons cascading downwards,

$$\Delta E_{ion} = -W_{rad}\Delta t \qquad (3)$$

and there is no change in the free electron thermal energy.

The Model

The modelling of the hydrodynamics assumes a self similar expansion of a well mixed plasma of lasant and impurity ions. The number of ions is fixed and the impurity is seeded as a percentage of this number. The hydrogen and helium-like ion stages of the lasant are treated with the collisional radiative model. The remaining stages are treated with a simpler two level model which considers the ground and first excited state. A time dependent average atom approach, based on the model of Locke and Grassberger (9), is employed for the impurity ions. Additional radiative losses due to $\Delta n = 0$ transitions are calculated assuming a Fermi-Dirac population distribution within the subshells with the energy separations calculated with Perrots formulation of the screened hydrogenic model with l-splitting(10). The atomic physics and hydrodynamics are calculated self consistently through a solution of the free electron energy balance, subject to an energy limit (11).

Simulations of Carbon

impurity	$E_{tot}(ergs)$	$E_{rad1}(ergs)$	$E_{rad2}(ergs)$
$N(Z=7)$	3.7×10^7	5.5×10^5	1.4×10^6
$Al(Z=13)$	5.6×10^7	1.5×10^6	2.7×10^7
$Ge(Z=32)$	1.3×10^8	1.6×10^7	3.1×10^7
$Mo(Z=42)$	1.4×10^8	2.6×10^7	6.1×10^7
$Au(Z=79)$	4.7×10^8	1.7×10^8	3.0×10^8

Table 1: *Carbon energy requirements.*

These simulations refer to a fibre of radius $0.9\mu m$ doped with a number of heavy ion impurities, irradiated with a $70ps$ optical pumping pulse. A summary of the peak gains for 50/50 mixtures is shown in figure 1. It is apparent that the gain falls as the mass of the impurity ion increases and this is also found to be the case as the percentage of the impurity is increased. Table 1 shows the energy requirements of the 50/50 mixtures. E_{tot} is the energy required to optimise the peak gain, which is found to increase for the heavier impurities. E_{rad1} is the energy radiated before $100ps$, approximately when the electron temperature peaks. This increases with the mass of the impurity and largely accounts for the increased pumping requirements. The energy delivered by the pumping laser is being radiated out by

78 Investigation of Radiative Cooling

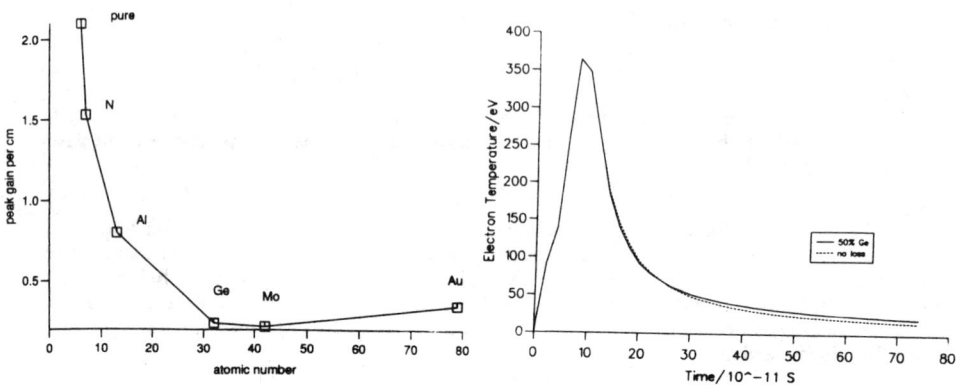

Figure 1: Optimised Peak gains for carbon.   Figure 2: Electron Temperatures.

the impurity, reducing the efficiency of the system. E_{rad2} refers to the total energy radiated before 0.5ns, approximately the time for peak gain. Despite increasing for the heavier impurities, this does not result in a faster cooling rate. Figure 3 shows the electron temperature profile for the 50 % Ge impurity. In order to assess the net effect of the impurity during the cooling, the calculation has been repeated with all atomic physics removed from the solution of the energy balance (referred to as 'no loss'). It apparent that removing the atomic physics energy components reduces the electron temperatures indicating the plasma is being reheated in the full simulation. The increased radiative losses are due to electrons which have recombined, radiatively cascading through the bound levels, but not modifying the electron temperature. An additional difficulty which should also be noted, is the slower expansion cooling which is a consequence of the increased inertia of the mixed plasma.

Aluminium Simulations

impurity	$E_{tot}(ergs)$	$E_{rad1}(ergs)$	$E_{rad2}(ergs)$
$Si(Z=14)$	2.6×10^8	1.9×10^7	2.6×10^7
$Ca(Z=20)$	3.6×10^8	3.6×10^7	4.3×10^7
$Ge(Z=32)$	6.7×10^8	1.2×10^8	1.6×10^8
$Mo(Z=42)$	9.2×10^8	2.8×10^8	3.5×10^8
$Sn(Z=50)$	1.2×10^9	3.9×10^8	4.8×10^8
$Au(Z=79)$	2.6×10^9	1.2×10^9	1.6×10^9

Table 2: *Aluminium energy requirements.*

Analogous simulations for a hydrogenic aluminium recombination laser have also been performed because these systems operate at much higher temperature and densities than the carbon. These results refer to a 1μm mixed fibre, irradiated with a 5ps optical pumping pulse. A summary of the optimised peak gains is shown in figure 3. The increase shown for the Si and Ca impurities is not due to radiative cooling. In these cases, the increase in gain is due to a reduction in trapping of the L_α line, depopulating the lower lasing level. However, in order to accurately assess the benefits of 'light impurities', relativistic splitting of the lasing levels must be considered (12,13). The energy requirements shown in table 2

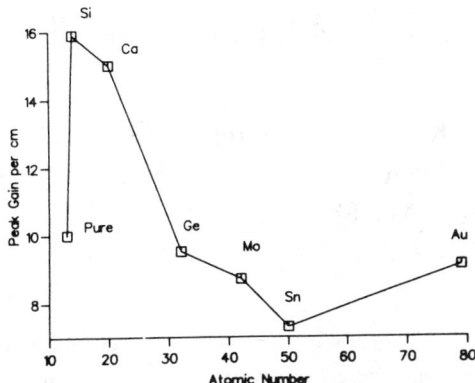

Figure 3: Summary of optimised peak gains for aluminium mixtures.

indicate trends analogous to the carbon simulations. More energy is required to heat the plasma as a significant fraction is radiated out during the pumping pulse. As with carbon, the simulations indicate that the mixed plasmas cool on a slower timescale than that due to the expansion cooling of a pure fibre.

Conclusions

The aim of this work has been to investigate the possibility of increasing the cooling rates in freely expanding recombination lasers by doping the lasant with heavy ion impurities. However, the simulations indicate that this is difficult for the following reasons. Firstly, the energy radiated out during the pumping pulse reduces the overall efficiency of the systems. The increased inertia of the mixture reduces the expansion cooling rate. In addition, when considering whether the impurity is cooling the plasma, the importance recombination reheating must be considered. Finally, it should be emphasised that radiative losses in a recombining plasma do not cool the free electrons as they result in a reduction of the potential energy stored in excitation/ionisation.

References

1. Pert, G.J., J.Phys. B 9 9 3301 (1976).
2. Jaeglé, P., et al., J.Opt.Soc.Am. B 4 563 (1987).
3. Suckewer, S., and Fishman, H., J.Appl.Phys. 51 1922 (1980).
4. Suckewer, S., et al., Phys.Rev.Lett. 55 1753 (1985).
5. Seely, J.F., et al., Opt.Comm. 54 289 (1985).
6. Nam, C.H., et al., J.Opt.Soc.Am. B 3 1199 (1986).
7. Epstein R., Phys.Fluids B 1 214 (1989).
8. Thornhill, J.W., et al., J Appl.Phys. 68 33 (1990).
9. Locke, W.A., and Grassberger, W.H., LLNL Report UCRL 52276 (1977).
10. Perrot, F., Physica Scripta 39 332 (1989).
11. Pert, G.J.,J.Comp.Phys. 39 251 (1981).
12. Borovskiy, A.V., et al., J.Phys.B 25 4991 (1992).
13. Eder, D.C.,Phys.Fluids B 2 3086 (1990).

Experiments of High Gain C VI X-Ray Lasing in Rapidly Recombining Plasmas

J Zhang[1], M H Key[1,2], P A Norreys[2], G J Tallents[3], A Behjat[3],
C Danson[2], A Demir[3], L Dwivedi[3], M Holden[3], P B Holden[4],
C L S Lewis[5], A G MacPhee[5], D Neely[2], G J Pert[4], S A Ramsden[4],
S J Rose[2], Y F Shao[6], O Thomas[2], F Walsh[2], Y L You[7]

[1]*Department of Atomic and Laser Physics, University of Oxford, Oxford, OX1 3PU, UK*
[2]*Central Laser Facility, Rutherford Appleton Laboratory, Chilton, OX11 0QX, UK*
[3]*Department of Physics, University of Essex, Colchester, CO4 3SQ, UK*
[4]*Department of Physics, University of York, York, YO1 5DD, UK*
[5]*Department of Pure and Applied Physics, Queen's University, Belfast, BT7 1NN, UK*
[6]*Scientific visitor from Institute of Applied Physics and Computational Mathematics, China*
[7]*Scientific visitor from Institute of Nuclear Physics and Chemistry, Chengdu, China*

Abstract. Recent experimental results of high gain C VI x-ray lasing in rapidly recombining plasmas are described. 7 μm diameter carbon fibre targets of up to 5 mm length were irradiated at intensities between $3 \times 10^{15} \sim 1 \times 10^{16}$ W/cm^2 by a 2 ps, 20 TW chirped pulse amplification (CPA) beam from the VULCAN Nd-glass laser ($\lambda = 1.053$ μm) at RAL. The gain length product on the 18.2 nm Balmer α transition of C VI ions was measured to be 6.5 ± 1. The ratio of intensities of resonance lines of H-like and He-like ions in the rapidly recombining plasmas was used as a useful diagnostic of initial conditions for high gain operation of the C VI recombination x-ray lasing.

1. INTRODUCTION

The quest for development of x-ray laser applications over the past few years has motivated research aimed at demonstrating saturated and diffraction limited laser operation in the water window spectral region in order to produce coherent sources suitable for applications in fields such as biological microscopy [1] and holography [2]. Adiabatically cooled recombination x-ray lasers are a promising scheme to achieve this because, in principle, they require much lower driving energy for the same lasing wavelength than collisional excited lasers [3-9]. A serious problem for recombination lasers, however, has been the failure so far to produce a sufficient gain length product by simply increasing target length to reach laser saturation where stimulated emission becomes a significant de-populating mechanism for the lasing population inversion.

One way to tackle this problem is to seek to increase the gain coefficient while remaining a relatively short target length [10]. Plasmas driven by picosecond laser pulses have a sufficient small scale length that heating occurs close to solid density, thus enhancing the volumetric expansion and adiabatic cooling of the plasma. Theory has predicted much higher gain coefficient on the C VI Balmer α transition at 18.2 nm when 7 µm diameter carbon fibre targets are irradiated by 2 picosecond laser pulses [10]. A newly available 2 psec, 20 TW chirped pulse amplification (CPA) beam of the VULCAN laser facility [11] ($\lambda=1.053$ µm) has offered the possibility of experimental verification of these ideas. High gain operation of the C VI recombination x-ray laser requires an electron temperature and density of $T_e \sim 10$ eV and $N_e \sim 1\times10^{19}$ cm^{-3} respectively during the peak of the population inversion. It is of great interest to study the conditions under which high gain recombination laser operate in order to optimise their performance. We report here experimental results of the first observation of a high gain C VI recombination x-ray laser driven by 2 psec laser pulses and also experimental measurements of absorbed energy, electron temperature, electron density and the fraction of fully stripped nuclei obtained by analysis of the ratio of the H-like to He-like resonance lines in the rapidly recombining plasmas.

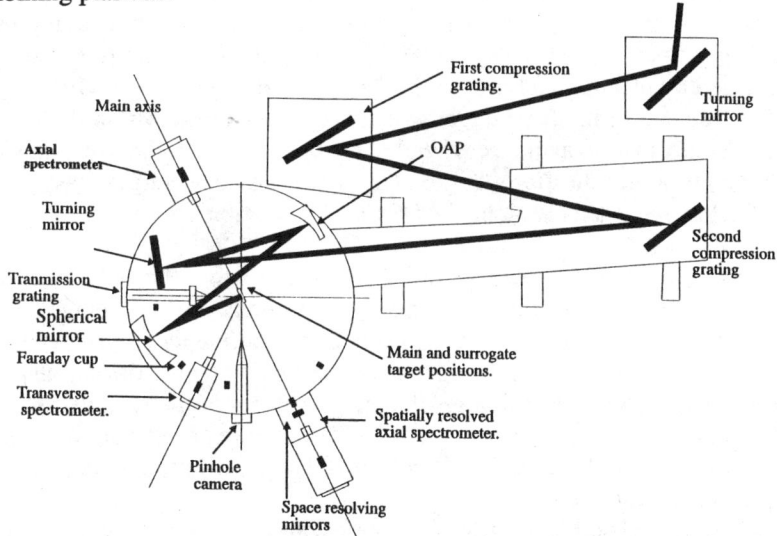

Figure 1. Experimental set-up showing various diagnostics in the experiment.

2. EXPERIMENTAL SET-UP

A schematic diagram of the experimental set-up is shown in figure 1. A 2 psec, 20 TW CPA laser beam at 1.053 µm was generated using the Nd:glass laser VULCAN which was equipped with a chirped pulse amplification (CPA) system [11]. The final pulse recompression grating was located inside the target chamber under vacuum. The recompressed beam of diameter 140x88 mm was incident onto a partially reflecting turning mirror which transmitted 5% of the recompressed pulse for measurements of the laser pulse length, prepulse (contrast ratio), spectral distribution and divergence. An off-axis parabolic mirror reflected the beam to a spot focus which was imaged by an off-axis spherical mirror to produce a line focus 7 mm long and 20 μm wide.

It is important to minimise pre-pulse and post-pulse to avoid pre-forming a plasma and post-heating of the plasma respectively. Pre-forming a plasma may reduce the laser gain and both pre- and post-pulses can be expected to increase the plasma x-ray emission which is not associated with x-ray lasing. For this experiment, a large background contrast ratio of between 10^6 to 10^7 was measured [11]. The pulse length was monitored during the experiment by an auto-correlator and was typically in the range of 2 ± 0.6 ps. 10 to 30 J of laser energy was incident on the line focus resulting in the spatially averaged incident irradiance in the line focus varying from 3×10^{15} W/cm^2 to 1.0×10^{16} W/cm^2, but only shots with laser irradiance in the range of $5 - 7 \times 10^{15}$ W/cm^2 are included in the data analysis.

The targets used in the experiment were 7 μm diameter carbon fibres supported at one end. The fibres remain straight for up to 1 cm. The fibre targets were positioned with better than ± 2 μm spatial accuracy and ± 1 mrad angular accuracy using a modified split-field microscope system [6,8]. The irradiation of the fibre targets was accomplished by placing the fibres in the 20 μm wide focus using a CW alignment laser. The free end of the fibre target was always placed well within the irradiated line focus length so as to avoid a cold end of the line plasma, which could absorb the x-ray radiation along the fibre length. The irradiated fibre length was varied by moving the line focus axially along the fibre so that the fibre free end position within the line focus varied.

The primary diagnostic viewing along the fibre axis was a flat-field spectrometer with a 1200 l/mm aperiodically ruled grating recording the spectral range 5.0 nm to 30.0 nm onto a KODAK 101-04 film. A grazing-incidence optical system consisting of two orthogonal cylindrical mirrors imaged the fibre end onto the entrance slit of the spectrometer to enable spatial resolution of ~30 μm of the recorded soft x-ray emission emitted from the target end [12]. Results presented here are spatially integrated as the spatial images produced show that the gain region is smaller than the 30 μm spatial resolution.

In order to infer absorbed energy in the plasma and to diagnose the characteristics of plasma which could effect the soft x-ray lasing performance during adiabatic expansion, a 2400 l/mm flat-field grating spectrometer was used at 45 degree off the fibre axis to record the Lyman series resonance Lines (np - $1s$) in the hydrogen-like ionisation stage and the $1snp$ - $1s^2$ resonance transitions in the helium-like ionisation stage in the spectral range of 2.0 nm - 10.0 nm. The intensity of an individual line was measured by subtracting the background continuum level and integrating over the line profile. The size and uniformity of the plasma were monitored using an X-ray pinhole camera. A group of Faraday cups were used to estimate the absorbed energy in the plasma. Using several cups, the angular variation of ion emission from the hot plasma was estimated and the Faraday cup signal converted to provide ion energy spectra, which were then integrated to give the total plasma expansion energy and hence an estimate of the absorbed laser energy.

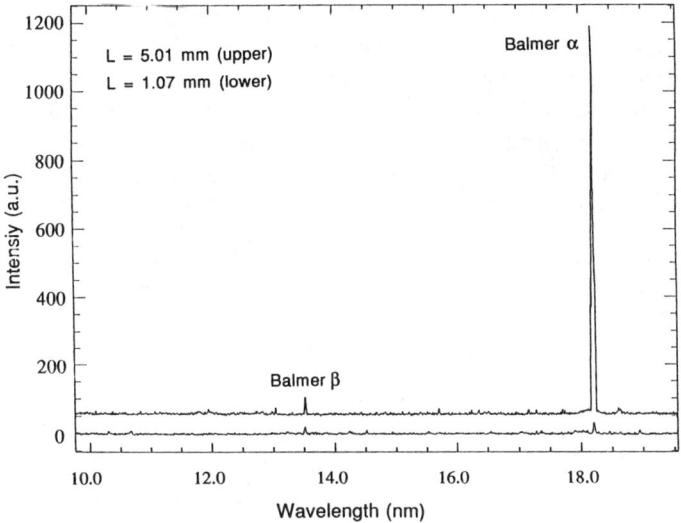

Figure 2. On-axis spectra for a 1.1 mm long carbon fibre plasma (lower spectrum) and a 5.0 mm long plasma (upper spectrum). The Balmer α line is 70 times more intense from the longer plasma which implies a gain length product of 6.5.

3. GAIN MEASUREMENTS

Two spectra from the on-axis flat-field spectrometer are shown in Fig. 2. The lower spectrum is from a 1 mm carbon fibre irradiated at an intensity of 5.6×10^{15} W/cm^2. It shows the major features of the spontaneous emission spectrum from carbon plasmas. Above it is the spectrum from a 5 mm long carbon fibre target irradiated at a similar intensity of 6.0×10^{15} W/cm^2. The Balmer α transition at 18.2 nm and Balmer β transition at 13.5 nm in the H-like ionisation stage are labelled on the figure. Comparison between the two spectra in figure 2 demonstrates the large enhancement in the on-axis signal for the Balmer α lasing transition as the plasma is lengthened. This is a strong indication of stimulated emission. During the experiment, a 1.6 μm thick Al foil filter was used near the film to differentially attenuate by a factor of 25 the strong Balmer α emission relative to the Balmer β emission in order to avoid film saturation of the Balmer α line. Such differential filtering allows an accurate measurement of the intensity of the Balmer α line relative to the Balmer β emission even though the intensity difference is large. The position of the filter midway between the Balmer α and β lines on the film could be seen in the continuum emission recorded by the spectrometer.

The photographic spectra were scanned by a digitising densitometer and the resultant film density values converted to intensity with a calibration curve [13]. The intensity of an individual line was then measured by subtracting the background continuum level and integrating spectrally and spatially (as discussed above) over the line emission. The results of these measurements are plotted in Fig. 3 for the Balmer α and Balmer β lines. The Balmer α data have been fitted by the Linford formula [14]. Least-square fitting for the data gives a gain value of 12.5 ± 1.5 cm^{-1}. This procedure underestimates the highest gain which may be present in the best spatial and temporal regions of the plasma. The uncertainty in the gain was calculated from the measuring error in the measurement of relative intensity values. The Balmer β line intensity increased linearly as the plasma length increased. This suggests the Balmer β line intensity is optically thin spontaneous emission.

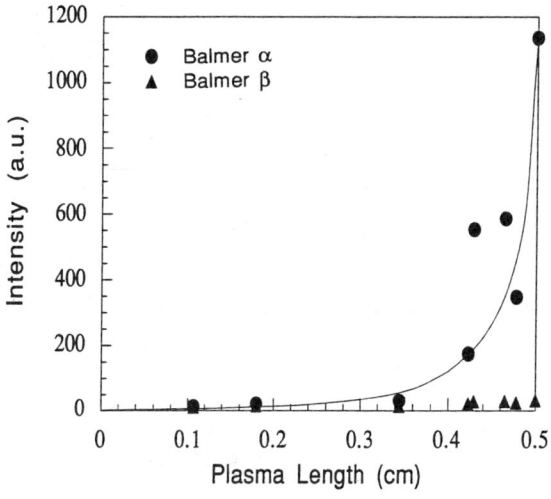

Figure 3. Intensity of the Balmer α lasing transition and the Balmer β spontaneous emission as a function of the plasma length. The solid dots represent the data for the Balmer α transition and the solid triangles the data for the Balmer β transition. The Balmer α data was fitted by a Linford formula with a gain coefficient of 12.5 cm^{-1}.

In order to obtain information on the uniformity, size and distribution of the gain region in the plasma, the exit plane of the recombining plasma was imaged onto an XUV sensitive phosphor that was coupled to a computer controlled CCD detector. A 250 mm focal length concave x-ray multi-layer mirror (central wavelength at 18.2 nm, bandwidth of 2.0 nm) [15] situated at a distance 320 mm axially from the fibre free end, imaged a plane from the exit face of the fibre target onto the phosphor/CCD detector at a magnification of 3.57. A 1.5 μm thick Al filter placed 15 mm in front of the CCD detector provided rejection of optical emission. Figure 4 shows the intensity distribution at the free end of the fibre target irradiated by the 2 ps driving laser beam. The peak intensity region has a horizontal and vertical FWHM (full width at half maximum) of 20 μm and 15 μm respectively and is approximately 100 μm from the target surface. Modelling [10, 16] suggests that the high gain region is

50 μm from the target surface over a 10 - 15 μm spatial range. However, the peak intensity region is likely to be slightly further from the target and larger because of refraction. The vertical section of the image shows a reasonable symmetrical profile, whilst the horizontal section shows a steep rise and a more gradual decay with distance away from the target surface.

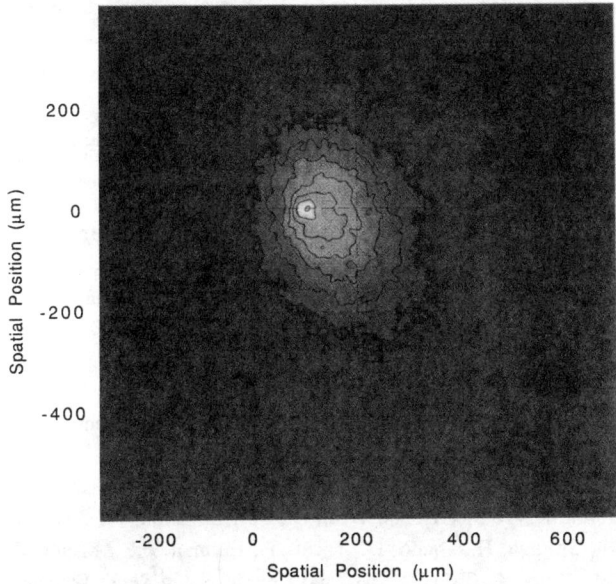

Figure 4. Image showing the intensity distribution of the Balmer α lasing beam at the exit plane of the recombining plasma. The positions are relative to the initial fibre axis. The plot shows contours representing 3, 5, 9, 14 , 28 and 32 relative Balmer α flux with the most intense central region occupying only ~ 20 μm.

4. CONCLUSION

In conclusion, we have demonstrated that high gain recombination x-ray lasers can be achieved by irradiating fibre targets with ultra-short laser pulses. Our results give one of the most unequivocal measurements of a high ratio of the Balmer α to Balmer β emission and also the highest gain coefficient observed for the C VI x-ray laser. The high value of achieved gain length product means that saturated and potentially useful x-ray laser output at 18.2 nm could in principle be produced with a carbon fibre length of only 10 mm and 80 J/ 40 TW output driving laser.

ACKNOWLEDGEMENTS

We would like to thank the VULCAN laser operations, the target preparation and engineering groups of the Central Laser Facility for their help and cooperation.

REFERENCES

[1] DaSilva, L.B., Trebes, J.E., Balhorn, R., Mrowka, S., Anderson, E., Attwood, D.T., Barbee, T.W., Brase, Jr., J., Corzett, M., Gray, J., Koch, J.A., Lee, C., Kern, D., London, R.A., MacGowan, B.J., Matthews, D.L., and Stone, G., *Science* **258**, 269-271 (1992).

[2] London, R.A., Rosen, M.D., and Trebes, J.E., *Appl. Opt.* **28**, 3397-3404 (1989).

[3] Suckewer, S., Skinner, C.H., Milchberg, H., Keane, C., and Voorhees, D., *Phys. Rev. Lett.* **55**, 1753-1755 (1985); Suckewer, S., Skinner, C.H., Kim, D., Valeo, E., Voorhees, D., and Wouters, A., *Phys. Rev. Lett.* **57**, 1004-1007 (1986).

[4] Jamelot, G., Klisnick, A., Carillon, A., Guennou, H., Sureau, A., and Jaegle, P., *J. Phys. B* **18**, 4647-4653 (1985).

[5] Kato, Y., Azuma, H., Murai, K., Yamakawa, K., Shiraga, H., Pert, G.J., Ramsden, S.A., and Key, M.H., in *X-Ray Lasers 1990*, edited by Tallents, G.J., Bristol: IOP Publishing Ltd., 1991, pp. 1-8.

[6] Chenais-Popovics, C., Corbett, R., Hooker, C.J., Key, M.H., Kiehn, G.P., Lewis, C.L.S., Pert, G.J., Regan, C., Rose, S.J., Sadaat, S., Smith, R., Tomie, T., and Willi, O., *Phys. Rev. Lett.* **59**, 2161-2164 (1987).

[7] Xu, Z.Z., Fan, P.Z., Zhang, Z.Q., Chen, S.S., Lin, L.H., Lu, P.X., Sun, L., Wang, X.F., Yu, J.J., and Qian, A.D., in *X-Ray Lasers 1990*, edited by Tallents, G.J., Bristol: IOP Publishing Ltd., 1991, pp. 151-158.

[8] Grande, M., Key, M.H., Kiehn, G., Lewis, C.L.S., Pert, G.J., Ramsden, S.A., Regan, C., Rose, S.J., Smith, R., Tomie, T., and Willi, O., *Opt. Commun.* **74**, 309-312 (1990).

[9] Nishimura, H., Shiraga, H., Daido, H., Tachi, T., Herman, P.R., Miura, E., Takabe, H., Yamanaka, M., Kato, Y., Tallents, G.J., and Key, M.H., in *Short Wavelength Coherent Radiation: Generation and Applications, Proceedings of OSA*, **20**, Washington: Optical Society of America, 1988, pp. 137-145.

[10] Zhang, J., and Key, M.H., *Appl. Phys.* B**58**, 13 (1994).

[11] Danson, C.D., Barzanti, L.J., Chang, Z., Damerall, A.E., Edwards, C.B., Hancock, S., Hutchinson, M.H.R., Key, M.H., Luan, S., Mahadeo, R.R., Mercer, I.P., Norreys, P.A., Pepler, D.A., Rodkiss, D.A., Ross, I.N., Smith, M.A., Smith, R.A., Taday, P., Toner, W.T., Wigmore, K.W.M., Winstone, T.B., Wyatt, R.W.W., and Zhou, F., *Opt. Commun.* **103**, 392-397 (1993).

[12] Fan, P.Z., Zhang, Z.Q., Zhou, J.Z., Jin, R.S., Xu, Z.Z., and Guo, X., *Appl. Optics* **31**, 6720-6723 (1992).

[13] Krishnan, J., Neely, D., Danson, C., Dwivedi, L., Lewis C.L.S., and Tallents, G.J., *RAL Annual Report* **RAL-92-020**, 22-23 (1992).

[14] Linford, G.J., Peressini, E.R., Sooy W.R., and Spaeth, M.L., *Appl. Opt.* **13**, 379-387 (1974).

[15] The mirror was manufactured by Changchun Institute of Optics and Fine Mechanics, Academia Sinica, Changchun, People's Republic of China.

[16] Borovskiy, A.V., Holden, P.B., Lightbody M.T.M., and Pert, G.J., *J. Phys.* B **25**, 4991-5003 (1992).

Theory of Recombination X-Ray Lasers Based on Optical-Field Ionization

D. C. Eder*, P. Amendt*, L. B. DaSilva*, T. D. Donnelly[†],
R. W. Falcone[†], R. A. London*, M. D. Rosen*, and S. C. Wilks*

*Lawrence Livermore National Laboratory, Livermore, CA 94550
† University of California at Berkeley, Berkeley, CA 94720

Abstract. Ultrashort-pulse, high-intensity laser drivers have the potential for creating tabletop-size x-ray lasers by ionizing the target gas via the electric field of the laser pulse. For appropriate plasma conditions following ionization, lasing can occur during the subsequent rapid recombination. A review of the theory and modeling for these optical-field-ionized x-ray lasers is presented. Particular attention is given to the issues of electron heating and ionization-induced refraction. We summarize modeling in support of experiments where evidence of lasing in H-like Li at 135 Å was obtained. In addition, we present modeling results for lasing in Li-like N at 247 Å. We briefly discuss new applications appropriate for tabletop-size high-repetition-rate x-ray lasers.

I. INTRODUCTION

Since the last conference in this series in Schliersee, Germany,[1] the major development in recombination x-ray lasers based on optical-field ionization has been the observation of faster than linear growth of the intensity with length for the L_α line in H-like Li at 135 Å. The initial observations[2] were at the RIKEN laboratory in Japan with similar results being obtained in a UC Berkeley/LLNL collaboration.[3] Other observations that are indicative of lasing in H-like Li are the directionality measured at RIKEN and the short period of emission ($\Delta t \approx 20$ psec) seen first in the UC Berkeley/LLNL experiment and subsequently also observed in the RIKEN experiment.[4] In addition to the experiments in Li, there was an observation of enhanced intensity for a transition between excited states in Li-like N (5f-3d line at 512 Å).[5] However, for all these experiments the measured gain-length product is relatively small ($gL \leq 4$). While lasing has most likely been observed, a definitive demonstration and sufficient output for applications awaits experiments with significantly larger gain-length products. In the RIKEN experiments, the lasing length is limited by the power available from their field ionizing laser.[4] Ionization-induced refraction appears to be the factor limiting the length in the Berkeley/LLNL experiments.[3] Recent experiments showing plasma waveguiding are very relevant to solving the refraction problem.[6,7]

There has also been significant progress in the theory and modeling of recombination x-ray lasers based on optical-field ionization. The importance of collisional heating at higher densities has been calculated by a number of authors,[8-9] but there are still some unresolved issues that would benefit from additional theory and experiments. There has been significant progress in understanding the importance of ionization-induced refraction.[10-12] While using an additional laser to create a plasma waveguide is one potential solution to the problem of refraction,[6,7] we are also investigating optimum placement of the gas jet or ablating vapor with respect to the location of vacuum best focus and the relative benefit of larger focal spots that give shallower density gradients as compared to using higher intensities that give narrow but nearly flat-top density profiles. Recent modeling for H-like Li shows the importance of recombination and collisional cascade heating and has shown the effects of two temperature electron distributions on the calculated gain coefficient.[13] Modeling for Li-like N has shown that a significant gain-length product is possible by using an element with lower ionization potential which allows for a larger spot size.[14]

Lasing during rapid recombination following field ionization is only one of the approaches being considered for tabletop-size x-ray lasing. A number of different tabletop-size x-ray lasers, all with relatively small gain-length products, have been demonstrated in the past few years.[15] In this proceedings Rocca, *et al.*, present the first demonstration of a tabletop system based on a capillary discharge that achieved a gain-length product of 7 in Ne-like Ar at 469 Å.[16] In addition, a variation of the OFI x-ray scheme, discussed in this paper, using circularly polarized laser was recently used to achieve a gain-length product of 11 in Pd-like Xe at 418 Å.[17] The basic difference in this latter scheme, as compared to the one discussed here, is that the electrons ionized by the circularly polarized field are left with enough energy to collisional populate the upper-laser state as compared to having cold electrons that populate an upper-laser state via rapid recombination. Since both schemes are based on forming a plasma using a confocal geometry, problems associated with ionization-induced refraction that affect the scheme discussed in this paper will also be important for the other OFI approach for shorter wavelengths where higher densities are required.

Following an overview of theoretical issues for OFI x-ray lasing discussed in Sec. II, we summarize modeling results for H-like Li and Li-like N in Secs. III and IV. In Sec. V, we briefly discuss some appropriate applications for table-top size x-ray lasers having relatively high repetition rates. We conclude with some comments on future prospects in Sec. VI.

II. THEORETICAL ISSUES FOR OFI X-RAY LASING

During rapid recombination following field ionization, lasing can occur between two excited states and between an excited state and the ground state. Most of the attention has been on lasing down to the ground state because of the relatively shorter wavelengths that can be achieved. For applications requiring less energetic photons, lasing between excited states is a potential x-ray source. There are some initial results for transitions between excited states of Li-like N, e.g., 5f-3d line at 512 Å.[5] In the rest of this paper we will restrict our attention to lasing to the ground state, although many of the results can be applied to lasing between excited states.

The basic requirement for achieving a significant gain coefficient in lasing down to the ground state, e.g., n=2 to n=1 or the L_α transition in H-like ions, is that the recombination to the upper Rydberg levels and subsequent collisional cascade down to the upper-laser level occur on a faster time scale than the radiative and collisional filling of the lower-laser (ground) level. This requirement places constraints on the density and temperature following field ionization. A lower density requires a lower temperature to ensure sufficiently fast 3-body collisional recombination. Except at very high intensities where Raman heating can be important,[18] the dominate heating mechanisms are above-threshold-ionization (ATI) and electron collisions. ATI heating is associated with the phase mismatch between the time of ionization and the peak of the oscillatory electric field.[19,20] Since the amount of energy given to the electron for a given phase mismatch depends on the quiver velocity, there is a benefit in using a shorter wavelength ionizing laser. Collisions by the rapidly oscillating electrons can give significant heating at higher densities.[8,9,20] There are differences in the choice of the Coulomb logarithm used in the different calculations that result in significant differences in the predicted heating.[14] For the H-like Li experiments, the densities are low enough that collisional heating is not be important for any of the choices used for the Coulomb logarithm. For the proposed Li-like N scheme discussed in Sec. IV., the collisional heating contribution varies from being very minor to being very significant depending on the model. Additional theoretical work is required to accurately treat collisional heating for OFI plasma parameters.

In most conventional x-ray lasing schemes a plasma is heated with a laser using a line focus geometry and the x-ray lasing axis is perpendicular to the driving laser axis. However, in OFI lasing schemes a confocal geometry is used and refraction of the driving laser as a result of field ionization is a major issue. In general, laser pulses have maximum intensity on axis which results in the electron density also having a maximum on axis giving the potential for refraction. One approach to reducing refraction is to use a short wavelength driving laser and operate at a low density. This is the case for the RIKEN experiments where 0.25 μm light is used and the electron density was measured to be of order 10^{17} cm^{-3}.[2] (As discussed below, modeling results for H-like Li predict very little gain for this low value of density.) Even if the density electron density is a factor of ten higher, refraction should not be a major problem. The UC Berkeley/LLNL experiments use a longer wavelength of 0.4 μm, and a simple calculation shows that the fall off in gain for lengths longer than 1.5 mm could be the result of refraction.[3] While there has been significant progress in modeling refraction,[10-12] a detailed study that self-consistently calculates transverse intensity and electron density profiles has not been completed. We have started such a study with one issue being the optimum placement of the gas jet or ablating vapor with respect to the location of vacuum best focus. If more energy is available by using an element with a lower ionization potential or by increasing the driving laser energy, there is a question if it is better to increase the size of the focal spot and thus produce a shallower density gradient or to use higher intensity to give a narrow but nearly flat-top density profile. We take the first approach in our discussion of potential lasing in Li-like N in Sec. IV. For some situations, the second approach might be better. We believe that the refraction study in progress will help answer these questions.

III. SUMMARY OF MODELING FOR H-LIKE LI AT 135 Å

Motivated by the recent experiments in H-like Li, there has been interesting modeling that has led to an unexplained discrepancy with experimental measurements of density. The electron density measured in the RIKEN experiment of order 10^{17} cm^{-3} is too low to produce any significant gain for an arbitrary cold electron distribution.[13] (Our understanding of the H-like Li experiments is based on the assumption of a two-component electron distribution, with the colder component arising from the outer electron and the hotter component arising from field ionization of the two inner electrons.) The key factor that has to be included in the modeling, to show the discrepancy with the density measurement, is heating of the free electrons by collisions associated with 3-body recombination and cascade to the lower levels. While a constant temperature of a fraction of 1 eV could produce significant gain for an electron density of 10^{17} cm^{-3}, the free electrons rapidly heat to temperatures above 1 eV thus slowing down recombination and inhibiting gain. Since the density measurement is based on time integrated data of He-like lines, there is a potential that the density could be higher for some time in a region that could have significant gain. Additional data on densities would be very useful.

The recent modeling of H-like Li also addressed the question of the effect of two electron components with different temperatures on the calculated gain coefficient.[13] Our initial calculations neglected the effect of the hotter distribution.[3] The modeling shows that the hotter component has the largest reduction on the gain when its temperature is of order 100 eV for a cold component having a temperature of 1 eV. For very high temperatures, the hot component no longer slows down the recombination and cascade because collisions with the hot component are very infrequent. However, an interesting effect results from having the same number of recombining cold electrons as ions. The electrons rapidly recombine to the upper Rydberg levels but this results in a large reduction in the free electron density which slows down the collisional cascade to the lower levels. In this case, one is better off having a "hot" component having a temperature of order 10 eV or less than having a hot component that does not interact at all because the temperature is too high. For details of these calculations see the proceedings in this collection by Donnelly.[13]

IV. SUMMARY OF MODELING FOR LI-LIKE N at 247 Å

One solution for problems associated with refraction is to use an element with a lower ionization potential which thus requires a lower laser intensity to field ionize. This allows for the use of a larger focal spot which gives shallower density gradients and less refraction. For this reason we have done calculations for lasing in Li-like N at 247 Å.[14] The required intensity to ionize N to the required He-like ionization stage is 3×10^{16} Wcm^{-2}. For a pulse duration of 100 fsec and an energy of 0.5 J, a focal spot with a radius of 70 μm is possible. A very rough estimate of the lasing length as a result of refraction is given by $L = 2b[n_c/(n_e \ln 2)]^{1/2}$, where n_c is the critical density, n_e is the electron density, and b is the radius of the focal spot.[3] For a wavelength of 0.4 μm, an electron density of 2.5×10^{19} cm^{-3}, and a focal spot radius of 70 μm, a lasing length of 3 mm is estimated. The density assumed here results in this length being only a factor of 2 greater than the 1.5 mm length in the UC Berkeley/LLNL experiments where refraction appears to be limiting the length. The need for the higher density is because a cold electron component, with a temperature significantly below that obtained from ATI heating, can not be justified in this case. To obtain a significant gain-length product ($gL > 5$) with only a lasing length of 3 mm, one requires a gain coefficient greater than 15 cm^{-1}. Based on our calculations of gain this implies that the temperature must be 12 eV or less for an electron density of 2.5×10^{19} cm^{-3} as shown in Fig. 1. Our estimates of ATI and collisional heating indicate that a temperature as low as 10 eV are possible for this system.[14]

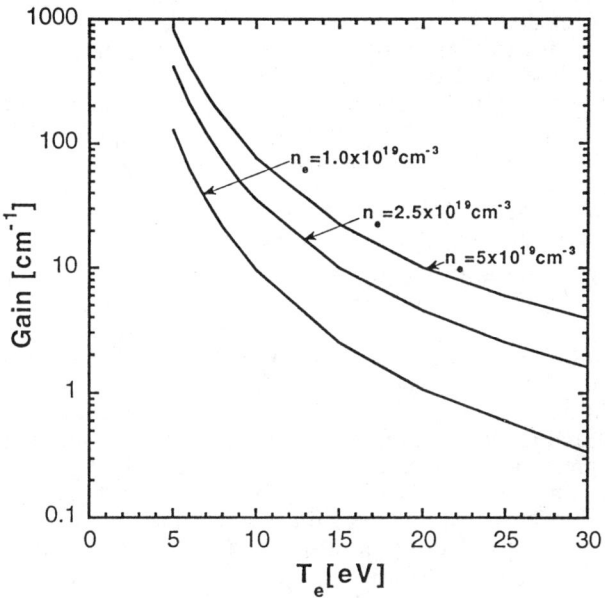

FIGURE 1. Calculations of the gain for the $3d_{5/2}$ to $2p_{3/2}$ transition in Li-like N at 247 Å.

V. APPLICATIONS FOR TABLETOP-SIZE X-RAY LASERS

An important aspect of most existing and proposed tabletop-size x-ray lasers is a relatively high repetition rate. The current repetition rate for the OFI Pd-like Xe x-ray laser operating at 418 Å is 10 Hz,[17] and the laser used to pump the OFI H-like Li scheme at 135 Å could also have this repetition rate. This allows one to have quite high average powers, with relatively small amount of energy per pulse, that can exceed the average power available from current 3rd generation synchrotron facilities. For example, the Advance Light Source (ALS) synchrotron facility is quoted to have an average flux of 2×10^{13} photons/sec at 150 Å for a 10^{-4} bandwidth.[21] An x-ray laser operating at 10 Hz would require 25 µJ per pulse to have the same average power. Most tabletop-size x-ray laser schemes are calculated to have a saturated energy per pulse of this order or greater. The big advantage of tabletop x-ray lasers is that they would be low cost and small enough to be in individual laboratories. A large number of the applications currently being studied using synchrotrons are not well suited to tabletop x-ray lasers because of the low bandwidth required (synchrotron flux levels are much higher for a 10^{-2} bandwidth) or the desire to have continuous tunability. However, there are many applications currently being studied with synchrotrons and other devices, as well as applications that can not be studied with synchrotrons, that are very well matched to the output properties of tabletop x-ray lasers.[22]

Photoelectron spectroscopy of core electrons to study dynamical processes on surfaces is one example where tabletop x-ray lasers could make an important contribution. Continuous tunability is not required because the measurement of the energy of the ejected electron combined with a well know photon energy (i.e., the narrow bandwidth x-ray laser source) determines the time dependence of the binding energy of the core level. The shifts in the binding energy of the core levels give important information on the dynamics of impurities, chemical processes, etc. In many case these shifts are relatively small which places the requirement on having a narrow bandwidth source. The quasi-continuous output of synchrotrons (tens of psec pulses separated by time intervals in the nsec range) makes the study of dynamical processes and the use of pump/probe techniques very difficult. In photoelectron spectroscopy a very wide range of photons wavelengths are of interest including the relatively long wavelengths ($\lambda > 400$ Å) of current tabletop x-ray lasers that have obtained significant gain-length products.

VI. CONCLUSIONS AND FUTURE PROSPECTS

Experiments in H-like Li at 135 Å have provided evidence that is very indicative of lasing during rapid recombination following optical-field ionization. In addition, substantial progress has been made in the theory and modeling of this new type of x-ray laser. Continued work in electron heating and refraction is clearly desired. Modeling has shown that significant gains are possible for the H-like Li scheme provide there is a cold electron component and the density is approximately 10^{18} cm^{-3}. There is a conflict between modeling and current experiments in that the measured density in the RIKEN experiment is significantly less than 10^{18} cm^{-3}. Additional measurements of density are clearly needed. The modeling results for Li-like N at 247 Å show that a significant gain-length product ($gL > 5$) is possible provided the electron temperature is of order 10 eV for electron densities of order 10^{19} cm^{-3}.

The future prospects for tabletop x-ray lasers, in general, are very good with many applications to explore. The outlook for x-ray lasers based on rapid recombination following field ionization appears to be very promising provided that work continues on propagation problems. Modeling for these schemes would benefit greatly from more measurements of electron density and temperature. Hopefully by the time of the next conference in this series these goals will be met.

Acknowledgments

Work performed under the auspices of the U. S. Department of Energy by Lawrence Livermore National Laboratory under Contract W-7405-ENG-48.

References

1. *X-Ray Lasers* 1992, Proceedings of the 3rd International Colloquium on X-Ray Lasers, IOP Conference Series 125, edited by E. E. Fill (Institute of Physics, Bristol, England, 1992).
2. Y. Nagata, K. Midorikawa, M. Obara, H. Tashiro, and K. Toyoda, Phys. Rev. Lett. **71**, 3774 (1993).
3. D. C. Eder, P. Amendt, L. B. DaSilva, R. A. London, B. J. MacGowan, D. L. Matthews, B. M. Penetrante, M. D. Rosen, S. C. Wilks, T. D. Donnelly, R. W. Falcone, and G. L. Strobel, Phys. Plasmas **1**, 1744 (1994).
4. K. Midorikawa, Y. Nagata, M. Obara, H. Tashiro, and K. Toyoda, these proceedings.
5. S. Borgström, E. Fill, J. Larsson, T. Starczewski, S. Svanberg, and C.-G. Wahlström, these proceedings.
6. C. G. Durfee III and H. M. Milchberg, Phys. Rev. Lett. **71**, 2409 (1993).
7. H. M. Milchberg, C. G. Durfeee III, and J. Lynch, these proceedings.
8. S. C. Rae and K. Burnett, Phys. Rev. A **46**, 2077 (1992).
9. P. Pulsifer, J. P. Apruzese, J. Davis, and P. Kepple, Phys. Rev. A **49**, 3958 (1994).
10. W. P. Leemans, C. E. Clayton, W. B. Mori, K. A. Marsh, P. K. Kaw, A. Dyson, and C. Joshi, Phys. Rev. A **46** 2077 (1992).
11. S. C. Rae, Opt. Commun. **97**, 25 (1993).
12. X. Liu and D. Umstadter, OSA Proceedings on Shortwavelength V: Physics with Intense Laser Pulses, edited by M. D. Perry and P. B. Corkum (Optical Society of America, Washington, DC, 1993), Vol. 17, p. 45.
13. T. D. Donnelly, T. E. Glover, M. Hofer, E. A. Lipman, R. W. Lee, L. Da SIlva, D. C. Eder, S. Mrowka, and R. W. Falcone, these proceedings.
14. P. Amendt, D. C. Eder, R. A. London, B. M. Penetrante, and M. D. Rosen, SPIE Proceedings, Conference on Short-Pulse High-Intensity Lasers and Applications, Edited by H. A. Baldis (SPIE, Bellingham. WA, 1993), Vol. 1860, p. 140.
15. See reference 1, SPIE Proceedings, Ultrashort-Wavelength Lasers II. edited by S. Suckewer (SPIE, Bellingham, WA, 1994), Vol. 2012, or these proceedings.
16. J. J. Rocca, F. G. Tomasel, V. N. Shlyaptsev, O. D. Cortazar, J. L. A. Chilla, and G. Giudice, these proceedings and J. J. Rocca, V. Shlyaptsev, F. G. Tomasel, O. D. Cortazar, D. Hartshorn, and J. L. A. Chilla, Phys. Rev. Lett. **73**, 2192 (1994).
17. B. E. Lemoff, G. Y. Yin, C. L. Gordon, III, C. P. J. Barty, and S. E. Harris, submitted to Phys. Rev. Lett.
18. D. C. Eder, P. Amendt, and S. C. Wilks, Phys. Rev. A **45**, 6761 (1992).
19. P. B. Corkum, N. H. Burnett, and F. Brunel, Phys. Rev. Lett. **62**, 1259 (1989).
20. B. M. Penetrante and J. N. Bardsley, Phys. Rev. A **43**, 3100 (1991).
21. *ALS Beamlines* Lawrence Berkeley Laboratory PUB-3104 (1992).
22. D. C. Eder, P. Amendt, C. B. Dane, L. B. DaSilva, L. A. Hackel, M. R. Hermann, R. A. London, B. J. MacGowan, D. L. Matthews, M. D. Rosen, and S. C. Wilks, Proceedings of the International Conference on LASERS '92, edited by C. P. Wang (Press, McLean, VA, 1993), p. 67.

Optical-Field-Induced Ionization X-Ray Laser Using Preformed Li+ Plasma

Katsumi Midorikawa, Yutaka Nagata*, Minoru Obara*,
Hideo Tashiro, and Koichi Toyoda

The Institute of Physical and Chemical Research (RIKEN)
Wako-shi, Saitama 351-01, Japan

Abstract. We report the production of an extremely cold lithium plasma by optical-field-induced ionization. Such a cold plasma was realized by irradiating a 0.5-ps KrF laser focused into a singly-ionized lithium plasma at an intensity of 10^{17} Wcm^{-2}. An average electron temperature of less than 1 eV was estimated by measuring the intensity ratio of the Lyman series line emission in a hydrogen-like lithium plasma. The time history of the Lyman-α laser (13.5 nm) in a lithium plasma strongly ensures the production of an extremely cold plasma.

1. INTRODUCTION

Cold and dense multiply ionized plasmas have a potential for generating soft x-ray lasers [1,2]. Recent researches show that an optical-field-induced ionization (OFI) process is an appropriate candidate to produce such plasmas [3-7]. In the OFI process, a high-intensity laser produces a plasma consisting of fully stripped ions and cold free electrons on a time scale much shorter than their recombination time. Because of a sufficiently low temperature of such a plasma, rapid three-body recombination could lead to a population inversion with respect to the ground state of hydrogen-like ion. The lasing to the ground state will make a soft x-ray laser wavelength much shorter compared with that between excited states. A high laser intensity required to produce desired ionization states in such a plasma is now achievable by the recent progress of table-top high power lasers with ultrashort pulse widths. The OFI scheme is, therefore, particularly attractive to produce short wavelength soft x-ray lasers with a moderate pump laser.

In order to realize an OFI soft x-ray laser, requirements for conditions of a pump laser as well as for an initial laser medium are investigated theoretically [5-7]. As for a wavelength of a pump laser, shorter wavelength is preferable. Since above threshold ionization (ATI) heating is proportional to the square of a pump laser wavelength, the use of UV wavelength can reduce the ATI heating. Secondly, the temperature of electrons produced by OFI becomes lower by using a linearly polarized laser pulse compared with by using a circularly polarized one [8,9]. Thirdly, as for a pump laser pulse, shorter pulse duration is not always preferable. An adequate pump laser pulse width makes the required laser intensity decrease, resulting for the decrease of the ATI heating [7]. However, a longer pulse than picoseconds would introduce additional heating.

Our previous paper [10] reported the observation of stimulated emission and a gain on the Lyman-α transition at 13.5 nm in H-like Li ions by a novel OFI scheme. Our OFI scheme is unique modification of the originally proposed method [4, 6] in which neutral gases are used as initial media. While, in our scheme, we proposed the use of a singly ionized lithium plasma instead of neutral gases. This modification would drastically change the final electron temperate and pump beam propagation in the plasma. As a result, a small signal gain coefficient of 20 cm^{-1} was observed on the Lyman-α transition at 13.5 nm by using a subpicosecond, high-intensity KrF laser.

In this paper, we describe the measurement of electron temperature of the plasma produced by our OFI scheme using a preformed Li$^+$ plasma. The electron temperature should be extremely low enough to produce a gain on the Lyman-α transition in Li^{2+}.

2. ADVANTAGES OF A PREFORMED Li$^+$ PLASMA

In our scheme, the use of subpicosecond KrF excimer laser as a pump source is essential because its high frequency effectively decreases the ATI heating. Furthermore, in the OFI scheme, ionization should occur in tunneling regime. That is, Keldysh parameter by which ionization mechanism is separated into two regimes [11]: multiphoton ionization and tunneling ionization, should be much smaller than unity. This requirement indicates that atom or ion having large ionization potential is preferable. The large ionization potential also allows high intensity interaction, avoiding pump beam defocusing.

Figure 1 shows the Keldysh parameters which are calculated from the ionization potentials and threshold intensities [12] of rare gases and rare-gas-like ions for KrF laser pumping. For neutral atoms having low ionization potentials, the Keldysh parameter becomes larger than unity, indicating that multiphoton ionization occurs dominantly over tunneling ionization. While the Keldysh parameter becomes smaller than unity for a singly-ionized Li because of its larger ionization potential of 75 eV. The ionization of Li$^+$ thus occurs in the tunneling regime. The high ionization potential of Li$^+$ also plays a role to decrease the

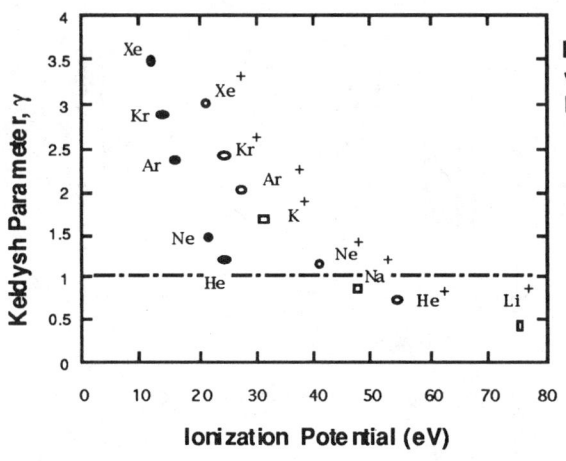

Figure1. Keldysh parameters for various rare gases and rare-gas-like alkali ions.

ATI heating because the electron is released near the crest of the optical cycle, and thus the phase mismatch becomes small [9]. Furthermore, this high ionization potential of Li^+ minimizes the defocusing effect of a subpicosecond KrF laser pulse because the leading edge of the subpicosecond pulse cannot further ionize the plasma medium. Therefore, the uniformity of the medium will be maintained until a high-intensity part of the subpicosecond KrF laser pulse reaches the plasma.

3. EXPERIMENT

The experimental setup for measurement of time-integrated spectra has been described in detail elsewhere [10]. In order to produce the initial Li^+ plasma, a 20-ns KrF laser pulse with an output energy of 200 mJ was line-focused onto a rotating lithium target located in a vacuum chamber ($<10^{-6}$ Torr). The focused laser intensity was 10^9 Wcm^{-2}. After a certain delay with respect to the 20-ns KrF laser pulse, a linearly polarized KrF laser pulse with a pulse width of 500 fs and an energy of 50 mJ was focused inside the Li plasma by using an f = 30 cm lens. The focused position was 0.5 mm from the lithium target. A spot size and a confocal length of this focused beam were about 10 μm and 2 mm, respectively. The maximum focused laser intensity was thus 10^{17} Wcm^{-2}.

The Lyman-series emission spectra from a H-like Li plasma were detected by using a flat-field grazing-incidence XUV spectrograph equipped with a microchannel plate and CCD camera. A thin film filter made of a carbon alloy was placed behind an entrance slit of the spectrograph to eliminate a stray light of the subpicosecond KrF laser pulse.

Time-resolved measurements of the Lyman-α transition in H-like Li ions were made with an x-ray streak camera (Hamamatsu model C4575, magnetic focusing for high temporal resolution). The time resolution of the streak camera has been measured to be 2 ps. The photocathode of the streak camera was made of thin

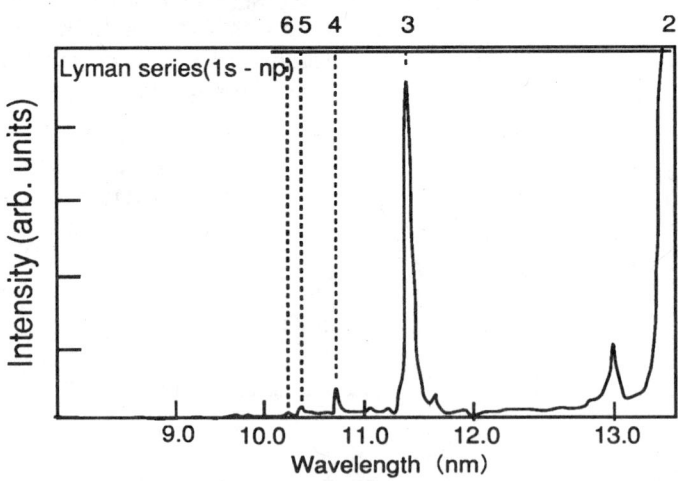

Figure 2.   The Lyman-series spectrum in H-like Li ion.

layers of parylene (500 nm thick), aluminum, and fluffy cesium iodide. The fluffy cesium iodide material was chosen as a photocathode because of its high quantum efficiency for the relevant soft x-ray spectral region [13].

Figure 2 shows a time-integrated Lyman-series spectrum in a H-like Li plasma. When an upper level of the transition is located higher than a level of the local thermal equilibrium (LTE) boundary, the electron temperature can be estimated from the intensity ratio of transitions [14]. The distribution of electron population densities is predominantly determined by collisional processes in the LTE limit. The intensities of the spontaneous radiation from upper states of $n = i$ and j ($i > j$) to the ground state $n = 1$, I_{i1} and I_{j1}, should thus obey the Saha-Boltzmann distribution:

$$\frac{I_{i1}}{I_{j1}} = \frac{\hbar \omega_{i1} N_i A_{i1}}{\hbar \omega_{j1} N_j A_{j1}} = \frac{\hbar \omega_{i1} A_{i1} g_i}{\hbar \omega_{j1} A_{j1} g_j} \exp\left(-\frac{E_i - E_j}{k T_e}\right) \quad (1)$$

where N_i is the population density of $n = i$. $\hbar w_{i1}$ is the photon energy between $n = i$ and 1. A_{i1} is the spontaneous transition probability from $n = i$ to the ground state, g_i is the statistical weight of the level i, E_i is the energy of the level i, and T_e is the electron temperature. The principal quantum number of the lowest LTE boundary, p, is described by Eq. (2) [14]:

$$p = \left(\frac{7 \times 10^{18}}{N_e} \left(\frac{k T_e}{z^2 \chi_H}\right)^{\frac{1}{2}}\right)^{\frac{2}{17}} \quad (2)$$

where N_e is the electron density (cm^{-3}), z is the ionic charge, and χ_H is one Rydberg (13.6 eV). In our experimental condition with the electron density of 10^{17} cm^{-3} [10], the LTE holds above $p = 3$.

The intensity ratio of 9 : 1 : 0.25 for the I_{41}, I_{51}, and I_{61} was obtained from Fig. 2. The electron temperature of the plasma was thus evaluated to be approximately 1.5 eV. This value gives an upper limit for the electron temperature because the population distribution of the lower levels slightly departs from that predicted by the Eq. (1) due to the faster relaxation rates of these levels [15]. Appearance of gain on the Lyman-α transition was, therefore, expected for this electron temperature which was much lower than the ionization potential of Li^{2+}.

Figure 3 gives a typical time-resolved streak image of the Lyman-α transition. In this measurement, the time resolution was limited to approximately 15 ps due to weaknee of the signal intensity. A thin-line curve signal peaked at t = 0 ps indicating a pulse shape of the KrF pump laser was intentionally introduced as a reference. On the other hand, a solid curveis a temporal evolution of the Lyman-α radiation at 13.5 nm. The peak intensity of the Lyman-α transition appears after 20 ps with respect to the pump laser irradiation. Time convolution gives that a typical pulse duration of the 13.5 nm radiation was less than 20 ps (FWHM). The observed pulse width indicates that the Lyman-α radiation is self-terminated since the radiative life time of the upper level of the transition is 20-ps.

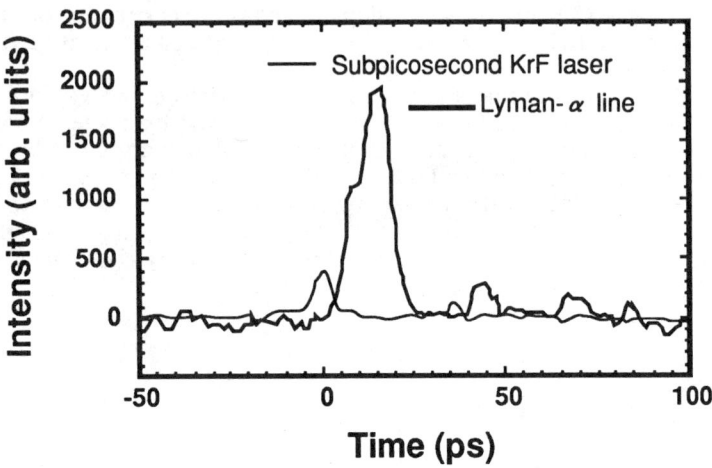

Figure 3. Time-resolved trace of the 13.5-nm Lyman-α emission in an OFI lithium plasma (bold curve), together with that of the subpicosecond KrF pump laser (thin curve).

4. DISCUSSION

The observed time-resolved signal of the Lyman-α transition is well reproduced by the numerical model under conditions of the electron temperature of 0.2 - 0.9 eV and of the electron density of 5×10^{17} - 10^{18} cm^{-3}. The results of the numerical calculations show that the three-body recombination completes within the order of 1 ps and the signal terminates within 20 ps under these conditions. The model also predicts the small signal gain coefficient of 20 cm^{-1} at peak. On the other hand, the pulse width of the spontaneous emission of the Lyman-α transition is predicted to be longer than 50 ps (FWHM) with a long decay tail. When the electron temperature is assumed slightly higher than 1 eV in the model, the numerical result predicts much smaller and unmeasurable gain.

Our experimental data and numerical results ensure that we have produced a plasma with a significantly lower temperature than what is expected from the ATI model [7, 16]. As described previously, our experimental scheme using a preformed Li$^+$ plasma qualitatively supports the optimum conditions for the Lyman-α lasing by minimizing undesirable heating. Acceleration of free electrons by the ponderomotive potential is also neglected because the Lyman-α emission originates from near the center of the pump beam focus where the laser field becomes maximum. The electron temperature of 1 eV, however, could not be produced even by taking these effects into account.

Additional effects should be considered in order to explain the observed low temperature quantitatively. One possible explanation is that a part of initial electrons produced prior to the field ionization do not recombine after 700 ns delay and the temperature of these electrons is kept low enough during the OFI. If it is true, a fast three body recombination leads to population inversion with respect to the ground state in spite of relatively high temperature of the OFI plasma. This is because the rate of three-body recombination becomes faster than that of electron-electron collision thermarization at an electron density of the order of 10^{18} cm^{-3} [17].

5. CONCLUSION

In summary we have demonstrated the production of fully stripped lithium ions by OFI, together with extremely cold electrons. The measurements suggested the electron temperature was approximately 1.0 eV, while the ionization potential of Li^{2+} is 122 eV. A streak camera measurement of the 13.5-nm Lyman-α emission showed a rise time of 20 ps and a pulse duration of 20 ps. The results of the numerical model indicate that these values support the extremely low temperature of the electrons produced by OFI. The low temperature of the plasma and the time-resolved Lyman-α signal provide evidences that the plasma conditions produced by our OFI scheme are appropriate to realize the Lyman-α laser at 13.5 nm.

ACKNOWLEDGMENTS

We would like to thank Prof. Y. Kato and Dr. S. Kubodera for their help and fruitful discussions and Dr. H. Shiraga, and Dr. K. Murai for their technical support of the x-ray streak camera measurement.

*Permanent address: Department of Electrical Engineering, Keio University, 3-14-1 Hiyoshi, Kohoku-ku, Yokohama 223, Japan.

REFERENCES

[1] J. Peyraud and N. Peyraud, J. Appl. Phys. **43**, 2993 (1972).
[2] W. W. Jones and A. W. Ali, Appl. Phys. Lett. **26**, 450 (1975).
[3] N. H. Burnett and G. D. Enright, IEEE J. Quantum Electron. **26**, 2580 (1991).
[4] P. Amendt, D. C. Eder, and S. C. Wilks, Phys. Rev. Lett. **66**, 2589 (1991).
[5] D. C. Eder, P. Amendt, and S. C. Wilks, Phys. Rev. A **45**, 6761 (1992).
[6] N. H. Burnett and P. B. Corkum, J. Opt. Soc. Am. B **6**, 1195 (1989).
[7] B. M. Penetrante and J. N. Bardsley, Phys. Rev. A **43**, 3100 (1991).
[8] W. P. Leemans, C. E. Clayton, W. B. Mori, K. A. Marsh, A. Dyson, and C. Joshi, Phys. Rev. Lett. **68**, 321 (1992).
[9] T. J. McIlrath, P. H. Bucksbaum, R. R. Freeman, and M. Bashkansky, Phys. Rev. A **35**, 4611 (1987).
[10] Y. Nagata,, K. Midorikawa, S. Kubodera,M. Obara, H. Tashiro, and K. Toyoda, Phys. Rev. Lett. **71**, 3774 (1993).
[11] L. V. Keldysh, Sov. Phys. JETP, **20**, 1307(1965).

[12] S. Kubodera, Y. Nagata, Y. Akiyama, K. Midorikawa, M. Obara, H. Tashiro, and K. Toyoda, Phys. Rev. A **48**, 4576 (1993).
[13] M. P. Kowalski, G. G. Gritz, R. G. Cruddance, A. E. Unzicker, and N. Swanson, Appl. Opt. **25**, 2440 (1986).
[14] H. R. Griem, Plasma Spectroscopy (McGraw-Hill, New York, 1964).
[15] R. W. P. McWhirter and A. G. Hearn, Proc. Phys. Soc. **82**, 641 (1963).
[16] S. C. Rae and K. Burnet, Phys. Rev. A **46**, 2077 (1992).
[17] L. Spitzer, Physics of Fully Ionized Gases (John Wiley and Sons, New York, 1962) p.133.

X-ray Spectroscopic Investigation of Optical-field Ionized Plasmas

S. Borgström[1], E. Fill[2], J. Larsson[1], T. Starczewski[1], S. Svanberg[1], and C.-G. Wahlström[1]

[1]*Department of Physics, Lund Institute of Technology, S-22100 Lund, Sweden*

[2]*Max-Planck-Institut für Quantenoptik, D-85748 Garching, Germany*

Abstract. Soft X-ray spectra of N, C, O, S, He and Ar ions were generated by optical field ionization with fs laser pulses. The experiments were carried out by focusing pulses of a high-power Ti:sapphire laser below a pulsed nozzle, using He, N_2, CO_2, O_2, SF_6, and Ar as parent gases. Gain measurements were made by comparing axial and transverse spectra. Gain on lines connecting to the ground state or to a quasi-ground state was not observed. However, strong indications of gain on lines between excited states are reported.

INTRODUCTION

It is well known that an atom can be stripped by many electrons in the field of a high-intensity laser pulse. This effect, known as "optical field ionization" has been suggested as a mechanism for generating ions in a relatively cold electron gas, thus creating favourable conditions for a recombination laser (1-4).

The simple classical barrier suppression model (4,5) predicts that at intensities well above 10^{16} W/cm^2 it should be possible to strip all electrons from the outermost shell of helium, carbon, nitrogen and oxygen as well as of sulphur and argon. The predicted X-ray spectra are single electron spectra from the next lower ionization stage. In this work we investigate experimentally the validity of this assertion.

EXPERIMENTS

The experiments were carried out with the high-power Ti:sapphire laser of the Lund Institute of Technology (6). Linearly polarized laser pulses (λ = 794 nm, pulse duration 150 fs, pulse energy 150 mJ) were focused with a lens of 50 cm focal length to a 50 µm diameter spot below a pulsed nozzle, thus generating an intensity of 5 x 10^{16} W/cm^2. The gas density 250 µm below the nozzle tip, where the laser pulse was focused, was about 10^{18} cm^{-3}, as measured by means of an interferometric method (7). To allow comparison of transverse and longitudinal spectra, the beam could be passed around the target chamber by means of a movable mirror and was then focused in a direction 90° to the direction of observation by a similar lens.

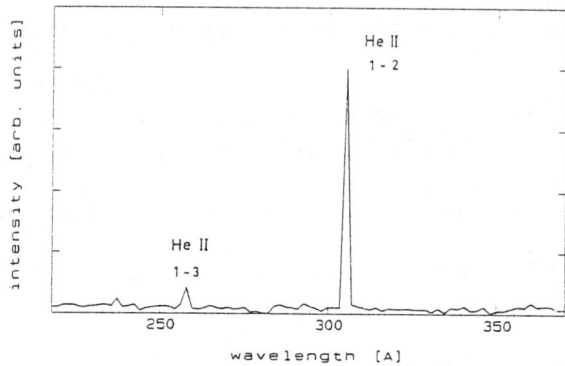

FIGURE 1. Axial spectrum of helium in the region of Lyman lines.

Time-integrated soft X-ray spectra were recorded by a 1 m grazing incidence spectrometer equipped with a photomultiplier.

It was found that the generation of high ionic charges depended quite critically on the adjustment of the focus into the gas jet. If the beam was not well focused below the nozzle tip, only odd harmonics of the laser frequency and lines from low ionization stages could be detected. With a well focused beam, however, reproducible spectra of hydrogenic helium, lithium-like carbon, nitrogen and oxygen as well as sodium-like sulphur and argon were seen.

A longitudinal spectrum of helium is shown in fig. 1. The most striking feature in this spectrum is the strong Lyman-α emission in relation to Lyman-β. The spectrum strongly suggests gain on the Lyman-α transition. However, it was found that the transverse spectra looked very similar.

Strong emission on resonance lines was also seen in the spectra of the other gases. In contrast to helium, however, the resonance lines from higher states were quite prominent. In nitrogen (see fig. 2) the lithium-like 2p-3d line is the strongest one, but 2p-4d has almost the same intensity. Observations similar to the case of nitrogen were made for carbon, oxygen, argon and sulphur.

In determining the dependence of He II Lyman-α and N V 2p-3d on laser intensity it was found that both lines could be observed down to an intensity a factor of five below our maximum intensity, in accordance with the intensities predicted by the barrier suppression model. The increase of both lines above threshold was nonlinear, which, nevertheless can be explained by a volume effect.

A razor blade scan of the axial emission showed that the emission originated from a region somewhat larger than the focal spot, about 100 μm in diameter in the case of nitrogen and about 150 μm in diameter in the case of helium. This finding can be explained by considering the expansion of the plasma, which occurs at a velocity of approximately $(3 q T_e / m)^{1/2}$, where q is the ionic charge, T_e the electron temperature and m the ion mass. Using estimated electron temperatures

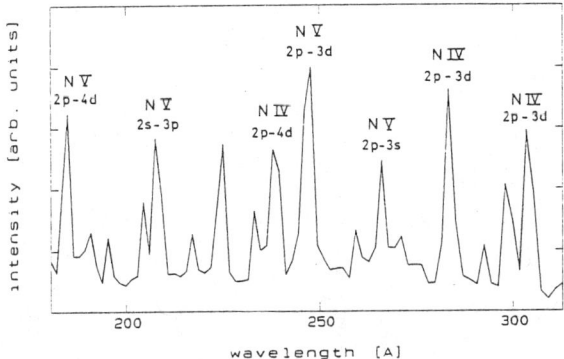

FIGURE 2. Transverse spectrum of nitrogen in the spectral region of resonance transitions, showing lithium-like and beryllium-like lines.

(see next section) one obtains expansion velocities of 7.5×10^6 cm/s for helium and 5×10^6 cm/s for nitrogen., The observed emission region thus corresponds roughly to the expansion during 2 ns.

ANALYSIS

To explain the striking dissimilarity between the helium spectra and the spectra of the other gases, numerical simulations were made, concentrating on helium and nitrogen. The plasma was assumed to be in a specified initial state after the laser pulse and then its evolution was simulated including ion kinetics and hydrodynamic expansion. The initial electron temperature was assumed to be due to ATI heating (4). Taking the spatial and temporal averages, a temperature of 40 eV was estimated for helium, whereas the corresponding temperature for nitrogen was 20 eV. The low electron temperature in the case of nitrogen is due to the fact that in nitrogen the main part of the electrons comes from comparatively low ionization stages.

In the case of helium the computer-generated spectrum can be brought into agreement with the experimental one by assuming that 0.5 % of the total ion population is in the hydrogenic ground state. The spectra are thus excitation spectra and the low intensity of the Lyman-β line results from a low excitation rate of the n = 3 level due to the low electron density of about 2×10^{18} cm^{-3}.

For nitrogen the computer-generated spectrum could not be matched to the experimental one. For any initial electron temperature or ground state population the 2p-4d emission did not exhibit the high intensity relative to the 2p-3d line. The discrepancy between the simulated and the experimental spectra is even more pronounced for the spectral region of transitions between excited states. The main feature there is the anomalously strong emission on the 3d-6f line in the experimental spectra, a feature not reproduced in the simulations. A mechanism other than three-

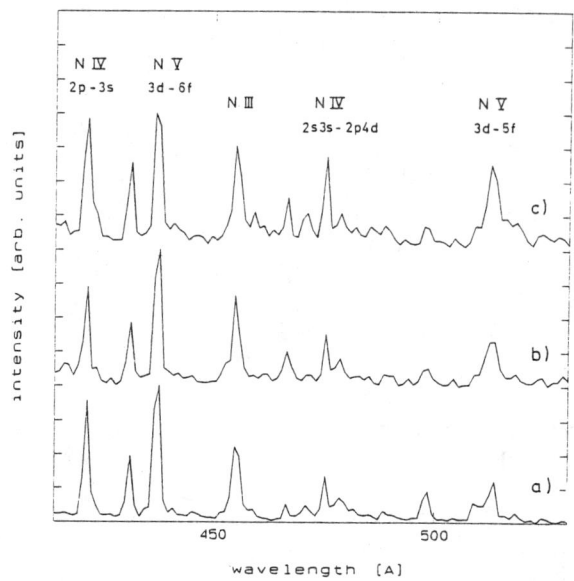

FIGURE 3. Plot comparing nitrogen spectra taken a) transversally, b) longitudinally along the short and c) longitudinally along the long dimension of the gas jet. Intensities are normalized to the N V 3d-6f line. Note the increase of the 3d-5f line relative to the 3d-6f line, which is optically thin.

body recombination must be populating excited levels, such as Rydberg stabilization (8) or state-selective charge transfer (9).

GAIN

A comparison of transverse and axial spectra did not reveal gain on any line to a ground state or quasi-ground state. This is not surprising considering the time-integrated nature of the spectra, which would require a amplification factor of several hundred for measurable gain.

However, strong indications of gain on lines between excited states could be observed. In comparing transverse and longitudinal spectra an increase in the emission in the axial direction was consistently observed for several XUV lines. One of these was identified as the NV 3d-5f transition at 511.9 Å. In order to perform a gain measurement a tube was mounted to the end of the nozzle tip. By squeezing the end of the tube a gas jet with long and short dimensions of 800 and 400 μm could be generated. In fig. 3 a transverse spectrum is compared with axial spectra taken along the short and long dimensions of the gas jet. It is seen that in the axial spectrum along the 400 μm dimension the emission of NV 3d-5f relative to other lines is slightly increased. The spectrum along the long axis shows a significant increase of the intensity at this line. Comparing the 3d-5f and 3d-6f lines (which have a common lower level) and fitting the data to a Linford plot a gain coefficient of 30 cm^{-1} is determined for the 3d-5f line. A detailed report of these findings is in preparation.

CONCLUSION

By focusing terawatt Ti:sapphire laser pulses into a gas jet of He, N_2, CO_2, O_2, Ar and SF_6 X-ray emission on lines of hydrogenic helium, lithium-like nitrogen, carbon and oxygen as well as of sodium-like argon and sulphur could be generated. This result is in accordance with the classical barrier suppression model, which predicts the corresponding parent ions to be readily generated at the intensities of up to 5×10^{16} W/cm^2 used in the experiment. A striking dissimilarity between the helium spectra and the spectra of the other gases was observed with regard to the relative strength of the first and second resonance lines. This finding is partially explained by the fact that the predicted electron temperature in helium is close to the ionization potential of hydrogenic helium, whereas in nitrogen it is considerably lower than the ionization potential of helium-like nitrogen. Simulations confirm that the helium spectra can be explained as excitation spectra, whereas the nitrogen spectra cannot be reproduced at all by the simulations. Anomalies in the nitrogen spectra need to be explained. Finally, gain on the 3d-5f transition in lithium-like nitrogen at a wavelength of 511.9 Å has been measured.

ACKNOWLEDGMENTS

Valuable discussions with D. Eder are gratefully acknowledged. We thank A. Persson for operating the laser and L. Engström for the loan of the spectrometer. The work was supported by the Swedish National Science Research Council and was performed within the framework of the European X-Ray Laser Network.

REFERENCES

1. Peyraud, J., and Peyraud, N., J. Appl. Phys. **43**, 2993–2996 (1972).
2. Burnett, N.H., and Corkum, P.B., J. Opt. Soc. Am. **B6**. 1195–1199 (1989).
3. Amendt, P., Eder, D.C., and Wilks, S.C., Phys. Rev. Lett. **66**, 2589–2592 (1991).
4. Penetrante, B.M., and Bardsley, J.N., Phys. Rev. **A 43**, 3100–3113 (1991).
5. Augst, S., Strickland, D., Meyerhofer, D.D., Chin, S.L., and Eberly, J.H., Phys. Rev. Lett. **63**, 2212– 2215 (1989).
6. Svanberg, S., Larsson, J., Persson, A., and Wahlström, C.-G., Physica Scripta **49**, 187–197 (1994).
7. Faris, G.W., and Hertz, H.M., Appl. Opt. **28**, 4662–4667 (1989).
8. Bucksbaum, P.H., and Jones, R.R., "Stability of atoms in intense laser fields", AIP Conference Proceedings 275, H. Walther, T.W. Hänsch and B. Neizert eds., New York 1992, pp 485–498.
9. Vinogradov, A.V., and Sobelman, I.I., Sov. Phys. JETP **36**, 1115–1119 (1973).

Plasmas for Short-Wavelength Lasers Driven by Ultra-Short, High-Intensity Laser Pulses

T. D. Donnelly, T. E. Glover, M. Hofer, E. A. Lipman,
R. W. Lee,[1] L. Da Silva,[1] D. C. Eder,[1] S. Mrowka,[1] and R. W. Falcone

Physics Department
University of California at Berkeley
Berkeley, CA 94720

A short-pulse, terawatt laser was used to generate and characterize plasmas appropriate for recombination pumped x-ray lasers. Thomson scattering was used to determine the electron and ion temperatures on a sub-picosecond time-scale. A recombination-pumped x-ray laser on the Ly-α transition of H-like Li was studied.

Ultra-short pulse, high-intensity lasers can be used to create plasmas with unique electron energy distributions. Here we report experiments and analysis that exploit and critically examine that capability.

In the first experiment, Thomson scattering was used to determine electron and ion temperatures of an under-dense plasma produced by a short-pulse laser.[1] We use a two-pulse technique to create and probe the plasma on a time-scale short enough that no significant plasma cooling occurs. Experiments were performed in a regime where ionization heating determined the electron energy. These experiments tested the accuracy of the tunneling model at densities and on time-scales relevant to recombination lasers.

Our experiments used an initial laser pulse[2] (125 fs, 30 mJ, 800 nm) to ionize helium gas and a second, co-linear pulse at 400 nm to probe the pre-formed plasma. The probe pulse is produced by frequency doubling the 800 nm pulse in KDP. We calculate that the wavelength dependent index of refraction of KDP results in a maximum pump-probe delay of 80 fs. We vary this delay up to 1 ps by passing both pulses through a glass window. This two-pulse technique avoids

© 1994 American Institute of Physics

complications that could arise from modification of the laser spectrum due to ionization induced blue-shifting.[3] Our ionizing laser intensity (2×10^{17} W/cm^2) is a factor of twenty in excess of the intensity necessary[4] to produce He^{2+}, which insures that He is fully ionized over the spatial dimension of our focused probe pulse.

Thomson-scattered spectra were used to determine electron and ion temperatures in the He plasma; the results are shown in Fig. 1. We find an experimental electron temperature of approximately 40 eV at all pressures studied. This measurement agrees with the prediction of tunneling theory for our conditions.[1] Ion temperature is observed to increases with pressure. This may result from ion-ion collisions in the laser and local space-charge fields.

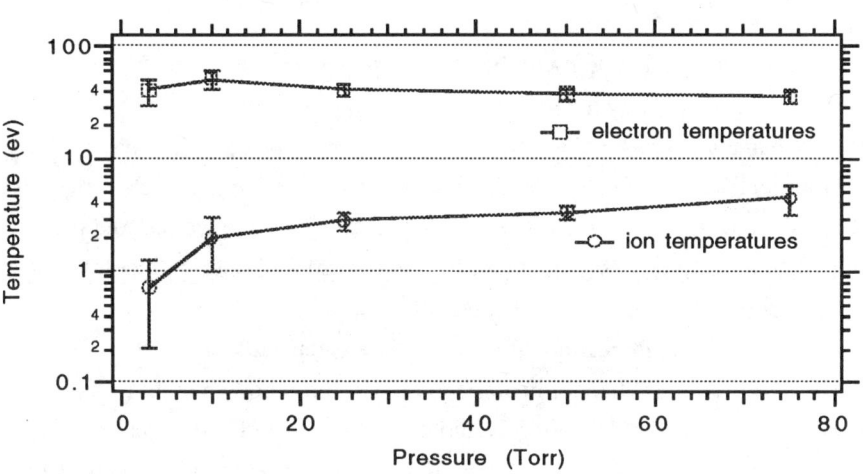

Figure 1. Measured temperatures as functions of pressure.

In a second experiment we generated a plasma appropriate for a recombination pumped x-ray laser on the Ly-α transition of H-like Li at 13.5 nm.[5,6] We carried out a gain-length study of the Ly-α transition, and the results are shown in Fig. 2. The data suggests non-linear growth of the 135 Å signal for short (< 2 mm) plasma lengths; however, the signal level saturates at longer plasma lengths. We believe that this apparent saturation is the result of breakup of the pump beam following ionization defocusing in the Li plasma.[7]

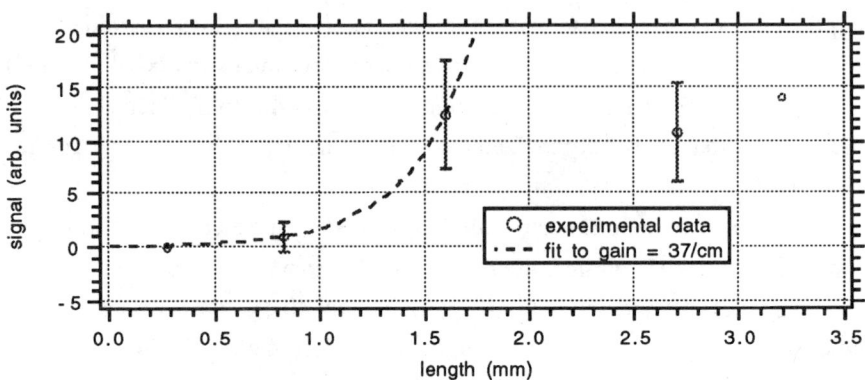

Figure 2. Ly-α emission in H-like Li as a function of plasma length.

Gain in the Li system will be strongly dependent on the electron distribution generated by the high-intensity pump laser. In order to study this effect, we analyzed how gain is affected by a multiply peaked electron distribution function. Such a distribution can result from optical-field ionization (OFI) of sequentially ionized atoms.[5] We further investigated the role of collisional heating processes that modify electron energies during recombination and the accompanying collisional cascade.

Multiphoton ionization of loosely bound electrons yields a cold distribution, T_e(cold), that dominates subsequent three-body recombination. However, collisional heating of this initially cold distribution can abruptly limit gain in these systems. Additionally, collisional excitation of the lasing ions by higher-energy, OFI electrons can inhibit the cascade of ion population to the lower energy levels and also limit gain. We quantified these effects using the atomic and plasma physics kinetics code FLY.[8]

We first consider the effect of the OFI electrons on the value of the gain that can be achieved in the Li system. Figure 3 shows calculated peak gain (assuming quasi-static Stark broadening[5]) on Ly-α as a function of the OFI electron temperature, T_e(OFI).

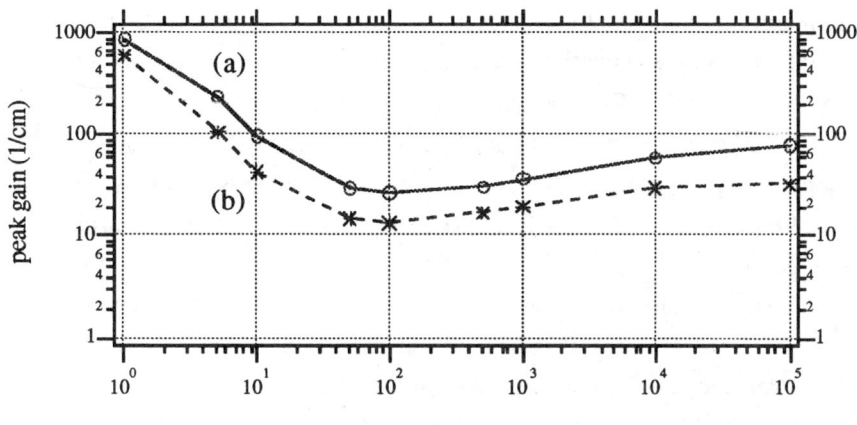

Figure 3. Peak gain on Ly-α as a function of the OFI electron temperature. T_e(cold) is held constant at 1 eV for both curves. Curve (a) represents a total electron density n_e(total) = 3×10^{19} cm^{-3}, where n_e(cold) = 1×10^{19} cm^{-3} and n_e(OFI) = 2×10^{19} cm^{-3}; curve (b) represents n_e(total) = 1.5×10^{19} cm^{-3}, where n_e(cold) = 5×10^{18} cm^{-3} and n_e(OFI) = 1×10^{19} cm^{-3}.

We neglect hydrodynamic effects in order to isolate the kinetic effects of the OFI electrons in the Li system. This assumption is reasonable for plasma columns with diameters > 100 μm.

The three-body recombination rate to all levels is given as: [11]

$$R_{3-body} = \frac{1.6 \times 10^{-9} Z^3 n_e^2 \ln(\Lambda)}{T_e^{9/2}} \; \text{sec}^{-1} \; . \quad (1)$$

It depends strongly on electron temperature T_e (K) and density n_e (cm^{-3}); Z is the ion charge state and $\ln(\Lambda)$ is the Coulomb logarithm. Low OFI electron temperatures (1 to 10 eV) result in collisional de-excitation of excited states of Li^{2+} to n = 2.[10] The result is that cold OFI electron temperatures will generate a high value of peak gain, as shown in Fig. 3.

If the OFI electrons are hotter, both three-body recombination and gain are reduced. Hotter OFI electrons do not contribute to the collisional cascade of population to n = 2. In fact, they result in a distribution of population over the entire manifold of excited Li^{2+} states (characterized by an effectively high-temperature Boltzmann distribution) and therefore a reduction of gain on Ly-α compared to the cold OFI situation.

If OFI electrons have energies that are significantly larger than the transition energies in the Li^{2+} ion, electron-ion collision cross-sections decrease and the OFI electrons decouple from the ions. Therefore, hot OFI electrons will neither participate in three-body recombination nor collisionally excite Li^{2+} ions. This allows the cold electrons that recombine to form Li^{2+} to cascade to n = 2 efficiently. This effect is reflected in an increased value of the peak gain as the OFI electron temperature increases beyond 100 eV.

Hotter OFI electrons also have a decreasing electron-electron collision frequency, and thus are decoupled from the cold electron distribution. However, if the OFI electrons have energies on the order of 10-100 eV, electron equilibration between the OFI and cold electron distributions will rapidly heat the cold electrons. Spitzer [9] gives the electron-electron equilibration time t_{eq}, between two electron temperature distributions T_1 and T_2, as:

$$t_{eq} = 0.137 \frac{(T_1+T_2)^{3/2}}{n_e \ln(\Lambda)} \text{ sec.} \qquad (2)$$

Thus, for a Li plasma where $n_e(\text{cold}) = 1 \times 10^{19}$ cm^{-3}, $T_e(\text{cold}) = 1$ eV, and the first OFI electron is described by $n_e(\text{OFI}) = 1 \times 10^{19}$ cm^{-3} and $T_e(\text{OFI}) = 30$ eV, we calculate an equilibration time on the order of 1 ps. This equilibration rate is fast relative to the time scale of the gain ($\approx$ 10 ps for our conditions) and would be the dominant gain limiting mechanism for the system.

Recombination pumped systems must therefore operate with densities and temperatures where electron equilibration with hot OFI electrons will not disturb a deliberately established cold electron distribution function. At a density of 1×10^{19} cm^{-3}, electron temperature distributions of 1 eV and 1 keV would equilibrate on the (relatively long) time scale of approximately 50 ps. OFI electron temperatures on the order of 1 keV could be generated by using circularly polarized pulses focused to an intensity of 2×10^{17} W/cm^2. This would leave the energy of the coldest portion of the electron distribution (first to be ionized) unaltered because it is generated in the multi-photon ionization regime. This scheme of producing keV OFI electrons would further enhance the gain by increasing the collisional decoupling the OFI electron from the lasing ions.

Another gain limiting mechanism in the Li system is collisional heating of the low energy portion of the electron distribution. Conservation of energy requires heating of the electron bath following three-body recombination and collisional cascade. We calculate these effects in the Li system using FLY to track the population of each of the Li^{2+} energy levels as a function of time. This procedure involves maintaining a self-consistent free-electron temperature during the simulation of plasma dynamics.

Figure 4 shows the resulting gain-time history; peak gain drops by about a factor of 3 when the effects of collisional heating are included.

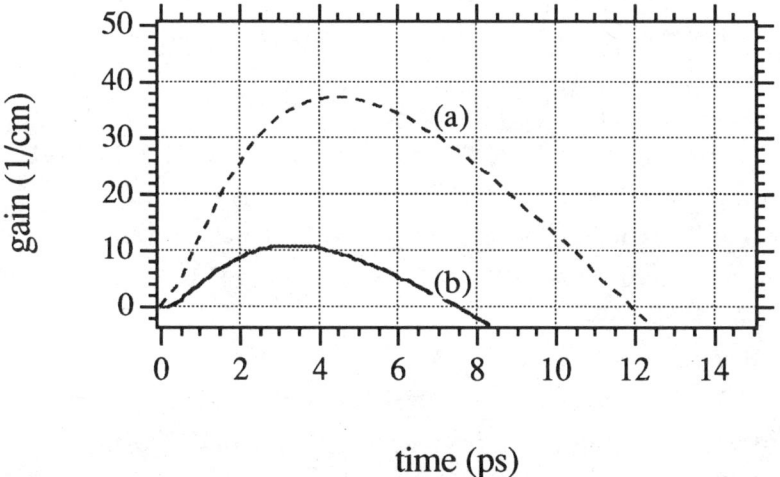

Figure 4. Time history of gain on Ly-α in Li^{2+} for the case of (a) no collisional electron heating, and (b) collisional electron heating. The curves were generated for $n_{ion} = 1\times10^{19}$ cm^{-3}, T_e(cold) = 1 eV, n_e(cold) = 1×10^{19}cm^{-3}, T_e(OFI) = 1 keV and n_e(OFI) = 2×10^{19} cm^{-3}.

The implication of the heating of initially cold electrons, by either collisional processes or through electron equilibration if linearly polarized light is used for the pump pulse, is an increase in the required electron density for the Li^{2+} laser in order to achieve a particular value of gain.

In conclusion, we find that collisional processes involving OFI electrons and collisional heating of low-energy electrons play important roles in determining gain in the Li^{2+} recombination pumped x-ray laser system. We believe that using circularly polarized light to ionize the Li plasma will decouple OFI electrons from the system and allow higher gains to be achieved.

This work was supported by the US Air Force Office of Scientific Research and through a collaboration with Lawrence Livermore National Laboratory under contract W-7405-ENG-48.

[1] Lawrence Livermore National Laboratory, Livermore, CA 94550

REFERENCES

1. T. E. Glover, T. D. Donnelly, E. A. Lipman, A. Sullivan, R. W. Falcone, "Sub-Picosecond Thomson Measurements of Optically Ionized Helium Plasmas," (to be published in Phys. Rev. Lett.).
2. A. Sullivan *et al*, Opt. Lett. **16**, 1406 (1991).
3. W. M. Wood, C. W. Siders, M. C. Downer, Phys. Rev. Lett. **67**, 3523 (1991).
4. S. Augst *et al*, Phys. Rev. Lett. **63**, 2212 (1989).
5. D. Eder *et al*, Phys. Rev. A. **45**, 6761 (1992); N. H. Burnett, G. D. Enright, IEEE J. Quantum Electron. **26,** 1797 (1990); N. H. Burnett, P. B. Corkum, J. Opt. Soc. Am. B **6**, 1195 (1989).
6. Y. Nagata *et al*, Phys. Rev. Lett. **71**, 3774 (1993).
7. A. Sullivan *et al*, "Propagation of Intense, Ultrashort Laser Pulses in Plasmas", in Shortwavelength V: Physics with Intense Laser Pulses, M. D. Perry and P. B. Corkum, eds. (OSA, Washington, DC, 1993), p. 40.
8. R. W. Lee, User Manual for RATION (unpublished).
9. L. Spitzer, *Physics of Fully Ionized Gases*, (Interscience, New York, 1962).
10. G. J. Pert, J. Phys. B **23**, 619 (1990).
11. Y. Zel'Dovich, Y. Raizer, *Physics of shock waves and high-temperature hydrodynamic phenomena*, W. Hayes, R. Probestein, eds. (Academic Press, New York, 1966).

Application of a Plasma Waveguide to X-ray Lasers

H.M. Milchberg, C.G. Durfee III, and J. Lynch

Institute for Physical Science and Technology
University of Maryland, College Park, MD 20742

Abstract. A recently developed laser-produced plasma channel is shown to be a promising means to produce a transversely coherent table-top x-ray laser. The channel is ideal for efficient high power laser pumping through optical waveguiding of the pump pulse and also acts as a waveguide for generated x-rays. It is shown that channel creation and guided pulses of moderate duration and energy can be highly effective in producing nonequilibrium plasmas and candidate ions appropriate for soft x-ray lasers. Examples are given for recombination-pumped and collisionally pumped systems.

INTRODUCTION

In the ten years since the first demonstrations of stimulated emission of soft x-rays [1,2], a major concern has been the large optical laser pumping requirements (> 1 kJ) to produce population inversions resulting in substantial (more than a few microjoules) x-ray output. The need for large optical pump sources in these early experiments has impeded the broad scientific impact that these soft x-ray lasers may otherwise have had. The large pump laser energy requirement in previous work has been driven by a need to produce an abundance of particular species of highly charged ions from slab or foil targets with pulses no shorter than 100 ps, and more commonly longer than 1 ns in duration. In the case of recombination pumping schemes, it is the recombination of these ions (typically fully stripped or He-like) to the next lower ion stage which results in population inversions. In collisional pumping schemes, the ground state of these ions (typically a closed shell structure such as Ne-like or Ni-like) is pumped by electron collisions. In these cases, almost all of the pump laser energy incident on target is conductively dissipated, radiated away as line and continuum emission, or directed into plasma hydrodynamic motion. In fact, the plasma densities at which gain has been generated is limited by thermalizing collisions and x-ray refraction to $N_e \lesssim 10^{20}$ cm^{-3}, down at least 3 orders of magnitude from the density of the initial solid targets. As a result, typical soft x-ray laser output to pump laser input efficiencies have been ~ 10^{-6}. Moreover, the presence of a high density boundary condition associated with solid or slab targets results in large electron density gradients which refract any soft x-ray laser beams produced. Refraction currently limits the gain-length product in electron collision pumped systems [3].

It would seem that gas targets might have been more promising: their initial density is closer to the final density at which gain is generated (1 atm of gas at single ionization yields $N_e \sim 10^{19}$ cm^{-3}) and the radiative, hydrodynamic, and conductive losses are lower than those of solid targets (although the absorbed fraction is reduced). However, the advantages of solid targets compared to gases are their reduced intensity threshold for ionization to stages of interest and high absorption (since the ionization and heating of solid targets is strongly collision dominated), making possible the use of long line focus geometries, and the less crucial role of gaseous plasma propagation issues such as filamentation and beam breakup [4] of the pump pulse as it approaches the focus.

In the last five years, progress in the development of high energy, high repetition rate (> 10 Hz) subpicosecond pulse laser systems [5] has led to suggestions for both field ionization induced recombination lasers [6] and collisionally excited systems [7]. The preformed channels described here could successfully guide intense subpicosecond pulses and eliminate beam propagation issues such as refraction from plasma creation [8] and Raman instabilities [9]. To create ions for transitions at the shortest wavelengths, however, field ionization cannot compete with collisional ionization, which can be much more effective at lower intensities. In addition, for shorter desired wavelengths and higher densities, Raman heating becomes significant [10], reducing the effectiveness of short pulse driving of recombination lasers. We show below that pulses of moderate width ($\lesssim$100ps) and energy (<100 mJ) can be very effective in producing candidate x-ray laser ions.

HIGH INTENSITY PLASMA WAVEGUIDE

Recently we demonstrated a technique for optically guiding intense laser pulses, in which a plasma refractive index channel was produced through the hydrodynamic evolution of a laser breakdown spark in an ambient gas [11]. The breakdown spark drives a radial shock wave in the ion density, leaving an ion density minimum on axis. Since strong electrostatic forces keep the plasma neutral (to within a Debye length of charge separation), the electron density also develops an on-axis minimum, appropriate for guiding a laser pulse. Figure 1 shows a calculation of gas response to irradiation by a typical pulse used in our experiments. The plasmas produced by the conditions of our experiment can be quite nonequilibrium in terms of ionization level for a given temperature. The important hydrodynamic timescales of experimental relevance are the shock generation timescale, $\tau_s = \lambda_{ii}/c_s$, and the plasma column evolution timescale $\tau_c = w_0/c_s$. Here, λ_{ii} is the ion-ion collisional mean free path, c_s is the local sound

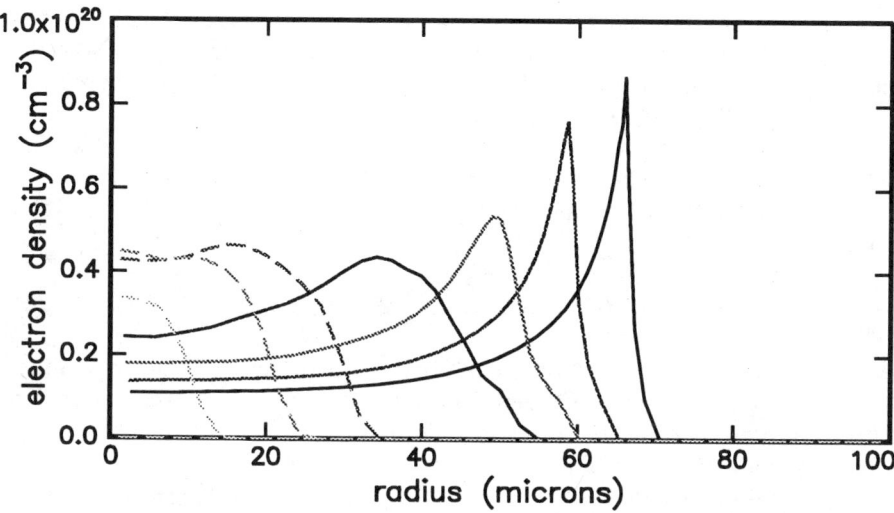

FIGURE 1. Calculation of response of 150 torr N_2 to peak laser intensity 5×10^{13} W/cm^2, $\lambda=1.064\mu$m, pulsewidth 100 ps. The dashed lines (laser on) are separated by 50 ps and the solid lines (laser off) by 450 ps. The laser pulse peaks at 120ps.

speed, and w_0 is the laser spot size. At gas densities less than approximately 1 atm, both of these times are comparable to or shorter than the plasma recombination timescale, while ionization, depending on whether field or collision dominated, may be faster or slower. The electron temperature increases on the time scale of the laser pulse and decreases according to the cooling loss channels available, such as thermal conduction and expansion. For our range of gas pressure, 5-300 torr, and $w_0=3-10\mu$m we have $\tau_s \sim 100$ ps and $\tau_c \sim 1$ ns.

In our experiments thusfar, plasma channels as long as 2.2 cm (70 Rayleigh lengths) have been generated and intensities up to 2×10^{14} W/cm^2 (40 mJ, 100ps, $w_0=10\mu$m) have been guided, with throughputs up to 75%. Figure 3 of ref. 11 shows our experimental setup for the guiding experiments. A 10 Hz Nd:YAG regenerative amplifier system [13] produces 250 mJ, $\tau=100$ps pulses at $\lambda=1.064\mu$m. The pulses are split, with 100-200 mJ directed to an axicon lens, and up to 50 mJ directed to a coupling lens. A corner cube mounted on an optical delay rail provides -1 to 15 ns delay between the axicon pulse and the lens pulse. The axicons which we use are glass cones, with base angles $\alpha = 20°-35°$. An incident beam refracts at the cone surface, forming a conical wave which interferes with itself, resulting in a Bessel function field profile $E(r,z)=|E_0(z)|^2 J_0^2(kr\sin\gamma)$, where $|E_0(z)|^2$ depends on the transverse field distribution incident on the axicon and γ is the angle of the conical wave normal with the optical axis.. An axial hole is present to allow passage of the guided beam.

A condition on the modes that will be supported by a plasma refractive index channel can be estimated by assuming an index variation $n(r)= 1 - 1/2\, N_e(r)/N_{cr}$ along with a parabolic electron density profile $N_e(r)=N_e(0)+N_{cr}(r/a)^2$, where a is a curvature parameter. If these expressions hold for $0<r<\infty$, there is an infinite number of bound modes (Laguerre-Gaussian). For a channel with $n(r)=n_{min}$ (or $N_e(r) = N(r_m)=N_e^{max}$) for $r>r_m$, there is a finite number of bound modes. The eigenvalue condition on the infinite channel propagation wavenumber can be used to obtain approximate cutoff and guiding conditions, respectively for the finite channel:

$$N_e^{max} - N_e^{min}=\Delta N_e(r_m) \geq (2p+m+1)^2/\pi r_e r_m^2 \; ; \; \Delta N_e(w_{ch})=(2p+m+1)^2/\pi r_e w_{ch}^2 \quad (1)$$

Here r_e is the classical electron radius, $N_e^{min}=N_e(0)$, $\Delta N_e(w_{ch})=N_e(w_{ch})-N_e(0)$, $p \geq 0$ and $m \geq 0$ are radial and azimuthal mode indices, and w_{ch} is the 1/e field radius of the p=0, m=0 mode. For a non-parabolic finite channel, these conditions are still a very good approximation.

It is useful to review some of our previous results [11,12] relevant for x-ray laser applications. Among the guiding regimes observed is single mode propagation. As indicated by Eqn. (1), reduced gas fill pressure (and thus reduced ΔN_e) can restrict propagation to the lowest order mode.

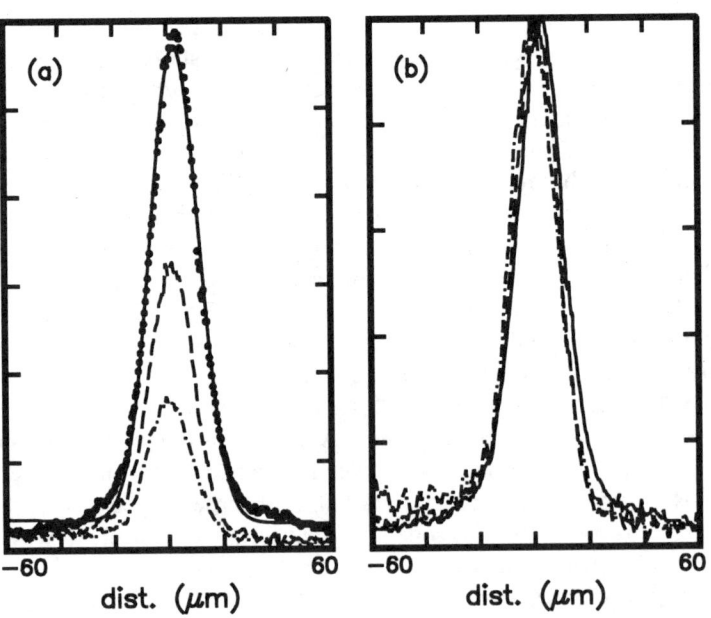

FIGURE 2. (a) Exit mode image lineouts for coupling beam waist moved away from channel entrance in 0.625 mm increments. One lineout is shown fit to a Gaussian. **(b)** normalized plots of (a).

Figure 2(a) shows lineouts of exit mode images for a sequence of axial coupling lens positions, with the plots of (b) normalized. Similar exit modes result if the coupling lens is moved sideways, although the coupling decrease compared to 2(a) is much more pronounced. This is consistent with single mode waveguide behaviour. Note the very good fit to a Gaussian of one of the curves in (a), indicating that the lowest order mode shape is somewhat insensitive to the index curvature [12].

APPLICATION TO X-RAY LASERS

For X-ray laser development, the plasma waveguide created in gas targets offers great advantages over slab or foil target geometries. The X-ray refraction problem characteristic of slab and foil targets is eliminated. In fact, as will be shown below, *the plasma waveguide acts as a waveguide for X-rays*. The plasmas produced can be highly non-equilibrium, of importance to both recombination and collisional pumping schemes. Gas target densities are closer to the ultimate operating densities of X-ray laser plasmas so that pump laser energy is used more efficiently. The guided pulse pump configuration delivers higher intensity at considerably lower pulse energy than line focus geometry. By taking advantage of collisional ionization of gases, candidate x-ray laser ions can be produced at moderate pulse duration and energy.

X-ray guiding by plasma waveguides

Equation (1) shows that the requirement on the electron density difference for guiding is wavelength independent, so that an x-ray beam with spot size w_0 is guided as well as an optical beam of the same spot size. This is due to the fact that the decreasing change in the plasma refractive index with light frequency, $-\omega_p^2/\omega^2$, is balanced by reduced diffraction at higher frequency for a given spot size. This can be shown simply as follows. The optical path difference induced by the plasma waveguide between the centre of the beam and the beam edge (taken at $\sim w_0$) is approximately $\Delta s = \Delta n \, d$, where d is the propagation distance along the waveguide and $\Delta n = \Delta(\omega_p^2)/\omega^2$ is proportional to the electron density difference across the beam, with $\omega_p^2 = 4\pi N_e e^2/m$. The effective angle of convergence of the wavefront due to the phase lag at beam centre with respect to the edge is then given by $\theta_{conv} = 2\Delta s/w$. The angle of diffraction is $\theta_{diff} = w_0/z_0 = \lambda/\pi w_0$, where $z_0 = \pi w_0^2/\lambda$ is the Rayleigh length. Setting the guiding condition as $\theta_{conv} = \theta_{diff}$ with $d = z_0$ gives $\Delta N_e = 1/\pi r_e w_0^2$, independent of wavelength and agreeing with Eqn. (1) for the case p=m=0. The implication of this result is that the plasma waveguide is as good a guiding structure for x-rays as it is for the optical pump pulse, and the possibility for transversely coherent x-ray output is evident.

Nonequilibrium plasmas

The gas density range appropriate for guiding and useful for x-ray lasers, a few torr through ~ 1 atm, is set at the low end by the requirement that the shock thickness (~λ_{ii}) be smaller than the laser spot size (in order that a distinct channel be recognized by the beam to be guided) and is limited at the high end by deleterious collisional mixing of the upper and lower levels of candidate laser transitions. In this density range, the plasma hydrodynamic and heating/cooling times may be shorter than the excitation/recombination times for the plasma ions. An example of this is shown in Fig. 3, a calculation of (a) Temperature profiles vs. time for 150 torr of N_2 irradiated by a $\lambda=1.064\mu m$, $\tau=100ps$, $w_0=10\mu m$, 5×10^{13} W/cm^2 peak intensity pulse and (b) <Z>, the corresponding average ionization level. The model includes tunneling ionization in the laser field [14], inverse bremsstrahlung heating, thermal conduction, and collision-based ionization and recombination. It is seen that peak <Z> ~5, characteristic of an abundance of He-like nitrogen, and this remains long after the peak temperature has dropped to ~10-20 eV. The temperature has declined rapidly through thermal conduction to the plasma and weakly ionized gas at the periphery of the channel, decreasing much faster than a recombination time (a few nanoseconds here). These are appropriate conditions for a recombination laser in Li-like nitrogen. The intensity used in this calculation is typical of intensities produced on axis by our axicons, so that a Li-like recombination laser need not have a guided heating pulse injected into the channel.

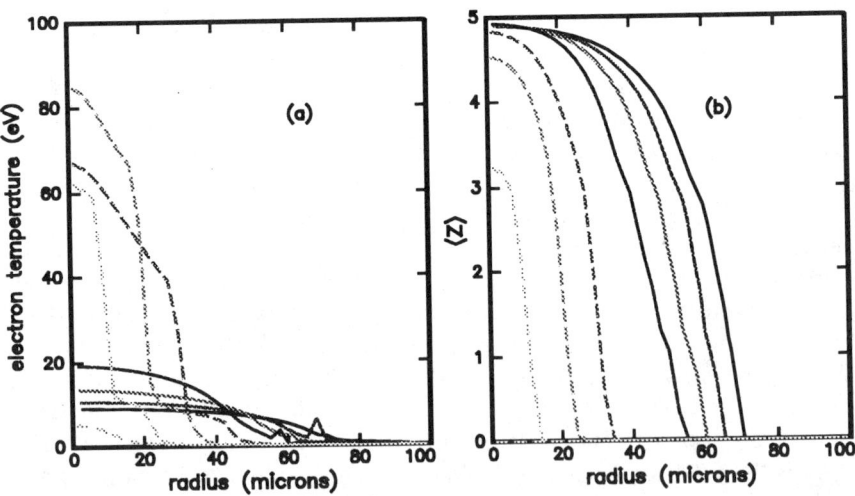

FIGURE 3. (a) T_e (electron temperature) profile vs. time for 150 torr N_2 irradiated by 5×10^{13} W/cm^2 peak intensity, $w_0=10\mu m$, $\lambda=1.064\mu m$, 100ps pulse. (b) corresponding <Z> profile vs. time. Dashed lines (laser on) are separated by 50 ps, solid (laser off) by 450 ps. The last curve is at t=2ns.

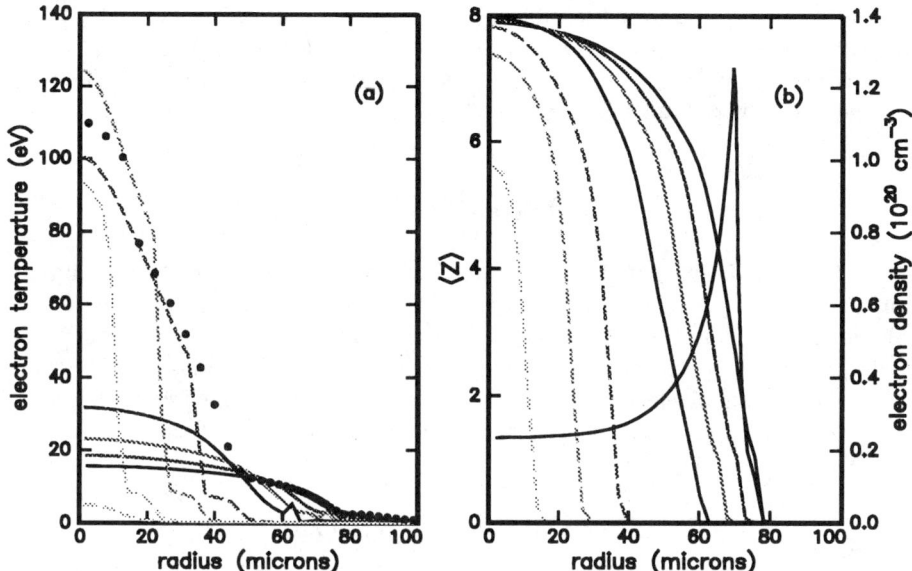

FIGURE 4. (a) T_e profile vs. time for 300 torr Ar irradiated by peak intensity 5×10^{13} W/cm^2, $\lambda = 1.064 \mu$m, $w_0 = 10 \mu$m, pulsewidth 100ps. The peak temperature induced by a 10^{14} W/cm^2, 100ps pulse injected at t=2ns is overlaid as dots. **(b)** left scale: corresponding <Z> profile vs. time; right scale: N_e profile at t=2.2ns, unchanged by the second pulse. Dashed lines (laser on) are separated by 50ps, solid (laser off) by 450ps. The laser peaks at 120ps.

For collisional pumping, however, a guided pulse is necessary to generate a heated electron distribution. Figure 4 shows a calculation of (a) electron temperature vs. time for 300 torr of Ar for the same pulse conditions as above and (b) corresponding <Z> profiles, where the peak abundance is that of the Ne-like species. The electron density profile at t=2ns overlays this plot. At t=2ns, the electron temperature is too low to adequately excite the 2p-3p transition in Ne-like Ar. The effect on the electron temperature of injecting a secondary $\lambda = 1.064 \mu$m, $\tau = 100$ps, $w_0 = 10 \mu$m, 10^{14} W/cm^2 peak intensity pulse is shown overlaid in (a). While high enough to effect ionization of Ne-like ions, this peak temperature rises and falls much faster than the Ne-like to F-like ionization timescale. The second pulse leaves <Z> and $N_e(r)$ unaffected, but serves as a very effective pump for 2p-3p excitations.

Pumping efficiency

We use as a requirement the delivery of 10^{14} W/cm^2 peak intensity to a 10μm radius x 1cm long volume in a 100ps pulse. This is adequate for a number of collisional schemes, among them Ne-like Ar, as shown above. If a cylindrical lens is used, ~13 J must be provided. With a solid or slab target, the focal width must be larger to avoid strong refraction of resulting x-rays, further increasing this energy requirement. In early Ne-like Se experiments [1], with focal widths of ~100μm and 1 ns pulses, ~1 kJ was used. By comparison, for a 100ps, 10^{14} W/cm^2 pulse guided at w_0=10μm by a preformed channel, only 20 mJ is required.

CONCLUSIONS

The plasma waveguide shows great promise as a vehicle for producing a compact, efficient, transversely coherent soft x-ray laser. The plasma waveguide will guide x-rays as effectively as the optical pump pulse. A promising and efficient approach for producing candidate x-ray laser ions is the use of moderate pulsewidth and energy in the collisional ionization regime. Guided end-pumping is shown to be far more efficient than side pumping with a line focus.

ACKNOWLEDGEMENTS

The authors thank T.J. McIlrath for useful discussions. This research is supported by the NSF(ECS-8858062 and ECS-9224520), the AFOSR(F49620-92-J-0059), and BMDO-T-IS.

REFERENCES

1. D.L. Matthews et al., Phys. Rev. Lett., **54**, 110 (1985).
2. S. Suckewer, C. H. Skinner, H.M. Milchberg, C. Keane, and D. Voorhees, Phys. Rev. Lett. **55**, 1753 (1985).
3. R. A. London, Phys. Fluids **31**, 184 (1988).
4. P.E. Young, H.A. Baldis, R.P. Drake, E.M. Campbell, and K.G. Estabrook, Phys. Rev. Lett. **61**, 2336 (1988).
5. C.P.J. Barty, B.E. Lemoff, and C.L. Gordon III, in *Ultrafast Phenomena 1994 Technical Digest*, vol. 7, p. 324 (Optical Society of America, 1994); J. Zhou et al., ibid, p. 327.
6. N.H. Burnett and P.B. Corkum, J. Opt. Soc. Am. B **6**, 1995 (1989).
7. B.E. Lemoff, C.P.J. Barty, and S.E. Harris, Opt. Lett. (1994).
8. P. Monot, T. Auguste, L.A. Lompre', G. Mainfray, and C. Manus, J. Opt. Soc. Am. B **9**, 1579 (1992).
9. T. M. Antonsen and P. Mora, Phys. Fluids B **5**, 1440 (1993).
10. D.C. Eder, P. Amendt, and S.C. Wilks, Phys. Rev. A **45**, 6761 (1992).
11. C.G. Durfee III and H.M. Milchberg, Phys. Rev. Lett. **71**, 2409 (1993)
12. C.G. Durfee III, J. Lynch and H.M. Milchberg, submitted for publication.
13. C.G. Durfee III and H.M. Milchberg, Opt. Lett. **17**, 37 (1992).
14. M.V. Ammosov, N.B. Delone, and V.P. Krainov, Sov. Phys. JETP **64**, 1191 (1986).

Short-Pulse Laser-Produced Plasma from C_{60} Molecules

Cornelius Wülker, Wolfgang Theobald, Donald Ouw, and Fritz P. Schäfer

Max-Planck-Institut für biophysikalische Chemie,
Postfach 2841, D-37018 Göttingen, Germany

Boris N. Chichkov

P.N. Lebedev Physical Institute,
Leninsky prospect 53, Moscow 117924, Russia

Abstract. The first experimental observations of a plasma produced in a vapor of C_{60} molecules with a high-intensity subpicosecond KrF laser (6×10^{15} W/cm^2) are reported (1). It differs from a plasma created in an ordinary carbon preplasma by reaching much higher ionization stages under the same experimental conditions. This remarkable property of C_{60} molecules (and other clusters (2)) opens new prospects for short-pulse driven X-ray lasers.

Recently it was suggested that a short-pulse laser-produced plasma in a vapor of C_{60} molecules can be an interesting object for X-ray laser research (3). If we directly apply the optical field ionization concept to obtain X-ray lasing with H-like carbon ions, we would need a short-pulse laser intensity of $I \geq 3 \times 10^{18}$ W/cm^2. There is practically no hope that electrons can remain cold under such intensities. The idea to use C_{60} clusters instead of carbon atoms has the following reasons: a) the energy deposition into C_{60} molecules can be much more efficient (a strong reduction of the necessary short pulse laser intensity), b) the spherical expansion of the resulting molecular microplasmas created due to rapid ionization of C_{60} molecules can result in strong electron cooling.

The aim of this paper is to study the interaction of intense short-pulse ultraviolet laser radiation with a vapor of C_{60} molecules. We have investigated the soft X-ray spectra emitted by a plasma created in a vapor of C_{60} molecules. The experimental set-up (1) is shown in Fig. 1. The C_{60} vapor is created by gently heating a C_{60} layer with a standard long-pulse (25 ns) KrF-excimer laser. The width of the line focus is 150 μm, and the maximum length is 3.5 mm. For the C_{60} measurements, the energy of the long pulse is 10 mJ which

© 1994 American Institute of Physics

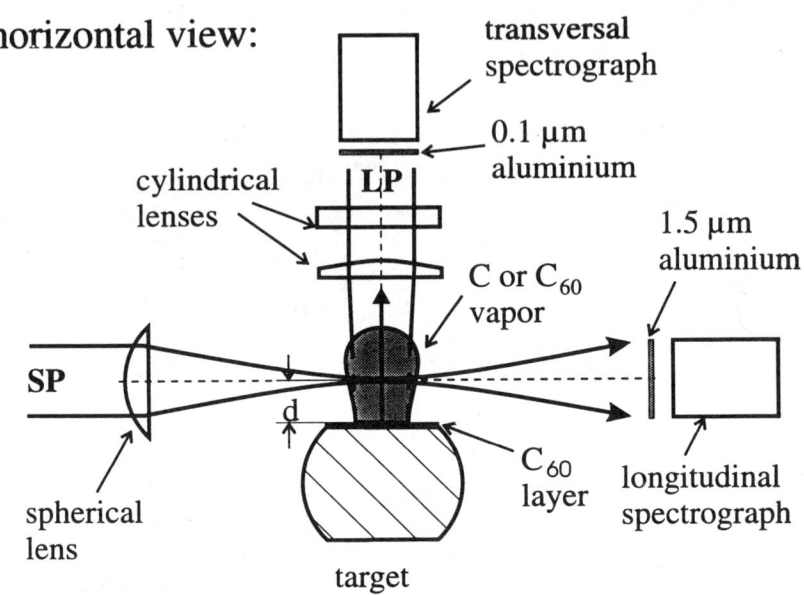

Figure 1: Schematic of the experimental set-up.

corresponds to 4×10^7 W/cm². For comparison, the measurements with glassy carbon are done with an intensity of 4×10^8 W/cm² (an ablation intensity of 4×10^7 W/cm² does not give a sufficient signal in this case). At a certain distance d from the plane surface, a subpicosecond (0.7 ps) high-intensity KrF laser beam (SP) parallel to the surface is focussed into the C_{60} (or carbon) vapor to a spot of 25 µm diameter with a confocal parameter of 2.5 mm, which was directly measured. The wavelength of both lasers is 248 nm. The distance d and the delay time τ, which is defined as the time between the leading edge of the long laser pulse and the short pulse at the interaction volume, can be varied. Two similarly constructed soft X-ray spectrographs ("longitudinal" and "transversal") record (in a single shot) the radiation which is emitted by the plasma generated in the vapor of C_{60} molecules (or ordinary carbon) by the short laser pulse.

In Fig. 2(a,b) the "longitudinal" spectra of plasmas produced in a vapor of C_{60} molecules (a) and in an ordinary carbon vapor (b) are compared. The spectra are recorded for different time delays between the short and the long laser pulses and $d = 500$ µm. Fig. 3 reproduces a part of Fig. 2(a) and shows the location of various spectral lines. The difference in the spectra is really striking. In the case of carbon vapor preplasma the spectrum consists

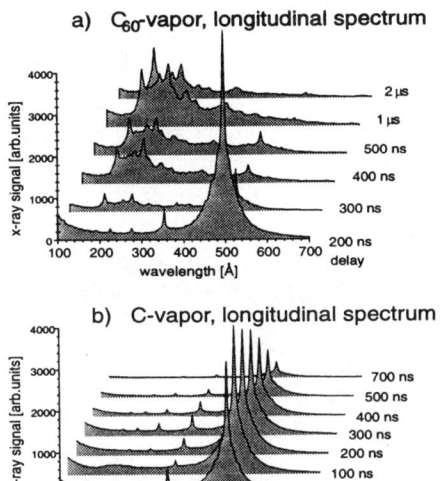
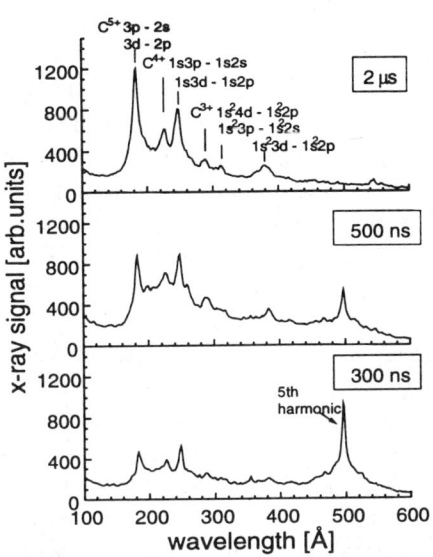

Figure 2: Comparison of "longitudinal" spectra of plasmas produced in a vapor of C_{60}-molecules (a) and in an ordinary Carbon vapor (b). The long pulse laser intensity was $I_{LP}=4\times10^7$ W/cm^2 for (a) and $I_{LP}=4\times10^8$ W/cm^2 for (b). The Intensity of the short pulse was $I_{SP}=6\times10^5$ W/cm^2 in both cases. The spectra are shown for different time delays between the short and long laser pulses. The short laser pulse was focused 500 μm away from the target surface.

Figure 3: Three selected spectra from Fig. 2(a) with line identifications. The parameter is the delay between the long and the short laser pulse (see text).

of odd harmonics (5-13) of the short-pulse KrF laser. The plasma signal is very weak. In contrast to this, with the C_{60}-vapor the plasma signal is very strong and the harmonics are weak. Ionic lines of all ionic states up to the H-like carbon can be seen. Note that the 182 Å line (C^{5+} $3d-2p$) increases for longer delays (see Fig. 3). In all our previous experiments with normal carbon targets (preplasma or solid state), this line was never observed so intense (4). It is most remarkable that the H-like ionic lines are still observed, when the short-pulse laser intensity is as low as 10^{14} W/cm^2 !

The origin of this difference becomes clear if we compare the ionization mechanisms of carbon atoms and C_{60} molecules. The carbon vapor 500 μm away from the target has a relatively low density ($\leq 10^{18}$ cm^{-3}). Therefore during the interaction with the short laser pulse the plasma can be created only due to tunnel ionization of carbon atoms (ions). And only tunnel ionization can compete with the process of high-order harmonic generation. The strong harmonic signal means that tunnel ionization is negligible.

The disappearance of harmonics in the case of C_{60} vapors is explained by the presence of more effective ionization mechanism (see discussion below). These molecules, which can be considered as a small (10.1 Å size) dense solid balls, are rapidly ionized due to electron impact ionization. Note that for a 200 ns delay, see Fig. 2(a), the "longitudinal" spectrum is analogous to the spectra shown in Fig. 2(b), since for this delay only single carbon atoms (ions) are able to reach the interaction region with the short laser pulse. More heavy and slow molecules reach this region later in time.

As can be seen in Fig. 4(a,b), where "transversal" spectra are compared, the Li- and Be-like lines are clearly dominating in the spectrum of the carbon plasma. In the case of plasma created in C_{60} molecules, the H-, He-, Li- and Be-like carbon lines have comparable intensities. This means that for our experimental conditions the C_{60} plasma has a very broad distribution over the

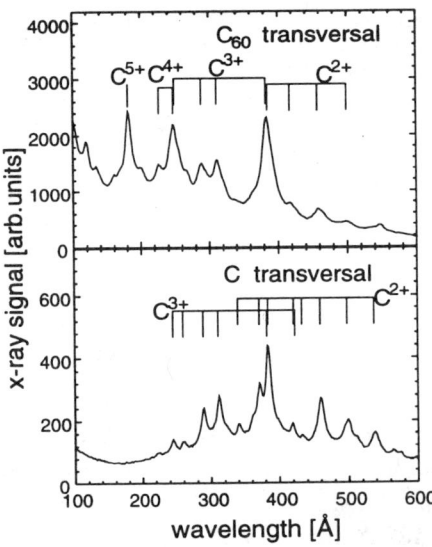
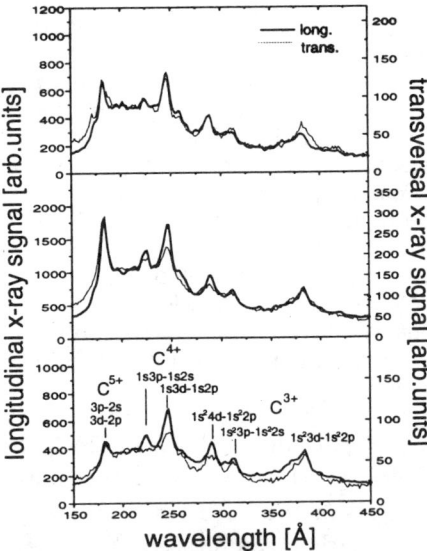

Figure 4: Comparison of "transversal" spectra for the C_{60}-target (a) and the C-target (b). The ionic states of the corresponding lines are shown. The delay is 2 μs for the C_{60}-spectra and 200 ns for the C-spectra. The Al-filter thickness is 0.1 μm.

Figure 5: Comparison of "longitudinal" and "transversal" spectra for different shots with the C_{60}-target. The Al-filter thickness is 1.5 μm for all spectra.

charge states ($C^{2+} \div C^{6+}$). This is the signature of a strongly nonequilibrium plasma state.

In Fig. 5 "longitudinal" and "transversal" spectra of a plasma produced in a vapor of C_{60} molecules are compared. They are recorded with the same, 1.5μm thick, aluminium filters. For approximately a half of laser shots the "longitudinal" and "transversal" spectra are identical. For another half there is some difference in spectra illustrated in Fig. 5. The He-like lines $1s3p(^3P) - 1s2s(^3S)$ and $1s3d(^3D) - 1s2p(^3P)$ are stronger in the "longitudinal" spectra.

The ionization dynamics of C_{60} molecules can be rationalized in the following way. The first few electrons of C_{60} molecules are ionized due to multiphoton ionization. They and the rest of weakly bound π-electrons experience an oscillatory motion in the strong laser field with an average energy of $E_{osc} = e^2 E_L^2/4m\omega_L^2 \simeq 34.6$ eV, and a maximum amplitude of $a = eE_L/m\omega_L^2 \simeq 6.5$ Å, where E_L and ω_L are the electric field strength and the frequency of the short laser pulse. During their oscillations these electrons remain in the vicinity of their parent molecular ion. In their collisions with the C_{60} ion skeleton they absorb energy from the laser field and become able to ionize (excite) carbon atoms to higher ionization states. Eventually follows the explosion of the whole C_{60} microplasma. Due to the possibility of a three-dimensional expansion the electrons may be cooled very efficiently—therefore these plasmas promise to be advantageous for recombination X-ray laser schemes with transient inversions on: $2s - 3p$, $2p - 3d$ transitions of Li-like carbon ions; $1s2s - 1s3p$, $1s2p - 1s3d$, $1s^2 - 1s2p$ transitions of He-like ions; and $2 - 3$, $1 - 2$ transitions of H-like carbon ions.

In conclusion, we would like to note that the ionization mechanism of other large clusters in an intense laser field is expected to be similar to the above discussed collisional ionization of C_{60} molecules. Therefore studies of short-pulse laser-created plasmas in a dense vapor of Ar, Xe or Kr clusters (2) are very interesting from the point of realization of standard, but more efficient, Ne-like and Ni-like collisional X-ray laser schemes.

We would like to thank Michael Stuke for providing us with the C_{60} material.

REFERENCES

1. Wülker C., Theobald W., Ouw D., Schäfer F.P., and Chichkov B.N., *Opt. Commun.*, (to be published).
2. McPherson A., Luk T.C., Thompson B.D., Boyer K., and Rhodes C.K., *Appl. Phys. B* **57**, 337 (1993).
3. Chichkov B. N., and Kato Y., *Research Report of Institute of Laser Engineering Osaka University*, Japan, ILE9302P (1993).
4. Theobald W., Wülker C., Szatmári S, Schäfer F. P., and Bakos J. S., *Appl. Phys. B*, (to be published).

Ultrashort Pulse Laser Produced Plasmas

J. Davis, R. W. Clark and J. L. Giuliani

Plasma Physics Division
Naval Research Laboratory, Washington, DC 20375

ABSTRACT. The interaction of an ultrashort pulse laser with planar aluminum targets and with layered aluminum/silicon targets is investigated with a non-LTE radiation hydrodynamics model. The energy deposition for an obliquely incident P-polarized laser beam is calculated with a Helmholtz wave equation. A fraction of the absorbed energy is expended in the production of fast electrons, which are transported and deposited in the cold target. These electrons produce K-shell vacancies which produce characteristic K_α line radiation. The atomic models include the ground states and an extensive manifold of excited states for each of the materials in the target. The ionization dynamics is calculated with a time dependent collisional radiative model self-consistently coupled to a probabilistic radiation transport scheme. The emitted x-rays, including the K_α radiation, provide information about the energy deposition in the target, the energetic electron spectrum, and the time dependence of the local ionization in the target. The focus of the investigation is directed towards characterizing the radiative properties of the plasma, as well as determining whether the plasma can support population inversions and lasing as a result of recombination into highly ionized excited states in the rapidly cooling blowoff plasma.

1. INTRODUCTION

The interaction of an intense ultrashort pulse laser with a solid target produces a bright source of soft x-rays. The x-rays from the laser produced plasma has attracted substantial interest because of the potential applications in such areas as laser fusion, x-ray lasers, strongly coupled plasmas, atomic physics, spectroscopy and microlithography. The interaction of these intense lasers with planar targets makes it feasible to investigate solid state plasmas as a potential source of intense x-ray radiation in a relatively quiescent environment, since the target plasma remains nearly stationary during the initial burst of x-rays. Before some of these potential applications can become a reality, a better understanding of the laser/plasma interaction must be obtained. An issue of major concern is the coupling of the laser energy to the target (1). Even though there is a wealth of experimental data on the reflectivity and x-ray spectra, the database is obscured because the results have been obtained with a variety of lasers and target materials, complicating a systematic assessment of the data.

In addition to the thermal plasma reservoir, there are groups of energetic electrons created in the absorption process that deposit their energy in the colder dense regions of the target producing inner shell transitions on a time scale of a few picoseconds. The mechanisms for the production and deposition of energetic electrons have recently been studied (2) by making use of the resulting inner shell radiation. For example, a series of K_α lines is produced by resonance fluorescence in various ions of the target material. The relative strengths of these lines can be used to estimate ionic abundances in the vicinity of the energetic electron deposition as well as the characteristics of the fast electrons.

In this paper, we investigate the interaction of an ultrashort pulse laser with planar aluminum and silicon/aluminum targets. The interaction is simulated with a 1D non-LTE radiation hydrodynamics numerical model self-consistently coupled to a Helmholtz wave equation describing the absorption of laser energy in the target material. The incident laser pulse is represented by a P-polarized wave obliquely incident on the target. Also, a fraction of the incident laser energy goes into the production of hot electrons which are slowed down and deposited in the cold dense material. The radiative response of the target is represented by the L- and K-shell line and continuum spectra along with the series of K_α inner shell transitions.

2. MODEL

2.1 Hydrodynamic and Laser Deposition Model

The basic hydrodynamic variables consisting of mass, momentum and total energy are transported in one dimension using a Lagrangian flux-corrected transport numerical scheme. The momentum and energy equations include the ponderomotive force, and the energy equation includes the rates of energy gain (or loss) due to radiation, thermal conduction, laser deposition and fast electron deposition.

The absorption of a laser pulse obliquely incident on a planar target is calculated by solving the Helmholtz wave equation. In this way, the incident, reflected and transmitted waves are self-consistently characterized in each spatial zone in the medium. The conductivity is obtained from the Cauble-Rozmus model (3). The conductivity is complex, with

$$\sigma = \frac{n_e e^2}{m_e} \frac{1}{\nu - i\omega} = \left(\frac{\omega_{pe}^2}{\nu^2 + \omega^2}\right) \frac{\nu + i\omega}{4\pi} = \frac{i\omega}{4\pi}(1 - \epsilon')$$

where ν is the collision frequency and ω is the laser frequency. ω_{pe} is the electron plasma frequency and n_e is the local electron number density. The Helmholtz equations for the electric and magnetic fields can be written

$$\nabla^2 \vec{E} - \nabla(\nabla \cdot \vec{E}) + \frac{\omega^2}{c^2}\left(1 + i\frac{4\pi\sigma}{\omega}\right)\vec{E} = \nabla^2 \vec{E} - \nabla(\nabla \cdot \vec{E}) + \frac{\omega^2}{c^2}\epsilon'\vec{E} = 0$$

$$\nabla^2 \vec{B} - \frac{\nabla \epsilon'}{\epsilon'} \times (\nabla \times \vec{B}) + \frac{\omega^2}{c^2}\epsilon'\vec{B} = 0$$

where ϵ' is the plasma dielectric function. For P-wave polarization, it is convenient to solve for the magnetic field, $\vec{B} = B_z(x,y)\hat{e}_z$, and the corresponding Helmholtz equation can be written

$$\frac{\partial^2 B_z(x)}{\partial x^2} - \frac{\partial \ln \epsilon'}{\partial x} \frac{\partial B_z(x)}{\partial x} + \frac{\omega^2}{c^2}[\epsilon' - \sin^2\theta_o] B_z(x) = 0.$$

A Runga-Kutta numerical scheme is employed to solve the appropriate Helmholtz equation. An adaptive gridding is used to resolve the fields in the vicinity of the critical surface. The computational mesh used for the solution of the fields is much finer than that used for the hydrodynamics and radiation transport; logarithmic interpolation is employed to obtain the electron density and collision frequency from the hydrodynamic computational mesh.

2.2 Atomic Model

The ionic populations in the plasma are characterized by a set of atomic rate equations, one for each of the atomic levels included in the model. The rate coefficients that are used to calculate the populating and depopulating processes are calculated using various scattering techniques and the methods used in calculating the corresponding rate coefficients are summarized elsewhere (4).

Radiation emission from the plasma and its opacity are dependent on the local atomic-level population densities. Except for optically thin plasmas, however, the level populations depend on the radiation field, since optical pumping via photoionization and photoexcitation can produce significant population redistribution. Thus, the ionization and radiation transport processes are strongly coupled and must be solved self-consistently. In this model, an iterative procedure (5) is used, where level populations are calculated using the radiation field from the previous iteration, then using these populations to calculate a new radiation field until convergence is reached.

The prescription for multiphoton ionization follows the tunneling ionization theory of Ammosov, Delone and Krainov (6).

The atomic model for aluminum used in this calculation consists of the ground states and 118 excited levels through neutral aluminum, including structure up to N=10 for H-like, and to N=9 for He-like and Li-like aluminum. The atomic model for silicon consists of the ground states and 65 excited levels. Ionization lowering is accounted for by means of an ion sphere model; the bound-free radiation is limited from states which merge with the continuum. Radiation transport is carried out using a probability-of-escape formalism which is described elsewhere (7,8).

2.3 Energetic electron transport

A fixed fraction of the laser energy deposited in each spatial zone (10 percent in the present calculation) is assumed to be expended in the production of energetic electrons. These fast electrons are assumed to be monoenergetic (at an energy between 4.0 and 10.0 keV). This assumption is made to aid in the interpretation of the resulting data; the high energy tail of a Maxwellian (or an arbitrary energy distribution) could be employed in the model. The energetic electrons free stream, are slowed down, and subsequently deposit their energy in the plasma, giving rise to local heating and inner-shell vacancies.

At sufficiently high electron energies, the Bethe theory is adequate. Then the fast electron stopping power can be written in the Bethe-Bloch (9) form

$$\frac{\partial E}{\partial x} = -\frac{2\pi e^4 n_i}{E} \left[Z_B \ln\left(\frac{1.16E}{\chi_{av}}\right) + Z_F \ln\left(\frac{1.16E}{\chi_o}\right) \right]$$

where χ_{av} is an average ionization potential, χ_o is an effective potential related to the electron plasma frequency, and Z_B and Z_F are the effective number of bound and free electrons per ion. The continuous slowing down model breaks down when quantum transitions are of the same order in energy as the beam (near the end of the trajectory).

2.4 K_α radiation model

The K-shell vacancies which are created are rapidly filled by either Auger decay (radiationless cascade) or resonance fluorescence (valence electron fills vacancy with the emission of K_α radiation). The energy-dependent K_α production rate is calculated for each aluminum ionization stage in each spatial zone, from Al I (neutral aluminum) through Al IX (boron-like aluminum), and for the corresponding silicon ionization stages. The K-shell vacancy cross sections for aluminum and silicon were obtained from the calculations of Blaha (10), and represent a small fraction of the total stopping power. From threshold to about 10keV, the energy dependence of these cross sections can be approximated by a simple slowing-down approximation

$$\sigma(E) = \frac{C}{(E\ \chi_k)} \ln\left(\frac{E}{\chi_k}\right)$$

where C for aluminum is equal to $9.5 \times 10^{-14} cm^2/eV^2$ and χ_k is calculated for each ionized state of aluminum. The fluorescence/Auger decay branching ratios (taken from Duston et al. (5)) vary with ionization stage, and are typically about 5 percent. The K_α line widths are determined from the inverse lifetimes. The K_α lines for Al I-IV (and for Si I-V) lie so close together in energy that they appear spectroscopically as a single line, whereas those of Al V-IX (and Si VI-X) are sufficiently separated that they can be resolved.

3. RESULTS

3.1 Energy Deposition

A series of numerical experiments was carried out involving an ultrashort-pulse laser (0.22 psec FWHM duration Gaussian with no prepulse) obliquely incident on a solid density aluminum target. The peak power was 1.0×10^{16} – 1.0×10^{17} W/cm² at a wavelength of 0.26 μ. Figure 1 shows the dependence of absorption fraction (absorbed energy relative to the total incident energy) on the angle of incidence for P-wave deposition. The dependence is similar to that calculated by Milchberg et al.(1) for $\nu/\omega_0 < 1$ in the solid target and $L/\lambda \sim 1$, where L is the plasma scale length in the vicinity of the critical surface and λ is the laser wavelength.

Figure 2 shows the electric and magnetic fields and the local energy deposition in the target near the peak of the laser pulse (0.30 picoseconds) for a peak intensity of 1.0×10^{16} W/cm² at 45° oblique incidence. The laser pulse is assumed incident from right to left, and only the rightmost portion of the plasma (from 0.8μ to the outer edge) is shown. The fields are averaged over an optical cycle, and normalized with respect to the amplitudes of the incident electric and magnetic fields. The location of the critical surface, i.e. where the incident laser frequency is equal to the plasma frequency ($\omega = \omega_p$) is represented as R_c. Mass density ($\log\rho$) is plotted in a stepwise manner to show the spatial resolution of the hydro mesh. Note the resonance in the electric field in the vicinity of the critical surface. The plasma heating from the laser deposition ($\log[dE_{Laser}/dt] \times 10^{-24}$ ergs/cm^3-sec) also peaks near R_c.

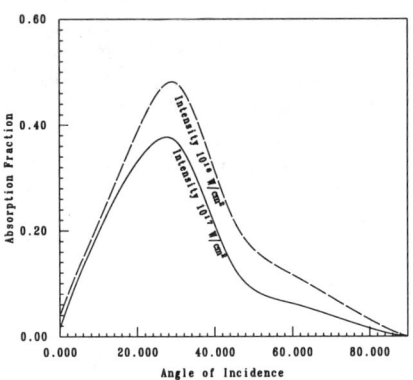 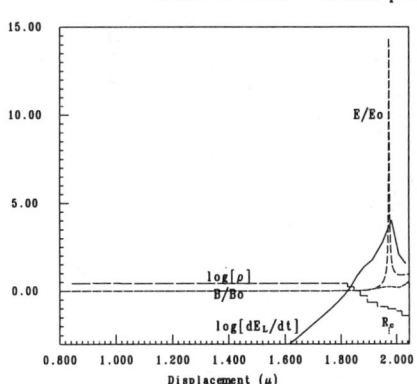

Figure 1. Absorption fraction (integrated over the laser pulse) as a function of incident angle for P-wave deposition.

Figure 2. Electric field for a 1.0×10^{16} W/cm^2 pulse at $45°$ oblique incidence. The mass density ρ and the location of the critical surface R_c are indicated.

3.2 Hydrodynamic response

Since most of the incident laser energy is absorbed in the low density blowoff plasma, and since thermal conduction and radiation do not efficiently transport energy into the cold dense target plasma, fast electrons, to the extent that they are generated, can be the principal dense plasma heating mechanism. This can be seen from figures 3 and 4, which show the hydrodynamic profiles at 0.30 nanoseconds, near the peak of the laser pulse, for 10keV and 4keV fast electrons, respectively. In both cases, P-wave deposition of a 1.0×10^{16} W/cm^2 pulse at $45°$ oblique incidence is assumed. Temperature, mass density, pressure and the radiative cooling rate are shown as functions of position in the target. The energetic electrons are generated near the critical surface, and for an initial energy of 10 keV, they have a range of about a micron into the dense plasma. The temperature profile in figure 3 reflects the primary laser heating and the heating from the fast electrons deep in the target. The temperature in the blowoff plasma reaches about 2 keV, and is about 30 eV near the end of the range of the fast electrons (at about 1.0 μ).

In the absence of an energetic electron beam, the temperature profile would be mostly dominated by thermal conduction from the laser deposition region and exhibit a rapid decay into the cold aluminum. The density is nearly constant behind the blowoff, although a small disturbance (a rarefaction and a weak shock) can be seen near x = 1.0 μ in Figure 3 due to the fast electron heating. Radiative losses are the greatest near x = 2.0 μ where the plasma is hot. A secondary cooling peak occurs at about 0.90 μ. The temperature profile in figure 4 shows a secondary peak anout 0.20 μ behind the critical surface, due to the reduced range of the 4 kcV fast electrons. The density profile is essentially unchanged. The temperature in the blowoff plasma reaches almost 3 keV, and is slightly greater than 30 eV in the secondary peak.

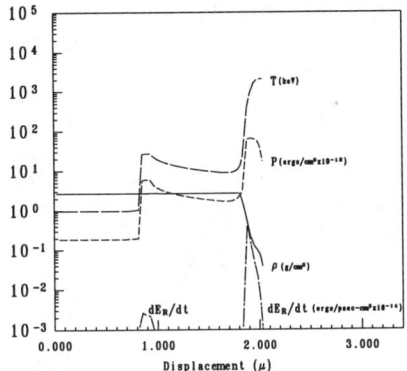

Figure 3. Hydrodynamic profiles for 10 keV fast electrons at 0.30 picoseconds. The mass density ρ [g/cm^3], temperature T [eV], pressure P [ergs/cm^3×10^{-12}], & cooling dE$_R$/dt [ergs/psec-cm^3×10^{-14}] from radiation are plotted as functions of position in the target.

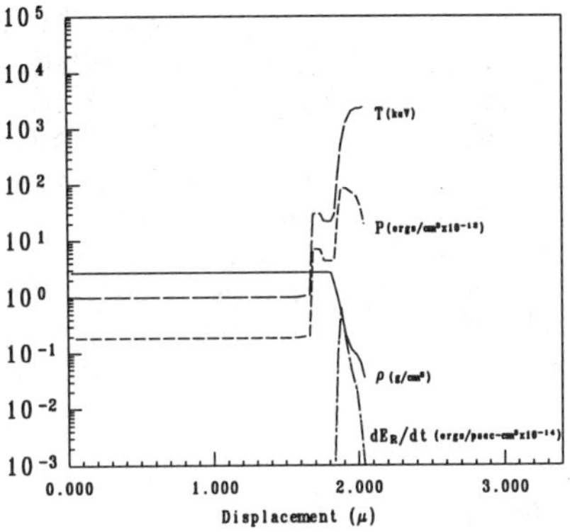

Figure 4. Hydrodynamic profiles for 4keV fast electrons. The mass density ρ, temperature T, pressure P and radiative cooling dE$_R$/dt are plotted as functions of position in the target.

The predicted frontside temperatures shown in Figs. 3 and 4 are somewhat higher than those inferred from recent experiments under similar conditions (11, 12). However, temperature estimates are generally based on spatially and temporally integrated spectra. Although we predict temperatures in excess of a keV in the blowoff region, most of the radiation comes from cooler, denser plasma deeper in the target. Also, much of the radiation is emitted after the laser pulse, when the plasma has cooled and recombined. Finally, there is a lag in the ionization of the plasma, due to its nonequilibrium nature (particularly at the lower densities encountered in the blowoff).

3.3 Radiation from Layered Targets

While the relative intensities of spectral lines can provide detailed information about atomic populations, it can be difficult to infer thermal (and density) gradients and fast electron effects from this data alone. However, the use of multiple materials in the target with layers of varying thicknesses provides additional spectral information. The range of energetic electrons in the target can be determined directly from such experiments, and the electron energy distribution function can thus be inferred. In addition, it may be possible to make quantitative measurements of laser absorption as a function of position in the target, and to verify assumptions regarding multiphoton processes, etc.

A series of layered target simulations was carried out, where the initial conditions were similar to the aluminum slab target described above (0.22 psec Gaussian pulse obliquely incident at 30° with peak intensity 1.0×10^{17} W/cm²). The target consisted of a 0.10 μ silicon layer coating a solid density aluminum substrate. Instantaneous spatially integrated x-ray emission spectra for selected K-shell and K_α lines are plotted as a function of photon energy at 0.30 picoseconds (near the peak of the laser pulse) in Figures 5 and 6 for the cases of 10 and 4 keV fast electrons, respectively. The spectral lines are identified on the figures. Two groups of K_α lines, from the various ionized stages of aluminum and silicon, are seen in the plots. In Fig. 5, very little K_α radiation from highly ionized aluminum is in evidence. This is because the 10 keV electrons are stopped deep (up to a micron) in the cold aluminum substrate. However, Fig. 6 shows substantial K_α radiation from hot aluminum. As in Figs. 3 and 4, the shorter range of the 4 keV electrons creates a warm layer of aluminum near the Al/Si interface. There is substantially more total silicon K_α radiation in the latter case, although the distribution by ion is not very different. Most of the 10 keV fast electrons produced near the critical surface escape through the silicon layer without creating K-shell vacancies.

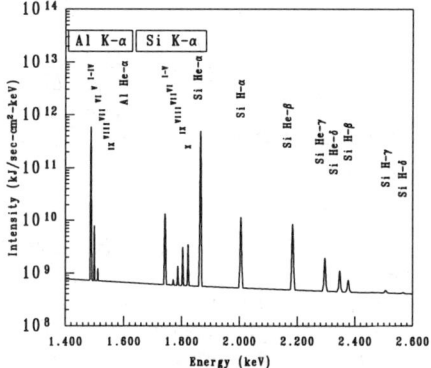 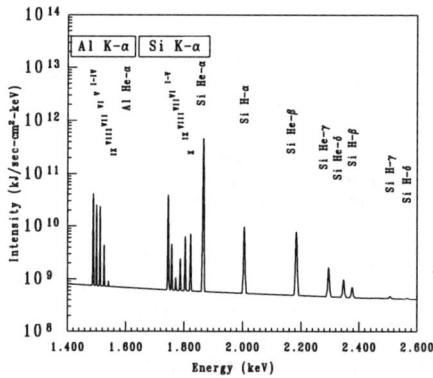

Figure 5. Emission spectrum at 0.30 picoseconds for 10 keV fast electrons. Selected K-shell and K_α lines from silicon and aluminum are identified.

Figure 6. Emission spectrum at 0.30 picoseconds for 4 keV fast electrons. K_α lines from highly ionized aluminum are in evidence.

4. CONCLUSIONS

The purpose of this investigation was to simulate the radiation hydrodynamics of an ultrashort pulse laser produced plasma and to study how the emitted x-rays could provide information about the energy deposition in the target, the production and transport of energetic electrons, and the resulting ionization of the plasma. The model which was used included a relatively detailed treatment of the atomic populations and radiation. The laser deposition was performed in a self-consistent manner by solving the appropriate Helmholtz equation. The treatment of the energetic electrons incorporated a number of prescriptions (production rate and energy distribution), but it provided for the self consistent transport of the electrons and the production of K-shell vacancies. Since K_α radiation is produced almost entirely by the energetic electrons, these inner shell K-lines are particularly useful in providing information on fast electron energy transport. However, the ratios of the K_α lines reflect the local atomic populations where the electrons produce K-shell vacancies. It is challenging to accurately determine the electron energy spectrum from this data alone. The details of the laser deposition and other factors which can alter the density and temperature profiles (and the local ionization of the plasma) can also affect the line ratios. The deposition is controlled by the local electrical conductivity, and progress continues to be made by a number of researchers (3, 13) in developing a satisfactory conductivity model for the conditions encountered in these plasmas. The use of very thin multiple material layers (2) is one way to provide additional information about the ionization of the plasma as a function of position in the target, and thus distinguish between these factors.

ACKNOWLEDGMENTS

We would like to thank Dr. P. Kepple for making the atomic data base available, and Dr. R. Cauble for providing assistance with the electrical resistivity. We would also like to acknowledge useful discussions with Drs. H. Milchberg and D. Meyerhofer.

REFERENCES

1. H. M. Milchberg and R. R. Freeman, *J. Opt. Soc. Am. B* **6**, 1351 (1989).
2. H. Chen, B. Soom, B. Yaakobi, S. Uchida and D. D. Meyerhofer,
    Phys. Rev. Lett. **70**, 3431 (1993).
3. R. Cauble and W. Rozmus, *J. Plas. Phys.* **37**, 405 (1987),
    R. Cauble, F. J. Rogers and W. Rozmus, *SPIE* **1229**, 211 (1990).
4. D. Duston, R. W. Clark, J. Davis and J. P. Apruzese,
    Phys. Rev. A **27**, 1441 (1983).
5. J. P. Apruzese, J. Davis, D. Duston and R. W. Clark,
    Phys. Rev. A **29**, 246 (1984).
6. M. V. Ammosov, N. B. Delone and V. P. Krainov,
    Zh. Eksp. Teor. Fiz. **91**, 2008 (1986)
    [*Sov. Phys. - JETP* **64**, 1191 (1986)].
7. J. P. Apruzese, *J. Quant. Spect. Rad. Transf.* **25**, 419 (1981).
8. J. P. Apruzese, *J. Quant. Spect. Rad. Transf.* **34**, 447 (1985).
9. C. J. Powell, *Rev. Mod. Phys.* **48**, (1) 33 (1976).
10. M. Blaha, *Private communication* (1992).
11. A. Mysyrowicz, J. P. Chambaret, and A. Antonetti,
    J. Phys. B: At. Mol. Opt. Phys. **27**, 1671 (1994).
12. S. C. Wilks, *Phys. Fluids B* **5**, 2603 (1993).
13. F. Perrot, *Phys. Rev. E* **47**, 570 (1993).

Controlled Power Compression in Materials for X-Ray Amplification

A. B. Borisov, A. McPherson, K. Boyer, and C. K. Rhodes

Laboratory for Atomic, Molecular, and Radiation Physics
Department of Physics, M/C 273, University of Illinois at Chicago
845 W. Taylor, Room 2136, Chicago, IL 60607-7059
Phone: (312) 996-4868
Facsimile: (312) 996-8824

ABSTRACT. New nonlinear phenomena involving (1) multiphoton excited X-ray emission from clusters and (2) stable channeled electromagnetic propagation in plasmas have combined scaling properties highly conducive for X-ray amplification in the kilovolt range.

Recent experimental and theoretical studies have led to fundamental developments concerning the amplification of X-rays in the kilovolt region. These results affect our ability to <u>controllably</u> apply power densities at or above the thermonuclear level in materials, the basic issue for the creation of bright and efficient sources of radiation in the X-ray range. Specifically, these findings concern (1) the multiphoton excitation of clusters [1-5] and (2) high-intensity stable confined modes of propagation in underdense plasmas [6-9]. In combination, these two new physical phenomena are being used to produce the necessary power compression; the first serves to establish the condition on power density locally, the second provides the required spatial organization.

Information on the former has come from studies of multiphoton excitation of Kr and Xe clusters. The principal findings, comprised of the results obtained from five cases [Xe(N), Kr(M), Xe(M), Kr(L), and Xe(L)] spanning spectrally from ~ 80 eV to ~ 5 keV, have (1) confirmed the important role of cluster formation on the X-ray emission [1], (2) established the scaling of this new phenomenon into the kilovolt spectral region [2], (3) demonstrated the production of hollow atoms (ions) [3], an essential feature of the proposed model [1] of the interaction, (4) provided evidence that coherent electron motions induced by the external field [4] play an important role in the interaction, (5) led to the conclusion that a regime of strong-coupling exists in which processes involving multi-electron ejection from an inner-shell can occur with high probability [5], and (6) revealed the scaling in atomic number which strongly favors heavy atoms [3-5]. The outcome of these studies is the finding that specific power densities of ~ 1 W/atom can be obtained in heavy systems.

Experimental studies with intense ($\sim 10^{19}$ W/cm^2) subpicosecond ($\sim$ 300 fs) ultraviolet (248 nm) radiation [10] are demonstrating how the channeled propagation involving relativistic and charge-displacement nonlinearities [6-9] can provide the necessary spatial organization of the deposited energy, the second phenomenon being used to produce conditions conducive to amplification. Analysis [11] shows that this spatial control is a key factor in establishing a favorable scaling relationship between the magnitude of the gain produced and excitation power needed for X-ray amplification. An essential result is that the radiative conditions required for the strong multiphoton production of the X-rays from the clusters are _identical_ to those producing the channeled propagation. The two processes appear to be exceptionally compatible. Moreover, recent experiments [12] have conclusively demonstrated the first _combined_ expression of these two highly nonlinear phenomena with the production of X-ray images of the self-trapped channels. These measurements [12] have provided additional supporting evidence for a highly ordered electronic motion induced in the clusters by the external field [4] and have revealed a radial dependence of the spatial pattern of X-ray emission which serves as a sensitive probe of the dynamics of the confined propagation. The resulting knowledge of scaling relations governing these phenomena enables the optimum conditions for amplification to be specified. Importantly, preliminary analysis indicates that the optimum parameters fall within the stable domain [9] found for the channeled propagation.

In conclusion, the key to efficient X-ray amplification lies in the ability to control properly the deposition of extremely high power densities ($\sim$ 1 W/atom) in matter. The results of preliminary studies involving the multiphoton excitation of clusters and new stable confined modes of electromagnetic propagation indicate that the necessary power density and control can both be simultaneously achieved through the combined use of these two new highly nonlinear mechanisms.

ACKNOWLEDGEMENTS

The expert technical assistance of P. Noel and J. Wright is acknowledged. The support for this research was provided by DoE, AFOSR, ARO, NSF, AND NRL.

REFERENCES

1. McPherson, A., Luk, T. S., Thompson, B. D., Boyer, K., and Rhodes, C. K., *Appl. Phys. B* **57**, 337-347 (1993).

2. McPherson, A., Luk, T. S., Thompson, B. D., Borisov, A. B., Shiryaev, O. B., Chen, X., Boyer, K., and Rhodes, C. K., *Phys. Rev. Lett.* **72**, 1810-1813 (1994).
3. McPherson, A., Thompson, B. D., Borisov, A. B., Boyer, K., and Rhodes, C. K., "4-5 keV Xe(L) Multiphoton-Induced X-Ray Emission from Multiply Core-Excited Xe Atoms," *Nature*, in press.
4. Boyer, K., Thompson, B. D., McPherson, A., and Rhodes, C. K., "Evidence for Coherent Electron Motions in Multiphoton X-Ray Production from Kr and Xe Clusters," *J. Phys. B*, in press.
5. Thompson, B. D., McPherson, A., Boyer, K., and Rhodes, C. K., "Multi-Electron Ejection of Inner-Shell Electrons through Multiphoton Excitation of Clusters," *J. Phys. B*, in press.
6. Borisov, A. B., Borovskiy, A. V., Korobkin, V. V., Prokhorov, A. M., Shiryaev, O. B., Shi, X. M., Luk, T. S., McPherson, A., Solem, J. C., Boyer, K., and Rhodes, C. K., *Phys. Rev. Lett.* **68**, 2309-2312 (1992).
7. Borisov, A. B., Borovskiy, A. V., Shiryaev, O. B., Korobkin, V. V., Prokhorov, A. M., Solem, J. C., Luk, T. S., Boyer, K., and Rhodes, C. K., *Phys. Rev. A* **45**, 5830-5845 (1992).
8. Borisov, A. B., Shi, X., Karpov, V. B., Korobkin, V. V., Shiryaev, O. B., Solem, J. C., McPherson, A., Boyer, K., and Rhodes, C. K., "Stable Self-Channeling of Intense Ultraviolet Pulses in Underdense Plasma Producing Channels Exceeding 100 Rayleigh Lengths," *J. Opt. Soc. B*, in press.
9. Borisov, A. B., Shiryaev, O. B., McPherson, A., Boyer, K., and Rhodes, C. K., edited by Corkum, P. B., and Perry, M. D., in *Short Wavelength V*, Washington, D. C.: Optical Society of America, 1993, Vol. 17, pp. 58-61.
10. Bouma, B., Luk, T. S., Boyer, K., and Rhodes, C. K., *J. Opt. Soc. Am. B* **10**, 1180-1184 (1993).
11. Boyer, K., Borisov, A. B., Borovskiy, A. V., Shiryaev, O. B., Tate, D., Bouma, B. E., Shi, X., McPherson, A., Luk, T. S., and Rhodes, C. K., *Appl. Opt.* **31**, 3433-3437 (1992).
12. Borisov, A. B., McPherson, A., Thompson, B. D., Boyer, K., and Rhodes, C. K., "Imaging of Xe(M) Emission from Xe Clusters in Stable Self-Trapped Channels," to be published.

Simulations of X-UV Gain in a Plasma Pumped by a Pulse-Train Laser

A. Klisnick, J. Virmont[#], N. Grandjouan[#],
H. Guennou, A. Sureau.

Laboratoire de Spectroscopie Atomique et Ionique, Université Paris-Sud, Bât. 350, 91405 Orsay Cédex, FRANCE and [#]Laboratoire de Physique des Milieux Ionisés, Ecole Polytechnique, 91128 Palaiseau Cédex, FRANCE,

Abstract. We present numerical simulations of X-UV gain in plasmas produced by irradiating a target with a train of 100 ps- pulses. Our predictions based on a 1.5D hydrocode do not account for the experimental gains reported on the AlXI 3d-5f by Hara and his co-workers. We suggest that the simulation of multiple-pulse irradiated plasmas may require a bidimensional description and propose an experimental verification of this assumption.

INTRODUCTION

A method for pumping a lithium-like aluminium X-ray laser with a very low driving laser energy was proposed by Hara et al. (1). This method is based on the use of multiple-pulse laser irradiation of the target. Experiments were performed in which a 10 mm-long aluminium plasma was created by a train of 2 to 16 identical pulses having each a duration of 100 ps and separated by 200 ps to 400 ps. A gain coefficient of 2.5 cm^{-1} was reported on the 3d-5f AlXI line situated at 105.7Å. These results are of great interest in the prospect of the future practical utilization of X-ray laser sources because they could allow to realize a compact source with very low pumping energy requirements.

However the reason why multiple-pulse irradiation should increase the efficiency of recombination pumping is still unclear. On the other hand only very few is known yet about the structure and the evolution of plasmas created by multiple-pulse irradiation. The purpose of the simulation work presented in this paper was to investigate the influence of such an irradiation on the calculated gain in lithium-like aluminium. Rather than restricting the conditions of the simulations to Hara's experimental ones we systematically explore the role of the number of incident identical pulses and of the temporal separation between them. Other parameters, such as the relative peak power for non-identical pulses, should also have an influence on the plasma evolution and on the calculated gain, the study of which is left for future work.

It should be noted that multiple- (or double-) pulse irradiation has also been proposed as a way of increasing the efficiency of collisionally excited X-ray lasers (2, 3). In this case the expected effect is more likely to lower transverse density gradients hence reducing refraction of the amplified beam propagating along the plasma column. For the aluminium recombination laser considered in this paper refraction of the amplified beam is negligible because amplification takes place in the plasma at relatively low electron density ($N_e \sim 10^{19}$ cm^{-3}) in a region where the density gradient is not very steep. Thus the effect expected in this case in using multiple pulse irradiation is to enhance recombination pumping and gain coefficients.

SIMULATION OF LITHIUM-LIKE RECOMBINATION X-RAY LASERS

The time- and space- evolution of gain on AlXI lasing transitions is calculated in two steps. First we use the 1.5D lagrangian hydrocode FILM to calculate the evolution of the line-shaped plasma. The 1.5D description allows to account in a simplified way (4) for the lateral expansion (i.e. perpendicular to both the incident laser axis and the focal line) of a plasma produced by irradiating a planar massive target with a cylindrically focused laser beam. The hydrocode provides the variation of quantities such as electron density and temperature or fractional abundances among the different ionic species. These data are then fed in a collisional-radiative model POPEXC which calculates the populations of the lithium-like singly-excited states at quasi-steady state (QSS) equilibrium (5). Population inversion densities and Doppler-gain coefficients (at line center) are then derived for the transitions of interest.

In the present work we have focused on the 3d-4f transition, instead of the 3d-5f which was studied in Hara's experiment. In fact the predicted gain on those two transitions behave very similarly with respect to variations on the laser pulse parameters, but the 3d-5f gain is always approximately 10 times lower than the 3d-4f one. This feature of the simulations is in contradiction with experimental observations, for reasons which are not yet understood (6) (5). The results presented below would thus be similar for the 3d-5f transitions except for the absolute value of the calculated gain.

The calculations were done for a set of different pulse-trains, composed of identical pulses with the same duration (gaussian, 100 ps FWHM) and the same peak power. The number of pulses, N_{pulse}, considered was varied between 1, 2, 4 and 8. The temporal separation, Δt, between the peaks of two successive pulses considered were 100 ps, 200 ps or 400 ps. In each case (i.e at fixed N_{pulse} and Δt) we performed several calculations to find the optimal value of the total driving energy for which the 3d-4f calculated gain was maximum. For all the cases considered the irradiance per pulse was in the range 0.7-2. TW/cm^2. The laser wavelength was 1.06µm and the width of the focal line was chosen to be 40µm, according to Hara's experimental one.

PREDICTED 3D-4F GAIN AND GL/E EFFICIENCY

The absolute maximum (in time and space) gain calculated on the 3d-4f line for each set of parameters explicited above is shown in Figure 1. One can see that whatever the time separation Δt between pulses is, gain decreases dramatically when the number of pulses N_{pulse} is increased from 1 to 8. The larger the time separation the stronger the drop of gain is. For a single pulse irradiation the predicted gain is 2.2 cm^{-1}, while it is as low as 0.1 cm^{-1} when 8 pulses separated by 400 ps are used.

The effect of multiple-pulse irradiation was then studied in terms of the gain-length efficiency GL/E. Here GL was calculated for a one centimetre long plasma and E is the optimal total energy found in each irradiation case, the values of which are listed in table 1. One can see in figure 2 that except for the case of 2 pulses separated by 100 ps the efficiency GL/E decreases as a function of the number of pulses.

For a single-pulse irradiation the optimal driving energy is of about 1.7 Joules and the efficiency GL/E reaches 1.3 J^{-1}. It should be noted that such a GL/E value is much larger than the typical values that have been reported for collisional excitation lasers, although they yielded much larger GL values. For example at LULI we have demonstrated a gain-length of approximately 16 at 212Å in a 2 cm zinc plasma placed in a half-cavity (6) with a corresponding GL/E of typically 0.04 J^{-1}.

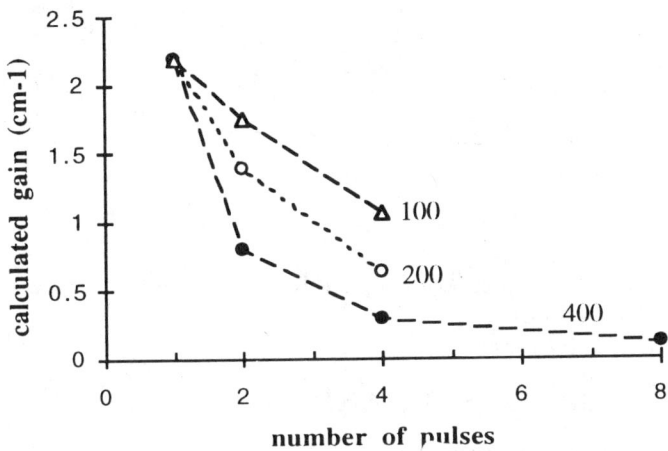

FIGURE 1. Calculated 3d-4f gain as a function of the number of incident pulses, for three different temporal separation between pulses.

The GL/E efficiency for a double-pulse irradiation with $\Delta t = 100$ ps was found to be similar to the single-pulse one, within the accuracy of our calculations (the optimal driving energy was determined with a ± 0.2 Joules accuracy). Since the corresponding gain was slightly higher for the single pulse case, this means that the optimal driving energy is lower for the double-pulse case. This must be due to the fact that the energy coupling between the driving laser and the plasma is better in the latter case, where the two pulses separated by 100 ps are in fact equivalent to a single pulse with about 200 ps duration.

TABLE 1. Optimal peak power and total driving energy for the different irradiations considered

number of pulses	Δt (ps)	optimal peak power per pulse (GW/cm)	optimal total driving energy (J/cm)
1		8	1.7
2	100	3	1.3
2	200	4	1.7
2	400	5	2.1
4	100	3	2.6
4	200	3	2.5
4	400	5	4.3
8	400	7	11.9

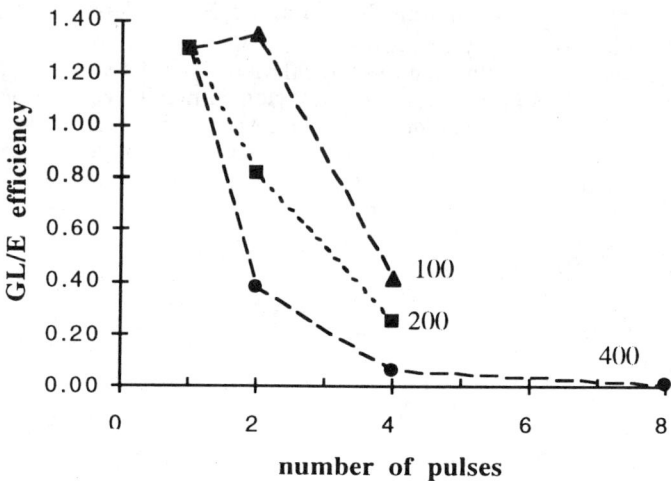

FIGURE 2. Calculated gain-length efficiency GL/E for the same irradiation parameters as in Fig. 1.

DISCUSSION AND CONCLUSIONS

The reason for which the calculated gain decreases when the number is increased (Fig. 1) can be understood by considering the quantity t_{gain}. This quantity is the temporal delay between the time of peak of the last driving laser pulse and the time of peak gain. The variation of this quantity for the different irradiations considered is plotted in Figure 3.

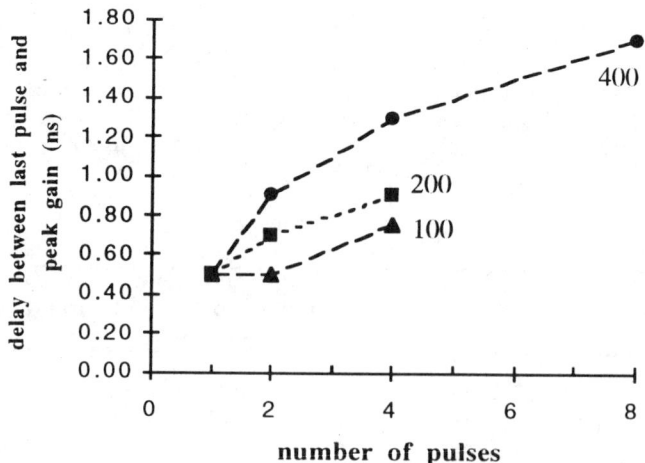

FIGURE 3. Temporal delay t_{gain} between the peak of the last driving pulse and the peak of gain.

It should be first recalled that gain in recombination X-ray lasers takes place during the cooling of the plasma after the extinction of the driving laser. We have shown (7) that the electron temperature must have decreased below a critical value T_C which is around 50eV for lithium-like aluminium. On the other hand the drop in temperature is mainly due to plasma expansion which yields a reduction of electron and ion densities. The value of the maximum gain reached during the expansion/cooling phase depends on the value of density at the time when the temperature is below T_C. It hence depends on the cooling rate. The faster the cooling the larger the density at T_C and the larger the gain on the lithium like transitions. The temporal delay t_{gain} between the peak of the driving laser pulse and the peak gain is thus inversely proportional to the cooling rate.

If we now examine the results plotted in Figure 3 we can see that the temporal delay t_{gain} increases with the number of pulses for the three separations Δt considered. The calculated gain is maximum 500 ps after the peak of a single pulse whereas in the case of a 8 pulse-train with $\Delta t = 400$ ps a delay of 1.7 ns after the peak of the 8th pulse is necessary to reach the maximum of gain. Actually a multiple-pulse irradiation produces a bigger volume of plasma which will cool more slowly than a plasma produced by a single, 100 ps- pulse.

In the frame of our modelling description we thus find that a short, single pulse is more efficient for recombination pumping than a pulse-train, in contradiction with the observations reported by Hara and co-workers. However we do not conclude that Hara's results could be questionable. It should be kept in mind that we use a monodimensional hydrocode to describe the plasma expansion. Now the successive laser pulses interact with a plasma which has already expanded both axially (i.e. toward the incident laser) and transversely (i.e. perpendicular to both the laser incidence axis and the plasma column axis). The laser focal spot is thus likely to be smaller than the dimensions of the plasma, in particular in the transverse direction. One could then expect two-dimensional lateral cooling to take place in a small restricted plasma zone heated by the successive pulses. This additional cooling process could be responsible for the enhancement of recombination pumping observed in Hara's experiments. This assumption could be verified experimentally by spatially resolving the gain zone in the directions transverse to the amplification axis: as a matter of fact the gain zone should be restricted to a small portion of the plasma column, or more precisely its transverse dimension should be smaller than the transverse size of the plasma at the time and axial position of gain.

We then conclude that more detailed, bidimensional simulations are necessary to investigate the role of multiple-pulse irradiation in plasma heating and cooling and in recombination pumping.

REFERENCES

(1) Hara T. et al., in *Proceedings of the 3rd X-Ray Lasers Conference*, IOP Conf. Series N°125, E.E. Fill ed., IOP Publishing, 1992, pp. 97-100
(2) Hagelstein P. et al., This Conference
(3) Jacquemot S. (private communication)
(4) Klisnick A. et al., in *Proceedings of the 3rd X-Ray Lasers Conference*, IOP Conf. Series N°125, E.E. Fill ed., IOP Publishing, 1992, pp.305-308
(5) Klisnick A. et al., *Appl. Phys. B* **50**, 153, 1990
(6) Makhrov V. et al., *J. Phys. B* **27**, 1899, 1994
(6) Rus B. et al., This Conference
(7) Klisnick A. et al., in *Proceedings of the 2nd X-Ray Lasers Conference*, IOP Conf. Series N°116, G.J Pert and G.J. Tallents eds., IOP Publishing, 1990, pp.17-20

IONIZATION STATE OF AN ALUMINIUM X-RAY LASER PLASMA BY Kα SPECTROSCOPY

A. Klisnick[*], C. Chenais-Popovics[#], C. A. Back[$], P. Zeitoun[*], P. Renaudin[#], O. Rancu[#], J.C. Gauthier[#], P. Jaeglé[*]

[*] *Laboratoire de Spectroscopie Atomique et Ionique, Université Paris-Sud, Bât. 350, 91405 Orsay Cedex, France;*
[#] *LULI: CNRS National Facility, Laboratoire pour l'Utilisation des Lasers Intenses, Ecole Polytechnique, 91128 Palaiseau Cedex, France;*
[$] *Lawrence Livermore National Laboratory, PO Box 808, Livermore, CA 94550, USA*

Abstract. An aluminium recombining X-ray laser plasma was probed by keV radiation emitted by an auxiliary plasma source to yield absorption structures on Kα transitions. From this technique we obtain the spatial distribution of the relative abundances of He- to B-like ions at different times during plasma cooling. He-like and Li-like ions are found to exist further from the target than predicted by the hydrocode, explaining previous gain measurements.

INTRODUCTION

Progress in the understanding of recombination X-ray lasers requires that we obtain measurements of plasma quantities other than gain coefficients on lasing transitions. In particular it is important to know the actual evolution of the plasma ionization state during the recombination phase since this evolution is the key to the production of population inversions.

We have applied the technique of absorption spectroscopy on Kα transitions to an X-ray laser aluminium plasma column produced by a 600 ps pulse, in conditions close to those used for the X-ray laser experiments performed at the LULI laser facility (1). This technique provides a direct probe of the ground state populations of the different ionic species present in the plasma. Under the conditions studied here the ground state population accounts for the bulk of the total population of the ionic species involved in the lasing process.

Absorption spectroscopy on Kα transitions was previously developed for point plasmas (2) as well as for X-ray laser plasmas produced by short (100 ps) pulses (3). We present an original method designed to improve the visibility of the absorption structures from X-ray laser plasma columns which have a diameter d smaller than the characteristic absorption length of the Kα transitions.

Absorption spectra have been obtained at three different times after the peak of the driving pulse, that is during plasma cooling and gain build-up. The experimental data were compared to synthetic spectra which were obtained from the predictions of a hydrocode. This comparison is discussed at the end of the paper.

EXPERIMENTAL ARRANGEMENT

These experiments were performed using two beams of the LULI laser facility. Figure 1 shows the experimental set-up. The primary beam

(λ=1.06μm, 1ω_0, pulsewidth 600 ps) was cylindrically focused onto a foil target with (1-5)x 10^{12} W/cm^2 to produce an Al plasma similar to those used in previous X-ray laser experiments that have demonstrated gain. This plasma was probed by an X-ray backlight source that was generated by the second laser beam. This beam (λ=0.53μm, 2ω_0, pulsewidth 600 ps) was spherically focused to ~5x10^{14} W/cm^2 onto a samarium coated foil target and was delayed relative to the 1ω_0 pulse. For this paper, t=0 is defined as peak of the 1ω_0 pulse.

The primary diagnostic was a Bragg crystal (ADP) X-ray spectrometer that dispersed X-rays onto direct X-ray film (Kodak SB392). The obtained spectral resolution of ~1000 allowed to resolve the structure of the multiple transitions in the Kα lines of different ionic species.

This set-up uses a point projection backlight, therefore the film records data that is a function of wavelength in one direction and space in the perpendicular direction. Absorption lines are formed when the backlight X-rays having energies that correspond to Al Kα transitions are absorbed by the plasma being probed. The wavelength of the absorption lines identify the ion that is in absorption, while the position and length along the spatial axis of the film correspond to the distance from the target surface where the ion species is located. The backlight source is only present for the duration of the 2ω_0 laser pulse and therefore absorption is recorded only during this instant in time.

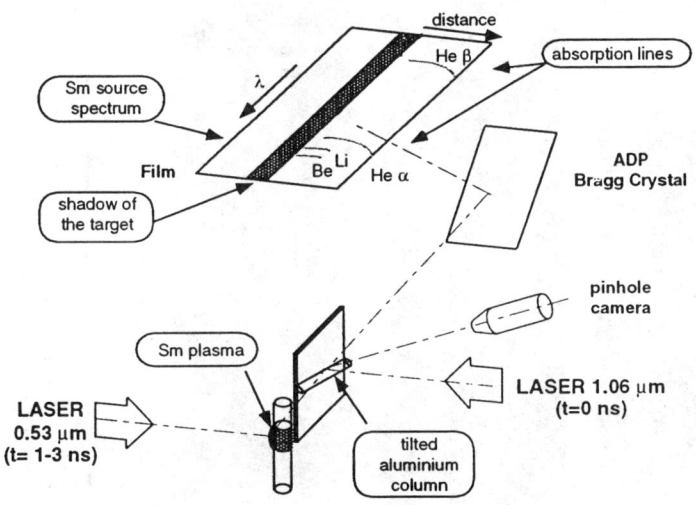

FIGURE 1. Experimental set-up showing the inclined position of the plasma column

The size of the probed plasma along the line of sight is critical to obtaining data that can be analysed. The application of the the Kα absorption technique to a line-focus plasma thus required special adaptations. The plasma column must be probed in a direction perpendicular to the plasma expansion which suggest that the plasma be probed along its diameter. Unfortunately, preliminary experiments indicated that the absorption length along this line of sight was not sufficient to produce a detectable absorption signal. To avoid modifying the plasma conditions of importance to X-ray lasing, the plasma column was tilted to increase the absorption length. In this geometry, which is shown in figure 1, the pathlength of the C-like to Li-like Kα transitions can be increased more than that of the He-like K-shell resonance lines which have much stronger oscillator strengths. The angle used in these experiments was 18.8° and was accurately set by orienting the strip target though a microscope objective.

EXPERIMENTAL ABSORPTION SPECTRA

The experimental spectra of figure 2 are typical of what was obtained for three different times (t= 1, 2, and 3 ns) after the laser pulse that created the aluminium plasma with an irradiance of approximately 1.6×10^{12} W/cm^2. The light area in the center of each spectrum corresponds to the shadow of the aluminium target. The spectrum in the upper part corresponds to unabsorbed samarium radiation that travelled behind the target. Finally the lower part shows the spectrum of the samarium radiation transmitted through the aluminium plasma. The thin and white structures which appear on the dark background show the variation of absorption on Kα transitions along the direction perpendicular to the target surface. Four groups of absorption lines belonging to He-like, Li-like, Be-like and B-like ionic species are observed.

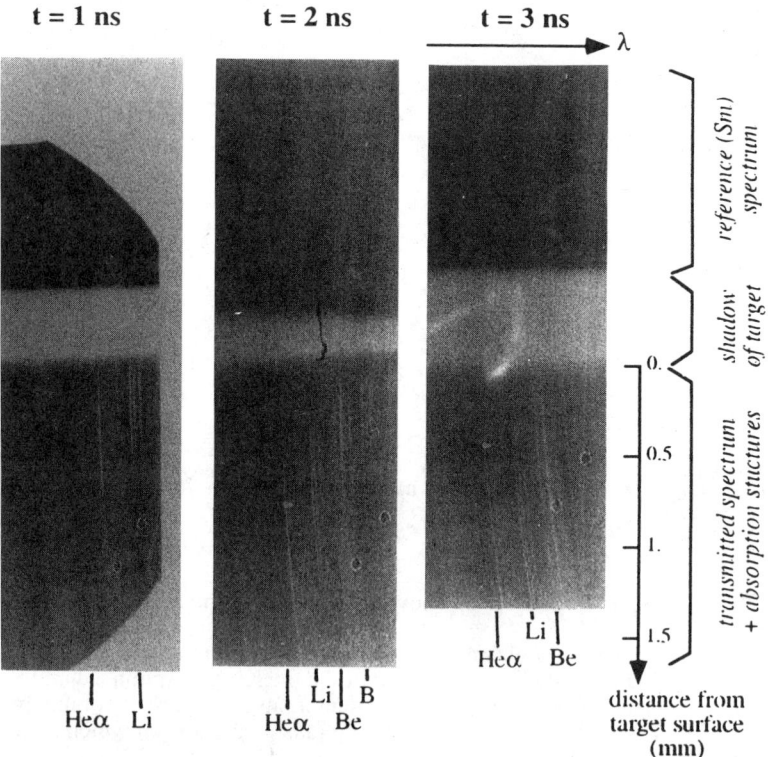

FIGURE 2. Absorption spectra in the 7.7-8.2Å range, taken at t = 1, 2, and 3 ns respectively after the peak of the pulse that created the probed aluminium plasma.

At t=1ns the absorption lines belonging to He- and Li-like ions extend from almost the target surface up to 600-700 µm. For this particular shot the spectral region observed did not include the positions of lines from lower ionic species. However other shots performed in similar conditions showed that at that time only He-like and Li-like lines are visible.

Later on, at t = 2 ns, the He- and Li-like lines appear further from the target surface whereas the lines belonging to Be- and B-like ions are visible in the region close to the target. The fact that the He-like line is relatively strong at that late stage of plasma expansion denotes that He-like ions are still abundant in a spatial region extending from 500µm up to 1.5 mm from the target. This feature is the signature of a frozen ionization which is known to prevail in plasma corona during cooling and which allows population inversions to exist. Lower ionization stages are present closer to the target where the electron density is high enough for recombination to be efficient.

Finally at t =3 ns the He-like line is barely visible while the strongest structures belong to Be-like ions. This means that He-like ions have now recombined down to Li-like and Be-like ions. Compared to earlier time (t = 2ns) the Li-like and Be-like lines turn on at approximately the same distance from the target but they extend further.

COMPARISON WITH SYNTHETIC SPECTRA

We performed a numerical simulation of the evolution of the aluminium plasma using conditions of irradiation similar to the experiments. We used the hydrocode FILM in the 1.5D geometry. This 1D geometry includes a self-similar description of the lateral expansion of a plasma column produced by iradiating a slab target with a line-focused laser beam (4). The width of the line focus entered in the simulation was 100µm according to the experimental one. The code FILM provides the temporal and spatial variation of several plasma quantities, such as electron density and temperature, as well as of the fractional abundances of the ionic species present in the plasma. The calculation of the ionization dynamics is performed on a time-dependent basis and the contribution of the He-like and Li-like excited levels is included via QSS effective rates.

The transverse size D_{lat} of the aluminium plasma and the distribution of ionic fractional abundances along the direction of plasma expansion at the times of interest are fed into an absorption model. At the wavelengths of the Kα transitions for the ionic species present in the plasma, the absorption model calculates the transmission I/I_0, i.e. the ratio of transmitted over incident samarium intensity, following the formula

$$I/I_0 (\lambda) = \exp (- K(\lambda) \cdot N_i \cdot l_{abs}(\lambda))$$

where $K(\lambda)$ is the absorption coefficient for each absorption line, calculated assuming a Doppler broadened profile with T_i = 70 eV. The ground state population density N_i for each ionic species is provided by FILM. Finally the absorption length l_{abs} is calculated from the transverse size D_{lat} of the plasma, taking into account the fact that the value of l_{abs} depends on the wavelength of the probing radiation through the inclined configuration of the aluminium column used in this experiment (see previous section).

Figure 3 shows the reconstructed transmission spectra at the three times of interest, t= 1, 2, and 3 ns. A 1 eV intrumental broadening is included and the quantity I/I_0 is averaged over a time interval of 600 ps to account for the experimental temporal resolution given by the duration of the samarium emission. Again four groups of absorption lines belonging to He-like, Li-like, Be-like and B-like ionic species can be observed.

One can see that the general main features of the experimental spectra of figure 2 are well reproduced by the synthetic spectra of figure 3. The zones of presence of the He-like and Li-like species extend up to the outer limit of the plasma whereas the less ionized species exist in a region closer to the target surface. As time goes on the presence zone of He-like ions moves outwards in the plasma.

A more detailed comparison of the location and extension of each ionic species reveals that the distances calculated by the simulations are generally smaller than those observed in the experimental spectra. For example at t=2ns He-like ions extend from 200μm to 1 mm on the synthetic spectrum whereas the observed positions are 500μm to more than 1.5 mm on the experimental spectrum. A similar disagreement is found for Li-like ions. This disagreement is coherent with the fact that the position of the maximum gain on the Li-like 5f-3d line was previously observed (1) to be around 400μm from the target, whereas the predicted value is smaller, typically 250μm.

Further analysis of the experimental data is required in order to extract ionic density numbers for each ionic species and to allow for a detailed quantitative comparison with the predictions of our hydrocode.

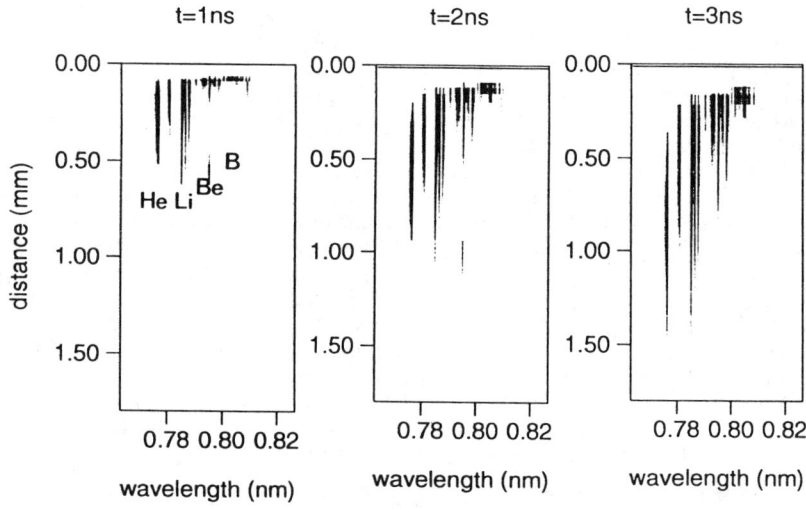

FIGURE 3. Synthetic spectra showing the transmission on the Kα lines as a function of the distance to the target surface.

REFERENCES

(1) Jamelot, G., et al., in *Proceedings of the 3rd X-Ray Lasers Conference*, IOP Conf. Series N°125, E.E. Fill ed., IOP Publishing, 1992, pp.89-95
(2) Back, C.A., Chenais-Popovics, C., Renaudin, P., Geindre, J.P., Audebert, P., and Gauthier, J.C., *Phys. Rev. A*, **46**, 3405, 1992
(3) O'Neill, D., Lewis, C.L.S., Neely, D., Davidson, S.J., *Phys. Rev. A* **44**, 2641, 1991
(4) Klisnick A. et al., in *Proceedings of the 3rd X-Ray Lasers Conference*, IOP Conf. Series N°125, E.E. Fill ed., IOP Publishing, 1992, pp.305-308

A SELF-CONSISTENT MODEL FOR LINE TRAPPING EFFECT IN X-RAY LASERS

Djamel E. Benredjem,[*] Alain Sureau, Clary Möller, and Hélène Guennou

Laboratoire de Spectroscopie Atomique et Ionique
Université Paris-Sud
Centre d'Orsay, Bât. 350
91405 Orsay Cedex (France)

[*]on leave from Université d'Annaba, Institut de Physique, BP 12
23000 Annaba (Algeria)

Abstract.- Line trapping appropriate for cylindrical geometry in recombining plasmas has been examined. Calculations are self-consistent and use the results of a 1D Lagrangian simulation as inputs. Resonant photons are assumed to travel a maximum distance which is given by the inverse of the absorption coefficient.

Introduction

The line trapping effect is due to the reabsorption of photons which are, in general, emitted non locally. Trapping is known to reduce gain of X-ray lasers by feeding the lower level of the lasing transition. Following Brunner *et al* and generalizing their work, we have developed a self-consistent model to account for line trapping in the gain calculation of Li-like recombination lasers. Other calculations in this scheme (Pert, Shestakov and Eder) as well as in the collisional excitation (Eder *et al*) have shown an important reduction of gain.

We apply our model to resonant photons (lower level $l \rightarrow$ ground level g). In this case, Einstein coefficients are large and we expect an important trapping effect.

Model

The resonant transition $l \rightarrow g$ is characterized by large emission (A_{lg}, B_{lg}) and absorption (B_{gl}) Einstein coefficients. Furthermore, the ground level of lasing ions is the most crowded. Then, absorption overcomes induced emission, resulting in a positive absorption coefficient:

$$K_{gl}(v|r) = \frac{h v \Phi(v)}{c} \left[N(Li, g|r) B_{gl} - N(Li, l|r) B_{lg} \right], \quad (1)$$

where $\Phi(v)$ is the Doppler profile and the N's the population densities of Li-like ions. The quasi-steady state (QSS) assumption (Klisnick et al) is used for Li-like excited levels and we obtain the following set of equations:

$$-N(Li,i|r)\Gamma_i(r) + R_i(r) + c\delta_{il} \iint dv \, d\Omega \, K_{gl}(v|r)n(v|r,u) = 0. \quad (2)$$

The integral involves the two processes named above and $n(v|r,u)$ designates the density of photons of wavelength λ_{lg} propagating in the direction of a unit vector u, while Γ_i and R_i are given by

$$\Gamma_i(r) = \sum_{j<i}\left[A_{ij} + N_e(r)Y_{ij}(r)\right] + N_e(r)\sum_{k>i} X_{ik}(r) + N_e(r)S_c(Li,i;He,g|r) \quad (3)$$

$$R_i(r) = N_e(r)\sum_{j<i} N(Li,j|r)X_{ji}(r) + \sum_{k>i} N(Li,k|r)\left[A_{ki} + N_e(r)Y_{ki}\right] + \quad (4)$$

$$N(He,g|r)N_e(r)\left[A_r(He,g;Li,i|r) + N_e(r)A_c(He,g;Li,i|r)\right]$$

where the different collision rates are thoroughly discussed by Klisnick et al.

The population density of the lower level can be written as

$$N(Li,i|r) = R_l(r)/\Gamma_l(r) + H(Li,i|r), \quad (5)$$

where the line trapping correction H is given by

$$H(Li,l|r) = \frac{c}{\Gamma_l(r)} \iint dv \, d\Omega \, K_{gl}(v|r)n(v|r,u). \quad (6)$$

After some mathematical manipulation in which we set $r' = r - s'u$ and approximate the integration over frequency by terms involving the centre value, at frequency v_0, we can show that H is solution of the following equation:

$$H(Li,l|r) - \frac{A_{lg}}{\Gamma_l(r)}\overline{N}(Li,l|r)$$

$$= \frac{A_{lg}}{\Gamma_l(r)} K_{gl}(v_0|r)\frac{1}{4\pi}\int_0^{s_m} ds \int d\Omega \, H(Li,l|r-su) \exp\left[-\int_0^s ds' K_{gl}(v_0|r-s'u)\right], \quad (7)$$

where

$$\overline{N}(Li,l|r) = \int dr' \, N^{(0)}(Li,l|r')G(r,r'), \quad (8)$$

$$G(r,r') = \frac{K_{gl}(v_0|r)}{4\pi|r-r'|^2} \exp\left[-\int_0^{|r-r'|} ds' K_{gl}(v_0|r-s'\frac{r-r'}{|r-r'|})\right]. \quad (9)$$

$N^{(0)}(Li,l)$ is the population density of the lower level, without trapping; s_m designates the surface of the plasma, while $s=0$ corresponds to the position of the lasing ion.

In Eq. 7 H and K_{gl} are replaced by Taylor expansions in the vicinity or r, with respect to su and $s'u$, to the second order. A rapid convergence is obtained if s and s' remain small, i. e. $|r-r'|$ is of the same order of magnitude as r. We examine line trapping in cylindrical symmetry, which is the most appropriate to the case of a thin fibre target. It is reasonable to assume that H, K_{gl} and the population densities depend only on the radial variable r, i. e. on the distance from the plasma axis. Calculation of the integral on the right-hand side member of Eq. 8 is performed by defining a volume for the variable r'. If r is in a plasma region where K_{gl} is large/small the distance of flight of resonant photons will be small/large. Then, we define the distance of flight by $\alpha/K_{gl}(r)$, where α is a constant ~ 1. The volume of integration is taken to be a cylinder centered on the position of the lasing ion and whose radius and length are equal to $\alpha/K_{gl}(r)$. This definition precludes photons emitted outside the cylinder from reaching the lasing ion. In our simulations, the characteristic dimension of the cylinder vary from a few μm near the plasma axis to a few tens of μm in the outer regions.

Eqs 7-9 lead to the final equation for H which involves only the radial coordinate r

$$H(Li,l|r) = 3K_{gl}(v_0|r)\int_r^R dr' K_{gl}(v_0|r')/r' \int_0^{r'} dr'' r''[H(Li,l|r'')(1-\Gamma_l(r'')/A_{lg}) + \overline{N}(Li,l|r'')], \quad (10)$$

where

$$R = r + \alpha/K_{gl}(v_0|r). \quad (11)$$

Equation 10 is resolved by a self-consistent procedure. In a first step we calculate K_{gl}, where the population densities are obtained from a simulation with the code FILM (without trapping). An initial guess for H and the calculated $\overline{N}(Li,l)$ are introduced in the right-hand side member of Eq.10 and a new correction is obtained. The population density of the lower level and that of the ground level, which is assumed to be insensitive to trapping, in regard to its relative importance, are used in Eq. 1 to obtain a new K_{gl} and so on. The variation of population for the lower level obtained after convergence induces a variation $H(Li,u|r)$ of the population density of the upper level, due to the collisional-excitation and de-excitation. By considering Eqs. 2-4 we can calculate $H(Li,u)$. Finally, we evaluate the correction $\Delta G(u,l|r)$ to the gain of the lasing transition by assuming a Doppler profile for the lasing transition.

Results and discussion

Figures 1 and 2 show the gain coefficient (cm^{-1}) for lasing transitions $3d_{5/2}-4f_{7/2}$ ($\lambda = 154.7 Å$) and $3d_{5/2}-5f_{7/2}$ ($\lambda = 105.7 Å$) of Li-like aluminium. Our calculation models the experiment of Carillon *et al* in which an incident laser of wavelength $\lambda = 0.53$ µm deposites an energy of *4 J/cm* over a *120 ps* pulse. The target consists of a thin fibre of diameter *7* µm coated with aluminium. Besides the peak gain reduction, which equals *25 %* for the transition *3 – 4* and *10 %* for the transition *3 – 5*, we notice a space-shift of the maximum gain which was noted by Makhrov *et al*. The two curves converge in the outer region of the plasma where the coefficient of absorption is small, due to the low population densities of the relevant levels.

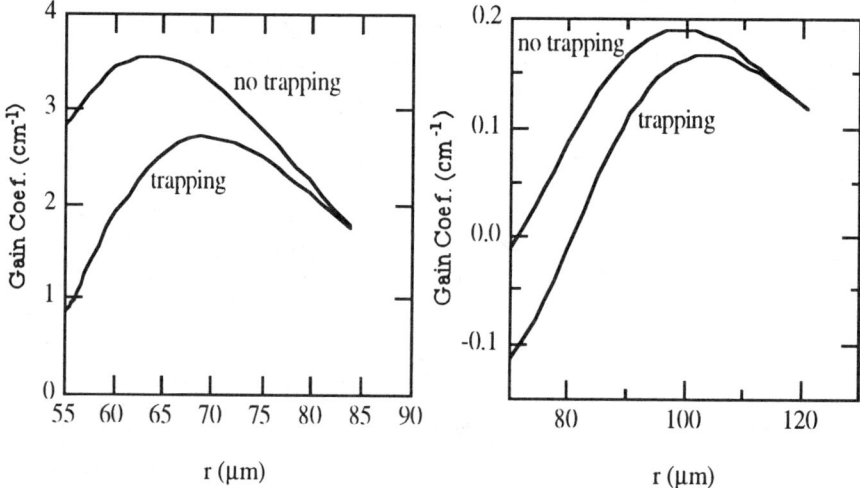

Figure 1. Gain coefficient vs distance from the plasma axis for the lasing line $3d_{5/2} - 4f_{7/2}$ of Li-like Al, at time of peak gain. Laser and target specifications are given in the text.

Figure 2. Same as in Fig. 1 for the lasing line $3d_{5/2} - 5f_{7/2}$.

Figure 3 represents the ratio of the gain coefficient with trapping included to that without trapping as a function of r, at different time steps. It is clear that the trapping effect decreases for increasing time, due to the expansion of the plasma.

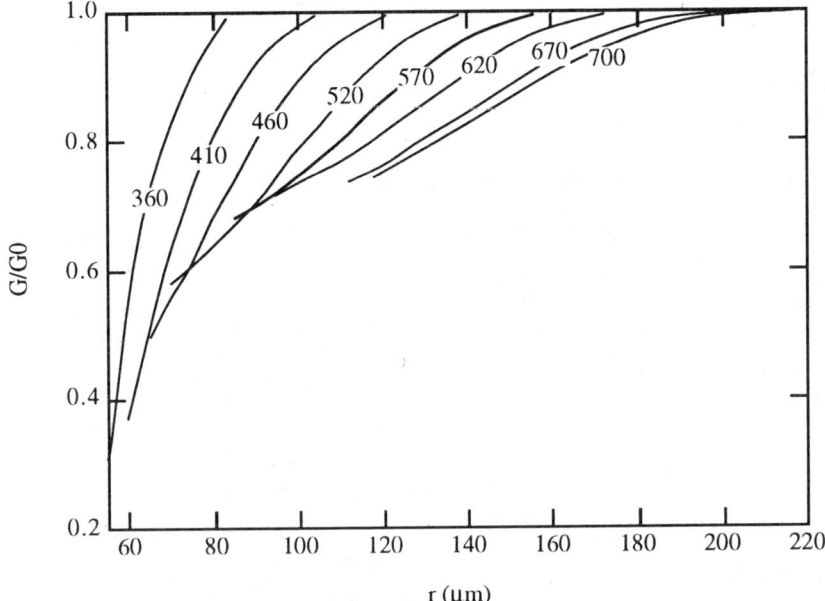

Figure 3. Ratio of the gain coefficients (with/without trapping) vs distance from the plasma axis, at various time steps, for the lasing transition $3d_{5/2} - 4f_{7/2}$. Laser and target specifications are given in the text. 360 and 460 ps are the times of peak gain for $3-4$ and $3-5$ transitions of Li-like Al, respectively.

In conclusion, trapping has been examined via a self-consistent method in the recombination scheme. Gain is calculated in the case of a thin fibre by using a monodimensional Lagrangian hydrocode. The main results are that gain is reduced but by a lesser amount than predicted by Makhrov et al by using a sophisticated expression for escape probability. Nevertheless, the space shift of the maximum gain reported by these authors is confirmed by our calculations.

References
Brunner W., John R. W., Paul H., and Steudel H., *Laser and Particle Beams* **6**, 723 (1988).
Carillon *et al*, *J. Phys. B* **23**, 147 (1990).
Eder D. C., Scott H. A., Maxon S., and London R. A., *Appl. Opt.* **31**, 4962 (1992).
Klisnick A., Sureau A., Guennou H., Möller C. and Virmont J., *Appl. Phys. B* **50**, 153 (1990).
Makhrov V., Roerich V., Starostin A., Stepanov A., Klisnick A., Sureau A. and Guennou H., *J. Phys. B.* **27**, 1899 (1994).
Pert G. J. and Rose S. J., *Appl. Phys. B* **50**, 307 (1990).
Shestakov A. I. and Eder D. C., *JQSRT* **42**, 483 (1989).

Experimental Study of Neonlike Zinc J=0-1 Soft X-ray Lasing at 21.2 nm

B. Rus[#], A. Carillon, P. Dhez, B. Gauthé, P. Goedtkindt, P. Jaeglé, G. Jamelot, A. Klisnick, M. Nantel, A. Sureau, and P. Zeitoun

Laboratoire de Spectroscopie Atomique et Ionique, URA 775 CNRS, Université Paris-Sud, Bâtiment 350, 91405 Orsay Cedex, France
[#]*on leave from Department of Gas Lasers, Institute of Physics, 18040 Prague 8, Czech Republic*

Abstract. Since our first observation of intense J=0-1 lasing in neonlike zinc in early 1993, we have investigated various issues of this X-ray laser. The results of these experiments, performed at Laboratoire pour l'Utilisation des Lasers Intenses (LULI), are briefly reviewed. Gain coefficients of ~5 cm^{-1} at 21.2 nm are now routinely obtained with ≈350 J delivered onto a Zn slab in ~650 ps pulses - the net irradiance is ~1.4×10^{13} Wcm^{-2}. The 21.2 nm emission appears in a ≈100 ps burst and precedes the lasing on the J=2-1 transitions at 26.2 and 26.7 nm which have a duration ≈350 ps. We have directly measured the gain regions of the J=0-1 line at 21.2 nm and the J=2-1 line at 26.7 nm and confirmed the initial claim that they are located at different distances from the target. We have also measured the refraction angles and divergences of the XRL beams of these two transitions. The beam at 21.2 nm was observed to be remarkably narrow with a divergence of ~2.6 mrad, peaking ~7 mrad off-axis. This contrasts to the beam at 26.7 nm which has a divergence of ≈10 mrad and peaks at ~15 mrad. The most recent finding is that the J=0-1 lasing exists only when the main pump pulse is preceded by a pulse train of mJ level, which is a remnant from the laser oscillator. An important achievement in developing the XRL at 21.2 nm was the demonstration of half-cavity operation: with a gain-length product of ~16.5, the emission exhibits saturation behaviour. The absolute energy measurements indicate ~400 µJ in the near-saturated beam.

INTRODUCTION

More than a year ago, we observed strong X-ray lasing on the J=0-1 transition in collisionally pumped Ne-like Zn (1). The line at 21.2 nm with a gain coefficient of 4.9 cm^{-1} was observed to entirely dominate the emitted spectrum for 2 cm long targets, being 20-30 more intense than the J=2-1 lines at 26.2 and 26.7 nm whose measured gains were 2.3 and 2.6 cm^{-1} respectively. In this paper as well as in Reference (2) we present results of our recent experimental efforts.

The simplified Grotrian diagram for Ne-like Zn is in Figure 1. Allowing for the fact that the transitions at 21.2 and 26.7 nm share the same lower level $(1/2,1/2)_{J=1}$, a detailed analysis of the first experimental results indicated that these two lines are emitted from distinct plasma regions (1). This was to be confirmed by a direct measurement. Furthermore, we carried out time-resolved measurements of beam refraction and divergence angles for the above lines to supply data for comparison with results of the modelling, and, in the case of the 21.2 nm line, to be able to implement a half-cavity. A question connected with the latter issue was whether the duration of gain at 21.2 nm is able to support the half-cavity roundtrip time for 2 cm plasmas, as suggested by the previous analysis indicating the gain duration exceeding perhaps two times the ≈100 ps duration of the X-ray laser (XRL) emission. The need

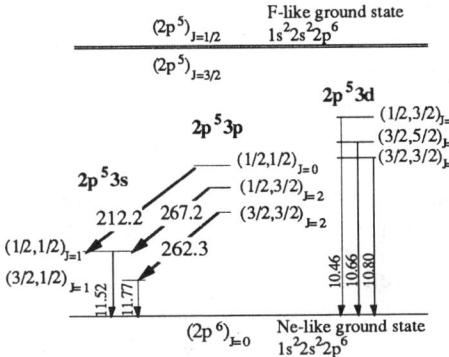

FIGURE 1. Simplified Grotrian diagram of Ne-like Zn (Z=30). Among 3s and 3p levels, we indicate only those involved in the observed lasing (wavelengths are in Å).

to minimize the roundtrip time by approaching the half-cavity mirror as close to the plasma as possible posed an additional question about the survival of the mirror reflectivity during the duration of the XRL pulse. Having successfully operated the half-cavity, we have measured the energy in the XRL beam.

Motivated by recent observations in Ne-like Ge (3) that even a very low-level prepulse can significantly enhance the J=0-1 output, we carried out a careful test whether the pulse train remnant from the laser oscillator, which is present in the laser shots, has a role in the strong J=0-1 lasing observed.

EXPERIMENTAL ARRANGEMENT

The setup is illustrated in Figure 2. Six LULI laser beams delivering a total of ≈350 J of net energy at 1.06 µm in nearly Gaussian pulses of ~650 ps FWHM are line focused onto a Zn slab by a combination of two perpendicular cylindrical lenses, creating a net irradiance of ≈1.4×10^{13} Wcm^{-2}. The plasma column is monitored for its width and uniformity by a multipinhole X-ray camera with a CCD detector.

The axial XRL output is detected by the 'Focal' grating spectrometer coupled to a streak camera, with a single viewing field ~20 Å. The spectrometer uses Wadsworth geometry and focuses parallel emission onto a point on the focal circle. It was adapted for collisional XRL experiments by adding beam collection optics. In the first configuration a plane mirror is employed to reflect the off-axis refracted XRL beam towards the spectrometer entrance slit. As the slit is open, the spectrometer works in spectrally unresolved mode but allows the analysis of the angular profile of the XRL beam intensity as each point at the detector corresponds to a particular ray direction. The angular coverage in a typical setup for 21.2 nm beam measurements is ~5.5-10 mrad, using a 20 cm plane mirror placed 10 cm from

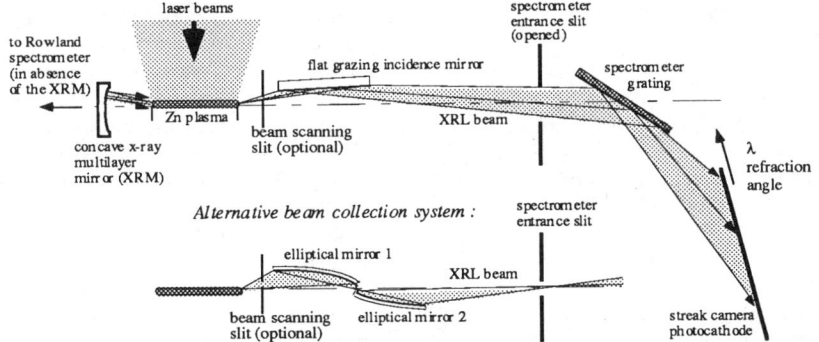

FIGURE 2. Experimental setup for investigation of collisionally-pumped XRLs at LULI. The Focal spectrometer is converted into an analyzer of XRL beam divergence with the help of plane mirror. The elliptical mirror system collects the total XRL emission produced.

the plasma. In the second configuration the slit has its normal width and the plane mirror is replaced by a system of two elliptical mirrors which focus the emission from the plasma exit plane onto the slit plane. As the divergence of the focussed beam at the slit plane is ~10-15 less than the divergence of the emitted beam, this setup provides both wavelength and angular resolution. However the latter is rather poor so that this device with an angular coverage of 0-15 mrad is used as a collector of the total XRL beam emitted. A 65 µm slit located at 85 mm from the plasma can be used to scan the beam to precisely obtain the angular distribution of its intensity.

To localise gain regions of the lasing lines we completed the above arrangement by a simple device consisting of a precisely positioned stainless cut edge placed at 1 mm from plasma end. The edge was advanced in each shot by steps of 25 µm away from the target surface to gradually screen the emitting region. The drive beams were carefully apertured so that no laser light hits and ablates the edge.

The half-cavity experiments at 21.2 nm were carried out with a Mo:Si multilayer mirror (XRM) of 13 cm curvature radius (4), whose reflectance was ~30%. The mirror was positioned 9 mm from the plasma end, making the roundtrip time for 2 cm plasmas to be 195 ps. The mirror was protected by an aperture so that only a spot of ≈2 mm in diameter was damaged during each shot and one mirror was usable many times. Several shots were also taken with a distance mirror-plasma of 6 mm, with the aim to increase the XRL output by increasing the feedback from the mirror.

The absolute energy contained in the XRL beam was deduced from "footprint" shots recorded on a thinned backside-illuminated CCD camera Thomson TSC 7395A of calibrated spectral sensitivity. To bandpass the emission near 21.2 nm, a 45° degree Mo:Si multilayer flat mirror designed at this wavelength (measured reflectance 28%) was used to relay the XRL beam onto a CCD back equipped by Al filters of various thickness (0.8-2.5 µm) to protect it against visible light and to properly attenuate the beam intensity so as to fall within the CCD dynamic range.

EXPERIMENTAL RESULTS

The time history of the lasing lines is plotted in Figure 3. The J=0-1 emission peaks ~100(±50) ps before the J=2-1 lines; its duration ≈100 ps is noticeably shorter than ≈300 ps duration of the J=2-1 lines. This behaviour was seen to be weakly sensitive on the driving laser energy.

The analysis of the above characteristics by unfolding the emissivity and gain time histories resulted into a conclusion (1) that the 21.2 nm J=0-1 line comes from a different plasma region than the 26.7 nm J=2-1 line. Direct measurements, whose results are shown in Figure 4, confirmed this hypothesis. We see that the 21.2 nm line is emitted from a narrow ≈30 µm zone

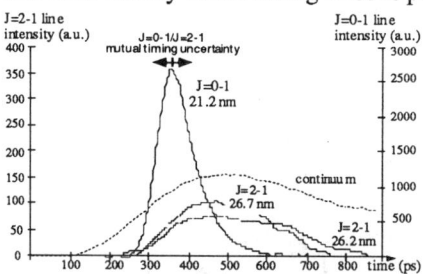

FIGURE 3. Temporal evolution of the lasing lines. Two streak records (J=0-1 and J=2-1) are mutually timed with the help of the continuum (dotted curve, not to scale).

located ≈30 µm from the target surface, in contrast to the 26.7 nm emission which peaks at ≈100 µm and is emitted over a ≈200 µm zone. Clearly the two lines are produced at different plasma conditions in space and time, which points at different processes predominantly involved in populating the corresponding upper 3p levels.

Another key parameter for understanding the XRL investigated and optimizing its performance are the X-ray beam characteristics. We carried out an extensive set of

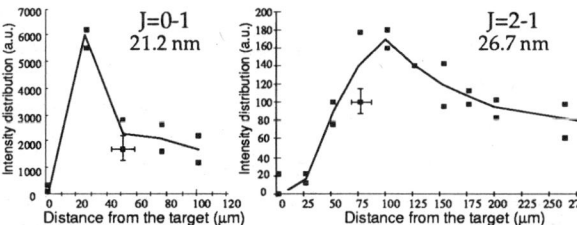

FIGURE 4. Distribution of intensity of the 21.2 and 26.7 nm lasing lines at a plane 1 mm away from the exit of a 2 cm long plasma. Data were obtained by scanning this plane by a fine stainless cut edge whose position was carefully measured by a telescope.

measurements of refraction angle and divergence at 21.2 nm, and limited comparative measurements of the same parameters at 26.7 nm. The results for 21.2 nm are in Figures 5 and 6. By increasing the plasma length the beam divergence strongly decreases down to ~2.6 mrad for a 2 cm plasma, although the refraction angle remains almost unchanged. This is consistent with high gain over a small region of plasma. The refraction angle quickly diminishes early in time and stabilizes at ~7.2 mrad starting from the emission intensity peak. The beam divergence, on the contrary, undergoes few evolution and rather broadens as time progresses. Both these variations suggest that the gain zone flattens and/or the gain decreases in time.

The measurements at 26.7 nm provided a fairly flat beam profile with a divergence of ≈10 mrad, peaking approximately at 15 mrad. The large beam divergence seems to be consistent with the wide and relatively uniform 26.7 nm gain region observed.

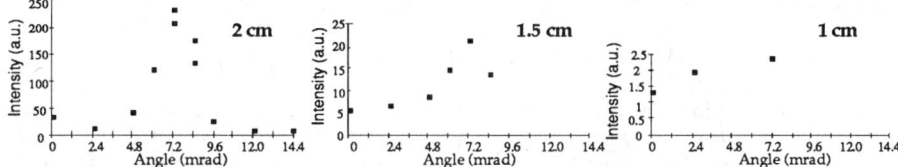

FIGURE 5. Time-integrated angular profiles of the beam at 21.2 nm for three plasma lengths. The data were obtained using the elliptical mirror system and the scanning slit.

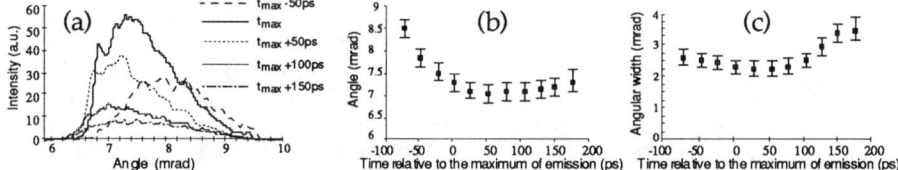

FIGURE 6. (a) 21.2 nm beam temporal profiles at different times relative to the instant of the emission maximum for a plasma of 2 cm; (b) temporal evolution of the peak intensity of the angular profile (refraction angle); (c) temporal evolution of the beam divergence.

All experiments presented were obtained with a nominal LULI laser pulse (Fig. 7). To establish the effect of the train of ≈10 mJ pulses on the J=0-1 lasing, we carried out a test in which the train was reduced by a factor of ~700, using additional Pockels cell in the laser chain. Under these conditions no 21.2 nm emission was seen, which suggests that Zn does not make exception from the elements where the J=0-1 line has been made to lase using a prepulse. However, further investigation is needed: compared to prepulses of joule level (5), the nature of effects due to tens-of-mJ prepulses, beneficial to the J=0-1 lasing, is not currently understood.

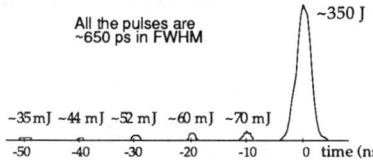

FIGURE 7. LULI laser pulse: the main heating pulse is preceded by a pulse train remnant from the laser oscillator.

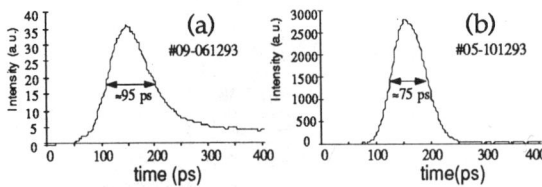

FIGURE 8. Temporal evolution of emission at 21.2 nm from a single pass XRL (a), and from the half-cavity (b) with the XRM placed 9 mm from the plasma end. The curve (b) clearly indicates that the XRM is efficient during the whole duration of the XRL emission.

The double-pass experiments were carried out with the XRM aligned to return the beam refracted at 7.2 mrad back into the amplifying region. Compared to single-pass, typically 60-fold enhancement of the output signal was observed for 2 cm plasmas with the XRM operating at the distance of 9 mm. The time evolution of a half-cavity generated pulse is similar to that of the single-pass (Fig.8), suggesting that no sudden destruction of the mirror took place within the pulse duration. With gain 4.9 cm^{-1} and with the calculated feedback from the XRM to be 5($\pm$3)$\times$10^{-2}, based on the knowledge of the mirror reflectance and the gain region width, the effective gain-length product attained for 2 cm plasma is gl_{eff}=16.6($^{+0.5}/_{-0.9}$). Should the ASE output intensity maintain the exponential behaviour, one would expect a ~900-fold enhancement compared to single-pass, about 15 times superior to the observed value. Apparently either the feedback from the XRM is much lower than the calculated value or the emission is not amplified during the second pass with the anticipated gain. However half-cavity shots performed in other configurations suggest that the above feedback estimation is basically correct. Positioned at 14 mm, the XRM increased the output from a 1.5 cm plasma by a factor of ~30 which is consistent with the expected ~60$\times$ enhancement corresponding to gl_{eff}=11.4($^{+0.5}/_{-0.9}$) calculated, once again accounting for gain 4.9 cm^{-1}. Moreover, as the roundtrip time is the same as for the shots with 2 cm plasmas, this consistency also virtually rules out a hypothesis about temporal diminution of the gain as the responsible effect for the lower-than-anticipated amplification during the second pass in that case. Another indication that our feedback estimations are realistic and that the XRM reflectivity does not dramatically differ for the distances used was obtained with the XRM operated at 6 mm from a 2 cm plasma. The ~80-fold output enhancement observed, compared to the factor 60 seen in the other shots, roughly equals to ratio of the calculated feedbacks in these cases.

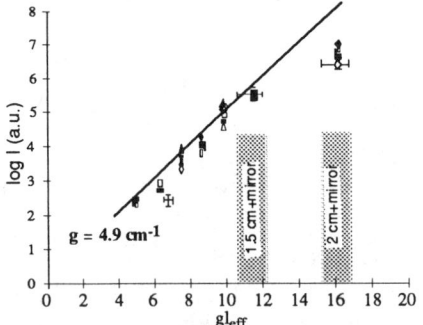

FIGURE 9. Measured intensity of emission at 21.2 nm as a function of effective gain-length product gl_{eff}.

From all these facts we conclude that the roll-off of the measured intensity from the exponential character, for gl_{eff} as large as $\approx$16.5, occurs due to onset of saturation.

Finally, the absolute measurements of energy provided 400 µJ in a half-cavity pulse, which combined with the $\approx$75 ps pulse duration gives a peak output power of $\approx$5 MW.

REFERENCES

1. Rus, B, Carillon, A, Gauthé, B, Goedtkindt, P, Jaeglé, P, Jamelot, G, Klisnick, A, Sureau, A, Zeitoun, P, JOSA B **11**, 564-573 (1994)
2. Jaeglé, P., et al., this Proceedings
3. Cairns, G, et al., this Proceedings
4. Chauvineau, JP, Institut d'Optique, Bâtiment 503, 91405 Orsay Cedex, France (1993)
5. Nilsen, J, Moreno, JC, MacGowan, BJ, Koch, JA, Appl. Phys. B, 57, 309-311 (1993)

Dielectronic spectra for Ne-like ions from F-like low-lying states

M. Cornille* and S. Jacquemot[+]

*Observatoire de Paris-Meudon DARC, 5 place J. Janssen, 92195 Meudon Cedex FRANCE
[+]Centre d'Etudes de Limeil-Valenton 94195 Villeneuve Saint-Georges Cedex FRANCE

Abstract. Energy levels, radiative decay and Auger rates have been calculated for dielectronic recombination of 7 Ne-like ions, ranging from Fe^{17+} to Ag^{38+}, from the $n=2$ F-like shell into the $n=3$ one, by use of the multiconfiguration AUTOLSJ code in intermediate coupling. Rate coefficients have then been computed versus the electronic temperature, the nuclear charge of the involved element, the final Ne-like state and the population distribution among the initial 3 F-like levels. Dielectronic spectra have also been obtained.

INTRODUCTION

Dielectronic recombination [DR] has been found to be an important process in non-LTE (1) and collisionally pumped Ne-like x-ray laser (2) plasmas. Then it must be taken into account in the determination of ionization balance and kinetics. In early modelings of high-temperature plasmas, the DR mechanism was often represented by semi-empirical analytic rate coefficients (3), but, as a significant way for selectively populating the Ne-like $3s$ and $3p$ lasing levels, the dielectronic recombination, from the F-like low-lying $n=2$ states to the Ne-like $n=3$ manifold, has to be carefully detailed (Fig. 1).

MODEL

The general picture implies an initial state $\underline{i}$ [in our case a singly-excited Ne-like $\left(1s^2 2s 2p^5\right) 2s 2p$, $2s 3l$ or $2p 3l$ ($l = s, p, d$) one], a doubly-excited autoionizing state $\underline{a}$ [belonging to the same sequence as $\underline{i}$ and to the manifold $\left(1s^2 2p^4\right) 2s^2 3l 3l' + 2s 2p 3l 3l' + 2p^6 3l 3l'$] and a final state $\underline{j}$, in the sequence more highly stripped by one electron than $\underline{i}$ [a low-lying F-like $\left(1s^2 2s 2p^5\right) 2s$ or $2p$ state]. The DR proceeds then through an initial capture from $\underline{j}$ to $\underline{a}$ followed by radiative stabilization from $\underline{a}$ to $\underline{i}$: $\underline{j} \overset{n_e C_a}{\to} \underline{a} \overset{A_r}{\to} \underline{i}$.

The DR rate coefficient can be written $\alpha_d^{jai} = \frac{1}{2}\left(\frac{(hc)^2}{2\pi mc^2}\right)^{3/2} T_e^{-3/2} F_s e^{-\Delta E_{aj}/T_e}$ where g denote level statistical weight factors, ΔE transition energies, A_a {A_r} Auger {radiative} decay rates and T_e {n_e} involved electronic temperature {density}. $F_s = \frac{g_a}{g_j} \frac{A_a^{aj} A_r^{ai}}{\sum_{j'} A_a^{aj'} + \sum_{i'} A_r^{ai'}}$ is the dielectronic line factor.

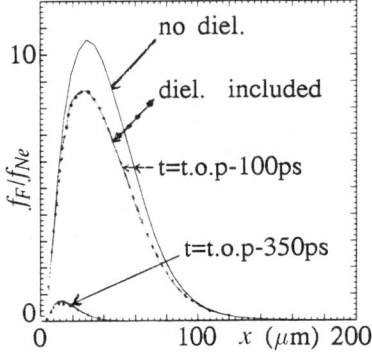 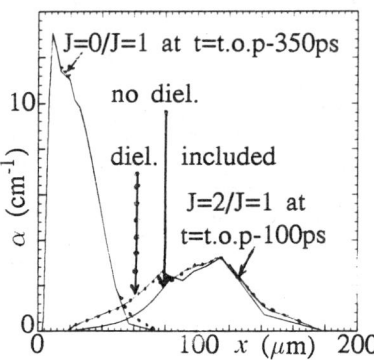

Figure 1. DR influence on local kinetic calculations (see S. Jacquemot): the ratio of the F- to the Ne-like fractional abundances [f_F/f_{Ne}] decreases when the system tends to be over-ionized which increases the "2/1" local gain coefficient [α]; the "0/1" one is not directly influenced: first, the low statistical weight of the upper lasing level disadvantages this populating mechanism; second the inversion occurs earlier, when only a few F-like ions are present, and near the critical surface, where the DR coefficient is reduced by a strong density dependence (4).

COMPUTATIONAL METHOD

Energy levels and radiative decay rates, for the whole Ne-like manifold [843 i and a levels], are provided in LSJ coupling by SUPERSTRUCTURE (5). First a set of non-relativistic wavefunctions is determined by diagonalization of the non-relativistic Hamiltonian, using orbitals calculated in a scaled Thomas-Fermi-Dirac potential. The required scaling parameters are obtained by a self-consistent minimization procedure, based upon the first four Ne-like $(1s^2 2s^2 2p^5) 2p-3l$ configurations. Second the program diagonalizes the Breit-Pauli Hamiltonian on this basis and therefore the final multiconfigurational wavefunctions include relativistic effects in the configuration mixing coefficients. Term coupling coefficients for the N and N+1 electron systems [N=9 here], also given by SUPERSTRUCTURE, are supplied as inputs to DWMDUB (6), whose spirit is very similar to the collisional codes DW and JAJOM (7), to compute bound–free transition matrix elements, in a distorted wave aproximation, and then autoionization probabilities.

DISCUSSION

The $\underline{a} \to \underline{i}$ lines are in fact satellites of F-like resonances, especially of the reference $(1s^2 2s^2 2p^4)3d\ ^2P_{1/2} - 2p\ ^2P_{1/2}$ one (Fig. 2), and are purely formed by DR. Their intensities are thus directly connected to α_d^{jai} according to: $I_s = \sum_j f_j n_e \alpha_d^{jai}$, and depend on the fractional abundances f_j of the F-like low-lying states [$\sum_j f_j = 1$]. LTE being assumed among the F-like $n=2$ shell, $f_j = g_j e^{-E_j/T_e}/\sum_j g_j e^{-E_j/T_e}$. If observed, these satellites can provide helpfull plasma diagnostics. No simple scaling laws can be derived from the study of the effective [$\alpha_d^i = \sum_j f_j \sum_a \alpha_d^{jai}$] DR rate coefficient, neither on nuclear charge Z_N [except at high temperature where $\alpha_d^i \propto (Z_N - 9)^{2.75}$] (Fig. 3) nor on total orbital momentum J of the final singly-excited level $\underline{i}$, configuration structures being apparently of some dominating influence (Fig. 4). In x-ray laser kinetic simulations, a statistical redistribution, among the Ne-like states, of semi-empirical total DR rate coefficients is then quite questionable.

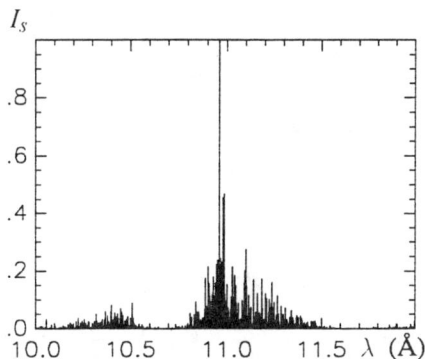

Figure 2. LTE Cu 3/3/ DR spectrum: the satellite intensities are calculated at $T_e = 10^{5/2}$ eV and normalized to $2 \max I_s$; the vertical line represents the F-like reference.

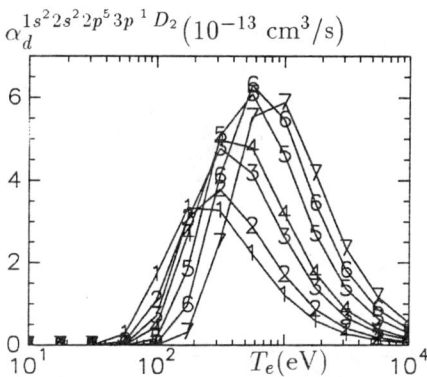

Figure 3. LTE effective DR rate coefficient [into $1s^2 2s^2 2p^5 3p\ ^1D_2$] versus T_e and Z_N: the curves are labeled according to 1≡Fe, 2≡Cu, 2≡Ge, 4≡Se, 5≡Sr, 6≡Mo, 7≡Ag.

α_d^i is in fact fairly sensitive to the F-like population distribution: assuming coronal equilibrium [$f_j = 0$ except for the ground $1s^2 2s^2 2p^5\ ^2P_{3/2}$ level] leads only to a 10% decrease of its maximum value (Fig. 5).

No detailed [A_a, α_d^{jai}] calculations have been found in the literature apart from for Ti^{13+} (8). Isoelectronic sequence studies of total DR rates have been reported (9) but don't allow direct comparisons. The only one relevant result, for the single case of Se^{25+} (10), is in basic agreement with this work (Fig. 6).

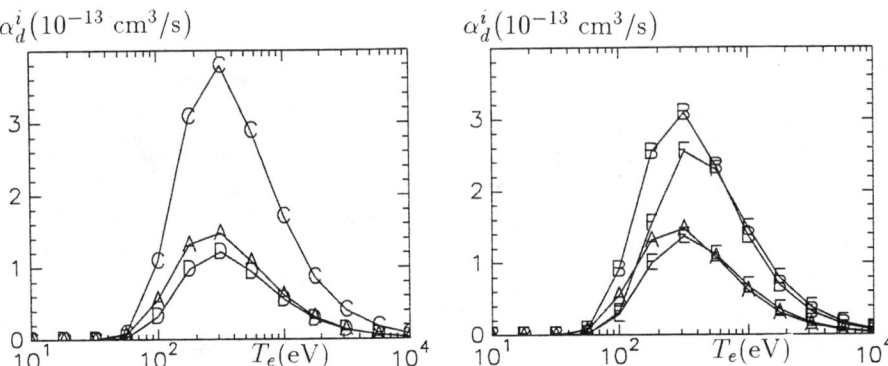

Figure 4. LTE Cu effective DR rate coefficient versus T_e and i [A≡ $2s3s\ ^3P_1$, B≡ $2s3p\ ^3P_1$, C≡ $2s3p\ ^1D_2$, D≡ $2s3p\ ^1S_0$, E≡ $2s3d\ ^3P_1$, F≡ $2p3p\ ^3P_1 - 1s^22s2p^5$ is omitted].

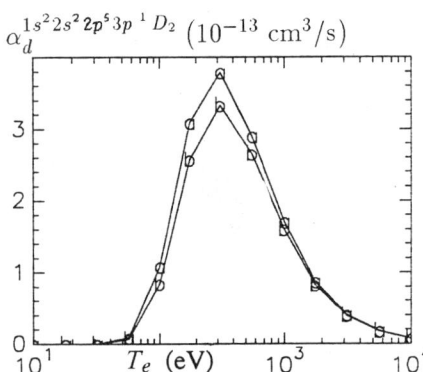

Figure 5. Cu effective DR rate coefficient [into $1s^22s^22p^53p\ ^1D_2$] versus T_e and $\{f_j\}$ [a≡LTE and b≡coronal].

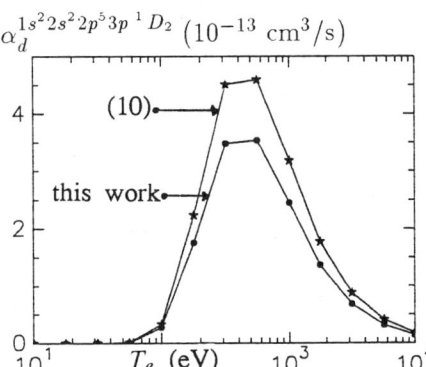

Figure 6. Se effective DR rate coefficient [from $1s^22s^22p^5\ ^2P_{3/2}$ and into $1s^22s^22p^53p\ ^1D_2$] versus T_e.

REFERENCES

1. Dubau, J., and Volonté, S., *Rep. Prog. Phys.* **43**, 199 (1980)
2. Whitten, B.L., et al., *Phys. Rev. A* **33**, 2171 (1986)
3. Burgess, A., *Astrophys. J.* **141**, 1589 (1965) — Merts, A.L., et al., *LASL Report* **LA-6220–MS** (1976) — Zhdanov, V.P., *Sov. J. Plasma Phys.* **5**, 320 (1979)
4. Hagelstein, P.L., et al., *Phys. Rev. A* **34**, 1931 (1967)
5. Eissner, W., et al., *Comput. Phys. Commun.* **8**, 270 (1974)
6. Dubau, J., and Loulergue, M., *Physica Scripta* **23**, 136 (1981)
7. Eissner, W., and Seaton, M., *J. Phys. B* **5**, 2187 (1972) — Saraph, H.E., *Comput. Phys. Commun.* **3**, 246 (1972)
8. Qui, Y., et al., *At. Data & Nucl. Dat. Tables* **55**, 1 (1993)
9. Roszman, L.J., *Phys. Rev. A* **35**, 2138 (1987) — Chen, M.H., *Phys. Rev. A* **38**, 2332 (1988)
10. Hagelstein, P.L., *J. Phys. B* **20**, 5785 (1987)

Hydrodynamic Simulation of a Laser-Produced Magnesium Recombination Plasma and Soft X-Ray Lasing

H. Kitazawa[1], S. Morinaga[1], H. Takakuwa[1], K. Shimizu[2], and S. Karashima[3]

1. Faculty of Engineering, Tokyo Institute of Technology, 2-12-1 O-okayama, Meguro-ku, Tokyo 152, Japan
2. Interdisciplinary Graduate School of Science and Engineering, Tokyo Institute of Technology, 4259 Nagatsuta-cho, Midori-ku, Yokohama 227, Japan
3. Faculty of Engineering, Science University of Tokyo, 1-3 Kagurazaka, Shinjuku-ku, Tokyo 162, Japan

Abstract. a laser-produced magnesium recombination plasma is simulated for the purpose of developing a soft X-ray laser operating in wavelengths within the water window. The plasma is assumed to be cylindrically symmetric and described by compressible fluid equations, which are numerically solved in the Lagrangian frame. The present paper concentrates chiefly on investigating the dependence of the spatial and temporal behavior of the plasma upon its initial hydrodynamic parameters.

INTRODUCTION

The emerging technology of soft X-ray lasers has noble applications to microscopy, lithography, and other fields. Especially, a soft X-ray laser operating in the biological water-window spectral region between 23.2Å and 43.7Å would be optimal in terms of the penetration, contrast, and resolution for a microscopy of biological specimens.

Recent work has demonstrated a significant amplification of soft X-ray radiation in the above spectral region, using collisional excitation (1) and recombination (2) schemes. The recombination X-ray lasers operating with the adiabatic cooling of plasma electron temperature are of continuing special interest because of their potential for the development of soft X-ray lasers requiring much lower driver energy, but are still unproved to be able to obtain a sufficient gain-length product for saturated laser action.

Also, it has been theoretically shown that an efficient amplification of soft X-ray radiation can be obtained in the plasma produced by high-power laser pulses with several ten picosecond pulse width (3). However, there seems to be some discrepancies between experimental and theoretical results. Further experimental investigations of the spatial and temporal behavior of the ionization degree and the radiation properties of the plasma would be needed to confirm those theoretical results.

In the present paper, a hydrodynamic simulation of the laser-produced magnesium recombination plasma is presented with calculations of the atomic-level population inversion in the plasma. Particularly, we will focus our discussion on the dependence of the spatial and temporal behavior of the plasma upon initial plasma parameters.

MODEL

A high power, ultra-short pulse laser illuminates a thin magnesium foil in an elongated line focus. Then it can be pictured that the laser radiation is absorbed by electrons, and that a highly ionized plasma is formed in the steady state. We assume that this plasma is cylindrically symmetric. The dynamic behavior of the plasma is described by the hydrodynamic equations for mass, momentum, and energy conservation, which are numerically solved in the Lagrangian frame.

The population densities of excited levels of ions are calculated at each time step, using the quasi-equilibrium and LTE models. The ℓ-splitting of those atomic levels and the Lyman-α trapping are not taken into account.

SIMULATIONS AND DISCUSSION

In order to simulate an expansion of the laser-produced recombination plasma, we assume an initial plasma at $t=0$:

1) The plasma consists of an ablation phase with a radius of 20 μm and an outer expansion phase. The ablation phase is a thermal equilibrium plasma with an electron temperature of 800 eV and an electron density of 4×10^{21} cm^{-3}. Here, the critical density plasma for incidence of the second harmonic of the neodymium-glass laser is supposed as an initial plasma. Also, it can be confirmed that the lasing for the Balmer-α transition in hydrogenic magnesium ions peaks at this electron temperature.

2) The electron density distribution in the expansion phase is taken to be adiabatic, and the velocity distribution of the plasma fluid to be proportional to the distance from the ablation phase, in the assumption of a self-similar expansion of the plasma. The maximum velocity is taken to be a sound velocity of 3.2×10^{7} cm.sec^{-1}.

The gain coefficient of a soft X-ray laser is highly sensitive to the ion temperature. However, it is in general difficult to estimate the ion temperature of the initial plasma. Therefore, we have investigated the temporal behavior of the plasma fluid for the initial ion temperatures of 10 eV, 100 eV, and 300 eV, and the dependence of the ion temperature upon those initial conditions. As is seen from Fig. 1, the ion temperatures increase rapidly for 50 ps, but at 100 ps the temperatures take the same value irrespctive of those initial temperatures, due to the equalization of ion and electron temperature. The gain coefficient for the Balmer-α transition peaks at ~140 ps (see Fig. 3). Therefore, the uncertainty of the initial ion temperature scarcely affects the gain coefficient. Hereafter, the initial ion temperature is taken to be 100 eV.

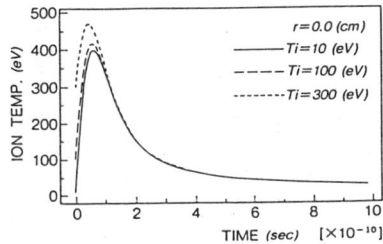

FIGURE 1. Temporal behavior of the ion temperature for the initial ion temperatures of 10 eV, 100eV, and 300 eV.

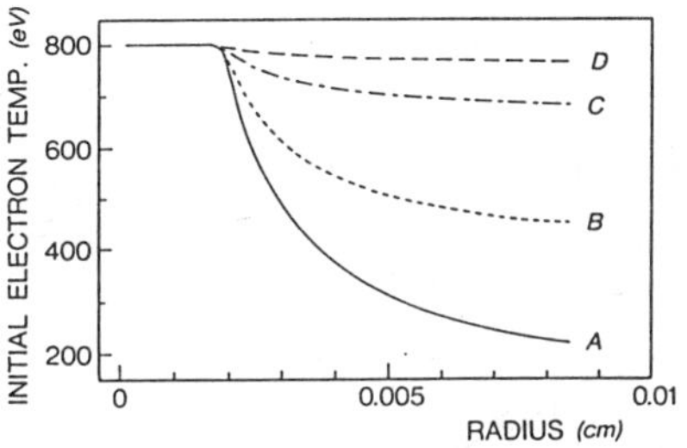

FIGURE 2. Assumed spatial distributions (A,B,C,D) of the initial electron temperature.

Moreover, we obtained, as is shown in Fig. 3, the spatial and temporal behavior of the electron and ion densities, fluid velocity, electron and ion temperatures, and gain coefficient for the Balmer-α transition in hydrogenic ions, using four types of initial electron temperature distributions (A, B, C, D) given in Fig. 2. In consequence it is found that the lasing for the Balmer-α transition in hydrogenic magnesium ions begins at ~100 ps after laser-pulse illumination, and culminates at ~140 ps, when the maximum gain coefficient near the center of plasma is 13–20 cm^{-1}. The temporal behavior is in reasonable agreement with recent experiments (2).

The larger gain coefficient is obtained for the lower electron temperature distribution of the ablation phase. The reason is that while the decrease of the electron density near the center of plasma is rather small, the plasma is rapidly cooled by a heat flow from the ablation phase to the expansion phase. However, the C or D plasma may be realized for incidence of the laser pulse beam with a pulse width of several ten picoseconds, because the root of a product of the thermal conductivity coefficient κ and the pulse width Δt is a degree of the radius of the initial plasma assumed. Probably, the initial plasma temperature is spatially equalized within Δt

Also, Fig. 3 shows that the electron temperature is spatially equalized in ~100 ps due to the large thermal conductivity of the plasma. It should be noted here that the maximum lasing for the Balmer-α transition is realized at ~140 ps, when the spatial electron density distribution has a steep gradient. This fact would make it more difficult to obtain a large gain-length product for the soft X-ray amplification.

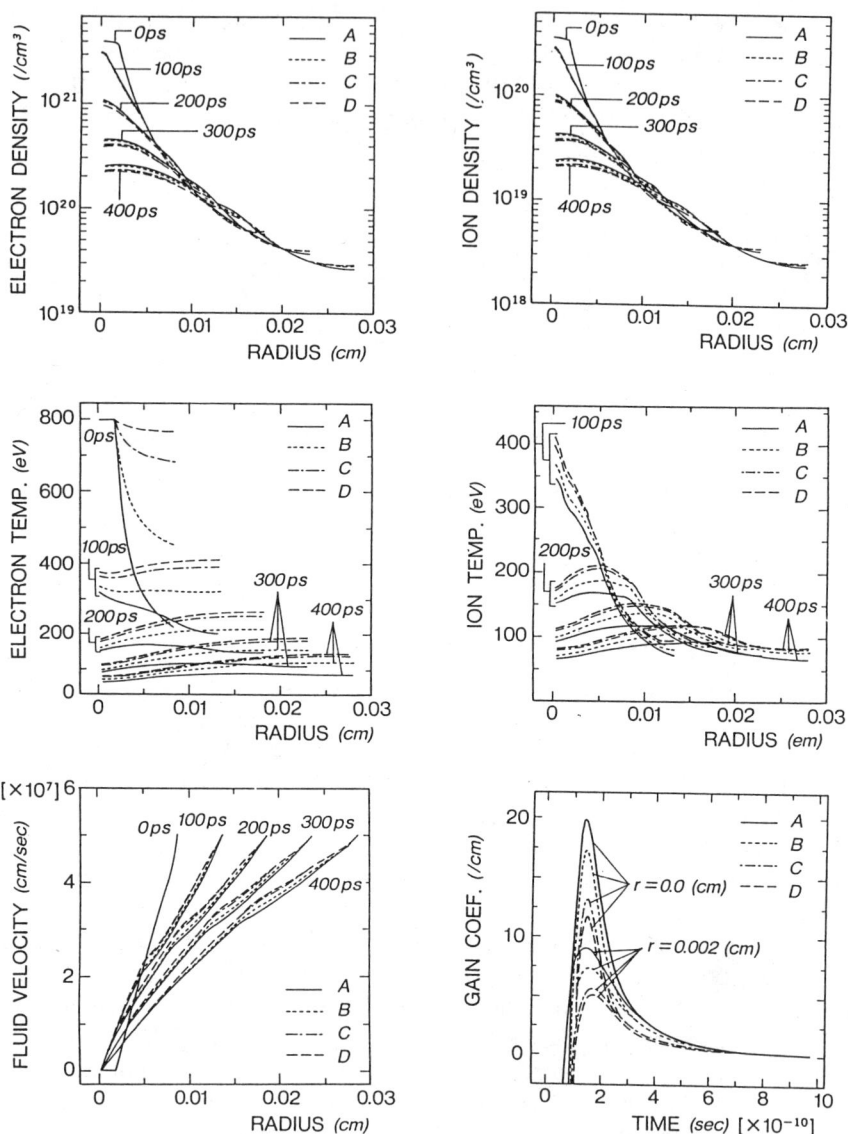

FIGURE 3. Spatial and temporal behavior of the electron and ion densities, fluid velocity, electron and ion temperatures, and gain coefficient for the Balmer-α transition in hydrogenic ions, calculated for four types of the initial electron temperature distribution(A,B,C,D).

SUMMARY

The dynamic behavior of a laser-produced magnesium recombination plasma has been simulated using the hydrodynamic equations for a cylindrically-symmetric plasma. As a result, we found that the temporal behavior of the gain coefficient for the Balmer-α transition in hydrogenic ions is in reasonable agreement with recent experiments. However, the absolute value of the gain coefficient is strongly affected by the electron temperature distribution in the expansion phase of the initially-assumed plasma. Therefore, the initial plasma parameters should be derived from a detailed investigation of the laser-plasma interaction. Also, the reabsorption of emitted soft X rays in the plasma which is not included in the present work, especially the Lyman-α trapping should be taken into account in order to obtain the absolute value of the gain coefficient.

ACKNOWLEDGMENTS

This work has been performed as one of the research projects of Tokyo Electric Power Company's Endorsed Chair of Highly Sophisticated Energy System at the Department of Electrical and Electronic Engineering in the Tokyo Institute of Technology. The authors gratefully acknowledge a financial support from Tokyo Electric Power Company.

REFERENCES

1. MacGowan, B. J. *et al.*, "Progress Towards a 44-Å X-Ray Laser," in *OSA Proceedings on Short Wavelength Coherent Radiation*, 1988, pp. 2–10.
2. Kato, Y. *et al.*, *Appl. Phys.* **B50**, 247–256 (1990).
3. Pert, G. J., *J. Phys.* **B9**, 3301–3315 (1976).

Effect of plasma nonuniformity in recombination pumping soft x-ray lasers

Akira Sasaki, Hitoki Yoneda, Ken-ichi Ueda, and Hiroshi Takuma

Institute for Laser Science, University of Electro-Communications
1-5-1 Chofugaoka, Chofu, Tokyo, JAPAN 182

ABSTRACT: We have developed a complete 2 dimensional model of recombination pumping soft x-ray lasers in short pulse intense UV laser produced plasmas. We have calculated spatial and temporal evolution of Al Hα gain by an atomic kinetics code as a postprocessor of a hydrodynamics code. We have included radiation trapping of H-like resonance lines using probability of escape which obtained from three dimensional ray trace calculation, in a plasma column produced by focusing the laser light into a narrow line. Using these codes, requirements to relax the reduction of soft x-ray gain due to line trapping, and pumping condition to achieve high gain is discussed.

1. INTRODUCTION

Recombination pumping soft x-ray lasers are expected to be more efficient than electron collisional pumping lasers, because the temperature required to produce upper laser level population is lower for typical H-like Balmer-α system rather than for Ne-like 3p-3s system[1] for a same x-ray laser wavelength. However, in recombination lasers the soft x-ray gain usually occurs while plasma undergoes rapid adiabatic cooling, after heated up to high temperature (>1keV) to fully ionize the lasant ion. Density, temperature, and velocity distribution in the plasma depend on laser intensity, spot size, pulse duration, target geometry, etc. The profile is highly nonuniform and this predominates the atomic kinetics. In particular, radiation trapping of H-like resonance lines changes population inversion significantly, which is also sensitive to velocity gradient[2,3]. Furthermore, the refraction of x-ray due to density gradient determines the effective amplification length of the x-ray laser beam[4]. Higher gain is expected in ultra short pulse laser irradiated narrow fiber target[5], in this case the plasma has steeper density gradient. Therefore, it is essential to develop a proper multi-dimensional model to analyze and to optimize the recombination pumping soft x-ray lasers.

2. THE SIMULATION CODES

We have developed a set of simulation code consists of a 2D hydrodynamics code and a 2D atomic postprocessor. We used postprocessing method assuming direct coupling between hydrodynamics and atomic physics was less pronounced in low-Z plasmas. Major part of radiant energy from Al plasma is emitted into K-shell line emissions. In this case, change of population via line trapping can be treated by introducing probability of escape in the atomic code for a few lines.

The 2D hydrodynamics code was based on CIP algorithm(6,7) and nonuniform Eulerian mesh system was employed to resolve steep density gradient near the target surface. Laser-plasma physics involved in the code were, heat conduction denoted by flux-limited Spitzer-Härm formula, radiation loss, and collisional absorption and critical dumping of incident laser. Since we are mainly interested in plasma of blow off region, simple ideal gas equation of state was considered. The atomic postprocessor calculated population of Al ions and soft x-ray gain on pseudo Lagrangian mesh. At the beginning of hydrodynamics calculation, we placed pseudo Lagrangian mesh near inside the target surface. Then, by tracking the trajectory of each mesh point time after time using local velocity distribution, temporal development of density and temperature were sampled, and were used as input data of atomic calculation.

The atomic code is divided into two parts, a rate equation solver and a line trapping solver. The rate equation solver is based on a non-LTE collisional radiative model(8,9), which calculates time-dependent population. Ground state of Al^{1+} to Al^{13+}, and excited states denoted by principal quantum number(≤ 8) of He-like and H-like ions were explicitly considered. Energy levels and rate coefficients were obtained from the screened hydrogenic model. For collisional ionization, we used simple Seaton's formula(10),

$$I_{z,n} = (2.15 \times 10^{-6}) T^{-3/2} \frac{\exp(-u)}{u^2}, \quad (1)$$

where,

$$u = E_{z,n} / T.$$

T is plasma temperature in eV, and $E_{z,n}$ is the ionization of the level. Rate coefficients for radiative recombination, collisional excitation, and spontaneous emission are given by,

$$R^r_{z,n} = (5.20 \times 10^{-14}) q u^{3/2} \exp(u) Ei(u), \quad (2)$$

$$C^U_{mn} = (1.58 \times 10^{-5}) f_{mn} T^{-1/2} \frac{\exp(-E_{mn}/T)}{E_{mn}}, \quad (3)$$

$$A_{nm} = (4.315 \times 10^7) f_{mn} E^2_{mn}, \quad (4)$$

where E_{mn}, f_{mn} are excitation energy and absorption oscillator strength from level n to m, respectively. Rate coefficients for inverse processes were determined from detailed balance. The soft x-ray gain was calculated after population of upper level N_m and lower level N_n are determined, assuming Doppler line profile as,

$$G = (2.657 \times 10^{-2}) f_{mn} \rho g_n \left(\frac{N_m}{g_m} - \frac{N_n}{g_n} \right), \quad (5)$$

where g is statistical weight of corresponding level, and ρ is a Doppler broadening function as,

$$\rho = (v/c)\sqrt{M/2kT}.$$

We considered trapping of H-like Lyman-α and β line, which are the only and the most important lines to cause significant change of population of H-like ions. The line trapping solver calculates probability of escape for typical arrangement of soft x-ray lasers schematically shown in Fig.1., from population of H-like levels (n=1,2,3) considering Doppler decoupling due to velocity gradient. The plasma is produced by a line focused laser, and is assumed to be uniform along the line focus.

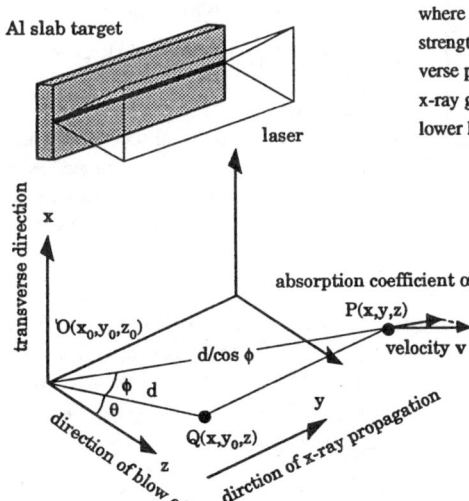

Fig.1 Arrangement of the x-ray laser plasma for calculation of escape probability of Lyman-α and β lines.

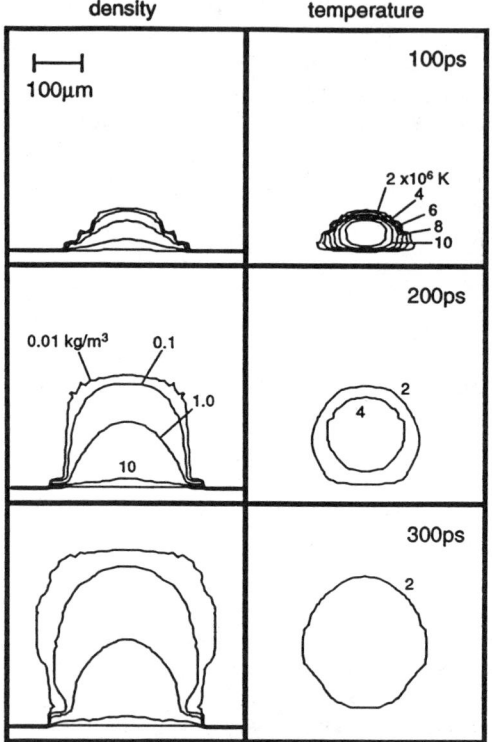

Fig.2 Typical density and temperature of the plasma for laser irradiation intensity of 5×10^{14} W/cm² (λ=249nm), pulse duration of 50ps.

Hydrodynamics and atomic physics calculations were carried out on the plane at $y=y_0$. The line center absorption coefficient α_p at arbitrary point $P(x,y,z)$ seen from the observation point $O(x_0,y_0,z_0)$ was obtained from the product of absorption coefficient at a corresponding point Q on $y=y_0$ plane, and decoupling factor ρ_p as,

$$\alpha_p = \alpha_Q \rho_p. \quad (6)$$

This was calculated from the relative velocity v_p at P towards or away from O as,

$$\rho_p = \frac{0.782}{\Delta v_D} \exp\left[-\frac{2.77(v_p/c)^2}{(\Delta v_D)^2}\right], \quad (7)$$

where Δv_D is a Doppler line width at Q. The optical depth for the specific direction from O was obtained by integrating αl to the edge of the plasma, and the line center escape probability was calculated by averaging it all over the solid angle as,

$$P_e = \iint P_e(\theta,\phi)d\theta d\phi$$
$$= \int_0^{2\pi} d\theta \int_{-\pi/2}^{\pi/2} \left[\int_0^\infty \exp(-\alpha dl)\right]\cos\phi d\phi. \quad (8)$$

In practice, ray trace calculations were carried out from O to discrete direction of θ and ϕ, for total of about hundred rays. Each ray was further divided into small ray element much shorter than the scale length of velocity gradient. Absorption length αdl was calculated for each element.

The rate equation solver and the line trapping solver were executed alternately to hold the consistency between kinetics and line transfer. In addition, we have been developing a x-ray laser propagation code. Preliminary calculation showed that even higher gain coefficient was obtained for shorter irradiation pulse duration and reduced focus width, effective gain length product was limited because x-ray was refracted away from the gain region. Detailed description of the code will be given elsewhere.

3. RESULTS AND DISCUSSION

Fig.2 shows typical hydrodynamics profile of 50ps, UV(λ=249nm) pulse produced Al plasma. Peak of the laser pulse is located at 50ps from the beginning of calculation. It is shown that temperature is as high as 1.2keV at the peak of incident laser pulse, which decreases rapidly below 300eV after 250ps of laser peak. Firstly, we have carried out kinetics calculation without line trapping. It was found that this optically thin calculation gave gain of around 0.25cm⁻¹ for n=3 to 2(Hα)

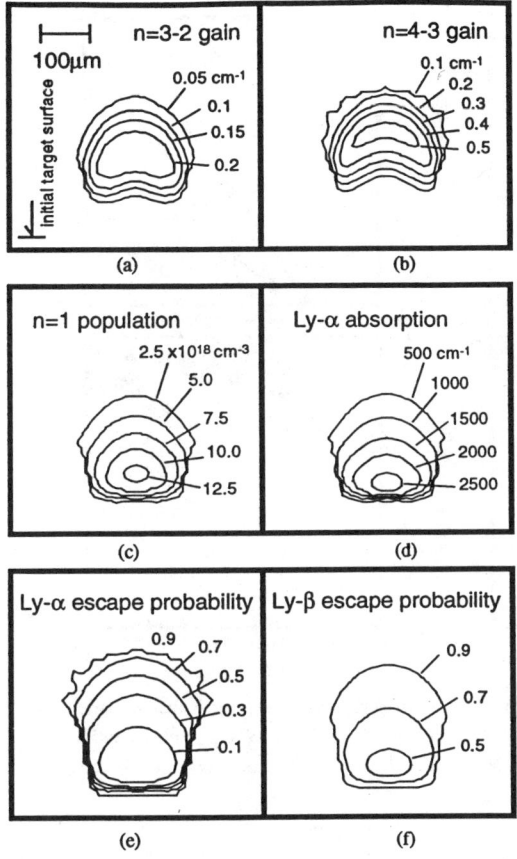

Fig.3 Results of kinetics calculation for 50ps, 5×10^{14} W/cm² UV laser irradiated Al plasma at 300ps from the beginning of calculation, showing "optically thin" gain of n=3-2(a), and n=4-3(b) transition, (c) population of H-like ground level, (d) line center absorption coefficient of Lyman-α line, escape probability of Lyman-α(e), and b(f), respectively.

and 0.5cm^{-1} for n=4 to 3 transition as shown in Fig.3(a) and (b). Secondly, calculations including trapping of Lyman-α and β lines were carried out, and their effects on population and soft x-ray gain were examined. At 300ps from the beginning of calculation, population of H-like ground state was of the order of 10^{19}cm^{-3}. This large population led to large absorption coefficient of Lyman-α line as high as 2500cm^{-1} at the center of plasma. In this case, it is clear that the absorption length was determined not by the radius of plasma (≈200μm) but by scale length of velocity gradient for corresponding temperature(≈20μm). However over the gain region, escape probability was found to be below 0.7. Due to this low escape probability population of n=2 was increased up to more than ten times, so that Hα gain was eliminated.

Finally, we have repeated calculations for shorter irradiation pulse duration and reduced focus width. Fig.4 shows absorption coefficient and escape probability of Lyman-α and β at the center of plasma as functions of time. The absorption coefficient decreased exponentially with time, and those for reduced focus width decreased faster. At 300ps, the escape probability was one order of magnitude greater than that for standard condition(d=100μm). This implies the scale length of velocity gradient also reduced proportionally as the focus width. For Lyman-β line, the escape probability became more than 0.9 so that late time gain of n=4 to 3 transition was found to be comparable to that of optically thin calculation. However, for Hα transition, even escape probability increased up to 0.5 at the end of calculation, enhancement of lower level was found to be still significant and we could hardly observe gain.

4. CONCLUSION

We have developed a complete set of 2 dimensional simulation codes for recombination pumping soft x-ray lasers. Absorption coefficient and escape probability and their effect to Hα gain have been quantitatively examined. From the results, in order to obtain Hα gain, it should be necessary to reduce focus width as narrow as 10μm. Using the present code, we can calculate gain for realistic situation, which is needed for comparison with experiments. The present code will be

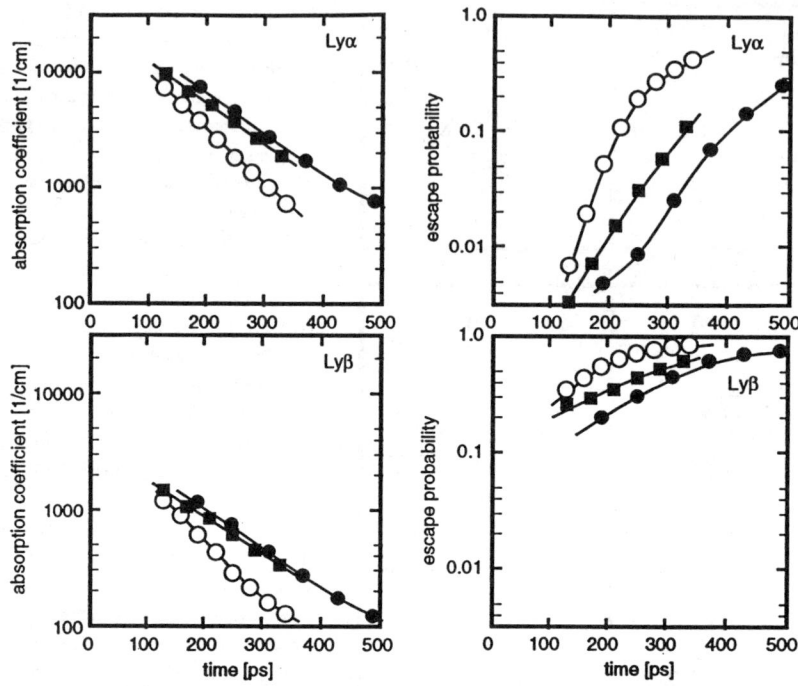

Fig.4 Calculated peak absorption coefficient and escape probability of Lyman-α and β line for shorter pulse duration and for reduced focus width; (●) peak intensity $I=5 \times 10^{14}$ W/cm^2, pulse duration $t_w=50$ps, focus width $d=100$μm. (■) $I=5 \times 10^{14}$ W/cm^2, $t_w=25$ps, $d=100$μm, (○) $I=1 \times 10^{15}$ W/cm^2, $t_w=50$ps, $d=40$μm.

useful for further optimization of gain in water window wavelength.

REFERENCES
1) R.C.Elton, X-ray Lasers, Academic Press, New York(1990).
2) G.J.Pert and S.J.Rose, Appl. Phys. B50, 307(1990).
3) J.S.Wark, et al., Phys. Rev. Lett., 72,1826(1994).
4) K.Murai, et al. Rev. Laser. Eng. 21, 625(1993).
5) J.Zhang, et al. Phys. Rev. A49, 4024(1994).
6) T.Yabe, et al., Comput. Phys. Commun. 66, 219(1991)
7) T.Yabe, et al., Comput. Phys. Commun. 66, 233(1991).
8) D.Duston and J.Davis, Phys. Rev. A21, 1664(1980).
9) M.Itoh, et al., Phys. Rev. A35, 233(1987).
10) M.J.Seaton, Atomic and Molecular Processes, edited by D.R.Bates, Academic Press, New York(1962).

Fluorine Recombination X-ray Laser Pumped by 10 ps KrF Laser Pulse

T. Tomie, E. Miura, I. Okuda, and Y. Owadano

Electrotechnical Laboratory
1-1-4, Umezono, Tsukuba, Ibaraki, Japan 305

Abstract Reported is a search for the x-ray amplification in a recombining fluorine plasma produced by the irradiation of a 10 ps KrF laser pulse. The pulse energy was 1 J and the maximum line-focus length was 1 mm with 20 μm width. Fluorine Hα (81Å) intensity at 50 μm showed exponential growth of gl ≈ 2, and gl ≈ 1.5 for ^{1}1s2p-1s3d line (103.8Å) at 100 μm. From the spatial distributions of He-like and Li-like ions, we discuss the detail of the plasma produced.

1. Introduction

The recombination scheme is considered to be the most efficient and the best scheme for realizing short wavelength x-ray lasers. The cost for this benefit is the complicatedness of the process for achieving population inversion; rapid change from the ionizing phase to the recombining gain producing phase is required.

For the rapid phase transition, very fast ionization and fast cooling of the plasma are required. For the rapid ionization, electron density should be very high. For this purpose, a short wavelength short pulse laser is needed for pumping. We have studied the amplification of ps KrF laser pulses to joule level (1) using the ASHURA system at ETL . We have so far reported the amplification of a 10 ps pulse to 4 J by an electron-beam-pumped amplifier of 29 cm diameter, and the amplification of 6 beams through the amplifier with the total energy of 23 J.(2)

Before challenging x-ray amplification with line focus irradiation, we have done some basic experiments with point focus. Results obtained are; electron density as high as 10^{23} /cm^2 can be generated if the prepulse is suppressed to lower than 10^{10} W/cm^2, to produce fully ionized Al ions the irradiation flux of 10^{15} W/cm^2 or higher is required on a slab target (3), smaller spot size irradiation allows faster expansion of the plasma(4), C-Hα line will not be a good candidate for us to try amplification because of too high density of the initial plasma produced by a 10 ps KrF laser pulse on a slab target (5), significant reduction of the plasma opacity can be expected near the surface (6), and so on.

These knowledges suggest us that it will be possible to observe gains in H-like F (Z=8) to Al (Z=13) ions using our ps KrF laser system. In this paper, we report our first try of observing x-ray amplification in KrF laser produced plasmas.

2. Experimental configuration

The size of beams on the chamber window is 10 cm diameter. For the creation of line-foci, we adopt Ross's method (7) in which point source is changed to a line focus caused by astigmatism of a spherical mirror, the width being determined by the size of the point source. We are going to use 6 line-foci, which rejects the use of off-axis parabola mirrors for focusing the beams to small spots because of the geometrical constraint. We have designed aspherical CaF_2 lenses which do not suffer from nonlinear loss of high power ps KrF laser pulses.

For the alignment of line focus, an alignment fiber was first set on the optical axis defined by a transit telescope, and then the beam was aligned by observing the obscuration of the beam by the fiber with a CCD camera. The width of the line focus observed by placing the CCD camera on the optical axis, was around 20 μm.

In the experiment reported below, the laser beam energy was about 1 J. Teflon (CF_2) slab target was shot with the maximum line focus length of 1 mm. The beam was shaped to produce line-focus of constant intensity distribution. The irradiating power density was estimated to be around 3×10^{14} W/cm^2.

The spectra were observed along the line-focus axis with a flat-field grating and Kodak 101-07 film. Space resolved observation along the plasma blow-off direction was performed with the use of 50 μm slit.

X-ray pinhole images were observed with a CCD camera to confirm that the plasma of designed length was produced.

3. Results and Discussions
spatial distribution of emission intensity

Fig.1 shows the spatial distribution of emission intensities of Li-like fluorine ions. All line intensities showed the peak at 100 μm from the target surface. High-

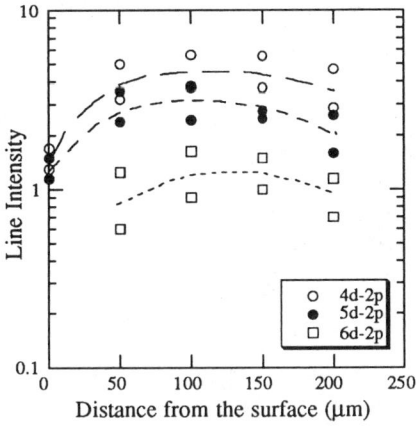
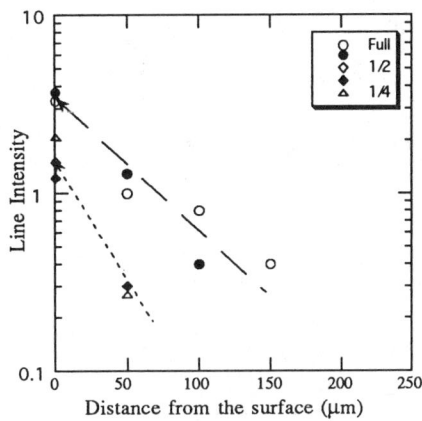

Fig.1 Spatial distribution of Li-like lines   Fig.2 Spatial distribution of Hα line

lying levels of Li-like ion are considered to be in thermal-equilibrium with the ground state of He-like ion. Hence, Fig. 1 suggests that, in our plasma, most of fluorine ions were ionized at least to H-like ions and recombined to He-like ions at around 100 μm from the surface. Therefore, we expect most of He-like lines were emitted during the recombining phase.

As seen in Fig.2, F-Hα line intensity decayed very quickly with the scale length of around 30 μm.

On the surface, intensities of Hα line (81Å) and ^{1}1s2p-1s3d line (103.8Å) were comparable in the present experiment. In our previous experiment with 300 ps, 0.53 μm laser, Hα line intensity was three times stronger than ^{1}1s2p-1s3d line intensity. Therefore, the degree of fully ionization of fluorine ions may not have been so high in the present experiment. We are not sure how large portion of the observed Hα line emission was emitted during the recombining phase.

gain and absorption of lines

In Figs. 3 and 4, intensities of 5d-2p line of Li-like ions and ^{3}1s3d-^{3}1s2p line of He-like ions are shown as a function of plasma length. Both lines increased linearly at 50 μm and 100 μm.

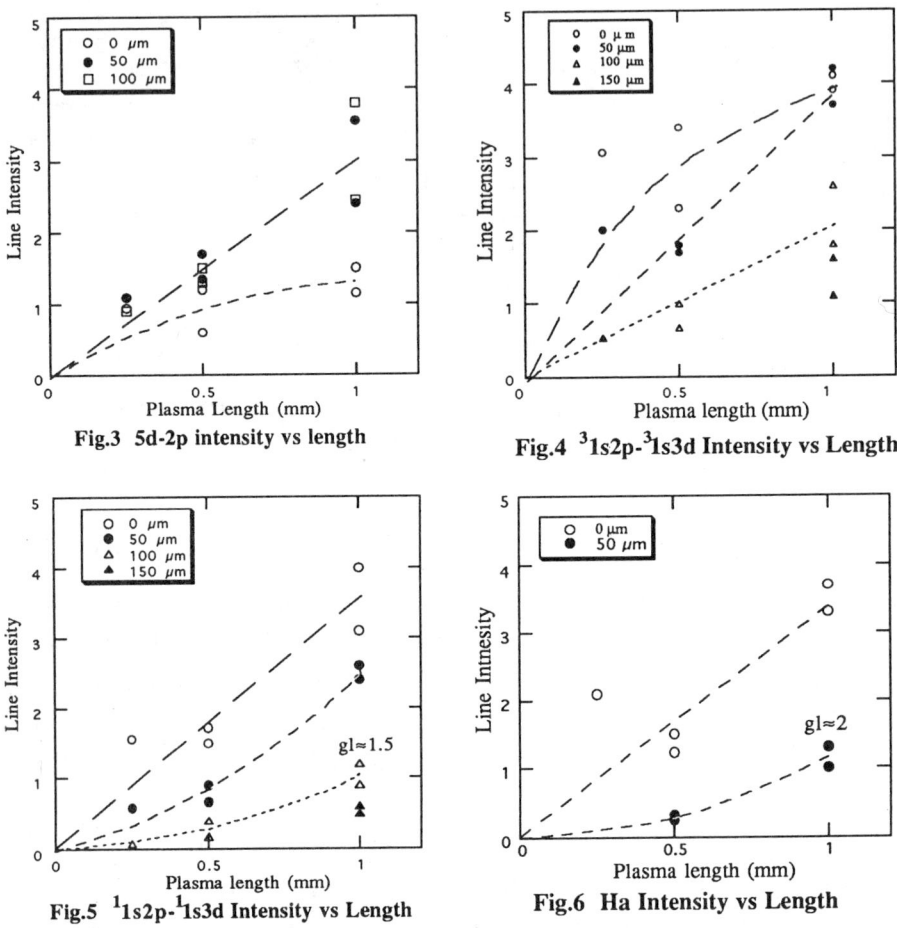

Fig.3 5d-2p intensity vs length

Fig.4 ^{3}1s2p-^{3}1s3d Intensity vs Length

Fig.5 ^{1}1s2p-^{1}1s3d Intensity vs Length

Fig.6 Hα Intensity vs Length

Figures 5 and 6 show intensities of n=3 to 2 emissions of H-like and He-like ions as a function of plasma length. On the surface, both intensities increased linearly. Away from the target surface, both lines showed some nonlinear increase. Gain length products are estimated to be 1.5 for ^{1}1s3d-^{1}1s2p at 100 μm and 2.2 for Hα at 50 μm.

We think linearity of the plasma emission is confirmed by the linear increase of Hα and ^{1}1s3d-^{1}1s2p intensities on the target surface, and 5d-2p line and ^{3}1s3d-^{3}1s2p line at 50 μm and 100 μm. Hence, the observed x-ray amplification is considered to be real one although the gl values were not so large. To be more confident, we want to have larger gl values. In next experiment, we are going to use multi-beams to create plasmas of longer length for realizing large gl values.

In summary, our first try of observing amplification in plasmas produced by the irradiation of one beam 10 ps KrF laser pulse on a teflon slab target was reported. The pulse energy was 1 J and the line focus length was 1 mm. Gain length products of 2.2 for Hα at 50 μm and 1.5 for ^{1}1s3d-^{1}1s2p at 100 μm from the target surface were observed. Linearity of the plasma emissions was confirmed from the linear increase of Hα and ^{1}1s3d-^{1}1s2p intensities on the target surface, and ^{3}1s3d-^{3}1s2p line and Li-like 5d-2p line at 50 μm and 100 μm. To increase the gl values, multi-beam irradiation experiment is scheduled in near future.

References

1. Tomie,T., Okuda,I., Owadano,Y., Tanimoto,M., Matsumoto,Y., Komeiji,S., Yaoita,A., and Yano,.M., "Picosecond high power KrF laser system for X-ray laser research", Laser Part. Beams, **8**, 299-306 (1990)

2. Tomie,T., Okuda,I., Owadano,Y., and Yano,M., "High power picosecond KrF laser system for water window x-ray laser", *X-ray lasers 1990,* ed. by G.J.Tallents, (Inst. Phys. Conf. Ser. No 116, IOP Pub., 1990) pp. 109-110

3. Tomie,T., Okuda,I., Matsushima,I., and Yano,M., "Generation of high density aluminum plasma by a laser pulse and the effect of prepulse ASE", *ibid.* pp. 243-246,
Tomie,T., Okuda,I., Matsushima,I., Owadano,Y., Staffin,R., and Yano,M., "High density aluminum plasma produced by a laser pulse", Proc. SPIE, **1551**, 213-223 (1991)

4. Tomie,T., Miura,E., Okuda,I., and Owadano,Y., "Focus diameter dependence of expansion of a plasma generated by a ps KrF laser pulse" presented in *40th Meet. Japan Soc. Appl. Phys.* (Tokyo, March 1993)

5. Miura,E.,Tomie,T., Okuda,I., and Owadano,Y., "Basic study of carbon x-ray laser pumped by a ps KrF laser pulse", presented in *48th Annu. Meet. Phys. Soc. Japan* (Sendai, April 1993)

6. Miura,E.,Tomie,T.,Okuda,I., and Owadano,Y., "Observation of Motional Doppler Decoupling Effect in a 10 ps KrF Laser Produced Plasma", presented in *4th Int. Conf. X-ray Lasers*, (Williamsburg, May 1994)

7. Ross,I.N., and Hodgson,E.M., "Some optical designs for the generation of high quality line foci", J.Phys. E :Sci. Instrum., **18** 169 -173(1985)

Observation of Motional Doppler Decoupling Effect in a 10ps KrF Laser Produced Plasma

E.Miura, T.Tomie, I.Okuda and Y.Owadano

Electrotechnical Laboratory
1-1-4 Umezono, Tsukuba, Ibaraki, 305 JAPAN

Abstract Observation of reduction of opacity in a 10ps KrF laser produced plasma is reported. The opacity of Lyβ line is estimated from the line intensity ratio to Hα line. The opacity nearer to the target surface was lower. The reduction of the opacity is considered to have been caused by a velocity gradient of the plasma blow-off.

1. Introduction

Recombination pumping scheme is the most promising candidate to achieve highly efficient short wavelength X-ray lasers, and there are many reports on gain observation[1] and numerical simulation[2]. However, there are some arguments about the discrepancy between experimental results and numerical simulations.[2] In numerical simulation, experimentally observed gain coefficient can be obtained, if the opacity of the plasma is neglected, which can not be validated theoretically. Especially in the recombination scheme, understanding of the opacity of the plasma is crucially important.

In most of X-ray laser schemes, population inversions are produced by a fast radiative decay of the lower lasing level, but the population inversions will be destroyed, if the escape of the radiation from the gain medium plasma is suppressed. Opacity can be significantly reduced in an expanding plasma owing to a large velocity gradient, which is called motional Doppler decoupling effect. There are many efforts of estimating this effect theoretically.[3] Experimentally, however, reduction of the opacity has not been directly measured, although there are some recent reports on the observation of modification of line profiles caused by motional Doppler shift.[4,5]

In this paper, we report an experiment to observe reduction of opacity in a 10ps KrF laser produced plasma. Spatial distribution of the opacity along the plasma blow-off direction was observed by recording spatially resolved spectrum at 45 degrees to the plasma blow-off direction.

2. Experimental Setup

The experimental setup is shown in Fig.1. A few joule energy pulses of 10ps duration are obtained from the ASHURA KrF laser system composed of one discharge and two e-beam amplifiers.[6] A 10ps KrF laser pulse of 1.5J energy was focused on plastic slab or 25 µm diameter carbon fiber targets with spherical (f=100cm) or aspherical (f=25cm) CaF$_2$ lenses. The spot diameters for the spherical and aspherical lenses were 150µm and 30µm, respectively. An X-ray spectrum from 60Å to 250Å was recorded on a Kodak 101-07 film with a flat field type grazing incidence spectrometer. The spectrum was spatially resolved

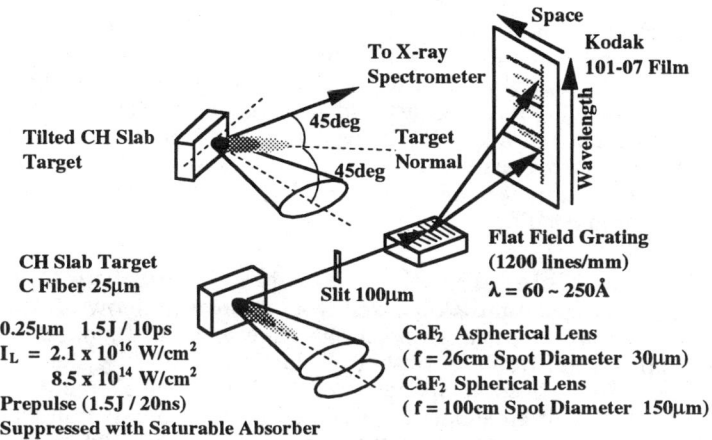

Figure 1 Experimental setup.

along the plasma blow-off direction and the spatial resolution was 100μm. It was found that a characteristic curve for long wavelength X-rays around 200Å was different from that for short wavelength X-rays around 30Å.[7] Optical density of the film was converted to the intensity using our characteristic curve.

Opacity of Lyβ line was estimated from the intensity ratio to Hα line. Deviation of the ratio from that in an optically thin plasma should be brought about by the absorption of Lyβ line, because Lyβ and Hα lines have the common upper level(n=3) and absorption or amplification of Hα line is considered to be very small. Many higher orders of Lyβ line were recorded on film. In this work, the 5th order of Lyβ line was used to evaluate the opacity, because the intensity was the largest among the higher orders and no other lines overlapped on this order.

3. Experimental Results
3.1 Intensity ratio under optically thin condition

Figure 2 shows spatial distributions of the intensity ratio of Lyβ line to Hα line under various irradiation conditions, when X-ray spectra were observed perpendicular to the plasma blow-off direction. If all relevant data such as the absolute

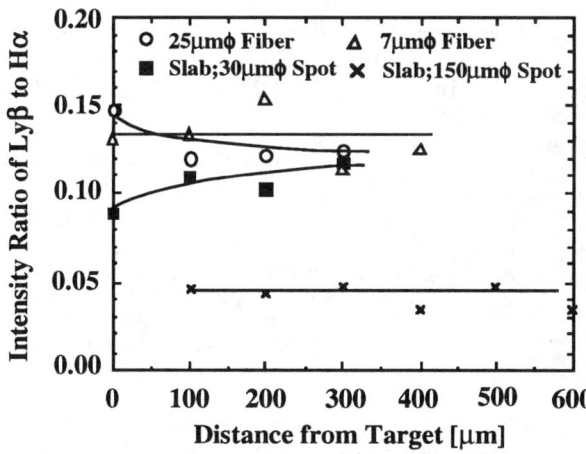

Figure 2 Spatial distribution of intensity ratio of Lyβ line to Hα line. All data were observed perpendicular to plasma blow-off direction.

sensitivity of the X-ray spectrometer were known, we could obtain theoretical intensity ratio of two lines in an optically thin plasma. However, in this work we estimated the ratio in the optically thin plasma experimentally as follows. In Fig.2, open circles represent data for a fiber target of 25μm diameter. Open triangles represent data for a fiber target of 7μm diameter obtained in our previous experiment using a 0.53μm laser of 300ps duration. Fiber targets are expected to produce optically thin plasmas, and both results gave nearly the same value. Therefore, we assume that intensity ratio of Lyβ line to Hα line on film is 0.14 under optically thin condition.

3.2 Observation of motional Doppler decoupling effect

If an X-ray spectrum is observed from the plasma blow-off direction as reported in ref. 4, all motional Doppler shifts are integrated along the blow-off direction, and it is difficult to observe reduction of opacity due to velocity gradient. In order to observe spatially resolved motional Doppler decoupling effect, the X-ray spectrum should be observed from an oblique direction to the plasma blow-off. In our experiment, the target was tilted by 45 degrees as shown in Fig.1.

The highest opacity is expected, when the X-ray spectrum was observed perpendicular to the plasma blow-off direction. As shown in Fig.2, large reduction of the intensity ratio was observed, only when a slab target was irradiated with a large spot diameter of 150μm. Therefore, spectrum observation at 45 degrees was performed only with a large spot diameter of 150μm. In Fig.3, closed circles represent data observed at 45 degrees to the plasma blow-off direction, and closed triangles represent data observed perpendicular to the plasma blow-off direction. When the X-ray spectrum was observed at 45 degrees, the opacity of Lyβ line was significantly reduced near the target surface.

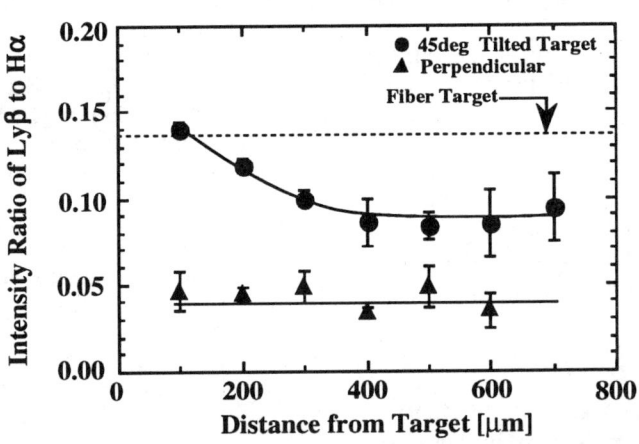

Figure 3 Spatial distribution of intensity ratio of Lyβ line to Hα line. A plastic slab target was irradiated with a laser spot diameter of 150μm. Reduction of opacity of Lyβ line is observed near the target surface, when the X-ray spectrum was observed at 45 degrees to the plasma blow-off direction.

4. Discussion

One would expect higher opacity nearer to the target surface because of expected higher ion density. Therefore, at first sight, it is quite strange that the observed opacity nearer to the target surface was lower. However, if one reminds of the motional Doppler decoupling effect, it will be found that the result in Fig.3 is quite reasonable.

Figure 4 is a schematic view of plasma expansion to understand the experimental result shown in Fig.3. In the lower part of Fig.4, length of solid arrows indicates the size of blow-off velocity at each small section of the plasma. We expect a large velocity gradient near the target surface. When an X-ray spectrum is observed at 45 degrees to the plasma blow-off direction, the X-ray emitted near the target surface suffers from motional Doppler shift of different amount at each small section, and each small section is optically decoupled each other if a spectral width is small enough.

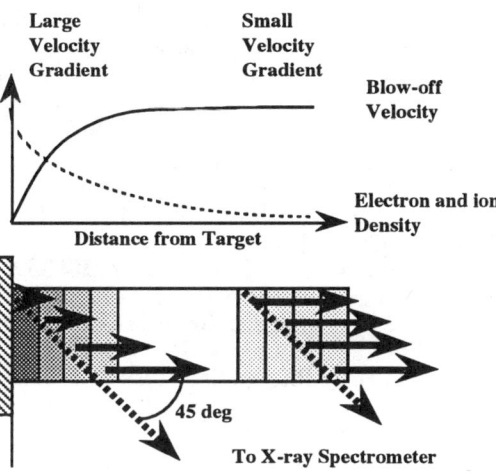

Figure 4 Schematic view of plasma expansion to understand the experimental result in Figure 3.

On the other hand, away from the target surface, the blow-off velocity becomes saturated, and the difference of motional Doppler shifts at each small section becomes very small. So, each small section of the plasma is optically coupled. Thus, opacity will be significantly reduced near the target surface owing to a large velocity gradient and will become higher away from the target surface. Figure 3 indicates that, under the present experimental condition, reduction of opacity due to motional Doppler decoupling effect was significant up to 400μm away from the target surface.

Reduction of opacity can be caused not only by motional Doppler shift but also by spectral broadening. Near the target surface, Stark broadening[8] due to high electron density may significantly reduce the opacity. The electron density was estimated to be $3 \times 10^{21} cm^{-3}$ on the target surface and $8 \times 10^{19} cm^{-3}$ at 200μm from Stark broadening of Hα line and He line. For Lyβ line, the Stark width for electron density of $3 \times 10^{21} cm^{-3}$ is much larger than the thermal Doppler width for electron temperature of several hundreds eV. Therefore, the reduction of the opacity on the target surface will include both contributions of velocity gradient and Stark broadening. It will be partially explained by Stark broadening effect that the opacity observed perpendicular was nearly constant along the plasma blow-off direction as seen in Fig.3.

At 200μm, for Lyβ line the Stark width for electron density of $8 \times 10^{19} cm^{-3}$ is nearly equal to the thermal Doppler width for electron temperature of 100eV, and motional Doppler shift is much larger than thermal Doppler width. Therefore, the observed reduction of opacity around 200μm is considered to have been caused mainly by velocity gradient.

5. Summary

Spatially resolved motional Doppler decoupling effect was observed. In a 10ps KrF laser produced plasma, significant opacity reduction of Lyβ line was observed near the target surface. On the target surface, the reduction is considered to have been caused partially by Stark broadening. At 200μm and further, the reduction of opacity was caused mainly by velocity gradient.

References

[1] For example, Chenais-Popovics C. et al., Phys.Rev.Lett.**59** 2161(1987).
[2] For example, Pert G.J. and Rose S.J. Appl.Phys.B**50** 307(1990).
[3] For example, Shestakov A.I. and Eder D.C. J.Quant. Spectrosc.Radiant. Transfer **42** 438(1989).
[4] Moreno J.C. et al. J.Opt.Soc.Am.**B9** 339(1992).
[5] Wark J.S. et al., Phys.Rev.Lett.**72** 1826(1994)
[6] Tomie T. et al., Proc. SPIE **1551** 213(1991).
[7] Miura E. and Tomie T., Rev.Laser .Eng.**21** 1011(1993).
[8] Griem H.R., Plasma Spectroscopy, McGraw-Hill, 1964, ch.4.

Compact Soft X-Ray Laser Pumped by a Pulse-Train Laser

Tamio Hara, Kozo Ando and Yoshinobu Aoyagi

The Institute of Physical and Chemical Research (Riken)
Wako-shi, Saitama 351-01, Japan

Abstract. Using a pulse-train laser with only 2.3 J/cm pumping energy, ASE's from two Li-like Al lines (105.7 Å and 154.7 Å) were observed. Gain coefficient of 105.7 Å was improved up to over 4 cm^{-1} when laser intensity of the second half of a pulse-train laser was reduced to 25 % of the first one. The time duration of the gain was about 1 ns within the 3 ns total time duration of pulse-train laser. Also a experiment on double-pass amplification of Al^{10+} 154.7 Å line was carried out by using an X-ray multi-layer mirror and increase of the line intensity by a factor of 4.4 was observed.

INTRODUCTION

For manufacturing microelectronic devices such as 1 Gbit dynamic random access memory (DRAM), the semiconductor industry needs X-ray projection lithography technology. As an X-ray source of this application, a table-size soft X-ray laser around 100 Å with high repetition rate must be developed. Recently we have observed soft X-ray amplified spontaneous emissions (ASE) in recombining Al and Si plasmas produced by a low power driving laser which has a pulse-train-like pulse form.[1-4] In the recombining plasma scheme, the control of plasma production and recombination is important for improvements of pumping efficiency and development of a compact X-ray laser. By use of a pulse-train laser, efficient heating and rapid cooling of plasmas have been achieved simultaneously. Highly charged ions such as Al^{10+} and Al^{11+} were produced efficiently by this laser. The other advantage of the pulse-train laser is that the electron temperature drops rapidly as soon as laser irradiation ceases, because the fall time of the train laser is the same as that of the last pulse. Therefore the use of a pulse-train laser instead of a long pulse laser is a powerful method to achieve high gain through recombination process. However, it is important for achieving large population inversions not only to cool rapidly the electron temperature but also to control the recombination process of plasmas. In the present paper, it is shown that the control of the envelop of the laser pulse-train leads to enhance the gain coefficients of soft X-ray lines due to optimizing the recombining process of plasmas. Double-pass experiment of soft X-ray line is carried out using a Mo/Si multi-layer flat mirror. We have also started to study a water window X-ray laser using a Ca target.

EXPERIMENTAL

A pulse-train glass laser system was constructed as a pump source. Here a 100-ps laser pulse from a mode locked oscillator injects into an optical pulse stacker[5] to get an 8-pulse train. This is doubled by a Michelson-type system to make 16 pulses.[4,6,7] Each pulse has almost the same peak power and the same interpulse time of 200 ps. Total time duration of pulse-train laser is 3 ns. The pulse-train was amplified by glass amplifiers and line-focused on an Al slab target by a lens system. For rapid cooling of the plasma, the width of the focus line was set at less than 40 μm. Temporal and spatial behavior of the emitted soft X-ray spectra were observed along the axis of a line plasma by an XUV flat-field spectrograph connected to a streak camera. On the opposite side, the Mo/Si multi-layer X-ray mirror was set at 5 cm apart from plasma end. A shutter was equipped between the X-ray mirror and the plasma. To control the envelop of the pulse-train, the neutral density filter with transparence of 50 % was inserted into the longer optical pass of a Michelson-type interferometer in order to reduce the laser intensity of the second half of the pulse-train to 25 % of the first one.

ASE EXPERIMENT

High temperature plasmas produced by irradiation of the pulse-train laser are rapidly cooling through adiabatic expansion into vacuum. The plasma over 0.6 mm from the target is in a strong recombining state. We have succeeded in observing the ASE signal of two Li-like Al lines (3d-5f 105.7 Å and 3d-4f 154.7 Å) with only 2.3 J/cm pumping energy of a pulse-train glass laser. The observation was made at the distance of 700 μm from the target surface. The duration of gain is about 1 ns.

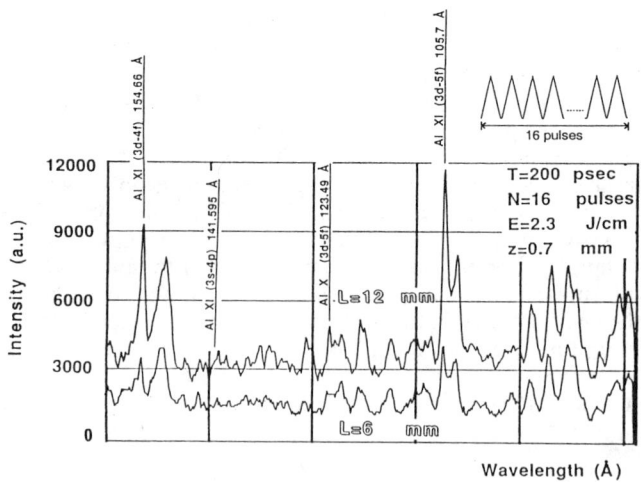

FIGURE 1. Time integrated soft X-ray spectra during 1 ns for the plasma lengths of 6 mm and 12 mm, which were observed at the distance of 700 μm from the target.

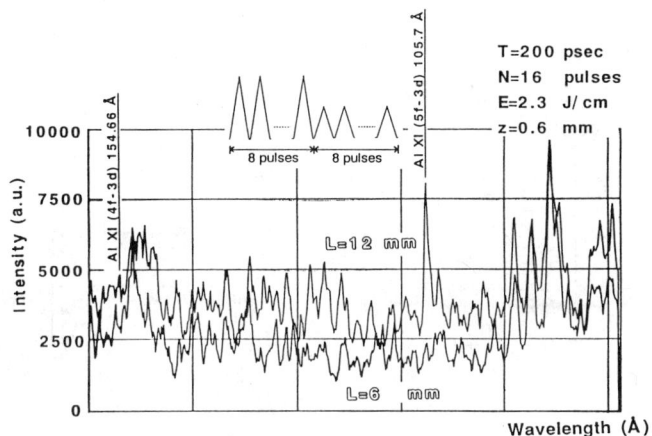

FIGURE 2. Time integrated spectra for the plasma lengths of 6 mm and 12 mm, which were observed at the distance of 600 μm from the target. The laser intensity of the second half of the pulse-train was reduced to 25 % of the first one.

Time integrated soft X-ray spectra during 1 ns are shown in Fig. 1. As shown in Fig. 1, the intensity emitted from L = 12 mm plasma was higher than two times of that from L = 6 mm plasma. After subtracting the background level from the observed line intensities, the gains estimated from the intensity ratios for 105.7 Å and 154.7 Å were 2.0 cm^{-1} and 1.5 cm^{-1}, respectively.

The form of the pump laser pulse gives a large influence to recombination processes of plasmas. We controlled the laser intensity of the second half of the pulse-train to increase gain coefficients of Al^{10+} lines. The laser intensity of the second half is reduced to 25 % of the first one, and the observed time integrated soft X-ray spectra at 600 μm from the target are shown in Fig. 2. The gain coefficient for 105.7 Å is improved up to over 4 cm^{-1} compared to 2.0 cm^{-1} of Fig.1. The line intensity of 154.7 Å is too weak to estimate the gain coefficient.

The experiment of double-pass amplification was carried out at the same position using a Mo/Si multi-layer mirror with the reflectivity of 34 % at 154.7 Å. Time histories of 154.7 Å line is shown in Fig. 3. When the shutter was open, the line intensity increased up to 3 times because the soft X-ray reflected by the mirror was amplified again in the plasma and came into the spectrograph. The time duration of the gain was also about 1 ns.

Fig. 4 shows time integrated soft X-ray spectra from 1.4 ns to 1.9 ns after laser irradiation, which corresponds to the time region of intensity enhance in Fig. 3. The time integrated line intensity of 154.7 Å became 4.4 times. This result means that this laser medium has the gain of 1.9 cm^{-1} for the 154.7 Å line. If an X-ray cavity is constructed with multi-layer X-ray mirrors, a first X-ray oscillator will be realized, which leads to great improvement of coherence, pumping efficiency and output energy of X-ray lasers.

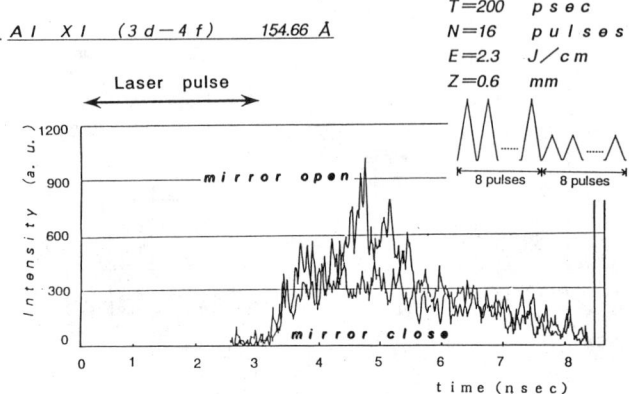

FIGURE 3. Time histories of the Al XI 154.7 Å line observed in the double-pass experiment using a Mo/Si multi-layer mirror.

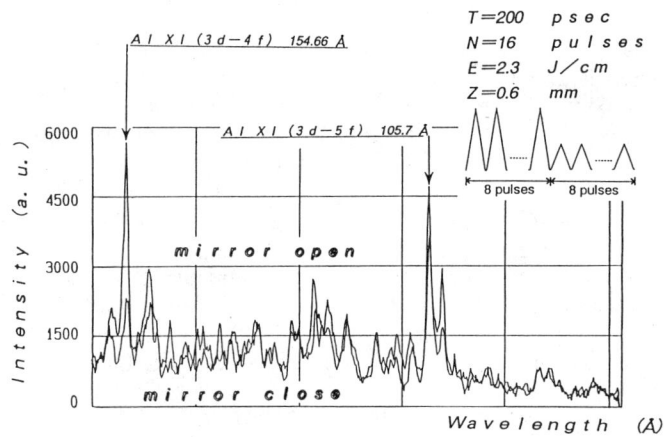

FIGURE 4. Time integrated soft X-ray spectra from 1.4 ns to 1.9 ns after laser irradiation.

We have started to study a water window X-ray laser using a Ca target. Up to now, some Li-like Ca lines (3d-6f 33.74 Å, 3d-5f 39.50 Å, 3d-4f 57.81 Å, etc.) are observed at 100 μm from the target surface, when the input energy is 50 J. This means that a compact water window X-ray laser is possible with low pumping energy less than 100 J.

REFERENCES

1. Hara, T., Ando, K., Kusakabe, N., Yashiro, H., and Aoyagi, Y., Jpn. J. Appl. Phys. **28**, L1010-L1012 (1989).
2. Hara, T., Ando, K., Yashiro, H., and Aoyagi, Y., "Compact soft X-ray laser," *Proc. 11th Int. Conf. on X-Ray and Inner-Shell Processes* ed T. A. Carlson, M. O. Krause and S. T. Manson (American Institute of Physics, 1990) pp. 197-207.
3. Yashiro, H., Hara, T., Ando, K., Negishi, F., Ido, S., and Aoyagi, Y., Jpn. J. Appl. Phys. **31**, L92-L94 (1992).
4. Hara, T., Hirose, H., Ando, K., Negishi, F., and Aoyagi, Y., *X-Ray Lasers 1992* ed E. E. Fill (IOP Publishing Ltd, Bristol, 1992) pp. 97-100.
5. Danson, C. N., Edwards, C. B., and Ross, I. N., Optics and Laser Technology **17**, 99-101 (1985).
6. Hirose, H., Hara, T., Ando, K., Negishi, F., and Aoyagi, Y., Jpn. J. Appl. Phys. **32**, L1538-L1541 (1993).
7. Hara, T., Ando, K., Negishi, F., Yashiro, H., and Aoyagi, Y., *X-Ray Lasers 1990* ed G. J. Tallents (IOP Publishing Ltd, Bristol, 1990) pp. 263-266.

Near Field Beam Characteristics of the beam from the Amplifier of an Injector/Amplifier Germanium XXIII XUV Laser System.

C G Smith[1], M H Key[1,2], G Cairns[3], C Lewis[3], D Neely[2], A Mac Phee[3]

1 Clarendon Laboratory, Parks Road, Oxford, OX1 3PU, England.
2 Rutherford-Appleton Laboratory, Chilton, Didcot, OX11 0QX, England.
3 Dept of Pure and Applied Physics, Queen's University, Belfast, BT7 1NN, N Ireland.

Abstract: A Ge XXIII XUV laser beam was generated from a double plasma source and used to inject a beam into a single plasma amplifier. The amplifier with a gain of 70 introduced refractive bending and positive lensing in the plane perpendicular to the target surface and negative lensing in the plane parallel to the target surface. Measurements were made of the beam in planes in the amplifier and upstream and downstream of it using a spherical mirror imaging system. The extent of astigmatism due to the crossed positive and negative lensing was measured.

INTRODUCTION

Previous Ge XXIII XUV laser experiments at the Rutherford- Appleton laboratory have used a double plasma arrangement to achieve high gain[1]. Using an XUV mirror to double pass the ASE through the plasmas has driven the laser to saturation[2]. Injecting a small fraction of the output of the double plasma into a separate amplifier plasma has increased the coherence and decreased the divergence of the XUV beam[3]. However the amplifier introduces refractive bending, lensing and non-correctable aberrations[4] to the beam. The refractive deviation can be compensated for by bending the target[5]. The positive lensing perpendicular to the target and negative lensing parallel to it can also, in principle, be corrected.

This experiment measured the magnitude of the amplifier astigmatism for the first time.

EXPERIMENTAL METHOD

The experimental system is shown in Fig 1.

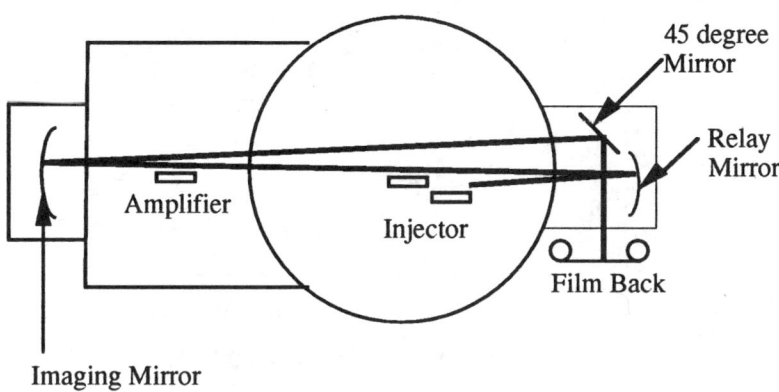

FIGURE 1. Schematic of the imaging apparatus.

The injector was composed of two 18mm Germanium stripe targets of width 100 μm. Each target was illuminated by three overlapping 22mm line foci of combined intensity $1.5 \ 10^{13}$ Wcm^{-2} in 650 ps at 1.06 μm; the width of the line focus was matched to target stripe width. The output of the injector was imaged to the input face of the amplifier target at a magnification of 3. The amplifier was a 150 μm wide 14mm long Germanium stripe target which was also irradiated at $1.5 \ 10^{13}$ Wcm^{-2} in 650 ps. ASE travelling directly from the injector to the amplifier was blocked.

The XUV beam at the amplifier was imaged on Kodak 101 film back via a 508 mm focal length XUV mirror, at a magnification of 4.2 and at a measured resolution of 2.5 μm. The imaging mirror was moved axially to produce images of different planes in a range +/- 30 mm about the amplifier input face. A series of images of the beam in different planes was recorded. The beam shape appeared to vary in keeping with the astigmatism of the amplifier. The beam patterns were digitised and converted to intensity using the response characteristics of Kodak 101 film.

RESULTS

The most striking observations were the minimum in the full width, half maximum (fwhm) in the beam profile about 10 mm downstream of the amplifier face for measurements perpendicular to the target face (fig 2A) and the converse minimum in the fwhm about 15 mm upstream of the amplifier face for measurements parallel to the target face (fig 2B). A typical intensity profile, that relating to figure 3C, is reproduced as fig 3A. Pictures of beam profiles around these two positions are shown in fig 3B and 3C respectively.

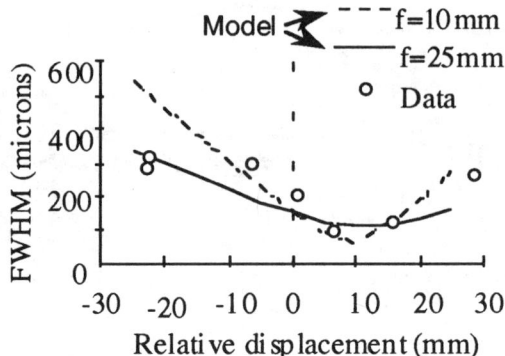

FIGURE 2A. FWHM of the beam image measured perpendicular to the target face and modelling for positive lensing with focal lengths 10 cm and 25 cm.

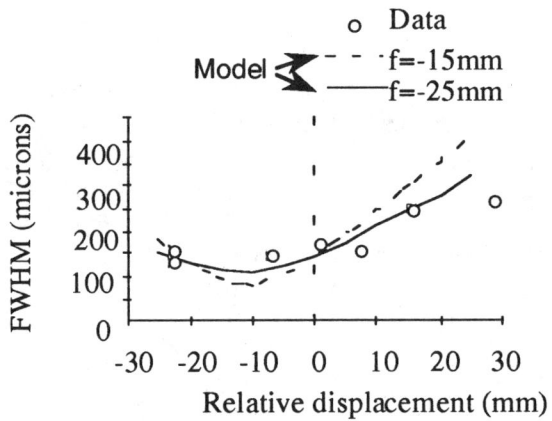

FIGURE 2B. FWHM of the beam image measured perpendicular to the target face and modelling for positive lensing with focal lengths 10 cm and 25 cm.

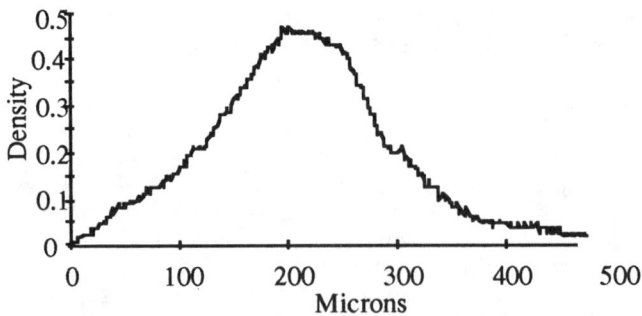

FIGURE 3A. An intensity profile of figure 3C in the direction indicated on that figure.

FIGURE 3B. The beam profile 7mm downstream of the amplifier injection face.

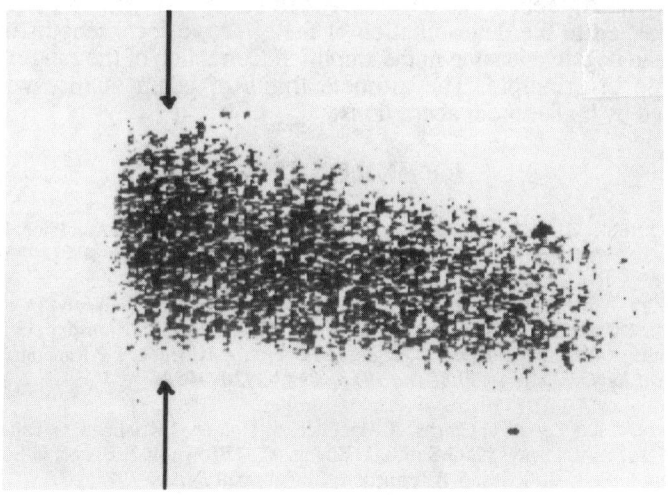

FIGURE 3C. The beam profile 7mm upstream of the amplifier injection face.

Model

The experimental system was modelled in a simple way based on an earlier analysis[3] of the amplifier's optical characteristics. The amplifier was represented by distributed positive and negative lenses in perpendicular planes. The injector was modelled as an incoherent disc of diameter 50 µm and at the furthest point of the injector. Rays were traced through the system to determine the fwhm of the beam pattern at the detector plane. Agreement with the positions of the minima in image width in the experimental data was obtained assuming positive focal lengths of 10 mm to 25 mm and negative focal lengths of -15 mm to -25 mm.

While the coupling system reduced beam divergence and increased coherence it did not inject a fully coherent beam. The minimum diameter of the beam was computed to be ~100 µm, comparable to the experimental result. To assess the higher order aberrations of the amplifier and the ultimate limit of the coherence of the beam produced from the amplifier would require further experiments. The reported experiment was sufficient to measure the astigmatism and determine the nature of optical correction required to eliminate it.

CONCLUSION

The astigmatic behaviour of an X-ray laser amplifier has been studied for the first time. Agreement achieved between the experiment and a simple model has led to the determination of the effective focal lengths of the positive and negative lensing in the amplifier. Correction of the astigmatism is possible in principle. The ultimate limits of beam quality will be determined by higher order aberrations.

REFERENCES

1. D M O'Neill, C L S Lewis, D Neely, J Uhomoibhi, M H Key, A MacPhee, G J Tallents, S A Ramsden, A Rogoyski, E A McLean, *Opt Comm 75 406 (1990).*

2. M H Key, A Kidd, P Norreys, R Kodama, H Z Chen, C Lewis, D Neely, D O'Neill, J Uhomoibhi, L Dwivedi, J Krishnan, G J Tallents, S Ramsden, G J Pert, J Zhang, A Carillon, P Dhez, P Jaegle, G Jamelot, A Klisnick, J P Raucourt., *Physical Review Letters Vol 68 (no 19) pp 2917-2920 (1992)*

3. D Neely, C L S Lewis, G Cairns, A MacPhee, M Holden, J Krishnan, G Tallents, M H Key, P A Norreys, C G Smith, J Zhang, M T Brown, R E Burge, G Slark, P Holden, G Pert, J Ploues, S A Ramsden. *CLF Annual Report 1993, p3.*

4. Optimisation of Brightness and Coherence in XUV Laser Amplifiers, M Key, C Smith, *CLF Annual Report, 1993, p64*

5. Y Kato, H Daido, R Kodama, K Murai, G Yuan, M Schulz, M Yamanaka, M Takagi, T Kanabe, S Nakai, D Neely, A MacPhee, C L S Lewis, G Slark, M Niibe, M Tsukamoto, Y Fukuda, H Tsunemi, S Nomoto, I Kodama, T Honda, K Shinohara, H Iwasaki, T Yoshinobu, *SPIE Vol 2012 Ultrashort Wavelength Lasers II (1993)*

Coupling between remote plasmas in an 'Injector-Amplifier' XUV laser system

GF Cairns, MJ Lamb, CLS Lewis, AG MacPhee, D Neely, C Pichler*
Department of Pure and Applied Physics, Queen's University, Belfast BT7 1NN

M Holden, J Krishnan, GJ Tallents
Department of Physics, University of Essex, Colchester CO4 3SQ, UK.

CG Smith, J Zhang
Clarendon Laboratory, University of Oxford, Oxford OX1 3PU, UK.

MH Key, PN Norreys
Central Laser Facility, Rutherford Appleton Laboratory, Chilton Oxon OX11 OQX, UK.

PB Holden, GJ Pert, SA Ramsden
Department of Physics, University of York, York YO1 5DD, UK.

RE Burge, MT Browne, GE Slark
Wheatstone Physics Laboratory, King's College, London WC2R 2LS, UK.

Introduction

Many applications of XUV lasers such as microscopy and holography require bright, highly coherent and monochromatic beams. We describe a geometry used for enhancing some of these characteristic properties of the neon-like germanium XUV laser. It involved coupling the output from one plasma column, an 'injector', via an image relay mirror to another remotely located 'amplifier' plasma(~0.8m away). Significant enhancement of the spatial coherence and brightness of the final output was observed.

The J=2-1(3p-3s) neon-like transitions(1), at 23.2nm and 23.6nm, were used to demonstrate the system. A refraction compensating double plasma(2,3) acting as the injector provided the ASE source for these lines. An X-ray multi-layer concave mirror allowed for control of the intensity and divergence, and thereby the coherence, of the injector beam into the amplifier.

Experiment

The setups for a number of configurations of the system are shown in figure 1. Six beams from the VULCAN glass laser, each 110mm diameter at 1.05µm, were used in a standard off-axis illumination geometry to drive a double slab target consisting of 18mm long germanium stripes coated onto glass substrates: the stripes were typically 115 and 160µm wide. A Gaussian pulse of 1.1ns FWHM was provided which delivered an average irradiance of 10^{13}Wcm^{-2}

*On student exchange from Inst. für Halbleiterphysik, Johannes Kepler Universität, A-4040 Linz, Austria.

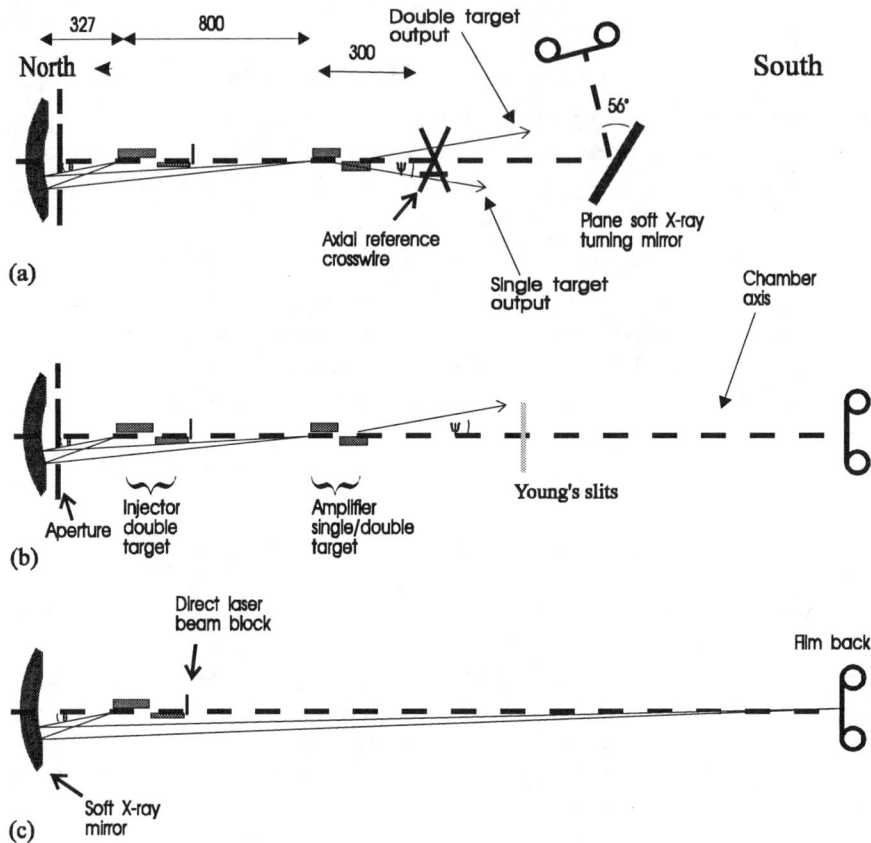

Figure 1 Different experimental configurations, a) coupling into amplifier and recording footprint images, b) using Young's slits coherence diagnostic, c) capturing nearfield images of the output from the injector.

on target. The output from one end (north) of the injector was image relayed to a similar amplifier target, whilst the output from the other (south) was stopped with a direct light block. Two 150mm diameter beams from VULCAN were used to drive the amplifier targets, also at an average irradiance of 1×10^{13} Wcm^{-2}. The amplifier had either single or double 14mm long and 160μm wide stripe targets. A combination of two rotatable plano-concave cylindrical lenses of focal length ~3m and a main focussing aspheric doublet of focal length ~40cm, were used to generate the line foci in the amplifier case. The X-ray mirror imaged the injector output with a magnification of ~3.3x to a plane ~60mm in front of the amplifier. A time delay of 4.8ns was introduced for the drive pulses at the amplifier to allow for the transit of the coupled XUV laser beam. Space resolving crystal spectrometers were used to monitor the resonance line emission in the 8-10Å region; this allowed for monitoring the plasma uniformity in the axial direction. A streaked crystal spectrometer was used to monitor the time variation of the neon-like and fluorine-like resonance lines under different drive and target conditions.

The soft XUV lasing emission was monitored using a flatfield grazing incidence grating spectrometer. Initially the injector and amplifier were tested separately for normal output of lasing emission before using the combination. However, the flatfield provided only 1D angular information about the beam and it was difficult to interpret the output from the injector-amplifier combination as alignment and beam directions were more uncertain from shot to shot. To acquire 2D information, a plane soft X-ray mirror, see 1(a), again preferentially reflecting at 23.2 and 23.6nm, was used to record a 'footprint' of the amplifier's output onto a film back. The mirror was operated at an angle of incidence of 34° and the film back contained Kodak 10402 film with a 0.8µm Al filter to exclude optical light. The total path length from amplifier to film was ≈1.3m. Various reference wires, used for the mirror alignment, also produced Fresnel interference fringes on the film which allowed us to infer changes in the spatial coherence of the beam(8). The spatial coherence was diagnosed more rigorously using a Young's slits pattern combined with diffracting elements to give wavelength isolation of the different laser lines(7). A series of slit pairs of different spacing were used simultaneously across the beam.

Results and Discussion

A typical footprint image for a single amplifier target is shown in figure 2. The target and plasma imprint are on the left whilst the coupled beam is refracted to the right. A feature of particular note were the fine striations observed towards the edges of the images: these are believed to be due to inhomogeneities in the gain zone. Other results for double as well as single amplifier targets have been reported elsewhere(4). For single targets, the divergence of the output beam was reduced in the vertical plane from a nominal input of ≈8mrad to ≈3mrad, whilst in the horizontal plane it was increased from a nominal value of ≈3mrad to ≈11mrad, the latter divergence being dominated by the refractive spreading of the beam's angular distribution. For double targets the divergence increased in both planes. The separation of the double targets controlled the integrated output beam energy as well as its pointing direction.

Typically ~2-3% of the relayed injector beam was missing in the shadow region of the plasma; this was consistent with what one might expect from the geometry of the system assuming a gain zone of scale-length ~100µm. By integrating under the missing portion of the relayed beam and comparing this with the total energy recorded on film due to the amplification of the injected beam in a single 14mm target, then typically the observed energy ratio, amplified/input, was between ~20-70x. Assuming that the output energy is given by the input energy times the exponential of the gain-length product of the lasing lines then this corresponds to a gain coefficient of $2.5\pm0.5\text{cm}^{-1}$.

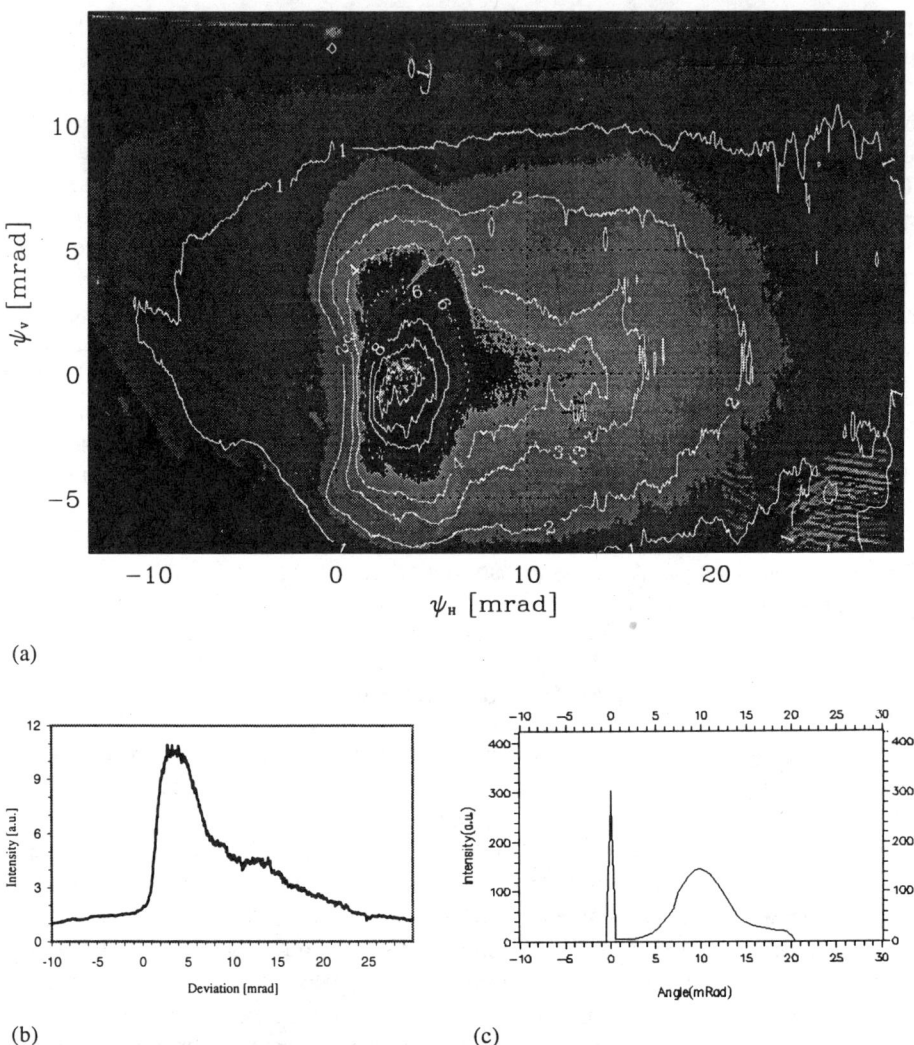

Figure 2 A footprint image of a typical single amplifier target output is shown in (a); the beam has been refracted to the right. (b) shows a section in the horizontal plane of the output for comparison with output from a simple semi-numerical model shown in (c).

A simple semi-numerical model was constructed to help assess the effects of refraction. The modelled output for a uniform collimated input beam is shown in 2(c); the parameters defining the plasma were typical of what we expect in the experimental conditions for a quasi-equilibrium state. The profile is on axis in the horizontal plane. For comparison a horizontal section through the ouput beam in 2(a) is shown in 2(b). Whilst the model is giving only qualitative information, some of the general features are worth noting such as the dominant deviated peak with its elongated shoulder to the right; these are discernible on both profiles. Note the peak at '0' on the model profile corresponds to that portion of the input beam which passes to either side of the amplifier plasma without interacting with it and is not an important feature. On the experimental profile the deviation cannot be taken as a reliable absolute measure but rather is relative; it was difficult to locate the fiducials for this shot. On other shots the peak occurred within an 8-12mrad range. Also the input beam was being focused and coming in at angle to the axis unlike the model input.

The model was based on a 1D expansion out from the slab target surface following similar analyses as outlined by London(5) and Boswell et al.(6), except applying to slab rather than foil targets. The density profile was based on a quadratic function and a parallel beam of ray packets of uniform intensity across their width were 'injected' into the plasma column. Using a semi-numerical approach similar to the technique as outlined in (6), the 'amplifier' mode, the output beam profile was obtained as shown in 2(c). The gain was based on a quadratic function with peak gain of $3cm^{-1}$, centred at 100µm from the target surface and having a width of ~80µm. Further work is being carried out to develop the model for double target amplifiers.

Conclusion

Successful coupling using a relay X-ray mirror has been demonstrated. Significant enhancement of the coherence and brightness of the output beam from the amplifier compared to that injected into it has been observed. However, it is important to minimise the destructive refractory effects in these amplifiers. We hope to demonstrate a fully coherent and saturated output beam in the near future.

References

(1) DM O'Neill, CLS Lewis, D Neely, J Uhoimoibhi, MH Key, A MacPhee, GJ Tallents, SA Ramsden, A Rogowski, A McLean, GJ Pert: Optics Comm. 75, 406,(1990)
(2) CLS Lewis, D Neely, DM O'Neill, J Uhoimoibhi, MH Key, Y AlHadithi, GJ Tallents, SA Ramsden: Optics Comm. 91, 71(1992)
(3) A Carillon, HZ Chen, P Dhez, L Dwivedi, J Jacobi, P Jaegle, G Jamelot, J Zhang, MH Key, A Kidd, A Klisnick, R Kodama, J Krishnan, CLS Lewis, D Neely, P Norreys, DM O'Neill, GJ Pert, SA Ramsden, JP Raucourt, GJ Tallents, J Uhoimoibhi: Phys. Rev. Lett. 68, 2917(1992)
(4) G Cairns, CLS Lewis, AG MacPhee, D Neely, M Holden, J Krishnan, GJ Tallents, MH Key, PN Norreys, CG Smith, J Zhang, PB Holden, GJ Pert, J Plowes, SA Ramsden: Appl.Phys. B 58, 51-56(1994)
(5) RA London: Phys. Fluids 31(1), 184-192(1988)
(6) B Boswell, D Shvarts, T Boehly, B Yaakobi: Phys. Fluids B 2 (2), 436-444(1990)
(7) CLS Lewis: These proceedings
(8) J Krishnan et al: These proceedings

Enhanced Output of Soft X-ray Lasers using Double Slab Targets

J. C. Moreno, J. Nilsen, and E. Chandler

*Lawrence Livermore National Laboratory
P. O. Box 808, Livermore, CA 94550*

Abstract. Double slab neon-like niobium soft x-ray laser experiments have been performed using the Nova laser. The two slabs have their front surfaces facing in opposite directions with either a 300 µm or 600 µm planar separation between them. Separate laser beams irradiate each slab with an intensity on target of 1.3×10^{14} W/cm^2. Best coupling was observed using a 300 µm separation. The angular divergence of the laser is measured for single slab and double slab configurations. Comparisons to numerical models are discussed.

INTRODUCTION

Neon-like collisional x-ray lasers have typically used either thin foils irradiated from both sides with a driving laser or massive slab targets irradiated on only one side (1). While thin foil targets have produced impressive results for both Ne-like and Ni-like x-ray laser systems, there are several advantages to using massive slab targets. Slab targets are generally easier to manufacture and can be more efficient due to a larger gain region. However at longer lengths refraction becomes a problem which limits the gain-length product. This can be compensated for by using double slab target configurations which couple the rays from one plasma into another. The planar separation of the two slabs with respect to each other must be optimized for best coupling. This type of double target configuration has been used effectively for Ne-like Ge to produce large gain-length products and recently has produced a saturated laser at 236 Å using a multilayer mirror for double-pass amplification (2-4).

Here we have used single and double targets of niobium. The intensity of the J=2-1 lines increased approximately four orders of magnitude by going from a single slab target of length 2.52 cm to a double slab target of 5.04 cm. Our strongest line, the J=2-1 line at 138.6 Å is most likely in the saturated laser regime for the double length targets. We have also measured a shift in the angular divergence of the x-ray laser lines when changing the planar separation between the slabs, in qualitative agreement with our numerical simulation codes.

EXPERIMENTAL RESULTS AND DISCUSSION

The double slab configuration is illustrated in Fig. 1. Two beams from the Nova laser pointed in opposite directions irradiate the two sides of the double slab target. The slabs are premounted on a rigid frame with the front surfaces all parallel to each other. There is a 5 mm gap between the two opposing slab targets and a planar separation d between the slabs to compensate for refraction. We used separations of d = 300 µm and 600 µm. A single 600 ps square pulse irradiated each slab target with an intensity of 1.3×10^{14} W/cm^2. Note that each single slab target actually consists of two slabs with a 16% gap between the slabs because of the corresponding gap in the laser beam. The effective length of each 3 cm target was therefore 2.52 cm. We also

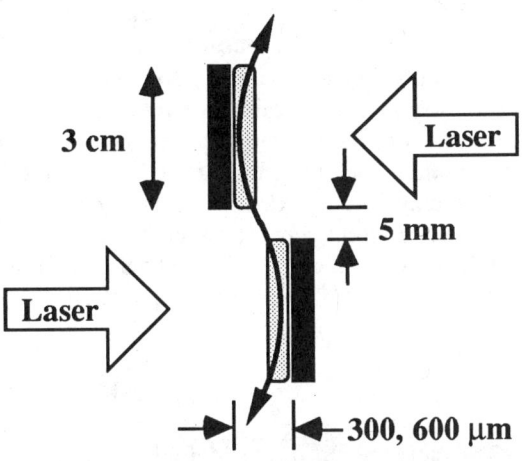

FIGURE 1. Schematic showing the double slab configuration and the refracted ray path.

employed combination double slab targets in which one slab was Nb and the other slab was Zr. This allowed us to measure and compare the intensity of two single slab lasers on a single shot with the same irradiance. The Nb x-ray laser is an interesting case to study because of the strong hyperfine effect that is observed in several lasing transitions (5,6).

Spectra containing the two J=2-1 lines at 138.6 Å and 140.4 Å of Ne-like Nb are shown in Fig. 2 for single slab and double slab configurations. These are time-integrated spectra obtained from a 1-meter grazing incidence spectrograph with a microchannel plate detector (McPigs). There is roughly four orders of magnitude increase in the intensity of these two lines when the length is doubled to 5.04 cm using a double slab target with a 300 µm planar

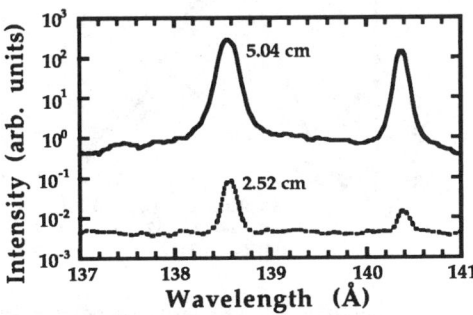

FIGURE 2. Comparison of Ne-like Nb spectra from single slab (dashed curve) and double slab (solid curve) targets with d = 300 µm.

separation. This separation gave better coupling and produced a brighter x-ray laser than using a 600 μm separation. Using the relative intensities and neglecting saturation effects we obtain a gain-length product of GL ≈ 17.5 (G=3.5 cm^{-1}) for the J=2-1 line at 138.6 Å, GL ≈ 19.1 (G=3.8 cm^{-1}) for the J=2-1 line at 140.4 Å, and GL ≈ 16.1 (G=3.2 cm^{-1}) for the J=1-1 line at 147.6 Å. These gain values are actually a lower limit on the small signal gain since it is expected that the measured intensities for the double slab targets are at or approaching the saturation regime. Comparing these values to earlier measured gains from Nb foils (6), one finds generally good agreement except for the 138.6 Å line which has a lower measured gain here. However this is the strongest line and assuming it is saturated at the double target length, as would be expected theoretically, then one would expect to measure a lower effective gain. More measurements at shorter lengths are needed to accurately measure the gain.

Time resolved spectral measurements were made using a flat field spectrograph with a streak camera detector (SFFS). The duration of the J=2-1 laser lines was measured to be < 200 psec for the 5.04 cm length double slab targets.

ANGULAR DIVERGENCE

The entrance slit of the McPigs spectrograph was aligned perpendicular to the target surface so that the angular distribution of the x-ray lines in the expansion direction could be measured. A comparison of the angular divergence of the 138.6 Å line for the two double slab cases is shown in Fig. 3a. We see that d = 300 μm gives the best coupling. The 600 μm separation is too large and allows mainly the large angle rays to couple between the two plasmas from each slab.

A Lagrangian hydrodynamics code (LASNEX) was used to model the plasma

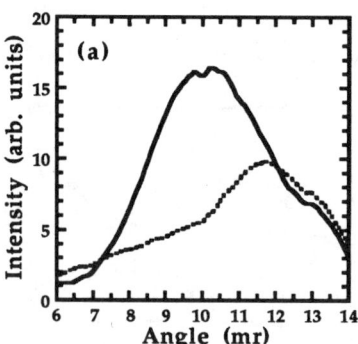 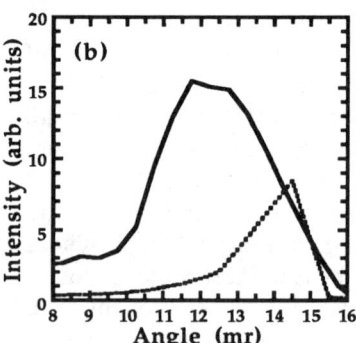

FIGURE 3. (a) Experimental measurement of angular divergence of 138.6 Å line for 300 μm separation (solid curve) and 600 μm separation (dashed curve). (b) Numerical simulation of angular divergence.

produced by line focus irradiation of Nb slab targets. We then used a post processor to model the gain profiles of the lasing transitions. Finally, the gain and electron density profiles were input into a 3-dimensional laser transport code (Beam3). A simulation of the angular distribution of the 138.6 Å Nb line is shown in Fig. 3b. We find good qualitative agreement with experimentally measured values. These results are also consistent with Ge double slab experiments in which the optimum separation was found to be ~ 200 - 300 µm (2,3).

CONCLUSION

The effects of refraction have been reduced by using a double slab configuration and we have significantly enhanced the brightness of Ne-like Nb x-ray lasers. We have measured GL $\geq$ 17 for the two J=2-1 lines at 138.6 Å and 140.4 Å and GL $\approx$ 16 for the J=1-1 line at 147.6 Å. Two separations were tried for the double slabs and the best coupling was observed for a 300 µm separation between the slabs. The angular distribution of the laser lines is observed to change with the separation. While we observe good qualitative agreement with our numerical codes, future experiments and more detailed analysis are planned to improve our understanding of these double slab x-ray lasers.

ACKNOWLEDGMENTS

The authors would like to thank S. Alvarez, T. Demiris, H. Louis, J. Ticehurst and the Nova operations crew for providing support for these experiments. This work was performed under the auspices of the U. S. Department of Energy by Lawrence Livermore National Laboratory under contract No. W-7405-ENG-48.

REFERENCES

1. Elton, R. C., *X-ray Lasers*, San Diego: Academic Press, 1990.
2. Lewis, C. L. S., et al, *Opt. Comm.* **91**, 71-76 (1992).
3. Wang, S., et al, *J. Opt. Soc. Am. B* **9**, 360-368 (1992).
4. Carillon, A., et al, *Phys. Rev. Lett.* **68**, 2917-2920 (1992).
5. Nilsen, J., Koch, J., Scofield, J., MacGowan, B. J., Moreno, J. C., and Da Silva, L. B., *Phys. Rev. Lett.* **70**, 3713-3715 (1993).
6. Moreno, J. C., Nilsen J., Koch, J. A., MacGowan, B.J., Scofield, J . H., and Da Silva, L. B., *Appl. Phys. B* **58**, 3-5 (1994).

Analytic Models for Beam Propagation and Far-Field Patterns in Slab and Bow-Tie X-Ray Lasers

Elaine A. Chandler

Lawrence Livermore National Laboratory,
PO Box 808, Livermore, CA 94550

Abstract. Simplified analytic models for beam propagation in slab and bow-tie x-ray lasers yield convenient expressions that provide both a framework for guidance in computer modeling and useful approximates for experimenters. In unrefracted bow-tie lasers, the laser shape in conjunction with the nearly-exponential weighting of rays according to their length produces a small effective aperture for the signal. We develop an analytic expression for the aperture and the properties of the far-field signal. Similarly, we develop the view that the far-field pattern of refractive slab lasers is the result of effective apertures that are created by the interplay of refraction and exponential amplification. We present expressions for the size of this aperture as a function of laser parameters as well as for the intensity and position of the far-field lineout. This analysis also yields conditions for the refraction limit in slab lasers and an estimate for the signal loss due to refraction.

INTRODUCTION

This work develops a general understanding of the origins of the far-field beam pattern from within the laser. We have used a very simple analytic model as an approximation to investigate the optics of slab and bow-tie x-ray lasers. While this approach is not a substitute for detailed computer models of these systems, the author has found it useful to have a set of approximate analytic expressions to guide her in computer modeling. These expressions may also be useful in roughing-out experimental designs.

In slab lasers, there is a large variation of the index of refraction in the ablation direction (see Fig. 1) that causes refraction of the rays in this direction. In the other directions, the propagation and "vertical" directions, the measured refraction is small and one assumes that the index of refraction is relatively invariant. Recently, in an effort to improve coherence, investigators have changed the shape of the laser in the vertical direction to produce active apertures in that direction; one shape under study is the "bow-tie" (1). We begin by analyzing the aperture for the unsaturated bow-tie and the shape and strength of the signal resulting from this geometry.

Similar calculations for a refracting slab reveal that only part of the slab laser participates in the production of the far-field signal at any given angle near the signal peak, resulting in an internal aperture for this signal. How the aperture changes as the laser is lengthened or as the slope of the index of refraction changes is a question we address analytically. Related to this question is the issue of how much signal is lost from a slab laser because of refraction.

SLAB AND BOW-TIE LASERS WITH NO REFRACTION

Let us consider first the simplest slab and bow-tie models, with no refraction. To make the analytic problem tractable, we assume that the laser is unsaturated with a small-signal gain g_0 that is uniform throughout. Since we are interested in the far-field image, we calculate the intensity of all rays leaving the laser at a particular angle. For each incremental ray tube comprising the signal at this angle, the strength is just the integral of the equation

$$\frac{dI}{dl} = g_0 I + \varepsilon$$

where l is the length along the ray and the source term ε accounts for spontaneous emission. The initial value is also taken to be ε. The integrated intensity I for a ray tube of length Λ in the laser is then

$$I_\Lambda = \varepsilon \frac{e^{g_0 \Lambda} - 1}{g_0} \quad (1)$$

The two geometries are shown in Fig. 1. The x-direction is taken as the direction of ablation while the z-direction is the propagation direction. We study the pattern produced in the far-field by calculating the signal along a line-out at the mid plane of the pattern and parallel to x for the slab. When we neglect refraction in the slab, the problem is symmetrical for the two non-propagating directions; but we also wish to study refraction, and its effect is to deflect the pattern in the x-direction. For the bow-tie, we are interested in the effect of the waist in the vertical direction, so the appropriate line-out is calculated perpendicular to x. We sum all

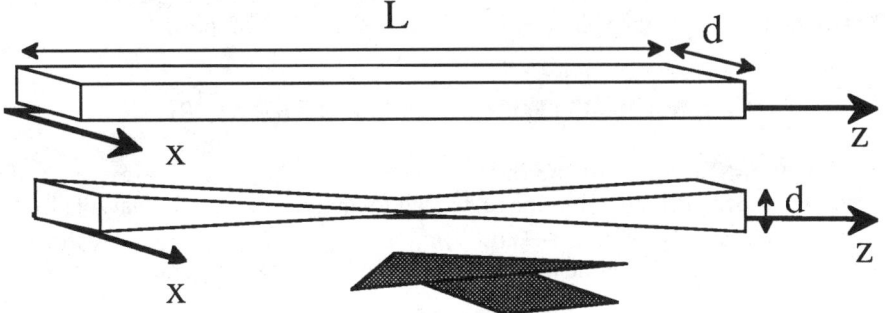

FIGURE 1. The geometries of a slab (upper sketch) and bow-tie laser (lower sketch) are simplified for this analytic study. The large arrow indicates the direction of the drive laser, x is the ablation direction, and z is the propagation direction.

rays emerging at each angle along the lineout direction with zero inclination in the perpendicular direction using a length-dependent weight given by Eq. (1). The geometric angle ϑ_g is defined by $\tan \vartheta_g = d/L$. Typically, the lasing region is 100–200 μm wide, while the length is several cm, so that $d/L \cong 3 \, mrad$. Clearly, with no refraction, ϑ_g is a natural measure for the limit of the pattern in the far-field. In fact, the calculations show that the intensity of the pattern $I(\vartheta)$ peaks at $\vartheta = 0$ and, in the slab, falls almost linearly to $I(\vartheta_g)$ where

Slab: $$I(\vartheta_g) = \frac{2}{g_0 L} I(0), \qquad (2)$$

so that the signal is less peaked for low-gain-length lasers. The bow-tie signal is flatter at angles smaller than ϑ_g and falls precipitously at ϑ_g:

Bow-tie: $$I(\vartheta_g) = \frac{1}{g_0 L} I(0). \qquad (3)$$

The ratio of the peaks of the bow-tie and the slab indicates the degree of signal loss in the bow-tie:

$$I(0)_{bow-tie} = \frac{1}{g_0 L} I(0)_{slab}. \qquad (4)$$

The loss is expected to be offset by an improvement in the coherence of the signal as the source size is decreased to an effective aperture at the waist. From Eq. (4) we estimate the size of the aperture to be $\sim d/g_o L$ full-width. In fact, calculations show that 70% of the 0° signal passes through this aperture. The aperture decreases in size as the gain-length increases.

Similarly, in the unrefracted slab at ϑ_g there is an effective aperture of full width $\sim 2d/g_o L$ which accounts for ~70% of the signal at this angle. Of course, at 0° with no refraction, the signal comes from the full width of the slab laser.

SLAB GEOMETRY WITH REFRACTION

We employ a model of the index of refraction n(x), $n(x) = n_o \exp(bx)$, that allows us to perform the integrations needed to analyze the signal analytically (2). Since b is typically of size $\sim 7x10^{-3} cm^{-1}$ and n_O is close to 1.0 in value, this expression is essentially a linear approximation of the index of refraction profile,

$$n(x) = n_o(1 + bx). \qquad (5)$$

Computer simulations of the ablating process made using the LASNEX code show a variation that is better fit to $n(x) = n_0 - n_1 \exp(-cx)$ which, while being an

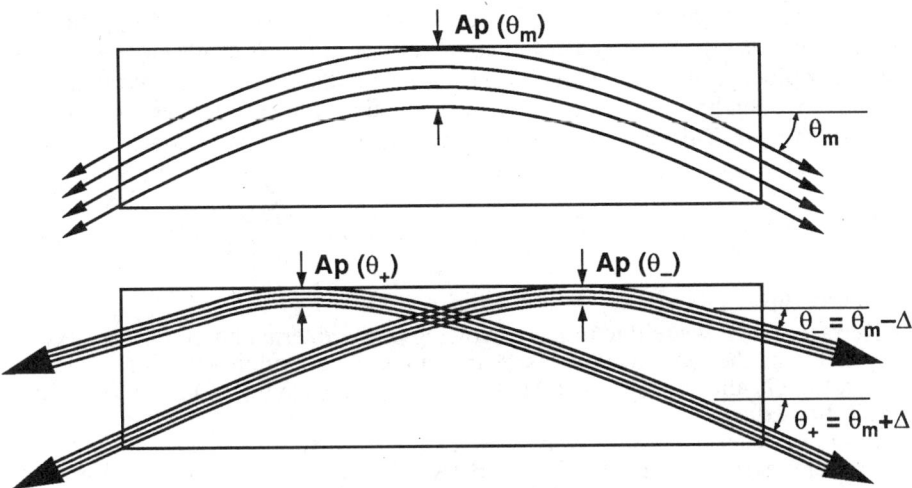

Figure 2. Schematics of the paths of rays within the refracting slab laser. In our approximations all rays that exit the laser at a particular angle are parallel throughout. The rays that comprise the peak of the far-field pattern are shown in the upper sketch; they are limited to an aperture $A_p(\vartheta_m)$ which is larger than the apertures for rays at greater or lesser angles. Note the symmetries of these rays.

increasing function of x in the lasing region, has a noticeable negative second derivative. However, n_o and b in Eq. (5) can be chosen to span the total variation in the index of refraction; by the mean-value theorem (3), b will match the actual derivative at some place along the x axis within the laser, which we take to be representative of the laser as a whole. To further simplify the calculations in the refracting laser, we initially ignore all rays but those traversing the full length of the laser.

The variation of the index of refraction shifts the far-field maximum to ϑ_m,

$$\vartheta_m = bL/2, \qquad (6)$$

which increases linearly with both b and the laser length, L. For $b = 7 \times 10^{-3}$ cm^{-1} and $L = 3$ cm, Eq. (6) gives $\vartheta_m = 10$ mrad which is a typical experimental result.

The paths of full-length rays that contribute to the signal at ϑ_m are symmetrical about the midpoint of the laser as shown in Fig. 2. They are parallel because of the linear index of refraction profile, and they have and internal aperture $A_p(\vartheta_m)$:

$$A_p(\vartheta_m) = d(1 - bL^2/8d). \qquad (7)$$

Full-length rays with other exit angles have smaller apertures.

The term in parentheses in Eq. (7) decreases quadratically as L is increased. For $b = .007\ cm^{-1}$, $d = .015\ cm$, and $L = 3\ cm$, $A_p(\vartheta_m) = .007\ cm$ or 48% of the width d. Because the rays are parallel, $A_p(\vartheta_m)$ is the size of the exit aperture as well.

The refraction limit is reached when the internal aperture of the rays contributing to ϑ_m disappears. This condition defines a critical angle ϑ_{cr} and critical length L_{cr} determined by Eqs. (6) and (7),

$$\vartheta_{cr} = \sqrt{2bd} \quad \text{and} \quad L_{cr} = \sqrt{8d/b}. \qquad (8)$$

For the conditions above, $\vartheta_{cr} = 14.5\ mrad$ and $L_{cr} = 4.14\ cm$.

Refractive losses are due to the narrowing of the internal apertures for rays of every angle as the length increases or the index of refraction variation becomes steeper. Eq. (7) allows us to calculate the fractional loss to the peak amplitude due to refraction effects.

As a check on our calculations, we have performed a series of Monte Carlo simulations with the LLNL code BEAM3 (1) which generates photons spontaneously throughout the laser and follows them as they pick up stimulated emission. In these simulations, g_0 and L are held constant while b is changed. As b increases, the signal moves out to higher angle and the peak power falls. We scale the peak signal from these simulations with the aperture formula in Eq. (7). The scaling works well until near the refraction limit where rays of almost-full length contribute appreciably to the weakened signal. By including rays of almost-full length in a modification to this calculation near ϑ_{cr}, we can account for the difference.

ACKNOWLEDGMENTS

The author wishes to thank Prof. M. Key, The Central Laser Facility of Rutherford Appleton Laboratory, and the University of Oxford for their hospitality while much of this work was done. In addition she wishes to acknowledge Dr. George Maenchen of Lawrence Livermore National Laboratory for the development of the BEAM3 code. This work was performed under the auspices of the US Department of Energy by Lawrence Livermore National Laboratory under contract No. W-7405-ENG-48.

REFERENCES

1. Wan, A. S., Da Silva, L. B., Moreno, J. C., Mayle, R. W., Cauble, R. C., Chandler, E. A., Dalhed, H. E., Libby, S. B., Nilsen, J., Ratowsky, R. P., Scott, H. A., and Van Wonterghem, B., "Development of X-ray Laser Architectural Components", in this Proceedings.
2. Snyder, A. W. and Love, J. D., *Optical Waveguide Theory*, New York: Chapman and Hall, 1991, ch. 1–3.
3. Apostol, T. M., *Calculus*, New York: Blaisdell Publishing Co., 1961, ch.7.

Light Scattering Measurement in a Ge X-ray Laser Plasma

C. H. Nam[1], I. W. Choi[1], Y. Oshikane[2], K. Murai[2], G. Yuan[2],
R. Kodama[2], H. Daido[2], Y. Kato[2], and S. Nakai[2]

1 Department of Physics, Korea Advanced Institute of Science Technology, Kusong-dong,
Yusong-gu, Taejon 305-701, Korea
2 Institute of Laser Engineering, Osaka University, 2-6 Yamada-oka, Suita, Osaka 565, Japan

Abstract. The light scattering at $3\omega/2$ was measured from X-ray laser plasmas produced using Gekko XII at ILE. Two kinds of pumping laser conditions were tried: traditional 1 ns pumping and 0.1 ns multiple-pulse pumping. The spectral intensity profile near $3\omega/2$ and the spatial intensity distribution of the $3\omega/2$ emission from line-focused plasmas were obtained using a visible spectrometer coupled with a streak camera. The spectral measurement showed two components of the $3\omega/2$ emission - strong red and weak blue components, and the spatial measurement revealed a series of localized $3\omega/2$ emissions from the line-focused plasma.

INTRODUCTION

The collisionally pumped X-ray lasers such as Se, Ge, and Y lasers (1-3) have been very successful in recent years. The X-ray laser output of these lasers reached saturated amplification either in single pass amplification or in double pass amplification with a multilayer X-ray mirror. The anomaly in the 0-1 transition of Ne-like ion over the 2-1 transition has triggered theoretical investigations on the detailed pumping mechanism of the collisional pumped X-ray laser. Even though this anomaly has been partially resolved by taking into account the contribution from the recombination of F-like ions to Ne-like ions, the theoretically predicted gain for the 0-1 transition well exceeds the experimentally observed value. For a complete understanding of the experimental result, a thorough diagnostics of X-ray laser (XRL) plasma is necessary. There are also some theoretical investigations on the effect of hot electrons to the excitation of the collisionally pumped XRL.

As a diagnostics to observe the plasma activity in an X-ray laser plasma, which is also linked to the hot electron generation, the optical emission, especially at $3\omega/2$, from X-ray laser plasma was monitored to characterize plasmas in the collisionally pumped XRL.

EXPERIMENT

The X-ray laser experiments based on the collisional excitation scheme (using either Ne-like or Ni-like ions) were performed on slab targets irradiated by one beam from Gekko XII. In the case of Ge X-ray laser experiment, a 1 µm thick and 1 mm wide Ge target was prepared on the top of 0.5 µm thick CH coated on a glass substrate. The pumping of X-ray laser was alternatively done either by G11 beam or by E08 beam of Gekko XII. Both beams were focused using a cylinderical lens and an aspherical lens to obtain the line focusing of 6 cm (G11) and 3 cm (E08), and the target lengths were set to less than 4 cm and 3 cm, respectively. The average focused intensity was 2.3×10^{13} W/cm^2 and 6.0×10^{13} W/cm^2 for the case of 6 cm and 3 cm line focusing, respectively. Two kinds of pumping laser condition were tried: traditional 1 ns single-pulse pumping (G11) and 0.1 ns multiple-pulse pumping with variable separation (G11, E08).

In order to observe the optical emission at $3\omega/2$ from both sides two similar setups were prepared. Each set consists of a visible spectrometer coupled with a streak camera and imaging optics. The line-focused plasma is imaged along the entrance slit of the spectrometer and the dispersed image by the spectrometer is relayed to the slit of streak camera.

Spectral Measurement

The $3\omega/2$ emission is usually generated through the two-plasmon decay (TPD) instability near the quarter-critical density region and the coupling of the plasmon from the TPD instability and the incident or scattered laser light (4,5). One characteristics of the $3\omega/2$ emission is the two-component spectral composition, i.e. red-shifted and blue-shifted spectra from the exact $3\omega/2$. The wavelength shift, especially the red component, calculated from the TPD instability is proportional to the electron temperature (6); consequently, it has long been studied for the application to the electron temperature diagnostics (7).

The temporally-resolved measurement showed that the $3\omega/2$ emission came after the peak of pumping laser in the case of 1 ns pumping and after the second pulse pumping in the case of 0.1 ns multiple pulse pumping. Even though the plasma heating can be significant at the first pulse in the case of multiple pulse pumping, the plasma density will be sharply decreased away from the target surface, generating too short a density scale length to generate the TPD instability. The density scale length at the quarter-critical density region obtained

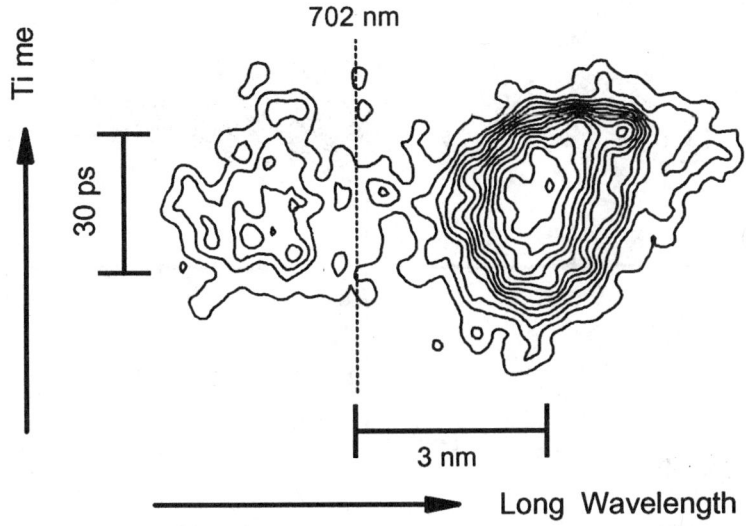

FIGURE 1. Contour plot of temporally-resolved $3\omega/2$ emission from the center of a 4 cm Ge target in the case of 0.1 ns double pulse pumping with 0.4 ns interval.

from 1-D simulation changes from 15 μm at the first pulse to 50 μm at the second pulse and to 100 μm at the third pulse with 0.3 ns interval between pulses. The output of X-ray laser is also sensitive to the density scale length, especially for the 0-1 transition of Ge X-ray laser, and it is also observed after a sufficiently long density scale length is developed to overcome the refraction from large density gradient.

Figure 1 shows a typical $3\omega/2$ emission from the center of a 4 cm Ge target at the second pulse of double pulse pumping with 0.4 ns interval. The pumping energy of each pulse was 140 J and 160 J. The wavelength shift of the red component shows that the wavelength shift by the E08 pumping beam with higher pumping intensity is larger than the G11 beam, and the wavelength shift by the third pulse is larger than that by the second pulse. The wavelength shift obtained from the same pumping beam, however, does not give a sharp intensity dependence. The wavelength shifts of the $3\omega/2$ emission from the tightest focusing region even showed nearly no correlation with the pumping intensity. This kind behavior of wavelength shift has diminished the diagnostic value of the $3\omega/2$ emission. One counter measure to restrict the wavelength shift at high temperature is the Landau damping effect which limits allowed plasmon wavenumber (6-9). The diagnostic value of the $3\omega/2$ emission is still controversial and a good modeling work is still required for a better understanding of this two-step nonlinear process.

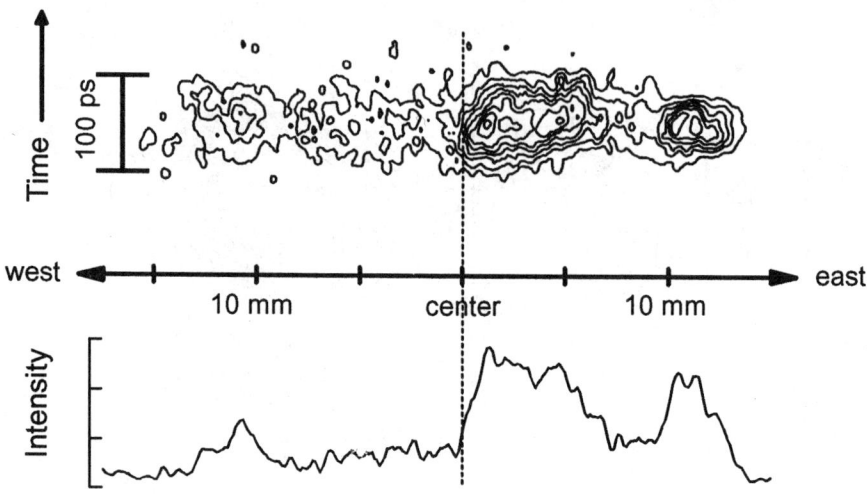

FIGURE 2. Contour plot of spatial intensity distribution of $3\omega/2$ emission.

$3\omega/2$ Emission from a Line-Focused Plasma

For the measurement of the spatial intensity distribution, the line imaging of the $3\omega/2$ emission was obtained. The line focusing achieved using a cylinderical lens and an aspheric lens generates a nonuniform intensity profile on the target. The tightest focusing position was located at 1.4 cm from the target center in the case of 6 cm line focusing (the G11 beam) and 1 cm in the case of 3 cm line focusing (the E08 beam). Even though the intensity difference of a factor 3 exists between the target center and the tightest focusing, the observed $3\omega/2$ emission was much more nonuniform than expected and even localized in the emission region.

Figure 2 shows the intensity distribution of the red component of the $3\omega/2$ emission. Two strong emissions with about 5 mm gap in the east side and scattered weak emissions in the west side were observed. The $3\omega/2$ emission generated by the stronger pumping (the E08 beam) showed a series of localized emissions throughout the line-focused plasma with strongest emissions at both tight focusing spots. One possibility for this kind of global modulation can be a nonuniformity in the pumping intensity distribution, such as filamentation which causes a nonuniform heating of plasma. Another possibility of the localized emission may come from a globally modulated TPD instability. For the thorough understanding of this $3\omega/2$ emission a more fundamental investigation of the TPD instability in a line-focused plasma is necessary.

CONCLUSION

The temporally-resolved $3\omega/2$ emissions from line-focused X-ray laser plasmas were obtained. In the case of multiple pulse pumping the $3\omega/2$ emission was observed with the second pulse pumping after a sufficient density scale length is developed. As with the change of experimental condition the spectral distribution of the $3\omega/2$ emission is varied, but the wavelength shift from the exact $3\omega/2$ did not show a close correlation with pumping intensity. The spatial imaging of the $3\omega/2$ emission from the line-focused plasma showed several localized emissions, which might come from a globally modulated TPD instability. The detailed analysis of the $3\omega/2$ emission will help to resolve the difference still existing between the predicted gain, especially for the 0-1 transition, and the observed value of the collisionally pumped X-ray laser.

REFERENCE

1. Koch, J. A., MacGowan, B. J., Da Silva, L. B., Matthews, D. L., Underwood, J. H., Batson, P. J., and Mrowka, S., *Phys. Rev. Lett.* **68**, 3291-3294 (1992).
2. Carillon, A. Chen, H. Z., Dhez, P., Dwivedi, L., Jacoby, J., Jaegle, P., Jamelot, G., Zhang, J., Key, M. H., Kidd, A., Krisnick, A., Kodama, R., Krishnan, J., Lewis, C. L. S., Neely, D., Norreys, P., O'Neill, D., Pert, G. J., Ramsden, S. A., Raucourt, J. P., Tallents, G. J., and Uhomoibhi, J., *Phys. Rev. Lett.* **68**, 2917-2920 (1992).
3. Da Silva, L. B., MacGowan, B. J., Mrowka, S., Koch, J. A., London, R. A., Matthews, D. L., and Underwood, J. H., *Opt. Lett.* **18**, 1174-1176 (1993).
4. Liu, C. S. and Rosenbluth, M. N., *Phys. Fluids* **19**, 967-971 (1976).
5. Berger, R. L. and Powers, L. V., *Phys. Fluids* **28**, 2895-2909 (1985).
6. Karttunen, S. J., *Laser Part. Beams* **3**, 157-172 (1985).
7. Amiranoff, F., Briand, F., Labaune, C., *Phys. Fluids* **30**, 2221-2225 (1987).
8. Powers, L. V. and Schroeder, R. J., *Phys. Rev. A* **29**, 2298-2300 (1984).
9. Seka, W., Bahr, R. E., Short, R. W., Simon, A., Craxton, R. S., Montgomery, D. S., and Rubenchik, A. E., *Phys. Fluids B* **4**, 2232-2240 (1992).

Refractive Compensation in Slab Target Geometries

D. Neely[1], Y Kato[2], R Kodama[2], H Daido[2], K. Murai[2], G Yuan[2], C. L. S. Lewis[3], A. MacPhee[3], M. J. Lamb[3], P. Holden[4], G. J. Pert[4] and A. Djaoui[1].

1) Rutherford Appleton Laboratory, Chilton, Didcot, Oxon, England, OX11 0QX
2) Institute of Laser Engineering, Osaka University, Suita, 565 Japan
3) Dept. of Pure and Applied Physics, Queens University of Belfast, N. Ireland, BT7 1NN
4) Department of Physics, University of Hull, Hull, England, Y01 5DD

Abstract. Results showing the enhancement effect on soft X-ray lasing emission intensity with target substrate curvature of up to 1 mrad per mm, which compensates for refraction due to electron density gradients present in the gain region of a Ne-like Ge expanding plasma are shown. Time integrated relative intensity and pointing data are presented for the Ne-like Ge 19.6, 23.6 and 28.6 nm 3p-3s transitions. A proposed simple illumination geometry for compensating for refraction due to lateral electron density gradients is also presented.

INTRODUCTION

Since the first conclusive demonstration of high gain length soft X-ray laser action (1) refractive effects caused by electron density gradients across the gain region have dominated the output characteristics of the lasers (2). Saturation has only been achieved by using target designs which minimise (3) or compensate (4,5) for the refractive effects encountered as the soft X-ray beam propagates (along the target axis) through the gain media. Lunney (6) proposed compensating for the transverse refractive density gradients using a curved target. This paper presents results obtained which validate this approach. Lunney also proposed using a toroidal shaped target to additionally compensate for the density gradients caused by the lateral expansion of the plasma. In this paper a much simpler scheme is proposed. Modelling indicates that by using a non Gaussian shaped lateral drive profile it is possible to create a near flat topped lateral density profile across the majority of the plasma. By combining this with a curved target the benefits of refraction compensation both axially and laterally could be obtained using a much simpler target design.

EXPERIMENT

The experiment was carried out using a single 1.053 μm beam of the GEKKO XII glass laser. Ge stripes deposited on 1.5 mm thick glass substrates were irradiated using a combination of cylindrical and F/3 aspheric lenses to generate a line focus. The targets were irradiated with a 1 ns FWHM Gaussian shaped pulse, which delivered up to 1.1 KJ to the focal plane. The average focal width was 100 μm giving an average target intensity of 1.7×10^{13} Wcm^{-2}.

A flat-field spectrometer with collection optics was used to examine the horizontal angular intensity distribution from each target which were recorded on Q-plates. The Q-plates were densitometered using a MK6 Applied Imaging densitometer with 16x 0.3 N.A. influx and 16x 0.35 N.A. efflux optics and a 2x400 μm sampling slit. Density to intensity conversion was carried out using a calibration curve. After background subtraction integration under the spectral profile was performed.

The Ge stripe was initially deposited onto a flat glass substrate which was subsequently bent to the required average curvature for each shot. This was achieved by tightly clamping the substrate between two circular metal formers. To maintain diagnostic and drive beam access to the stripe the metal formers were kept at least 4 mm below the stripe. The surface profile P(y) of each target was measured (at the location of the stripe) immediately prior to it being placed in the vacuum chamber using the following method. The deflection S(y) of a reflected laser beam was measured (at a distance d = 1.91 m from the target surface) as a function of position y along each target. For small angular deflections ($\theta(y) \ll 0.1$ rads) of the reflected beam, the approximation $dP(y)/dy \approx S(y)/2d$ is reasonably accurate. Figure 1a shows two deflection curves, typical of those obtained during

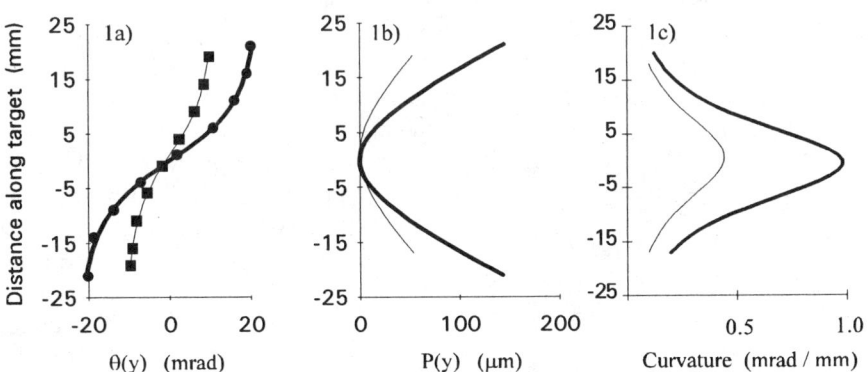

FIGURE 1. Graph 1a showing the measured deflection curve and fit assuming a hyperbolic surface shape for two 40 mm long targets with average curvatures of 0.5 and 0.25 mrad/mm, 1b Graph showing the surface profile of the two targets and 1c Graph showing the distribution of curvature as a function of position along the two targets.

the experiment and a least squares fit assuming a hyperbolic target surface profile, which was found to adequately describe the general shape of the curved targets used. Figure 1b shows the calculated surface profile and Figure 1c displays the angular curvature C (mrad mm^{-1}). Figure 1c clearly shows that the distribution of curvature is not constant but tends to be greatest in the centre and smallest at the ends. This is a consequence of not being able to clamp the targets both above and below the stripe but such an approach was not practical during the run.

RESULTS / DISCUSSION

The time integrated peak emission intensities for the 19.6, 23.6 and 28.6 nm transitions from 40 mm targets are plotted as a function of average target curvature on Figure 2. The results clearly show a strong enhancement ($\approx$4x) for the 19.6 and ($\approx$2x) 28.6 nm lines, peaking at an average curvature of 0.7 mrad mm^{-1}. The 23.2 and 23.6 nm lines did not show a significant enhancement for this data set (for clarity the 23.2 nm transition is not shown here but its output intensity closely followed that of the 23.6 nm line). Time resolved data (7,8) for the 19.6 and 23.6 nm lines shows the most marked effect on the output lasing intensity for curved targets when compared to flat targets. The 23.6 nm line showed a general increase ($\approx$2x) in output intensity. The 19.6 nm transition displayed a much shorter duration pulse occurring 400 ps before the centre of the driving pulse which was $\approx$5x the level of signal from a flat target of the same length. Although no streaked data was taken on the 28.6 nm transition, modelling would suggest that high levels of gain are present over a large volume of the plasma late in time when the plasma has significantly expanded. This may account for the enhancement observed on this line in the time integrated data set whilst, within the scatter of the data set no strong trends were observed for the 23.2 and 23.6 nm transitions. This was not the case with the time resolved data set, and may be due to shot to shot uncertainties.

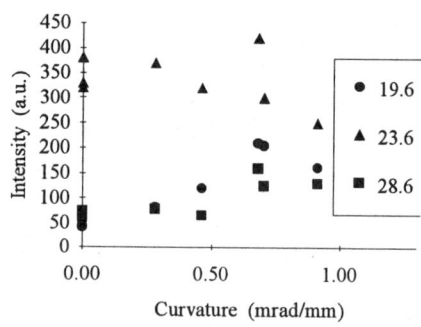

FIGURE 2. Graph showing the relative output intensities as a function of average target curvature for 40 mm targets.

The non-uniform axial drive intensity distribution, caused by the bisection of the curved focal plane with the target, has two peaks occurring near the ends of the plasma and falls towards the centre. Modelling indicates that at the lower intensities present in the centre of the target the peak gain occurs slightly later in time further from the target surface in a lower density gradient. Further modelling including a ray trace where account of the varying axial intensity distribution and curvature along the line focus will be necessary to

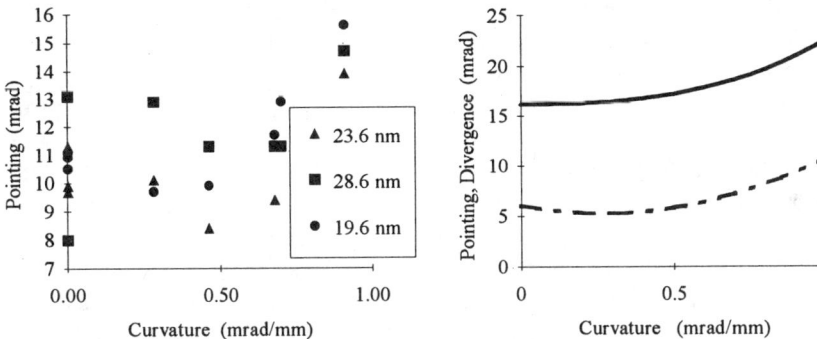

FIGURE 3. Graph showing the beam angular pointing direction for the 19.6, 23.6 and 28.6 nm transitions as a function of target curvature for 40 mm long targets.

FIGURE 4. Graph showing the predicted angular pointing direction (solid line) and beam divergence (dashed line) for 40 mm long targets as a function of target curvature

fully explain the results obtained. The early peak for the 19.6 nm transition is in good agreement with modelling (9) which predicts that this transition optimises early in time in a higher density region than the 23.2 and 23.6 nm lines.

The measured output beam pointing direction Ψ for three of the lasing transitions is plotted as a function of target curvature in Figure 3. For a target of length L, Ψ is defined by the relationships between the density and gain profiles. For small target curvatures (C $\ll$ $2\Psi_{Flat}/L$) the output pointing direction from a curved target remains almost constant. The main improvement is that the ray spends more time in the high gain region of the plasma so the output signal intensity increases and the divergence reduces slightly as shown by the modelling results in figure 4. For C $\geq$ $2\Psi_{Flat}/L$ the effect of curvature is likely to be more appreciable and $\Psi \approx CL/2$ because the optimum ray can actually be guided through the region of gain (10). This can be seen for the high average curvature data shots where the pointing direction of the lines starts to increase in agreement with theory (although code results tend to slightly overestimate Ψ) and the output intensity increases. As the curvature increases significantly beyond this point the optimum ray path moves to higher density gradients where the corresponding densities become so large that the lasing levels equilibrate.

REDUCTION OF LATERAL DENSITY GRADIENTS

By placing an appropriate grating-like phase plate in the optical line focus illumination system (11), it is possible to significantly modify the lateral intensity distribution at the focal plane. If a flat topped lateral density profile can be obtained then by combining this with a curved target (possibly also with pre-pulse (7)) a high quality soft X-ray amplifier would be produced. Such an amplifier would be ideal for obtaining a highly coherent saturated beam suitable for holographic imaging. In an attempt to characterise the benefits of such a scheme, modelling was carried out using the 2D hydro code Pollux for 100 ps pulses. The density profiles obtained using a Gaussian and a modified drive pulse are shown in figure 5. The double peaked intensity distribution clearly produces an almost flat distribution across 60% of the line focus width which lasts for $\approx$ 40 ps and would prove ideal as a coherence preserving amplifier. Modelling was also carried out for 1 ns drive pulses which did produce flatter density profiles than using a Gaussian shaped profile, but the lateral gradients were not as ideal as in the 100 ps driver case.

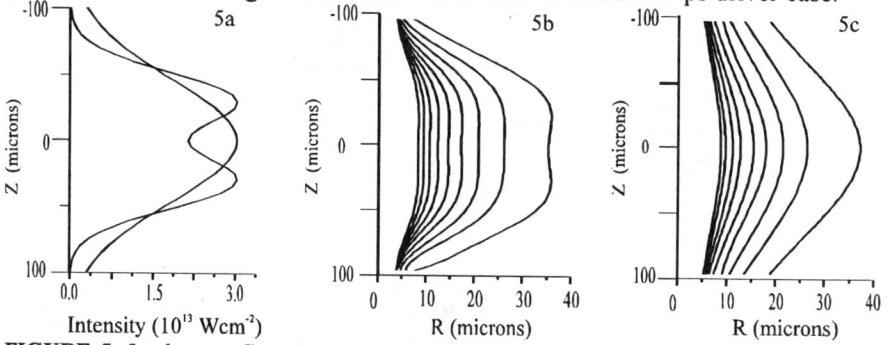

FIGURE 5. 5a shows a Gaussian and a double peaked input drive pulse. The electron density contours at a time 20 ps before the peak of the driving pulse for the double peaked pulse are shown in 5b and 5c shows the results obtained at the same time but using the Gaussian pulse.

CONCLUSION

This experiment has demonstrated the benefits of using curved targets to compensate for refractive effects caused by plasma expansion. The significant increase of the 19.6 nm J=0-1 transition with target curvature brings experimental results into closer agreement with theoretical modelling. In the future we hope to combine the effect of using a laterally modulated driving pulse (possibly also with pre-pulse) and curved target to produce a high quality soft X-ray amplifier.

1 Matthews, D. L. et al, Phys. Rev. Lett **54** 2, p110-113 (1985).
2 London, R. A. and Rosen, M.D., Phys. Fluids **29**, 11 p3813-3822, (1986).
3 MacGowan, B.J., et al, Phys. Fluids, B **4** p2326-2337 (1992).
4 Carillon, A. et al, Phys. Rev. Lett, **68** 19, p2917-2920, (1992).
5 Wang, S., et al, Chinese Phys. Lett. **8** 12 p618-621 (1991).
6 Lunney, J., Appl. Phys. Lett. **48**, p891 (1986).
7 Kodama, R., et al, Phys. Rev. Lett **5** 13 (1994).
8 Daido, H. et al, These proceedings (1994).
9 Holden, P. B., et al, J. Phys. B, **27**, p341 (1994).
10 Kato, Y., et al, These proceedings (1994).
11 Neely, D., et al,. Submitted to ECLIM (1994).

Na-like Autoionizing Levels: Plasma Diagnostics and Prospects for Photopumped Soft X-Ray Lasers

A. L. Osterheld, J. Dunn, B.K.F. Young, S.B. Libby, A. Szoke, R.S. Walling, W.H. Goldstein and R.E. Stewart

Lawrence Livermore National Laboratory, L-59
P.O. Box 808, Livermore, CA 94550

A. Ya. Faenov, I. Yu. Skobelev and S. Ya. Khakhalin

Multicharged Ions Spectral Data Center, National Scientific-Research Institute for Physical-Technical and Radio-Technical Measurements (VNIIFTRI)
Mendeleevo, Moscow Region, 141570 Russia

Abstract. We have investigated the kinetics of doubly excited sodium-like copper states with 3l3l' and 3l4l' configurations, and the feasibility of photopumped x-ray lasing schemes and plasma diagnostics using transitions from these levels. These x-ray lasing schemes, like similar lasing schemes in Li-like ions (1–3), have high quantum efficiency and potentially reduced radiative trapping effects. In addition, the satellite transitions emitted by these sodium-like levels may provide valuable diagnostics of x-ray laser plasmas.

INTRODUCTION

Sodium-like autoionizing levels have interesting applications in x-ray laser research. The n=3–2 and n=4–2 satellite spectra from these levels provide valuable information about the electron temperature, density and ionization distribution. In addition, transitions between these levels may be inverted, if excited by resonant photopumping. These schemes are analogous to proposed lithium-like photopumped x-ray lasing schemes (1–3).

While the plasma conditions in neon-like x-ray laser plasmas have been diagnosed by seeding a low-Z tracer into the plasma (4), it may be advantageous to use the L-shell radiation produced by the lasing element. The use of n=3–2 sodium-like satellite transitions for plasma diagnostics has previously been investigated (5–7). The n=4–2 satellite transitions offer several advantages including the lower optical depth of the corresponding neon-like resonance lines, the better isolation of the 4–2 satellites, and the smaller contributions from resonance excitation and cascades to the resonance and satellite lines. Because

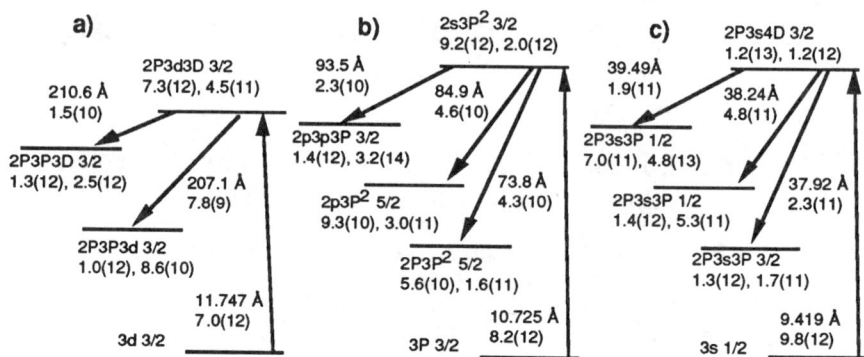

FIGURE 1. Energy level diagrams of the important levels for examples of a) n=3–3, b) inner shell 2s–2p, and c) n=4–3 Na-like copper photopumped x-ray lasing schemes

of lack of space, we refer the reader to references 8 and 9 for a discussion of the application of n=4–2 satellite lines to plasma diagnostics.

Resonantly photopumped x-ray laser schemes between autoionizing levels potentially have high quantum efficiency and reduced trapping problems. In most conventional lasing schemes, the lower laser level is depleted by fast radiative decays to low-lying ground levels. As the transverse size of the x-ray laser is increased, the reabsorption of this radiation increases the effective lifetime of the lower level, reducing the inversion density. The conventional collisional x-ray lasers have low quantum efficiency, as they lase on $\Delta n=0$ transitions between highly excited levels. This is because the levels which are strongly excited by electron collisions typically do not have $\Delta n > 0$ decays to levels with shorter lifetimes. The use of transitions between autoionizing levels opens up the possibility of lasing transitions whose lower levels decay non-radiatively. This potentially ameliorates the trapping problem, and allows $\Delta n > 0$ lasing transitions.

THEORETICAL MODEL AND RESULTS

We investigate the kinetics of the autoionizing levels using a collisional-radiative model. The level populations are determined by solving a set of linear rate equations which describe the various excitation, de-excitation, ionization and recombination processes which transfer population between the energy levels. The collisional-radiative model consists of detailed atomic data embedded in an average-level model. Further details may be found in ref. 10.

The detailed atomic data come from the Hebrew University–Lawrence Livermore Atomic Code package (HULLAC). This code is used to calculate wave functions, energy levels, autoionization rates, and all important radiative transition rates from the relativistic, multi-configuration parametric potential model with full configuration interaction (11–13). Relativistic, distorted wave electron excitation cross-sections are calculated for all transitions (14), including all collisions between doubly excited states. The cross-section calculations use a powerful angular factorization and an empirically motivated interpolation scheme for calculating collisional rates. This code package uniquely combines

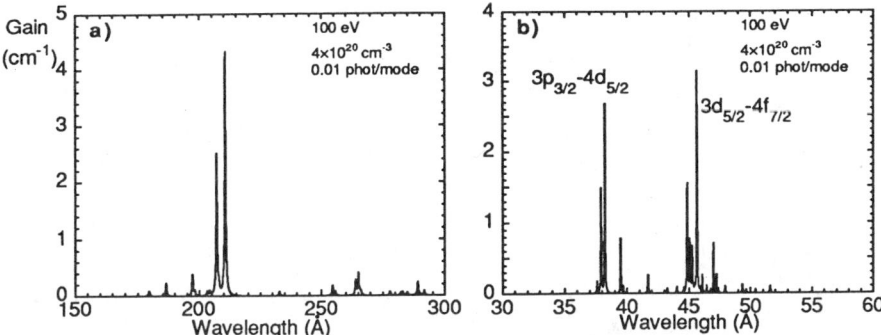

FIGURE 2. Gain calculations for the a) n=3–3, and b) n=4–3 sodium-like copper photopumped x-ray lasing schemes.

accuracy and highly efficient computational techniques and enables practical calculations of even very complex atomic systems.

In the present work the collisional-radiative model explicitly includes the doubly excited sodium-like levels with configurations $2s^2 2p^5 3l3l'$ and $2s^1 2p^6 3l3l'$ $l,l' = 0$–2, and $2s^2 2p^5 3l4l'$ and $2s^1 2p^6 3l4l'$ $l = 0$–2, $l' = 0$–3. There are slightly more than 1000 of these levels. This full treatment of the kinetics consistently describes the mechanisms which populate the doubly excited states and allows the gain calculations to include the effects of cascades between autoionizing levels, collisional quenching, and collisional broadening.

The main goal of the present work is to investigate the kinetics to identify promising lasing schemes in a single ion (copper), rather than to search for resonant line pairs in sodium-like ions of many elements. The large number of sodium-like autoionizing levels gives rise to many potential lasing transitions, most of which can not be pumped to practical inversion densities. Initially, an analytic three level formula was used to select transitions for further study with the full collisional-radiative model.

Because of the complex level structure of a many-electron ion, the photopumped sodium-like lasing schemes are more varied than the corresponding lithium-like case. The sodium-like lasing schemes fall into three qualitatively different classes. These include $\Delta n=0$, n=3–3 (and n=4–4) lasing transitions, inner shell 2s–2p transitions pumped by 2s–3p (or 2s–4p) excitations, and n=4–3 transitions produced by pumping the $3l4l'$ doubly excited levels. The n=3–3 schemes lase at wavelengths similar to the normal neon-like copper x-ray lasers, but will be the easiest to achieve experimentally. The inner-shell 2s–2p schemes have no analogy in lithium-like ions. These are at relatively short wavelength even in mid-Z ions, and are analogous to collisional (14) and photopumped (15) neon-like 2s–2p lasing schemes previously investigated. The n=4–3 schemes are directly analogous to the lithium-like schemes, and are at quite short wavelength. A favorable example of each type of lasing scheme is discussed below.

Examples of each type of lasing scheme are shown in Figure 1, along with the relevant lifetimes and transition rates. The levels are labeled with their total angular momentum and a relativistic configuration name formed from the orbitals of the n=2 hole and the outer two electrons (where 2p=2p1/2, 2P=2p3/2,

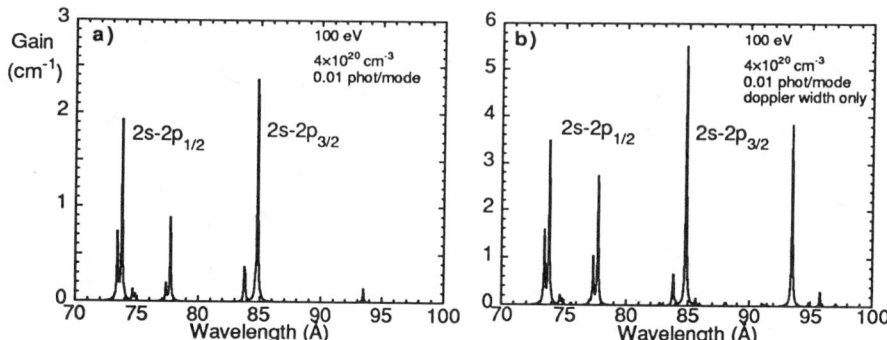

FIGURE 3. Gain calculations for the 2s-2p lasing scheme a) including autoionization and collisional widths, and b) with Doppler broadening only.

etc.). The total radiative and autoionization decays rates (sec^{-1}) are also indicated, as well as transition wavelengths and rates, where 1.5(10) is shorthand for 1.5×10^{10}.

Steady state gain calculations are shown in Figs. 2 and 3. The calculations are for electron and ion temperatures of 100 and 75 eV, and an electron density of 4×10^{20} cm^{-3}. For these initial calculations, the ionization balance was not calculated consistently with the kinetics, and the total abundance of sodium-like ions was taken to be 50%. In each case, the photopumped lines was pumped with an intensity of 0.01 photons/mode. This corresponds to equivalent radiation temperatures of roughly 228, 250, and 285 eV for the schemes in Fig. 1. The gain transitions in the n=4-3 case in Fig. 2b include both 4d-3p transitions directly pumped and 4f-3d gain lines produced by collisional transfer. This effect is most prominent for this case because of the strong 3d-4f collision rates. The gains in all cases are affected by decreased lifetimes owing to fast autoionizations or collisional mixing. This is illustrated in Fig. 3, which compares calculated 2s-2p gains including the auger and collisional lifetimes to values calculated from Doppler broadening alone. In particular, the line at 93.5 Å is drastically reduced owing to the rapid auger decay of the lower level (see Fig. 1b). Even larger lifetime effects occur for the longer wavelength n=3-3 gain transitions. The gains in Fig. 2a would increase by a factor of four if only Doppler broadening were considered.

CONCLUSIONS

We have investigated the prospects for soft x-ray lasing between autoionizing sodium-like copper levels pumped by photo-resonant excitation and have identified a number of promising gain transitions on $\Delta n=0$ and $\Delta n=1$ transitions, including relatively short wavelength transitions in medium-Z ions. The potential lasing transitions fall into three categories. These include long wavelength n=3-3 transitions which are the easiest to realize experimentally; inner-shell 2s-2p transitions which are much shorter in wavelength that the normal neon-like lasing transitions; and $\Delta n=1$, n=4-3 transitions which are at short wavelengths, but will be very difficult to experimentally achieve.

A number of general conclusions may be drawn. All of these schemes require fairly intense pumping transitions. This is because the existence of multiple sodium-like ground states reduces the ion density in any given ground level of a three-level scheme, and because the lifetimes of the highly excited upper lasing states are not truly "metastable". In addition, the gain coefficients are strongly reduced by both the homogenous broadening from autoionization and collisional decays, and the quenching of the inversions from collisional mixing. Finally, these photopumped lasing schemes require separate pump and lasing plasmas. This results from the conflicting requirements of maintaining a large population of sodium-like ions and simultaneously producing an intense pump line whose transition energy is large compared to the sodium-like ionization potential. This requirement of separate plasmas reduces the application of this method for aperture scaling. The amelioration of the trapping problem is offset by the problem of propagating the externally produced pumping radiation into the lasing region.

ACKNOWLEDGMENTS

Work performed under the auspices of the U.S. Dept. of Energy by the Lawrence Livermore National Laboratory under Contract No. W-7405-Eng-48.

REFERENCES

1. Elton, R.C., *NRL Report 9103* (1988); Elton, R.C., *X-ray Lasers*, New York: Academic Press, 1990.
2. Lunney, J.G., *Optics Comm.* **53**, 235 (1985).
3. Libby, S.B., Osterheld, A.L., Szoke, A., Walling, R.S., Young, B.K.F., "Progress toward X-ray lasing between autoionizing transitions, in *Proceedings of the 3rd International Conference on X-Ray Lasers, IOP Conference Series* **125**, 1992, pp. 163–166.
4. Young, B.K.F., Osterheld, A.L. Shimkaveg, G.M., Shepherd, R.L., Walling, R.S., Goldstein, W.H., and Stewart, R.E., *J.Q.S.R.T.* **51**, 417 (1994).
5. Goldstein, W.H., Walling, R.S., Bailey, J., et al., *Phys. Rev. Lett.* **58**, 2300 (1987).
6. Peyrusse, O., Combis, P., Louis-Jacquet, M., et. al., *J. Appl. Phys.* **65**, 3802 (1989).
7. Khakhalin, S.Ya., Bryunetkin, B.A., Skobelev, I.Yu., Faenov, A.Ya., Nilsen, J., Osterheld, A., Pikuz, S.A., *ZhETF* **105**, 1181 (1994), and references therein.
8. Osterheld, A.L., Dunn, J., Young, B.K.F., Goldstein, W.H., Faenov, A. Ya., Skobelev, I. Yu., Khakhalin, S. Ya., submitted to *Phys. Rev. A*, (1994).
9. Khakhalin S.Ya., Dyakin, V.M., Faenov, A.Ya., Fiedorowicz, H., Bartnik, A., Parys, P., Nilsen, J., and Osterheld, A., submitted to *JOSA B*, (1994).
10. Osterheld, A.L., Walling, R.S., Young, B.K.F., Goldstein, W.H., Shimkaveg, G., MacGowan, B.J., DaSilva, L., London, R., Matthews, D. and Stewart, R.E., *J. Quant. Spectrosc. Radiat. Transfer*, **51**, 263 (1994).
11. Klapisch, M., *Comput. Phys. Commun.* **2**, 239 (1971).
12. Klapisch, M., Schwob, J.L., Fraenkel, B.S., and Oreg, J., *J . Opt. Soc. Am.* **61**, 148 (1977).
13. Oreg, J., Goldstein, W.H., Klapisch, K., Bar-Shalom, A., *Phys. Rev. A* **44**, 1750 (1991).
14. Bar-Shalom, A., Klapisch, M. and Oreg, J., *Phys. Rev. A* **38**, 1773 (1988).
15. Enright, G.D., Baldis, H.A., Dunn, J., La Fontaine, B. and Villeneuve, D.M., "A search for gain on 2p–2s transitions in a collisionally excited Ge plasma" in *Proc. OSA Conf. on Short Wavelength Coherent Radiation: Generation and Application* **11**, 87 (1991).
16. Politov, V.Yu., Loboda, P.A., Lykov, V.A. and Nilsen, J., *Optics Comm* **108**, 283 (1994).

Recombination X-Ray Lasers Driven by Ultra-Short Pulses in Hydrogen-Like Ions: Theoretical Analysis

Yuelin Li[†] and Jie Zhang[‡]

[†]*Shanghai Institute of Optics and Fine Mechanics, Academia Sinica*
P. O. Box 800-211, Shanghai 201800, China

[‡]*Department of Atomic and Laser Physics, University of Oxford, Oxford, OX1 3PU, UK*
SERC Rutherford Appleton Laboratory, Chilton, Didcot, Oxon, OX11 0QX, UK

Abstract. We have analysed the physics of recombination x-ray lasers in H-like ions produced in fibre targets irradiated with ultra-short pulses with a simplified theoretical model. The experimental parameter dependencies of the gain coefficient are discussed.

1. INTRODUCTION

It is recently of great interests to develop x-ray lasers with ultra-short driving pulses. In the traditional recombination pumping scheme, shorter driving pulses are favourable for producing higher initial density, therefore can enhance the recombination processes and produce higher gain-length product. In this paper, we analyse the physics of the recombination gain production in ultra-short pulse irradiated fibres in the light of last advances in the physics of ultra-short pulse laser-matter interactions(1-4). The gain dependence on the experimental parameters is discussed.

2. THE THEORETICAL MODEL

In ultra-short pulse cases, the laser-matter interaction is localised in the so called "skin layer"(3). We assume here that the plasmas produced in this layer play the most important role for lasing as they expand first and are cooled down fastest. During an ultra-short heating, the electron temperature of the surface

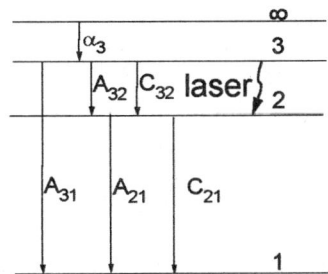

FIGURE 1. The diagram of the level structure used in the theoretical calculation. The electrons cascade to upper level of the laser transition (n=3), and the lower laser level (n=2) is evacuated due to the fast spontaneous transition. Only "down flux" were considered.

solid-density plasma can be calculated by assuming a constant absorption(1, 2), here we use the formula in reference(1),

$$T_{eo}=1.33\times I_{abs}^{4/9}\tau^{2/9}C_v^{-2/9}\kappa_o^{-2/9} \text{ K} \quad (1)$$

where C_v, κ_o, I_{abs} and τ are the heat capacity at constant volume, heat conductivity, the absorbed laser intensity, and the driving pulse duration, respectively. The plasma density at the end of the laser irradiation is obtained with an assumption of a sonic isothermal expansion during the heating(1, 4), and can be approximated by taking the volume ratio

$$N_{eo}=Z_oN_o[R_o^2-(R_o-l_s)^2]/[(R_o+c_s\tau)^2-(R_o-l_s)^2] \quad (2)$$

where N_o, Z_o, and N_{eo} are, respectively, the solid ion density, the average ionisation stage and electron density at the end of the laser pulse. l_s is the skin depth of the solid density plasma and c_s the ionic sound speed. R_o is the fibre radius. In the calculation, the fibre targets are assumed to be illuminated by laser beam from all directions. Equation (1-2) are solved together with collision-radiation rate equations self-consistently.

The expansion of the ring-shaped plasma is described by a self-similar model(5-7). With the adiabatic exponential adopted from numerical simulation, the electron density and temperature can be calculated. The atomic physics is described by a steady-down-flux model(6, 7), and for the Balmer-α transition in H-like ions(see Fig. 1 for the level structure) the populations of the upper and lower levels are $N_3=\alpha_3N_e^2N_b/(C_{32}+A_{32}+A_{31})$ and $N_2=(A_{32}+C_{32})/(C_{21}+A_{21})$, where N_b is the number density of the bare nuclei, α_3 the three-body recombination rate coefficient, C and A are the collision de-excitation and spontaneous transition rates between the corresponding levels. The gain coefficient can be calculated with the standard formula with the known N_3, N_2. Optically thin condition is assumed.

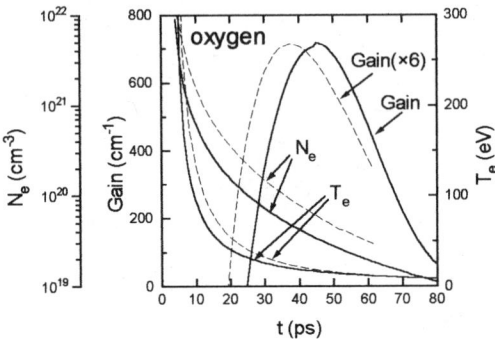

FIGURE 2. Theoretical and simulated results for the time history of T_e, N_e, and the gain for H-like oxygen Balmer-α transition. Driving pulse duration $\tau=2$ ps, fibre radius $R_0=3.5$ μm. Solid lines: theory calculation with optimised laser intensity $I_{abs}=5.6\times10^{14}$ Wcm^{-2}. Dashed lines: simulation of the 1-D hydrodynamic code MEDUSA with optimised $I_{abs}=2.5\times10^{14}$ Wcm^{-2} (from the Lagrangian cell showing the highest gain).

3. RESULTS AND DISCUSSION

Fig. 2 gives the comparison of our model results with those of the numerical simulation by the 1-D Lagrangian hydrodynamic code MEDUSA(8, 9). With the optimised absorbed laser intensity, the model calculated temporal histories of the Balmer-α transition gain, the temperature and the density agree with the simulated results roughly. The higher gain coefficient from the model is probably the result of the incomplete consideration of atomic physics.

For searching the optimum experimental parameters, we have calculated the Balmer-α transition gain under different conditions. In Fig. 3(a), the optimal atomic number is found between 7~8 with 2 ps pulse duration and 7 μm fibre diameter, and the optimised absorbed laser intensities found for carbon, oxygen and sodium fibres are also comparable to the MEDUSA simulation results of 0.4, 2.5, and 12.5×10^{14} Wcm^{-2}. In Fig. 3(b) we give the gain dependence upon the fibre diameter and pulse duration. We can see that the gain always pass through a maximum with changes in the parameter, which is very different from the longer driving pulse situation(5-7) and could be attributed to the non-equilibrium nature of the dynamics of the population inversion(10). Analysis shows that shorter pulses are favourable for high Z target, but thinner fibres are favourable for lower Z medium. This can also be understood by the adiabatic trajectories in $\eta=N_e/Z^7$, $\theta=T_e/Z^2$ the phase diagram in Fig. 4. The solid lines in Fig. 4 correspond to the optimum condition in Fig. 3(a). The starting points of the trajectories at the higher density represent the conditions at the end of the heating, and at the other end the conditions under which the maximum gain appears. It is seen that the trajectory of

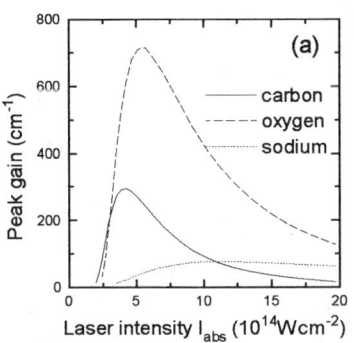
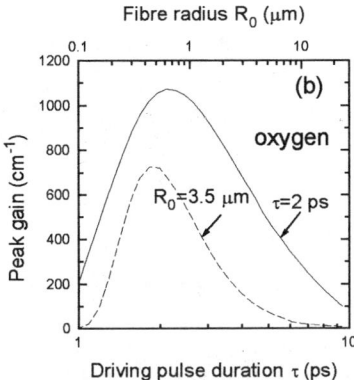

FIGURE 3. (a) Calculated laser intensity dependencies of peak gain coefficient for the Balmer-α transition in carbon, oxygen and sodium ions produced in fibre targets. Driving pulse duration τ=2 ps, fibre radius R_0=3.5 μm. Optimal absorbed laser intensities were found to be 4.6, 5.6, and 11×10^{14}Wcm^{-2} for carbon, oxygen and sodium targets, respectively. (b) Gain coefficient as functions of driving pulse duration τ and target radius R_0. The absorbed laser intensity was 5.6×10^{14} Wcm^{-2}. Dashed line: gain dependence on τ with fibre radius R_0=3.5 μm. Solid line: gain as function of R_0 with pulse duration τ=2 ps.

sodium plasma is shorter and that of the carbon is longer than that of oxygen plasma, indicating that the initial density of sodium plasma is too low and that of the carbon is too high. Another set of calculation with shorter pulse duration (τ=0.5 ps) for sodium and thinner fibre (R_0=1 μm) for carbon, which are conditions expected for higher gains, are also given as dashed lines in Fig. 4. The trajectory of sodium becomes obviously longer, and both dashed trajectories are closer to that of the oxygen, for which the condition has been relatively optimised. This may imply the existence of a universal adiabatic trajectory for optimum gain operation. The peak gain coefficients also become higher for the dashed cases.

Our analysis is based on two major assumptions, i. e., the skin layer assumption and the isothermal pre-expansion during the heating. It should be noticed that the condition for both of the assumptions to be valid is that the density scale length of the plasma does not exceed the heat conduction depth during the laser heating. Therefore the model may be more favourable for much shorter drive pulses other than 2 ps.

4. CONCLUSION

We have analysed the physics of the recombination x-ray lasers in hydrogen like ions in fibre targets irradiated by ultra-short laser pulses with a compact theoretical model. Considering the Balmer-α transition in low Z target, a qualitative agreement has been observed between our results and the simulation results by the MEDUSA. The dependence of the gain coefficient upon the laser

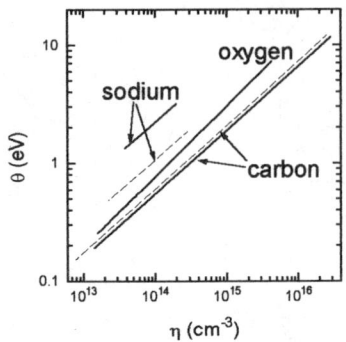

FIGURE 4. Trajectories of adiabatic expansion for carbon, oxygen and sodium plasmas on $\eta=N_e/Z^7$-$\theta=T_e/Z^2$ diagram. Solid lines: $\tau=2$ ps, $R_0=3.5$ μm, and optimum intensities from Fig. 3(a)(The peak gain for carbon, oxygen and sodium are respectively 200, 700 and 70 cm^{-1}). Dashed lines: $\tau=2$ ps, $R_0=1$ μm, $I_{abs}=4.6\times10^{14}$ Wcm^{-2} for carbon; and $\tau=0.5$ ps, $R_0=3.5$ μm, $I_{abs}=22\times10^{14}$ Wcm^{-2} for sodium(The peak gain are 400 and 490 cm^{-1} for carbon and sodium)

pulse duration, fibre radius and laser intensity, as well as atomic numbers have been obtained and discussed. The methods we used may be also very useful for analysis for "table top" x-ray laser performance in ultra-short pulse drive range.

ACKNOWLEDGEMENTS

This work is within the framework of the cooperation between the Royal Society, UK, and the Academia Sinica, P. R. China. One of the author (Li) would like to thank Professor M. H. Key for his hospitality during Li's stay at the Rutherford Appleton Laboratory.

REFERENCES

1 Fedosejevs, R. et al., *Appl. Phys.* **B50**, 79-98(1990).
2 Murnane, M. M. et al., *Science*, **251**, 531-536(1991).
3 Rozmus, R. and Tikhonchuk, V. T., *Phys. Rev.* **A42**, 7401-7412(1990).
4 Landen, O. L. et. al., *Phys. Rev. Lett.* **63**, 1474-1478(1989).
5 Pert, G. J., *J. Opt. Soc. Am.* **B4**, 602-608(1987).
6 Borovskii, A. V. et al., *Sov. J. Quantum Electron* **17**, 1447-1457(1987).
7 Solovev, N. A. and Fedotov, M. A., *Opt. Spectrosc.* **65**, 409-501(1989).
8 Zhang, J. and Key, M. H., *Appl. Phys. Lett.* **74**, 7606-7608(1993).
9 Zhang, J. and Key, M. H., *Appl. Phys.* **B58**, 13-18(1994).
10 Li, Y. and Zhang, J., *J. Phys.* **D27**, 707-713(1994).

Soft X-ray Laser of He-like Aluminum ions

Baifei Shen, Zhizhan Xu and Shensheng Han

Shanghai Institute of Optics and Fine Mechanics, Academia Sinica
P.O.box 800-211, Shanghai, 201800

Abstract. With numerical calculations, the plasma conditions, in which the laser gains of 1s3d $^1D_2$ - 1s2p $^1P_1$ transition of He-like aluminum ions can be produced, have been obtained. The method of radiative cooling used to produce the necessary plasma conditions has also been proposed and simulated.

1. INTRODUCTION

One of the most important objectives of current x-ray laser research is the attainment of signification gain in the wavelength region known as the "water window" (2.33nm-4.37nm) [1,2]. Due to its high quantum efficiency compared with the electron collisional excitation scheme, the recombination scheme has attracted much attention. For He-like aluminum ions investigated in this article, the wavelength of the laser transition 1s3d $^1D_2$ - 1s2p $^1P_1$ is 4.5nm, at the long wavelength edge of the "water window". For He-like ion scheme, owing to similar ionization energy for He-like and H-like ion, when the driving laser is turned off, there are not as many "mother ions" as for the H-like or Li-like ion scheme. However, usually we can obtain 50 per cent of H-like ions and 30 per cent of He-like ions [3] and that's enough because of large transition energy between n=2 and n=1 of He-like ions. in comparison with Li-like ion scheme, the necessary abundance is relatively small for He-like ion scheme, . Large emptying rate of the transition from the lower laser level to ground state is also an attractive advantage.

2. REQUIREMENT TO PLASMA PARAMETER

The necessary plasma parameter for lasing of transition 1s3d $^1D_2$ - 1s2p $^1P_1$ of He-like aluminium is investigated with collisional radiative model. In our model, the population of n=5,6 of He-like ions and the ground state of H-like ions are assumed to satisfy the Saha-Boltzmann equilibrium. In this model, the atomic parameters, such as rate of spontaneous transition, wavelength and oscillator strength are calculated with Hartree-Fock method by using Cowan's code [4] and the other atomic parameters are calculated with ex-

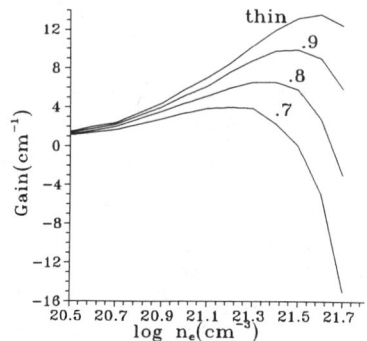

Fig.1 Gains of transition 1s3d $^1D_2$ − 1s2p $^1P_1$ of He−like ions as a function of electron density. The electron temperature T_e=170eV, ion density N_i=2x10^{19}cm^{-3} and escape factor are 1,.9,.8,.7, respectively.

periential formula.

With the model described above, the gain coefficients are calculated by assuming that the electron temperature $T_e = 175eV$, there are 50 per cent of H-like ions and 20 per cent of He-like ions, and the ion density of aluminium is $2 \times 10^{19} cm^{-3}$. It is due to serious opacity we do not assume higher ion density. Fig.1 is the gain coefficients of transition 1s3d $^1D_2$ - 1s2p $^1P_1$ of He-like aluminium ions as a function of electron density. It shows that if the plasma is optically thin, the maximum gain is about 13 cm^{-1} at the electron density $6 \times 10^{21} cm^{-3}$. When opacity is included, the gain coefficients decrease and the most suitable electron density decreases, too.

3. EXPERIMENTAL RESULT OF INVERSION

Experiment was carried out at the Six Beam Laser Facility at SIOFM. The energy of driving laser was 10J, the pulse width was 250ps and the intensity on target surface was about $2 \times 10^{14} W/cm^2$. Fig.2 is one typical spectrum of aluminium heated by the laser described. It shows that there is inversion between level 1s2p and 1s3p of He-like aluminium ions.

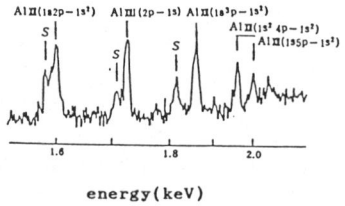

Fig. 2 Microdensitomrter trace of He-like Al ions

4. METHOD OF EXPERIMENT

From the discussion above, we know that the necessary requirement to obtain the gain of transition 1s3d $^1D_2$ - 1s2p $^1P_1$ of He- like Aluminium ions is that the electron density is as high as about $5 \times 10^{21} cm^{-3}$, the electron temperature is lower than about 175eV and at the same time there must be enough H-like ions in plasma. In order to obtain such a plasma, the plasma must be cooled rapidly. Supposing the initial electron temperature are 2000eV or 1200eV, and the lasing electron temperature is 175eV, the cooling time must be shorter than 500ps and 300ps, respectively, as shown by calculation.

There are three possible cooling method: thermal conduction, expansion and radiative. In this article, radiative cooling is investigated. We consider a two -level model consisting of a lower level g and excited level i as the method of and Keane [5]. Electrons are cooled by electron collisional excitation, followed by radiative decay.

The population excited from the lower state is given by

$$N_i = N_g N_e S \qquad (1)$$

Here N_g is the lower level population, N_e the electron density, S is the collisional excitation rate. The population losing by radiative decay can be written

$$N_i' = N_i \frac{A}{A + N_e S exp(\Delta E/T_e) + N_e S'}, \qquad (2)$$

where A is the radiative decay rate, S' is the collisional excitation rate from the excited level. The electron cooling power per electron is given by

$$K_e = N_i' \Delta E / N_e. \qquad (3)$$

Due to the optimum energy level separations X_{opt} for electron temperature $T_e = 175eV$ and electron density $N_e = 10^{21} cm^{-3}$ is about 1.5 [5], copper was selected as radiative cooling material because its character energy level is about 300eV. In order to decrease the influence of opacity, 90 percent of copper and 10 percent of Aluminium was assumed in our calculation. The code CASTOR 2 [6] is used to calculate the variation of plasma parameter. CASTOR 2 was a two-dimensional and point-focused model. In our calculation, the plasma is still at the direction r (rectangular to the direction of pumping laser) so that the code become one-dimensional and planar. The pulse width of driving laser is assumed 110ps, the intensity is $1.1 \times 10^{14}/cm^2$, and the wavelength is $0.53 \mu m$. The step in space is $10 \mu m$ and the step in time is 100fs. With the plasma parameters calculated by CASTOR 2, the laser gain coefficients of transition 1s3d $^1D_2$ - 1s2p $^1P_1$ are calculated with collisional radiative model. Owing to low ion density of He-like aluminium in our calculation, opacity is neglected. The result is shown in Fig.3. The large laser gains appear at the place near to the target surface as expected, the time is about 250ps later, and the maximum gain is above $20 cm^{-1}$.

5.RESULT

It is shown by calculation that the necessary plasma parameters for laser gain of transition $1s3d^1D_2$ - $1s2p^1P_1$ of He-like aluminium are that the electron temperature is lower than 175eV, and the electron density is about $1 - 7 \times 10^{21} cm^{-3}$ and it is possible to obtain larger laser gain of transition $1s3d^1D_2$ - $1s2p^1P_1$ of He-like aluminium by radiative cooling. The method of making use of the mixed target of copper and aluminium is proposed.

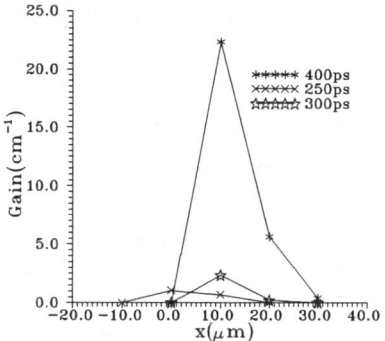

Fig.3 Gains of transition $1s3d\ ^1D_2$–$1s2p\ ^1P_1$ of He-like Al ions as a function of distance from the target surface, at the time 250ps, 300ps, 400ps, respectively.

REFERENCES

1. Rosen,M.D. et al.,Phys.Rev.Lett.,Vol.54,p106,1985
2. Matthews,D.L. et al.,Phys.Rev.Lett.,Vol.54,p110,1985
3. Apruzese,J.P. et al., IEEE Trans. on Plasma Sci.,Vol.16 p529,1988
4. Cowan,R.D.,Theory of atomic structure and spectra(Berkely: University of California Press, 1981)
5. Keane,C. et al.,Phys.Rev.A,Vol.33,p4179,1986
6. Christianen,J.P. et al.,Comput.Commun.,Vol.17,p397,1979

A Numerical Method for Obtaining the Fine Structure of X-Ray Spectra

Ling-qing Zhang, Shen-sheng Han, Chun-hong Jiang, Zhi-zhan Xu, Zheng-quan Zhang, and Lan Sun

Shanghai Institute of Optics and Fine Mechanics,
P.O.Box 800-211, Shanghai 201800, P.R.China

Abstract. A numerical method based on techniques of inverse Fourier convolution and nonlinear least square algorithm, etc. is presented for obtaining high-resolved X-ray spectra of laser plasmas, which can eliminate the line-broadening induced by radiation sources and spectrographs, and can improve the spectral resolving power of spectrographs. The code ESDAP has now been successfully applied to the analysis of spectra from line-shaped Mg and CaF_2 laser plasmas.

I. INTRODUCTION

Data processing techniques are very important in X-ray spectrium data analysis as most laser plasma(LP) x-ray spectra are still recorded with film as detector. In fact, the spectra obtained by spectrographs are usually not the real spectra radiated by the laser plasmas, but mixed with broadenings, distortions and noises from the radiation source, optical system and x-ray films. In this paper we present a procedure which employed two principal techniques of inverse Fourier convolution and nonlinear least square algorithm in data processing with the aim to obtain the real emission and the fine structures of laser-produce plasma. The procedure has now been successfully applied to the analysis of space-resolved spectra from line-shaped Mg laser plasmas recorded by a pinhole crystal spectrograph(PCS)[1]. Data analysis of CaF_2 is now being in progress.

II. THEORY

Any spectral line recorded by a spectrograph can be expressed as a convolution of a lineshape function $G(z)$ and a real spectra[2]

$$F(\lambda) = \int_{-\infty}^{+\infty} G(\Delta\lambda) f(\lambda - \Delta\lambda) d\Delta\lambda \qquad (1)$$

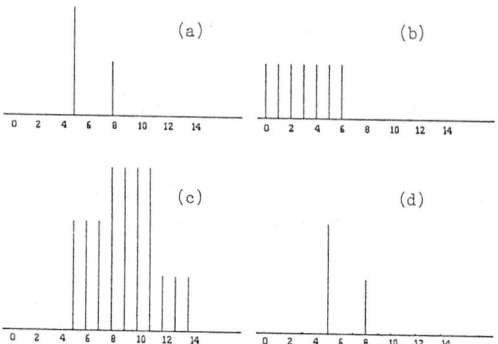

Figure 1: Simulated spectra of inverse Fourier convolution. (a) Real spectra, (b) Lineshape function, (c) Recorded spectrum, (d) Deconvolution spectrum.

where $F(\lambda)$ is a recorded spectrum, $f(\lambda)$ is a real spectrum. $G(\Delta\lambda)$ is a function represent the distortion of unit spectral intensity of the spectrograph. In order to deconvolute $G(\Delta\lambda)$ from $f(\lambda - \Delta\lambda)$, We take the inverse Fourier transform of both sides of Eq.(1), giving

$$I(X) = D(X) \times i(X) \qquad (2)$$

where $I(X) = F^{-1}[F(\lambda)]$, $i(X) = F^{-1}[f(\lambda)]$, $D(X) = F^{-1}[G(\lambda)]$ with $G(\Delta\lambda)$ being known, then

$$f(\lambda) = F[i(X)] = F\left[\frac{I(X)}{D(X)}\right] \qquad (3)$$

$G(\Delta\lambda)$ can be precisely known from spectrographic geometric parameters, if not, $G(\Delta\lambda)$ can be estimated empirically.

Data function F should multiply a rectangle window function W(i), since the data points is limited in number in practical data processing. With data point number N

$$f_i = F_i W(i)$$
$$W(i) = 1, \quad i = 0, 1, \cdots, N-1$$

Therefore, carrying FFT out of f_i is practically implementing convolution of data function F and window function W. The effect of window function on data processing is not only relevant to its width, but also its shape. Then Eq.(3) should modified as

$$f(\lambda) = F\left[W(X) \cdot \frac{I(X)}{D(X)}\right] \qquad (4)$$

The real spectra obtained through data processing mentioned above sometimes have partially overlapping spectral lines. In order to obtain fine spec-

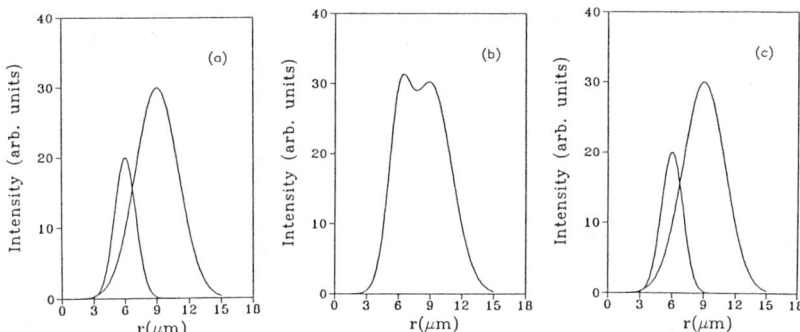

Figure 2: Simulated spectra of nonlinear least square fitting. (a) Real spectra, (b) Recorded spectrum, (c)Fitted spectrum.

tral structures, we adopted a method of Gauss-Newton or Marquardt algorithm[3] to fit the experimental data with a proper theoretical fitting function

$$f(\lambda) = f_B(\lambda) + \sum_{i=1}^{m} f_p(\lambda_{io}) \qquad (5)$$

Where f_p is the lineshape function of spectral lines, λ_{io} is the center position of lines, f_B is the background function of spectra. Shown in Fig.1 and Fig.2 are simulated spectra of inverse Fourier convolution and nonlinear least square fitting, respectively. The flow chart of the procedure is given below

III. APPLIATIONS

With $G(\Delta\lambda)$ precisely deduced[4] and Hamming window function used, we have the spectrum of Mg laser plasma resolved, the result is shown in Fig.3. While fitting the overlapping lines and the continuum edge in Fig.3, Gauss function was chosen as the line shape function[5][6].

Shown in Fig.4 and Fig.5 are the fitting results of the two parts of the spectrum in Fig.3. One part is the Ly-α line and its satellites of MgXII (polynomial is chosen as the background function), the other part is the continuum edge(only part of the edge was recorded, exponential is chosen as the background function). It is obvious in Fig.4 and Fig.5 that data processing can conveniently separate the lines from the background in the spectrum. The logarithmic slope of background curve in Fig.5(b) gives the value of the electron temperature T_e.

Two diagnostic approaches of electron temperature were adopted using the fitting results in order to prove the applicability of the spectral data processing method. One is the determination of T_e from the ratio of the intensities of the resonance and satellites lines of H-like magnesium ions, the other is from the logarithmic slope of the continuum edge. The space-resolved electron temperatures are shown in Fig.6 (they are slightly different

PROCEDURE FLOW CHART

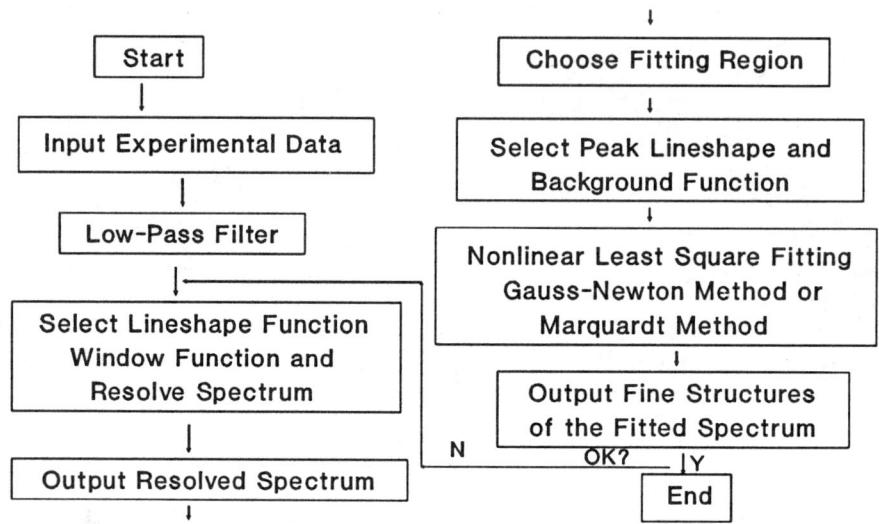

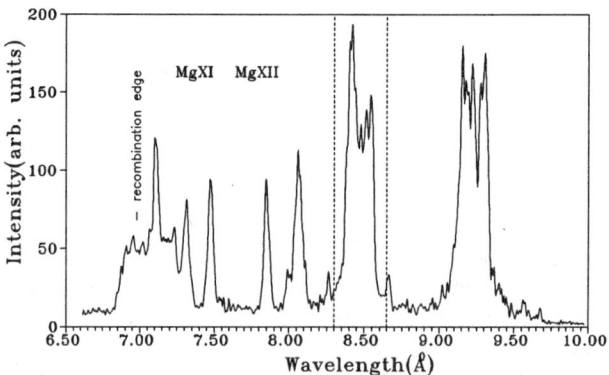

Figure 3: Resolved X-ray spectrum of Mg laser plasma.

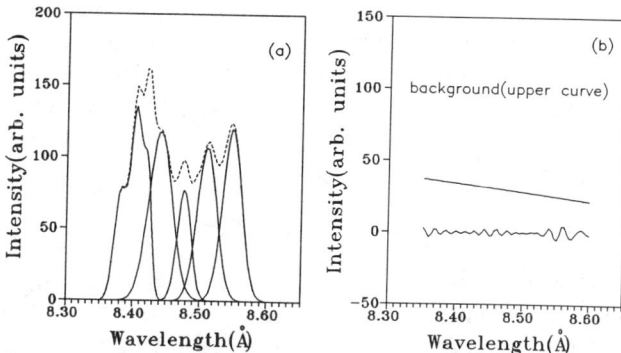

Figure 4: Fitting results of spectrum in Fig.3. (a) Fitted Ly-α line and its satellites, (b) Resolved background and fitting errors.

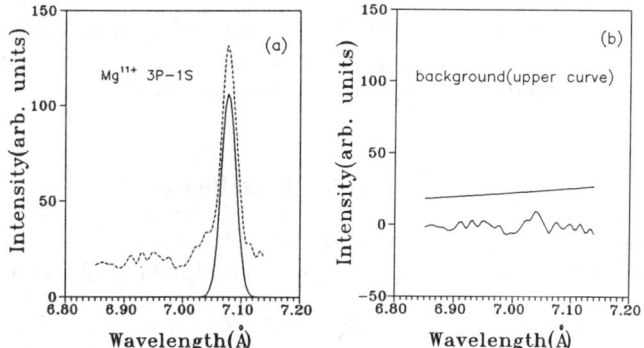

Figure 5: Fitting results of continuum edge in Fig.3. (a) Experimental line(3p-1s) and resolved peak, (b) Resolved background and fitting errors.

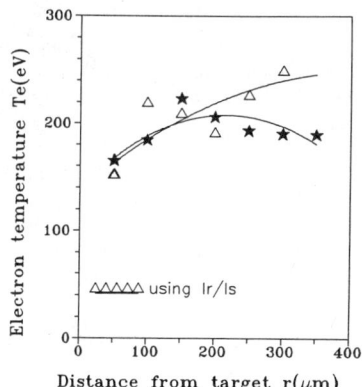

Figure 6: Electron temperature profiles obtained by method of I_r/I_s of MgXII and the slope of the continuum edge using the fitting results.

when far away from the target surface in the normal direction, the reason is that the satellite intensities of Ly-α line decrease gradually to zero while away from the target, so the error cause by the first approach will become significant). The similar results was obtained by V. A. Boiko et al in the spot-focused Mg laser plasmas[7]. The data processing of CaF_2 spectra is now being in progress.

IV. CONCLUSION

Two general techniques of inverse Fourier convolution and nonlinear least square algorithm were employed in the data processing of X-ray spectra radiated by laser-produced plasmas. The procedure-ESDAP which bases on the two techniques has two main functions. One is to get the real spectra by means of eliminating broadenings and distortions induced by radiation sources and spectrographs, the other is that it can resolve overlapped spectral lines, this equal to improve the resolving power of spectrographs through spectral data processing. The key point is how to precisely structure the fitting functions fit to different spectral lines when applied it to other wave length region.

ACKNOWLEDGMENTS

The authors gratefully acknowledge the LF12 Laser Facility Operation Group for their efforts, and Dr. Rong-qing Chen for helpful discussions.

REFERENCES

1. Han S. S. et al, 1992, Inst. Phys. Conf. Ser., No 125: Section 7, 383
2. Kauppinen J. K. et al,1981, Applied Spectroscopy, 35(3), 271
3. Marquardt D. W., 1963, J. Soc. Ind. Appl. Math., 11(2), 431
4. Zhang L. Q. et al, 1994, ACTA OPTICA SINICA, No.7
5. Espen P. VAN et al, 1977, Nuclear Instruments and Methods, 142, 243
6. Statham P. J., 1976, X-Ray Spectrometry, 5, 16
7. Boiko V. A. et al, 1979, J. Phys. B: Atom. Molec. Phys., 12(11), 1889

Possible approaches to the Recombination X-Ray Lasers with Large GL Value

Shen-sheng Han, Bai-fei shen, Zhi-zhan Xu, Zheng-quan Zhang, Hua-guo Teng, Wen-qi Zhang and Ling-qing Zhang

Shanghai Institute of Optics and Fine Mechanics
P.O.Box 800-211, Shanghai, 201800, P.R.China

Abstract. Two possible designs are considered in this paper for increasing the GL value of recombining X-ray lasers. A "fibers target" structure was proposed to get a rapid 3-dimensional adiabatic expanding at the initial state followed by a slower 2-dimensional plasma expansion. A "hybrid model" was used to calculate the plasma behavior of the target and the results are presented. Theoretical results about X-ray pumped recombination X-ray laser are also reported.

I. INTRODUCTION

Recombination X-ray lasers with wavelength near the "water window" range have been successfully demonstrated[1][2]. But regretfully, unlike the electron-collisional excitation X-ray lasers, the saturated output of recombining X-ray laser has not been achieved though the predicted pumping efficiency is much higher than that of collisional pumping mechanism.

For recombination pumping mechanism, the minimum electron temperature (and maximum pumping) during lasing is usually determined by how much cooling can be accomplished after the plasma is first heated to get the initial ionization state. In most gain experiments to date, the method of such cooling, at least for laser-produced plasmas, is adiabatic expansion. It is well known that the heating and ionization should be completed before the plasma has expanded freely to a significant degree, so short-pulsed lasers are desirable and that the higher the dimension of free expansion, the more rapid the cooling. Because of the strong dependence of collisional recombination rate coefficient on the electron density ($\propto N_e^3$) and temperature ($\propto T_e^{-2}$), the pumping is particularly sensitive to values of these variables, and the proper operating conditions are limited in a quite small region of plasma parameter space. In an adiabatic expanding cooling scheme, the faster the

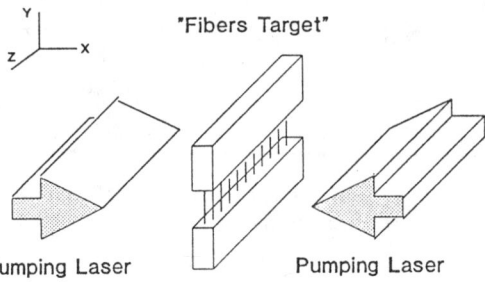

Figure 1: Schematic structure and pumping mode of "fibers target"

cooling, the narrower the "gain window" in time and space, which will result in a higher requirement for irradiation uniformity and maybe a shorter target length that could be used in the experiments. Short laser pulse (< 1ps) is weakly absorbed by the target and ionizes very little material, and the refraction may also cause a significant decrease of gain length in this case. So in conventional free adiabatic expansion cooling scheme, the experimental requirements for high gain and long target length are somewhat contradictory, this makes the achievement of large GL value in recombination X-ray lasers very difficult.

Two possible approaches toward the recombination X-ray laser with large GL value are considered in this paper. In Sec.II, a "fibers target" structure is proposed. The adiabatic expansion is controlled by proper target design to get a rapid 3-dimensional adiabatic expansion at the beginning in order to get a large gain, and a slower 2-dimensional plasma expansion after it reaches the optimum parameter space to increase the "gain window" in time and space. In Sec.III, the quasi-cw X-ray pumping He-like Al 3d-2p (λ=4.5nm)recombination X-ray laser scheme will be discussed, where the initial recombination ions will be produced by photoionization of a relatively cold plasma, and the recombination can proceed at a high rate without forced cooling. This makes the scheme possible to get a large gain and long target length at the same time. The summary and the conclusion will be presented in Sec.IV.

II. CONTROLLED EXPANSION LI-LIKE AL RECOMBINATION X-RAY LASER

"Fibers target" structure and the pumping mode are showed in Fig.1. The target possess a rapid 3-dimensional expansion after the initial shortpulse laser irradiation, and the distance between two adjacent fibers is determined by the requirement that the expansion approaches 2-dimensional when the plasma reaches the optimum parameter space. A hybrid model[3]

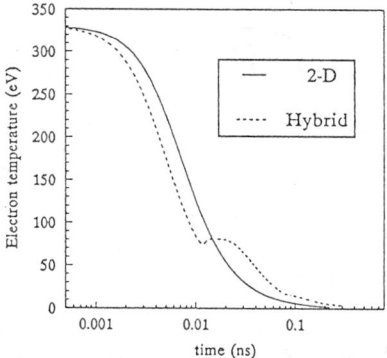

Figure 2: Time history of electron temperature of "fibers target" and 2-D adiabatic expansion with the same initial values. laser pulse duration: 30ps; heating rate: $1.2 \times 10^{25} ergg^{-1} sec^{-1}$; width of line focus: $3\mu m$; distance between two adjacent fiber: $6\mu m$.

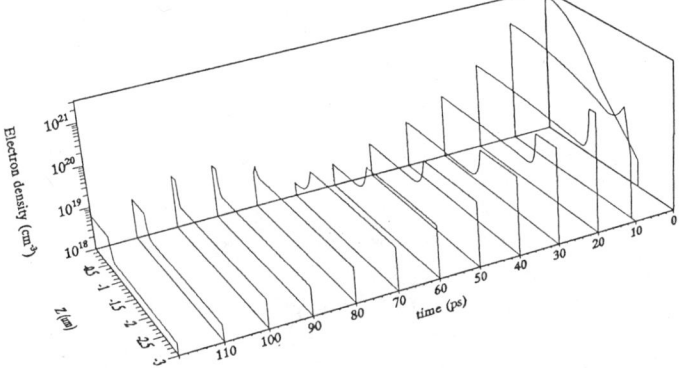

Figure 3: Time history of electron density profile distributed along the central Z-axis between two adjacent fibers.(Only half of the profile is shown in the figure)

of compressible fluid flow involving a separation of variables into a similarity solution in two dimensions and one-dimensional flow in the third was used to simulate the fluid dynamic behavior of the target expansion. The plasma possessing a similarity density profile in the (X,Y) plane, but varying in Z; and a uniform temperature over all fluid cells. The initial parameters after the laser-target interaction were estimated by the asymptotic 2-dimensional similarity formula of constant heating[4] and the mass conservation. The gain is obtained by a collisional-radiative post-processor under the quasi-steady-state (QSS) approximation, with 48 lithium-like levels and the helium-like ground state.

The time history of electron temperature and electron density distribution along the central Z axis between two adjacent fibers are presented in Fig.2 and Fig.3, where the width of the line focus is 3.0 μm. Compared with 2-dimensional "free" adiabatic expansion with the same initial values. They

238 Recombination X-Ray Lasers

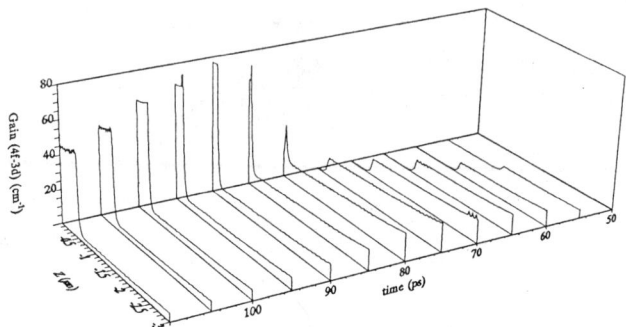

Figure 4: Time history of Al^{10+} 4f-3d gain distribution at the center of X-Y plane along Z axis.

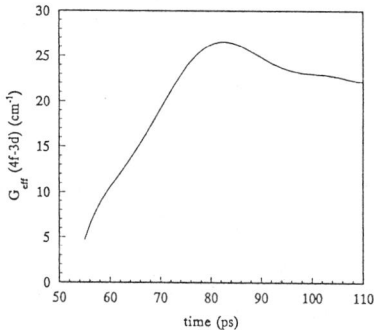

Figure 5: Time history of Al^{10+} 4f-3d gain at the center of X-Y plane

clearly demonstrate a more rapid cooling rate at the beginning for the "fibers target" structure. In Fig.4, we present the time history of Lithium-like Aluminum 4f-3d gain distribution at the center of X-Y plane along the Z axis. Also, the time history of "effective" gain deduced from

$$\frac{I_{st}}{I_{sp}} = \frac{(e^{G_{eff}L} - 1)^{3/2}}{G_{eff}L(G_{eff}Le^{G_{eff}L})^{1/2}} \tag{1}$$

is shown in Fig.5, where I_{st} and I_{sp} are stimulated and spontaneous emission respectively. Because there is no refraction and radial displacement at the center of X-Y plane, the simulation results show that it is possible to get a large GL value with this target structure. The experiment on the stagnation and interpenetration of laser-created colliding plasmas[5] also shown that better uniformity along Z-axis is possible in this controlled expanding plasma medium.

III. X-RAY PUMPED HELIUM-LIKE ALUMINUM 3d-2p RECOMBINATION X-RAY LASER

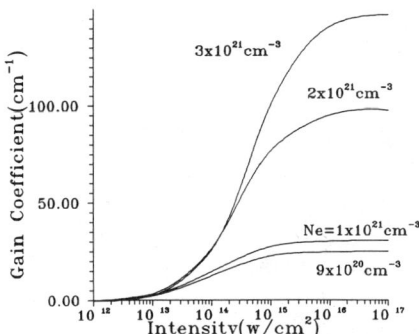

Figure 6: Electron density dependence of 3d-2p gain for He-like Al ions as a function of the incident X-ray intensity.

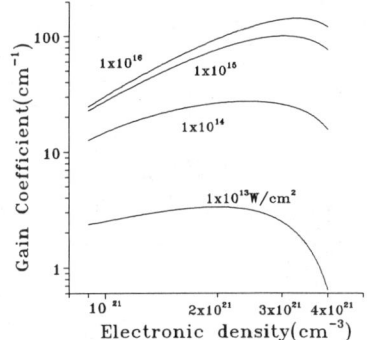

Figure 7: X-ray intensity dependence of 3d-2p gains for He-like Al ions as a function of the electron density at T_e=175eV.

A collisional-radiative model was used to investigate the atomic process in the laser produced plasma pumped by an intense X-ray beam. The radiation field effect is modeled by including photoionization and stimulated recombination. The ionization stages of H-like and He-like ions are treated and the excited states of He-like ions are included. The atomic processes included in the model are: spontaneous radiative transition; collisional excitation and deexcitation; collisional ionization and three-body recombination; photoionization from the ground state of He-like ions; spontaneous and stimulated recombination.

The 1s3d-1s2p transition of Al^{11+} was calculated with the model, suppose $T_i = T_e = 175$ eV and an escape factor for $1s^2 - 1s2p$ of 0.7 to include the opacity effect. The energy of pumping X-ray is supposed to be equal to 2.1 keV and the multi-photo ionization and the reduction of ionization energy caused by electron density has not been included.

The electron density dependence of the gain is shown in Fig.6, also the gain is shown in Fig.7 as a function of incident X-ray pumping intensity.

When the incident X-ray intensity is higher than $1 \times 10^{13} W/cm^2$, there are apparent 3d-2p gain and the maximum gain is at the electron density $N_e \simeq 3 \times 10^{21} cm^{-3}$.

According to the quasi-stationary model for determination of ablation parameters in soft-X-ray-driven low-to-medium-Z plasma ablation[6], the electron temperature and density of the plasma medium could be achieved by a blackbody radiation pumping source with the flux intensity of about $3 \times 10^{13} W/cm^2$, which according to Ref.7, corresponding to about $2 \times 10^4 J$ laser pumping energy for a 2cm-long cylindrical hohlraum heated by two ring irradiation sources. From this estimation, it seems that the X-ray pumped He-like Al X-ray laser is very difficult to be realized by soft-X-ray-driven ablation, a two-step pumping mode is necessary where the first pulse is to produce the plasma medium with proper electron temperature and density, and the second to generate a intense flash X-ray flux in the proper wavelength range to enhance the abundance of H-like ions.

IV. SUMMARY

Two possible approaches toward saturated output recombination X-ray lasers are discussed in this paper. The results show that, with "fibers target" structure, a table-top saturated output Li-like Al recombination X-ray laser is possible, and there is also possibility of "water window" He-like Al recombination X-ray laser pumped by flash X-ray source from intense short pulse laser produced plasmas.

ACKNOWLEDGMENTS

The author gratefully acknowledge Dr. Wei Yu for the enlightening discussions. This work was supported by the National High Technology Program of China.

REFERENCES

1. Zhi-zhan Xu, Zheng-quan Zhang, et al; "Recent progress on short wavelength recombination X-ray laser research at SIOFM", This colloquium
2. H.Daido, et al; X-ray Laers 1992, Inst. Phys. Conf. Ser. 1992, N0.125; Section 7, 111
3. G.J.Pert, J. Fluid Mech., Vol.131, 401 (1983)
4. James H.Hunter, Jr; Richard A London; Phys. Fluids, Vol.31, No.10, 3102 (1988)
5. S.M.Pollaine, R.L.Berger and C.J.Keane, Phys. Fluids B, Vol.4, No.4, 989 (1992)
6. T.Endo, H,Shiraga and Y.Kato, Phys. Rev. A, Vol.42, No.2, 918 (1990)
7. G.D.Tsakiris, Phys. Fluids B, Vol.4, No.4, 992 (1992)

Z-dependent Spatial Behavior of 4d-3p and 4f-3d Transitions of Li-like Ions in Laser-produced Plasmas and the Electron Density Measurement

Ru-xin Li, Bai-fei Shen, Zhi-zhan Xu, Pin-zhong Fan, Zheng-quan Zhang, Xiao-fang Wang, Pei-xiang Lu and Shen-sheng Han

Shanghai Institute of Optics and Fine Mechanics, Academia Sinica, P.O.Box.800-211, Shanghai, 201800, P.R.China

Abstract. We presented herein an observation on the difference of spatial behavior between 4d-3p and 4f-3d transitions of lithium-like ions in laser-produced plasma, by a flat-field grating spectrograph with a spatial resolution of about 35μm. The Z-dependent and space-dependent characteristics of intensities ratio of 4d-3p and 4f-3d transitions of Li-like ions was found and a new method for the electron density measurement was proposed.

INTRODUCTION

Although great progress has been achieved in the study of recombination X-ray laser, some problems still remain unresolved in understanding the lasing mechanism. The nd-3p and nf-3d (n=4,5,6) transitions of Lithium-like ions in line-focused laser-produced plasma are the most important lasing transitions of Lithium-like scheme recombination X-ray laser(1,2). Population inversions obtained around the nf-3d transitions are due to the strength of the 3d-2p decay channel. Although there is evidence for the existence of considerable gain on the nd-3p lines of Li-like ions, this is not understood and cannot be reproduced theoretically. Even if the assumption is made that collisions instantaneously equilibrate the 3d and 3p levels, thus allowing the 3p level an additional decay channel, no significant gain is predicted on these transitions(3). Then possible additional populating mechanism for the nd level (upper level of lasing) such as photo-excitation, must be considered. The studies on the difference between nd-3p and nf-3d (n=4,5,6) transitions of Lithium-like ions in laser-produced plasma are of great importance, because they will offer information for better understanding of the lasing scheme.

In this paper, we presented the preliminary results on the difference of spatial behavior between 4d-3p and 4f-3d transitions of lithium-like ions in laser-produced plasma, by a flat-field grating spectrograph with a spatial resolution of about 35μm. The Z-dependent and space-dependent characteristics of intensities ratio of 4d-3p and 4f-3d transitions of Li-like ions was found and a new method of the electron density measurement was proposed. We found that the line intensity ratio is very useful to determine the plasma characteristics.

Z-DEPENDENT CHARACTERISTICS

In order to find the Z-dependence, Si, KCl, CaF$_2$ and Ti polished slab targets were adopted, and the laser intensities on target surface were in the range from 3×10^{12} to 4.3×10^{13} Wcm^{-2}.

Experimental results showed that the ratio of intensities of these line pairs (i.e. nf-3d and nd-3p transitions of Li-like ions) changed with the distance from target surface and these characteristics become more apparent for higher atomic number, and the ratio for 4f-3d and 4d-3p were most sensitive among those line pairs. Plasma electron density effect can mainly account for these phenomena. At those places far away from the target surface, the population of 4f and 4d levels differs far from Saha-Boltzmann distribution because of low electron density. And because the requirement of electron density scales with atomic number by the form of Z^7, the Z-dependence characteristics of the ratio of the intensities of these two lines can be understood.

ELECTRON DENSITY DIAGNOSIS

If the density is high enough, the small energy gap between the 4d and 4f levels allows a much faster electron collision between both levels than radiative decay, thus the population densities of both levels are governed by Boltzmann distribution, and the theoretical intensity ratio of the two lines I_{4f-3d} and I_{4d-3p} is given by,

$$\frac{I_{4f-3d}}{I_{4d-3p}} = \frac{\lambda^3_{4d-3p}}{\lambda^3_{4f-3d}} \frac{g_{3d} f_{4f-3d}}{g_{3p} f_{4d-3p}} \exp\left(\frac{E_{4d} - E_{4f}}{KT_e}\right) \qquad (1)$$

where I, λ, g, f, and E are total intensity(integrated over the line profile), wavelength, statistical weight of the lower state of the line, absorption oscillator strength and excitation energy for the upper level of the transition, respectively. As can be seen, no density-dependence of intensity ratio exists, but a apparent dependence of Te (electron temperature) can be found. However in the region of moderate density region (e.g. 10^{17}-10^{19} cm^{-3} for Si^{11+} ion and 10^{18}-10^{20} cm^{-3} for Ca^{17+} ion), the ratio of the intensities is sensitive to the electron density, thus an approach for density diagnosis is available. The seeds of high Z element in the target of concerned element will help to measuring the density of high density region. Fig.1 shows the curves of the ratio of Si^{11+}4f-3d and 4d-3p line pair as a function of electron density, under two values of electron temperature, calculated with a C-R model provided that the abundance is large enough and optical thin approximation. A strongly Ne-dependent behaviour of line intensities ratio for the density region around 10^{18} cm^{-3} can be seen in this figure. The details of this method can be found in Ref.4.

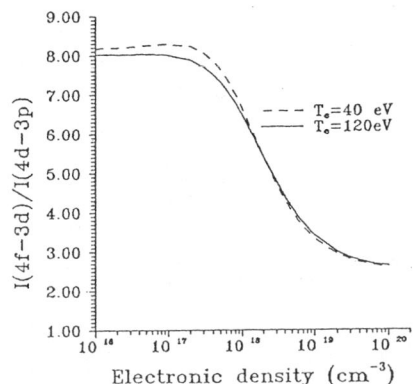 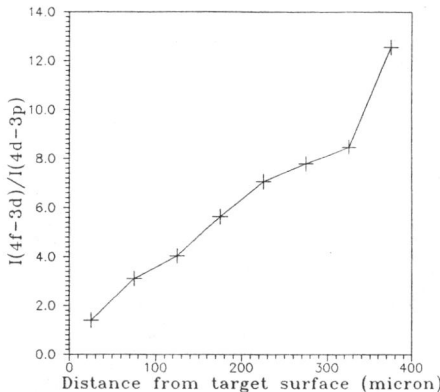

FIGURE 1. Curves of intensities ratio of Si^{11+} 4f-3d and 4d-3p line pair as a function of electron density.

FIGURE 2. Spatial distribution of the axial intensities ratio of Ti^{19+} 4f-3d and 4d-3p line pair.

AXIAL INTENSITIES RATIO OF Nf-3d AND Nd-3p TRANSITIONS

Fig.2 shows the spatial distribution of the axial intensities ratio of 4f-3d and 4d-3p transitions of Ti^{19+} ion for a 10mm long plasma, corresponding laser intensity was $4.3 \times 10^{13} W/cm^2$ with 110ps duration. The value of the intensities ratio of this line pair according to the equilibrium relation (Eq.1) is 4.4, then it can be approximately deduced from this curve that the region from target surface to the position of about 125μm is an absorption region for 4f-3d transition because the values of intensity ratio are lower than the value determined by Eq.1 in this region. This result is quite coincide with that obtained by comparing the intensities for different plasma lengths (5). It can also be seen that the value of ratio increases rapidly with distance from target surface and it is mainly caused by lower and lower density.

For better understanding the lasing of Si^{11+} 5f-3d and 5d-3p transitions, we considered the intensities ratio of this line pair. Fig.3 shows the curves of the intensities ratio of Si^{11+} 5f-3d and 5d-3p line pair as a function of electron density, under two values of electron temperature, calculated with the method mentioned above. Fig.4 shows the spatial distribution of the axial intensities ratio of 5f-3d and 5d-3p transitions of Si^{11+} ion for a 10mm long plasma, corresponding laser intensity was $6.5 \times 10^{12} W/cm^2$ with 116ps duration. Also shown in this graph is the electron density distribution obtained by measuring the Stark broadening of 5f-3d transition for case of dense region and by comparing the experimental values of intensities ratio with the calculated curves in Fig.3 for the case of less dense region. An absorption region near the target surface can also be found and an atomic physics model considering opacity, using the density data obtained from stark broadening can interpret this result. But those low values of ratio in the gain region around 200μm (5) had not been understood yet, and the consideration on abundance change and additional population of 5d level in modeling is of great importance and we will present the details elsewhere. Comparison of the distribution of axial intensities ratio from different plasma length is still being made.

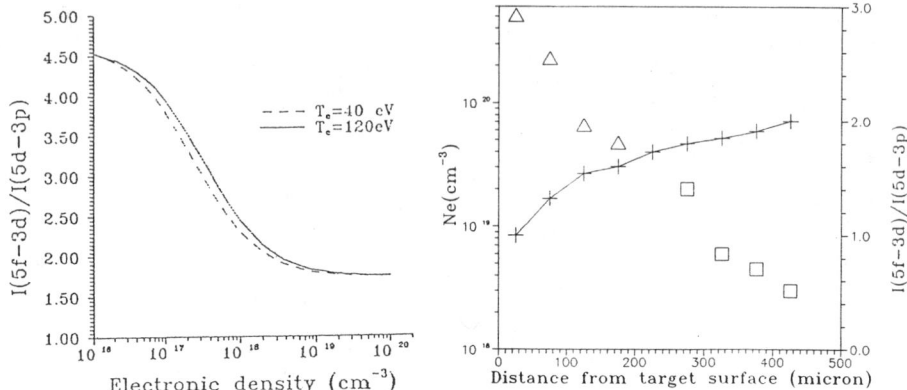

FIGURE 3. Curves of intensities ratio of Si^{11+} 5f-3d and 5d-3p line pair as a function of electron density.

FIGURE 4. Spatial distribution of the axial intensities ratio of Si^{11+} 5f-3d and 5d-3p line pair (+) and the electron density(▲ and □).

CONCLUSION

The Z-dependence and space-dependence characteristics of the ratio of the intensities of 4f-3d and 4d-3p line pairs were found and they can be interpreted by electron density effect. Based on the density effect on the intensities ratio of 4f-3d and 4d-3p lines, an approach for density measurement is proposed.

The spatial distribution of the axial intensities ratio of nf-3d and nd-3p transitions of Li-like ions in line-shaped plasmas were studied. And we found that these intensities ratio are very useful to determine the plasma characteristics and helpful to check the modeling.

REFERENCES

1. Z. Z. Xu, Z. Q. Zhang, P. Z. Fan, S. S. Chen, L. H. Lin, P. X. Lu, X. P. Feng, X. F. Wang, J. Z. Zhou, and A. D. Qian, Appl. Phys. **B50**, 147-156 (1990); Z. Z. Xu, P. Z. Fan, L. H. Lin, Y. L. Li, X. F. Wang, P. X. Lu, R. X. Li, S. S. Han, L. Sun, A. D. Qian, B. F. Shen, Z. M. Jing, Z. Q. Zhang, and J. Z. Zhou, Appl. Phys. Lett., **63**, 1023-1025 (1993)
2. see, for example, P. Jaegle, G. Jamelot, A Carillon, A. Klisnick, A. Sureau, and H. Guennou, J.Opt. Soc. Am. **B4**, 563-574 (1987)
3. P.B.Holden and G.J.Pert, J.Phys., **B25**, 3085-3092(1992)
4. B.F.Shen, Z.Z.Xu, and R.X.Li, Chinese Science Bulletin, 1994, (to be published)
5. Z.Z. Xu, Z.Q. Zhang, P.Z. Fan, R X. Li, X.F. Wang, P.X. Lu, S.S. Han, L.Q. Zhang, B.F. Shen, W.Q. Zhang, X.P. Feng, A.D. Qian, and H.Z. Xiang, this colloquium.

Simulation of X-ray Laser for Li-like Silicon

Baifei Shen, Xiaofang Wang, Zhizhan Xu and Huaguo Teng

Shanghai Institute of Optics and Fine Mechanics,Academia,Sinica
P.O.Box 800-211,Shanghai 201800,P.R.China

Abstract. The simulation of X-ray laser has been performed for Li-like silicon, pumped by the 116ps laser pulse, with the intensity of $6.5 \times 10^{12} W/cm^2$ and the wavelength of $1.06 \mu m$. The populations of energy level $4f$ and the gains of transition $4f - 3d$ of Li-like silicon ions have been obtained as a function of time. The results have been explained and compared with experiment.

1.INTRODUCTION

In recombination x-ray lasers, the Li-like ion scheme is one of the most important methods. It has large efficiency and can easily scale to "water window" region. At initial stage, the driving lasers of long pulse width($\sim$ 1ns), were used to heat the plasma. However, it's difficult to scale to short wavelength with the laser of such long pulse. The laser of relatively shorter pulse (1ps-100ps) is now used to pump the laser medium around the world. We have successfully completed the recombination x-ray laser experiment of Li-like silicon ions. The pulse width of the pumping laser is about 116ps. The time- and space- resolved line intensity and gain coefficients have been measured. Here the simulation of the experiment is performed, and the comparison of result between the simulation and experiment have also been made.

2.MODEL AND METHOD

The laser-plasma interaction is simulated with the code CASTOR 2 [1]. The CASTOR 2 used to be two dimensional and point-focused. In our calculation, the motion along the direction r is not permitted, so the code becomes to be one dimensional and planar. The code TRIP [2] and our own code calculating the distribution of ionic stage are included in the code CASTOR 2. The pulse width of pumping laser is 116ps and intensity is $6.5 \times 10^{12} W/cm^2$.

With the plasma parameter calculated with the code CASTOR 2 and the atomic parameter calculated with R.D Cowan's code [3], the population of $1s^2 4f$ and the gain coefficients of $1s^2 4f - 1s^2 3d$ of Li-like silicon are calculated with collisional-radiative model.

3. RESULT AND DISCUSSION

Fig.1 is population of energy level 4f of Li-like silicon ions, as a function of time at the place $60\mu m$, $120\mu m$ and $200\mu m$ far from the target surface, respectively. Because the plasma fluid moves to the place near the target surface earlier, the population of level 4f near the target surface appears earlier than the population at the place farther from the target surface. And, because the electron density is large at the place near the target surface and the recombination rate is large, too, so the population of level 4f at the place near to the target surface is large and disappears earlier.

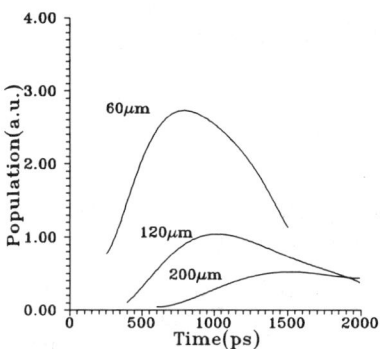

Fig.1 Population of transition 4f–3d of Li-like Si ions, as a function of time at place $60\mu m$, 120μ and $200\mu m$ far from the target surface, respectively.

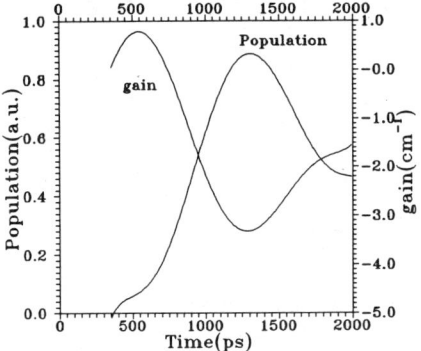

Fig.2 Population of 4f and gain of 4f–3d of Li-like Si ions as a function of time at the place $160\mu m$ from the target surface.

Fig.2 is population of level 4f and gain of transition 4f-3d of Li-like silicon ions as a function of time at the place $160\mu m$ far from the target surface. It shows that the largest gain coefficients of transition 4f-3d is at about 500ps and the largest population of 4f at about 1ns, which can be understand as follows. At earlier time, when the electron density is suitable and ion temperature is low, the Doppler broadening is small, there are relatively large gain, in spite of that the electronic temperature is not very low. While the electron density increases continually, there are still many He-like ions so that the population of 4f increases continually, too. Then, however, the populations of 4f and 3d come to Boltzmann equilibrium, the gain, therefore, becomes small or even negative. At the place about $300\mu m$ or larger than $300\mu m$ far from the target surface, the delay time between maximum gain and maximum population is short.

Fig.3 gives gains of transition 4f-3d of Li-like silicon ions as a function of time at the place $160\mu m$, $200\mu m$, $240\mu m$, $280\mu m$, $320\mu m$ and $380\mu m$ far from the target surface. It shows that at the place near to the target surface, the gain time is short and there are a long period of time in which the gain is negative, there should be no time integrated gain. At the place $300\mu m$ or ever farther away from the target surface, there are a long period of time the gain is positive, so there are time integrated gain.

Population of level 5f and gain of transition 5f-3d have also been calculated. The profiles of population and gain are similar to the transition 4f-3d, but the gains are smaller than the transition 4f-3d.

4. EXPERIMENT

The experiment was carried out at LF12 Laser Facility of SIOFM. The wavelength of the driving laser is $1.06\mu m$ and the pulse width is 116ps, the average laser intensity on target was about $6.5 \times 10^{12} W/cm^2$. Fig.4 is the gain of transition 4f-3d as a function of distance from the target surface [4]. The result is obtained by assuming the transition 4d-3p is optically thin. The result accords with the theoretical results shown in fig.3.

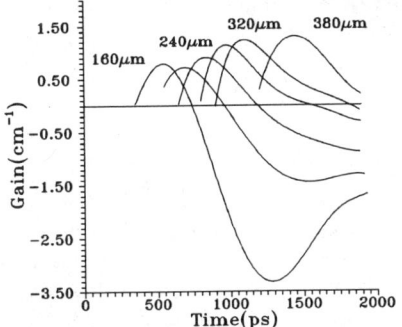
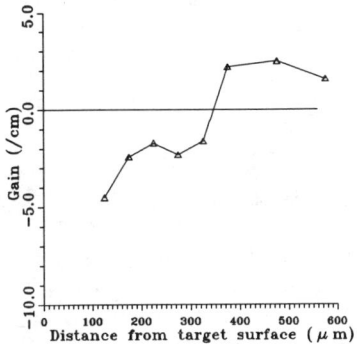

Fig.3 Gains of transition 4f-3d of Li-like Si ions, as a function of time at place $160\mu m, 200\mu m, 240\mu m, 280\mu m$, $320\mu m$ and $380\mu m$ far from the target surface, respectively.

Fig.4 Experimental gains of transition 4f-3d of silicon ions, as a function of distance from the target surface.

REFERENCES

1. Christianen,J.P. et al. 1979 Comput.Phys.Commun.17,397.
2. Cowan,R.D. 1981 Theory of atomoc structure and spectra(Berkely: University of California Press).
3. Magill,J. 1978 Comput.Phys.Commun.16,129.
4. Wang,X. et al.,in this proceeding.

Investigation on the homogeneity of line plasmas created with the aid of a cylindrical lens array

A. Glinz and J.E. Balmer

Institute of Applied Physics, University of Berne, CH-3012 Berne, Switzerland

Abstract: The homogeneity of plasma columns produced by line-focused laser radiation has been investigated. An array of cylindrical lenses in conjunction with a spherical lens was used to produce a line focus with a uniformity of the irradiance along the line of better then +/- 3 %. The uniformity of the resulting plasma radiation in the keV range was studied for copper and aluminum plasmas created with both 700-psec and 100-psec shots of a Nd:glass laser (λ = 1054 nm) on slab targets at an irradiance of about $1 \cdot 10^{12}$ W/cm^2 and $3 \cdot 10^{12}$ W/cm^2, respectively.

1. Introduction

Long narrow plasma columns are a basic requirement for X-ray laser studies. It was shown that uniform plasma conditions are crucial for the achievement of maximum gain, particularly in recombination-pumped schemes [1]. Plasma columns can be produced, for example, by focusing a high-power laser beam into a line. The uniformity of the plasma column depends on the uniformity of the irradiance along the line focus because plasma production depends on the irradiance. However a high uniformity of the irradiance along the line focus does not necessarily result in plasma conditions of the same uniformity. Plasma instabilities such as self-focusing can degrade the initial degree of uniformity [2]. A line focus is usually achieved by utilizing a pair of crossed cylindrical lenses, a spherical lens in conjunction with a cylindrical lens or a spherical lens together with an off-axis spherical mirror. A shortcoming of all these methods is that the intensity along the line focus is a continuous mapping of the intensity distribution of the pump laser beam.

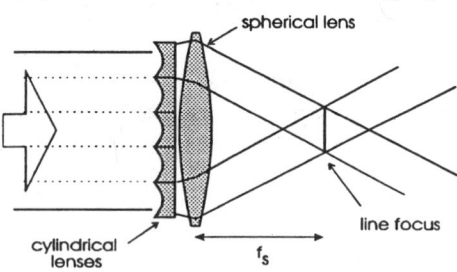

Fig. 1. *Schematic of the focusing configuration of the cylindrical lens array (CLA)*

In this way a circular beam of uniform intensity can not been brought to a line focus of uniform intensity. One possibility to solve this problem is to use a cylindrical lens array (CLA) as the focusing optics [3]. As shown in fig. 1, it consists of an array of cylindrical lenses in conjunction with a spherical lens. The CLA splits the incident pump laser beam into as many parts as there are cylindrical lens elements in the array. Each part of the split beam is then separately focused to a line and all these individual line foci overlap at the same place, forming a common line focus. Due to the superposition of the individual beamlets the intensity variations of the incident beam are averaged out leading to a smooth line focus. The goal of the work presented here is to investigate the homogeneity of copper and aluminum plasmas, created by 700-psec and 100-psec laser pulses (λ = 1054 nm), focused to a line of known uniformity. Both elements are known to have XUV lasing transitions in a recombination scheme

2. Experimental setup

The incident pump laser beam (diameter: 42 mm) of a Nd:glass laser system (λ = 1054 nm) is brought to a line focus by means of a CLA in combination with a spherical lens (f_s = 100 mm). The cylindrical lens array consists of 5 cylindrical lenses, each of them 9 mm wide and with a focal length f_c = -80 mm. According to geometrical optics the length L of the line focus is given by

$$L = d \cdot \frac{f_s}{f_c}$$

where d is the width of a cylindrical lens element. With the parameters given above, our line focus is calculated to be 11 mm long. The dimensions and the homogeneity of the line focus were measured by focusing the attenuated laser beam directly onto the chip of a CCD camera. The spatial resolution of this measurement is given by the pixel size of 23 x 23 μm^2. Both copper and aluminum slab targets were irradiated with 100 psec/1.5 J and 700 psec/4.7 J laser pulses (λ = 1054 nm), line focused to approximately $3 \cdot 10^{12}$ W/cm^2 and $1 \cdot 10^{12}$ W/cm^2, respectively. The plasma radiation was imaged with a double-slit camera onto a P20 phosphor screen coated with 200 nm of aluminum and followed by an image intensifier and a CCD camera. To restrict the investigated plasma radiation to the keV range, a 14-μm thick Be filter could optionally be inserted between the plasma and the phosphor screen. The corresponding spectral responsivity curves are shown in figure 2. The magnification of the double-slit camera was 20 x and 1 x in the directions perpendicular and parallel to the plasma column, respectively. The slits perpendicular and parallel to the plasma column were 110 μm and 50 μm wide, respectively.

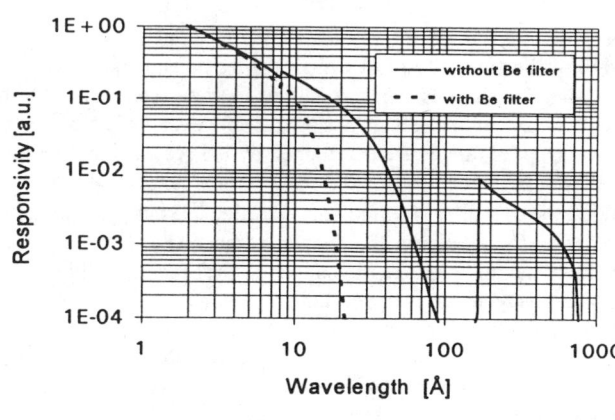

Fig. 2. *Responsivity curves of the P20 phosphor screen coated with 200 nm of aluminum. The optional Be filter is 14 μm thick.*

3. Experimental results

The measured width of the line focus was found to be less than 50 μm (FWHM). Figure 3 shows plots of the intensity along the line focus as recorded for a 100-psec shot at $1 \cdot 10^{13}$ W/cm^2. These intensity plots along the line focus are the average taken over a width of 4 pixels, i.e. 92 μm. A characteristic feature of the intensity distribution is the strong modulation observed at both ends. This may be shown in a straightforward manner to be due to Fresnel diffraction at each of the

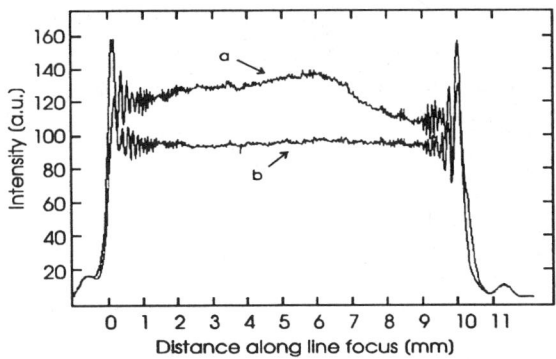

Fig. 3. Scans of the irradiance along the line focus.
Plot (a) shows the irradiance for the situation where the incident laser beam is smaller than the cylindrical lens array.
Plot (b) shows the intensity after the two outermoust cylindrical lenses of the array had been masked.

cylindrical lenses. Curve a in figure 3 shows the homogeneity between the modulations at the ends to be about +/- 8 %. It corresponds to the case where the incident laser beam with a diameter of 41 mm does not completely fill the 45-mm wide cylindrical lens array. Curve b in figure 3 shows the intensity distribution along the line focus after the two outermost cylindrical lenses of the array had been masked so that only the part of the laser beam falling onto the three cylindrical lenses in the center of the array was focused. With this masking about 25 % of the incident energy was lost but the homogeneity between the modulated ends was improuved to better than +/- 3 %. This clearly indicates that the incident laser beam should not be smaller than the CLA in order to obtain maximum uniformity along the line focus.

Figures 4 and 5 show the intensity distribution of the plasma radiation along the plasma line created by the laser beam focused with the two outermost cylindrical lens elements of the CLA masked. The intensity scale was the same for all plots. The two curves in these figures correspond to the plasma radiation in the keV X-ray region and in a spectral range extended to longer wavelengths, respectively. These two spectral ranges are determined by the responsivity curve of the phosphor screen coated with 100 nm of aluminum plus the transmission curve of the optional 14-μm thick Be filter (fig. 2). With the Be filter inserted the recorded plasma radiation was restricted to K-shell and L-shell radiation for aluminum and copper, respectively. For the present experiments the irradiance was not high enough to excite the copper K-shell radiation. Figures 4a and 4b show the measured radiation intensity along a copper and an aluminum plasma created by 700-psec shots at an irradiance in the line focus of about $1 \cdot 10^{12}$ W/cm^2. Figures 5a and 5b show the measured intensity along copper and an aluminum plasmas created by 100-psec shots at an irradiance of about $3 \cdot 10^{12}$ W/cm^2. (Where energy or irradiance values are explicitly given, the energy loss caused by the mask on the CLA had always been taken into account.)

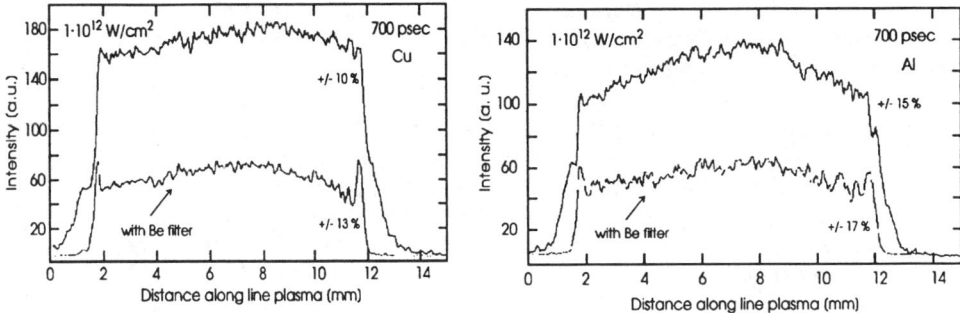

Fig. 4a and 4b. *4a) Intensity of copper plasma radiation along a line plasma created with a 700-psec laser pulse at an irradiance of about $1 \cdot 10^{12}$ W/cm^2. With the Be filter the recorded plasma radiation is restricted to wavelengths shorter than ~10 Å. The plasma homogeneity in the unfiltered and filtered radiation range is about +/- 10 % and +/- 13 %, respectively.*
4b) Same as 4a), but for an aluminum plasma. The plasma homogeneity in the unfiltered and filtered radiation range is about +/- 15 % and +/- 17 %, respectively.

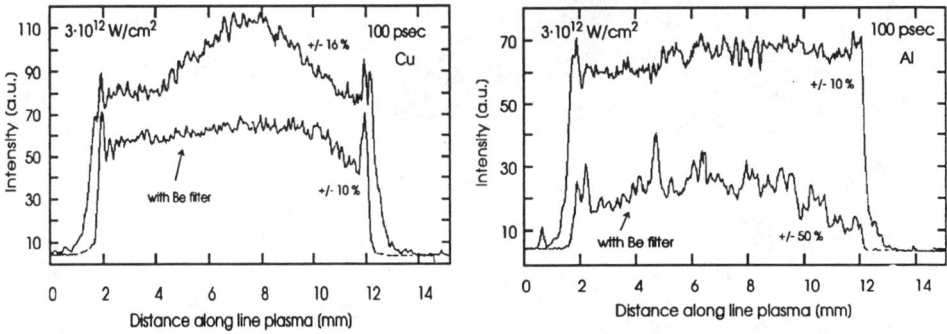

Fig. 5a and 5b. *5a) Intensity of copper plasma radiation along a line plasma created with a 100-psec laser pulse at an irradiance of about $3 \cdot 10^{12}$ W/cm^2. With the Be filter the recorded plasma radiation is restricted to wavelengths shorter than ~10 Å. The plasma homogeneity in the unfiltered and filtered radiation range is about +/- 16 % and +/- 10 %, respectively.*
5b) Same as 5a), but for an aluminum plasma. The plasma homogeneity in the unfiltered and filtered radiation range is about +/- 10 % and +/- 50 %, respectively.

Figure 6 is a scanning electron micrograph of part of an impact generated by a 100-psec shot on a copper slab at an irradiance of about $3 \cdot 10^{12}$ W/cm^2. It reveals a regular stripe pattern perpendicular to the line. This pattern is most likely caused by the interference between individual beam segments. The beat pattern has a theoretical period of $\Delta x = \lambda/\sin\theta$ for adjacent beam segments where θ is the angle between the beam segments. With $\theta = 5°$ this gives $\Delta x = 12.1$ μm, in good agreement with the measured value of 11.5 μm. Such small-scale structures could be seen neither in the plasma radiation recorded with the double-slit camera nor in the line-focus irradiance recorded with the CCD chip, because the spatial resolution of both the double-slit camera and the CCD chip was not sufficient. However, he width of the line focus measured by focusing the beam directly onto the chip of a CCD camera seems to be in agreement with the scanning electron micrograph of the impact.

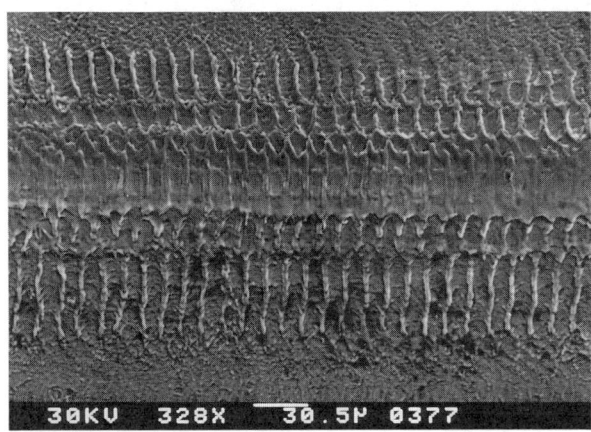

Fig. 6. *Scanning electron micrograph of part of an impact generated by a 100-psec shot on a copper slab at an irradiance of about $3 \cdot 10^{12}$ W/cm^2*

4. Summary

With the aid of a CLA we obtained a line focus with a uniformity of better than +/- 3% between the modulated ends. The copper and aluminum plasmas created by a line focus of the above-mentioned homogeneity proved to be also relatively homogeneous. The homogeneity in the keV radiation range was found to be better than +/- 17 % for both copper and aluminum plasmas created with long pulses as well as for copper plasmas created with short pulses.

5. References

[1] H. Dumont, J.-L. Bourgade, J. Bruneau, D. Desenne, A. Dulieu, M. Louis-Jacquet, L. Berthet, A. Decoster, S. Jacquemot, A. Carillon, P. Jaeglé, G. Jamelot, J.-P. Raucourt, and B. Gauthé, Opt. Commun. 96, 87 (1993).

[2] M.J. Herbst, J.A. Stamper, R.R. Whitelock, R.H. Lehmberg, and B.H. Ripin, Phys. Rev. Lett. 46, 328 (1981).

[3] W. Chen, S. Wang, B. Chen, A. Xu, and C. Mao, Chinese Physics 12, 403 (1992).

Observation of Strong Emission from Ne IX and NeX Transitions in a Laser-Driven Plasma

John K. Crane, Todd Ditmire, Hoang Nguyen, and Michael D. Perry

Lawrence Livermore National Laboratory, L-493, P. O. Box 808, Livermore, CA 94550, U. S. A.
Tel: 510-422-0420 / FAX: 510-422-1930
email: crane1@llnl.gov

Abstract. We observe strong emission from the $1s^2$-$1snp$ Rydberg series in He-like neon and from the Lyman-α transition in H-like neon. These emissions are observed when 1.05 µm light from a 650 femtosecond laser is focussed into the dense, localized output of a pulsed, supersonic nozzle. The maximum focal irradiance of our laser was measured at full power in a vaccum to be 2×10^{18} W/cm^2. Although emissions from lower charge states such as Ne^{6+} and Ne^{7+} closely follow rates predicted by tunneling theory, emissions from Ne^{8+} and Ne^{9+} are observed at irradiances two orders of magnitude below tunneling theory estimates (e.g. the He-α line appears at 2×10^{17} W/cm^2). We discuss the origins of these anomalously high charge states and the implications to recombination-pumped x-ray lasers.

INTRODUCTION

Recently McPherson et al.[1] reported strong, energetic emission(greater than 1 keV) from laser excited, krypton atoms in a pulsed gas source. Audebert et al.[2] observed Kα emission from argon in the interaction between a subpicosecond laser and the output of a gas jet. In both cases the emissions are anomalous in the sense that they derive from ions stripped to a higher charge state than predicted by tunneling theory. McPherson et al. postulate that this anomalous emission results from a new type of absorption in the clusters of Kr atoms, formed in the supersonic expansion of the gas nozzle. Other mechanisms that may cause sufficient excitation to produce these emissions include collisional ionization from the field-driven free electrons, or from the hot electrons produced by stimulated Raman scattering in the plasma. In any case the observation of such strong anomalous emission of a collisional origin will affect the design of ultrashort pulse, recombination pumped XUV lasers that depend on the existence of cold electron temperatures at electron densities of $\sim 10^{19}$ cm^{-3}. In this paper we present our observations of fluorescence from He-like and H-like neon. We compare our results with a model that combines a laser-driven collisional excitation term with tunneling ionization.

DESCRIPTION

The laser system used in our experiments is a Nd:glass system that can generate up to 10 Joules at a wavelength, λ=1.053 µm.[3] The system employs chirped pulse amplification to produce high contrast, time-bandwidth limited pulses that are 650 femtoseconds in duration at a power level of 15 TW. To determine the focussed irradiance of the laser in vacuum we simultaneously measure the pulse duration, pulse energy and the spatial beam

© 1994 American Institute of Physics

profile at the focus at full laser power. To make x-ray emitting plasmas we focus the laser at the output of a pulsed gas jet that is equipped with a supersonic nozzle designed for Mach 8 flow. To determine the neutral and electron densities in the interaction region we measure the frequency shift of the back-scattered Raman wave.[4] These measurements yield values greater than 10^{19} atoms/cm^3 for the neutrals and more than 10^{20} electrons/cm^3 for neon plasmas at our typical operating conditions.

The x-ray emission that is created in the bright plasma located at the gas jet output is sampled with a grazing incidence spectrometer, which reimages the dispersed light onto a microchannel plate/ CCD detector or x-ray streak camera. In our experiments we measured the line emission from different charge states as a function of laser irradiance and electron or neutral density. Figure 1 shows a typical spectrum in neon, containing lines from He and H-like ions. This particular spectrum was taken at a focal irradiance of 5×10^{17} W/cm^2; at 1×10^{18} W/cm^2 the He-like transitions and the Lyman-α line are sufficiently intense to saturate the detector; in this case we determine line strengths from the same line emissions seen in 2nd order.

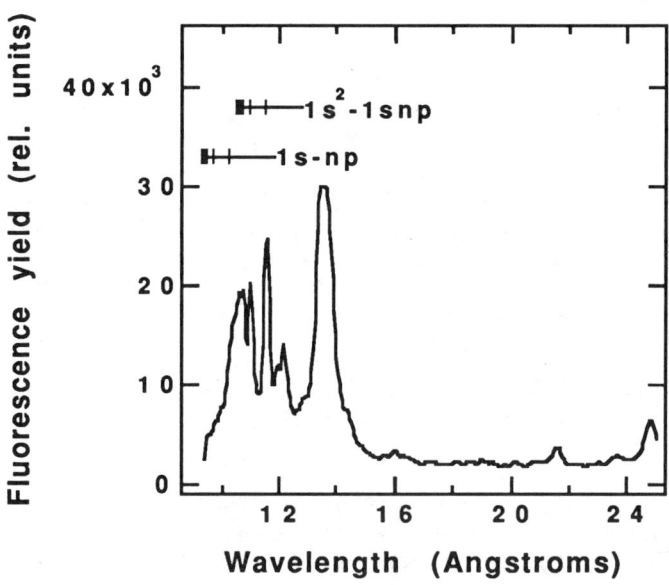

FIGURE 1. Neon spectrum at 5×10^{17} W/cm^2 and gas density, 2×10^{19} atoms/cm^3.

At the irradiances used in our experiments tunneling ionization is the dominant source of photoionization for most of the ionization stages in neon. The ADK model[5] has been shown to accurately predict experimental results for the production of high charge states in low density gases.[6] In Fig.2 we show the production of ions in neon by tunneling as a function of laser irradiance. The curves are generated by integrating the intensity dependent rate over time and space for a Gaussian shaped laser pulse. As can be seen from these curves, values for laser irradiance well above 10^{19} W/cm^2 are required to produce significant ionization to the the Ne^{9+} ground state, which is an order of magnitude greater than the peak irradiance that we measured in vacuum.

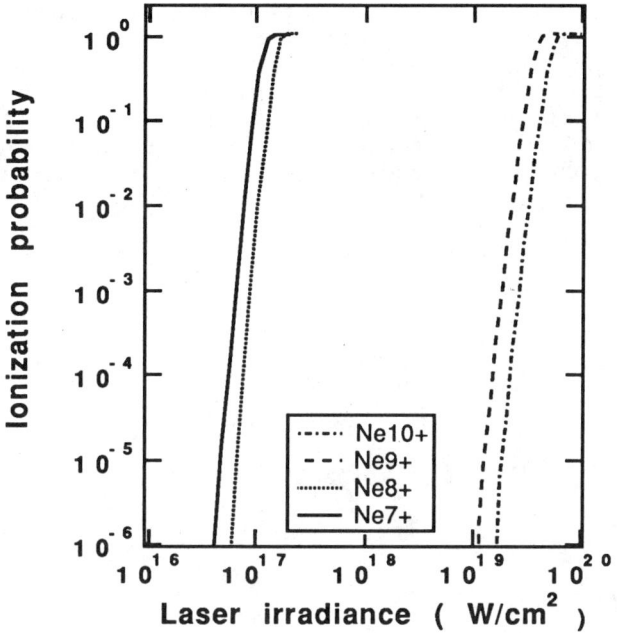

FIGURE 2. Ionization probability versus laser irradiance for higher charge states in neon.

To account for an enhanced rate of ionization necessary to yield x-ray line emission from transitions in Ne^{8+} and Ne^{9+}, we have considered laser field-driven collisional ionization. The number of ions of each charge species produced with the passage of the focused laser pulse is calculated by solving the coupled series of equations for sequential ionization as a function of both space and time. Laser assisted collisional ionization is modeled by including in the rate equations a cycle averaged collisional ionization rate. Only the quiver velocity of the free electrons is considered; all thermal motion is considered negligible compared to the very large quiver motion. If the classical expression for the collisional ionization of a bound electron is used, the cycle averaged collisional ionization rate for the ith charge species is given by:

$$\left\langle \frac{dn_{i+1}}{dt} \right\rangle_{coll} = \frac{q_i e^4 n_e n_i}{\sqrt{mU_p}} \left[\left(\frac{1}{I_p} + \frac{1}{4U_p} \right) \ln \left(\frac{1+\sqrt{1-I_p/2U_p}}{1-\sqrt{1-I_p/2U_p}} \right) - \frac{\sqrt{1-I_p/2U_p}}{I_p} \right]$$

where q_i is the number of electrons in the outermost subshell, n_e is the electron density, n_i is the ion density, I_p is the ionization potential, and U_p is the electron ponderomotive energy. Figure 3 shows a plot of emission yield for transitions from different charge states in neon versus laser irradiance. The emission yields are adjusted for spectrometer and detector efficiencies and transition oscillator strengths, so that they are proportional to the upper state population of the transitions. To compare the model results, which calculate ion densities, to the emission strengths measured with the detector, we multiplied the model results by a scaling factor that overlays the data for Ne VII emission with its corresponding model curve in the region of saturation of the photoionization. In this irradiance regime the ionization rate is the same with and without collisions, thus emission from the Ne VII line should be accurately predicted by pure tunneling.

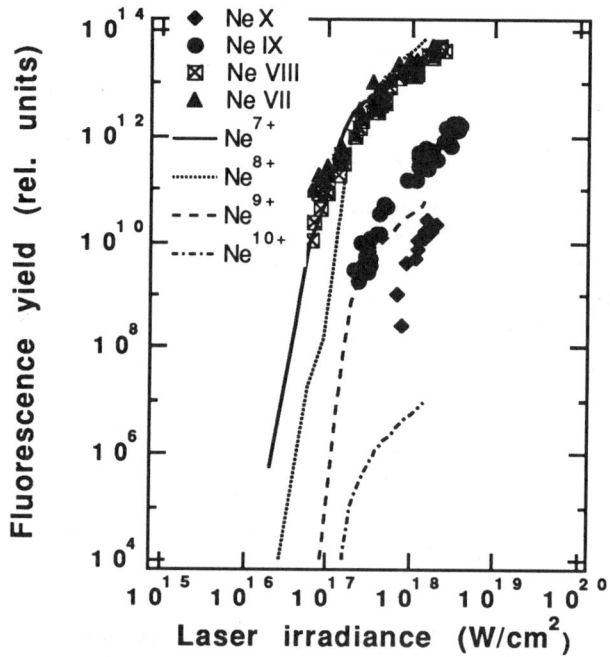

FIGURE 3. Line emission strength as a function of laser irradiance for transitions in neon: Ne X, 1s-2p, 12.1 Å; Ne IX, $1s^2$-1s2p, 13.6 Å; Ne VIII, 2s-5p, 60.8 Å; Ne VII, 2s-3d, 106.1 Å. Curves show model predictions. Neon density equals 2×10^{19} atoms/cm^3.

As can be seen in comparing the model results with the emission data, Ne VII and NeVIII emissions are accurately simulated by the model, however the deviation of the model predictions from the observed emissions for the Ne IX and Ne X transitions is significant- these emissions are much stronger than what our simple collisional model would indicate. Recent work at

Rutherford[7], where similar observations of H-like and He-like emissions in Ne are seen, suggests that Raman heating may produce the necessary density of sufficiently energetic electrons to yield the strong emissions that we observe. We also see strong Raman backscattering in our experiments and are currently developing a similar Raman heating model for use in understanding our results. Alternatively, similar anomalous emissions have been reported by McPherson in Kr and Xe and are attributed to collective electron effects in the large clusters that are generated by their supersonic nozzle. Under the conditions of operation of our pulsed supersonic nozzle (Mach 8 flow) we are also likely to be producing large clusters of neon atoms[8], however, until we can obtain more quantitative evidence for a cluster-based mechanism to explain our results we will continue to focus our attention on the Raman-based mechanism for producing anomalous x-ray emission.

SUMMARY

We observe anomalous x-ray emission from He and H-like transitions in neon plasmas produced from a gas jet. We see these emissions at focal irradiances that are two-orders of magnitude below the values that would be predicted by a purely field ionization mechanism. Although preliminary estimates for collisional excitation from field driven electrons yield a substantial enhancement of the rates over a pure tunneling mechanism they fail to predict the large yields that we observe experimentally. We continue in investigate other mechanisms to explain the bright line emissions from NeIX and NeX transitions in laser driven plasmas produced from gaseous targets.

References:
1. McPherson, A., Luk, T. S., Thompson, B. D., Borisov, A. B., Shiryaev, O. B., Chen, X., Boyer, K., and Rhodes, C. K., *Phys. Rev. Lett.* **72**, 1810-1813 (1994).
2. Audebert, P. et al., "X-ray emission from an Ar gas jet induced by an intense femtosecond laser pulse", presented at the Conference on Generation and Application of Ultrashort X-Ray Pulses, Salamanca, Spain (1994).
3. Patterson, F. G., Perry, M. D., and Hunt, J. T., *J. Opt. Soc. Am B* **8**, 2384-2390 (1991).
4. M. D. Perry, C. B. Darrow, C. Coverdale, and J. K. Crane, *Opt. Lett.* **17**, 523-525 (1992).
5. Ammosov, M. V., Delone, N. B., and Krainov, V. P., *Sov. Phys. JETP* **64**, 1191-1194 (1986).
6. Augst, S., Meyerhofer, D. D., Strickland, D., and Chin, S. L., *J. Opt. Soc. Am B* **8**, 858-867 (1991).
7. Keys, M. H.,"Reduction of driver energy for x-ray lasers," presented at the 4th International Conference on X-Ray Lasers, Williamsburg VA, May 16-20, 1994.
8. Hagena, O. F., Obert, W., *J. Chem. Phys* **56**, 1793-1802 (1972).

This work was performed under the auspices of the United States Department of Energy by the Lawrence Livermore National Laboratory under contract no. W-7405-Eng-48.

Towards a 38 Å X-Ray Laser

L. Bonnet, S. Jacquemot and A. Decoster

Centre d'Etudes de Limeil-Valenton 94195 Villeneuve Saint-Georges Cedex FRANCE

Abstract. The aim of this study is the design of experiments on the P102 facility at CEL-V showing evidence that a significant population inversion can be achieved on a very short wavelength transition with the help of a new generation of lasers. Apart from multiple theoretical interests, from dense and hot plasma study to short pulse laser-matter interaction, applications can already be considered: microlithography as well as plasma diagnostics.

INTRODUCTION

Traditional experiments on H-like recombination X-Ray Lasers seem to have reached a limit in the maximum value of the gain-length product. Using ultra-short (Δt_L), high peak irradiance (P_L) lasers should help overgoing it by creating, when interacting with a massive target, an initially totally ionized high density plasma that can, by very rapid cooling, reach suitable temperature conditions for lasing but remain quite dense (Fig. 1).

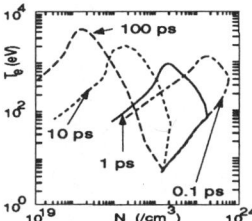

Figure 1. Thermodynamical paths of an outer cell in a 4 μm thick massive Al target for various pairs ($P_L, \Delta t_L$) at constant laser energy density of 10^6 J/cm².

Complete simulations (hydrodynamics+kinetics) confirm that the efficiency of the H-like recombination scheme could thereby be greatly increased in the case of the Balmer α transition of Al at 38.8 Å.

MODELING

A complete set of numerical tools has been developed at CEL-V to describe X-Ray Lasers (see S. Jacquemot), ranging from hydrodynamics and laser-matter interaction (CHIVAS) to plasma kinetics and population inversions (LASIX) and beam propagation (OPTIQX). Two major schemes, Ne-like collisional excitation and H-like recombination, have been successfully modeled: simulations of the H-like recombination C fiber experiment performed at RAL (1) have recently given theoretical gains in good agreement with experimental values (2). However, when dealing with high intensity short pulses, several improvements have to be done to CHIVAS in order to correctly describe laser-matter interaction.

Laser absorption model

The traditional WKB approximation, treating laser propagation by ray-tracing and absorption (3) by Inverse Bremsstrahlung from surface up to the critical density and back, is no longer valid (Fig. 2) due to high density gradients. Therefore, we have enhanced the hydrodynamics code CHIVAS by installing a subroutine that solves the Helmholtz wave equation for the electric field E of normally incident linearly polarized lasers:

$\frac{d^2E}{dz^2} + \epsilon' \frac{\omega_L^2}{c^2} E = 0$ (Eq. 1)

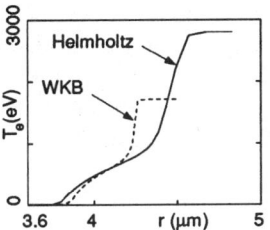

Figure 2. Influence of the laser absorption model on the spatial profile at t.o.p. of the electronic temperature T_e for a 4 μm thick massive Al target and a 1 ps–10^{18} W/cm² pulse.

where ω_L is the angular frequency and ϵ', the complex permittivity of the plasma. Its expression is taken from the Lee/More conductivity model (4) at zero frequency $\sigma(0)$ accounting for various properties of matter during the interaction (solid-state effects, electron degeneracy,...) and analytically extended for non zero frequencies according to: $1/\sigma(\omega) = (1 - i\omega_L\tau)/\sigma(0)$, where τ is the relaxation time.

Particular attention has been paid to the numerical resolution of Eq. 1. A variable step (h_n) Numerov-like method has been used. In each hydrodynamical (hyd.) cell n, subzoning is determined by both the laser wavelength and the cell thickness (whole number N_n of electromagnetic (em) cells in each hyd. cell). Physical initial conditions are taken in the critical part of the plasma accounting for the skin depth and the exponential decay of the field amplitude. The deposited power Q_n in a hyd. cell n is just obtained by summing for all its em sub-cells i_n the product of the current density vector by the electric field: $Q_n = \sum_{i_n=1}^{N_n} \vec{J}_{i_n} \cdot \vec{E}_{i_n}$; then, the total absorbed power Q in the whole plasma is equal to: $Q = \sum_n Q_n$. Unfortunately, two problems arise: first, the summation leads to a renormalization in order to restitute the exact difference between the incident and reflected energies and, secondly, the field amplitude is not rigorously conserved in vacuum. That's why we have developed a finite difference "Poynting" scheme specifically adapted to our problem. In this case, eq. 1 becomes:

$h_n E_{n-1} - (h_n + h_{n-1}) \frac{1 - f_n \frac{h_n + h_{n-1}}{4}}{1 + f_n \frac{h_n + h_{n-1}}{4}} E_n + h_{n-1} E_{n+1} = 0$. By first noticing that the Poynting vector temporal mean $\bar{S} = \frac{c}{8\pi} Re(\vec{E} \times \vec{B}^*)$, which represents the energy flux at the interface ($n+1/2$) between cells n and $n+1$, can be written $\bar{S}_{n+1/2} = \frac{c^2}{8\pi\omega} \frac{E_{n+1}^* E_n - E_{n+1} E_n^*}{2i h_n}$ and, secondly, that the deposited power Q can also be expressed by: $Q = -\frac{\omega}{8\pi} Im(\epsilon') |E|^2$, one can directly deduce, <u>from the wave equation itself</u>, the following result: $\mathrm{div}(\bar{S}) + Q = 0$: the deposited power in a given cell can be simply obtained from the difference between the Poynting vector temporal means at its two borders. This scheme has the advantage of requiring no renormalization and it also provides the exact reflexion coefficient at the interface between two media of different densities. Furthermore, it gives, without additional calculation, the photon flux at each interface needed for the evaluation of the light absorption by multi-photon processes.

Ionization

Time-dependent ionization models have been recently coupled to the hydrodynamical code CHIVAS. The first one, CORAT, is a pseudo-coronal model describing only the ground states of the different ionization stages which are coupled by collisional ionization, 3–body and radiative recombinations (5). It is solved by a GGGV algorithm (6) and is very fast. The second one is the non LTE screened hydrogenic model NOHEL (7). The time-dependent version runs only on line and is therefore time consuming. It is also generally coupled to radiative transfer.
At those high intensities ($>> 10^{16}$ W/cm^2), the multi-photon and tunnel ionization should be taken into account and lead to an enhancement of the mean ionization in the cold and dense part of the target, thereby increasing the coupling between laser and solid (8). This is in progress.

GAIN COEFFICIENT RESULTS

All the simulations treat a 4 μm thick massive Al target described by 40 cells in geometric progression from the rear side to the interaction surface. Different configurations have been studied by varying the intensity of the incident laser, from 10^{16} up to 10^{19} W/cm^2, and the pulse duration, from 0.1 ps to 1 ps. The ionization model is CORAT except in one case (10^{17} W/cm^2–0.3 ps) for which a comparison between the two ionization models CORAT and NOHEL has been done. Reabsorption along the resonance lines is always taken into account in LASIX (according to the well–known 1D Sobolev-Hummer model). Lasing lines are described by Voigt profiles. The gain coefficients reported here are those of the $3d_{5/2}$–$2p_{3/2}$ transition at 38.8 Å which is the most intense in our results.

Fig. 3 shows the time and space evolution of the <u>local gain coefficient</u> for the optimal pulse duration at peak intensity of 10^{17} W/cm^2: $\Delta t_L = 0.3$ ps, leading to a maximum gain of 231 cm^{-1}. On Fig. 4 are drawn the thermodynamical paths of the cells of maximum gain for various experimentally accessible Δt_L: 0.2 ps (dashed-dot), 0.3 ps (solid line) 0.5 ps (dots), 1 ps (dashed), at constant peak intensity of 10^{17} W/cm^2; the points show the time of maximum gains together with their values. It demonstrates that, at a given P_L, there is a choice of Δt_L which provides the optimal initial state of the plasma (hot enough but still dense) and the most efficient cooling. It should be noticed that the maximum local gain coefficient obtained in the case $(\Delta t_L = 0.3\text{ps}, P_L = 10^{17}\text{W/cm}^2)$ with NOHEL model in CHIVAS falls to 130/cm. However in neither cases are propagation effects taken into account.

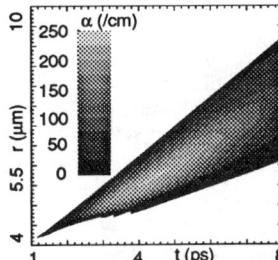

Figure 3. Time- and space- map of the $3d_{5/2}$–$2p_{3/2}$ gain coefficient for a 0.3. ps–10^{17} W/cm^2 pulse.

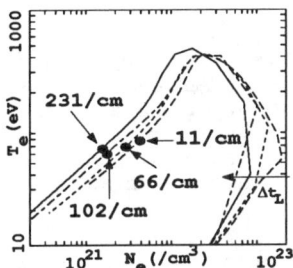

Figure 4. Thermodynamical paths of the optimal cells for various Δt_L at constant P_L of 10^{17} W/cm^2.

CONCLUSIONS

Several improvements are in progress in order to get more reliable simulations. In normal incidence, an automatic fit to the WKB solution in smooth gradients is necessary to easily treat large targets. Our "Poynting" modeling should be extended to obliquely incident lasers. Calculations with OPTIQX are underway to evaluate the influence of beam propagation on those so high but so localized gain coefficients. They will be followed by an optimization of the experimental conditions for the P102 facility at CEL-V.

1. Pert, G.J., and Rose, S.J., *Appl. Phys.* **B 50**, 307–311 (1990)
2. Bonnet, L., *not published*
3. Johnston, T.W., and Dawson, J.M., *Phys. Fluids* **16**, 722 (1973)
4. Lee, Y.T., and More, R.M., *Phys. Fluids* **27**, 1273–1286 (1984)
5. Bonnet, L., "Model calculations of H-like recombination schemes" in *Proceedings of NATO ASI on Laser Interaction with Atoms, Solids and Plasmas*, Cargese, 1993, in press
6. Gauthier, J.-C., et al, *J. Phys. D: Appl. Phys.* **16**, 321–331 (1983)
7. Decoster, A., Rapport des activités Laser CEA/CEL-V **4**, 3 (1994)
8. Perry, M.D., UCRL-53852 (1987)

Inner-Shell Photo-Ionized X-Ray Laser Schemes for Low-Z Elements

S. J. Moon*[†], D. C. Eder* and G. L. Strobel[‡]

Lawrence Livermore National Laboratory, Livermore, CA 94550

[†] *Department of Applied Science, University of California Davis-Livermore, Livermore, CA 94550*

[‡] *Department of Physics, University of Georgia, Athens, GA*

Abstract. Gain calculations for inner-shell photo-ionized lasing in C at 45 Å are performed. An incident x-ray source represented by a 150 eV blackbody with a rise time of 50 fsec gives a gain of order 10 cm^{-1}. The x-ray source and thus the driving optical laser requirements are significantly reduced as compared to what is needed for Ne at 15 Å. We expect that existing ultra-short pulse lasers can produce the required x-ray source and thus produce a table-top x-ray laser at 45 Å.

I. INTRODUCTION

Previous theoretical work in inner-shell photo-ionized (ISPI) laboratory x-ray laser schemes have mainly focused on the 5 to 15 Å wavelength regime, where laboratory x-ray lasing using any approach has not yet been obtained. This was investigated for Ne at 15 Å by Kapteyn[1] and extended by Strobel *et al.*[2] also treating Mg at 10 Å. The experimental validation at these short wavelengths is dependent on the development of an ultra-short pulse (100 fsec FWHM) optical laser with energy of order 10 J or greater. Current "table-top" size ultra-short pulse (USP) lasers with energy of order 1 J exist. We present results for C at 45 Å as a representative low-Z element where lasing can be tested experimentally using current high energy ultra-short pulse lasers. Carbon has a smaller Auger rate compared with Ne and a longer lasing wavelength thus requiring a less energetic pump source. An x-ray laser at 45 Å, just outside the water window, is optimal for many biological applications[3].

Although current x-ray lasers using Ni-like ions operate at above and below the wavelength considered in this paper, 45 Å, they require high energy (E > 1 kJ) driving lasers[4]. As a result of using a lower energy driving laser (E ≈ 1 J), an inner-shell x-ray laser would operate at a higher repetition rate, albeit with less energy in each x-ray pulse. Despite a very short lasing duration (Δt < 100 fsec) and small cross sectional area (A ≈ 10^{-6} cm^2), the large saturation intensity, I_{sat}, associated with the relatively large Auger rate out of the upper lasing state[5] results in significant energy per pulse yielding a high average energy.

In the ISPI scheme, lasing takes place between the L shell and the K shell. Neutral Ne having a closed L shell makes it a good candidate for ISPI lasing, yet current lasers can not provide the needed energy to produce a significant gain-length product[1]. Because of the open L shell structure in C, it is relatively easier to both collisionally- and photo-ionize the L shell thus destructive filling of the lower-lasing state is more severe for C than Ne. Due to the lower energy requirements for K-shell ionization of C and the smaller Auger rate the requirement on the intensity of the pump is reduced as compared to Ne; however, the rise time requirements are not changed. A blackbody source is chosen to represent the x-ray source yet by optimizing the target material it may be possible to use a line or band source which would give more efficient pumping.

In section II we discuss general details of the ISPI scheme first proposed 25 years ago by Duguay and Rentzepis[6]. In section III we report our results for C at 45 Å and in section IV discuss conclusions.

II. INNER-SHELL PHOTO-IONIZATION

An USP (100 fsec FWHM) optical laser with energy ≥ 1 J is used to produce a hot plasma at line focus. The plasma generates a broad-band x-ray spectrum with a rapid rise time. A low-Z filter is sandwiched between the target and lasant to stop a majority of the low energy x rays that can ionize outer-shell electrons and thus populate the lower-laser state. The remaining high energy x rays primarily photo-ionizes the inner-shell electrons of the lasant atoms. This produces a population inversion, and resulting positive gain for an allowed 2p-1s radiative transition in the singly charged ion for a sufficiently intense x-ray source. Rapid Auger decay of the 1s hole state competes with the lasing transition and produces a large number of energetic electrons into the lasant material. Electron induced ionization to the lower-laser state limits the magnitude and duration of positive gain. Ultra-short pulse x-ray lasing is inherent in this scheme.

A high intensity source of x rays is required to compete with the Auger rate and cause a significant upper-laser state population. To achieve a high absorption of the driving laser's energy a structured target, parallel grooves on a solid material, or a composite of clusters, *e.g.*, gold-black, can be used[7]. The cluster targets are relatively inexpensive to produce but difficult to model due to their fractal properties. A new inexpensive structured target consisting of vertical rods[8] has been shown to also have high absorption properties[9]. We are currently modeling this type of target, but in this paper we concentrate on gain calculations for an assumed x-ray source.

A time dependent single temperature blackbody is used to approximate the x-ray emission from the plasma. For work done using a Au target composed of parallel grooves this is shown to be a conservative assumption[10]. An ideal source would be a line source with the difference in energy with the lasant's K edge being within the L-shell energy. This provides maximum coupling of x-ray energy to the lasant atoms, because the cross-section is peaked at threshold. In addition, such a line source would effectively reduce electron ionizations of the L shell from photo-ionized electrons. In the lasing medium, electrons come from both photon ionization and from Auger decay. The energy spectrum of the photo-ionized electrons is dependent on the x-ray source. As stated above an optimized source can mitigate this problem. However, the negative effect of Auger electrons will not be affected. If the rise time of the x rays is rapid enough, lasing can be achieved before significant electron ionization can occur.

To achieve lasing a filter is needed in order to reduce the low energy x rays. A low-Z filter can be chosen to optimize the ratio of the x rays at K-shell energies to x rays at the L-shell energies in the lasant. Filtering is primarily through K-shell ionization of the low-Z filter element. For Ne it was found that 3.5 microns of Be with $E_K = 118.4$ eV yield maximum gain. In C, we find that 2 microns of Li with $E_K = 59.9$ eV is optimal. This thickness does result in a reduction of x rays at the K edge of C by 60%. However this is required to sufficiently reduce the amount of lower energy x rays. There are windows of high transmission below the filter's K-edge energy and a trade-off is made between filtering at the lasant's K edge to reduce the low energy photons enough for lasing to occur. Geometrical effects associated with the plasma being a line source of finite transverse extent and with the separation between the plasma and the lasant given by the filter thickness are included in our calculations.

III. RESULTS

Previous work[1, 2] has shown that for gains of order 10 cm-1 in Ne, a maximum blackbody temperature of order 500 eV with rise time of 50 fsec is required. We find that for C a much reduced blackbody temperature ($T_{bb} \approx 150$ eV), with the same 50 fsec rise time, gives comparable gains. Shown in fig. 1 are blackbody spectrums appropriate for Ne and C. The filtered spectrum is also shown with the K edges marked for reference. As can be seen for both Ne and C the peak of the filtered spectrum is to the right of the K edge allowing for the broad band nature of the filtered spectrum to be taken advantage of. However, the cross section decays rapidly from its maximum value at the K edge, for example, in C at the peak of the filtered spectrum the cross section is 1/4 of its K-edge value, where as the filtered spectrum only increases by a factor of 2 of its K-edge value. This results in the convolution of the intensity and the absorption cross-section having a peak very near the K edge and decreasing monotonically for higher energies. The replacement of the broad-band source with a line source near the K edge or a

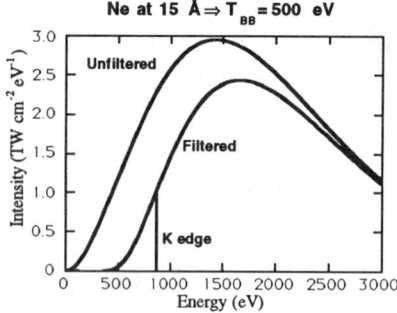 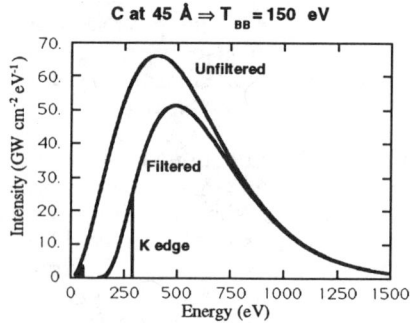

FIGURE 1. X-Ray source requirements for both Ne at 15 Å and C at 45 Å. The filtered source for Ne uses a 3.5 μm Be filter and for C a 2 μm Li filter is used.

band of emission above the K edge would reduce the requirements on the source. For a line source at the K edge the flux required for the x-ray source is approximately 1/6 that of the 150 eV broad-band source.

For the time dependence of the x-ray source we use a simple expression appropriate for a sech^2 driving pulse[1]. Expressed in terms of a blackbody temperature it is given by the equation,

$$T_{bb} = T_{Max}\left[0.02\int_{-\infty}^{t} \mathrm{sech}^2(1.76t'/\tau)dt'\right]^{4/9}$$

where τ is the FWHM of the driving laser and T_{max} is the maximum temperature (model assumes no cooling). This is shown in fig. 2 for $T_{max} = 150$ eV and $\tau = 50$ fsec which are appropriate parameters for C. The corresponding gain curve in fig. 2 is for a neutral C density of 4.0×10^{19} cm^{-3} mixed with 4 H atoms for every C atom. Molecular effects of CH_4 were not treated. The x-ray source is taken to have a traverse extent of 10 μm used in conjunction with a 2 μm Li filter. Results for C using a driving laser with $\tau = 100$ fsec show a reduction in gain by a factor of 3. As shown in fig. 2, the gain has a FWHM of ≈ 60 fsec, showing the ultra-short pulse nature of this scheme. In fig. 3, the populations of the upper- and lower-laser states are plotted with the filtered intensity of the x-ray source. From this plot we can see that the upper-laser state population follows the intensity which is expected given the fast Auger exit channel out of the upper state. This will be the case unless the intensity changes on a time scale faster than the inverse of the Auger rate which for C is 10.7 fsec. The lower-laser state population grows exponentially due to electron-ionizations. Since the degeneracy between the lower- and upper-laser states is 3 to 1, the gain goes to zero when the lower-state population reaches three times the upper-state population.

Given the calculated gain coefficient, a line source of x rays with a length of order 1 cm is required in order to have a gain-length product of order 10. (Gain-length products between 5 and 10 provide clear evidence of lasing.) Assuming a conversion efficiency to incoherent x rays of 20%, the energy required for the

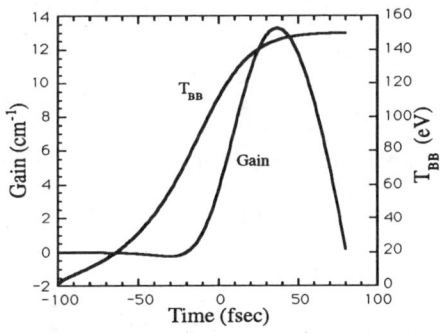
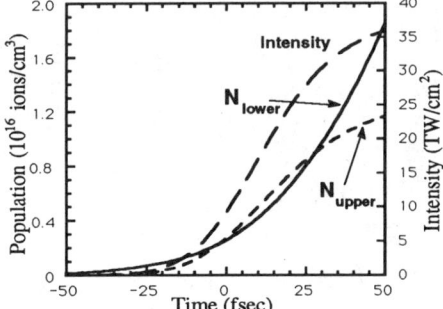

FIGURE 2. A gain coefficient of 13 cm^{-1} with FWHM = 58 fsec is shown for C with T_{max} = 150 eV time dependent blackbody source.

FIGURE 3. Time dependent plots of the upper- and lower-laser state populations leading up to max. gain are shown along with the filtered intensity of the source.

driving laser is 1.0 J. Lasers with this energy are currently available. The major issue is whether the rise time of the x rays is sufficiently rapid ($\tau \approx 50$ fsec) since this can not be currently measured.

IV. CONCLUSIONS

Theoretical work on inner-shell photo-ionized x-ray lasers in the 5 to 15 Å wavelength regime shows that the equivalent blackbody temperature of the x-ray source must be of the order 500 eV, requires a driving laser with energy of order 10 J or greater. Our preliminary results for C at 45 Å indicated a driving laser with energy of order 1 J is sufficient to produce a large gain-length product. Gains of over 10 cm^{-1} were found for C of a density of 4.0×10^{19} cm^{-3} using a 2 μ Li filter and a maximum black-body temperature of 150 eV pumped with 50 fsec rise time. Collisional ionization to the lower lasing levels limits the duration of lasing giving a pulse on the order of 60 fsec FWHM. Such short coherent x-ray emission is important for many applications involving fast dynamical processes.

Acknowledgments

Work performed under the auspices of the U. S. Department of Energy by Lawrence Livermore National Laboratory under Contract W-7405-ENG-48

References

1. H. C. Kapteyn, Applied Optics **31**, 4931 (1992).
2. G. L. Strobel, D. C. Eder, R. A. London, M. D. Rosen, R. W. Falcone, and S. P. Gordon, SPIE Proceedings, Short-Pulse High-Intensity Lasers and Applications II, L.A., CA, Jan. 1993, Vol. 1860 p. 157.
3. R. A. London, M. D. ROsen, and J. E. Trebes, Applied Optics **28**, 3397 (1989).
4. X-Ray Lasers 1992, Proceedings of the 3rd International Colloquium on X-Ray Lasers, IOP Conference Series 125, edited by E. E. Fill (Institute of Physics, Bristol, England, 1992).
5. G. L. Strobel, D. C. Eder, and P. Amendt, Appl. Phys. B, **58**, 45 (1994).
6. M. A. Duguay and P. M. Rentzepis, Appl. Phys. Lett. **10**, 350 (1967).
7. M. M. Murname, H. C. Kapteyn, S. P. Gordon, J. Bokor, E. N. Glytsis, R. W. Falcone, Appl. Phys. Lett. **62**, 1068 (1993).
8. D. Al-Mawlawi, C. Z. Liu, and Martin Moskovits, J. Mater. Res. **9**, p. 1014 (1994).
9. F. Budnik, G. Kulcsár, L. Zhao, R. Marjoribanks, P. Herman, D. Al-Mawlawi, M. Moskovits, 24th Anomalous Absorption Conference, Pacific Grove, CA, June 1994.
10. D. C. Eder, G. L. Strobel, R. A. London, and M. D. Rosen, *SPIE Proceedings*, Applications of Laser Plasma Radiation, San Diego, CA, 1993, edited by M. Richardson (SPIE, Bellingham, WA, 1994), Vol. 2015, p. 234.

The Monochromaticity and Intensity Scaling of the Neon-like Yttrium Laser

P B Holden, M Nantel, B Rus and A Sureau

Laboratoire de Spectroscopie Atomique et Ionique, Bâtiment 350, Université Paris Sud, 91405 Orsay Cedex, France.

Abstract. The two dominant lasing lines in the Ne-like yttrium laser exhibit an apparently anomalous behaviour. The line at 157Å demonstrates a marginally higher gain than that at 155Å despite the fact that the latter is observed to be up to 100 times brighter. We describe a convincing explanation for this behaviour in terms of the 155Å line being composed of two partially overlapping transitions - the existence of such an overlap (between a J=0-1 and a J=2-1 transition) has long been the suspected cause. We demonstrate that the intrinsic line shapes narrow along the path of the beam due to variations in electron density and ion temperature which arise as the beam refracts through the expanding medium, and that these variations are sufficient to partially decouple the overlapping transitions at long plasma lengths. The resulting effect is that the beam undergoes large amplification initially, producing the observed intensity ratio between the 155Å and 157Å emission, but that the gain at 155Å is reduced at the longer lengths examined experimentally.

    High gain length lasing is routinely obtained at LLNL in Ne-like yttrium [1]. In contrast to all other Ne-like lasers, a single line (at 155Å) dominates the output. The cause has long been a suspected overlap between a J=2-1 and a J=0-1 transition [2]. Assuming complete overlap, earlier work [3] reproduced the intensity dominance but achieved this only as a consequence of a higher gain on the combined line, in contrast to experimental measurements which indicate that the low intensity J=2-1 line at 157Å is, somewhat anomalously, more strongly amplified. The presence of an adequately high emissivity at 155Å is impossible to justify and the explanation we propose here is that the overlapping transitions decouple at longer lengths (as a result of the change in intrinsic linewidths as the beam refracts through an exploding foil) leading to reduced apparent amplification.

    A full treatment of the problem would require the self-consistent time dependent solution of the hydrodynamics, atomic physics and beam propagation (considering the beam as a collection of frequency packets). To investigate the wide parameter space of uncertainty in the relative wavelengths and gains on the two 155Å transitions, such an approach is intractable and we here simplify the problem through a parameterisation of the relevent quantities along an assumed ray path which is assumed to be the same for all three transitions. The temporal and spatial dependencies (at 400ps) of the electron density and ion temperature from a hydrodynamic EHYBRID calculation [4] of typical LLNL conditions - the irradiation of an yttrium foil with a 500ps flat topped pulse of 0.53micron light at an intensity of $1.4 \times 10^{14} Wcm^{-2}$ - are illustrated in figures 1 and 2. These are well represented by exponential (temporally) and gaussian (spatially) profiles as illustrated. If it is assumed that the beam path is parabolic with an exit angle of 10mrad at both ends of the plasma, the density and ion temperature along the path ($z=ct$) are given by

© 1994 American Institute of Physics

$$\rho(z) = \exp\{-5.56 \cdot z/3 - 2(z-l/2)^4/3l^2\} \text{ gcm}^{-3}$$
$$T_i(z) = \exp\{6.30 - 2z/15 - (z-l/2)^4/2l^2\} \text{ eV}$$

where l is the plasma length. We note that over a 3cm plasma (approximately that required for saturation) density variations are of the order of a factor 3.

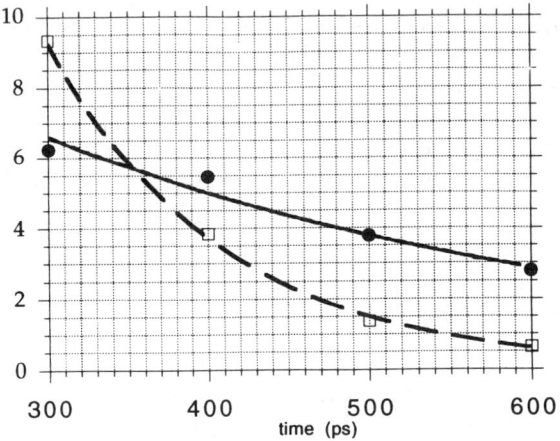

Figure 1. Time dependence of the mass density (mgcm^{-3}, dashed) and ion temperature (100eV, solid) at the foil centre [3] fitted to an exponential decay.

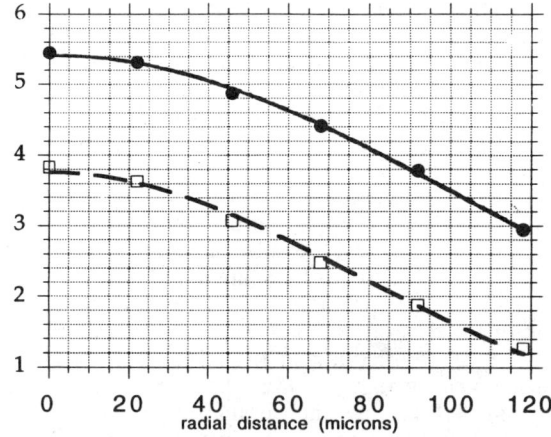

Figure 2. Spatial dependence of the mass density (mgcm^{-3}, dashed) and ion temperature (100eV, solid) at 400ps [3] fitted to a gaussian profile.

From these values we calculate the Doppler and Lorentz widths and the emissivity and gain as a function of frequency. The total emissivity is assumed to be proportional to the density and the line centre gain proportional to the density and inversely proportional to the linewidth. The line separation is assumed to be less than 50mÅ in accordance with the finding that the lines are unresolvable [5]. We neglect the effect of saturation to avoid complicating the interpretation but note that saturation is observed for gain lengths ~ 15. The details of the calculation and the assumptions made can be found in [6].

The intensity vs length scaling for the two wavelengths is illustrated in figure 3 for a configuration which gives an intensity ratio of ~100 for a 3cm plasma. The 157Å line was attributed an on-axis gain of 9cm^{-1} which was found to reproduce the LLNL data and the 155Å J=2-1 line a value of 7cm^{-1} reflecting the theoretical ratio of the two [3]. A virtually identical curve can be obtained for any degree of separation between 10mÅ and 50mÅ if a suitable J=0-1 gain coefficient is assumed. The lower bound approximately defines a separation which prevents decoupling and the upper a degree of separation which is resolvable. The J=0-1 gains required to fit this curve as a function of line separation are illustrated in figure 4.

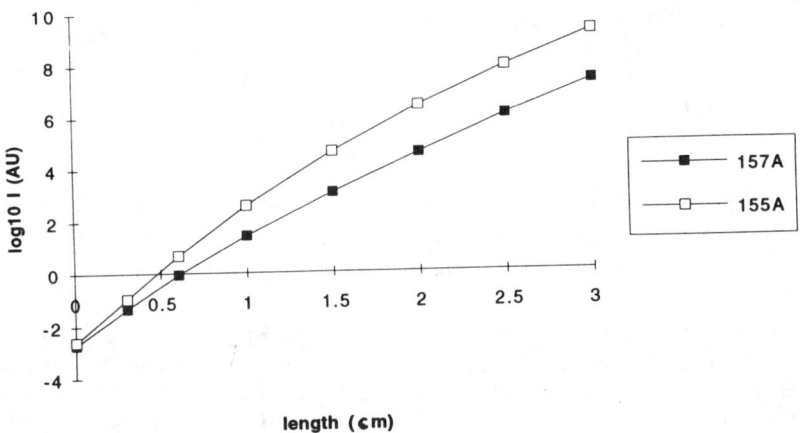

Figure 3. The intensity of the two wavelengths as a function of length. Apparent gain coefficients are ~5.5 per cm on both lines, although a large intensity ratio is exhibited.

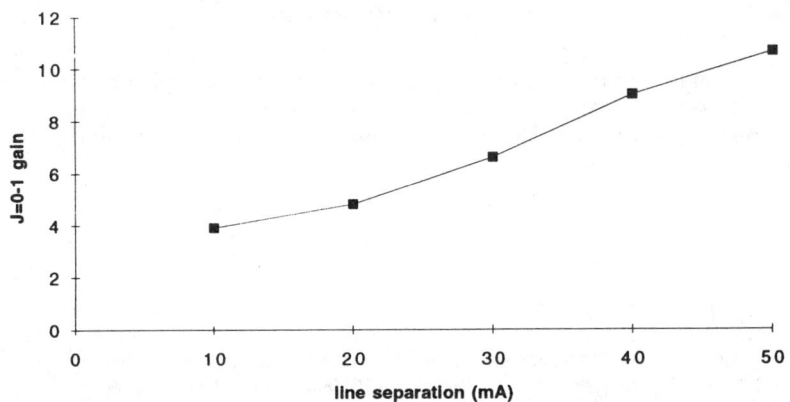

Figure 4. The on-axis J=0-1 gain coefficient required to reproduce figure 3.

Gain narrowing on the composite transition is illustrated in figure 5 for line separations of 20mÅ and 40mÅ. It is apparent that in the former case, the narrowing is indistinguishable from the well known √Gl dependence on a single line. The effect may consequently be very difficult to observe directly.

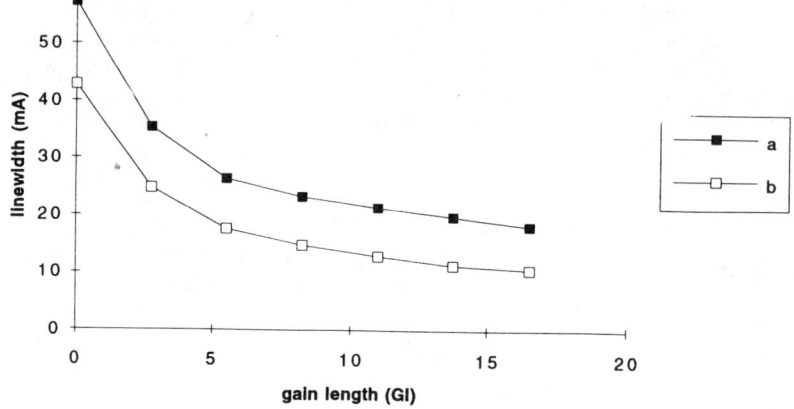

Figure 5. The line narrowing on the composite transition for a) 40mÅ and b) 20mÅ separation

Conclusions

We have used a simple description of the passage of a refracting laser beam through an expanding medium which is a good representation of the results of more detailed modelling. The variations in linewidth along the trajectory are shown to be sufficient to result in a significant decoupling of the two transitions which compose the 155Å line. The effect is that far higher output intensities are obtained than would be expected for the apparent degree of amplification. We note that the use of geometries which facilitate beam propagation through high density plasma (eg. curved slab targets) may be especially beneficial for yttrium as they would be expected to inhibit the degree of decoupling.

Acknowledgements

We are grateful to GJ Pert for providing the hydrodynamic code EHYBRID and to JA Koch for useful discussions. The work was carried out under the auspices of the CNRS. One author (PB Holden) is sponsored by the EEC capital mobility scheme.

References

1) Shimkaveg GM, Carter MR, Walling RS, Ticehurst JM, London RA and Stewart RE 1992, Xray lasers 1992, IOP Conf. Ser. Vol. 125 (IOP, Bristol)
2) Matthews DL et al 1986, *Journal de Physique*, Colloque C6, supplément no 10, Tome 478
3) Holden PB, Pert GJ, Kingston AE and Robertson E 1994, *Appl. Phys. B* **58** 23
4) Pert GJ 1994, personal communication
5) Keane CJ 1991, SPIE Proc. Vol. 1551 (SPIE, San Diego)
6) Holden PB, Nantel M, Rus B and Sureau A 1994, submitted to *J. Phys. B*

Hyperfine Splittings, Prepulse Technique, and other New Results for Collisional Excitation Neon-like X-ray Lasers

Joseph Nilsen, Juan C. Moreno, Jeffrey A. Koch, James H. Scofield, Brian J. MacGowan, and Luiz B. Da Silva

Lawrence Livermore National Laboratory, P. O. Box 808, Livermore, California 94550

Abstract. The observation of hyperfine splitting on an X-ray laser transition is presented and the impact on the laser gain is discussed. We measure the lineshape of the 3p → 3s(J = 0 → 1) transition in neon-like niobium and zirconium and observe a 28 mÅ splitting between the two largest hyperfine components in the niobium(Z=41) line at 145.9 Å, in good agreement with theory. In zirconium(Z=40), no splitting is predicted or observed since the hyperfine effect is proportional to the nuclear moment and this is present primarily in elements with odd Z. The hyperfine splitting is used to explain why the low-Z ions with odd Z have not lased. We discuss the use of a prepulse technique to achieve lasing in low-Z neon-like ions from Z = 22 to 32 on the 3p → 3s(J = 0 → 1) transition with wavelengths from 326 to 196 Å. Using this technique on selenium(Z=34) we show a large enhancement of the J = 0 → 1 transition at 182 Å. Using a series of short pulses to drive selenium we observe the 182 Å line to completely dominate the spectra. In an effort to reduce the large density gradients associated with hydrodynamic expansion, we discuss the use of low density foams for the laser target and present results which show lasing in zirconium aerogel with an initial density of 20 mg/cm^3. Finally we discuss recent double slab experiments with ruthenium(Z=44) targets in which we observe lasing at 117 and 118 Å for the first time.

Experimental Setup

Experiments were conducted at Lawrence Livermore National Laboratory (LLNL) on the Nova laser using λ = 0.53 μm. In a typical experiment on the low-Z materials the Nova laser illuminates a 125 μm thick, 4.5 cm long slab target of nickel. The actual target length is reduced to 3.8 cm by a 16% gap in the center. The pump laser beam was a 600 ps FWHM gaussian pulse with 1100 J of energy in a 120 μm wide (FWHM) by 5.4 cm long line focus, resulting in a peak intensity of 34 TW/cm^2. A 6 J prepulse (also 600 ps FWHM) preceded the main pulse by 7 ns. For the selenium experiments, a 1 μ thick coating of selenium on a nickel substrate was used as the target and the main pulse was increased to 2200 J while keeping the prepulse energy and delay the same as above. As an alternative to using a weak prepulse before the main pulse, we also illuminated targets with a series of three 100 ps FWHM gaussian pulses which were 400 ps apart and had 250 to 400 J of energy in each pulse. For the niobium and zirconium experiments 3.0 cm long slab targets were illuminated by a 500 ps square pulse with 2.4 kJ of energy in a 120 μm wide (FWHM) by 3.6 cm long line focus, resulting in a peak intensity of 130 TW/cm^2. For the zirconium aerogel experiment a 2.0 cm long target was illuminated by a 500 ps square pulse with 2.4 kJ of energy in a 120 μm wide (FWHM) by 2.4 cm long line focus, resulting in a peak intensity of 200 TW/cm^2 while the ruthenium experiments used 1.0 cm long targets illuminated with the same beam energy, and therefore twice the intensity, as the aerogel.

The principal instruments were a time-gated, microchannel plate intensified grazing-incidence grating spectrograph(MCPIGS) and a streaked flat field spectrograph(SFFS); both of these instruments observed the axial output of the X-ray laser. The MCPIGS provided angular resolution over 10 mrad near the X-ray laser axis, while the SFFS integrated over an angular acceptance of 10 mrad. The MCPIGS used a 600 line per mm grating and had spectral coverage of approximately 150 to 680 Å. For the zirconium, niobium, and ruthenium experiments a 1200 line per mm grating was used with spectral coverage of approximately 75 to 340 Å. When measuring the laser lineshape the SFFS was replaced with a high-resolution, grazing incidence grating spectrometer which recorded time-integrated but spatially resolved data using a Princeton Instruments camera with a backside illuminated EEV CCD. This spectrometer was centered on the laser axis with an angular acceptance of 12 mrad and had a measured spectral resolution of 20000 at 146 Å and spectral coverage of 2 Å. The angular resolution of all three instruments was perpendicular to the target surface.

Experimental Results and Analysis

Over the last few years we have observed rather different behavior for neon-like lasers which used materials with even Z as compared with odd Z. Since elements with odd Z have a nuclear spin and a nuclear moment and those with even Z tend to have no nuclear spin, one possible explanation for this anomalous behavior is that hyperfine splitting is playing an important role in the gain of the neon-like laser lines. Hyperfine splitting can affect the gain of the laser line by effectively increasing the linewidth. Since the gain is inversely proportional to linewidth the gain will decrease. If the splitting is large enough, a single line may be split into several weaker lines. The hyperfine effect is largest for the $J = 0 \rightarrow 1$ line which dominates the spectra of the low-Z neon-like ions. We did a series of experiments, described in a previous paper(1), to measure the lineshape of the $J = 0 \rightarrow 1$ laser line in neon-like niobium. Niobium was chosen because it has a very large nuclear spin, I = 9/2, and a large nuclear moment, μ = 6.167, its wavelength is in the range of the high resolution spectrometer which we had available, and it had been observed to lase(2). Figure 1 shows the measured(solid) and calculated(dotted) intensity versus wavelength for the niobium line. Two components are clearly visible with a separation of 28 mÅ, which is very close to the 32 mÅ prediction given the 7 mÅ resolution of the spectrometer. This is the shortest wavelength transition and most highly ionized plasma in which the hyperfine effect has been directly observed on a laser transition(1). If we consider vanadium, assuming an ion temperature of 50 eV based on calculations, the hyperfine splitting(3) reduces the gain coefficient of the $J = 0 \rightarrow 1$ laser line at 304 Å by 40%. Given a nominal gain coefficient of 2.6 cm^{-1} for titanium(4,5), this reduces the gain coefficient to 1.6 cm^{-1} for vanadium. For the 3.8 cm long targets tried with vanadium, this would make the vanadium fifty times weaker than the titanium. While this is still within the detectable range of the diagnostics, the hyperfine effect appears to play a major role in the non lasing of vanadium and scandium and the poor lasing of the other odd Z ions(3).

Using the prepulse technique described above we have done experiments(4-6) on all elements from calcium(Z=20) to germanium(Z=32) with the exception of gallium(Z=31). Different illumination conditions were used depending on the element but most were tried with the nominal conditions described above. Lasing

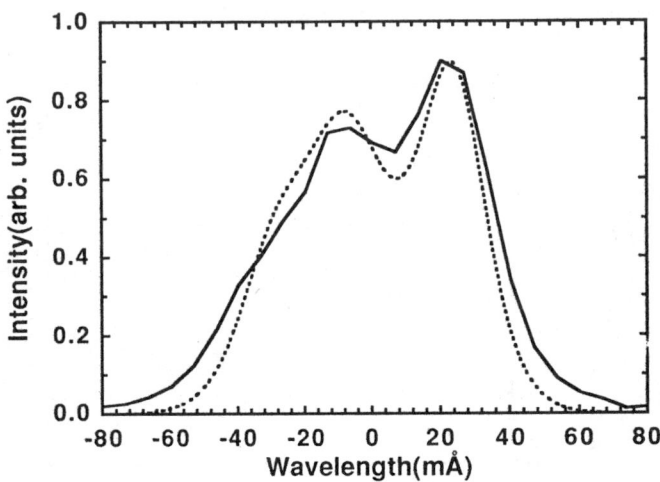

Fig. 1. Measured(solid line) and calculated(dotted line) intensity vs wavelength around line center for the $J = 0 \rightarrow 1$ laser line in neon-like niobium at 145.9 Å. The curves are normalized to the same intensity and a peak gain-length product of 2 is assumed for the calculated curve.

was determined by observing the high spectral brightness of the lasing lines relative to the strong emission lines on-axis, the absence of the lasing lines off-axis, the short time duration of the lasing relative to the optical drive pulse, and the exponential growth of the laser output as the length was increased. In the experiments(4-6) we saw strong lasing on the 3p → 3s(J = 0 → 1) lines at 326, 285, 255, 231, 212, and 196 Å in titanium, chromium, iron, nickel, zinc, and germanium, respectively while scandium(Z=21), vanadium(Z=23), and manganese(Z=25) do not lase. Weak lasing was observed in cobalt(Z=27) and copper(Z=29). Calcium did not lase but we attribute that to target handling difficulties. For the even Z ions we also observed some weak J = 2 → 1 laser lines. Figure 2 shows the spectrum from the MCPIGS spectrograph for a zinc target. The strong J = 0 → 1 laser line at 212 Å completely dominates the spectrum and is fifteen times more intense than the weak J = 2 → 1 laser lines at 262 and 267 Å. In recent experiments(7) on selenium using the prepulse technique, the J = 0 → 1 line at 182 Å suddenly jumps up and becomes a strong line as was originally predicted but never observed in standard X-ray laser experiments(8). However, unlike the lower-Z cases, the J = 2 → 1 lines at 206 and 209 Å still dominate the spectrum. Based on our calculations(4,5), we believe the prepulse is playing a key role in creating a larger, more uniform density plasma, at the densities required for lasing at these wavelengths. The combination of the small gain region with the inability to propagate the length of the laser is no doubt the reason the low-Z neon-like lasers have not worked without the prepulse.

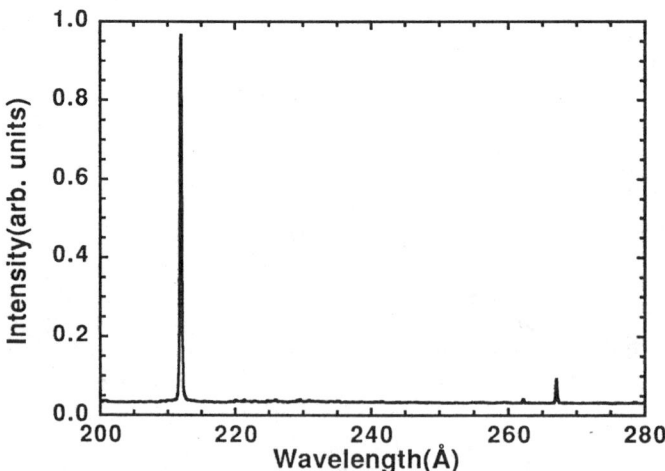

Fig. 2. MCPIGS spectrum of a 3.8 cm long target of zinc using the prepulse technique.

As another approach to preforming a plasma, we have done experiments on germanium and selenium targets using a multiple pulse technique. For the selenium experiments we illuminated 3.0 cm long targets with a series of three 100 ps FWHM gaussian pulses which were 400 ps apart and had 400 J of energy in each pulse. Figure 3 shows the selenium spectrum from the MCPIGS spectrograph. The 182 Å line completely dominates the output. The time resolved spectrum shows strong lasing on the second and third pulse. The first pulse provides the initial heating and expansion of the plasma, thereby preparing the plasma for lasing during the subsequent pulses.

To eliminate the violent hydrodynamics which takes place when a solid target is heated by an optical laser such as Nova we are pursuing the use of foam targets which could potentially be fabricated at the final density needed for lasing and be volume heated by the Nova laser, thereby eliminating the large density gradients in the plasma. These density gradients cause significant refraction of the X-ray laser as it propagates down the laser axis and limits the effective length of the plasma as well as the laser coherence. We have tried several experiments on different foams with modest success. To achieve very low density we have tried molybdenum-doped agar foam (a hydrocarbon foam) at 3 mg/cm^3 and selenium-doped agar foam at 8 mg/cm^3. Both foams were nominally 50% metal by weight and neither lased. The agar has the difficulty that the cell size is micron scale and the foam must be doped with the element of interest, either as small particles or a compound. A more promising route is using aerogels such as

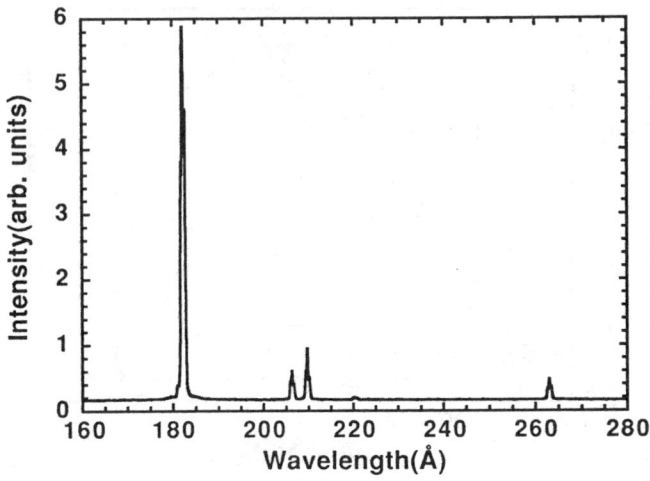

Fig. 3. MCPIGS spectrum of a 2.5 cm long target of selenium illuminated with multiple pulses.

zirconium aerogel which has a small cell size the order of 500 Å and is pure zirconium oxide, so the doping issue is avoided. We have tried zirconium aerogel with densities of 470, 90, and 20 mg/cm^3 and they have lased quite well, as shown in Fig. 4 which presents the intensity versus wavelength as measured with the MCPIGS spectrograph for the lowest density foam. The pair of J = 2 → 1 lines at 146 and 148 Å lase quite well. Presently we are trying to produce lower

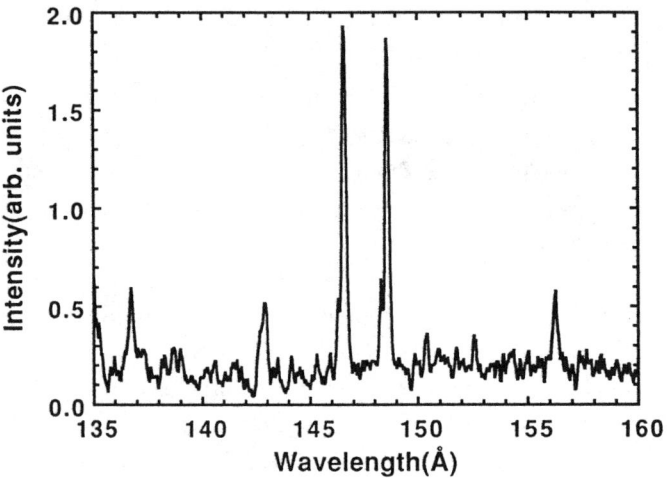

Fig. 4. MCPIGS spectrum of a 1.7 cm long target of zirconium aerogel.

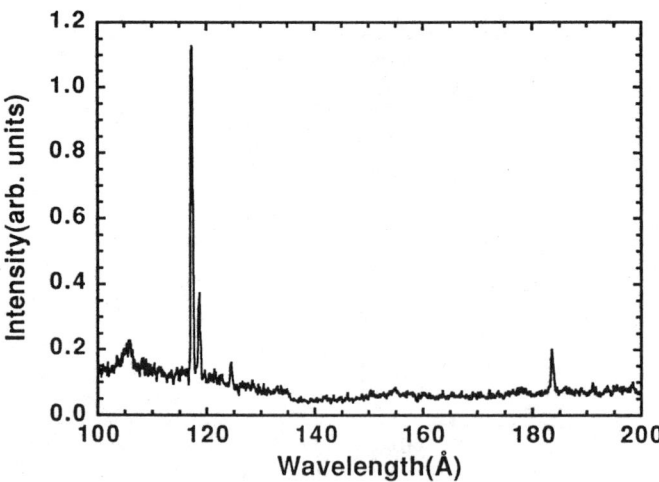

Fig. 5. MCPIGS spectrum of a double slab target of ruthenium.

density aerogel. Silicon aerogel is the most mature technology and can be produced down to 1 mg/cm^3 so we are optimistic that the zirconium aerogel density can be lowered to the 1 - 3 mg/cm^3 range appropriate for lasing.

In an effort to make brighter X-ray lasers with slab targets we have been doing experiments using double slab targets. We recently observed lasing in ruthenium(Z=44) at 117 and 118 Å, as shown in Fig. 5, using this approach. The target consisted of two 1.0 cm long slab targets which were separated by 300 μ in the transverse dimension and were aligned end to end with a separation of 0.5 cm. The slabs were illuminated from opposite sides by two beams of the Nova laser using 2.4 kJ of energy from each beam in a 500 ps square pulse.

Conclusions

We show that using the prepulse technique many low-Z neon-like ions from titanium to germanium lase well on the 3p → 3s(J = 0 → 1) transition. Using this technique on selenium caused the "missing" J = 0 → 1 line at 182 Å to lase quite strongly. New multiple pulse experiments on selenium show the 182 Å line completely dominating the spectrum. The hyperfine effect is shown to be the dominant line broadening mechanism for the 3p → 3s(J = 0 → 1) neon-like niobium laser line at 145.9 Å. We measured the lineshape of this transition and observed a 28 mÅ splitting between the two largest hyperfine components, in good agreement with theory. This is the largest hyperfine splittings ever measured on a laser transition. In the effort to produce more uniform plasma, lasing is observed for the first time using a foam target of zirconium aerogel. Finally, the double slab target was used successfully to demonstrate lasing in ruthenium at 117 and 118 Å for the first time.

Acknowledgements

The authors would like to thank Larry Hrubesh for providing the zirconium aerogel and Sharon Alvarez, Hedley Louis, Tony Demiris, Judy Ticehurst, and the Nova facilities crew for providing support for the experiments. The support of S. B. Libby, D. A. Nowak and D. L. Matthews is greatly appreciated. Work performed under the auspices of the U. S. Department of Energy by the Lawrence Livermore National Laboratory under contract No. W-7405-ENG-48.

References

1. J. Nilsen, J. A. Koch, J. H. Scofield, B. J. MacGowan, J. C. Moreno, and L. B. Da Silva, Phys. Rev. Lett. **70**, 3713-3715 (1993).

2. J. Nilsen, J. L. Porter, B. J. MacGowan, L. B. Da Silva, and J. C. Moreno, J. Phys. B **26**, L243-247 (1993).

3. J. H. Scofield and J. Nilsen, Phys. Rev. A **A 49**, 2381 - 2388 (1994).

4. T. Boehly, M. Russotto, R. S. Craxton, R. Epstein, B. Yaakobi, L. B. Da Silva, J. Nilsen, E. A. Chandler, D. J. Fields, B. J. MacGowan, D. L. Matthews, J. H. Scofield, and G. Shimkaveg, Phys. Rev. A **42**, 6962-6965 (1990).

5. J. Nilsen, B. J. MacGowan, L. B. Da Silva, and J. C. Moreno, Phys. Rev. A **48**, 4682 - 4885 (1993).

6. J. Nilsen, J. C. Moreno, B. J. MacGowan, and J. A. Koch, Applied Physics B **57**, 309 - 311 (1993).

7. J. Nilsen and J. C. Moreno, "Using the prepulse technique to enhance the weak 18.2 nm laser line in neon-like selenium," Opt. Lett. (in press, 1994).

8. R. C. Elton, *X-ray Lasers* (Academic Press, Inc., San Diego, 1990), pp. 99 - 126.

Theory of Ne-Like Collisional X-Ray Lasers

S. Jacquemot

Centre d'Etudes de Limeil-Valenton 94195 Villeneuve Saint-Georges Cedex FRANCE

Abstract. The modeling of collisionally pumped neon-like soft x-ray lasers is discussed in terms of hydrodynamics, atomic physics, kinetics and beam propagation. The influences of different mechanisms [bi-dimensional expansion, line transfer and reabsorption, unstationarity, refraction ...] are illustrated by examples from application to neon-like copper, which moreover show how these complex simulation procedures are crucial in explaining experimental results.

METHODOLOGY

The idea of a neon-like collisional excitation inversion scheme was suggested early in the 1970s by Zherikhin and co-workers (1). Since 1984 and the theoretical groundwork given by Rosen *et al.* for the first lasing action at Livermore (2), the physics used to design and analyze such soft x-ray laser experiments has been deeply improved. It can be now classified in four main areas: {i} laser heating and target hydrodynamics, {ii} detailed atomic physics, {iii} ionization-inversion kinetics and {iv} x-uv radiation propagation and amplification (Fig. 1) and each laboratory involved in this research has in fact developed its own set of numerical tools (3–6).

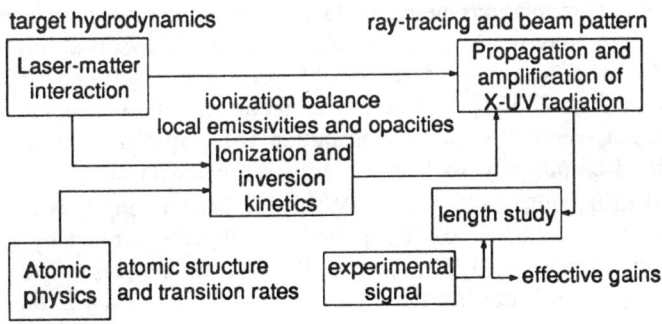

Figure 1. Steps used in modeling of neon-like collisional x-ray lasers.

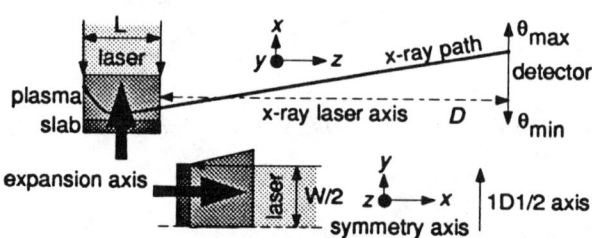

Figure 2. Test problem: a gaussian [Δt_L=600 ps FWHM] line-focused [W=150 μm × L (cm)] laser pulse at ω_0 irradiates a slab target of copper [nuclear charge Z=29] [the corresponding experiments being performed with zinc — Z=30 — (see P. Jaeglé et al), the usual $(Z-9)^6$ scaling law (4) has been used to determine the peak irradiance, I_L=1.2 10^{13} W/cm^2, reached at time $t=t_L=0$ — 5% error bars may be applied to all the following numerical results due to this "shift" in the Mendeleev table]; a detector, viewing a small solid angle [$\Delta\theta=\theta_{max}-\theta_{min}$, variable] about the x-ray laser axis, is located at large distance from the plasma [D=185 cm].

HYDRODYNAMICS

The hydrodynamics provides time and space profiles for electronic [T_e] and ionic [T_i] temperatures, mass densities and velocity gradients. Ns interaction being involved, the classical geometric optics approximation is valid and the exact resolution of the Helmholtz wave equation is not required, in opposite to ps schemes. Then are assumed laser deposition by inverse Bremsstrahlung and resonant absorption through a given energy dump at critical density [~30% deposition, the computations being in fact quite insensitive to this parameter]. Some corrections may be done to the usual Johnston-Dawson absorption coefficient (7): the most widely used are the Skupsky one describing strong ion-ion correlations in high-Z plasmas and the Langdon effect to take into account a non-Maxwellian electron velocity distribution (8). Spitzer thermal conduction (9) is modified empirically so that the diffusive electron heat flux is harmonically limited to $f_{lim}mn_e(k_BT_e/m)^{3/2}$. When the laser intensity is below 2 10^{13} W/cm^2, a classical value of 0.6 is applied. Stronger f_{lim} must be invoked for higher-Z experiments up to 0.03 above 10^{14} W/cm^2 (10,11). The parametric instabilities are usually not introduced in the simulations even if the stimulated Brillouin scattering is suspected to play some role in anomalous absorption and ion heating processes (6,12). Real 2D simulations [such as Livermore performes (13)] being quite expensive and difficult to post-process, so-called 1D1/2 codes have been developed, considering self-similar models for lateral expansion and thermal conduction (14), and suffice to restitute the drastic effects due to these two additional coolings (Fig. 3).

The Ne-like collisional X-ray laser experiments involving high-Z active materials, radiation transport can't be ignored in hydrodynamical simulations. If just radiative losses [through Bremsstrahlung, photo-recombination and line emission] are included, the electronic temperature is strongly underestimated (11) (Fig. 4). Opacities and emissivities required by the multi-angle and multi-group treatment of this radiative transfer are usually given by a non-LTE screened hydrogenic average-ion model (15) [except in the case of the EHYBRID code in York (6) and the new GLF code in Livermore (16), where hydrodynamics and kinetics are not decoupled] which also evaluates the ionization state of the plasma and thermodynamically consistent non-LTE equations of state. The amplifications happening thereabouts t_L, these calculations are steady-state; unstationary effects are seen only during the late recombination phase [t > +600 ps] which is, unlike the recombination schemes, not studied.

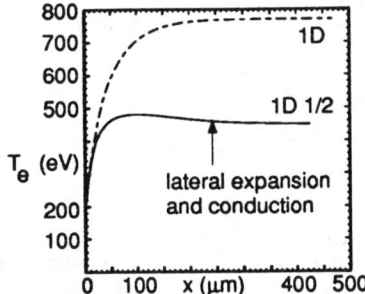

Figure 3. Influence of the lateral coolings on the temperature spatial profile [t ≃ +320 ps]: strong decrease of the peak T_e and limitation of the expansion [the amplification area, not shown here, is also affected, especially contracted in time and space].

Figure 4. Influence of the radiative transfer on the temperature temporal profile [in one specific "inner" Lagrangian cell]: T_e is directly increased by reabsorption in the inner part of the plasma of radiation emitted by the corona.

ATOMIC PHYSICS

The collisional-radiative models, used in the kinetic modeling, require an extensive data base: energy levels and transition rate coefficients, calculated by more or less sophisticated codes. A refined modeling of the excited-state level structure for the neon-like isoelectronic sequence includes the L-, M- and N-shell states in detail [i.e. 89 levels in intermediate coupling] and an additional set of 3 Rydberg series from n=5 to n=10 leading to the 3 F-like n=2 ground states. The Na-like M-shell may also be described. The kinetics must of course take into account all the ionic sequences and all the atomic processes. Then these detailed levels are surrounded by a simpler model, hydrogenic in Limeil or maybe more sophisticated elsewhere (6,17). Accurate collisional excitation and radiative decay rates into the detailed manifold must be obviously calculated, especially for the forbidden 2p/3p transitions (18). Hydrogenic or semi-empirical formulations are used to complete the data base, but two processes, from and to Ne-like n=2–3 levels, proved to be crucial and have to be correctly described: the collisional ionization (Fig. 5) and the dielectronic recombination (see M. Cornille and S. Jacquemot). The simulations presented below are restricted to only two lasing lines, the $(1s^2 2s^2 2p^5)3p\ ^1S_0/3s\ ^3P_1$ line [the well–known 182 Å in selenium] refered as "0/1" and corresponding [they share the same lower level] $3p\ ^1D_2/3s\ ^3P_1$ one ["2/1"].

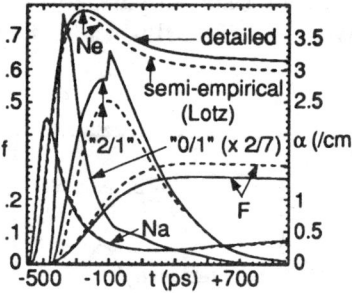

Figure 5. Influence of the formulation of the collisional ionization [in the "optimal" Lagrangian cell for the "2/1" inversion]: a non negligible 30% variation is observed on the "2/1" local gain coefficient α, directly due to very small – 3% – changes in F- and Ne-like fractional abundances f.

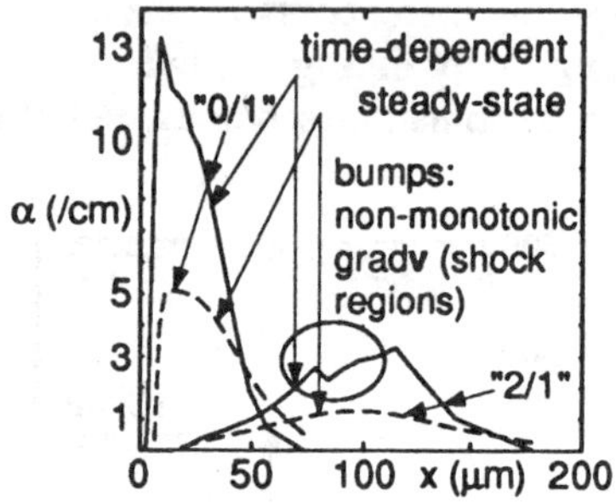

Figure 6. Time-dependent kinetics: a severe underestimation of the local gain coefficients α can follow from steady-state calculations [here, at "optimal" times: t ≃ −100 ps for the "2/1" inversion and −350 ps for the "0/1" one], due to fast hydrodynamical evolutions.

KINETICS

This major modeling step calculates the time-dependent ionization balance, level populations, electron density [n_e], emissivities [j] and opacities – or local gain coefficients – [k=-α] for all the required lines [especially the 3p/3s lasing ones], at specified positions in the plasma. To simplify, it can be considered as just the resolution of a set of coupled equations, but special attention has to be paid to some specific mechanisms. The spatial distributions of density and temperature change so significantly over the time scale of the x-ray laser pulse that gain must be treated in a time-dependent manner (Fig. 6). Even if steady-state calculations provide useful information on level populations and save computer-time, they definitely can't restitute accurate gain histories.

Since radiative decay to the ground state is the dominant mechanism for depopulating the lower lasing level and for maintaining the inversions, reabsorption of 2p/3s resonance lines has an important, and deleterious (Fig. 7), influence on the performances of the Ne-like x-ray laser. Trapping in the 2p/3d one is also of some interest as the excited states from which it issues contribute by cascades to the population of the upper lasing level. This process is well handled with escape probability methods which decouple the spatial dependence in the kinetic problem. This simplification has been validated by numerical solutions of the line transfer equations (19). The 1D planar geometry, high expansion velocity gradients and large Doppler shifts make the trapping rather local in space and then the Sobolev approximation can be used [equation 11 in (20)].

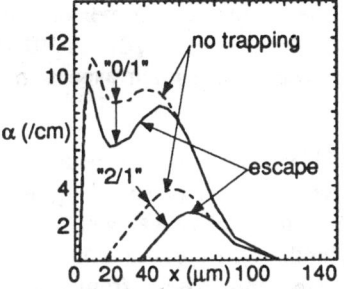

Figure 7. Resonance line trapping: the spatial evolutions of the local gain coefficients [at t ≃ −240 ps] in the inner part of the plasma are strongly influenced by reabsorption.

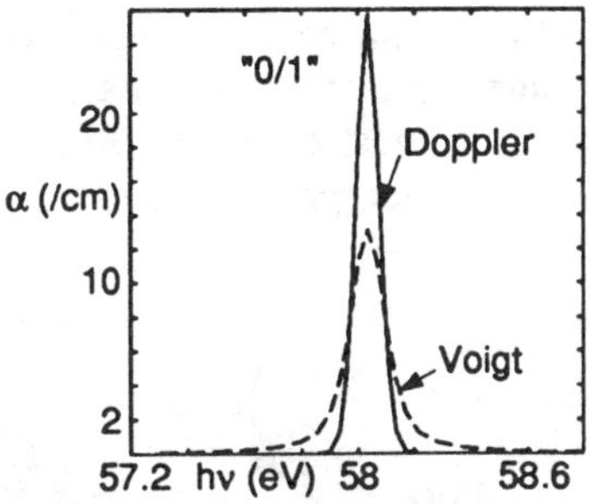

Figure 8. Line broadening: the reduction of gain due to the use of a more realistic line profile can reach 50% for the "0/1" line [at optimum conditions].

The collisional excitation inversions occuring, by definition, in high density regions, collision broadening can't be neglected. The radiation, or natural, and the collision damping being supposed completely uncorrelated, and the effects of the unavoidable Doppler broadening being taken into account, Voigt profiles are adopted to describe the x-ray laser line shapes. The total Lorentz width [Γ] is obtained by summing all outgoing kinetic rates from both the upper and the lower levels of the involved lasing line. The usual Doppler gain [α_D] is then multiplied by a complementary error function which is the exact expression of the Voigt function at line center: $\alpha = \alpha_D e^{a^2} \text{erfc}(a)$ [$a = \frac{\Gamma}{4\pi\Delta\nu_D}$ – refer to Chapter 9 in (21) – where $\Delta\nu_D$, the Doppler width, may involve an arbitrary corrected temperature, for example at Limeil $T_D = 0.5(T_e + T_i)$, to simulate additional ion heating mechanisms and match previous selenium line width measurements (22)]. Hyperfine splitting [$\Delta\nu_s$] must be included if larger than these broadenings, but can't be invoked in the studied copper lasing system: $\Delta\nu_s \lesssim 5$ meV while $\Delta\nu_D \simeq 100$ meV (23).

RAY-TRACING

Amplification is achieved as the x-rays propagate down a long narrow plasma column. Due to electron density gradients, refraction tends to curve them, according to the ray equation $\frac{d}{ds}\{\eta\frac{d\mathbf{r}}{ds}\} = \nabla\eta$, where η is the plasma index, and therefore to direct them out of gain regions, which shortens the effective propagation length and reduces the final gain value (24). The transport of the time- and frequency-dependent intensities $I(t,\nu)$ along such trajectories (Fig. 9) is found from the steady-state transfer equation $[\eta^2\frac{d}{ds}\{I/\eta^2\} = j - kI]$ in a refracting medium. An angular radiation pattern $F(t,\nu,\theta)$ is then mapped onto transverse position [for a given exit angle θ] at the plane of a virtual detector, located at large distance from the plasma. Integrations over desired combinations of frequency, angle and time can be performed, following experimental device performances, to determine the macroscopic nature of the output beam (Fig. 10). A 2D [x,z] ray-trace is, in most cases, accurate enough to describe the main aspects of the Ne-like x-ray laser propagation.

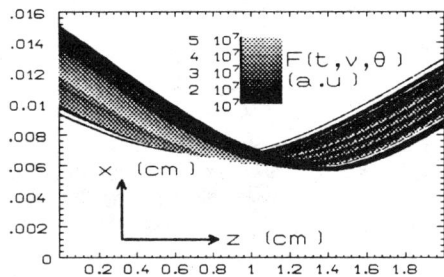

Figure 9. Families of plane ray trajectories for a 2 cm long copper plasma at "0/1" line center ν [θ = 7.5 mrad, t $\simeq$ +180 ps].

This procedure is repeated twice, for two different plasma lengths, and an effective theoretical gain is finally obtained, as the measured one to allow accurate comparisons, from a Linford formula (25). The refraction process and integrations [with an exacerbated sensibility of the final results to any variation in their limits] finally lead to great reductions of gain values. Furthermore gain narrowing and saturation can be naturally predicted from gain-length studies (5).

Some multiple pass issues may also be explored as, for example, the dependence of the x-ray laser output energy and coherence [using additional wave optics calculations (13)] on the duration of lasing and the characteristics of the required multilayer mirror [reflectivity, shape, damage threshold] (26).

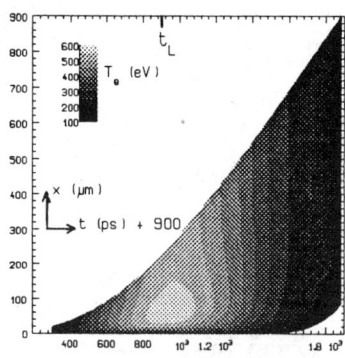
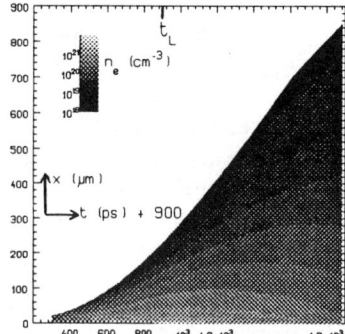

(Continued)...

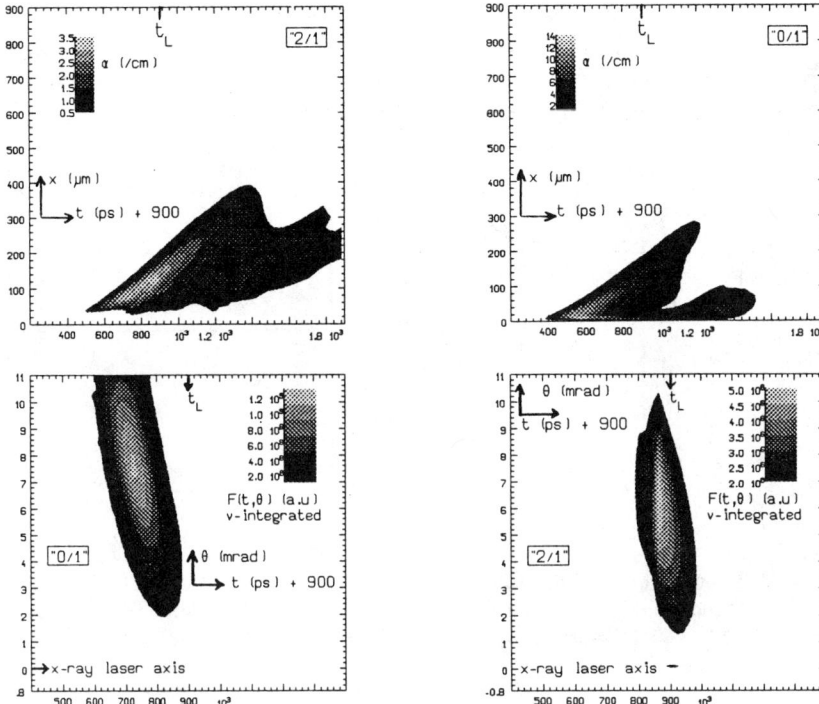

Figure 10. Complete simulation: {i} time- and space-maps for electronic temperature and density [first entry, last page] and local gain coefficients [second entry] – the localization of the inversions is clearly exhibited: the "2/1" occurs in the corona, approximately 100 μm from the target surface and thereabouts t_L, where the temperature reaches its maximum value, while the "0/1" takes place near the critical surface [$n_c \simeq 10^{21}$ cm^{-3}], 10 μm from the target at lower temperature and earlier times –; {ii} time- and angle-maps for beam patterns [third entry] – while the "2/1" flux profile is smoothed over, the "0/1" one strongly peaks at 7.5 mrad and, as density gradients relaxe, moves towards the axis, which is the signature of high refraction influenced amplification. The "0/1" [resp. "2/1"] gain decreases from locally 13.1 [resp. 3.3] cm^{-1} to peak 4.92 [resp. 2.57] cm^{-1}, if integrated over the line profile and a $[-0.8, 0.8] \cup [2, 11]$ (mrad) detector aperture, or 3.89 [resp. 1.72] cm^{-1} on the total experiment duration (see P. Jaeglé et al. for the theory/experiment agreement).

BACK TO THE POPULATION INVERSION SCHEME

To summarize, the "0/1" inversion appears to be purely collisional – it occurs during the ionization phase near the critical surface, which optimizes the collisional populating flux from the ground state to the upper lasing level – while the "2/1" one is, so to speak, hybrid – it happens during the primary recombination period at high temperature, when the probabilities of finding F-like ions and high Ne-like excited states are the largest, which favours the cascade processes and the dielectronic recombination (Fig. 11). The same interpretations can be given from temporal (Fig. 5) or spatial (Fig. 6) local gain profiles (27).

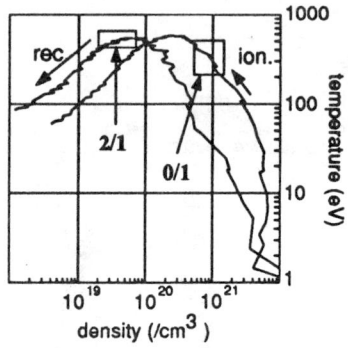
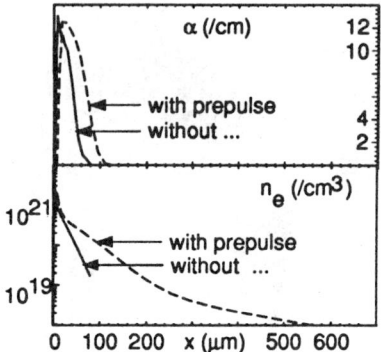

Figure 11. Thermodynamical paths of the "optimal" Lagrangian cells [in which the local gain coefficients reach their maxima].

Figure 12. Influence of a {−5 ns, 1% I_L} prepulse on density and "0/1" local gain spatial profiles [t ≃ −360 ps].

OPTIMIZATIONS

Some experiments have been recently performed to improve the efficiency of the Ne-like collisionally pumped x-ray laser, using {i} a low intensity prepulse (28) or {ii} multiple shorter pump pulses (see L.B. Da Silva *et al.*). Some interesting theoretical behaviors of the "0/1" line may then be stressed.

{i} A prepulse technique, which creates a larger, and more uniform, plasma and cancels density gradients, increases the inversion area (Fig. 12) [without changing the local optima for T_e, n_e and α] and makes the propagation of the x-ray radiation easier (Fig. 13). In addition, it may hydrodynamically smooth [especially in the case of very low intensities] any inhomogeneity in the laser energy deposition, whose role in the macroscopic evolution of the "0/1" beam pattern has to be properly defined.

{ii} A multiple pump technique is used to produce very short [≃ 40 ps] x-ray laser pulses, which makes them a good source for 2D imaging: the first incident laser pulse creates the lasing plasma while the second one heats and ionizes it, to produce optimal hydrodynamical and kinetic conditions suitable for high gain; a possible third one can reproduce them to extend the inversion.

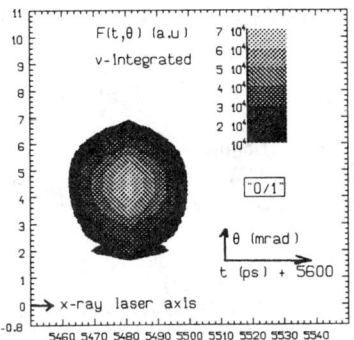 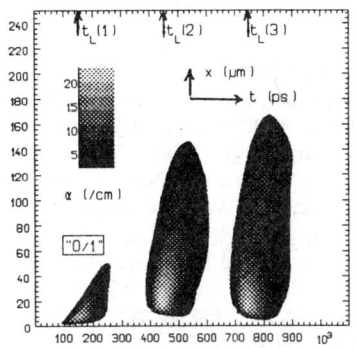

Figure 13. Influence of a 600 ps FWHM gaussian prepulse – a peak intensity I_l = 1% I_L is reached 5 ns before the main pulse – on the time- and angle-map for the "0/1" beam pattern: the deflection angle is reduced to 4.5 mrad, the gain not damaged by refraction [8.9 cm^{-1}] but the brightness greatly subdued [from 10^9 to 7 10^4 in arbitrary units].

Figure 14. Influence of the multiple pump technique – 3 {100 ps FWHM, I_L} laser pulses separated by 300 ps – on the "0/1" local gain: 3 amplification lobes can be observed, with increased efficiency [8.13 cm^{-1}, after propagation and integrations, and 8 10^{13} a.u brightness] and refraction of the second one [10.25 mrad, hence a necessary use of focus optical device].

CONCLUSION

The success of the current simulations shows that the mechanisms governing population inversions in Ne-like ions and x-uv radiation amplification in refracting plasma columns seem to be quite understood and numerically overcome. The "J=0 mystery" is even to be solved, without changing anything to the computation ingredients [especially in atomic physics]: experimental integrations [destroying strongly peaked gain profiles] and inhomogeneities of laser energy deposition onto the target surface [hydrodynamically smoothed by the prepulse technique] may, most certainly, be invoked.

ACKNOWLEDGMENTS

I am grateful to M. Cornille and J.C. Gauthier for their atomic physics calculations, to all members of the LSAI x-ray laser team for enlightening discussions and to L. Bonnet and A. Decoster for critical reading of this manuscript.

REFERENCES

1. Zherikhin, A.N., *et al.*, *Sov. J. Quantum. Electron.* **6**, 82 (1976)

2. Rosen, M.D., et al., *Phys. Rev. Lett.* **54**, 106 (1985) — Matthews, D.L., et al., *Phys. Rev. Lett.* **54**, 110 (1985)
3. London, R.A., *J. Phys. B* **22**, 3363 (1989) [LLNL]
4. Jacquemot, S., and Decoster, A., *Laser and Particle Beams* **9**, 517 (1991) [CEL-V]
5. Holden, P.B., et al., *J. Phys. B* **27**, 341 (1994) [York — RAL]
6. Guoping, Z., et al., *Inst. Phys. Conf. Ser.* **125**, 327 (1992) [Beijing — SIOFM]
7. Johnston, T.W., and Dawson, J.M., *Phys. Fluids* **16**, 722 (1973)
8. Skupsky, S., *Phys. Rev. A* **36**, 5701 (1987) — Langdon, A.B., *Phys. Rev. Lett.* **44**, 575 (1980)
9. Spitzer, L., and Härm, R., *Phys. Rev.* **89**, 977 (1953)
10. Rosen, M.D., *Comments Plasma Phys. Controlled Fusion* **8**, 165 (1984)
11. Jacquemot, S., and Decoster, A., *Rapport des activités laser CEA/CEL-V* **4**, 14 (1994)
12. Rosen, M.D., *Phys. Fluids B* **2**, 1461 (1991)
13. Ratowsky, R.P., et al, *Inst. Phys. Conf. Ser.* **125**, 315 (1992)
14. Jacquemot, S., et al, in *Ultrashort Wavelength Lasers II*, S. Suckewer, ed., Proc. **SPIE 2012**, 180 (1994) — Pert, G., *J. Fluid Mech.* **131**, 401 (1983)
15. Decoster, A., *Rapport des activités laser CEA/CEL-V* **4**, 3 (1994) — Lokke, W.A., and Grassberger, W.H., *LLNL Report* **UCRL-52276** (1977) — Pollak, G., *Los Alamos Report* **LA-UR-90-2423** (1990)
16. Wan, A.S., et al, *Inst. Phys. Conf. Ser.* **125**, 293 (1992) — Scott, H.A., and Mayle, R.W., *Appl. Phys. B* **58**, 35 (1994)
17. Osterheld, A.L., et al, *Inst. Phys. Conf. Ser.* **125**, 309 (1992)
18. Cornille, M., et al, *At. Data & Nucl. Dat. Tables*, in press
19. Lee, Y.T., et al, *Phys. Fluids B* **2**, 2731 (1991)
20. Shestakov, A.I., and London, R.A., *J. Quant. Spectrosc. Radiat. Transfer* **42**, 483 (1989)
21. Mihalas, D., *Stellar Atmospheres*, San Francisco: Freeman W.H. and co, 1978
22. Koch, J.A., et al, *Phys. Rev. A*, in press
23. Scofield, J.H., and Nilsen, J., *Phys. Rev. A* **49**, 2381 (1994)
24. London, R.A., *Phys. Fluids* **31**, 184 (1988)
25. Linford, G.J., et al, *Appl. Optics* **13**, 379 (1974)
26. Eder, D.C., et al, *LLNL Report* **UCRL-JC-105900** (1991)
27. Whitten, B.L., et al, *LLNL Report* **UCID-21152** (1987)
28. Nilsen, J., et al, *Appl. Phys. B* **57**, 309 (1993) — Nilsen, J., et al, *Phys. Rev. A* **48**, 4682 (1993) — Nilsen, J., and Moreno, J.C., *LLNL Report* **UCRL-JC-115290** (1993)

Enhancement of the J=0-1(19.6nm) Transition Relative to the J=2-1(23.6nm) One Using a Prepulse with the Ne-like Germanium XUV Laser System

GF Cairns[1], MJ Lamb[1], CLS Lewis[1], AG MacPhee[1], D Neely[4], P Norreys[4], MH Key[2,4], C Smith[2], SB Healy[3], PB Holden[3], G Pert[3], JA Ploues[3]

(1) Department of Pure and Applied Physics, Queen's University of Belfast, BT7 1NN.
(2) Clarendon Laboratory, Parks Road, Oxford, OX1 3PU.
(3) Department of Physics, York University, York, YO1 5DD.
(4) Central Laser Facility, Rutherford Appleton Laboratory, Chilton, OX11 0QX.

Introduction

We report here an experiment carried out using a very low energy prepulse(~0.02%) with the Ne-like germanium XUV laser. One of the anomalies in these schemes has been the relatively weak lasing observed on the J=0-1 transition compared to that observed for the J=2-1 transitions: the lasing observed on the J=2-1 lines at 23.6 and 23.2nm has been much stronger for the Ne-like germanium laser in all of our previous experiments. This has been contrary to the modelling which predicts that a higher gain should be expected on the J=0-1 transition. It had been suggested that this might be due to the adverse effects of refraction, which tend to bend the ASE radiation out of the gain region; the gain region for the J=0-1 line lies in a higher density region where the density gradients tends to be larger.

Recent modelling(1,2) suggests that the use of a prepulse to pre-form the plasma before driving it with the main pulse, should help to relax the density gradients in the higher density regions and thereby reduce the effects of refraction. Nilsen et al(1) have observed the J=0-1 lasing line to dominate in Cr, Fe and Ti when using the prepulse technique. This led them to conclude that resonant photo-pumping was not a major pumping mechanism in the operation of the Ti XUV laser as originally proposed(1) but rather that refraction out of the gain region was reduced. We have specifically looked at the germanium lasing transitions where modelling has suggested that a significant increase in the ASE for the J=0-1 transition can be expected with the use of the prepulse. This modelling has been carried out for prepulses of 1, 3 ,5 and 7%. Current modelling is taking place at York for very low prepulses of ~0.02% as was used in this experiment. Cohesive forces of the target material are highly significant in this regime and must be taken adequately into account.

Experiment

The experiment was carried out using 3 beams of VULCAN, the Nd-glass laser, which were arranged in the standard format, to give a line focus of ~25mm long and ~100μm wide(3). The targets consisted of germanium stripes, 100μm wide and 22mm long, coated onto glass substrates. Diagnostics included a streaked flat-field spectrometer at one end and a time integrated flat-field spectrometer at the other, used for examining the lasing emission. Resonance lines in the wavelength range 7.0-10.5Å were monitored on a spatially resolving time integrating bragg crystal spectrometer which used a CCD detector and on a streaked crystal spectrometer which gave time resolved spectra recorded onto HP5 film. The timing slit for the streaked lasing emission was offset at the peak of the normal lasing lines at 8±2 mRad from the axis.

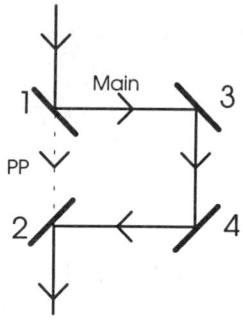

Figure 1 Mirror Dogleg

The prepulse was produced by means of a dogleg, using mirrors(figure 1), which was placed before the preamplifiers in the Nd-glass laser. Mirrors 1 and 2 had a reflectivity of ~97.7% and transmitted ~1.2% each. These were not ideal but were sufficient for this 'look and see' trial. Consequently, a very low energy prepulse of approximately 0.02% of the total energy at target was produced and delivered 5ns early. The drive pulses were ~650ps FWHM at 1.06μm giving irradiances of ~ $1.6 \times 10^{13} W/cm^2$ on target.

Results and Discussion

A spectral profile is given in figure 2 showing clearly the dominance of the 19.6nm line in comparison to the 23.6 and 23.2nm lines for a typical prepulse shot. This trace was produced from a time integrated spectrum measured at 3.0 mRad from the main axis; this corresponded to the peak in the spatial profile for the 19.6nm line. Even though the J=2-1 lines peaked almost on axis their intensity was still significantly below the peak for the J=0-1 line shown in the figure.

Figure 3 shows a time resolved profile of the spectrally integrated intensities for the 19.6, 23.6 and 23.2nm lines. Again the dominance of the 19.6nm line is evident. It tended to start lasing typically 150ps earlier than the J=2-1 lines and had a shorter FWHM pulse duration of about 160ps: the J=2-1 lines tended to lase for longer. It must be noted that the slit was no longer at the peak of the lasing line output which tended to be closer to the axis: this was believed to be responsible for the flattened shapes of the profiles, as the beam deviation tends to vary with time. The Ne-like and F-like 3s-2p resonance lines were also time

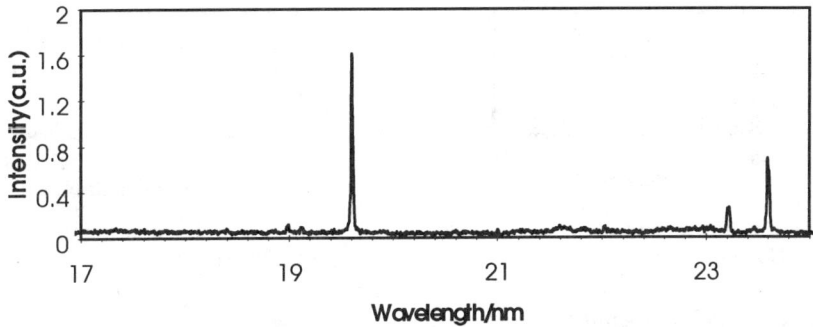

Figure 2 Spectral profile from time integrated data.

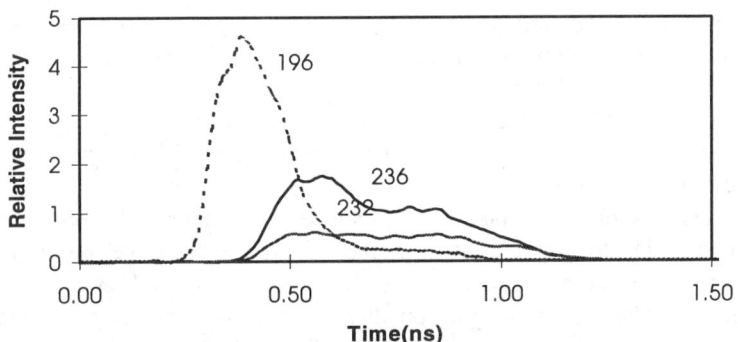

Figure 3 Temporal profile of lasing lines with prepulse

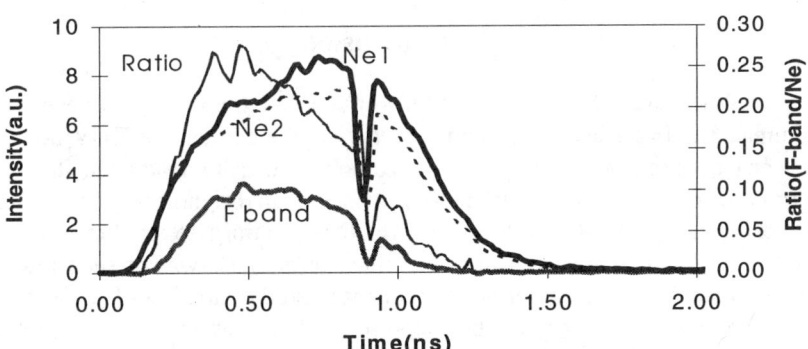

Figure 4 Temporal profile of resonance lines.

resolved. Figure 4 shows the resonance lines corresponding to the same shot as in figure 3. The noticeable dip in these lines is due to a fiducial marker on the detector. Also shown in figure 4 is the ratio of the F-like(3s-2p) band at 9.439-9.553Å and the Ne-like lines at 10.01(Ne1) and 9.762Å(Ne2). This data indicates that the temperature reaches a peak earlier in time relative to the drive pulse. These results suggest, as expected, that the conditions for the J=0-1 to lase

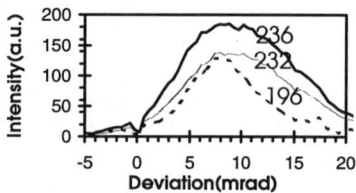

Figure 5 Deviation without the prepulse (#112710as)

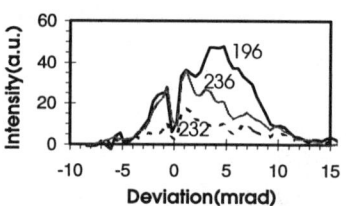

Figure 6 Deviation with prepulse (#132910as)

requires higher temperature and densities.

To assess the effects of refraction on the ASE output, the angular deviation from the main axis of the lasing beams was considered. Figures 5 and 6 show the time integrated profiles for the three lines. Without the prepulse the intensity peaked off axis, as expected(3), at typically 8-10mrad. With the prepulse the deviation is significantly reduced. The 19.6nm line peaked at ~3-4 mrad off-axis whilst the 23.2 and 23.6nm lines peaked almost on axis. This would be consistent with the density gradients having been relaxed in the gain regions with the use of the prepulse.

Three shots were taken under similar conditions with lengths 9, 14, & 22mm in order to provide an estimate of the gain on the lasing lines. The estimated gains were $g\sim3.0\text{cm}^{-1}$ for the 19.6nm line and $g\sim2.5\text{cm}^{-1}$ for the 23.6nm line. Values for the gain on these lines in previous experiments(3) have been measured as $2.7+/-0.1\text{cm}^{-1}$ and $3.8+/-0.3\text{cm}^{-1}$ respectively.

Conclusion

With a very low energy prepulse the J=0-1 transition has been shown to dominate the J=2-1 lasing transitions in the Ne-like germanium XUV laser. Whilst this can be explained in terms of relaxation of density gradients, it is still not certain with such low energy pulses. The overall brightness of the lines did decrease, especially the J=2-1 lines which is contrary to predictions from the modelling; this may be due to the fact that the prepulse was of such low energy. We are planning to carry out an experiment to study variation of parameters such as the timing and energy of the prepulse, with a view to optimising the output intensity of the 19.6nm line; this may provide an alternative route to a bright, monochromatic and highly coherent source.

References

(1) J Nilsen, BJ MacGowan, LB DaSilva, JC Moreno, Physical Review A, 48, 4682(1993)
(2) PB Holden, York, Private Communication
(3) DM O'Neill, CLS Lewis, D Neely, J Uhoimoibhi, MH Key, AG MacPhee, GJ Tallents, SA Ramsden, A Rogoyski, EA McLean, and GJ Pert, Optics Comm., 75, 406(1990)

Overview of Research on Ne-Like Ge Soft-X-Ray Laser in China

Shiji Wang[1], Gunalin Zhou[1], Guoping Zhang[2],
Jiatian Sheng[2] and Shutai Chunyu[3]

[1]Shanghai Institute of Laser plasma
P.O.Box 800-229, Shanghai, 201800, China

[2]Institute of Applied Physics and Computational Mathematics
P.O.Box 8009, Beijing, 100088, China

[3]Southwest Institute of Nuclear Physics and Chemistry
P.O.Box 525, Chengdu, 610003, China

Abstract. With the specially arranged four short targets in a total length of 56mm, a saturated output of soft-X-ray laser with gain length GL=16-17 and minimum divergence angle=(3-4)mrad was obtained for two lasing lines at 23.2 and 23.6nm wavelengths. In the experiment where the double-pass output of a set of targets enters into the other at a distance of ~1000mm, a sixth saturated output with a divergence angle 1.5mrad was achieved.

INTRODUCTION

Since the first successful demonstration of Ne-like Se X-ray laser in 1985 by Lawrence Livermore National Laboratory using exploding foil targets(1) and the first successful demonstration of Ne-like Ge X-ray laser in 1987 by Naval Research Laboratory using slab targets (2), six runs of Ne-like Ge X-ray laser experiments were performed in 1989-1993 on Shenguang laser facility at National Laboratory of High Power Laser and Physics (NLHPLP) in China(3-8) (See table 1). As space is limited, the four runs of experiments by SILP will be introduced mainly in this paper, and for convenience, they will be called experiments 1, 2, 3 and 4 below.

Experimental Conditions

I.Line focus system
Shown in Fig.1 is a cylindrical lens array developed by NLHPLP. D is diameter of array. d is width of cylindrical lens. n is number of cylindrical lens, generally, it is 4 or 6. Line focus system consists of a cylindrical lens array and an aspherical lens. Its operating principle is shown in Fig.2. Each cylindrical lens produces a focus line. A homogeneous illumination focus line can be obtained from superposition of n focus lines. Shown in

TABLE 1. Summary of Ne-like Ge experiments in China

No	Institution	Time	Target(mm)	Mirror	GL	Φ_d(mrad)
1	SILP	1990.4	22+18		14.3	8
2	"	1991.8	4×14		16.5	3
3	"	1992.9	4×14	spherical	17	6
4	"	1993.9	4×14	"		1.5
5	SINPC	1989.7	18		7.2	12.7
6	"	1991.3	20	plane	9.6	12

Note: 1. Only data of 23.2nm line are listed.
2. SILP represents Shanghai Institute of Laser Plasma.
3. SINPC represents Southwest Institute of Nuclear Physics and Chemistry.
4. In experiment 4, an output of X-ray laser was obtained with a sixth saturated intensity.
5. Φ_d represents horizontal divergence angle of X-ray laser beam.

Fig.3 is irradiance distribution along line focus measured by use of cw YAG laser. The irradiance fluctuation along the focus line is ±10% for n=6. The length of focus line can be calculated from the following formula

$$L = dF/f = DF/nf$$

where, F is focal length of aspherical lens, and f is focal length of cylindrical lens. The width of focus line was determined from experiments of passing through slit of laser. The measured width is ~120 μm.

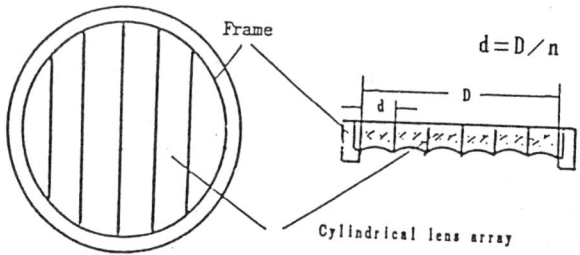

FIGURE 1. Cylindrical lens array

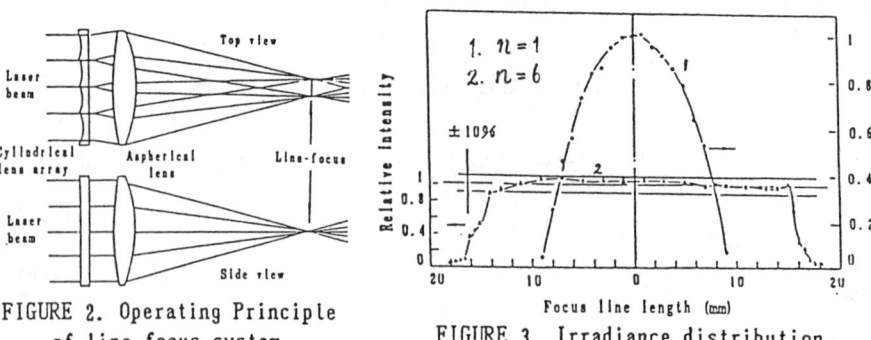

FIGURE 2. Operating Principle of line focus system

FIGURE 3. Irradiance distribution along focus line

II. Driving laser
  Two beams
  Laser parameters of one beam:
   Wavelength   1.054 μm
   Energy output   ~600J
   Pulse width   ~1ns
   Focus line   ~25mm×120μm, ~30mm×120μm
   Irradiance on target ~1×10^{13} w/cm^2
   Irradiance fluctuation
   along focus line   ±10%
III. Target (made by the Applied Physics Department of Tongji University)
  Thickness   2mm
  Width   6mm
  Length   22+18mm, 4×14mm
  Surface flatness   <10μm
IV. Mirror (made by Changchun Institute of Optics and Precision Mechanics)
  Number of Mo/Si-layer pair   30
  Nominal period   12.6nm
  Mo/Si ratio   ~1/2
  Measured reflectivity for normal
  incidence 23.4nm X-ray   ~20%
  Bandwidth   ~2nm
  Roughness of glass substrate   <1nm
  Size of substrate   φ10mm×3mm
  Radius of curvature   80mm
V. Diagnostic instruments
  The experimental arrangement is shown in Fig.4.
  1. Aspherical lens;
  2. Cylindrical lens array;
  3. Flat-field grating spectrometer;
  4. Rowland grating spectrometer;
  5. Space-resolved crystal spectrometer;
  6. Time-resolved crystal spectrometer;
  7. Double-slit camera.

The line intensity given in this paper is time-integrated one obtained from the flat-field grating spectrometer.

With a streak camera coupled to 3, the time-resolved X-ray laser spectrum was recorded. With a stigmatic system placed in front of 3, the spaceresolved

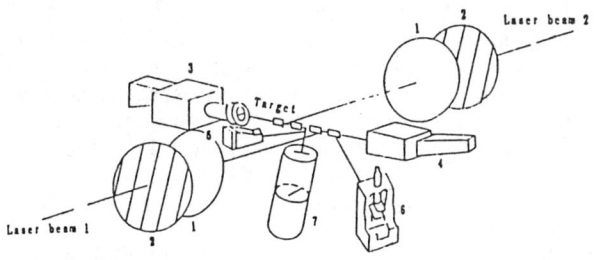

FIGURE 4. Experimental arrangement

X-ray laser intensity across the emission aperture was measured.
In different experiments, different diagonstic instruments were used.

Simulation and Experiment

Three codes have been developed for X-ray laser research since 1986 by Beijing Institute of Applied physics and Computational Mathematics.

I. One-dimensional nonequilibrium radiation hydrodynamic code (JB19).

It provides time- and space-dependent state parameters for a laser-produced plasma, such as electron temperature, ion temperature, electron density and so on. A two-direction pseudo-two-dimensional scheme, in which a plane target was assumed to be a cylindrical one with a radius of curvature equal to twice width of focus line, was used in this code, so that, two-dimensional effect can be simulated for line focus irradiation of exploding foil and slab targets.

II. Steady-rate-equation code on Ne-like ions and F-like ions (ALPHA).

It provides time- and space-dependent small-signal gain coefficient and spontaneous emission rate. Based on an escape probability approximation, trapping of seven lines from dipole transitions of $n=3$ levels to ground level was considered.

III. X-ray laser path code (XBY).

Based on geometrical optics approximation, it calculates propagation and amplification of X-ray laser line and provides intensity of X-ray laser in temporal, angular and spectral resolution.

The experimental principles are schematically shown in Figs.5-8. In experiment 1, when total length of targets is over $22+18$mm, X-ray laser intensity does not increases continuously. Based on this experience, we cut a long target into two short targets. They are arranged as targets 1, 2 or 3, 4 in Fig.6-8. Here targets 1, 2 or 3, 4 are irradiated by the same beam of laser, the directions of electron density gradients and so refractive index gradients in gain regions 1, 2 or 3, 4 are nearly the same. This kind of couplig between two targets is called following one. Targets 1, 2 in experiment 1 or targets 2, 3 in experiments 2, 3, 4 are irradiated by two opposite beams of laser, the directions of electron density gradients and so refractive index gradients in corresponding gain regions are nearly opposite. This kind of conpling between two targets is called opposing one.

In experiment 2, total effecive length of targets is 4×14mm.

In experiment 3, double-pass amplification output of targets 1 and 2 enters into gain regions 3 and 4 to amplify further.

The purpose of experiment 4 is to improve the quality of X-ray laser beam. From preliminary simulation, in order to realize single-mode high amplification, acceptance angle of gain region 3 should be equal to divergence angle of single-mode output, which is 0.3 and 0.6mrad for plane and spherical mirrors, respectively. The size of gain region 3 is $\sim 100 \mu$m, so that, the distance between targets 2 and 3, Y_{23}, should be chosen as (150-300)mm. But, under present condition of target chamber, Y_{23} is ~ 1000mm, acceptance angle of gain region 3 is ~ 0.1mrad.

In these experiments, critical problems are to select proper parameters of X_{23}(or X_{12}), $\theta i(i=1,2,3,4)$ and β. Simulation predicts these parameters correctly. Optimal parameters are shown in table 2.

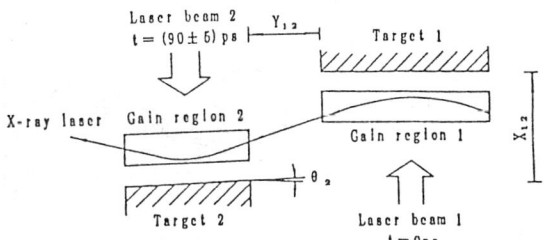

FIGURE 5. Experiment 1

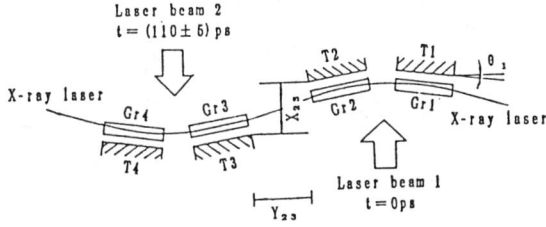

FIGURE 6. Experiment 2

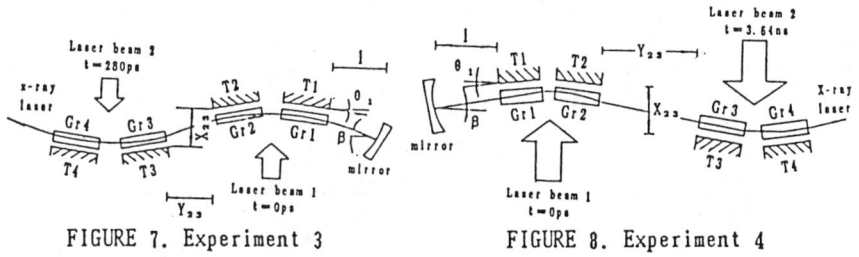

FIGURE 7. Experiment 3   FIGURE 8. Experiment 4

TABLE 2. Optimal parameters

	experiment 1	experiment 2	experiment 3	experiment 4
X_{12} (μm)	320			
X_{23} (μm)		250	230	10mm
θ_1 (mrad)	2-9	4.5	4.5	4.5
θ_2 (mrad)		″	″	″
θ_3 (mrad)		″	″	″
θ_4 (mrad)		0-2	″	″
β (mrad)			13.5	13.5

Note: $Y_{12}=4$mm in experiment 1. $Y_{12}=Y_{34}=1$mm and $Y_{23}=5$mm in experiment 2. $Y_{12}=Y_{34}=1$mm, $Y_{23}=5$mm and $l=30$mm in experiment 3. $Y_{12}=Y_{34}=1$mm, $Y_{23}\approx 1000$mm and $l=30$mm in experiment 4.

Key techniques of experiments:
1. Accurate alignment of two focus lines
2. Accurate adjustment of targets and mirror

Due to establishment of focus line collimation technique and adjusting and monitoring technique of targets and mirror, we obtain:
1. Alignment accuracies of two focus lines
   Pointing $<\pm 0.5$ mrad
   Positioning $<\pm 10\mu$m
2. Adjustment accuracies of X_{23}, θ_1 and β
   X_{23} $<\pm 5\mu$m
   θ_1 $<\pm 0.1$ mrad
   β $<\pm 0.1$ mrad

Results

The line intensity and divergence angle of X-ray laser beam are shown in Figs. 9-13 It is seen from these figures that
1. The rollover of intensity occurs at $GL\approx 14$, $GL<14$, $I_{23.2}<I_{23.6}$ and $GL>14$, $I_{23.2}>I_{23.6}$.(See Fig. 9).
2. A saturated output is achieved for 23.2 and 23.6nm lines at target length of 56mm(See Fig. 9).
3. The intensity of 19.6nm line increases exponentially up to L=56mm (See Fig. 9).
4. In two 14mm targets 1 and 2 of following coupling, divergence angle decreases fast. In 14mm target 3 which is oppositely coupled to target 2, divergence angle decreases slowly. Minimum divergence angle is (3-4)mrad at L=56mm (See Fig. 10).

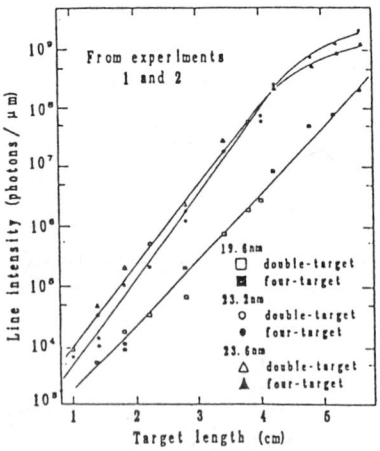

FIGURE. 9. Lasing line intensity vs target length

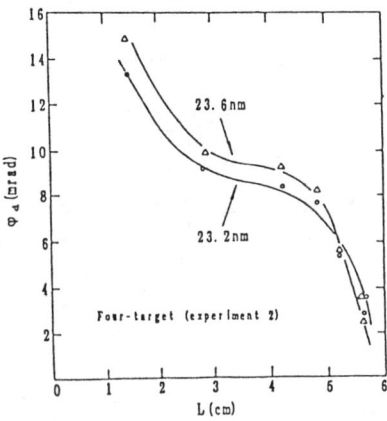

FIGURE 10. Horizontal divergence angle vs target length

5. For two 14mm targets of folloing coupling, double-pass amplification output is 40 times more intense than single-pass one (See Fig. 11), its divergence angle is ~3.7mrad, but with a broader region of less amplification(See Fig. 12). A fully saturated output is obtained for 23.2 and 23.6nm lines at L=56mm (See Fig. 11), but with divergence angle ~6mrad.

6. It is known from our experiments that $I_{23.2}/I_{23.6} \approx 1.5$-$2$ at saturation.

7. In experiment 4, a sixth saturated output with a minimum divergence angle ~1.5mrad was observed (See Fig. 13).

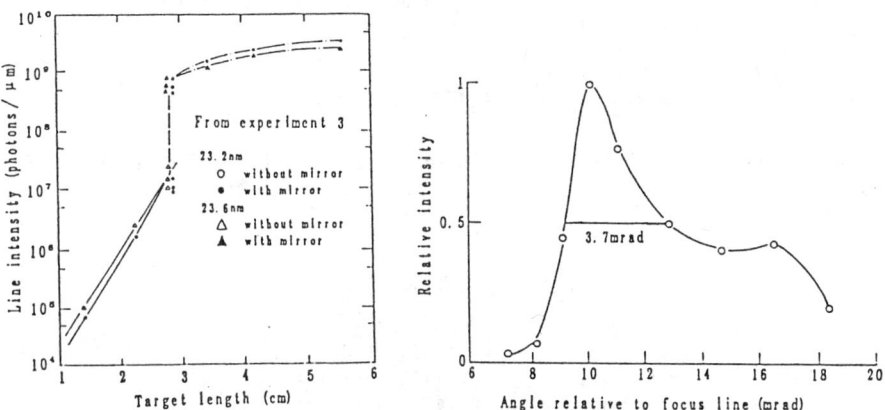

FIGURE 11. Lasing line intensity vs target length

FIGURE 12. Normalized angular distribution of double-pass output intensity

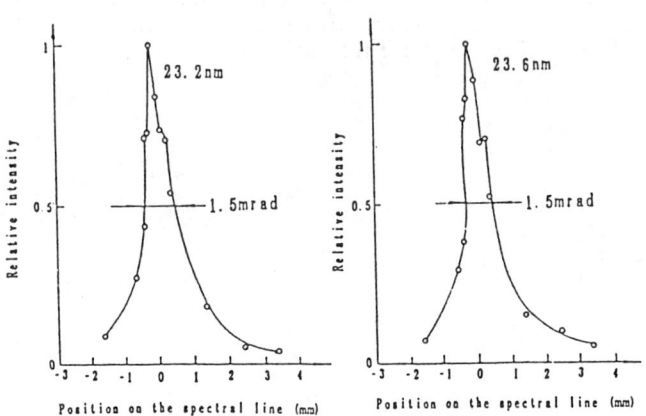

FIGURE 13. Normalized intensity profile along spectral line

Conclusion

Research of Ne-like Ge soft-X-ray laser is useful for exploring some of properties of X-ray laser.

With the specially arranged four short targets in a total length of 56mm, a saturated output with divergence angle $\sim(3-4)$mrad has been obtained using two laser beams of shenguang facility.

For two 14mm targets of following coupling, double-pass amplification with a third saturated intensity and a divergence angle ~ 3.7mrad has been achieved only using one laser beam of shengnang facility.

In experiment 4, a sixth saturated output with a minimum divergence angle ~ 1.5mrad has been observed.

ACKNOWLEDGEMENTS

The authors are grateful to professors Nengkuan Chen, Min Yu, Renyu Hu, Ximing Deng, Zucong Tao, Xiangwan Du and Xiantu He for their direction and support.

REFERENCES

1. Matthews D.L. et al, Phys. Rev. Lett, 54, 110 (1985)
2. Lee T.N. et al, Phys.Rev. Lett, 59, 1185 (1987)
3. Shiji Wang et al, J.Opt. Soc. Am.B, 9, 360 (1992)
4. Shiji wang et al, "Experimental study of a nearly saturated Ne-like Ge soft-X-ray laser by multi-target series coupling", in proceedings of the Colloquium on X-ray Lasers, 1992, PP49-52.
5. Shiji Wang et al, High Power Laser Part. Beams (China), 5, 557 (1993)
6. Shiji Wang et al, Chinese Journal of Laser, B2, 481-484 (1993)
7. Shutai Chunyu et al, High Power Laser Part. Beams(China), 2, 280-290 (1992)
8. Shutai Chunyu et al, Science in China (Series A), No 8, 875-879, 1992
9. An He et al, High Power Laser Part. Beams (China), 5, 88-92, 1993.

A Linearly Polarized Soft X-ray Laser

B. Rus[#,ø], G. F. Cairns[*], P. Dhez[#], P. Jaeglé[#], M. H. Key[¶,@],
C. L. S. Lewis[*], D. Neely[¶], A. G. MacPhee[*], S. A. Ramsden[€],
C. G. Smith[@], and A. Sureau[#]

[#]*Laboratoire de Spectroscopie Atomique et Ionique, URA 775 CNRS, Université Paris-Sud,
Bâtiment 350, 91405 Orsay Cedex, France*
[*]*Department of Pure and Applied Physics, Queen's University of Belfast, Belfast BT7 1NN, U.K.*
[¶]*Central Laser Facility, Rutherford Appleton Laboratory, Chilton, Oxon, OQ11 0QX, U.K.*
[@]*Clarendon Laboratory, University of Oxford, Oxford OX1 3PU, U.K.*
[€]*Department of Computational Physics, University of York, York YO1 5DD, U.K.*
[ø]*Department of Gas Lasers, Institute of Physics, 18040 Prague 8, Czech Republic*

Abstract. We report results of polarization experiments on the collisionally excited Ne-like Ge soft X-ray laser where we have used an injector-amplifier multistage geometry. The polarization state of the X-ray beam was analysed by two crossed 45° angle of incidence multilayer mirrors which act as linear polarizers. Results were evaluated by comparing intensities of time-integrated beam patterns behind each polarizer. The polarization state of the 23.2 and 23.6 nm ASE output of the injector plasma was systematically studied and, as expected, revealed no macroscopic degree of polarization observable within the precision of the experiment. When the injector output beam was linearly polarized and coupled into the amplifier plasma, the degree of polarization of the amplifier output was ~0.98, the total gain-length product attained for the polarized beam was ≈12 and the beam energy was ≈20 nJ.

INTRODUCTION

One of the parameters of the X-ray laser (XRL) beam, which is of importance for applications such as interferometry, nonlinear XUV optics and XUV induced processes in atomic or solid-state systems, is its polarization state.
A scheme consisting of injecting a polarized emission into an active medium and amplifying it to the desired intensity, has been demonstrated in this work. Its objectives were twofold: to verify that a simple non-saturated ASE is unpolarized, and to demonstrate that a polarized X-ray beam can be amplified without significant degradation of its degree of polarization. The details of the experiment and an investigation of polarization of ASE systems can be found in (1).
In the absence of works dealing with the macroscopic, time-dependent polarization properties of ASE systems, we base the assessment of the ASE polarization features on its coherence parameters. The interconnection of the coherence and polarization may be illustrated by considering collisions in the plasma. As the collisions randomly change the phase of the ion's dipole oscillations as well as the orientation of its dipole moment, they will manifest themselves both in the macroscopic coherence and the polarization. The frequency of electron-ion elastic collisions may be estimated as ≈10^{14} s^{-1} which is much larger than the

spontaneous and stimulated emission decay rates. Thus within the radiative lifetime of an excited ion the collisions occur so frequently that the phase and orientation of the dipole moment will be randomized.

The phase of quasimonochromatic radiation of wavelength λ is maintained during the coherence time t_{coh} which may be evaluated (2) via the spectral linewidth $\Delta\lambda$:

$$t_{coh} = \gamma \lambda^2 / (c \Delta\lambda) \quad (1)$$

where $\gamma = 0.32$ and 0.66 for Lorentzian and Gaussian profiles respectively. It is obvious that t_{coh} is a useful estimate for the characteristic time during which a given polarization state of the macroscopic field is maintained, especially in the case when homogeneous broadening is important. Taking $\Delta\lambda \sim 20$ mÅ as a linewidth estimate for ASE at $\lambda \approx 200$ Å under moderate gain-length product conditions (3), Equation 1 gives $t_{coh} \approx 0.3$ psec. A typical XRL pulse is hundreds of psec in duration and will accordingly consist of perhaps one thousand of "wavetrains" each with a definite polarization state. The polarization states of these wavetrains are mutually uncorrelated; therefore, in the absence of a monitor with subpicosecond resolution, an unsaturated XRL beam should appear to be almost completely unpolarized.

Different groups of atoms contribute to excite each of the transverse modes to a different extent. As a result, different modes will have different polarization. To obtain a uniform direction of polarization over the beam, a single-mode operation is necessary - the same requirement as to achieve perfect transverse coherence.

AMPLIFICATION OF POLARIZED EMISSION

We describe polarization properties of an XRL beam by the degree of polarization

$$D_p = I_p / (I_p + I_u) \quad (2)$$

where I_p and I_u are the intensities of the polarized and unpolarized components.

In the experimental arrangement investigated, a source plasma of length l_s delivers an intense beam of completely unpolarized radiation (i.e. $D_p = 0$) which is polarized by an X-ray optics having a total throughput R and then coupled into an amplifier plasma of length l_a with a coupling efficiency C. The intensity emerging from the amplifier in the non-saturated regime can be expressed as

$$I_{amp} = RCw \frac{j_s}{g_s} \frac{exp(g_s l_s + g_a l_a)}{\sqrt{g_s l_s + g_a l_a}} + w \frac{j_a}{g_a} \frac{exp(g_a l_a)}{\sqrt{g_a l_a}} \quad (3)$$

where j_s, g_s and j_a, g_a are peak spectral emissivities and gain coefficients of the source and amplifier plasmas respectively, and w treats the spectral line profile. In deriving Equation 3 it has been assumed that $exp(g_s l_s) \gg 1$, $exp(g_a l_a) \gg 1$.

Supposing that the emission injected into the amplifier maintains its polarization state during the amplification (the validity of this assumption is discussed later), the degree of polarization of the amplifier output D_p^{amp} may be written as:

$$\frac{1}{D_p^{amp}} = \frac{1}{D_p^{in}} + \frac{j_a}{j_s} \frac{g_s}{g_a} \sqrt{\frac{g_s l_s + g_a l_a}{g_a l_a}} exp\left(-g_s l_s - \ln(D_p^{in} R C)\right) \quad (4)$$

To ensure $D_p^{amp} \cong D_p^{in}$ the second term on the right side must be minimised; a requirement met even for moderate values of $g_s l_s + \ln(D_p^{in} RC)$.

X-RAY POLARIZERS

The polarizers used in this work were multilayer mirrors coated onto super-polished fused silica substrates. Mo:Si layers were produced by ion beam sputtering (4) and the device properties were tested with synchrotron radiation (5). A reflectance R_s of ~24 % was measured for s-polarized light (electric vector parallel to the polarizer surface) at the design wavelength for a 45° incidence angle. Calculations suggest (4) that the ratio s-reflectance/p-reflectance is typically ~7.4, i.e. the p-reflectance R_p ~3.25%; they also point at a weak dependence of this ratio on the surface roughness as well as on other fabrication parameters.

When a totally unpolarized beam of intensity I_{0u} is incident on the polarizer, the reflected beam will consist of an unpolarized component I_u and a linearly polarized component I_p. After two reflections the beam components are described as

$$I_p = 0.5\ (R_s^2 - R_p^2)\ I_{0u}$$
$$I_u = R_p^2\ I_{0u} \quad\quad\quad (5)$$
$$D_p = (R_s^2 - R_p^2) / (R_s^2 + R_p^2)$$

For R_s =0.24(±0.04) and R_p =0.03(±0.01), D_p achieved is 0.97($^{+0.02}/_{-0.05}$).

EXPERIMENTAL ARRANGEMENT

The experiment was carried out at RAL. The source plasma was a double target amplifier and consisted of two 22 mm × 100 μm Ge stripes coated onto glass substrates. Both were irradiated, from opposite directions, by a group of three VULCAN laser beams. The net irradiance on each target was ~1.6×10^{13} Wcm^{-2} with an expected gain coefficient on each of the J=2-1 lines at 23.2 and 23.6 nm of 3.5(±0.5) cm^{-1} (6). We refer henceforth to this target as 'injector'.

In the setup for injector ASE polarization studies, the North-going output is reflected from a near-normal incidence XRM (i.e. polarization insensitive) and returned along an axis which takes it close to but not through the source plasma.

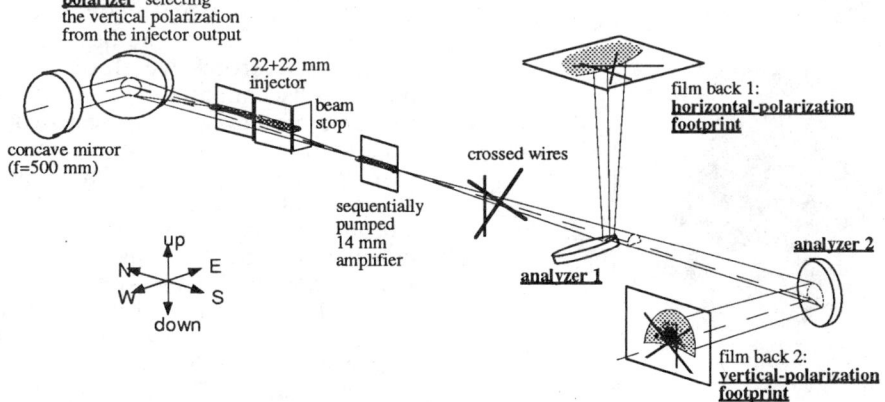

FIGURE 1. Experimental setup (not to scale). The North-going injector output is vertically polarized and relayed toward the amplifier. When polarization state of the injector is investigated, the concave mirror alone is used to relay the beam towards the analyzers.

The beam polarization state is then analyzed using two crossed polarizers. In the second experiment devoted to study amplification of the polarized beam, the source ASE beam is linearly-polarized to $D_p \sim 0.97$ by making it reflect twice off a 45° mirror polarizer before injection into the polarization analyzer system via an additional amplifier plasma. In the lab-frame the injected beam is vertically polarized.

The exit plane of the injector was image relayed to a plane ~60 mm in front of the amplifier input plane at an angle of ≈3 mrad and purposely offset ≈70 μm laterally from the amplifer surface to centre on the expected position of the gain zone in the amplifer plasma. The amplifier was a 14 mm long Ge stripe, otherwise similar to the injector components, and shot with a single VULCAN beam of 150 mm diameter, providing a net irradiance of $\approx 1.3 \times 10^{13}$ Wcm^{-2}. The gain in the amplifier plasma was estimated to be 3.0($\pm$0.5) cm^{-1}, hence the net amplification ≈70×.

The intensity of the vertically polarized emission extracted from the injector beam is given by Equation 5 and amounts to ~0.028 of the intensity leaving the concave mirror in the 'injector' shots. With C≈ 0.25, the total coupling RC is ≈0.25×0.028 ≈ 7×10^{-3} so that ≈1/140 of the injector beam energy is coupled into the amplifier.

The beam emerging from the amplifier was detected as a 'footprint' on two film backs containing Kodak 104-02. Since we did not have a soft X-ray beam splitter, we carefully arranged the first analyzer to intercept and reflect the lower half of the XRL beam; the top half of the beam is transmitted past the edge of the polarizer and then analyzed for a vertically polarized component by the second, crossed analyzer. Assuming a beam symmetry according to the horizontal plane, we were able to analyse the degree of polarization of the XRL beam produced.

EXPERIMENTAL RESULTS

In the first part of the experiment, devoted to the injector ASE output polarization state measurements, 11 shots were fired with ±7% variance in drive intensity. A typical footprint image of the injector output beam is shown in Figure 2 where the composite beam appears to have similar signal strength in each of the crossed polarizer arms. The "centre of gravity" of the beam relative to the first polarizer edge and to the spatial fiducial moved a little on shot-to-shot basis making it inappropriate to exactly compare top and bottom halves of the beam on all shots. Therefore we looked for intensity discontinuities across the join to signal the presence of macroscopic polarization in the beam. Intensity discontinuities were averaged over several of such traces in a given shot.

From the data thus collected we conclude that neither upper nor lower half of the footprints is systematically more intense, and thus that there is no preferred, systematic direction of polarization in the X-ray beam. Within the accuracy of the experiment, estimated to be ~5%, the time-integrated degree of polarization is zero.

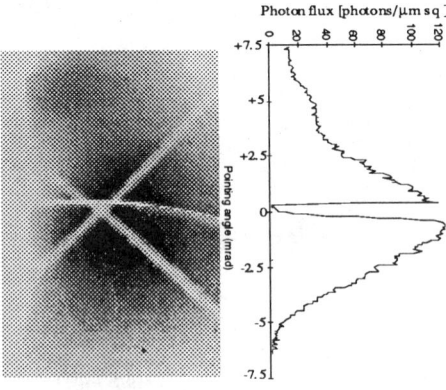

FIGURE 2. Composed footprint image of the injector output (#01-140993) and its densitometric trace in vertical direction.

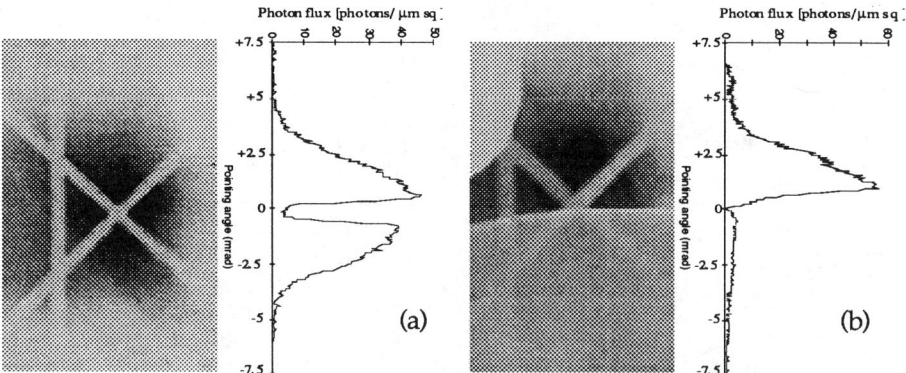

FIGURE 3. (a) Footprint of the polarized amplifier output (#02-121093) viewed by the vertical analyzer, and (b) by the tandem of crossed analyzers (#04-121093).

In the second part of the experiment where amplification of polarized light was studied, 8 shots were taken. The high quality of the amplified beam profile is shown in Figure 3a where the first analyzer was removed from the beam path allowing the second analyzer to reflect the vertically polarized component of the whole beam. A typical result with both analyzers in place is shown in Figure 3b. Here we see that the bottom half of the beam, analyzed for a horizontal polarization, contains virtually no signal, which indicates a high degree of vertical polarization of the amplified beam. Quantitative analysis, averaged over the performed shots, provides $D_p = 0.98(^{+0.02}/_{-0.05})$ which is essentially the same value as in the injected beam. This means that within the precision of the experiment and and under its conditions there is no depolarization of the amplifying beam.

From absolute calibration of the film we estimate ≈ 20 nJ in the polarized beam.

DISCUSSION AND CONCLUSIONS

Two aspects of the experiment should be noted. First, the footprints contain signals from both J=2-1 lasing transitions. We thus have an 'overlap' of two beams whose polarization states are mutually uncorrelated (both the upper and lower lasing levels of these transitions are different); this lessens the resulting degree of polarization of the whole ASE beam. Secondly, the Fresnel number of the injector is ~4 in the horizontal and ~16 in the vertical plane, assuming the gain region 75 μm × 150 μm (7); we are thus still far from a single-mode ASE.

In general, three processes are able to 'depolarize' the beam during the amplification.

First, the contribution of the amplifier own ASE to the output can be evaluated using Equation 3. Taking D_p^{in} of the injected beam 0.97 and RC=7×10^{-4} (accounting for the throughput of all the used X-ray optic elements), we obtain $D_p^{amp} = 0.969$. This is of course a non-detectable change compared to 0.97.

Secondly, under the presence of a magnetic field in the amplifying plasma the Faraday effect must be considered. The dominant source term of the spontaneously generated magnetic fields (8) is here grad(n_e) × grad(T_e). Considering the experimental geometry, the B-field is directed downstream from the X-ray beam propagation in the upper part of the amplifier plasma and counterstream in the lower part. In the upper plasma region, the beam E vector will be rotated counter-clockwise (viewed downstream from the X-ray beam propagation), whereas in the lower region

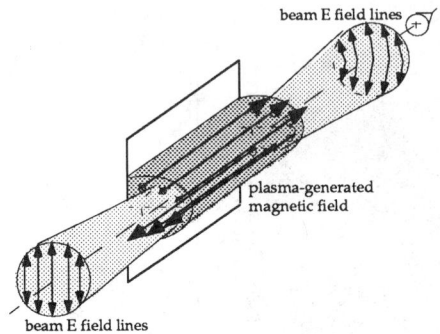

FIGURE 4. Effect of self-generated magnetic field in an XRL plasma amplifier on a linearly polarized beam.

it will be rotated clockwise. The amplifier plasma thus bends the beam field lines. The plane of polarization of an emission propagating along the magnetic field B over a distance l_a will locally rotate by $\Phi \approx (n_e/n_{cr})(qB/2m_e c)l_a$ where $n_{cr} \sim 2 \times 10^{24}$ cm^{-3} for 23.6 nm, and q and m_e are the electronic charge and mass. Taking $B \approx 20$ T, we obtain $\Phi \approx 0.7$ degrees. The effect of the plasma-generated magnetic fields is thus negligible under the experimental conditions encountered. However, it might be of importance for short driving pulses and/or for small-lateral-dimension plasmas.

For the third possible source of depolarization - collisions experienced by the amplifying ions, no quantitative assessment is available. However, the short-distance collisions which have the strongest 'depolarizing' effect influence predominantly the wings of the spectral profile, where gain is marginal. The amplification further reduces the spectral width of the lasing line by gain-narrowing and therefore the emission whose polarization state was collisionally changed will not be effectively amplified. As a result, the overall depolarization due to the collisions is expected to be small. This seems to be evidenced experimentally.

In summary, the polarization state of two J=2-1 lasing lines in Ne-like Ge has been investigated. According to expectations, the injector ASE output was observed neither to be preferentially polarized in any direction nor to show a measurable degree of polarization. To control the polarization state of the XRL beam, vertical polarization was selected from the injector beam prior to feeding it into the amplifier. The degree of polarization ~0.98 of the amplifier output has been found identical (within the experimental precision) to the degree of polarization of the injected beam. Producing a highly polarized X-ray beam of a good spatial quality is the major achievement of this work. Regarding the attained gain-length of ≈ 12, exploiting another amplifier would allow the polarized beam to reach the saturation.

REFERENCES

1. Rus, B., Cairns, G. F., Dhez, P., Jaeglé, P., Key, M. H., Lewis, C. L. S., Neely, D., MacPhee, A. G., Ramsden, S. A., Smith, C. G., and Sureau, A., submitted to Phys Rev A, 1994
2. Goodman, J. W., *Statistical Optics*, New York: John Wiley & Sons, 1985, Ch 5
3. Koch, J. A., MacGowan, B. J. , DaSilva, L. B., Matthews, D. L., Underwood, J. H., Batson, P. J., and Mrowka, S., Phys. Rev. Lett. **68**, 3291-3294 (1992)
4. Chauvineau, J. P., Institut d'Optique Théorique et Appliquée, Bât. 503, Centre Scientifique d'Orsay, 91403 Orsay Cedex, France, personal communication (1993)
5. NIST Physics Laboratory, Physics Bldg., Rm B160, Gaithersburg, MD 20899.
6. Neely, D., Lewis, C. L. S., O'Neill, D. M., Uhomoibhi, J. O., Key, M. H., Rose, S. J., Tallents, G. J., and Ramsden, S. A., Optics Comm. **87**, 231-236 (1992)
7. Cairns, G., Lewis, C. L. S., MacPhee, A. G., Neely, D., Holden, M., Krishnan, J., Tallents, G. J., Key, M. H., Norreys, P. N., Smith, C. G., Zhang, J., Holden, P. B., Pert, G. J., Plowes, J., and Ramsden, S. A., Appl. Phys. B **58**, 51-56 (1994)
8. Stamper, J. A., Laser and Particle Beams **9**, 841-862 (1991)

EBIT X-ray Spectroscopy Studies for Applications to Photo-Pumped X-ray Lasers

S. R. Elliott, P. Beiersdorfer, and J. Nilsen
Lawrence Livermore National Laboratory
Livermore, CA 94550

Introduction: Several pumping mechanisms have been suggested for x-ray lasers including collisional excitation, recombination, photo-ionization and photo-pumping [1]. The success of photo-pumping as an x-ray laser scheme hinges on sufficient overlap of the emission and absorption lines. For such a scheme to exhibit gain, the difference of the energies of the two lines must be within the line widths determined by the plasma dynamics, such as Doppler and opacity broadening. Typically, an overlap of a few parts in 10^4 is required. Due to correlation effects, high-n levels of multi-electron ions are difficult to calculate and are reliable to roughly a part in 10^3. These differences are large enough to preclude accurate predictions of successful overlaps. As a result, precise measurements of the overlaps are needed. The continued interest in photo-pumping schemes lies in its potential to improve the laser output. It also allows the excitation of lasing transitions not accessible to other mechanisms and thus to test laser kinetics from a different perspective. Figure 1 shows an example of a photo-pumped x-ray laser scheme.

We have studied several such photo-pumping schemes at the LLNL electron beam ion trap (EBIT)[2]. The Ni-like isoelectronic sequence 3d-5f and 3d-6f transitions were studied for photo-pumping by He-like ions, the Ne-like 2p-4d transitions were studied for photo-pumping by Ni-like 3d-4f transitions, and Ni-like $3d_{5/2}$-$6f_{7/2}$ transitions were studied for photo-pumping by H-like Ly-α transitions. A number of other chance coincidence pairs which do not follow an isoelectronic sequence were also studied. The data were taken with a flat-crystal vacuum spectrometer [3], a flat-crystal helium atmosphere spectrometer, or a curved-crystal spectrometer in the von Hamos geometry[4].

The advantage of EBIT over laser-produced or tokamak plasmas for such experiments is its ability to control the charge balance and the excitation process. By choosing the electron beam energy, we can select a dominant charge state. In particular, by operating below various ionization potentials, the contributions of various charge states to a spectrum can be deduced. Moreover, blends with satellite lines produced by dielectronic recombination can be avoided by proper choice of the beam energy. Thus, wavelength measurements are unambiguous, reliable, and precise.

Ni-like ions pump Ne like ions We have investigated a particular class of schemes whereby a $2p_{1/2}$-$4d_{3/2}$ transition in a Ne-like ion of atomic number Z is photo-pumped by a $3d_{5/2}$-$4f_{7/2}$ transition in a corresponding Ni-like ion of atomic number (2Z + 5) [5]. These potential resonances lie along an isoelectronic sequence. Therefore we measured a number of the transitions along the sequence to search for good overlap candidates. The theoretical predictions of the energies of these transitions, based on a multi-configuration Dirac-Fock calculation using

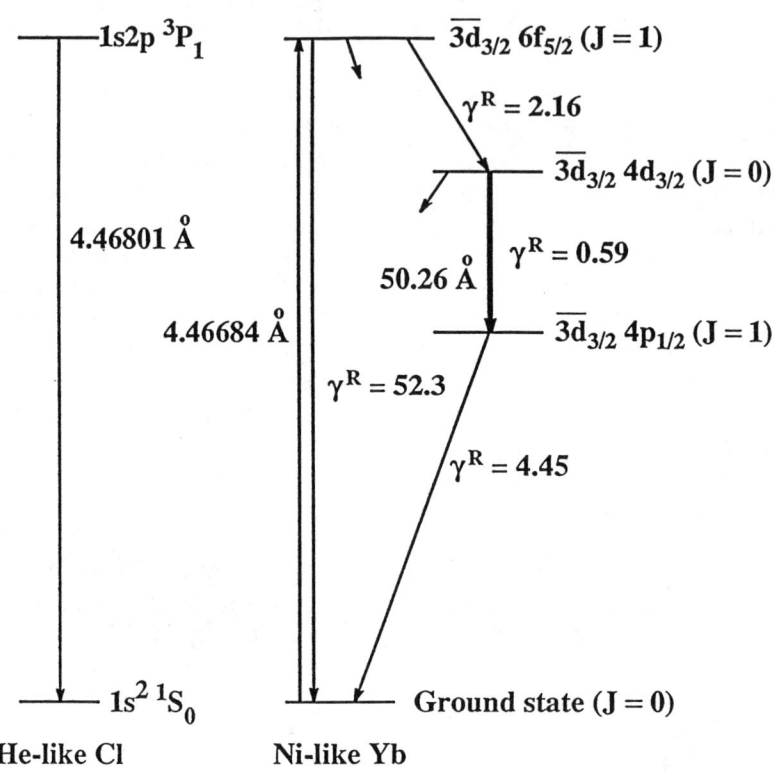

Figures 1: Level diagrams demonstrating an example laser scheme. Transition rates are indicated γ^R and are in units of psec^{-1}. The bar over the 3d indicates a vacancy in the closed M shell.

the code of Grant et al. [6], were found to be offset by 1.88 eV from the measured values. By having observed a family of these transitions, the determined offset improves future predictive power.

A similar class of schemes exists where the $2p_{1/2}$-$4d_{3/2}$ transition in a Ne-like ion of atomic number Z is photo-pumped by the $3d_{3/2}$-$4f_{5/2}$ transition in a corresponding Ni-like ion of atomic number (2Z + 4). Here the offset was determined to be 2.86 eV and a favorable resonance was identified with Ne-like Rb pumped by Ni-like Pt at 2512 eV. The energy difference was found to be 0.4 ± 0.1 eV or 160 ppm.

He-like ions pump Ni-like ions This class of schemes also follows an isoelectronic sequence. The He-like 2^1P_1-1^1S_0 transition may pump a 3d-6f or 3d-5f transition in a Ni-like ion. Which transition depends upon the ion pair in question. For this data we also measured an offset between theory and experiment for the Ni-like transitions and found that the $3d_{5/2}$-$6f_{7/2}$ transition differed from theory by 1.60 eV, the $3d_{3/2}$-$6f_{5/2}$ transition differs from theory by 2.01 eV, the $3d_{5/2}$-$5f_{7/2}$ transition differs by 1.06 eV, and the $3d_{3/2}$-$5f_{5/2}$ transition differs by 1.29 eV[7]. The most favorable overlap was found for the He-like F 2^1P_1-1^1S_0 transition pumping the Ni-like Ag $3d_{3/2}$-$6f_{5/2}$ transition. The two lines differ by 190 ppm at 738 eV.

H-like ions pump Ne-like and Ni-like ions The collection of possible lasant-pump pairs in this group do not follow an isoelectronic sequence but are chance coincidences between the H-like Ly-α line and various transitions in a number of Ne-like ions. The pairs we have studied to date are Ne and Fe [8], Na and Co [9], and Mg and Ge [10]. The H-like Ne Ly-α and $2p_{1/2}$-$4d_{3/2}$ Ne-like Fe transitions differ by 600 ppm at 1022 eV, the H-like Ly-α Na and $2s_{1/2}$-$4p_{3/2}$ Ne-like Co transitions differ by 100 ppm at 1237 eV, and the H-like Ly-α Mg and $2s_{1/2}$-$3p_{1/2}$ Ne-like Ge transitions differ by 340 ppm at 1472 eV. Figure 2 shows an example of the spectra of Na and Co elucidating the overlap.

A measurement was also done on the H-like Al and Ni-like Er pair. The H-like Al Ly-a and $3d_{3/2}$-$4f_{5/2}$ Ni-like Er transitions differ by 1000 ppm at 1726 eV [11].

An additional class of Ly-α pumped schemes uses a H-like ion of atomic number Z and the $3d_{5/2}$-$6f_{7/2}$ Ni-like transition in ions of atomic number (3Z + 21) [12]. The best candidate was determined to be H-like Ca pumping Ni-like Tl. The Ca Ly-α and Ni-like Tl $3d_{5/2}$-$6f_{7/2}$ transitions differ by 260 ppm at 4108 eV.

He-like ions pump Ne-like ions This collection of possible lasant-pump pairs does not follow an isoelectronic sequence. The pairs we have studied to date include Ar and Y [13] and Mg and Cu [8]. The He-like Ar 2^1P_1-1^1S_0 and Ne-like Y $3p_{1/2}$-$5d_{3/2}$ transitions differ by 150 ppm at 3140 eV and the He-like Mg 2^3P_1-1^1S_0 and Ne-like Cu $2p_{3/2}$-$4d_{5/2}$ transitions differ by 200 ppm at 1343 eV. Figure 3 shows an example of the spectra of Mg and Cu.

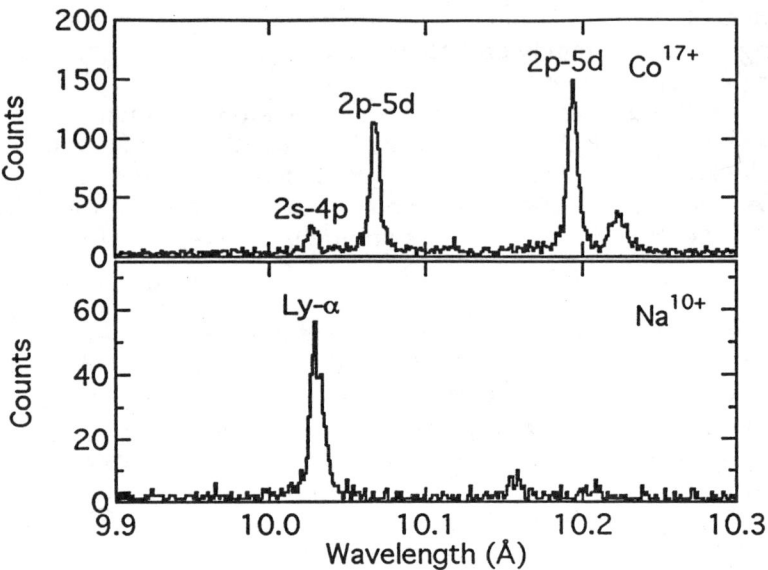

Figure 2: The spectra of H-like Na and Ne-like Co elucidating the potential overlap.

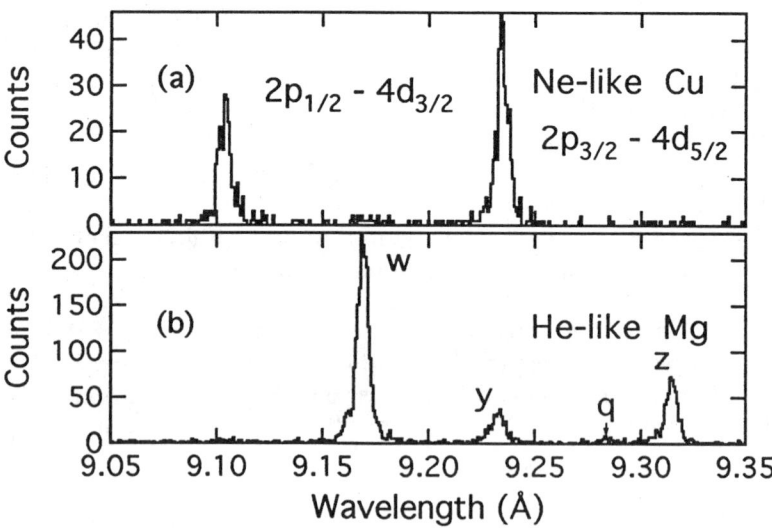

Figure 3: The spectra of He-like Mg and Ne-like Cu elucidating the potential overlap.

Conclusion: Several of the above pairs are good candidates for photo-pumping. For example, if the appropriate plasma conditions could be prepared, the Pt-Rb scheme described above would lase on a transition of about 165 Å. The Ar-Y scheme would lase on a transition near 155 Å, and the Mg-Cu scheme would lase near 235 Å.

Acknowledgments: This work was performed under the auspices of the U. S. department of Energy by Lawrence Livermore National Laboratory under contract W-7405-Eng-48.

References
[1] R. C. Elton, X ray Lasers, (Academic Press, Inc., San Diego, 1990) pp 99 - 198.
[2] M. A. Levine et al., Physica Scripta **T22**, 157 (1988); and R. E. Marrs et al., Phys. Rev. Lett. **60**, 1715 (1988).
[3] P. Beiersdorfer, and B. J. Wargelin, Rev. Sci. Instrum. **65** (1), 13 (1994).
[4] P. Beiersdorfer et al., Rev. Sci. Instrum. **61** (9), 2338 (1990).
[5] S. Elliott, P. Beiersdorfer, and J. Nilsen, Phys. Rev. **A47**, 1403 (1993).
[6] I. P. Grant et al., Comput. Phys. Commun. **21**, 207 (1980).
[7] S. Elliott, P. Beiersdorfer, and J. Nilsen, Physica Scripta, In Press.
[8] P. Beiersdorfer, S. R. Elliott, and J. Nilsen, Phys. Rev. **A49**, 3123 (1994).
[9] Joeseph Nilsen, S. R. Elliott, and P. Beiersdorfer, To Be Submitted.
[10] J. Nilsen, et al., Submitted to Phys. Rev. A.
[11] In Preparation.
[12] P. Beiersdorfer et al., Phys. Rev. **A46**, R25 (1992).
[13] J. Nilsen et al., Physica Scripta **47**, 42 (1993).

High-Order Harmonic Generation

C.-G. Wahlström

Department of Physics, Lund Institute of Technology
P.O. Box 118, S-221 00 Lund, Sweden

Abstract. A brief overview of high-order harmonic generation is given. Special emphasis is made on the use of this technique as a source of coherent radiation in the extreme ultraviolet. The experimental techniques as well as some characteristics of the generated radiation are presented. In particular, the wavelength range that can be covered and the number of photons per pulse that can be generated will be discussed and compared with other sources of radiation in this wavelength range. Finally some applications of radiation generated as high order harmonics will be briefly mentioned.

EXPERIMENTAL SETUP

High-order harmonic radiation is generated when an intense laser field interacts with free atoms, ions or plasma gradients. In this paper only interactions with free atoms and ions will be covered. The experimental setup can look like the one in Figure 1. Laser pulses from a short-pulse high-power laser are focused by a lens into a low density (10^{17}-10^{18} cm^{-3}) non-linear medium inside a vacuum system. In Figure 1 the medium is a jet of a rare gas emerging from a pulsed nozzle, but can also be, e.g., a plume of alkali-metal ions, produced by laser irradiation on a rotating solid target. In either case, the focused high-power laser interacts with

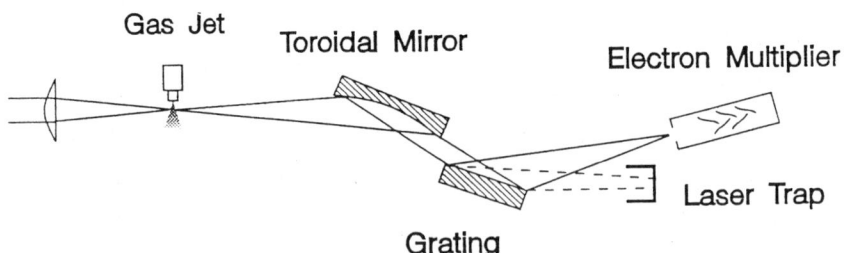

FIGURE 1. Example of experimental setup for high-order harmonic generation. From Ref [1].

free atoms or ions and only odd order harmonics are generated. The generated radiation and the laser beam propagate in the same direction after the interaction region and a VUV/XUV spectrometer is required to separate the laser radiation and the harmonics of different orders.

If a conventional spectrometer with an entrance slit is used, one must place the slit close to the interaction region to obtain a high collection efficiency. However, at this point the diameter of the diverging laser beam is still small, the intensity is very high, and there will be a severe risk for laser-plasma generation on the entrance slit. An alternative arrangement is to use the spectrometer without an entrance slit, placing the interaction region directly at the slit position (as in Fig. 1). With this arrangement, the grating is directly exposed to the diverging laser beam. To protect the grating from optical damage, the laser beam must diverge to a large enough diameter before reaching the grating. As will be discussed below, tight focusing, with a rapidly diverging laser beam after the interaction region, leads to much less efficient harmonic generation compared to loose focusing. The grating, therefore, must be placed at a relatively large distance from the interaction region. Finally, the harmonics can be detected, e.g. using an electron multiplier tube or a micro-channel-plate detector.

HARMONIC SPECTRA

The yield of the harmonics, according to lowest-order perturbation theory, is expected to drop rapidly with increasing order. This is indeed observed for the lowest orders, but from the 7th or the 9th harmonic, a *plateau* is found where the photon yields of different harmonic orders are roughly the same. The plateau extends to increasingly higher harmonics as the peak intensity of the laser field is increased, but it always ends with a rather sharp cut-off. From recent experiments, harmonic orders as high as the 135th [1] and wavelengths as short as about 7 nm [2] have been reported. Both the formation and the extent of the plateau have been investigated by performing systematic studies of the intensity dependence of individual harmonics [3]. For the high order harmonics, the cut-off energy, i.e. the photon energy of the highest harmonic in the plateau, is found to increase linearly with laser intensity.

Numerical calculations, valid in the regime of tunneling ionization, have shown that the cutoff-energy in the single-atom response, is well approximated by the simple formula I_p+3U_p [4]. I_p is the ionization energy and U_p is the ponderomotive energy, given by $U_p(I) = 9.33 \cdot 10^{-14} \cdot I\lambda^2$, where the energy is expressed in eV, I, the intensity, in W/cm^2, and λ, the wavelength, in µm. (The maximum value of I in this expression is the saturation intensity for ionization,

I_{sat}.) In the experimental spectra, the cutoff energy is somewhat reduced due to intensity-dependent phasematching effects [5]. The constant of proportionality in the above expression is reduced from 3 to about 2, depending on the experimental conditions.

Some important observations can be made directly from the above expression for the cutoff energy. First, to generate harmonic radiation with very short wavelength (high cutoff energy), one should choose as non-linear medium atoms or ions with high ionization potential. Secondly, since the ponderomotive potential is proportional to the square of the laser wavelength, lasers in the near-infrared region can generate harmonic radiation with shorter wavelengths than, e.g. excimer lasers operating in the ultraviolet region. Finally, since I_{sat} decreases with the duration of the laser pulses, very high harmonic orders (large U_p) require short pulse duration.

The most frequently used non-linear media for harmonic generation are the rare gases. The ionization potential as well as the saturation intensities for ionization for these gases decrease with atomic number. The harmonic conversion efficiency, on the other hand, increases with atomic number. For an application where radiation corresponding to a particular harmonic order is required, one should therefore choose as non-linear medium the heaviest rare gas possible for the given harmonic to be part of the plateau (which depend on the laser pulse duration). Some relevant parameters for the rare gases and estimates of the plateau extent in the single atom response (I_p+3U_p) and in a typical experiment (I_p+2U_p) are given in Table 1.

TABLE 1. Properties of the rare gases and estimated extents of the harmonic plateau cutoff when used as non-linear medium in high-order harmonic generation experiments.

	I_p (eV)	I_{sat} (1ps, 1μm) (10^{14} W/cm^2)	$I_p+3 \cdot U_p$ ($I=I_{sat}$) (eV)	$I_p+2 \cdot U_p$ ($I=I_{sat}$) (eV)
He	25	7	230	160
Ne	22	5	170	120
Ar	16	1.5	60	45
Xe	12	0.7	25	21

The ionization energies and saturation intensities for the rare-gas ions or rare-gas-like alkali-metal ions are substantially higher than for the corresponding neutral atoms. According to the I_p+3U_p formula, higher harmonic orders should therefore be possible to generate in these ions. A numerical calculation [6] for He$^+$, using a 527 nm laser at 5×10^{15} W/cm^2, predicts a cut-off energy well above 400 eV. This implies that coherent radiation in the biologically interesting "water window" (2.3 - 4.4 nm) might be possible to generate as high-order harmonics of a laser field in the visible. However, when producing ions through photo ionization, a significant electron densitiy is simultaneously produced. The dispersion of these free electrons severely reduce the macroscopic efficiency, in particular if long-wavelength lasers are used.

PHASE MATCHING

The existence of a macroscopic field requires that there be proper phase matching between the induced and the driving fields. There are essentially two reasons for the phases not to be matched: Firstly, the dispersion in the non-linear medium makes waves with different frequencies travel at different speeds in the medium and therefore get out of phase with each other. (For frequency doubling, etc. in the visible or infrared spectral regions, this effect is circumvented by the use of birefringent crystals.) In most of the recent high-harmonic generation experiments, using gas jets, the atomic density was low so the dispersion was small. Ionization of the medium, however, creates free electrons that can introduce a substantial phase mismatch between the generated harmonics and the driving polarization.

The second reason for the phases not to be matched is connected to the unavoidable phase shift of π, occurring in a light wave in passing through a focus. Using a loosely focused laser beam, and a short non-linear medium, only a small fraction of this phase shift occurs in the medium. However, even if the shift inside the medium is only a fraction of the total phase shift π, the difference in phase between the driving field and the generated field will be q times as large for a harmonic of order q. For sufficiently high harmonics, therefore, the phase lag will reach π, after a distance L_{coh} (the coherence length) in the medium, resulting in destructive interference.

The focusing geometry has a drastic effect on the macroscopic efficiency. A useful quantity that characterizes the focus is the confocal parameter, b. For a Gaussian beam, b is equal to twice the distance on the propagation axis over which the beam section increases by a factor of two. It has been found that in the tight focusing limit, the measured number of photons is proportional to b^3 [7].

CHARACTERISTICS OF THE RADIATION

There has been no direct measurements of the spatial coherence reported yet. However, measurements of the angular structures of the harmonic radiation have been reported [8]. Harmonic orders that are in the cut-off region show smooth, narrow angular profiles. Harmonics in the plateau, on the other hand, are often broader and can even exhibit ring structures. When preparing an experiment where XUV radiation with good focusability and corresponding to a particular harmonic order is needed, one should therefore choose the intensity so that the harmonic order in question is just at the end of the plateau or in the beginning of the cutoff region. Instead of decreasing the available laser energy, and thereby reducing the harmonic photon flux, one can change the focusing condition. By increasing the focal cross section (proportional to the confocal parameter, b), one can take advantage of the b^3 rule and increase the total flux at the same time as the spatial profile is improved.

Using peak intensities exceeding the saturation intensity for ionization leads to rapid ionization in the medium during the risetime of the laser pulse. The rapid temporal variation in the density of free electrons induces a time-dependent change in the refractive index, and consequently, a spectral blueshift of the fundamental field and of the generated harmonic field. The blueshift of the harmonic field is found to be due mainly to the shift of the fundamental. It is therefore approximately equal to the shift of the fundamental divided by the process order [3]. In Figure 2, the normalized spectral profiles of the 15th harmonic, obtained in xenon with a short-pulse laser are shown for a fixed peak intensity and different gas pressures.

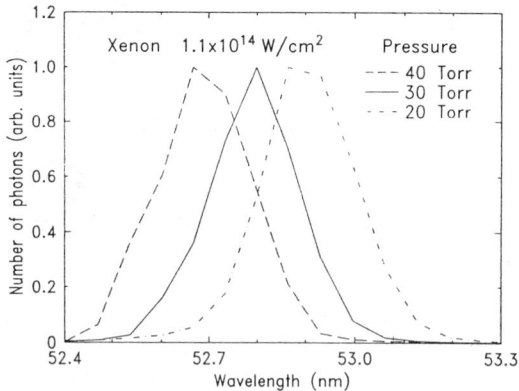

FIGURE 2. Normalized spectral profiles in Xe, obtained with a fixed peak intensity and different gas pressures. From Ref [3].

In order to get an idea of the characteristics of the radiation, generated as high order harmonics from a short-pulse laser, various parameters are presented in Table 2 together with the corresponding numbers for an undulator. The numbers for the harmonic radiation are estimates based on experiments performed with the 150 fs, 150 mJ Ti:sapphire laser of the Lund High-Power Laser Facility [9], operating around 800 nm. (10^8 photons per pulse, given in the table, corresponds to a pulse energy of about 1 nJ.) The undulator data are corresponding estimates for the new undulator under construction at the MAX II, 1.5 GeV synchrotron facility in Lund [10]. The comparison is made at 100 eV, corresponding to about the 63rd harmonic. If the comparison would be instead made at, say, 30 eV, the number of photons per pulse in the harmonic generation case would be at least two orders of magnitude higher since a heavier gas could be used. Making the comparison at much higher photon energies, on the other hand, would greatly favor the synchrotron/undulator source. (Above about 170 eV harmonic radiation has not yet been reported experimentally while synchrotron radiation with much higher photon energies are readily generated.) The first observations to be made from Table 2 is that the average power obtained from an undulator is considerably higher due to the very high repetition rate. The peak power, on the other hand, is several orders of magnitude higher in the case of harmonic generation. The two sources therefore complement each other and are best suited for different applications. Another, very important difference between these sources is the

TABLE 2. Estimated characteristics for radiation at 100 eV, generated as high-order harmonics of a 150 fs Ti:sapphire laser or in an undulator at a 1.5 GeV storage ring.

	Harmonics	**Undulator**
	150 mJ @ 1.5eV 150 fs, $\eta = 10^{-8}$	Max II 200 mA 1.5 GeV
Photons /pulse	10^8	10^5
Repetition rate	10 Hz	500 Mhz
P_{ave}	15 nW	1 mW
Pulse width	0.1 ps	20 ps
P_{peak}	15 kW	0.1 W
Bandwidth	0.01%	0.1%
Coherence cycles	3000	100

pulse duration. With a short-pulse laser, the harmonic pulse is short as well. (However, great care must be taken in order to maintain the short pulse duration when separating the harmonics of different orders from each other and from the fundamental radiation.) Many potential applications of high-order harmonic radiation depend on the ultra-short pulse duration. Experiments based on pump-probe technique for dynamic studies, e.g., molecular kinetics and surface studies, frequently require sub-picosecond resolution. These experiments cannot be performed with synchrotron radiation, but harmonic generation might be the ideal radiation source.

Comparing harmonic generation as a source of XUV radiation with existing x-ray lasers, the most striking differences are tunability, repetition rate and pulse energy. The wavelength of the harmonic radiation follows directly the laser wavelength, so a tunable high-power laser can be used to generate tunable radiation in the XUV. Even if the laser is tunable over only a limited wavelength range, the coverage in the XUV becomes complete by the possibility of choosing harmonics of successive orders. Ti-sapphire terawatt lasers with pulse durations in the 100 fs range, suitable for efficient high-order harmonic generation, can frequently be operated at repetition rates of 10 Hz or more. Most conventional x-ray laser schemes, on the other hand, require driving lasers with considerably higher pulse energies and correspondingly lower repetition rates. Therefore, x-ray lasers usually have considerably lower repetition rates than harmonic-generation sources and are best suited for applications requiring single-shot exposure. (X-ray lasers based on optical-field ionization might, one day, prove to be exceptions.) The pulse energies of conventional x-ray lasers, on the other hand, are in many cases several orders of magnitude higher than the most efficient harmonic generation source existing today. This might strongly favor x-ray lasers in applications such as x-ray holography. Some x-ray lasers have also been made to operate at shorter wavelengths than so far obtained with harmonic generation. X-ray lasers and harmonic generation sources therefore complement each other and might be found to be useful in completely different applications.

APPLICATIONS

In a recent experiment harmonic radiation in the 10 - 120 eV range was used to measure photo-ionization cross sections in various rare gases [11]. This is one example of an experiment that could have been performed using synchrotron radiation. However, the experiment proved that "synchrotron experiments" can be brought into normal sized laboratories with table-top equipment. In another experiment time-resolved photoemission studies on surfaces was performed in a pump-probe experiment with picosecond resolution [12]. This experiment, in the VUV/XUV, could not have been done with any other conventional source due to the high temporal resolution required. This experiment thus illustrates how the harmonic generation source can open up new fields of investigations. The very high peak intensities expected in the XUV by focusing the harmonic radiation by suitable x-ray optics might open up the field of multi-photon processes and non-linear optics in the XUV.

REFERENCES

1. L'Huillier, A. and Balcou Ph., *Phys. Rev. Lett.* **70**, 774 (1993).
2. Macklin J.J., Kmetec J.D. and Gordon III C.L., *Phys. Rev. Lett.* **70**, 766 (1993).
3. Wahlström C.-G., Larsson J., Persson A., Starczewski T., Svanberg S., Salières P., Balcou Ph. and L'Huillier A., *Phys. Rev. A* **48**, 4709 (1993).
4. Krause J.L., Schafer K.J., and Kulander K.C., *Phys. Rev. Lett.* **68**, 3535 (1992).
5. L'Huillier A., Lewenstein M, Salières P., Balcou Ph, Ivanov M. Yu., Larsson J. and Wahlström C.-G. *Phys. Rev. A* **48**, R3433 (1993).
6. Xu H., Tang X. and Lambropoulos P., *Phys. Rev. A* **46**, R2225 (1992).
7. Lompré L.-A., L'Huillier A., Monot P., Ferray M., Mainfray G. and Manus C., *J. Opt. Soc. Am.* **7**, 754 (1990).
   Balcou Ph., Cornaggia C., Gomes A. S. L., Lompré L.-A. and L'Huillier A., *J. Phys. B* **25**, 4467 (1992).
8. Tisch J.G.W., Smith R.A., Ciarrocca M., Muffett J.E., Marangos J.P. and Hutchinson M.H.R. *Phys. Rev A* **49**, 28 (1994).
   Salières P., Ditmire T., Budil K.S., Perry M.D. and L'Huillier A., *J.Phys B Letters*,
   Peatross J. and Meyerhofer D.D., to appear in *the Proceedings of the Int. Conf. on Multiphoton Processes*, Quebeck, Canada, July 1993.
9. Svanberg S., Larsson J., Persson A. and Wahlström C.-G., *Physica Scripta* **49**, 187 (1994).
10. Werin S. (1994), Private Communications.
11. Balcou Ph., Budil K.S., Ditmire T., Perry M.D., Salières P. and L'Huillier A., *Preprint* (1994).
12. Haight R. and Peale D.R., *Phys. Rev. Lett.* **70**, 3979 (1993).

Prospects for High Power Linac Coherent Light Source (LCLS) Development in the 1000Å-1Å Wavelength Range

R. Tatchyn*, K. Bane*, R. Boyce*, G. Loew*, R. Miller*,
H.-D. Nuhn*, D. Palmer*, J. Paterson*, T. Raubenheimer*,
J. Seeman*, H. Winick*, D. Yeremian*, C. Pellegrini[†],
J. Rosenzweig[†], G. Travish[†], D. Prosnitz[◊], E. T. Scharlemann[◊],
S. Caspi[‡], W. Fawley[‡], K. Halbach[‡], K.-J. Kim[‡], R. Schlueter[‡],
M. Xie[‡], R. Bonifacio[°], L. De Salvo[°], P. Pierini[°]

*Stanford Linear Accelerator Center, Stanford University, Stanford, CA 94309
[†]Department of Physics, University of California (UCLA), Los Angeles, CA 90024
[◊]Lawrence Livermore National Laboratory, Livermore, CA 94550
[‡]Lawrence Berkeley Laboratory, University of California, Berkeley, CA 94720
[°]Istituto Nazionale di Fisica Nucleare, Sezione di Milano, 20133 Milano, Italy

Abstract. Electron bunch requirements for single-pass saturation of a Free-Electron Laser (FEL) operating at full transverse coherence in the Self-Amplified Spontaneous Emission (SASE) mode include: 1) a high peak current, 2) a sufficiently low relative energy spread, and 3) a transverse emittance ε [r-m] satisfying the condition $\varepsilon \leq \lambda/4\pi$, where λ [m] is the output wavelength of the FEL. In the insertion device that induces the coherent amplification, the prepared electron bunch must be kept on a trajectory sufficiently collinear with the amplified photons without significant dilution of its transverse density. In this paper we discuss a Linac Coherent Light Source (LCLS) based on a high energy accelerator such as, e.g., the 3km S-band structure at the Stanford Linear Accelerator Center (SLAC), followed by a long high-precision undulator with superimposed quadrupole (FODO) focusing, to fulfill the given requirements for SASE operation in the 1000Å-1Å range. The electron source for the linac, an RF gun with a laser-excited photocathode featuring a normalized emittance in the 1-3 mm-mrad range, a longitudinal bunch duration of the order of 3 ps, and approximately 10^{-9} C/bunch, is a primary determinant of the required low transverse and longitudinal emittances. Acceleration of the injected bunch to energies in the 5 - 25 GeV range is used to reduce the relative longitudinal energy spread in the bunch, as well as to reduce the transverse emittance to values consistent with the cited wavelength regime. Two longitudinal compression stages are employed to increase the peak bunch current to the 2 - 5 kA levels required for sufficiently rapid saturation. The output radiation is delivered, via a grazing-incidence mirror bank, to optical instrumentation and a multi-user beam line system. Technological requirements for LCLS operation at 40Å, 4.5Å, and 1.5Å are examined.

1. INTRODUCTION

The basic physical mechanisms underlying the onset and exponential amplification of FEL gain have been discussed in prior literature by numerous authors (1,2,3,4,5). From among the many different types and classes of FELs it has proved useful to distinguish those that operate: 1) *with* vs. *without* an external optical cavity, and 2) at beam energies and particle densities at which Coulomb effects within the beam are *negligible* vs. *non-negligible*. Further distinctions may be drawn depending on whether or not an external radiation train is introduced at the beginning of the amplification process (viz., seed light amplification vs. SASE); with respect to the details of the structures employed to guide and enhance the growth of gain (e.g., the Optical Klystron (OK), harmonic generation schemes, etc.); and the source of the particle beam driving the FEL (e.g., storage ring vs. linac). Since the 1970s a wide variety of FELs included in a number of these categories have been developed and operated in the infrared (IR) through the ultraviolet (UV) regimes (6). Heretofore, extension of FEL operation to increasingly higher photon energies has been inhibited by a number of factors critical to the gain process. First, significantly increased particle densities, or beam currents, are required to maintain the interparticle interactions underlying the growth of gain. Second, the energy spread of the particles in the beam must be kept sufficiently small so that the radiation field that they generate and through which they interact remains sufficiently monochromatic (coherent). Third, most of the radiation generated by the particles must be physically confined to the lasing phase-space volume, a condition expressible by the criterion $\varepsilon \leq \lambda/4\pi$, viz., the emittance of the particle beam must be substantially dominated by the wavelength of the amplified light.

In recent years, a number of technical developments has made it possible to consider FEL schemes wherein these and other factors could be satisfied down to x-ray wavelengths of the order of 1Å (7). First, RF electron guns with laser-driven photocathodes and normalized emittances in the 1-3 mm-mrad range have been designed and operated. Assuming particle beam energies E as high as 50 GeV, viz., γ factors ($\gamma = E/m_e c^2$) approaching 10^5, the possible fulfillment of the emittance criterion with the use of, e.g., a high energy linac is evident. Second, assuming that the energy spread, σ_E, of the particle bunch emitted from the cathode stays relatively constant, the multi-GeV acceleration following the gun can be used to significantly reduce the relative energy spread, σ_E/E, of the beam to values required for efficient gain amplification. Third, longitudinal compression stages in the acceleration cycle, such as have been developed and utilized, e.g., at the Stanford Linear Collider (SLC) at SLAC (8), can be employed to increase the peak current (i.e., the particle density within the bunch) to values required for efficient gain stimulation. Finally, progress in undulator and focusing lattice technologies, in particular at 3rd Generation Synchrotron Radiation

Sources (9), can now yield insertion devices of sufficient length and quality to maintain the cited beam conditions during the FEL amplification process.

Given these developments, a multi-institutional collaboration has been established to study the feasibility of utilizing a portion of SLAC's 3 km linac as a driver for an x-ray FEL. Since highly efficient reflector arrays are presently unavailable in this spectral regime, the basic operating mode assumed for this FEL has been either SASE or seed-light amplification designed to saturate the gain in a single pass through the undulator. The design of this x-ray source, referred to as the LCLS ("Linac Coherent Light Source"), has been structured as an R&D facility for the development and utilization of x-ray lasers (see Fig. 1).

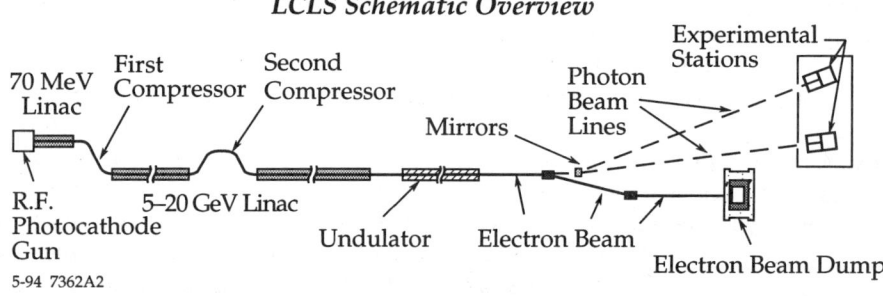

FIGURE 1. Proposed component layout of the LCLS experimental facility at SLAC.

The activities of the research group have focused on a number of areas associated with the various systems indicated in the layout. These include theoretical and design studies of: 1) the FEL gain process in the insertion device; 2) emittance control and beam transport in the RF gun, the longitudinal bunch compressors, and the linac acceleration sections; 3) the insertion device and its auxiliary beam focusing and monitoring systems; and 4) the optical beam line instrumentation. In the following sections we will review selected aspects of these studies.

2. PHYSICAL CHARACTERISTICS OF THE LCLS

The basic energy transfer mechanism in the SASE FEL is dependent on the interaction of the free electrons with the combined undulator and radiation fields. At the proper phase the kinetic energy of electron motion induced by the undulator is transferred to the radiation field. Once this energy exchange begins favoring a particular wave field (out of the large aggregate of waves generated by the randomly distributed electrons) the process becomes regenerative; viz., an initially weak bunching fluctuation coherently enhances the field, which in turn induces stronger bunching, etc., until limiting effects lead to saturation. In the multi-GeV energy regime of the SLAC LCLS, Coulomb effects play a minor role in this process. For an N-period undulator of length L_u and a peak bunch current I_p, an analysis of gain amplification in the bunch (5) leads to an estimated total output power of $P_{tot}[GW] \cong 1000 E I_p / N$, from which the potential performance of the SASE FEL can be assessed.

Theoretical and numerical studies of the LCLS have focused on the various factors and mechanisms underlying the growth of the bunching process. Using the codes FRED and GINGER (10), FRED3D and TDA3D (11), and NUTMEG (12), gain in both transverse and helical undulators, including the effects of external focusing, beam phase space distributions, and field errors, has been simulated for various FEL schemes. These include: 1) SASE on the 1st and 3rd harmonics of a single undulator, 2) SASE on the 1st harmonic of an Optical Klystron (OK) configuration, and 3) harmonic amplification of seeded light and SASE through a series of harmonically tuned undulators. Theoretical analyses and code development for increasingly refined simulations of gain startup from noise, field tapering, and the effects of undulator field harmonics on gain are also in progress. Basic parameter sets from three single-undulator case studies are listed in Table 1.

Table 1. Parameters for three LCLS cases:

	FEL1 (40Å)	FEL2 (4.5Å)	FEL3 (1.5Å)
Normalized emittance $\gamma\varepsilon$ [mm-mrad]	3.5	1	1
Peak current I_p [kA] *	2.5	5	5
Electron beam energy E [GeV]	7	15	25
σ_E / E [%]	0.02	0.02	0.02
Pulse duration $\sqrt{2\pi}\sigma_\tau$ [fs]	300	150	150
Repetition rate [Hz]	120	120	120
Undulator period λ_u [cm]	8.3	4.0	4.0
Peak field B_u [T]	0.76	1.6	1.6
Saturation length L_u [m]	60	40	70
Peak coherent power [GW]	10	100	40
Average coherent power [W]	0.4	1.4	1.6
Energy/pulse [mJ]	3	12	5
Coherent photons/pulse (x10^{13})	6.6	3.3	0.5
Approximate Bandwidth (BW) [%]	0.1	0.1	0.1
Peak brightness** (x10^{31})	5	500	500
Average brightness** (x10^{21})	2	100	100
Transverse size [microns, FWHM]***	80	30	20
Divergence angle [μrad, FWHM]***	25	10	5

*Bunch charge in all cases is 1 nC; **Photons/s/mm^2/mrad^2/0.1%BW; ***At exit of undulator

3. RF GUN AND BEAM TRANSPORT

A typical RF gun/injector design under present study at SLAC consists of a UV laser-excited photocathode (e.g., Cs_2Te (13)) followed by a 1.6 cell S-band cavity and acceleration of the bunch up to 50MeV. The development of a working prototype using codes such as SUPERFISH (14), ITACA, and PARMELA (15) to study the effects of cavity geometry on beam transport is in progress in collaboration with groups at the University of California at Los Angeles (UCLA), the Los Alamos National Laboratory (LANL), and the Brookhaven National Laboratory (BNL). Emission parameters anticipated for the final structure include a 1nC charge, a 3ps (FWHM) length, and a 1mm-mrad normalized emittance.

Following injection, the bunch passes through a first (2.5x) compressor at 70MeV, an acceleration up to about 7GeV, a second (4x) compressor, and a final acceleration up to the LCLS operating energy. The number of compressions, their energies, and the compression factors are selected to minimize the bunch's energy spread, emittance growth, and time and intensity jitters (16). In simulations of the beam acceleration and compression, a number of known emittance degradation effects and possible methods for their control are under continuing study. These include: 1) transverse and longitudinal wakefield effects associated with the RF and passive accelerator components, 2) optical dispersion effects, and 3) the effects of random misalignments of the transport lattice components.

4. THE LCLS INSERTION DEVICE

Use of a linac allows for a minimal aperture in the LCLS insertion device, in principle down to a few millimeters. Limiting factors include: 1) the proximity of the magnetic material and vacuum duct walls to the e-beam's bremsstrahlung and synchrotron radiation cones; 2) complications associated with Beam Position Monitor (BPM) and vacuum engineering; and 3) undulator type (helical vs. transverse). For the undulator parameter ranges studied for various LCLS cases, viz., $2 \leq K \leq 6$ and $2.5cm \leq \lambda_u \leq 10cm$, the available gap range and the sub-KHz repetition rate allow a wide range of candidate technologies to be considered (see Fig. 2, left). Practical issues associated with the various technologies include: 1) cost factors such as, e.g., undulator materials and device length; 2) the attainability of 0.1%-0.2% field errors within a single gain length; and 3) constraints on the external focusing lattice design. In recent studies, e.g., it has been found that: 1) helical superconducting configurations (bifilar (17) or linear (18)) lead to the shortest structures; and that 2) a transverse pure permanent magnet (PM) design with adequate field quality and sufficiently strong focusing (19,20) could be developed from proven 3rd generation storage ring technology (Fig. 2, right). Substantial experience in the design and operation of transverse FEL undulators based on electromagnetic (E&M) DC technology has been acquired at LLNL (21). Other important factors in the construction and operation of LCLS insertion devices include field control, metrology, and alignment. These issues are being studied and are likely to entail the development of novel or improved techniques (22,23) for reliable operation. A vital aspect of the LCLS as an x-ray source is the tunability of its spectral and polarization properties. In

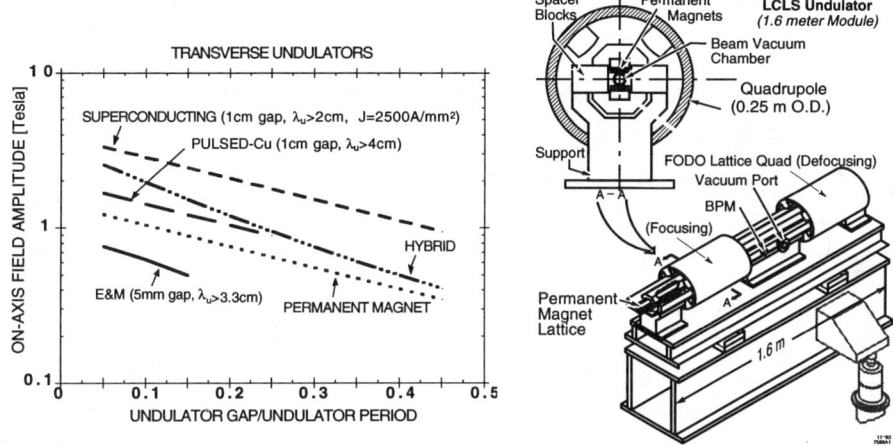

FIGURE 2. Candidate LCLS undulator technologies and parameter regimes (left) and a section of a 60m PM insertion device (gap=1.5cm; period=8cm) for a 40Å LCLS (right).

addition to being tunable with the linac energy, the lasing frequency, as well as the polarization, can be controlled by the undulator. A possible mode of operation would be to induce bunching in an initially transverse insertion device, and then pass the beam through the final few gain lengths in a field-controllable structure such as, e.g., a pulsed-Cu field synthesizer (18).

5. LCLS BEAM LINE OPTICS

The temporal, peak power, brightness, and spectral-angular parameters of the LCLS are all critical to its beam line and end-use experimental designs (24). A graphical display of both the spontaneous and coherent (angle-integrated) photon flux distributions generated by the three LCLS cases of Table 1 is shown in Fig. 3. The opening angle of the coherent FEL peaks ($[(1+K^2)/N_i]^{1/2}/\gamma_i$ for the 1st harmonic) is at least a factor $\sqrt{N_i}$ smaller than that of the spontaneous background. For $K = 6$ the total power in the spontaneous spectrum, which is seen to extend out to the vicinity of the 100th harmonic, can be comparable to or even greater than the power under the coherent peaks. Due to the low average linac current, the time-averaged LCLS power is relatively small, of the order of 1W.

From the output parameters in Table 1 it is evident that a coherent photon pulse at normal incidence can deposit of the order of 1eV per atom for absorptivities (25) and penetration depths typical of solid state materials in the x-ray range. This level of energy loading, which can be shown to lead to the enhanced probability of lattice damage, can be reduced by decreasing the angle of incidence, θ_i, on the optical surface (see Fig. 4, left). This leads to the notion of

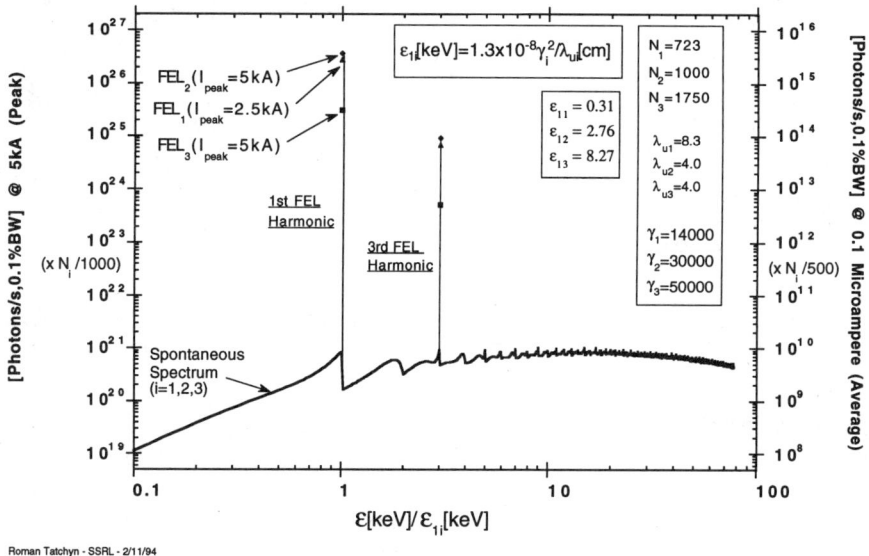

FIGURE 3. Energy-normalized spectral flux curves for three LCLS undulators.

multiple reflections at grazing incidence to deflect the LCLS beam by a total angle θ_T. For m reflectors with equal reflectivities R ($R \cong 1$), indices of refraction $n = 1 - \delta + ik$, atomic densities #[cm^{-3}], and vertical penetration depths δ_p [cm] of the order of $\lambda/4\pi k$, an absorbed-energy parameter, η_A [eV/atom], can be defined for a given peak power, P_{peak} [W], and incident beam diameter D_w [cm]:

$$\eta_A = \frac{P_{peak} \sqrt{2\pi}\sigma_\tau}{q} \left[\frac{\theta_i}{D_w^2} \right] \left[\frac{1-R}{\delta_p \#} \right] \ll 1. \quad (1)$$

Selecting, e.g., $\eta_A \leq 0.01$, a criterion suggested by earlier experimental work at SSRL (26), parameter studies of eq. (1) indicate that grazing incidence arrays (Fig. 4, right) of practical size and economy can be designed for the coherent peaks of the LCLS down to 1Å wavelengths.

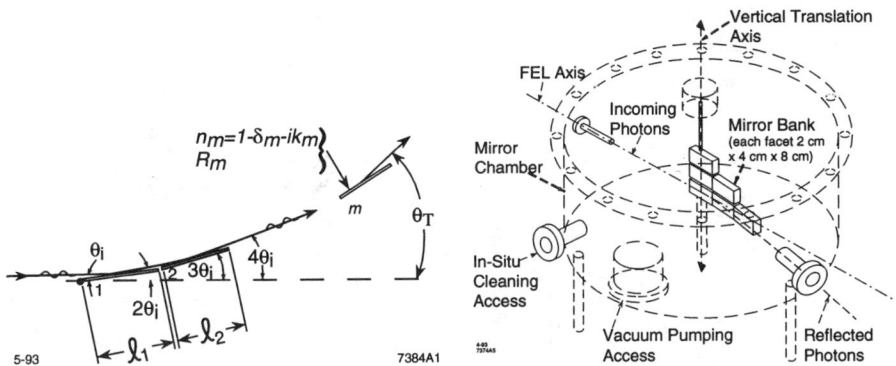

FIGURE 4. Multiple grazing incidence geometry (left) and practical configuration (right).

The design of beam line instrumentation for experimental applications (e.g., monochromators, beam splitters, delay lines, etc.) to further control the spectral-angular, temporal, and coherence properties of the LCLS beam is expected to present both significant challenges and opportunities. As an example, the diffraction-limited source volume of the LCLS should allow the use of efficient monochromator configurations in which the beam itself is the entrance aperture. Due to potential damage effects and the extreme brevity of the radiation pulses, however, special techniques such as beam expansion and compression, or novel elements such as multi-phase or dynamical optics, may need to be developed to attain the desired spectral profiles, resolving powers, and efficiencies (27).

6. SUMMARY

As indicated in Fig. 5, the peak power, brightness, and temporal properties of the LCLS are expected to open important new regimes for x-ray science and technology. Two workshops for exploring such possibilities have recently been held at SLAC (27, 28). The first focused on an LCLS and various applications in the 40Å range, with an emphasis on imaging techniques such as, e.g., single-shot holography of biological samples. The subsequent workshop, emphasizing LCLS sources and applications in the 4.5Å-1.5Å regime, has generated significant interest in areas such as surface and liquid-phase chemistry, materials science, structural biology, and non-linear physics. New techniques and extended parameter ranges in time-resolved, structural, and coherence studies have been considered. These include the possibility of real-time studies of fast chemical reactions and phase transitions; real-time lattice dynamics studies using speckle interferometry; structural analysis using the spontaneous LCLS spectrum for Laue diffraction; and multiple-beam techniques for holographic tomography. Studies

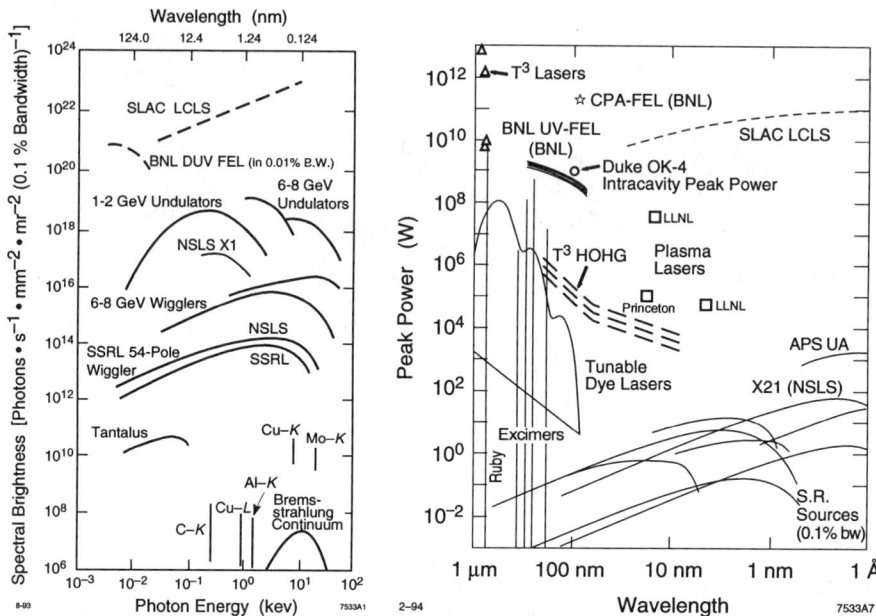

FIGURE 5. Comparison curves contrasting the average spectral brightness (left) and the peak coherent output power (right) of the LCLS with alternative coherent and quasi-coherent sources spanning the visible through the sub-100keV x-ray regimes.

of these and other possibilities, including issues related to sample damage, are currently in progress.

ACKNOWLEDGMENTS

Work supported in part by the Department of Energy Offices of Basic Energy Sciences and High Energy and Nuclear Physics and Department of Energy Contract DE-AC03-76SF0015.

REFERENCES

1. Madey, J., "The Development of the Free Electron Laser," in *Near Zero: New Frontiers of Physics*, Fairbank, J. D., Deaver, Jr., B. S., Everitt, C. W. F., and Michelson, P. F., eds., New York: W. H. Freeman and Company, 1988, pp. 431-441.
2. Murphy, J. B., and Pellegrini, C., "Introduction to the Physics of the Free Electron Laser," in *Lecture Notes in Physics No. 296*, Month, M., and Turner, S., eds., Berlin: Springer-Verlag, 1988, pp. 163-219.
3. Colson, W. B., *SPIE Proceedings* **738**, 2-27(1988).
4. Pellegrini, C., *Nuclear Instruments and Methods* **A272**, 364-367(1988).
5. Kim, K.-J., *Physical Review Letters* **57(13)**, 1871-1874(1986).

6. Van Amersfoort, P. W., Van Der Slot, P. J. M., and Witteman, W. J., eds., *Nuclear Instruments and Methods* **A341,** 1994, Part II, pp. ABS3-ABS140.
7. Winick, H., Bane, K., Boyce, R., Cobb, J., Loew, G., Morton, P., Nuhn, H.-D., Paterson, J., Pianetta, P., Raubenheimer, T., Seeman, J., Tatchyn, R., Vylet, V., Pellegrini, C., Rosenzweig, J., Travish, G., Prosnitz, D., Scharlemann, E. T., Halbach, K., Kim, K.-J., Schlueter, R., Xie, M., Bonifacio, R., DeSalvo, L., Pierini, P., "Short wavelength FELs using the SLAC linac," presented at the 8th Synchrotron Radiation Instrumentation Conference, Gaithersburg, MD, August 23-26, 1993.
8. Raubenheimer, T., Emma , P., and Kheifets, S., "Chicane and Wiggler Based Bunch Compressors for Future Linear Colliders," in *Proceedings of the 1993 Particle Accelerator Conference*, Washington D.C., May 17-20, 1993, pp. 635-637.
9. Cornacchia, M., and Winick, H., eds., *Proceedings of the Workshop on Fourth Generation Light Sources*, SSRL Pub. 92/02, pp. 385-616.
10. Scharlemann, E. T., and Fawley, W. M., *SPIE Proceedings* **642**, 2-9(1986).
11. Kim, K.-J., Xie, M., Scharlemann, E. T., Pellegrini, C., and Travish, G., "Performance Characteristics, Optimization, and Error Tolerances of a 4-nm FEL Based on the SLAC Linac," in *Proceedings of the 1993 Particle Accelerator Conference*, Washington D.C., May 17-20, 1993, pp. 1533-1535.
12. Bonifacio, R., DeSalvo, L., Pierini, P, and Scharlemann, E. T., *Nuclear Instruments and Methods* **A296**, 787-790(1990).
13. Chevallay, E., Durand, J., Hutchins, S., Suberlacq, G. , and Wurgel, M., *Nuclear Instruments and Methods* **A340**, 146-149(1994).
14. Menzel, M. T., and Stokes, H. K., "User's guide for the POISSON/SUPERFISH group of codes," January, 1987, LANL Report LA-UR-87-15.
15. Rosenzweig, J., Smolin, J., and Serafini, L., "Design of a High Brightness RF Photoinjector for the SLAC X-Ray Linear Coherent Light source," in *Proceedings of the 1993 Particle Accelerator Conference*, Washington D.C., May 17-20, 1993, pp. 3024-3026.
16. Bane, K., Raubenheimer, T., and Seeman, J., "Electron Transport of a Linac Coherent Light Source (LCLS) Using the SLAC Linac," ibid., pp. 596-598.
17. Brau, C. A., *Free-Electron Lasers*, Boston: Academic Press, Inc., 1990, ch. 6, pp. 269-271.
18. Tatchyn, R., and Cremer, T., *IEEE Transactions on Magnetics* **26(6)**, 3102-3123(1990).
19. Tatchyn, R., Boyce, R., Halbach, K., Nuhn, H.-D., Seeman, J., Winick, H., and Pellegrini, C., "Design Considerations for a 60 Meter Pure Permanent Magnet Undulator for the SLAC Linac Coherent Light Source (LCLS)," in *Proceedings of the 1993 Particle Accelerator Conference*, Washington D.C., May 17-20, 1993, pp. 1608-1616.
20. Tatchyn, R., *Nuclear Instruments and Methods* **A341**, 449-453(1994).
21. Deis, G. A., Burns, M. J., Christensen, T. C., Coffield, F. E., Kulke, B., Prosnitz, D., Scharlemann, E. T., and Halbach, K., *IEEE Transactions on Magnetics* **24(2)**, 986-989(1990).
22. Warren, R. W., *Nuclear Instruments and Methods* **A272**, 257-260(1988).
23. Ben-Zvi, I., and Qiu, X. Z., *SPIE Proceedings* **2013**, 44-53(1993).
24. Tatchyn, R., and Pianetta, P., "X-Ray Beam Lines and Beam Line Components for the SLAC Linac Coherent Light Source (LCLS)," in *Proceedings of the 1993 Particle Accelerator Conference*, Washington D.C., May 17-20, 1993, pp. 1536-1538.
25. Henke, B. L., Lee, P., Tanaka, T. J., Shimabakuro, R. L., and Fujikawa, B. K., *American Institute of Physics Proceedings* **75**, 340-388(1982).
26. Tatchyn, R., Csonka, P., Kilic, H., Watanabe, H., Fuller, A., Beck, M., Toor, A., Underwood, J., and Catura, R., *SPIE Proceedings* **733**, 368-376(1986).
27. Spicer, W., Arthur, J., and Winick, H., eds., *Workshop on Scientific Applications of Short Wavelength Coherent Light Sources*, Stanford, CA, October 21, 1992, SLAC Report 414.
28. Arthur, J., Materlick, G., and Winick, H., eds., *Workshop on Scientific Applications of Coherent X-Rays*, Stanford, CA, February 12, 1994, SLAC Report 437.

Numerical Simulation of X-ray Zone Plates with High Aspect Ratio

Yuri V. Kopylov, Alexei V. Popov, and Alexander V. Vinogradov[*]

Institute of Terrestrial Magnetism, Ionosphere and Radio Wave Propagation

Russian Academy of Sciences, 142092 Troitsk, Moscow region, Russia

[*]Lebedev Physics Institute

Russian Academy of Sciences, 53 Leniskiy pr., 117924 Moscow, Russia,

1. Introduction.

The prospects of diffractive X-ray optics are mainly associated with design and manufacturing of high resolution and efficient Fresnel zone plates. An adequate wave theory of this fine multiscale object is still to be developed. Classical wave theory of the Fresnel zone plate is restricted to treating it as a plane amplitude or phase screen [1-2]. This approximation is not valid for actually high resolution and efficient zone plates which necessarily must be optically thick and having high aspect ratio (thickness over the zone width)

$$\frac{b}{\Delta} \geq \frac{1}{\epsilon - 1} \gg 1 \qquad (1)$$

Recently, a number of papers appeared considering realistic zone plate models beyond the limits of the plane screen Kirchhoff theory [3-5]. All of them use semianalytical approaches based on specific models of the scattering object. On the other hand, straightforward numerical methods of solving Maxwell's equations are extremely time consuming due to high ratio of all the geometrical scales (zone plate radius and thickness, focal length) to the wavelength λ.

We suggest an alternative computational approach based on the parabolic wave equation (PWE) [6]. It is applicable because the permittivity ϵ of all materials in X-ray range only slightly differs from its free-space value: $|\epsilon - 1| \ll 1$. Physically, this causes that all diffraction processes in an X-ray optical element have unidirectional and almost paraxial character. In point of fact, we deal rather not with diffraction but with wave propagation in a multiscale nonuniform medium composed of weak dielectric elements. The parabolic equation method has proven to be a very efficient computational tool for such problems. Using PWE crucially reduces computational time and permits to simulate numerically not only various types of Fresnel zone plates but many other elements and systems of X-ray transmission optics.

2. Parabolic Wave Equation (PWE).

The Leontovich-Fock parabolic wave equation has become a classical approximate approach to radio wave propagation [6] and underwater acoustics [7]. A variety of its modifications has been used in diffraction theory [8-10] and nonlinear optics [11]. Recently, an extensive use of the PWE takes place in many branches of computational electromagnetics which was anticipated earlier by Malyuzhinets [8]. We believe that X-ray imaging optics will become a new important application of this well-proven and efficient method because most of optical schemes work with paraxial wave beams and all the materials are almost transparent in this spectral region.

The parabolic approximation can be derived from the exact wave theory of a weak nonuniform medium with complex dielectric permittivity $\epsilon = 1 + \alpha(x,y,z)$, $|\alpha| \ll 1$ by substituting the paraxial Ansatz

$$E(x,y,z) = u(x,y,z)e^{ikz} \qquad (2)$$

into the wave equation

$$E_{xx} + E_{yy} + E_{zz} + k^2 \epsilon(x,y,z) E = 0 \qquad (3)$$

and neglecting the second derivative of the slow amplitude u_{zz}. The following approximate equation arises

$$2ik\frac{\partial u}{\partial z} + \Delta u + k^2 \alpha(x,y,z) u = 0 \qquad (4)$$

which describes adequately the main diffraction phenomena ("transversal diffusion" [8]) in the nonuniform dielectric medium as well as the free-space propagation and focusing outside the optical elements. Formally, the applicability conditions of the PWE (4) can be obtained by putting $\alpha(x,y,z) = \tilde{\alpha}(\frac{x}{a}, \frac{y}{a}, \frac{z}{b})$ where a and b are characteristic scales of transversal and longitudinal nonuniformity of the dielectric permittivity (for a Fresnel zone plate, a is the minimum zone width: $a \approx dr_N = r_N - r_{N-1} \approx \frac{1}{2}(\lambda f/N)^{1/2}$ [1], and b is the zone plate thickness). Estimation of the neglected terms yields:

$$kb \mid \alpha \mid \leq 1, \quad \frac{ka^2}{b} \geq 1 \tag{5}$$

These conditions are usually met for X-ray optical elements. In practice, however, the main averaged features of the wave field can be described correctly even under less restrictive assumptions.

If the Fresnel parameter ka^2/b proves to be large, it is possible to omit in (4) the diffusion term Δu which leads to the plane screen approximation

$$u(x,y,b) = u(x,y,0) \exp\left[i\frac{k}{2}\int_0^b \alpha(x,y,z)\,dz\right] \equiv T(x,y)u(x,y,0) \tag{6}$$

supplying initial values $u(x,y,b)$ for the Fresnel-Kirchhoff integral. However, as X-ray zone plates use to have high aspect ratio b/a and large number of zones N, the complete PWE (4) is to be used to accurately take into account the diffraction processes inside the plate body.

3. Computational Aspects of the PWE.

Transition from rigorous wave theory to the transversal diffusion approximation leads to the change of kind of the differential equations and to a new formulation of the boundary value problem. In contrast to the elliptic wave equation, the PWE (4) describes the evolution of the wave amplitude in process of almost unidirectional propagation along the optical axis. Physically, it means neglecting the backward reflections from the interfaces and the waves diffracted into large off-axis angles (such approximation is adequate for the variations of the refraction index are supposed to be small).

The marching type of the parabolic equation (4) makes correct the Cauchy problem with a given initial distribution $u(x,y,0) = u_0(x,y)$ and with some radiation condition excluding spurious waves coming from infinity. Analytically, the simplest way is to demand that the solution must be bounded at infinity when an arbitrarily small absorption $Im\,k > 0$ is introduced (Malyuzhinets' "principle of extinguishing" [8]). Unfortunately, the extinguishing condition is difficult to implement in the process of the numerical solution of the problem which demands truncation of the spatial domain. It makes one to seek a way of transferring the radiation condition from infinity to a certain finite surface surrounding all the sources and diffractive elements. Since any spurious reflection from this artificial boundary would change the wave field structure, such boundary condition has to provide full transparency for arbitrary radiation from inside.

In a model two-dimensional case, an exact form of the transparency condition

$$\frac{\partial u}{\partial x}(x, \pm A) = \mp e^{-i\pi/4}\sqrt{2k\frac{\partial}{\partial z}}u(x, \pm A) \tag{7}$$

has been found in [15,16]. It can be derived formally by factorization of the differential operator

$$2ik\frac{\partial}{\partial z} + \frac{\partial^2}{\partial x^2} = \left(\frac{\partial}{\partial x} + e^{-i\pi/4}\sqrt{2k\frac{\partial}{\partial z}}\right)\left(\frac{\partial}{\partial x} - e^{-i\pi/4}\sqrt{2k\frac{\partial}{\partial z}}\right) \tag{8}$$

corresponding to the free-space version of the PWE (4); here, the symbol of fractal derivative

$$\sqrt{\frac{\partial}{\partial z}}u(z) = \frac{1}{\sqrt{\pi}}\frac{\partial}{\partial z}\int_0^z u(\zeta)\frac{d\zeta}{\sqrt{z-\zeta}} \tag{9}$$

is used to describe the nonlocal surface admittance.

Quite similarly, in three dimensions the radiation condition can be transferred from infinity to an arbitrary cylindric surface $r = A$ surrounding all the dielectric elements. In the most interesting axial symmetric case, the following boundary condition arises

$$\frac{\partial u}{\partial r}(A, z) = \frac{\partial}{\partial z}\int_0^z u(A, \zeta)K(z - \zeta)\,d\zeta \tag{10}$$

with

$$K(z) = e^{-i\pi/4}\frac{\sqrt{k}}{\pi\sqrt{2}}\int_{c-i\infty}^{c+i\infty}\frac{H_0^{(1)'}(A\sqrt{2ikp})}{H_0^{(1)}(A\sqrt{2ikp})}e^{pz}\frac{dp}{\sqrt{p}} \tag{11}$$

Moreover, as usually the integration region is very wide compared with the wavelength ($kA \gg 1$), the Hankel function can be replaced by its far-field asymptotic expression $H_0^{(1)}(t) \sim -\sqrt{2k/\pi t}e^{i(t-\pi/4)}$. Then the integral in (11) can be calculated explicitly:

$$K(z) \approx -e^{-i\pi/4}\sqrt{2k/\pi z} \tag{12}$$

which reduces the boundary condition (10) to the form

$$\frac{\partial u}{\partial r}(A, z) = -e^{-i\pi/4}\sqrt{\frac{2k}{\pi}}\frac{\partial}{\partial z}\int_0^z u(A, \zeta)\frac{d\zeta}{\sqrt{z-\zeta}} \tag{13}$$

coinciding with (7). Its numerical implementation [15] provides, with high accuracy, full transparency of the artificial boundary $r = A$ in all practical examples.

Many aspects of the numerical solution of the parabolic equation have been thoroughly studied (e.g. [17-22]). Confining ourselves to the axial symmetric case, we use the following six-point implicit finite-difference scheme

$$2ik\frac{u_m^{n+1} - u_m^n}{\tau} + \frac{u_{m+1}^{n+1} - 2u_m^{n+1} + u_{m-1}^{n+1} + u_{m+1}^n - 2u_m^n + u_{m-1}^n}{2h^2}$$

$$+ \frac{u_{m+1}^{n+1} - u_{m-1}^{n+1} + u_{m+1}^n - u_{m-1}^n}{4mh^2} + k^2 \alpha_m^{n+1/2} \frac{u_m^{n+1} + u_m^n}{2} = 0 \qquad (14)$$

reducing the PWE (4) to a three-diagonal set of linear algebraic equations. Here, τ and h are the mesh steps in z and r directions, and $u_m^n \approx u(n\tau, mh)$. The linear equations (14) written for $m = 0, 1, \ldots, M = A/h$, along with the finite-difference approximation of the transparency condition (13)

$$\frac{u_{M+1}^{n+1} - u_{M-1}^{n+1}}{2h} + 2\sqrt{\frac{2k}{i\pi\tau}}\left(u_M^{n+1} - \sum_{s=1}^{n}\gamma_s u_M^{n+1-s}\right) = 0 \qquad (15)$$

form a complete set of equations which can be solved step by step, from n to $n+1$, by the marching method - cf. [15]. The energy balance law

$$\sum_{m=0}^{M} m \mid u_m^n \mid^2 \leq \sum_{m=0}^{M} m \mid u_m^0 \mid^2 \qquad (16)$$

grants convergence and stability of the scheme (14).

The finite difference method has some important advantages compared with other possible methods of solving the PWE: it is logically simple, weakly depending on specific geometry of the diffractive elements and especially efficient for calculation of the global field distribution throughout the optical system. Such global patterns, visualized by means of color graphics, are extremely useful for understanding all the details of diffraction and focusing processes.

4. Numerical Examples.

A. Thin zone plates.

We have calculated a series of global field intensity distributions produced by an amplitude Fresnel microlens composed of 75 opaque zones ($N = 150$) with rectangular transmission function

$$T(r) = \begin{cases} 0 & for \quad r_{2n-2} < r < r_{2n-1}, \quad n = 1, \ldots, N/2 \\ 1 & for \quad r_{2n-1} < r < r_{2n}, \quad n = 1, \ldots, N/2 \end{cases} \qquad (17)$$

Several sources of illumination have been examined: point source and ring sources of different radii. By choosing the external radius $r_N = 4243 nm$ we obtain the focal length $f = r_N^2/\lambda N = 50 \mu m$ for a typical wavelength $\lambda = 2.4 nm$ in the "water window" range. The outermost zone width $dr_N = r_N/2N$ equals $14 nm$, so we can expect maximum spatial resolution about $1.22 dr_N \approx 17 nm$ [1].

In our first model example, the ideal thin zone plate is illuminated by a point source placed at $Z = 75 \mu m$ apart, so the image must be expected at $Z' = 150 \mu m$. The calculated field intensity distribution shows a sharp focus at this point. The radial intensity distribution near the optical axis is similar to the "point spread function" of an equivalent ideal lens

$$T(r) = \begin{cases} \exp\left(ik\frac{R^2-r^2}{2f}\right), & r < R \\ 0, & r > R \end{cases} \tag{18}$$

(quadratic phase screen with the same external radius and focal length) calculated by the same finite-difference method. The only difference is a much lower peak intensity. The ratio of these peak intensities may be used as a measure of diffraction efficiency of the zone plate. In our example, this ratio is equal to 0.1 which coincides with the theoretical estimate $1/\pi^2$ [1].

The radius of the first dark ring in the image plane is about $Y' \approx 50$ nm which gives, after dividing by magnification factor $M = Z'/Z = 2$, the value $Y \approx 25 nm$ in the object plane. This agrees with the well-known Rayleigh criterion [23]

In order to estimate numerically the field of view of this model Fresnel zone plate, we calculated a series of images of extended objects. As a model, we used a fine coherent ring source. Its position was chosen at $Z = 75 \mu m$ which gave twofold magnification $M = 2$.

Calculations performed for the ring radii $a = 200 nm$ and $a = 600 nm$ show a clearly seen image of the source as a bright ring with radius $a' = 2a$ at the range $Z' \approx 150 \mu m$. It is interesting to note that the width of the main intensity peak does not decrease with increasing radius a and the spatial resolution calculated according to the Rayleigh criterion remains to be about $20 nm$. What happens is a fast increase of the background intensity level near the optical axis which becomes to exceed in brightness the ring image when the radius $a \geq 500$ nm. This proves to be the main factor limiting the field of view of the Fresnel zone plate.

B. Thick zone plates.

We have estimated numerically the performance of a series of thick Fresnel zone plates. The external radius r_N and the number of zones have been chosen coinciding with the above model examples. The optical constants of the opaque zone material (germanium) in the complex refraction index representation

$$n = \sqrt{\epsilon} = 1 - \delta + i\beta \qquad (19)$$

are: $\delta = 0.0026, \beta = 0.00094$ for the chosen wavelength $\lambda = 2.4 nm$.

As the refraction and absorption factors δ and β are of the same order of magnitude, such realistic zone plate works unlike both amplitude and phase idealized Fresnel zone plates. In a rough approximation, the plane screen transmission function (6) can be used to choose its optimal thickness. After substituting $\alpha = \epsilon - 1 \approx 2(i\beta - \delta)$ into the formula (6) we get an estimate of absorption and phase shift caused by the opaque zones:

$$\log |T| = -kb\beta \approx -b/400, \qquad \arg T = -kb\delta \approx -\pi b/460 \qquad (20)$$

where b is the zone plate thickness in nanometers.

We see that phase shift is about π for $b = 460 nm$. If it were not for absorption, this thickness would provide a very good performance close to that of an ideal phase zone plate [1]. For greater thicknesses, due to absorption, the amplitude contrast plays the main role and the zone plate should behave like a classical thin Fresnel lens. However, diffraction in the zone plate body may modify the output field considerably. Only numerical calculations allow to take this effect into account.

First we obtained the field distribution on the back side of the thick zone plate for two different thicknesses $b = 460 nm$ and $b = 920 nm$ by the numerical integration of the PWE inside the zone plate body. The plane incident wave has been taken in order to obtain directly the effective transmission function of the zone plate. In both cases, the output field differs considerably from the "meander" transmission function

$$T(r) = \begin{cases} \exp\left(ik\frac{\alpha}{2}b\right) & for \quad r_{2n-2} < r < r_{2n-1}, \quad n = 1, \ldots, N \\ 1 & for \quad r_{2n-1} < r < r_{2n}, \quad n = 1, \ldots, N \end{cases} \qquad (21)$$

predicted by the plane screen approximation (6).

Then we studied the imaging performance of these realistic Fresnel zone plates illuminated by a ring source of radius $a = 200 nm$. As it has been expected, a more bright image is produced by the zone plate with the resonant thickness $b = 460 nm$. The efficiencies estimated by comparison with the equivalent ideal lens (18) are 0.23 for $b = 460 nm$ and 0.15 for $b = 920 nm$. Moreover, it is easily seen that the zone plate of thickness $920 nm$ behaves almost like the classical (amplitude) thin Fresnel zone plate. The only difference reveals in some additional splitting of the ring image.

Thus, our calculations show that realistic zone plates with high aspect ratio $A = b/dr_N \sim 30-60$ retain most of the imaging properties of a classical Fresnel zone plate providing at the same time a higher diffraction efficiency. Some illustrations are given in Figs.1-4.

5. Conclusion.

A new method to calculate imaging properties of X-ray Fresnel zone plates is suggested and implemented numerically. The mathematical basis of our approach is the parabolic wave equation solved by finite differences inside the zone plate body as well as in a free space region surrounded by an artificial transparent boundary. The computational procedure allows for fast calculation and visualization of the global field distribution. Resolution, diffraction efficiency and field of view of the zone plate are easily derived from the numerical solution.

The numerical examples include comparison with planar Fresnel-Kirchhoff diffraction theory and simulation of some realistic high aspect ratio X-ray zone plates designed for the "water window" spectral range.

7. References

1. A.G.Michette, Optical Systems for Soft X-Rays (Plenum Press, New York and London, 1986).

2. A.G.Michette, C.J.Buckley, X-Ray Science and Technology (Institute of Physics, Bristol, 1993).

3. J.Maser, G.Schmahl, Coupled Wave Description of the Diffraction by Zone Plates with High Aspect Ratios, Optics Comm., V 89., (1992), pp.355-362.

4. A.Sammar, J.-M.Andre, JOSA, 10, (1993), 2324.

5. V.E.Levashov, A.V.Vinogradov, Analytical theory of the zone plates efficiency, Physical Review E, V.49, N 5, (May 1994).

6. V.A.Fock. Electromagnetic diffraction and propagation problems, (Pergamon Press, Oxford, 1965).

7. F.D.Tappert, The parabolic approximation method, Lectures Notes in Physics, 70, in: Wave propagation and underwater acoustics, eds. by J.B.Keller and J.S.Papadakis, Springer, (New York, 1977) pp.224-287.

8. G.D.Malyuzhinets, Progress in understanding diffraction phenomena (in Russian), Soviet Physics (Uspekhi), 69, No 2 (1959) pp.312-334.

9. L.A.Vainstein, Open resonators and open waveguides (in Russian), *Soviet Radio*, (Moscow, 1966).

10. V.M.Babič and V.S.Buldyrev, Short - wavelength diffraction theory (Asymptotic methods),(Springer New York, 1991).

11. V.E.Zakharov, A.B.Shabat, Exact theory of two-dimensional self-focusing and unidimensional self-modulation of waves in nonlinear media (in Russian), JETP, 61, No 1(7), (1971) pp. 118-134.

12. J.Claerbout, Fundamentals of geophysical data processing with applications to petroleum prospecting, (McGraw - Hill, New York, 1976).

13. A.V.Popov, S.A.Hoziosky, On a generalization of the diffraction theory parabolic equation (in Russian), J.Comp.Math. and Math. Phys., 17, No 2 (1977), pp.527-533.

14. E.A.Polyansky, Method of correction of the parabolic equation solutions in a nonuniform waveguide (in Russian), (Nauka, Moscow, 1985).

15. V.A.Baskakov, A.V.Popov, Implementation of transparent boundaries for numerical solution of the Schrödinger equation, Wave Motion, v.14, No 1 (1991) pp.123-128.

16. S.W.Marcus, A generalized impedance method for application of the parabolic approximation to underwater acoustics, J.Acoust.Soc.Am., 90, No 1 (1990) pp. 391-398.

17. G.D.Malyuzhinets, A.V.Popov, Yu.N.Cherkashin. On development of a numerical method in diffraction theory (in Russian), Proc. of the 3rd Symposium on diffraction, Nauka, Moscow (1964) pp.176-178.

18. A.V.Popov, Solution of parabolic equation of diffraction theory by finite difference method (in Russian), J.Comp.Math.and Math.Phys., 8, No 5 (1968) pp.1140-1143.

19. G.Botseas, J.S.Papadakis, Finite-difference solution to the parabolic wave equation, J. Acoust. Soc. Amer., 70, No 3 (1981) pp.795-799.

20. W.L.Siegmann, D.Lee, Aspects of three-dimensional parabolic equation computations, Comput.and Math. with Appl., 11, No 7/8 (1985) pp.853-862.

21. V.Yu.Zavadsky, Modeling of wave processes (in Russian), (Nauka, Moscow, 1991).

22. A.A.Samarsky, A.V.Gulin. Numerical Methods (in Russian), (Nauka, Moscow, 1989).

23. M.Born, E.Wolf, Principles of Optics (Pergamon, Oxford, 1980).

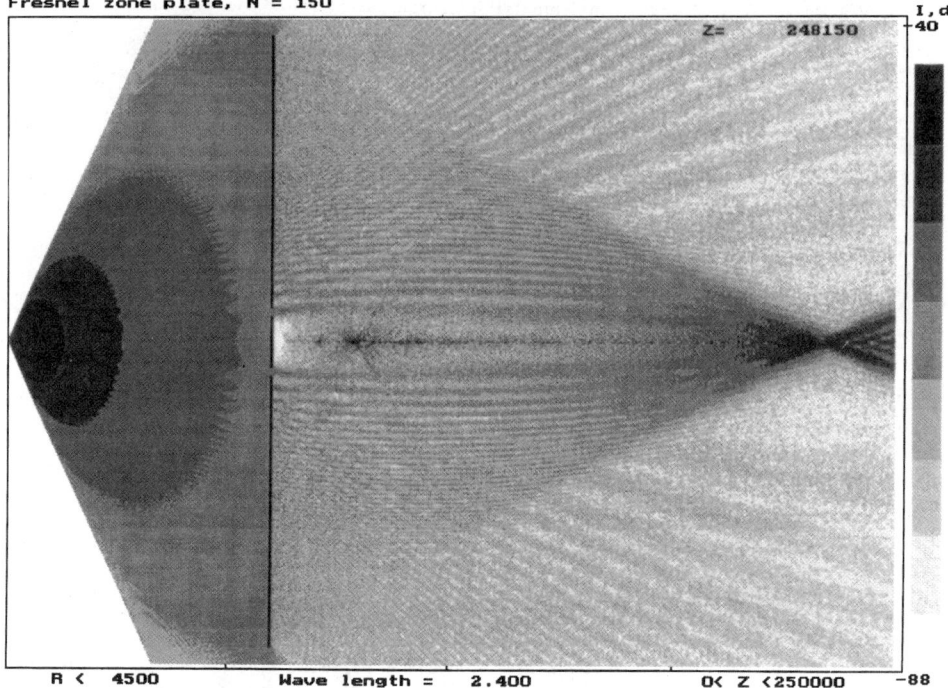

1. Global field intensity distribution displayed in a logarithmic graded grey. Thin Fresnel zone plate illuminated by "point" source (a circular hole of diameter 10 nm) is placed at $z = 75\,\mu m$; the image appears at $z \approx 225\,\mu m$ (all dimensions are given in nanometers). Parameters of the zone plate are: number of zones $N = 150$, external radius $r_N = 4.2\,\mu m$, focal length $f = 50\,\mu m$, wavelength $\lambda = 2.4\,nm$.

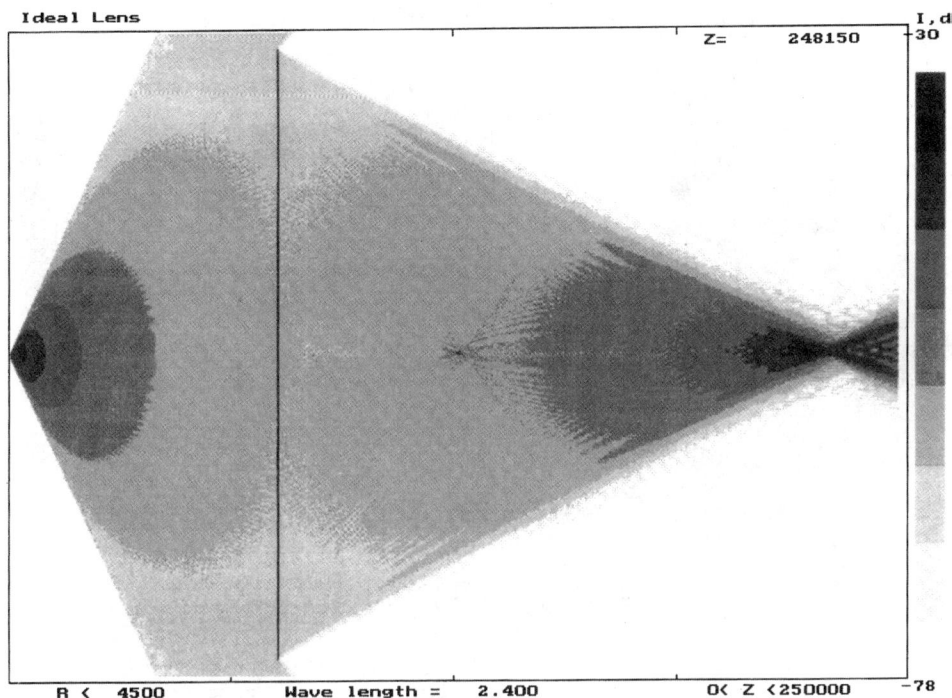

2. Field intensity distribution for an equivalent ideal thin lens (quadratic phase screen) calculated for the geometry of Fig.1

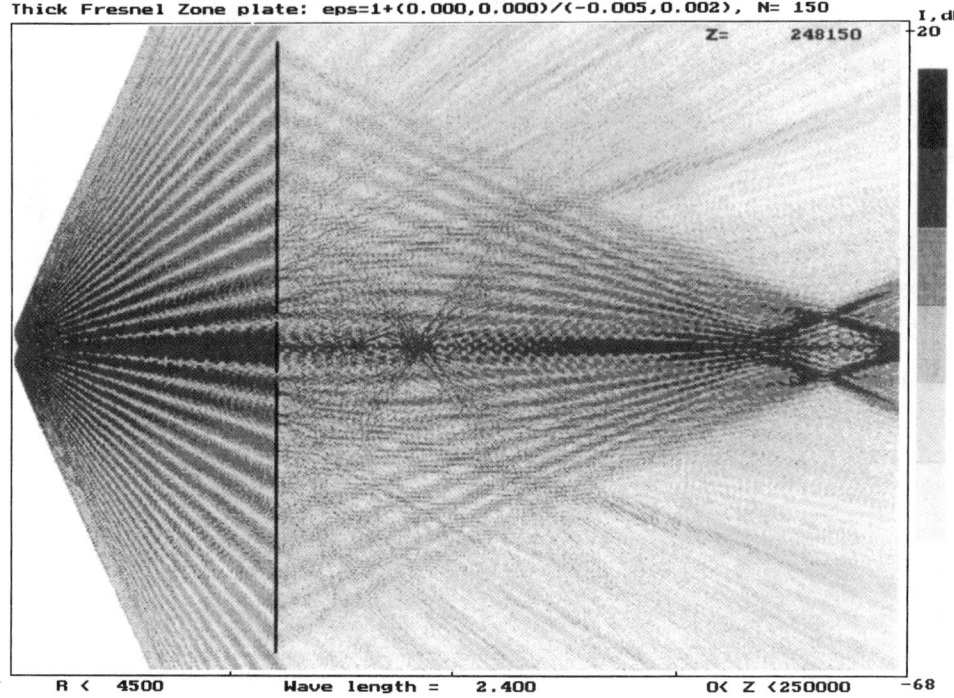

3. Field intensity produced by a thick zone plate (Ge-vacuum, number of zones $N = 150$, external radius $r_N = 4.2\mu m$, thickness $b = 460 nm$) illuminated by a coherent ring source of radius $a = 200 nm$.

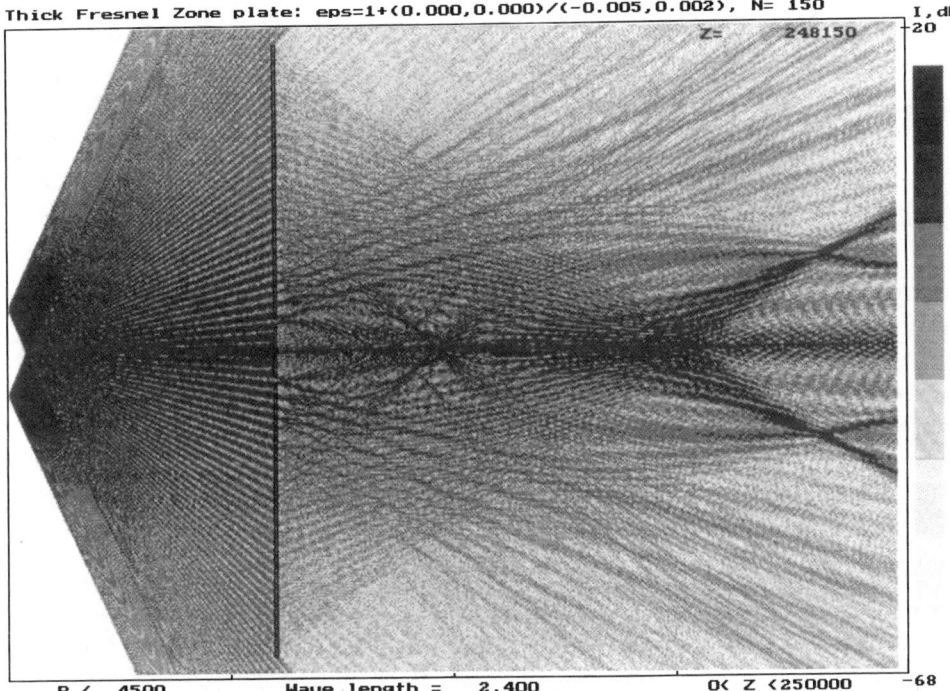

4. Field intensity produced by a thick zone plate (thickness $b = 920nm$) illuminated by a coherent ring source of radius $a = 600nm$.

Self-Induced Spatially-Hole-Burned Distributed Feedback Resonator for X-Ray Lasers

Ferenc Ráksi

Department of Chemistry, University of California, San Diego, La Jolla, CA 92093

Abstract. A new distributed feedback resonator is proposed for X-ray lasers. The laser does not have external mirrors but instead feedback is provided by scattering on periodic gain modulations. A self-inducing effect is investigated for the creation of the distributed feedback, where small initial spatial modulations in the gain medium are amplified due to the positive feedback for the growth of the field and the spatial modulation of the gain.

DISTRIBUTED FEEDBACK LASERS

Distributed feedback (DFB) resonators for X-ray lasers have been proposed for many years to overcome problems associated with mirrors (1-3). The schemes proposed utilize the periodic structure of crystals as resonators. However there are problems with using crystals as laser materials. The main difficulty is concerned with the transmission of X-rays through the crystal. But also of concern is the distortion and destruction of the periodic structure of the crystal due to the high powers involved in pumping the laser.

DFB lasers utilize Bragg scattering from periodic variation in the gain or refractive index of the gain medium for feedback. Depending on what property of the material we modulate, DFB lasers show different dynamics. When the real part of the complex refractive index is modulated, the feedback does not show a relevant dependence on the population of the laser levels. On the other hand, when the feedback happens on gain modulation, the laser show a rich nonlinear dynamics. The dinamics are due to the dependence of the coupling strengths of the Bragg grating on the population of the laser levels. In practice the laser can be pumped by interfering pump beams. By choosing the correct geometry, the Bragg resonance for feedback can be tuned to fall within the gain curve of the laser material. At the beginning of the pump the population and feedback are low. Then the feedback grows simultaneously with the gain, like in a passively Q-switched laser. As a result, a short intense laser pulse is emitted. As the laser field depletes the gain, the feedback is also reduced, releasing the laser pulse from the cavity. This nonlinear dynamics is utilized in gain coupled DFB lasers to generate visible laser pulses as short as several hundred femtoseconds (4-6). Due to the narrow bandwidth of the Bragg resonator, the gain coupled DFB laser normally operates in a single mode. The emitted short pulses are nearly transform limited.

Spatial-Hole-Burning

The dynamic behavior of single mode lasers is further modified by the spatial-hole-burning effect. The two counter propagating waves of the laser field form an interference pattern inside the gain material. The gain depletion is thus not homogeneous, but spatially modulated. The laser field is scattered on this spatially-hole-burned grating and provides additional feedback. Normally the phase of this spontaneous grating is such that it couples the fields to each other with a phase such that the scattered waves interfere destructively with the existing waves. The overall effect is the decrease of the coupling strengths (7-8).

Spatial-hole-burning can also be used to create the Bragg grating for a DFB laser by propagating interfering, depopulating laser beams through the gain medium, which are derived from an external laser (9-10). This scheme was implemented using dye lasers. Incoherent pulses of 200 ps duration and 340 nm wavelength were used to pump laser dyes in a spatially homogeneous way. Coherent interfering depopulating beams in various parts of the visible spectrum created gain modulation by gain depletion. The spatially hole-burned dye laser emitted pulse trains or single pulses of around 6 ps duration (9).

Stability of the gain modulation

In a gain coupled distributed feedback laser the peaks of the gain modulation coincide with the intensity peaks of the standing wave pattern. This provides maximum coupling between the electromagnetic field of the laser and the gain material. This also means that spatial hole-burning reduces the inversion at the peaks of the gain to a larger extent than at the minima. Thus during lasing, the modulation depths is reduced.

SELF INDUCED SPATIAL HOLE-BURNING

For an X-ray laser we can not use the schemes presented in refs. (9-10). We do not have a coherent X-ray laser with which to optically pump the gain material, or to use as a hole-burning laser. Instead we can pump the laser in a spatially homogeneous way. We investigate how to create spatial modulation spontaneously. Counter propagating amplified spontaneous emission can initiate spatial hole-burning or some small spatial gain modulation noise can initiate coupling between the right and left propagating optical waves. To obtain a macroscopic effect built up from noise, we need a positive feedback for the growth of the modulation and the field. We showed in the previous section that in simple gain materials gain modulation does not increase spontaneously. The same is true for the laser intensity. The feedback contribution from a spatially hole-burned grating decreases the growth of the field, due to the phase relation of the scattering and spatial hole-burning.

The situation is entirely different if we use a mixture of gain components and saturable absorbers in the gain material. The absorption cross section of the absorber atoms should be higher than the stimulated emission cross section of the gain atoms. To have net gain the concentration of the excited gain atoms should be higher than the concentration of the saturable absorber. In this case, when the increase of the laser field induces saturation, it happens first in the absorber atoms. The optical gain then increases instead of decreases, until we achieve full saturation of the absorber. If we consider the spatial hole burning of the counter propagating amplified spontaneous emission, the gain increases at the peaks of the interference fringes. This is opposite to the pure gain case.

The gain and laser intensity fringes are now in-phase, and so the scattered waves interfere constructively with the waves already present in the laser. The result is positive feedback for both the gain modulation and the growth of the field.

The DFB laser models (7, 8, 10-12) treat spatial hole-burning. Extending the field equations with new components and adding new material equations for the saturable absorber results in the following differential equation system. We assume that the laser works at exact resonance and at the peak of the gain profile, and perfectly overlapping gain and saturable absorber spectral profiles.

In these equations R and S are the optical fields propagating in opposite directions and W and W_{abs} are concentrations of excited gain and absorber atoms. The 0 and 1 indices refer to the order of the Fourier component of the modulation.

$$\left(\frac{\partial}{\partial z} + \frac{n}{c}\frac{\partial}{\partial t}\right) R = \left(W_0\alpha - W_{0abs}\alpha_{abs}\right) R + \frac{1}{2}\left(-W_1\alpha + W_{1abs}\right) S \quad (1)$$

$$\left(-\frac{\partial}{\partial z} + \frac{n}{c}\frac{\partial}{\partial t}\right) S = \left(W_0\alpha - W_{0abs}\alpha_{abs}\right) S + \frac{1}{2}\left(-W_1\alpha + W_{1abs}\right) R \quad (2)$$

$$\frac{d}{dt}W_0 = \lambda(N - W_0) - B\left(|R|^2 + |S|^2\right) W_0 \quad (3)$$

$$\frac{d}{dt}W_{0abs} = -B_{abs}\left(|R|^2 + |S|^2\right) W_{0abs} \quad (4)$$

$$\frac{d}{dt}W_1 = 2B|RS|W_0 - B\left(|R|^2 + |S|^2\right) W_1 \quad (5)$$

$$\frac{d}{dt}W_{1abs} = 2B_{abs}|RS|W_{0abs} - B_{abs}\left(|R|^2 + |S|^2\right) W_{1abs} \quad (6)$$

B is the coupling strength between the light and the atoms, α and α_{abs} are the small signal gain and absorption. A constant λ pump power was applied, beginning at time = 0. The unit of time is the light propagation time across the laser material. The intensity is measured in photons / stimulated emission cross section area.

With the saturable absorber terms removed, we obtain amplified spontaneous emission (Fig. 1).

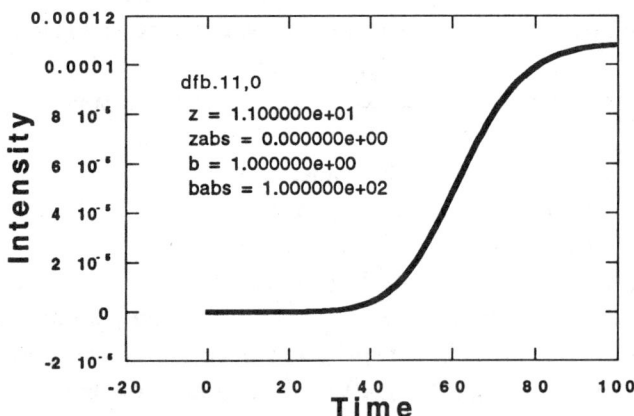

FIGURE 1. Laser intensity for gain material without absorber.

When saturable absorber is added, the intensity evolution is as shown in Fig. 2. The gain-length product is 11 for both examples, while the additional absorption-length product is 2 for the second case. The saturable absorber has 1000 times higher absorption cross section than the stimulated emission cross section of the gain atoms. The laser emits a transient before the amplified spontaneous emission saturation when absorber is added. We expect this spike to be highly coherent. In addition to gain narrowing of the laser line width, "feedback narrowing" occurs. This narrowing is due to the increasing feedback and sclectivity of the Bragg grating.

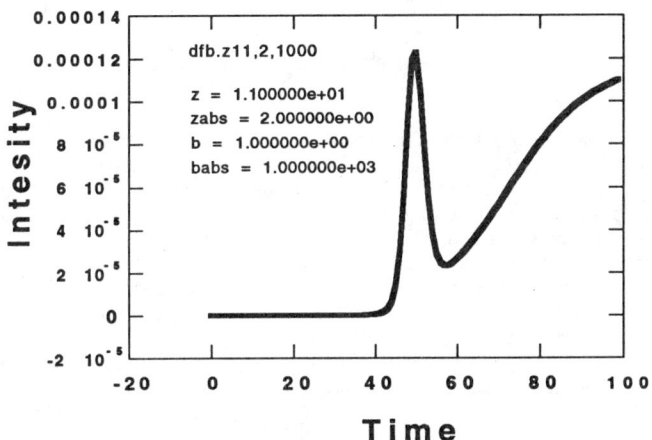

FIGURE 2. Temporal evolution of the laser intensity when the gain is mixed with a saturable absorber.

REFERENCES

1. Fisher, R. A., *Appl. Phys. Letters* **24**, 598-9 (1974).
2. Yariv, A., *Appl. Phys. Letters* **25**, 105-7 (1974).
3. Lyakhov, G. A., *Sov. J. Quant. Electron.* **6**, 456 (1976).
4. Bor, Z., *IEEE Journal of Quantum Electronics,* **16**, 517-24 (1980).
5. Szatmári, S., Schäfer, F. P., *Appl. Phys. B* **46**, 305-11 (1988).
6. Ráksi, F., Zacharias, H., Heuer, W., *Applied Physics B* **53**, 97 (1991).
7. Sargent, M. III., Swattner, W. H., Thomas, J. D., *IEEE Journal of Quantum Electron.* **4**, 465-72 (1980).
8. Rabinovich, W. S., Feldman, B. J., *IEEE Journal of Quantum Electron.* **25**, 20-30 (1989).
9. Ráksi, F., *Appl. Phys. B* **47**, 91-96 (1988).
10. Ráksi, F., *IEEE Journal of Quant. Electron.* **26**, 61 (1990).
11. Duling, I. N., Raymer, M. G., *IEEE Journal of Quantum Electron.* **20**, 1202-7 (1984).
12. Szczepansky, P., *IEEE Journal of Quant. Electron.* **24**, 1248-57 (1988).

Development of X-ray Laser Architectural Components

A. S. Wan, L. B. Da Silva, J. C. Moreno, R. C. Cauble,
E. A. Chandler, H. E. Dalhed, S. B. Libby, R. W. Mayle,
J. Nilsen, R. P. Ratowsky, H. A. Scott, B. Van Wonterghem

Lawrence Livermore National Laboratory, Livermore, CA 94550

Abstract. This paper describes our recent experimental and computational development of short-pulse, enhanced-coherence, and high-brilliance x-ray lasers (XRLs). We will describe the development of an XRL cavity by injecting laser photons back into an amplifying XRL plasma. Using a combination of LASNEX/GLF/SPECTRE-BEAM3 codes, we obtained good agreement with experimental results. We will describe the adaptive spatial filtering technique used to design small-aperture shaped XRLs with near diffraction-limited output. Finally we will discuss issues concerning the development of high-brilliance XRL architecture, with emphasis on scaling the XRL aperture. Combining these advances in XRL architectural components allows us to develop a short-pulse, high-brilliance, coherent XRL suitable for applications in areas such as biological holography, plasma interferometry, and nonlinear optics.

1. INTRODUCTION

With its short wavelength, controllable pulse duration, high brightness and coherence, the X-ray laser (XRL) is ideally suited as a diagnostic probe to image rapidly evolving (<1 ns) laser-driven plasmas with high electron density (10^{21} cm^{-3} < n_e < 10^{24} cm^{-3}).[1,2] Over the past several year, we have developed a number of XRL applications to probe laser-produced plasmas on the Nova 2-Beam facility at Lawrence Livermore National Laboratory (LLNL), such as radiography,[3] Moiré deflectometry,[3] and 2D interferometry.[2] This paper focuses on the experimental and theoretical development of short-pulse, enhanced-coherence, high-brilliance XRLs for various applications.

Experimentally we use one beam to set up the target plasma to be imaged, which leaves us only one beam to use to generate the XRL, severely limiting our experimental parameters. For our current plasma imaging applications we typically use a 3-cm long exploding-foil Y-XRL [4] with an output energy of ~8 mJ, 200-ps FWHM pulse, ~100 µm diameter source size, 10-mrad divergence, and bandwidth ($\lambda/\Delta\lambda$) of 10^{-4}, which corresponds to a brightness of ~10^{17} W/sr-Å-cm^2. This brightness is equivalent to a 6-GeV blackbody, which overwhelms the self-emission of the target plasma and allows us to use multilayer optics for imaging applications.

Although the Y-XRL has sufficient coherence and brightness for most current imaging applications, we still need to pursue a short-pulse, enhanced-coherence XRL to achieve better spatial resolution and to move toward 3-D holographic applications. Typical laser-produced plasmas have a blowoff velocity of ~10^7 cm/s. Thus, an XRL with a 200-ps FWHM pulse duration yields a spatial resolution of ~20 μm. To obtain higher spatial resolution images of the plasmas, we need to shorten the XRL pulse by another order of magnitude. At LLNL we are developing a short-pulsed XRL using two approaches: traveling wave, which is described in a separate paper in this proceeding,[5] and a short-pulse XRL cavity, which we will describe in detail in Sec. 2 of this paper.

The exploding foil Y-XRL has a temporal coherence of ~30–60 μm, which is sufficient for 2-D interferometry.[2] Beyond the 2-D interferometer, a diffraction-limited XRL can yield 3-D holographic images of ICF capsules. Amplified spontaneous emission becomes diffraction limited as the Fresnel number, N_F, approaches 1. N_F can be defined as $r^2/(\lambda L)$, where L is the XRL length, λ is the wavelength, and r is the radius of the source size. To improve laser coherence, for a fixed λ, we need to either increase L or reduce r. In a refractive medium, L is limited by the maximum refraction length,[6] which can be controlled by reducing ∇n_e,[7] or by curving the XRL target.[8] We can then increase L by using multiple-stage laser architecture [9] or by using multilayer optics.[10]

An alternative approach to obtaining a diffraction-limited XRL is to reduce the XRL aperture. Traditionally the XRL source size is defined by the driving laser optics, with a typical best focus configuration of order of 100 μm FWHM. By controlling the shape of XRLs during plasma expansion and reducing the effective XRL source size using the concept of adaptive spatial filtering,[11,12] we can obtain a diffraction-limited XRL with a high effective power. In Sec. 3 we will summarize our recent efforts in shaped XRLs.

As XRLs advance beyond plasma imaging applications and toward high deliverable power on target, we need to develop a high brilliance XRL. Brilliance is proportional to the product of intensity and XRL aperture, and inversely proportional to the square of the divergence. In Sec. 4 we will briefly discuss some issues on scaling of the XRL aperture as we march toward a high brilliance XRL architecture. We will summarize and outline our future research direction in Sec. 5.

2. DEVELOPMENT OF AN X-RAY LASER CAVITY

Given the limited experimental flexibility of the Nova 2-Beam facility, one possible technique for achieving a short-pulse yet high-brightness XRL is to use multilayer optics [13] in a cavity configuration. The formation of an XRL cavity relies on the reinjection of XRL photons back into a gain medium using multilayers. Although the half-cavity concept has been tried in the past,[14] mirror damages have limited the performance of the cavity. Our approach on cavity development is to place the injection mirror far away from the XRL to minimize multilayer damage and to use the multiple-pulse [15] configuration on Nova, timed such that the XRL photons produced in earlier pulses will be reinjected and propagate through the gain medium created by latter pulses. In the following sections we will describe our experimental and theoretical studies of the XRL cavity.

2.1 Experimental Setup and Results

To properly design and understand an XRL cavity we need to know the spatial gain and n_e profiles. The first step of our experimental efforts is to measure the near-field emission profile of the $J = 2-1$ Y laser line at 155 Å, for a 1-cm-long, 100-μm-thick Y-slab target using the XUV Imaging Diagnostic,[3] which employs an imaging multilayer mirror to look at the near-field spatial emission profile at the chosen wavelength. A spatially defining imaging slit limits the viewing of a 10-μm (in the transverse direction) slice of the plasma in the blow-off direction. With our current setup we have a spatial resolution of order of 2 μm.

In our experiments, we used both available beams to drive two sides (designated as east and west sides) of the Y-slab XRL. The injection mirror is aligned such that the east-side XRL can reinject while the west-side XRL will not couple. The driving Nova pulse shape is a multiple-pulse configuration with four 600-ps gaussian pulses at ~150 TW/cm^2 peak incident intensity at 2ω. The injection mirror is placed at 49.5 cm from the output end of the XRL. We observed little evidence of injection mirror damage during our experiments. The temporal peak-to-peak separation between these four pulses is 1.6 ns, which allows the photons generated by the first pulse to reinject into the gain medium created by the third pulse, and the second pulse to inject into the fourth pulse.

One of the major experimental uncertainties is the damage to the XUV imaging mirror, which is placed at ~30 cm away from the XRL target, caused by the intense optical and broad-band x-radiation from the XRL plasma.[16] This damage limits our ability to see the full effect of the injection. The damage to the imaging mirror is evident as the output intensity weakens as a function of time. But taking the west-side XRL as a reference of non-injected XRL intensity, the output intensity of the east-side XRL is a factor of six or more stronger than the reference intensity. This enhanced intensity agrees well with our estimates and demonstrates the successful coupling between the pulses due to reinjection.

Closer examination of the images of the near-field emission profiles reveal several clues to the properties of this multiple-pulse XRL configuration. The gain regions for all four pulses peak at ~100 μm off the target surface. Furthermore, the gain regions stay at approximately the same positions through all four pulses with no appreciable shifts in position. We also observe some residual gain after the pulses, with this small gain (of order of 1/10 of the peak gain) moving away from the target surface with increasing time. We will compare these features with our numerical simulation of these multi-pulsed XRL plasmas in the next section.

2.2 Modeling of Multiple-Pulse Laser Plasmas

The modeling of XRL plasmas is performed in stages. For collisionally dominated plasmas we can decouple the detailed level populations and line transfer physics from the rest of the problem. We use LASNEX[17] to carry out the laser-deposition

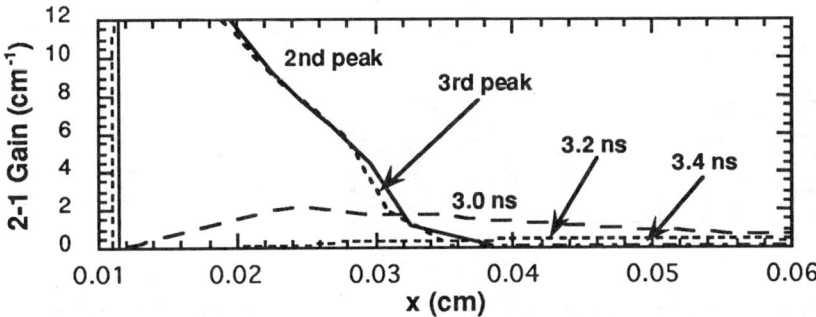

Fig. 1 Spatial gain profiles of Y-XRL peak near the slab surface (0.01 cm) at times corresponding to the peaks of the second and third pulses. Gains between pulses are weaker, with spatial peaks moving away from the slab surface.

and hydrodynamics simulations. From LASNEX we obtain plasma characteristics, such as T_e and n_e, and variables such as mesh positions and velocities. Using our atomic kinetic post-processing code, GLF,[18] we calculate the level populations and line transfer. With the spatial gain and n_e profiles, we can perform ray propagation calculations to match the simulations with actual observable quantities such as laser output intensities. For this purpose we can use SPECTRE, which calculates time-dependent ray trajectories, and BEAM3,[19] which is a 3-D Monte-Carlo ray propagation code to study the effect of refraction and gain saturation.

With the narrow line focus, we can't approximate the plasma expansion by limiting the plasma expansion to a straight 1-D geometry, in which case we calculated T_e ~2.5 keV at the peak of each pulse with an averaged ionization balance between O- and F-like. The plasma remains dense in this geometry. Between pulses the plasma cools to 1–1.5 keV with the majority of the ions at Ne-like. The critical surface moves out as a function of time, which results in a gain profile that moves away from the slab surface as a function of time. Between pulses we still observe large gain due to high T_e and n_e, which can collisionally pump the population inversion. These features clearly disagree with the measured near-field emission profiles.

A much better approximation of the hydrodynamics will be to use a 1-D wedge geometry to represent the 2-D expansion of the laser plasma. An on-axis image of the XRL plasma [20] shows a wedge-shaped plasma expansion of order of 20 degrees. This expansion is also consistent with 2-D LASNEX simulations of line-focused configuration on slab targets. Each driving laser pulses heats the plasma to a T_e of ~1.5–1.6 keV and burns through slightly beyond an averaged ionization state of Ne-like. Between pulses the plasma expands and cools to ~200-300 eV. The plasma cools and expands after each pulse. Given the long time (1.6 ns) between pulses, by the time the next pulse hits the target, we observe a large n_e reduction, and the majority of the energy will be deposited at a higher n_e region, near the critical surface. Therefore each pulse behaves as if it incidents a new slab target, without large perturbation by the pre-formed plasma in front of the slab, and the plasma behavior reproduces pulse after pulse, which is in agreement with our experimental observation with spatially fixed emission profiles.

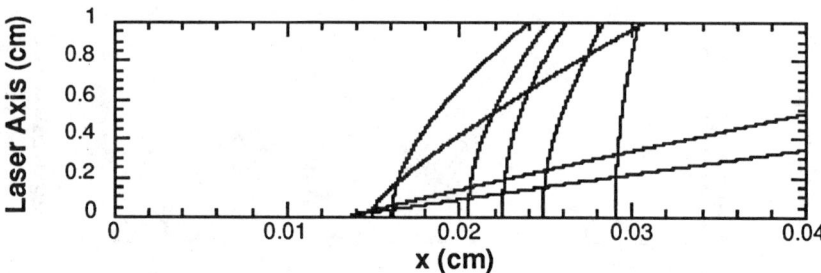

Fig. 2 Ray trajectories across a 1-cm-long Y-XRL, at the peak of the second pulse, showing large refraction for rays launched at the high gain region near the slab surface (0.01 cm), and ~100 μm refraction for the rest of the rays.

We can further postprocess the LASNEX output by using GLF to calculate detailed level populations (and therefore the small signal gains) and line transfer. Figure 1 plots spatial snapshots of the small signal gain profiles of the $J = 2-1$ line at 155 Å at times between the peaks of the second and third driving pulses. Here we observe strongly peaked gain profiles, with gain >20 cm^{-1} near the slab surface (0.01 cm in Fig. 1). With the reproducible plasma parameters for every pulse, the gain profiles also stay fixed at these positions, in agreement with observation.

With the low T_e between each driving pulse, the plasma recombines past the Ne-like configuration. At these temperatures, the collisional excitation rate is low and we observe a factor-of-2 increase in recombination flux from F-like ions to the $3p_{3/2}$ upper laser state. As shown in Fig. 1, the gains between pulses are weak, by about an order of magnitude, and the spatial gain profiles move away from the slab surface with increasing time due to changes in the ionization balance as the plasma cools and expands. These features are also in agreement with our experimental observation and further strengthen our model assumptions.

The one remaining puzzle is the location of the calculated gain, peaking close to the slab surface as compared to our measured emission profiles peaking at 100 μm off the target surface. Several possible mechanisms exist for altering the spatial gain profiles. The effect of radiation trapping, obtained by GLF line transfer calculation, reduces the peak gain, at the peak of the pulses, from ~22 cm^{-1} to ~13 cm^{-1}. But the spatial profiles still peak close to the target surface. We then use SPECTRE and BEAM3 to study the role of refractive ray propagation on the near-field emission profiles. Figure 2 plots the ray trajectories for a series of parallel rays launched at one end of a 1-cm Y-slab XRL. Rays near the slab traverse across large n_e gradients and are refracted out of the gain region, while rays launched further away from the slab surface face smaller n_e gradients and they refract by about 100 μm as they traverse across a 1-cm region. This 100-μm displacement is in excellent agreement with our experimental observations.

BEAM3 is a 3-D Monte Carlo code that includes the effect of saturation. We can specify the input gain, index of refraction, spontaneous emission rate, and geometries to simulate the Y-XRL. Without the effect of refraction we can recover much of the calculated local gain profile. However, as we turn on refraction, we observe a similar 100-μm shift in the near-field emission profile with a significant reduction in emission intensity. The combination of both SPECTRE and BEAM3 simulations is our final step in analyzing the Y-slab amplifier. Using all our code capabilities, we obtained a reasonable agreement with the experimental observation.

2.3 Future Research Direction for the XRL Cavity

To produce a short-pulse, enhanced-coherence XRL cavity, we have configured the pulse shape for 100-ps gaussian pulses. We found no evidence of locking with this setup. We also observed very small gain in the first pulse. This is consistent with observations of experiments done at Osaka.[21] Preliminary calculations support the experimental data, showing small gain due to low T_e's with few Ne-like ions. The first pulse, in this configuration, serves to pre-form the plasma to set up the latter pulses. With damage of our imaging mirror, we can't observe locking on the fourth pulse by the XRL photons generated by the second pulse. We will focus our effort on the continued development of the short-pulse XRL cavity in the next set of experiments. We are also planning to measure the change of coherence [22] of Y XRLs with and without reinjection.

3. ENHANCED COHERENCE: ADAPTIVE SPATIAL FILTERING

Instead of letting the driving laser optics define the XRL aperture, the concept of adaptive spatial filtering of an XRL uses geometric shaping to control the laser aperture. This could be in a conical or bowtie-shaped laser. In the unsaturated regime, rays that traverse the longest gain regions have the possibility of attaining the greatest gain lengths (GL) and highest output intensities. The neck of the bowtie laser acts as a narrow spatial filter where the high-GL rays must pass through. These high-GL rays have strong correlation between their angle of tilt and their transverse positions. Rays that travel outside of the bowtie neck achieve lower GLs and are also attenuated by surrounding materials. In a non-refractive and uniform-gain medium, the output of an unsaturated bowtie XRL would be weaker than that of a stripe XRL by the ratio of the neck area to the end area.

In the saturated regime, the coherent power of a bowtie XRL increases by extracting significant power into only a few modes and by eliminating parasitic modes. In saturation the loaded gain depends on the overall intensity pattern in a nonlinear and self-consistent way with the maximum gain located at the neck region, where the adaptive spatial filtering occurs. This effective spatial gain distribution maintains the correlation between ray tilt and transverse position at the laser output end that is needed to achieve high coherent power. Thus the energy content of a saturated XRL preferentially flows through the few Fresnel zones defined by the bowtie neck. In a non-refractive and uniform-gain medium, the output intensity of a bowtie XRL would be weaker than that of a stripe XRL by the ratio of active gain volumes between the two geometries. However the coherent power of a bowtie XRL would be much greater than that of a stripe XRL.

A key issue in the performance of bowtie XRLs is our ability to maintain the bowtie shape during the plasma expansion. One way to fabricate a bowtie XRL is to deposit a bowtie-shaped thin film coating on a thick substrate. The substrate is made of similar-Z material such that the substrate plasma provides a hydrodynamic tamper to the XRL plasma and maintains the bowtie shape during the lasing period. We have performed a series of numerical studies on the hydrodynamics, kinetics, and lasing properties of a simple stripe-shaped Ge XRL.[11] A series of 2-D LASNEX calculations simulating a line-focused slab XRL and a similar-width stripe XRL shows that a similar-Z hydrodynamic tamper, such as Cu on Ge, can effectively control the XRL plasma expansion to 1-D. Such one-dimensionally expanding plasmas should also provide a flatter transverse n_e profile in the lasing region and reduce the transverse laser divergence.

We have performed a set of experiments comparing a 120-μm line-focused Ge-slab XRL and a 120-μm Ge stripe and bowtie (120-μm ends with 25-μm necks) XRLs deposited on solid Cu substrates. Pinhole camera images contrasting the stripe and slab emission features and imaging of targets at 33.8 Å using a normal-incident multilayer mirror [20] qualitatively confirmed our LASNEX results on the hydrodynamic behaviors of these Ge/Cu plasmas. At or near saturation the stripe XRL intensities are comparable to a line-focused slab XRL. The comparable laser intensities indicate no significant differences in gains and refractive ray propagation between the two XRL configurations. The bowtie XRL intensities were factors of 10–15 lower than the intensities for the stripe and slab XRLs. This intensity difference is likely due to less gain medium of the bowtie configuration and the refractive nature of the plasma medium. X-ray emission spectra in the 8–11-Å range of slab, stripe, and bowtie targets found contributions dominated by Ne-like Ge lines, and a forest of Cu lines ranging from N- to Ne-like. Intensities of Cu and Ge emissions are in agreement with the types of XRLs fielded. The comparable temporal histories of the identified Ne-like Ge 3d and 3s → 2p emission lines indicate comparable ionization histories for the slab and stripe XRLs.

We are planning further characterizations of stripe and bowtie XRLs, such as measurements of small signal gains, plasma expansion characteristics, and coherence. We also have plans to further enhance the bowtie output by varying Nova pulse shapes, using new XRL target designs, and using multiple-staged oscillator-amplifier configurations.

4 SCALING THE XRL APERTURE

To develop a high-brilliance XRL we need to reduce the XRL divergence, increase the intensity, and scale the XRL aperture. Intensity is limited by saturation. In an idealized XRL architecture, we want to maximize the extraction of laser energy by using a small-source-size oscillator, control the divergence by collimating the XRL photons using multilayers or Fresnel zone plates,[23] and scale the aperture using a large-aperture, low-refraction amplifier. In the previous section we have discussed one possible approach of generating a diffraction-limited, small-source-size oscillator using the adaptive spatial filtering technique, and there are many efforts around the world [13,23] trying to develop optics components in the x-ray regime.

In designing an idealized large-aperture XRL amplifier we want to minimize the effect of refraction and maximize the spatial gain such that the amplifier plasma will enhance output intensity while maintaining the optical quality of the x-rays. When a

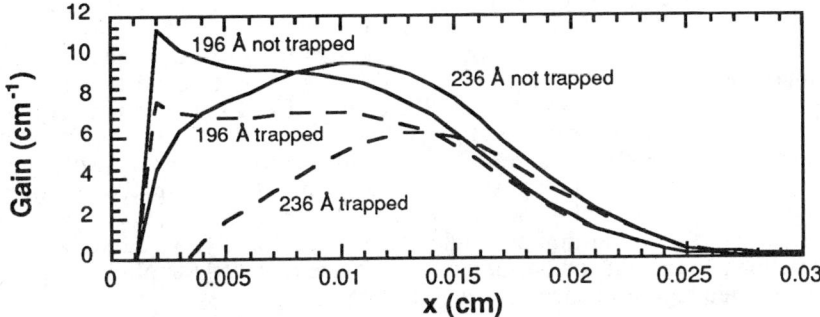

Fig. 3 Effect of trapping on the spatial gain profiles of the Ge 196- and 236-Å line at the center of a 500-μm line-focused slab XRL, as calculated by GLF line-transfer calculations.

prepulse technique is used to pre-form the plasma,[7,20] lasing occurs at region with flatter density gradient and the effect of refraction is reduced. We can also curve the XRL target to match the angle of curvature with the refraction angle.[8] But even if we can develop techniques to minimize the effect of refraction, we still need to address our ability to design a high-gain, large-aperture amplifier. In this section we will address the effect of trapping on the spatial gain profile.

Trapping can be described as a radiative excitation process that pumps the ground-state electrons to the lower laser state, thus reducing the inversion density. Unless the Doppler effect is large enough to shift the frequencies of the emitted photons far from the absorption profile, the effect of trapping depends largely on the ground-state population and the laser aperture. As the aperture increases, the optical depth of the trap line increases, resulting in the trap field buildup at the high-density region, where gain is normally largest for this collisionally excited scheme.

The population inversion, ΔN, is dependent on the populations of the upper and lower laser states, where the lower laser state is adjusted by the statistical weight, g (defined as $g \equiv 2J + 1$), of the upper and lower states:

$$\Delta N = N_{upper} - \left(g_{upper}/g_{lower}\right) N_{lower},$$

For the $J = 0$–1 line, $g_{upper}/g_{lower} = 1/3$, as compared to the $J = 2$–1 line with $g_{upper}/g_{lower} = 5/3$. Therefore trapping, which populates the lower laser state, favors the $J = 0$–1 line. To illustrate this trapping effect we simulate a large-aperture XRL using a 500-μm-FWHM line focus on a Ge-slab, with a 600-ps squared pulse at an intensity of 20 TW/cm^2 at 2ω. Figure 3 compares the gain profiles, at 400-ps into the pulse, of the Ge $J = 0$–1 (196-Å) and $J = 2$–1 (236-Å) lines at the center of this large-aperture XRL, for trapped and untrapped line-transfer calculations using GLF. Near the slab we observe a large gain reduction of the 236-Å line while the gain of the 196-Å line remains close to the untrapped case.

5. SUMMARY

This paper describes our experimental and theoretical studies of the development of various XRL architectural components for different experimental conditions. We have begun developing a short-pulse XRL cavity, and have obtained good agreement between the measured near-field emission profile and the simulation using a combination of LASNEX/GLF/SPECTRE-BEAM3. In our simulations we have demonstrated the importance of properly modeling the effect of hydrodynamic expansion, atomic physics, and refractive ray propagation. We are also developing a small-source-size oscillator using the concept of adaptive spatial filtering by geometric shaping. Using a bowtie-shaped XRL, we hope to eventually demonstrate a diffraction-limited single-stage XRL. Finally, we addressed issues of refraction and trapping on developing a large-aperture XRL amplifier as the final component of a high-brilliance XRL architecture.

ACKNOWLEDGMENT

Work performed under the auspices of the U. S. DOE by LLNL under contract number W-7405-ENG-48 and is partially supported by the Institute Sponsored Research Program.

REFERENCES

1. See papers in *Proc. of the Appl. of X-ray Lasers Workshop*, R. A. London, D. L. Matthews, S. Suckewer, Eds., LLNL Report CONF-9206170 (1992).
2. L. B. Da Silva et al., this proceeding.
3. L. B. Da Silva et al., *Proc. Ultrashort Wavelength II, SPIE* **2012**, 158, S. Suckewer, Ed. (1993).
4. L. B. Da Silva et al., *Opt. Lett.* **18**, 1174–1176 (1993).
5. J. C. Moreno et al., this proceeding.
6. R. A. London et al., in *X-Ray Laser 1990*, G. J. Tallents, Ed., 363–370 (IOP Publishing, Bristol and Philadelphia, 1991).
7. J. Nilsen et al., *Phys. Rev. A* **48**, 4682–4685 (1993).
8. Y. Kato et al., *Proc. Ultrashort Wavelength II, SPIE* **2012**, 12, S. Suckewer, Ed. (1993).
9. C. L. S. Lewis, et al., *Opt. Commun.* **91**, 71–76 (1992).
10. M. Key, *Proc. Ultrashort Wavelength II, SPIE* **2012**, 22, S. Suckewer, Ed. (1993).
11. A. S. Wan et al., to be published in *Optical Engineering*.
12. S. B. Libby, R. A. London, T. A. Weaver, to be published.
13. T. Barbee, this proceeding.
14. P. Jaeglé et al., in *X-Ray Laser 1992*, E. E. Fill, Ed., 1–7 (IOP Publishing, Bristol and Philadelphia, 1992).
15. P. L. Hagelstein, *Proc. Short Wavelength Coherent Radiation: Generation and Applications*, R. W. Falcone and J. Kirz, Eds. (1988)
16. B. J. MacGowan et al., *J. X-ray Science and Technol.* **3**, 231–282 (1992).
17. G. B. Zimmerman et al., *Com. Plasma Phys. and Cont. Fus.* **2**, 51 (1975).
18. H. A. Scott and R. W. Mayle, *Applied Physics B* **58**, 35–43 (1994).
19. G. Maenchen, private communications.
20. J. F. Seely et al., submitted to *Physics of Plasma*.
21. H. Daido et al., this proceeding.
22. J. E. Trebes et al., *Phys. Rev. Lett.* **68**, 588–591 (1992).
23. A. V. Vinogradov, this proceeding.

Soft-x-ray Amplification in a Capillary Discharge Plasma

J.J. Rocca, F.G. Tomasel, V.A. Shlyaptsev, O.D. Cortázar, J.L.A. Chilla, and G. Giudice

Electrical Engineering Department, Colorado State University, Fort Collins, Colorado 80523.

Abstract. We have realized the first demonstration of large soft-x-ray amplification in a discharge-created plasma. A gain-length product of g*l=7.2 was measured at 46.9 nm in the $J=0-1$ line of Ne-like Ar in a 12-cm-long plasma column generated by a compact, fast capillary discharge. The beam divergence was measured to be less than 9 mrad.

INTRODUCTION

In this paper we report the first demonstration of large amplification (g*l > 7) in a pulsed power driven plasma. Soft-x-ray lasing was first achieved almost ten years ago in plasmas generated by large laser facilities.[1,2] Subsequent experiments with laser-created plasmas have succeeded in expanding the number of x-ray laser transitions, reaching the spectral region of the water window, and in increasing the power of the x-ray laser transitions.[3] These laser sources have also been successfully utilized in proof-of-principle application experiments.[4] However, the widespread use of x-ray lasers in important applications has been limited by their size, cost and complexity. In these Proceedings, several groups report progress in the utilization of table-top laser drivers for the development of soft-x-ray lasers on a smaller scale.[5-7] Schemes based on both collisional excitation and plasma recombination, including optical-field-induced ionization followed by recombination[8], are reported to have achieved gain-length products up to 4.[5-7]

It was early recognized in the development of x-ray lasers that direct excitation of the gain medium by a pulsed discharge could result in a significant increase in laser efficiency. However, despite significant efforts[9], the for long pursued goal of demonstrating discharge-pumped lasing at wavelengths below 100 nm had remained unresolved. In pulsed power driven plasmas, a major obstacle has been the axial inhomogeneities in the plasma produced by non-symmetric compressions and instabilities, which result in severe distortions of the plasma column and the destruction of the amplification.[10] This plasma uniformity problem has brought attention to laser schemes which are less sensitive to the symmetry of the plasma, such as resonant photoexcitation and photoionization followed by recombination. The latter scheme has been successfully used in demonstrating population inversion in He-like Ne photopumped by a Na Z-pinch generated by Saturn, the largest pulsed power machine in the world.[11]

We have proposed to overcome the limitations associated with discharge-pumping by the use of fast discharge excitation of capillary channels.[12] These discharges provide the advantages of producing the necessary hot plasma columns of small diameter using only moderate radial compressions, and highly homogeneous initial plasma conditions, which can result in more stable plasma columns. Discharges through plastic capillaries have been previously studied as soft-x-ray sources for spectroscopy, microscopy and lithography.[13] Several experiments, two of which are reported in these Proceedings, have been recently conducted to explore for amplification following plasma recombination in capillary discharge plasmas[14-19] and in a gas liner pinch.[20] Gain[15,19,20] and anomalous line intensities indicative of amplification[21] have been reported, but only limited scaling with length has been observed.

At the previous International Colloquium on X-Ray Lasers, held at Schliersee in 1992, we reported the successful generation of high temperature ($T_e > 150$ eV), small diameter (200 μm) plasma columns, with aspect ratios up to 250:1, for the excitation of collisionally pumped lasers by fast capillary discharges.[22,23] Measurements conducted with a gated pinhole x-ray camera indicated that the rapidly rising current detaches the plasma from the walls to generate a compressed plasma column with a high degree of ionization, in agreement with hydrodynamic calculations.[23,24] These initial experiments, conducted in argon-filled capillaries, showed that the conditions of these plasma columns approach those necessary for amplification in transitions of Ne-like and Ni-like ions. Soft-x-ray spectra showed that these discharges require only modest currents to ionize Ar to the Ne-like and F-like states. The possibility of utilizing capillary discharges in the excitation of collisionally excited lasers was discussed at Schliersee[22], and in other publications.[23,25-27]

Herein we report the first clear demonstration of large soft-x-ray amplification in a discharge-created plasma. A gain of 0.6 cm^{-1}, determined by fitting the experimental data to the Linford formula[28], was obtained in the 46.9 nm line of Ne-like Ar in plasma columns up to 12 cm in length, resulting in a maximum gain-length product of 7.2.

MEASUREMENTS AND DISCUSSION OF THE RESULTS

The experiments were conducted using a fast capillary discharge, having half-cycle period of 60 ns, to excite plasma columns up to 12 cm in length in 4-mm-diam channels filled with either pure Ar, or with a mixture of Ar and H_2. The pulse generator and capillary discharge set up has been described in previous publications.[21,23] It consists of a 3 nF capacitor which is pulse-charged by a Marx generator, and then rapidly discharged through a capillary channel to produce a hot plasma column with a large length-to-diameter ratio. The soft-x-ray emission in the axial direction was measured with a 2.2-m grazing-incidence spectrograph having a 1200 l/mm gold-coated grating placed at 85.8°, and detected with an intensified array detector. The entire system was characterized with a ray tracing code to determine the angles of acceptance of the radiation by the spectrometer, and to correlate the source divergence to the image size in the focal plane of the instrument. The calculated results were verified by measurements, which yielded an acceptance angle of approximately 15 mrad in the direction perpendicular to the slit, and of approximately 1.5 mrad per millimeter in the detector, in the direction of the slit.

Our previous calculations predicted that the maximum gain in Ne-like Ar would be observed in the $J=0-1$ line, and that could reach 0.7 cm^{-1}.[24] Spectra obtained in the region around 47 nm identified the presence of this line, as well as strong resonance transitions of Mg-like Ar, and one 3d-3p transition of Ne-like Ar. Our measured wavelengths for these Ne-like Ar lines, 46.875 ± 0.015 nm and 48.50 ± 0.015 nm, respectively, agree well with previously calculated and measured values in Θ-pinch plasmas.[29,30] Spectra in other wavelength regions also identified the $J=2-1$ line of Ne-like Ar at 69.77 nm, but no gain experiments were conducted for the $J=2-1$ transitions. The intensity ratio of the $J=0-1$ laser line and the closely spaced 48.5 nm 3d-3p line, which can not have amplification, provides a convenient reference in the search for amplification. At low plasma densities (1×10^{16} cm^{-3}), and in an optically thin plasma, this intensity ratio has been measured to have a value of about 6,[29] and decreases in value to ~1 at higher (1×10^{19} cm^{-3}) plasma densities. We have performed line intensity calculations utilizing two independent sets of atomic data[31], with up to 157 levels to model the population distributions in Ne-like Ar as a function of the plasma conditions. These yielded intensity ratios between 1.5 and 3 in the absence of amplification for electron densities in the range of interest for this experiment (1×10^{18} to 1×10^{19} cm^{-3}).

Experiments with pure argon discharges at relatively low pressures between 0.2 and 0.32 Torr and discharge currents between 30 and 40 kA showed the intensity of the $J=0-1$ line at 46.9 nm to reach up to 5 times the value of that of the 48.5 nm line in 12-cm-long plasma columns. Calculations performed utilizing a ray tracing code as a postprocessor of hydrodynamic/atomic computations to synthesize axial spectra, suggest this line ratios are the result of small

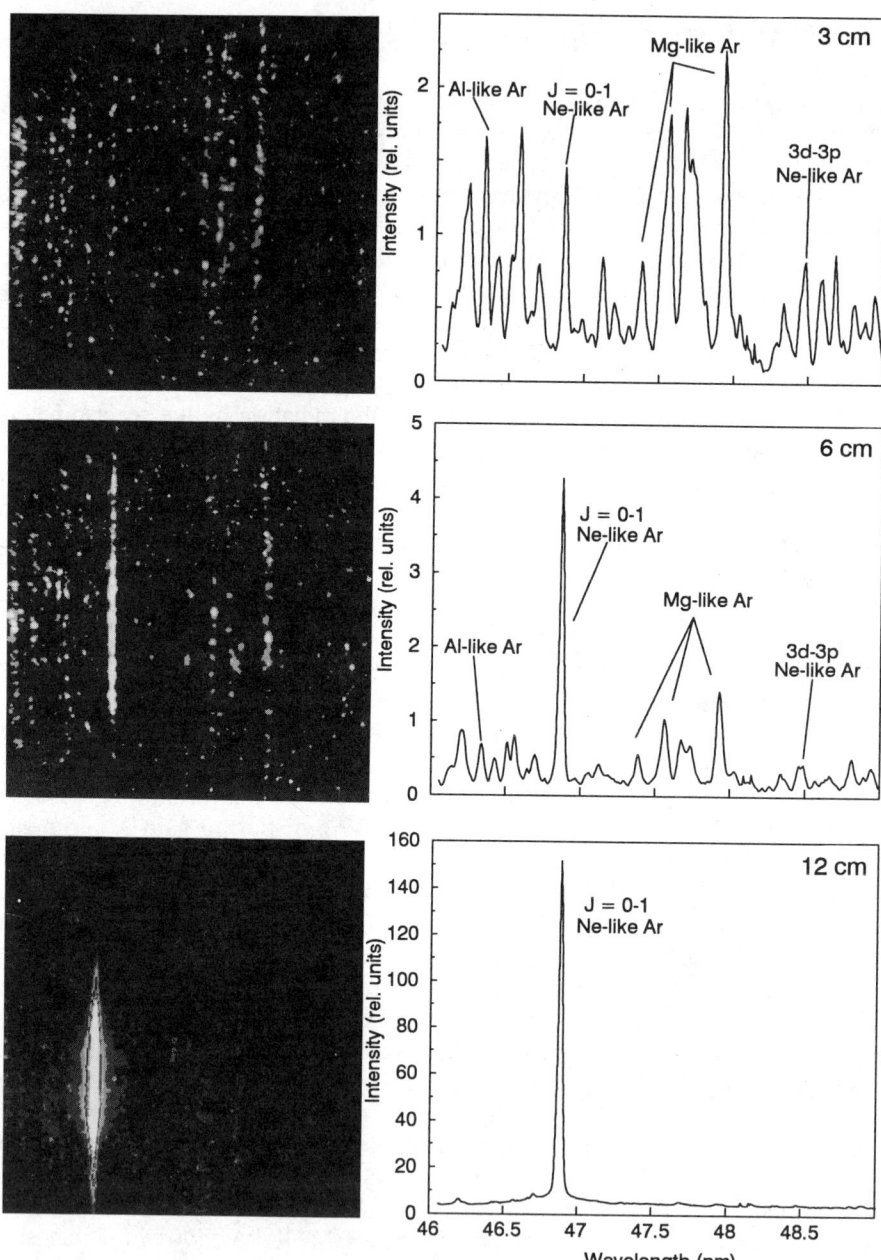

FIGURE 1. Variation of the intensity of the spectral lines in the neighborhood of 47 nm as a function of capillary length. For a 3-cm plasma column, the 46.9 nm $J=0-1$ line of Ne-like Ar is observed to be less intense than the surrounding lines from Mg-like Ar. In the 12-cm plasma column the $J=0-1$ line totally dominates the spectrum.

amplification, of the order of 0.1 - 0.2 cm^{-1}. This result is supported by time-resolved spectra corresponding to 6- and 12-cm capillaries, which show an intensity increase of approximately 2.6 for the longer plasma columns. Model calculations, discussed in a separate paper in these Proceedings[32], suggest that at these discharge conditions the maximum gain is usually obtained in a narrow annular region with small ion temperature ($\sim$ 50 eV), before the collapsing plasma column reaches the axis. At later times, the plasma compression is calculated to rapidly overionize the plasma and overheat the ions (to Ti > 1 KeV), thus quenching the gain both in the annular region and in the axis.

A large increase in the intensity of the $J=0$-1 line was detected when H_2 was added to the Ar in the capillary channel to raise the total pressure to 0.6 - 0.7 Torr. The $J=0$-1 line was measured to reach an intensity greater than 300 times that of the neighboring 48.5 nm Ne-like Ar line. These large intensity ratios are a clear indication of amplification.

The value of the gain was determined by measuring the increase of the line intensity as a function of capillary length. For this measurement, plasma columns of 3, 6, and 12 cm in length were generated. Special care was observed in maintaining the amplitude and pulsewidth of the current pulse constant. For this purpose, the inductance of the discharge circuit was kept approximately constant by adjusting the geometry of the capillary discharge electrodes for each capillary length, to compensate for the reduced inductance associated with the shorter capillaries. In this manner, the current half-cycle period was measured to be between 59.5 and 62 ns for all the capillary lengths. The peak amplitude of the current pulse was set to 38 $\pm$ 1 kA in all the shots by adjusting the charging voltage. Figure 1 shows typical spectra for each of the three plasma column lengths. In the spectrum from the 3-cm capillary, the intensity of the $J=0$-1 line of Ne-like Ar at 46.9 nm is observed to be smaller than the intensity of the surrounding resonant lines of Mg-like Ar ions, and to be only about twice the intensity of the neighboring 48.5 nm 3d-3p Ne-like Ar line. For the 6-cm capillary, the increase in the intensity of the $J=0$-1 line makes it significantly more intense than the neighboring lines, and in the 12-cm capillary the laser line totally dominates the spectrum, resembling the spectra of the emission from soft-x-ray lasers pumped by large laser facilities. Figure 2 is a plot of the integrated intensity of the $J=0$-1 line of Ne-like Ar as a function of capillary length. A least squares fit of the data to the Linford formula[27] results in a gain coefficient of 0.6 $\pm$ 0.04 cm^{-1}, corresponding to a gain-length product g*l=7.2. This is to our knowledge the largest amplification obtained in a "table-top" soft-x-ray laser device.

Another conclusive evidence of lasing results from the measurement of the divergence of the radiation of the $J=0$-1 Ne-like Ar line relative to all other transitions, and in particular to the neighboring 48.5 nm Ne-like line, which originates in the same region of the plasma. The magnitude of the divergence was determined from the image size in the direction of the slit, which is correlated to

FIGURE 2. Integrated line intensity of the $J=0$-1 line of ArIX as a function of the plasma column length. A fit to the Linford formula yields a gain coefficient of 0.6 ± 0.04 cm^{-1}, corresponding to $g*l = 7.2$.

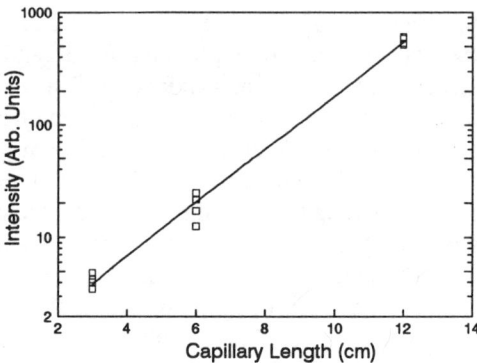

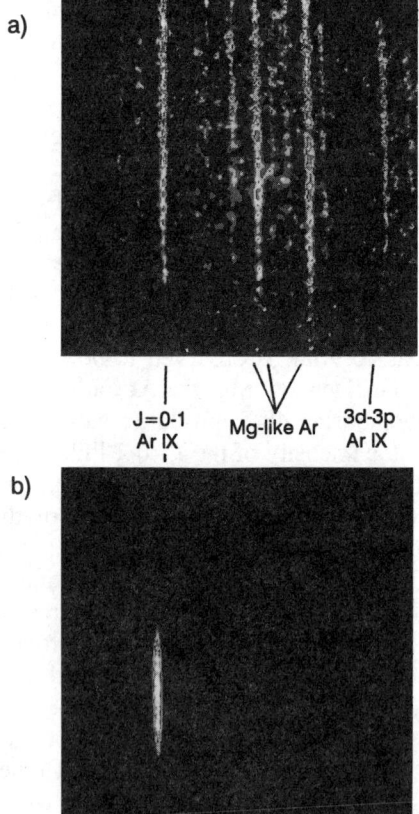

FIGURE 3. Spatial distribution of line intensities in the detector for discharge conditions corresponding to: a) Negligible amplification, and b) large amplification of the ArIX $J=0$-1 line. The divergence of the $J=0$-1 emission in the latter case is sharply reduced due to amplification, and corresponds to ≤ 8.7 mrad. In b), the detector sensitivity was reduced to avoid saturation by the 46.9 nm line.

the divergence and the size of the source. Figure 3(a) shows the intensity distribution, on the detector plane, of Ne-like and Mg-like Ar lines corresponding to a condition of negligible amplification. In this case all the lines, including the $J=0$-1 line, are observed to have similar intensity distributions, stretching over the majority of the detector. A sharp decrease in the divergence of the $J=0$-1 line was measured for the discharge conditions in which large amplification is obtained [Figure 3(b)]. The intensity distribution of the amplified 46.9 nm line in the spectrum shown in the latter figure (obtained by decreasing the sensitivity of the detector to avoid saturation), corresponds to a divergence of less than 8.7 mrad. This maximum divergence value results from assuming a point source. A slightly smaller divergence would result from considering a finite source size.

CONCLUSIONS

In summary, we have realized the first demonstration of large soft x-ray amplification using a discharge-created plasma. The observed amplification ($g^*l=7.2$) is also the largest obtained to date from a "table-top" soft-x-ray laser. The discharge excitation scheme demonstrated in the experiments reported herein is likely to result in amplification of soft-x-ray laser transitions in other ions, and has the potential for increasing the wall-plug efficiency of ultrashort wavelength lasers by two orders of magnitude.

ACKNOWLEDGEMENTS

We want to acknowledge the contributions of M.C. Marconi, B. T. Szapiro, K. Richardson, H. Mancini, K. Floyd, and A. Vinogradov. We also want to thank A. Osterheld for providing the atomic data utilized in some of the calculations and to gratefully acknowledge the technical support of B. Bach from Hyperfine Inc, Boulder, Co. Encouraging discussions with P. Hagelstein, W. Silfvast, J. Apruzese, and S. A. Lee are appreciated. This work was supported by the N.S.F. grant ECS -9013372, through its Quantum Electronics, Electromagnetics and Plasma Division, by the U.S. Dept. of Energy, grant DE-FG02-91ER12110 of the Office of Basic Energy Sciences, Division of Advanced Energy Projects, and by an award from the U.S National Research Council.

REFERENCES

1. Matthews, D.L., et al., *Phys. Rev. Lett.* **54**, 110 (1985).
2. Suckewer, S., et al., *Phys. Rev. Lett.* **55**, 1753 (1985).
3. Lee, T.N., McLean, E.A., and Elton, R.C., *Phys. Rev. Lett.* **59**, 1185 (1987); Jaeglé, P., et al., *J. Opt. Soc. Am. B* **4**, 563 (1987); O'Neil, D.M., et al., *Opt. Commun.* **75**, 406

(1990); Skinner, C.H., *Phys. Fluids B* **3**, 2420 (1991); MacGowan, B.J., et al., *Phys. Fluids B* **4**, 2326 (1992).
4. *Proceedings of a Workshop on Applications of X-Ray Lasers*, San Francisco, CA, 1992. London, R.A., Matthews, D.L., and Suckewer, S., Eds.
5. Hagelstein, P., et al., "Table-top XRLs at MIT", *in these Proceedings*.
6. Morozov, A., et al., "On Developing Table-Top Soft X-Ray Laser and Its Applications", *in these Proceedings*.
7. Hara, T., Ando, K., and Aoyagi, Y., "Compact soft X-Ray laser pumped by a pulse-train laser", *in these Proceedings*.
8. Midorikawa, K., et al., "An Optical-Field-Induced Ionization X-Ray Laser Using a Preformed Plasma", *in these Proceedings*.
9. Burkhalter, P., et al., *J. Phys. (Paris) Colloq., Suppl. 10*, **47**, C-247 (1986); Davis J. et al. *IEEE Trans. on Plasma Science* **16**, 482, (1988), Krishnan M. et al., *Phys. Intl. Rep. PITR-3411-01*, 1987; Davis J. et al., *NRL Memorandum Report 5771* (1986); Apruzese, J.P., "X-ray Laser Research Using Z Pinches", *in these Proceedings*.
10. Elton, R.C., *X-Ray Lasers*, Boston: Academic Press, 1990.
11. Porter, J.L., et al., *Phys. Rev. Lett.* **68**, 796 (1992).
12. Rocca, J.J., Beethe, D.C., and Marconi, M.C., *Opt. Lett.* **13**, 565 (1988).
13. Conrads, H., *Z. Phys.* **444**, 200 (1967); Bogen, P., et al., *J. Opt. Soc. Am.* **58**, 203 (1968); McCorkle, R.A., *Appl. Phys. A* **26**, 261 (1981), *Annals of the New York Acad. Sci.* **342**, 53 (1980); Zakharov, S.M., et al., *Sov. Tech. Phys. Lett.* **6**, 486 (1980).
14. Marconi, M.C., and Rocca, J.J., *Appl. Phys. Lett.* **54**, 2180 (1989).
15. Steden, C., and Kunze, H.-J., *Phys. Lett. A* **151**, 1534 (1990).
16. Rocca, J.J., Marconi, M.C., and Tomasel, F.G., *IEEE J. Quantum Electron.* **29**, 182 (1993).
17. Morgan, C.A., Griem, H.R., and Elton, R.C., *Phys. Rev. E* **49**, 2282 (1994).
18. Tomasel, F.G., et al., *Phys. Rev. E* **47**, 3590 (1993).
19. Shin, H.J., Kim, D.E., and Lee, T.N., *submitted for publication to Phys.Rev. E*.
20. Glenzer, S., and Kunze, H.-J., *Phys. Rev E* **49**, 1586 (1194); Kunze, H.-J., et al., "Lasing at short wavelengths in a capillary discharge and in a dense z-pinch", *in these Proceedings*.
21. Rocca, J.J., et al., *SPIE J.* **1551**, 275 (1991).
22. Rocca, J.J., et al., "Fast discharge excitation of small scale soft X-ray lasers", in *Proceedings of the Third International Colloquium on X-Ray Lasers*, 1992.
23. Rocca, J.J., et al., *Phys. Rev E* **47**, 1299 (1993).
24. Shlyaptsev, V.N., et al., *SPIE J.* **2012**, 99 (1993).
25. Rocca, J.J., et al., *Phys. Rev. E* **48**, R2378 (1993).
26. Rocca, J.J., et al., *SPIE J.* **2012**, 67 (1993).
27. Ivanov, L.N., Ivanova, E.P., and Knight, L.V., *SPIE J.* **2012**, 265 (1993).
28. Linford, G.J., et al., *Appl. Opt.* **13**, 379 (1974).
29. Elton, R.C., et al., *Phys. Rev A* **40**, 4142 (1989).
30. Preissing, N., et al., *Phys. Rev. E* **48**, 3867 (1993).
31. HULLAC (*H*ebrew *U*niversity-*L*awrence *L*ivermore *A*tomic *C*ode), Osterheld A.L. (private communication); Vainshtein, L.A., Sobelman, I.I., and Yukov, E.A., *Excitation of atoms and broadening of spectral lines*, Moskow: Nauka, 1979.
32. Shlyaptsev, V.N., Rocca, J.J., and Osterheld, A.L., "Modeling of a capillary discharge soft x-ray amplifier", *in these Proceedings*.

Soft X-ray Lasing in a Capillary Discharge

Tong-Nyong Lee, Hyun-Joon Shin, and Dong-Eon Kim

Department of Physics, Pohang University of Science and Technology, Pohang, 790-784, Korea

Abstract. Soft x-ray lasing in the C VI Balmer α transition is observed in a capillary discharge. The capillary is made of polyethylene with a bore diameter of 1.2 mm. Plasma radiation from the discharge is analyzed using a toroidal mirror and a two-meter grazing-incidence spectrograph-monochromator. The electron temperatures are measured at both the axial and the peripheral region close to the capillary wall, using space-resolved spectra. A comparison of the branching ratio in the hot (axial) and the cool (peripheral) plasma regions indicates that there is a large population inversion between n=3 and 2 states of C^{5+} ions in the cool ($T_e \sim 13$ eV) region of the capillary plasma. Relative line intensities of the C VI H_α and a number of non-lasing lines are compared in this cool region as a function of capillary length. The C VI H_α line intensity increases exponentially whereas those of non-lasing transitions increase linearly with an increase of the capillary length. The gain coefficient thus measured indicates 2.8 cm^{-1}. The lasing line intensity does not seem to increase exponentially beyond a capillary length of 16 mm and the gain-length product, gL, obtained here is 3.9, which is a typical value one would expect for a recombination soft x-ray laser. The photoelectric signals of the lasing line indicate that the lasing takes place about 40 ns after the current peak in the first half cycle of the capillary discharge, with a lasing pulse width of 60 ns in FWHM.

INTRODUCTION

In the most of existing x-ray lasers[1-7], the pumping power is delivered onto a suitable target by focusing a high-power driving-laser beam. In addition to the high pumping power required[8] for short wavelength lasers, since the efficiencies of both the driver itself and the coupling between the driver laser light and the plasma are poor, the overall energy efficiency of the x-ray laser is extremely low. A practical soft x-ray laser needs to be more energy efficient, compact, inexpensive, and easy to maintain. To this end, there have been a number of suggestions to utilize discharge plasmas as a lasing medium, including the capillary discharge.

Recently, there have been a number of reports[9-13] about capillary discharge experiments aiming at a possible x-ray lasing based on both the recombination as well as the electron collisional pumping scheme. Plasma parameters for given input discharge conditions have been studied extensively from time-resolved spectral data.

Here, we report unambiguous observation of the recombination lasing in the C VI H_α transition[14]. The result was obtained from space-resolved and time-resolved spectral data, using a polyethylene capillary discharge. The gain in the C VI H_α transition was directly measured by comparing spectral line intensities of the lasing and non-lasing lines by varying the capillary length. The photoelectric signals were obtained in order to observe the lasing pulse signal as well as shot-to-shot behavior.

At the initiation of the current rise, electric breakdown in a capillary discharge takes place initially along the inner wall, since this is the least inductive current path, as in many linear pinch devices[15,16]. The plasma current sheath which is formed by the ablated material from the surface breakdown will then implode radially toward the capillary axis. With this compression the plasma forms hot dense core in the axial region at or slightly before the current maximum. The plasma now expands radially from the stagnation and eventually collides with the capillary wall as the current decreases. There the plasma will experience a sudden cooling, mainly due to conduction loss to the wall, and subsequent recombination. If the plasma condition is right then the outer annulus region close to the wall is likely the place where the lasing will take place. It is therefore important to observe this region exclusively, because the mixing of the contribution from other region (from axial hot plasma) may obscure the net effect of the gain.

EXPERIMENTAL ARRANGEMENT

The schematic of the capillary discharge[17,18] used in our experiment is shown in Fig.1. Carbon plasmas were produced using a polyethylene [$(CH_2)_n$] capillary with a 1.2 mm bore diameter and their lengths were varied from 8 to 16 mm. The capillary discharge was powered by two low-inductance capacitors (60 nF each) connected in parallel to the carbon electrodes (A) via identical transmission lines (B). The discharge was initiated by a carbon trigger-pin (F) inserted near one of the electrodes. The charging voltage was varied from 17 kV to 27 kV; and the dicharge current was monitored for every discharge by a Rogowski coil.

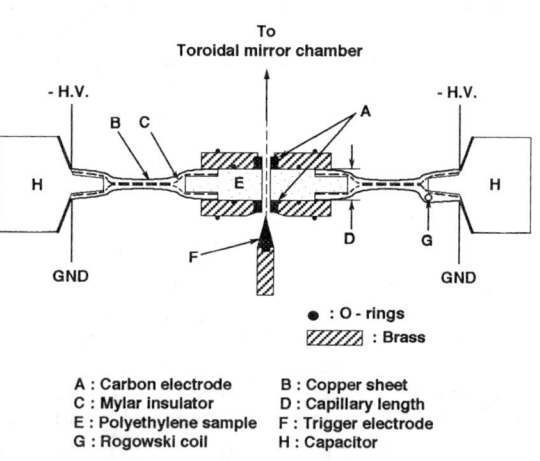

A : Carbon electrode   B : Copper sheet
C : Mylar insulator   D : Capillary length
E : Polyethylene sample F : Trigger electrode
G : Rogowski coil   H : Capacitor

● : O - rings
▨ : Brass

FIGURE 1. Capillary Discharge Device.

The half-cycle time of the discharge was typically 220 ns (145 ns FWHM) for the 14 mm long capillary. Throughout the experiment, the input power density (=stored energy / capillary volume / FWHM of the discharge current) of the capillary volume was kept the same (1.5×10^{10} W cm^{-3}) by adjusting the capacitor charging voltage for different capillary lengths.

The experimental layout is shown in Fig.2. A toroidal mirror is placed between the capillary and a two-meter grazing-incidence spectrograph-monochromator to collect light from the discharge into the spectrograph through the entrance slit. A spherical grating of 600 lines/mm which is set at a grazing angle of 1.5 degree was used with a 22-μm wide entrance slit. Time-integrated spectra from the discharge were recorded on Kodak 101-05 photographic plates. The optical density on each developed plate was compared against a calibrated diffuse density wedge, and the relative spectral intensities were then obtained using Henke's formula[19].

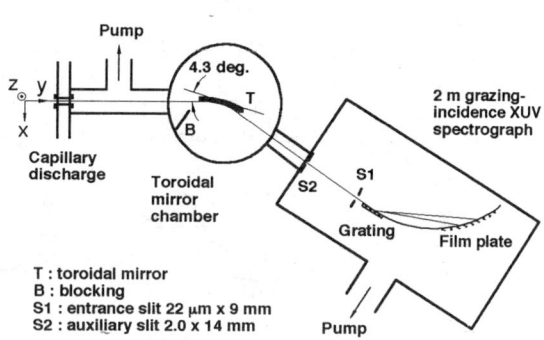

FIGURE 2. Experimental Layout.

Using the sagittal focus at the photographic plane formed by the toroidal mirror, a space-resolved spectrum in a direction perpendicular to the dispersion can be obtained[20]. However, this technique is valid only for the sagittal focal region of the dispersion and also when the magnification of the mirror is reasonably large.

In order to obtain spatial resolution in a wide spectral range with a single shot and also to exclusively observe cool peripheral region, the capillary axis is transversely moved away from the line of sight of the spectrograph-toroidal mirror system. Let the line of sight be the y-axis and the perpendicular axis in the meridional plane be the x-axis (see Fig.2). The capillary axis is moved 0.4 mm toward the + x-axis, keeping it parallel to the line of sight, in order to admit the plasma radiation only from a peripheral region. This is possible because in this case the slit only intersects the peripheral part of the astigmatic image formed at the entrance slit, as confirmed by the ray-tracing. With this method, it is possible to obtain spatial resolution of less than 200 μm in the entire wavelength of 2 to 20 nm[21].

RESULTS

Figure 3(a) and (b) show the microdensitometer scans of a single-shot spectra (in wavelengths of 2.5 to 4.2 nm) which are obtained from the axial (a) and the peripheral (b) regions of the capillary cross section. In the case of Fig. 3 (a), there are strong Lyman series lines of H-like C VI and n --> 1 resonance lines of He-like C V ions. There is a relatively strong continuum on the short wavelength side of the C V series limit. On the other hand, in the case of (b), which represents the radiation exclusively from a part of the outer annulus region, the spectrum is much different from that of (a); the intensities of the high n Lyman series lines are negligible, and the C VI recombination continuum is weak relative to those of the C VI and CV resonance lines. The electron temperatures at both the axial and the peripheral regions estimated from the slopes of the recombination continuum of the C^{5+} ions[22] are 25 eV (a) and 13 eV (b), respectively.

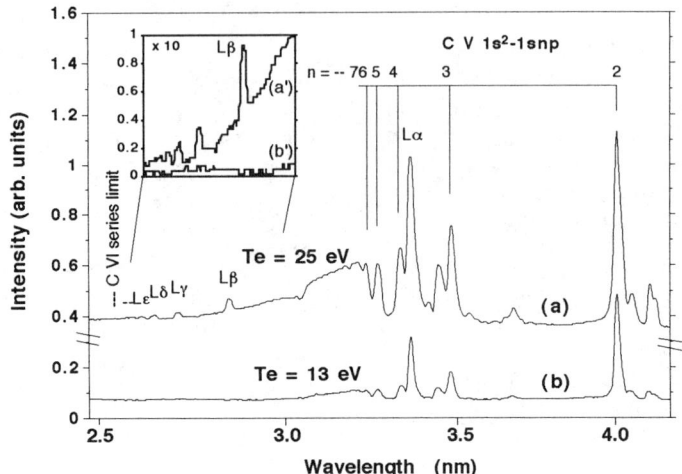

FIGURE 3. Time-integrated spectra representing plasma radiation from the axial (a) and from the peripheral (b) region of the capillary.

One of the features of the spectra is that the intensity of the L_β line in the peripheral region is drastically decreased to a value less than 1/15 times that of axial region as can be seen in the insert (x 10 version) of the figure, while that of H_α line (not shown in the figure) decreased only to the value of 1/1.2 ~ 1/1.5 times that of axial region. The branching ratios of $I(H_\alpha)$ (n = 3-->2) to $I(L_\beta)$ (n = 3-->1) for hot and cool regions were obtained by comparing the line intensities. The result shows that $[I(H_\alpha)/I(L_\beta)]_{hot} \ll [I(H_\alpha)/I(L_\beta)]_{cool}$. According to a calculation[24], the present trend can only be explained by the existance of population inversion between n = 3 and 2 states of the C^{5+} ions in the cool plasma.

In order to verify the gain in the cooler plasma, relative intensities of the C VI H_α and other spectral lines nearby were measured by varying the capillary length. In this series of experiment special care is paid to exclude impurities in the discharge as much as possible and the input power density (1.5 x 10^{10} W/cm^3) was kept the same for all capillary lengths. Figure 4 shows the microdensitometer scans of time-integrated spectra in the wavelength region of 10 - 23 nm, which are obtained with two different capillary lengths, 14 and 10 mm. Each spectrum is obtained with a single-shot exposure, right after two low-voltage (~ 10 kV) cleaning shots with a newly prepared capillary.

The intense line radiation at 18.2 nm is the Balmer H_α line of the C^{5+} ion. The other members of the Balmer series lines are blended with either the high order lines of the Lyman series lines or the third order line of the C V resonance line. There are others which include the CV lines of 1s2s - 1snp, 1s2p - 1snd transitions, and some impurity O VI lines in the spectra. The spectra shown in Fig. 4 indicate that the C VI H_α line intensity increases more rapidly than any other lines when the capillary length increased from 10 mm to 14 mm.

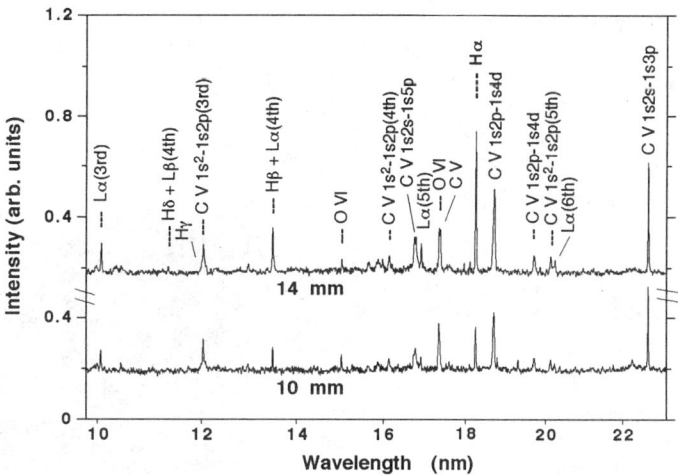

FIGURE 4. Time-integrated spectra obtained using capillary lengths of 14 and 10 mm. Each spectrum is obtained with a single discharge with input power density of 1.5×10^{10} Wcm^{-3}.

In order to check the tendency, intensities of the major spectral lines in this wavelength region are plotted as a function of capillary length and these are shown in Fig. 5(a) and (b). Each data point is obtained using a newly prepared capillary. The line intensities of the C VI L$_\alpha$ (3rd and 5th order), the C V 1s2p - 1s4d, the 1s2s - 1s3p, and the 1s2s - 1s5p transitions increase linearly, whereas the C VI H$_\alpha$ line increases exponentially with the increase of capillary length. The solid line in Fig. 5 (b) represents a gain curve with a gain coefficient of $g = 2.8$ cm^{-1}, obtained from the formula given by Linford et al[23]. Also plotted is the intensity of the C VI H$_\beta$ / L$_\alpha$ (4th order) blend line, indicating some evidence of gain for the C VI H$_\beta$ transition. The C VI H$_\alpha$ line intensity for a 16 mm long capillary does not

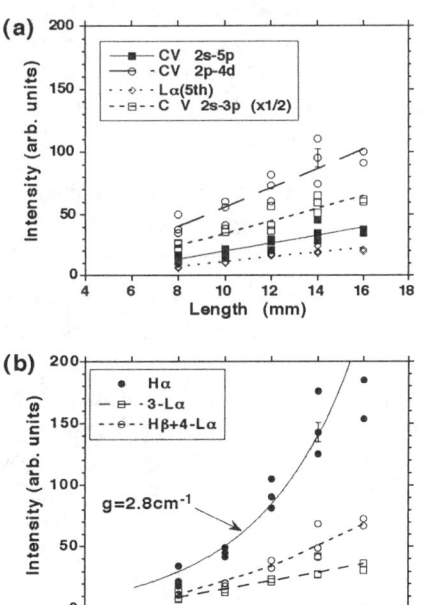

FIGURE 5. Relative intensities of the C V 1s2s-1s5p, 1s2p-1s4d, 1s2s-1s3p, and C VI L$_\alpha$ (5th order) lines (a) and C VI H$_\alpha$, H$_\beta$ / L$_\alpha$ (4th order) blend and the L$_\alpha$ (3rd order) lines (b), as a function of capillary length.

show exponential increase, but tends to level off, and this gives the gain-length product, gL value of 3.9 in the present experiment. This is a typical tendency in a recombination-pumped soft x-ray laser where the gL values are much smaller[3] (typically, gL $\leq$ 4) than those in the electron-collisionally pumped scheme [4-7]. Such experiment has been repeated allowing some impurities to the discharge. However, the general feature was the same and the measured gain coefficient of the C VI H_α line was again 2.8 cm^{-1}.

The temporal behavior of the spectral radiation is investigated by operating the spectrograph in a monochromator mode. A thin (100 μm) plastic scintillator (NE102) was placed behind the 39 μm wide exit slit and this was coupled to a photomultiplier tube (Hamamatsu H3284) outside the vacuum chamber via fiber optics cable. Figure 6 (a), (b), (c) and (d), respectively show oscillograms of the discharge-current waveform and the photoelectric signals of the C VI H_α, the C V 1s2p - 1s4d line (18.7 nm), and the C VI H_β / L_α (4th order) blend line, superposed with those of continuum in the vicinity of each spectral line. The C VI H_α line signal peaks about 35 ns after the current maximum and has a pulse width of 60 ns FWHM. The C V line signal peaks much later (~40 ns) than that of C VI H_α signal, indicating a recombination of the C^{5+} ions. The H_β / L_α (4th order) blend line also peaks near the peak of the H_α signal but the duration is a little longer due to the L_α transition. These results show the capillary plasma of strong recombination characteristics.

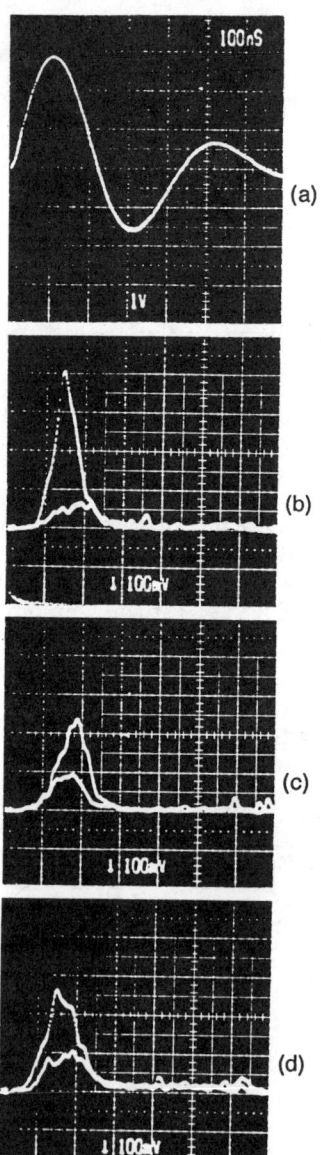

FIGURE 6. Oscillograms of the discharge current waveform (a), and the photoelectric signals of the C VI H_α (b), the C V 1s2p-1s4d (c) and the C VI H_β / L_α (4th order) blend lines (d).

Shot-to-shot intensity variation of the C VI H$_\alpha$ and the C V 1s2p - 1s4d lines are examined from the spectral signals and this is shown in Fig. 7(a) and (b) for 14 and 10 mm capillary lengths respectively. The number of shots are made starting from the very first discharge (after two cleaning shots) using a same capillary. In the case of the 14 mm long capillary, the C VI H$_\alpha$ line intensity decreases very rapidly during the first four shots, whereas non-lasing lines do not change appreciably. In the case (b) of the shorter (10 mm) capillary, however, there is no such rapid decrease in signal height for either the C VI H$_\alpha$ or the C V line signals with an increase of number of shot. Also, it is found that an increase of the discharge power density to the 10 mm capillary neither enhances the intensity appreciably from the normal power density operation nor does it cause a rapid decrease of intensity during the first few shots. An examination of the capillary bore diameter after five shots indicates neither noticeable damage to the wall nor appreciable increase in diameter (corresponding to about 10 % increase in volume in 14 mm capillary).

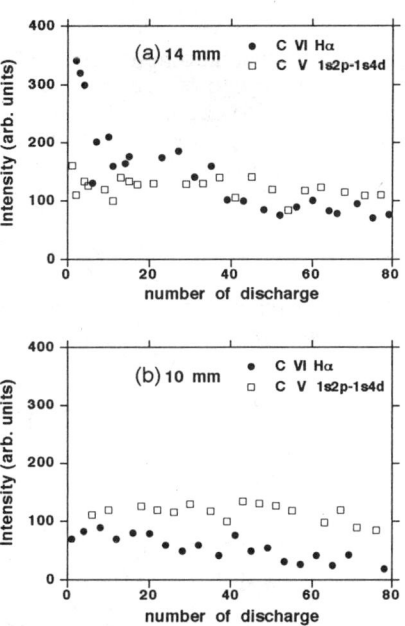

FIGURE 7. Shot-to-shot intensity variation of the C VI H$_\alpha$ and C V 1s2p-1s4d (λ=18.7nm) lines.

These results based on the photoelectric signals are in good agreement with those obtained from the time-integrated spectra described above, i.e., 1) the gain exists only for the first few shots in a newly prepared capillary, 2) an increase in the power density does not increase the gain coefficient, and 3) the C VI H$_\alpha$ line intensity from a capillary longer than 16 mm does not increase and tends to level off.

SUMMARY

The carbon-plasma which is produced in a polyethylene capillary discharge is investigated using a two-meter grazing-incidence spectrograph-monochromator. Space-resolved spectra are obtained using a toroidal mirror system and the electron temperatures are measured at both the axial and the peripheral region close to the capillary wall. The branching ratios in the cool and the hot region of the capillary indicate that there exists a population inversion between n = 3 and 2 states of C^{5+} ionic species. Relative line intensities of the C VI H$_\alpha$ and a number of non-lasing transitions are compared in this cool region as a function of capillary length. The C VI H$_\alpha$ line intensity increases exponentially whereas those of non-lasing transitions increase linearly with an increase of the capillary length. The gain coefficient measured indicates 2.8 cm^{-1}. The photoelectric signal of the C VI H$_\alpha$ line indicates that the amplification takes place about 40 ns after the current peak in the first half cycle of the capillary discharge, with a pulse width of 60 ns in FWHM. It is also important to note that there seems to be an optimum power input to the discharge for a given capillary geometry, and that the amplification lasts for only a few discharges with a freshly prepared capillary.

ACKNOWLEGEMENTS

This research was partially supported by POSCO basic research grant and also by the Korea Science and Engineering Foundation.

REFERENCES

1. Rosen, M. D., et al., *Phys. Rev. Lett.* **54**, 106 (1985).
   Matthews, D. L., et al., *Phys. Rev. Lett.* **54**, 110 (1985).
2. Suckewer, S., et al., *Phys. Rev. Lett.* **55**, 1753 (1985).
3. Skinner, C. H., *Phys. Fluids B* **3**, 2420 (1991).
4. Carillon, A., et al., *Phys. Rev. Lett.* **68**, 2917 (1992).
5. MacGowan, B. J., et al., *Phys. Fluids B* **4**, 2326 (1992).
6. Koch, J. A., et al., *Phys. Rev. Lett.* **68**, 3291 (1992).
7. Da Silva, L. B., et al., *Opt. Lett.* **18**, 1174 (1993).
8. Elton, R. C., *X-ray Lasers*, San Diego: Academy Press, 1990.
9. Steden, C., and Kunze, H.-J., *Phys. Lett. A* **151**, 534 (1990).
10. Rocca, J. J., Marconi, M. C., and Tomasel, F. G., *IEEE J. Quantum Electron.* **29**, 182 (1993).
11. Tomasel, F. G., et al., *Phys. Rev. E* **47**, 3590 (1993).
12. Morgan, C. A., Griem, H. R., Elton, R. C., *Phys. Rev. E* **49**, 2282 (1994).
13. Rocca, J. J., et al., *Phys. Rev. E* **47**, 1299 (1993).
    Rocca, J. J., et al., *Proc. SPIE* **2012**, 67 (1993).
14. Shin, H. J., Kim, D. E., and Lee, T. N., to be published in *Phys. Rev. E*.
15. Glasstone, S. and Lovberg, R. H., *Controlled Thermonuclear Reactions*, New York: D. Van Nostrand Co., 1960.
16. Lee, T. N., *Annals New York Acad. of Sciences* **25**, 112 (1975).
17. Bogen, P., Conrads, H., Gatti, G., and Kohlhaas, W., *J. Opt. Soc. Am.* **58**, 203 (1968).
18. McCorkle, R. A., and Vollmer, H. J., *Rev. Sci. Instrum.* **48**, 1055 (1977).
19. Henke, B. L., et al., *J. Opt. Soc. Am. B* **1**, 828 (1984).
20. Tondello, G., *Optica Acta.* **26**, 357 (1979).
21. Shin, H. J., et al., to be published elsewhere.
22. Seely, J. F., Dixon, R. H., and Elton, R. C., *Phys. Rev. A* **23**, 1437 (1981).
23. Linford, G. J., Peressini, E. R., Sooy, W. R., and Spaeth, M. L., *Appl. Opt.* **13**, 379 (1974).
24. Calculation have been made with Ration code which is developed by Lee, R.W. and his group at Lawrence Livermore National Laboratory, including or excluding opacity effect. And we would like to express our thanks to them for allowing us to use the code.

Modeling of a Capillary Discharge Soft X-ray Amplifier

Vyacheslav N. Shlyaptsev[a], Jorge J. Rocca[b], and Albert L. Osterheld[c]

[a] *P.N. Lebedev Physics Institute, Leninsky pr.53, Moscow, Russia;*
[b] *Colorado State University, Electrical Engineering Dept., Fort Collins, CO 80523;*
[c] *Lawrence Livermore Natl. Laboratory, P.O.Box 808, Livermore, CA 94550*

Abstract. In this paper we report the results of numerical modeling of the first discharge pumped soft X-ray laser (1), recently experimentally demonstrated in a capillary discharge plasma. Interaction and comparison with experiment enabled us to develop an adequate numerical model that comprehensively describes the dynamic, ionization and spectral characteristics of the capillary discharge plasma. It allows to gain understanding of the complex nature of the Z-pinch capillary discharge plasma column utilized as gain medium, and facilitates the development and improvement of these new X-ray sources.

1. INTRODUCTION

Generally, a potentially higher efficiency and a deceptive simplicity were the main reasons that generated fascination and attracted attention to Z-pinches as a source of high temperature plasma for different applications during the last decades. But many years of studying Z-pinches for X-ray laser excitation have shown that this simplicity is illusory and misleading, as it is in the case of Z-pinch thermonuclear fusion. Unfortunately, severe distortions, disruptions, hot spots etc. in basically unstable MHD plasmas usually prevent stable pinch formation and confinement. These properties of pinches are not so crucial for raw X-ray production in the sub-keV region, but are a key obstacle in soft X-ray laser development. Consequently, in X-ray laser development, the main attention was devoted to the utilization of X-ray laser schemes that are less sensitive to compression symmetry and to the investigation of plasma stabilization methods (2,3 and refs. in 1).

The Fast Capillary Discharge (FCD) attracted attention as a potential driver for X-ray lasers due to the following important factors (4,5):
- Its perfect initial symmetry (this requirement is important to the same extent it is for laser ICF targets).
- A small initial radius of several millimeters is exactly what is necessary for stable compression to the final dimension of several hundred microns using only relatively moderate radial compression factors $r/R_0 \leq 5\text{-}10$, where r and R_0 are final and initial radii. At such values of compression most Z-pinches behave relatively predictable.
- Both experimental work (4) and numerical calculations (5) have shown that in fast capillary discharges, like in the classical Z-pinch, the plasma

detaches from the wall, and the pinch effect leads to plasma compression to a size ~ 200 μm. A high temperature of several hundred eV is reached, that is favorable for the collisional X-ray laser scheme. In our previous work (5) we reported results of experimental and theoretical investigations of the hydro-, ionization and radiation parameters of Ar FCD plasmas. It was shown that 1D theory relatively satisfactory reproduces the temperature, density, ion charge, and X-ray source size of the FCD.

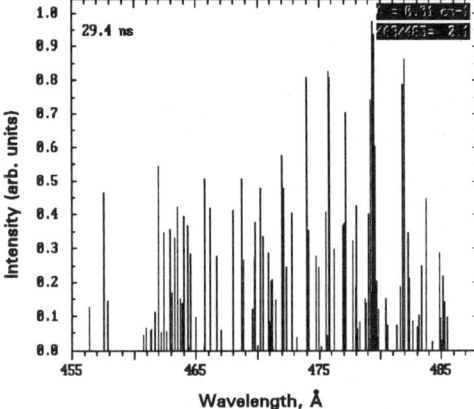

FIGURE 1. Relative intensities of Ar spectral lines as a function of wavelength. Each bar is an individual line.

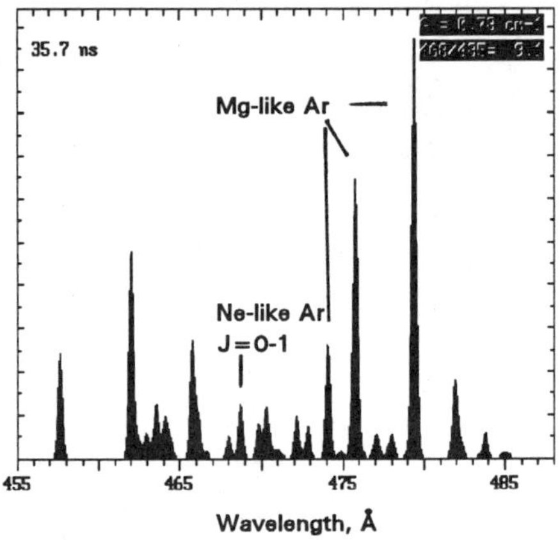

FIGURE 2. Example of a calculated time integrated spectra in the transverse (off-axis) direction (Ar 0.32 Torr).

In a separate paper (1), we report the results of the first measurement of large amplification on the J=0-1 line of Ne-like Ar in a fast capillary discharge plasma. In this work we present a theoretical analysis of a FCD plasma as an X-ray amplifier, and investigate the gain coefficients resulting from electron collisional excitation in Ne-like Ar in comparison with experiments.

2. RESULTS

The 1D MHD approximation used in the theoretical study is still in the limits of its validity for the capillary dimensions, discharge parameters, ions and transitions of interest. The 1D approximation enabled us to describe physical phenomena with substantial extent of complexity and detail, including boundary and wall ablation effects, the kinetics of Ne-like and others ions with time-dependent ionization and line radiation transport, refraction and ray tracing along the axial direction of the plasma, etc.

Atomic data for the detailed atomic kinetics of ions of $Z = 6\text{-}16$ was calculated using HULLAC (6) and the methods described in (7). Data for the Ne-like ion ArIX with n=2,3 (37 levels) (7), n=2,3,4 (89 levels) or with n=2,3,4,5 (157 levels) (6) was used in the calculations. For the Na-, Mg-, Al-like Ar ions, data for excited configurations lower than n=3 was used, for F-like ions n=2,3 was employed, and finally, for Li-like ions, data with n=2 was utilized (for example, 148 levels were used for the Al-like ions, 113 for F-like ion etc.). In summary, a detailed atomic model of the Ar ions including more than 500 levels for the Ne-like neighboring ions from ArVI to ArXVI was utilized for the accurate simulation of spectra that were compared with experiments.

To analyze features of the experimental line intensities, spectra in the 47 nm region were simulated for different discharge parameters and capillary lengths. Fig.1-3 show the results of spectra calculations for a 12 cm long, 4 mm diam. capillary, filled with pure Ar at 0.32 Torr and discharge current $I_0 = 37.5$ kA. Fig.1 is an example of an "identification" spectra, obtained neglecting all broadening mechanisms on the graph (but not in the calculations!), that illustrates the contribution to the spectrum of each individual line. Most of the lines here belong to the Al-like and Mg-like ions. Figs. 2 and 3 are off-axis and on-axis simulated spectra, respectively, and Fig.4 shows the corresponding on-axis experimental data. Note, that in the off-axis spectra (Fig.2), lines corresponding to Ne-line ions, including the J=0-1 line, are very weak. Gain values for the Ne-like Ar lines, computed utilizing two different sets of atomic data (6) and (7), were compared and found to be in reasonable agreement. Computed spatio-temporal profiles of the gain for the J=0-1 46.9 nm line show peak values of $G = 0.4 - 0.7$ cm^{-1} in pure Ar.

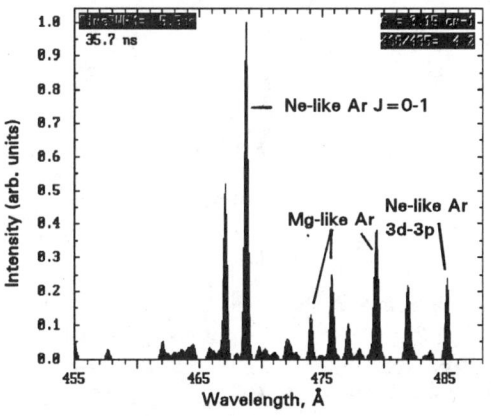
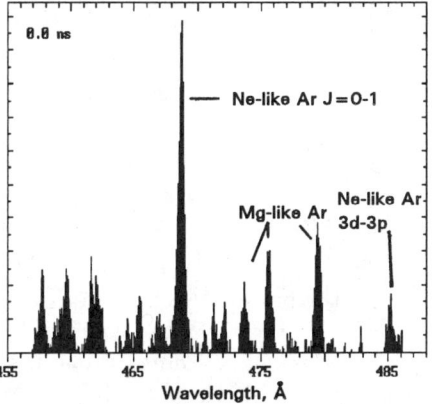

FIGURE 3. On-axis time resolved (5.5ns) spectra at the parameters of Fig.2. Strongest is J=0-1 line of ArIX, note artificial "phantom" line left of it.

FIGURE 4. Averaged (sum of 5 spectra) experimental time resolved spectrum at the same parameters of Fig.3

Usually at the pressures noted above the gain has the maximum in a thin ring region near the skin layer, in the front of the heat conductive shock waves. However, at these plasma conditions, spatial and temporal integration of the gain, required to have a valid comparison with measurements, results in reduced predicted gain values (according to Linford formula) in the neighborhood of $\sim 0.15 - 0.3$ cm^{-1} for spectrometer collection angles of $\sim 15\text{-}20$ mrad (note here the very small active solid angle of FCD).

For a clear identification of the amplification properties of the plasma we introduced in the computed spectra an artificial, "phantom" line to the left of the J=0-1 line, that represent the same J=0-1 line without amplification or absorption, i.e. in the radiation transport and ray tracing we used the condition $G = 0$ (Fig.3). Comparison of the intensity of the J=0-1 line and this artificial line clearly illustrates the contribution of spontaneous emission in the total line intensity. Particular attention was paid in modeling the absolute and relative intensities of the 0-1 gain line and the neighboring 3d-3p Ne-like transition at 48.5 nm, which does not amplify, as a gain diagnostics tool. The utilization of these theoretical tools in analyzing the features in the spectral data from capillary discharge experiments assisted us in determining the gain coefficients and in developing a better control of the amplification.

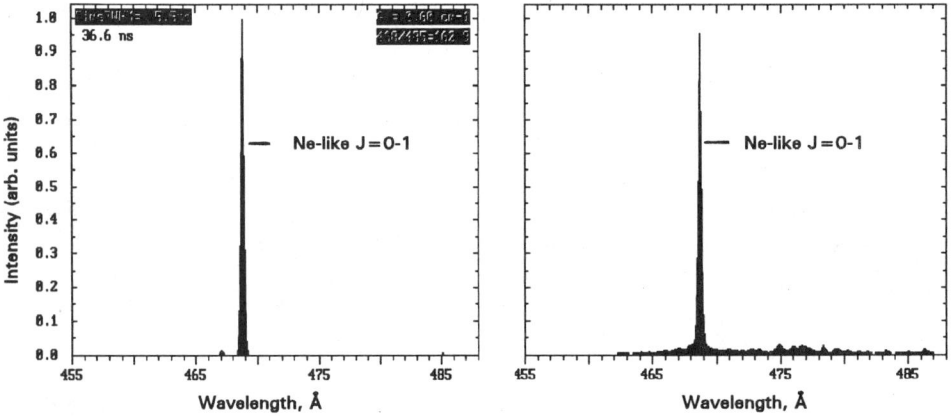

FIGURE 5,6 Calculated (left) and experimental(right) time resolved spectra for a FCD in Ar/H_2 at 0.6 Torr, 1:2 partial pressures, $I_0 = 37.5$ kA

At the parameters noted in Fig.5,6, an increase in the pressure by the addition of H_2 increases the gain and substantially boosts the intensity. Here the calculations show the gain maximum is achieved in the dense axial region of the Z-pinch. At these conditions, hydrogen works like a buffer gas, that beneficially changes the heat conductivity and resistivity, decreases the compression ratio, increases the electron-ion relaxation decreasing the ion temperature (5-10 times from 1-3 keV to 100-200 eV), and hence increases the gain. Also hydrogen possibly improves the stability of the pinch compression.

Simple estimations for plasma densities of $(2\text{-}5) \times 10^{18}$ cm^{-3} show that refraction effects restrict the maximun plasma column length for amplification of the J=0-1 ArIX line to $\sim$ 10-15 cm. However the result of a more precise numerical calculation, shown in Fig. 7, indicates that the intensity should continue to grow for lengths as large as twice the maximun (12 cm) used in these experiments.

A parameter that looks hardly achievable experimentally, is an active solid angle of amplifying region $2 \times r/L \sim 0.001$. The recent successful realization of this needle-like channel illustrates the excellent compression symmetry achieved with the fast capillary discharge, and emphasizes the advantage of this unique device in X-ray lasers design and, in Z-pinch practice.

Finally, Fig.8, which shows the computed relative intensities of the $\sim 10^4$ Ar lines used in the present calculations, illustrates the amplification obtained in the experiments reported in (1). The points marked by the arrow and the circles represent the ArIX J=0-1 46.9 nm line intensities corresponding to an Ar point source and to the Ar FCD X-ray laser respectively. The latter is several orders of magnitude more intense.

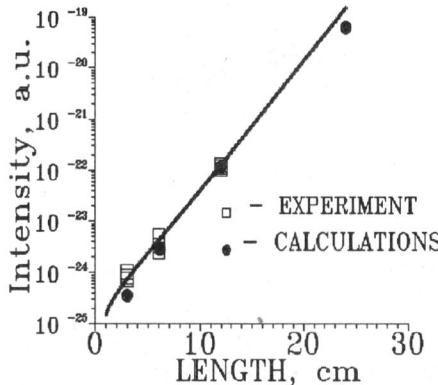 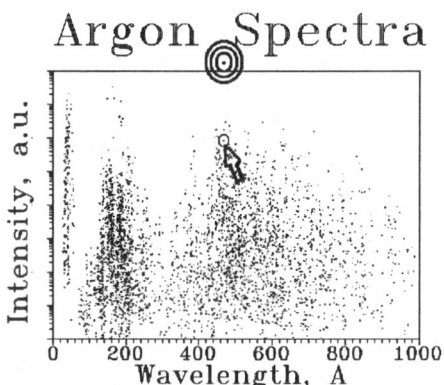

FIGURE 7. Output intensity of the 46.9 nm FCD ArIX X-ray Amplifier as a function active media length. The solid line is a Linford fit with G = 0.62

FIGURE 8. "Space" of Ar lines. The arrow and circle illustrates the intensity of the ArIX J=0-1 46.9 nm line for a point source and for the experimentally demonstrated FCD Ar X-ray laser

3. REFERENCES

1. J.J. Rocca, F.G. Tomasel, V.A. Shlyaptsev, D.O. Cortázar, J.L.A. Chilla, and G. Giudice, "Soft-x-ray Amplification in a Capillary Discharge Plasma", *in this Proc.* .
2. *Proc. 3rd Int. Conf. on Dense Z-pinches*, London, UK, 1993.
3. V.N.Shlyaptsev, A.V.Gerusov, "On two methods of table-top X-ray laser design", *Proc. 3rd Int. Colloq. on X-ray Lasers*, Ed. E. Fill, Schliersee, Germany,1992.
4. J.J.Rocca, O.D.Cortázar, B.T.Szapiro, K.Floyd and F.G.Tomasel, *Phys. Rev. E*, **47**, 1299, 1993 and J.J.Rocca, O.D.Cortázar, B.T.Szapiro, F.G.Tomasel, "Study of fast capillary discharge plasmas column for soft X-ray amplifiers", *SPIE J*, **2012**, 67, 1993.
5. V.N.Shlyaptsev, A.V.Gerusov, A.V.Vinogradov, J.J.Rocca, O.D. Cortázar, F.Tomasel, and B.Szapiro, "Modeling of fast capillary discharge for collisionally excited soft X-ray lasers; comparison with experiments", *SPIE J*, **2012**, 99, 1993.
6. A.L.Osterheld et al.," Atomic physics modelling of X-ray laser plasma", *Proc. of Third Int. Colloq. on X-ray lasers*, Schliersee, Germany, 1992.
7. A.V.Vinogradov, V.N.Shlyaptsev, *Sov. J. Quant. Electronics*, **10**, 754, 1980.

Lasing at Short Wavelength in a Capillary Discharge and in a Dense Z-Pinch

H.-J. Kunze*, S. Glenzer*, C. Steden*, H. T. Wieschebrink*,
K. N. Koshelev[†], and D. Uskov[††]

Institut für Experimentalphysik V, Ruhr-Universität, 44780 Bochum, Germany
[†] *Institute for Spectroscopy, Russian Academy of Sciences, Troitzk, Russia*
[††] *P. N. Lebedev Physical Institute, Russian Academy of Sciences, Moscow, Russia*

Abstract. Results on the emission of the CVI Balmer-α transition obtained with a fast capillary discharge are summarized, and a model is discussed, which explains the observations as result of fast ions produced by a m=0 instability and charge exchange with CIII ions in the cold plasma region. Plasmas of large dimensions were produced in the gas-liner pinch discharge, and the emission of the 4f-3d transition has been studied in CIV, NV, OVI and FVII. Amplification is seen on the transition in OVI and FVII.

CAPILLARY DISCHARGE

For a couple of years we study the emission from small fast capillary discharges. The plasmas are made up of ablated wall material and their parameters are similar to those reported already some decades ago (1). Since they are surprisingly close to those reported for laser produced plasmas which display lasing according to the recombination scheme (2), the expectation to achieve lasing also in such a small device initiated our investigations.

The capillary arrangement has been described previously (3,4). Electrical energies of the discharges are typically between 3 and 12 joules. Most of our observations were carried out with capillaries made of polyacetal since the carbon debris was low in comparison with capillaries of polyethylene. We start by summarizing previous and new results:

- The characteristic feature is a strong spike on the emission of the Balmer-α line of CVI at 18.22 nm which always occurs at the time of the second current maximum. This spike is extremely reproducible in time, the jitter being a few nanoseconds. Figure 1 shows an example of the emission recorded with a 1-m grazing incidence spectrometer and a fast scintillator-photomultiplier combination. The spike is not observed on other lines of carbon or oxygen ions in the vuv or visible spectral regions.

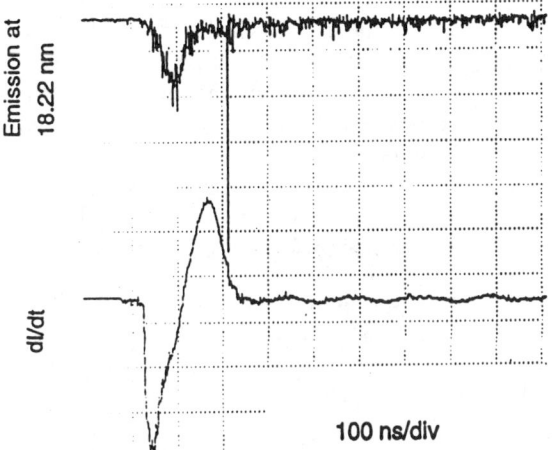

Figure 1. Emission of the Balmer-α line of CVI and dI/dt of the discharge.

- The intensity of the spike increases only weakly with the length of the capillary, i.e. less than expected for the case of a constant gain factor in the plasma. Gain-length products of 2.7, 3.0, and 3.6 have been deduced from the spike for capillaries of 1 cm, 2 cm, and 3cm in length (3).

- The narrower the capillary the better is the spiking. During a series of discharges the wall diameter increases due to ablation of wall material and the spike finally disappears even if the discharge current is increased correspondingly.

- A plane multilayer mirror (reflectivity 12 % at 18.22 nm) placed at one end of the capillaries increased the amplitude of the spike by factors 1.5 to 1.7 for capillaries of different length (4). This implied a gain-length product of GL = 2.7 for a 1 cm capillary and of GL = 2.5 for a 3 cm capillary. After a few shots this effect disappeared, the surface of the mirror was destroyed locally.

- Time integrated spectra recorded with a grazing incidence spectrograph reveal that the emission of the CVI Balmer-α transition is directed. The divergence decreases with the length of the capillary (4). This effect is not observed on any other line of carbon or oxygen ions.

- The divergence of the Balmer-α line decreases with the multilayer in place.

- The emission of the CIII ion at 229.7 nm and of the CV ion at 227.1 nm was scanned across the diameter. It became evident that the emission of CV peaks on the axis, CIII emission has a minimum there. Time integrated spectra recorded with a pinhole transmission grating support this observation. This suggests a hot plasma channel on the axis.

- Strong spectra of CVI and CV ions are observed for a capillary 1 cm long and 0.5 mm in diameter made of ABS polymer. The Balmer-α to Balmer-δ are clearly seen, even the resonance line of OVII shows up.

- Temperatures deduced in our device from the continuum emission at long wavelength and also obtained by other groups in similar arrangements (5,6) seem to contradict these observations: they are so low that it is impossible to ionize an appreciable fraction to the high ionization stages and to excite the heliumlike and hydrogenlike ions.

In order to explain our observations we advance therefore a model, which relies on an instability of the m = 0 type to produce hot regions in the plasma and on charge exchange between ions as the pumping mechanism for the population inversion (7). The ocurrence of an instability just at the time of the second current maximum certainly is not obvious. At this time the current is typically 10 kA and density and temperature are such that equilibrium by the Bennett relation is reached. It is usually assumed that such equilibria are stable if surrounded with high pressure gas or even a wall. Resistivity and viscosity should have further stabilizing effects. However, several observations indeed corroborate the occurrence of the instability.

First the current derivative dI/dt shows a perturbation indicative of an instability at the time of the spike (see Fig.1). For some discharges this is really distinct. We then cut the capillaries after use and stain the inside slightly using a lead pencil: the imprints of a m=0 instability on the wall are unambiguous, the periodicity appears very clearly. For capillaries of 1 cm length we obtain $ka_0 \approx 2.4$ for all diameters up to 1.4 mm, where a_0 is the radius of the capillary and k is the wavenumber of the instability. For capillaries of 2 cm, 3 cm, and 5 cm length the values of ka_0 are 1.6, 0.8, and 1.2, respectively. It is surprising that the wavelength of the instability increases with the length of the capillary. The growth time of such an instability is about a_0/c_s, where c_s is the ion sound speed. The parameters of our plasma predict a value of about 10 ns; the high-temperature phase of the instability will be much shorter and hence comparable to the jitter of the Balmer-α spikes and their duration.

It is well known that hot plasmas are produced in the neck regions of such an instability, and we advance that they are the origin of the emission from very highly ionized species observed in our capillary discharge, the bulk plasma still remaining cold. Constrictions by factors between 2 and 3 is all that is needed to reach a temperature (about 150 eV and higher) necessary to strip carbon atoms completely, which is the precondition for populating excited states of CVI by recombination.

Plasma jets of fully ionized carbon atoms flow from the neck regions and interact with the cold bulk plasma. The velocity is given approximately by the ion sound speed c_s, which is about 10^7 cm/s for a plasma of fully stripped carbon atoms at 150 eV; this corresponds to a kinetic energy of about 600 eV. The fast ions experience Coulomb collisions, collisional recombination and charge exchange recombination, and we now estimate the magnitude of these processes in the cold plasma. At second current maximum the electron density n_e of the capillary plasma is of the order of 10^{19} cm^{-3}, the temperature is between 10 and 20 eV, and the carbon atoms are mainly in the CIII and CIV ionization stage. In the case of capillaries made of polyacetal the carbon ion density should be 1.3×10^{18} cm^{-3}.

The slowing down probability and the energy loss probability by Coulomb collisions with CIII and CIV ions have the same magnitude of 10^{10} s^{-1} (8). This suggests a penetration length of the fast beam into the cold plasma of about 10 µm. This is certainly a crude approximation and has to be modelled in detail.

The probability for collisional recombination into all levels above the collision limit is about 10^9 s^{-1} (9). This is indeed low and we directed our attention therefore to charge exchange with carbon ions. Using a multilevel Landau-Zener model we calculated the cross-section for charge exchange with CIII(2s^2), and the over-barrier transition model yielded respective cross-sections for collisions with CIII(2s2p). The results were astonishing: cross-sections are larger than 10^{-15} cm^2, they remain large over some energy interval, and most importantly, they are selective into the n=3 level of CVI, i.e. the upper level of the Balmer-α transition. The pumping probability into this n=3 level is approaching nearly 10^{11} s^{-1}. This is about two orders of magnitude higher than collisional recombination into all levels n≥3. Taking into account radiative decay and collisions to the n=4 level, the relative population in the n=3 level can be estimated. Charge exchange into the n=2 level is several orders of magnitude slower and can be neglected. The assumption of an inversion factor F=1 thus is reasonable and the gain coefficient G may be estimated. If we take a beam density of the magnitude of the ion density in the cold bulk plasma, we obtain G≈400 cm^{-1} in the case of plasmas from polyacetal. For active layers of about 10 μm thickness and 14 plasma jets going into one direction, we arrive at GL≈5 for the 1 cm capillary. Given all uncertainties this compares favorably with the experimental values of about 2.7 and 2.5.

The wavelength of the instability increases with the length of the capillary resulting in less plasma jets per unit length. This explains the behavior of the gain-length product GL with length L: the total length of the active region simply does not increase linearly with the length of the capillary. For the 3 cm capillary our estimate gives GL≈5 compared with experimental values of 2.5 and 3.6. Larger diameters of the capillary also result in longer wavelength of the instability and hence in lower values of GL, again consistent with the observations.

Although crude our model thus explains reasonably well the observations in our capillary discharge, the major question remaining, why the instability occurs at all and why at the time of the second current maximum.

GAS-LINER PINCH

We now turn to studies carried out employing the gas-liner pinch device (10). It is specifically designed and suited for various spectroscopic investigations. In principle it is a large aspect ratio z-pinch, the diameter of the discharge chamber is 18 cm and the electrode separation is 5 cm. One important feature are two independent gas-inlet systems. The schematic arrangement is shown in Fig. 2. The main gas, usually called driver gas, is injected with a fast electromagnetic valve through an annular nozzle into the vacuum chamber. It forms initially a hollow gas cylinder which, after preionization, is imploded towards the axis by the main discharge. On the axis it forms a plasma cylinder of about 2 cm in diameter and 5 cm long. For

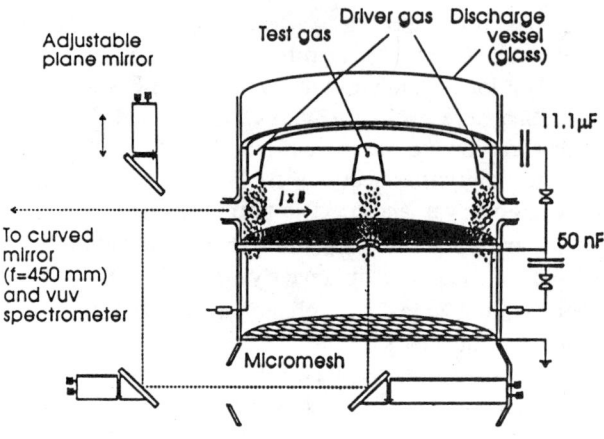

Figure 2. Schematic of the gas-liner pinch setup

the present studies we used hydrogen. The capacitance of the discharge is 11.1 µF, the charging voltage is between 25 and 35 kV. Depending on the fill pressure the implosion time is between 1 µs and 2 µs.

The second fast electromagnetic valve injects the so-called test gas through a nozzle in the center of the upper electrode along the axis. The concentration usually is adjusted to be less than 1% and can easily be varied to identify effects of radiative transport on the emission characteristics. The test gas ions are essentially confined to the central region of the plasma column where the plasma is rather homogeneous. No test gas ions are in cool boundary layers and influence spectral lines by self-reversal.

Plasma parameters are determined by collective Thomson scattering performed at the same time as the spectroscopic measurements: electron density and temperature, impurity density and temperature are obtained (11). More recently also drifts between the protons and the electrons are deduced (12). Spectroscopic observations are carried out side-on in the midplane of the discharge chamber through one of four ports and end-on through a hole of 1.8 cm diameter in the lower electrode. The spectrometer is a 1-m vuv normal-incidence instrument equipped with a microchannel plate in the exit plane. Gate time is 20 ns. The phosphor at the rear side is imaged onto an optical multichannel analyzer.

The present investigations were triggered by the idea, that the imploding plasma cylinder forms a high-density and high-temperature peak on the axis for a rather short time, which quickly relaxes and leads to population inversion by three-body recombination. We studied therefore the emission of the 4f-3d and 4d-3p transitions in the lithiumlike ions CIV, NV, OVI, and FVII. Test gases were CH_4, N_2, CO_2, and SF_6, however, high concentrations were chosen in order to improve the chances of observing single pass gain. The electron density was in the range from 1.5 to 6×10^{18} cm^{-3}, the electron temperature was typically between 7.5 and 45 eV. Discharges without test gas indicated possible impurities, and the continuum radiation gave the relative sensitivity calibration.

For CIV and NV the intensity ratio I(4f-3d)/I(4d-3p) was found to be independent of the plasma conditions and corresponded to an equilibrium population between the 4f and 4d levels without any indication of amplification. This was true for radial as well as for axial observation.

The results are different for OVI. At times around maximum pinch compression strong enhancement of the 4f-3d line at 51.97 nm with respect to 4d-3p transition at 49.83 nm was recorded in axial direction. The side-on ratio varied between 2.8 and 4.7, the end-on ratio averaged over 16 discharges was 25. The comparison of both ratios yields a gain-length product of GL$\approx$4.5. At earlier and later times during the discharge no enhancement is observed. The effect was confirmed by time resolved observations end-on. For this purpose the microchannel plate was replaced by a fast photomultiplier, and indeed emission bursts were seen on the 4f-3d transition at the same time when the enhancement was seen with the MCP-OMA-dectection system. No bursts were detectable on the 4d-3p or any other line. From the burst we derive GL$\approx$3.9.

In order to see any influence of the plasma parameters we carried out side-on spectroscopic observations and Thomson scattering at the same time. At high electron densities of about 5×10^{18} cm^{-3} the side-on intensity ratio corresponded to the equilibrium value of 2.7, at lower densities the weaker collisional coupling became evident and the 4f-3d transition was relatively stronger. This indicates some amplification already in radial direction at lower densities.

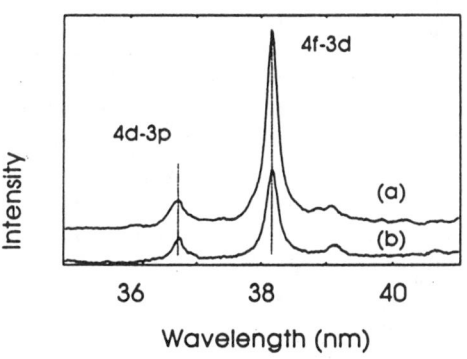

Figure 3. Side-on spectra of the 4f-3d and 4d-3p transitions of FVII from Ref. (10)

This effect was even more pronounced for the respective transitions in FVII. Figure 3 shows two examples of the 4f-3d and 4d-3p transitions at 38.12 nm and 36.77 nm recorded in radial direction at about 40 ns before maximum compression. The different ratio is evident, during a series of discharges it varied between 2.8 and 13, the lowest value corresponding again to an equilibrium population between the 4f and 4d levels. End-on observations were not possible yet because of the low reflectivity of the mirrors in the optical path, but they are planned for the near future.

In order to find an explanation of the observations, we set up a collisional-radiative model for the population of all levels up to the principal quantum number n=5 in OVI and FVII. Levels with n=6 to n=14 were assumed to be coupled by collisions, and their population densities were linked to the ground state of the heliumlike ion by the Saha-Boltzmann equation. The population densities were calculated for various electron densities around 1×10^{18} cm^{-3} and temperatures between 15 and 50 eV, and then it was assumed that the temperature relaxes quickly to values between 7 and 10 eV. Population inversion and gain coefficient certainly depend on the cooling time, but it became evident that the axial amplification observed for the transition in OVI can be explained in terms of three-body recombination if the cooling time is of the order of 1 ns.

The case of FVII, on the other hand, poses some problems, since three-body recombination alone probably cannot explain the experimental results. Additional observations, however, may point to a possible explanation. Pinhole pictures of the plasma column taken with a four-frame microchannel plate camera revealed the development of a Rayleigh-Taylor instability during the implosion if high amounts of test gas were used as in our present studies (10). From the neck regions of this instability ion beams will emerge and interact with the bulk plasma (7). Highly selective charge exchange as discussed in the case of the capillary discharge readily may make up for the missing population channels, although the specific processes between ions still have to be analyzed and cross-sections calculated. Spectra of the FVII lines recorded along the axis of the plasma column employing a CCD camera give further support to this possibility. Enhanced emission along the axis is seen in regions of about 0.8 mm.

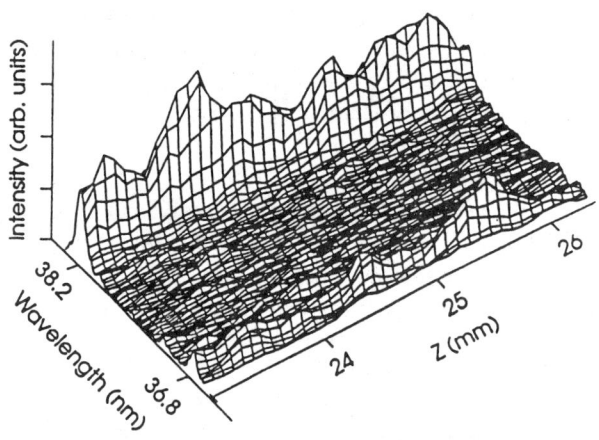

Figure 4. Emission of the FVII lines along the axis from Ref. (10)

THETA-PINCH

We finally conclude by referring to our recent investigations on neonlike ArIX (13) for which such successful lasing has been reported by Rocca *et al.* at this conference (14). We measured the relative intensities of all 3p-3s transitions in theta-pinch discharges at temperatures which were about 80% of the excitation energy of the 3p level. About 1% argon was added to the initial hydrogen or deuterium filling of the discharge. The plasma was heated so rapidly that the argon ions were in an ionizing regime and the concentration of ArX was still low, when AIX peaked. This has the advantage that possible contributions to the population of the excited states by recombination can be neglected in the analysis at this time during the discharge. Electron density and temperature were known and obtained by Thomson scattering, they were of the order 10^{16} cm^{-3} and 200 eV, respectively. The relative intensities were compared with theoretical calculations, which employed various collisional and radiative transitions given in the literature. Very good overall agreement was observed with a collisional-radiative model which included all levels up to n=5, i.e. it consisted of a total of 157 levels. The standard deviation of the relative intensities of all 15 transitions from their theoretical values was 15% in the hydrogen discharge and 21% in the deuterium discharge. The experimental results thus confirm the best available theoretical calculations and hence modelling of neonlike ArIX in plasmas should be possible with sufficient reliability.

ACKNOWLEDGMENTS

This work was supported by the Deutsche Forschungsgemeinschaft. We thank Dr. Danz for supplying the ABS polymer.

REFERENCES

1. Bogen, P., Conrads, H., Gatti, G., and Kohlhaas, W., *J. Opt. Soc. Am.* **58**, 203 (1968).
2. Suckewer, S., Skinner, C. H., Milchberg, H., Keane, C., and Voorhees, D., *Phys. Rev. Lett.* **55**, 1753 (1985).
3. Steden, C., and Kunze, H.-J., *Phys. Lett.* **151**, 534 (1990).
4. Steden, C., Wieschebrink, H. T., and Kunze, H.-J., "Line emission from carbon ions in a capillary discharge," in *X-Ray Lasers 1992*, ed. E. E. Fill, London: Inst. Phys. Conf. Ser. No. 125, 1992, pp. 423.
5. Tomasel, F. G., Rocca, J. J., Cartázar, O. D., Spiro, B. T., and Lee, R. W., *Phys. Rev. E* **47**, 3590 (1993).
6. Morgan, C. A., Griem, H. R., and Elton, R. C., *Phys. Rev. E* **49**, 2282 (1994).
7. Koshelev, K. N., and Kunze, H.-J., "Ion beams from axial discharges and the x-ray laser problem, in *Dense Z-Pinches*, eds. M. Haines and A.Knight, New York: AIP Conf. Proc. 299, York, 1994, pp. 231-235.
8. Trubnikov, D. A., in *Reviews of Plasma Physics*, Vol.1, ed. Leontovich, M. A., New York: Plenum, 1965, pp. 105.
9. Elton, R. C., *X-Ray Lasers*, San Diego: Academic Press 1990.
10. Glenzer, S., and Kunze, H.-J., *Phys. Rev. E* **49**, 1586 (1994).
11. DeSilva, A. W., Baig, T. J., Olivares, I., and Kunze, H.-J., *Physics Fluids* **4**, 458 (1993).
12. Glenzer, S., Wrubel, Th., Büscher, S., and Kunze, H.-J., "Measurements of electron-proton drifts with collective Thomson scattering" in *Proceedings of the XXI.. Int. Conf. Phen. Ionized Gases*, APP Bochum 1993, eds. Ecker, G., Arendt, U., Böseler, J., Vol. II, pp. 367.
13. Preissing, N., Campos, D. O., Kunze, H.-J., Osterheld, A. L., Walling, R. S., *Phys. Rev. E* **48**, 3867 (1993)
14. Rocca, J. J., Tomasel, F. G., Shlyaptsev, V. A., Cortazar, D. O., and Guidice, G., these proceedings.

NONSTATIONARY ARGON PLASMA, CONTAINING Ne-LIKE and Na-LIKE IONS. "FAST COMPRESSION" and POPULATION INVERSION.

L.N. Ivanov
Institute of Spectroscopy, Russian Academy of Science, Troitsk
Moscow Region, Russia 142092
L.V. Knight
296 ESC Brigham Young University, Provo, UT, 84602

ABSTRACT. Evolution of levels populations in Ar plasma with varying parameters is under theoretical investigation. The model imitates fast compression and expansion of the capillary plasma column. The role of the HYDROGEN admixture is discussed.

INTRODUCTION and METHOD. We study the argon plasma under conditions where Ne- like and Na- like ions are the most abundant. This includes a sufficiently large region of plasma electron density N_{el} and plasma electron temperature T_{el} to be realized and investigated experimentally. Under these conditions in the steady state plasma with electron density $N_{el} > 10^{17}$, the population inversion for all pairs of lasing candidates is small, not more than 0.001 of the total number of argon atoms. Recently J.J.Rocca et.al. [1,2], in there experiments with capillary discharge showed, that the preionized plasma column with a diameter 0.2 cm can be rapidly heated by magnetic field arising during the fast increasing of the discharge voltage. The radiating region, under the influence of the compressing magnetic field, shrinks and its temperature increases. During the whole cycle of compression - expansion, the hot dense plasma never contacts with the capillary walls.

In [3] the idea of the simultaneous confinement and heating of the thermonuclear plasma inside tubelike powerful laser beam, has been suggested. We believe, that the same idea would be useful in the capillary discharge problem. The tubelike laser beam, enveloping capillary additionally compress its plasma due to ponderomotive forces and serves as an additional source of heating energy, that moderates the requirements to the characteristics of the discharge. As it is noticed in [3], such laser beam protects plasma against sausage- type instability.

The nonstationarity features in the compressing plasma, drastically influence the inversion of "lasing" levels population. The maximum of inversion does not obligatory coincides with the maximum of compression or temperature. Thus an appreciable inversion can appear at the decompression phase with moderate plasma temperature and relatively high electron density.

Here we investigate this question theoretically. A definite time dependence of the diameter D(t) of the compressing column, is accepted. The model of the Maxwellian homogeneous plasma, characterized by the correlated functions of electron temperature $T_{el}(t)$ and electron density $N_{el}(t)$ is considered. The $T_{el} \sim D^{-4/3}$ and $N_{el} \sim D^{-2}$ time dependencies follow from the simplest model of the homogeneous compression of the plasma obeying the local neutrality condition. Certainly it is a very simplified model of the process. Experiment shows that energy transfer between electro- magnetic field and plasma proceeds much faster than it follows from this model, possibly due to other physical processes in the plasma column. Nevertheless, we believe,that the accurate calculation of the evolution of the states populations in the plasma with the phenomenological time dependence of parameters of the active media, is a realistic way to evaluate perspectives of lasing, even if we do not specialize real physical processes that manifest itself in such a manner. Thus, we use here the terminus "compression" and "expansion", by convention, to designate some phenomenological model of the correlated variation of the plasma parameters.

In our calculations, we account for all the elementary atomic processes of the states population, in the framework of the collisional - radiational model. The rate coefficients of the processes changing the state of the Ne - like ion or Ne - like residue in the Na - like ion, are calculated due to the relativistic many - body perturbation theory with the model bare potential [4]. All the other rate coefficients are calculated in the semi - classical approximation. [5,6]. 37 basic states of the Ne - like ion and 37 multitudes of adjacent Rydberg states of the Na - like ion are accounted for. The infinite number of states of each Rydberg series is included. We investigated the role of the diffusion - like migration of the state of the system "Ne - like ion plus electron" over the overlapping Rydberg multitudes on the states population kinetics. It proved that the contribution of this migration into the relative populations of the states of the Ne - like ion is important, even in the case when Na - like ion states are poorly populated themselves. It is due to intensive population fluxes through these states and immensely large number of reaction channels. The summary effect of populating of Rydberg levels through dielectron recombination followed by different ionization processes, is very selective in relation to final state of system; it is important for creation of inversion.

Time dependent population calculations are based on our code [5], that solves the system of kinetic equations for populations of 37 lowest Ne- like levels and 37 population distribution functions for the multitudes of Rydberg states. Reabsorption of the spontaneous radiation for all transitions is accounted for due to the standard routine [7]. We considered the case of plasma cylinder with diameter 0.10 - 0.02 cm., the ion temperature is supposed to be equal to T_{el}. We experimented with the ion temperature function other than T_{el}: its reasonable variation did not change qualitatively the results presented below.

RESULTS. As usual, we numerate the states of the Ne - like ion in accordance with increasing energies. Special attention is paid to the states 3,5 and 6,8-15. The first two are assumed to be lower, the other nine - upper levels in the collisional - radiational lasing scheme. Transitions 15-5 (3p,J=0 - 3s,J=1) and 13-5 (3p,J=2 - 3s,J=1) are usually under especial attention in experimental and theoretical investigations. For the "compressed" column, with high temperature $T_{el} > 100\,eV$, three other transitions show high inversion. Those are: transition 37-33 ($\lambda = 520.5$ Å and radiation transition probability $A = 3.2 + 09\,sec^{-1}$), 35-33 ($\lambda = 572.1$ Å, $A = 2.5 + 09\,sec^{-1}$), 34-33 ($\lambda = 572.4$ Å $A = 1.1 + 09\,sec^{-1}$). All wavelengths and transition probabilities are due to our calculations, see [5], and references therein.

Figure 1 shows evolution of the populations inversion for the transition 13-5 during the rather slow expansion starting from the steady state hot dense plasma. Two slightly different starting densities are considered. Line A: $N_{el} \rightarrow 35 + 16\,cm^{-3}$, while $t \rightarrow \infty$, $D \rightarrow 0,1$, cm.; line B: $N_{el} \rightarrow 30 + 16\,cm^{-3}$. The unambiguous transient effect manifests itself as increase of relative inversion

$$I(i-j) = \big(P(i)g(j) - P(j)g(i)\big)/P\,g(j),$$

where $P(i)$, $P(j)$ are levels populations, $g(i)$, $g(j)$ statistical weights, P- total number of Ne- like and Na- like ions that is close to the total number of Ar ions, for the processes here considered.

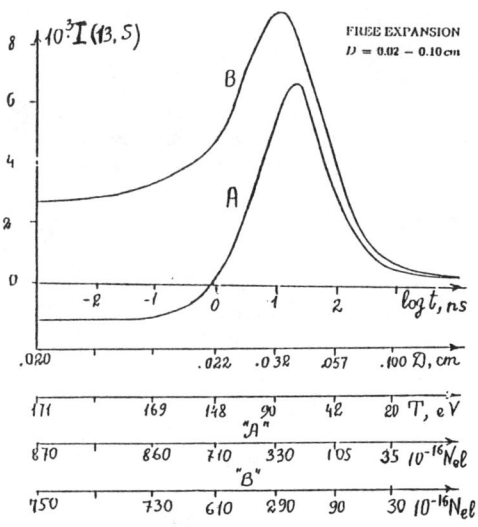

Figure 1

We see, that considerable inversion appears even if there is no asymptotical inversion for both steady states initial and final. This transient effect is due to thermodynamical relaxation of the ensemble of ions with overheated inner degrees of freedom. This effect is of the opposite sign, as compared with that predicted in [8] for "Ni - like Xe plasma". There, a short lasting (1 to 10 ps) increase of gain was predicted during the relaxation of the overcooled ions in the "strongly overheated plasma".

Figures 2 and 3 show evolution of the population inversion for some 3p-3s - transitions during the total cycle with moderate velocity compression and two different, moderate velocity, regimes of expansion: free expansion and delayed expansion, with $D(t)$ function symmetrical relatively to the point of the maximal compression.

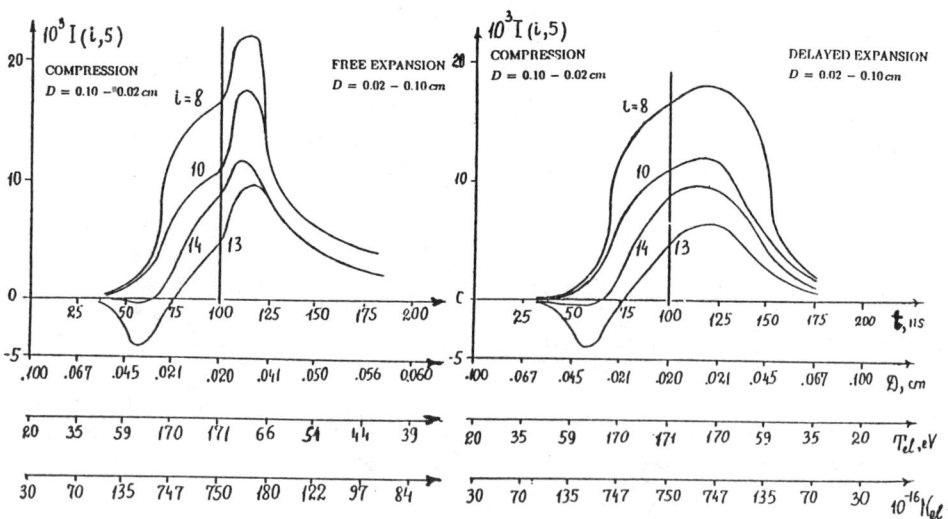

Figure 2..Figure 3.

In both cases, the inversion reaches its maximum during the expansion phase: the faster expansion the larger inversion. Figures 4 and 5 display the cases with the same as in figures 2 and 3, moderate velocity compression and fast (800 ps and 400 ps) expansion. Only decompression phase is presented. We do not discuss here how to realize such a fast changing of the parameters of the working media in the real device. The faster process seems to be impractical for the long (about 10 cm) capillary.

One can see a large transient inversion for two 3s-3p transitions, that remains sonsiderable even at the totally "decompressed" phase. Three transitions between the highest levels have large inversion at the "hot" phase, but expeience no transient phenomena. The same is for the 15-3 transition: no trannsient increasing of inversion for any regimes have been observed. It is due to the special mechanism of population of the level 15 as compared with other 3p - levels. It was discussed in [5].

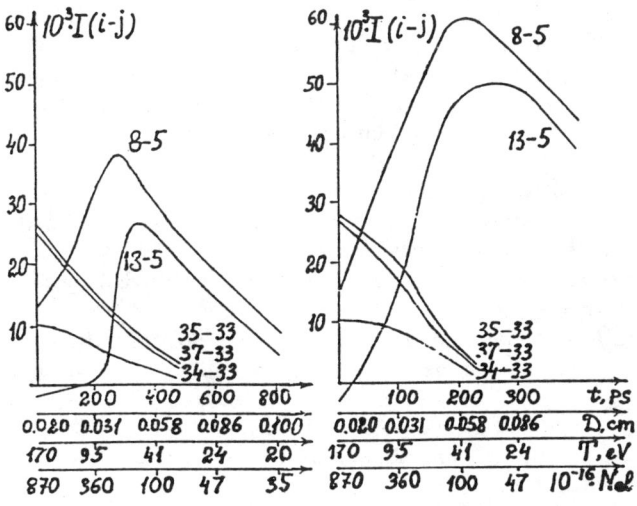

..........Figure 4.................................Figure 5.

Hydrogen in Argon plasma. During the whole process, the average velocity of Ar ions never exceeds $2.0 + 06\, cm/sec$, electrons move 250-400 times faster. Because of the local neutrality condition, the rate of the matter compression or expansion is restricted by the low mobility of the heavy positively charged component. In our case this time is not less than 50 ns. Our calculations show that the transient effects are small in this case.

The rate of matter compression or expansion can be increased by introducing into the plasma the light positively charged component, with equivalent amount of electrons. This has been realized in [2] by adding a certain amount of H_2, that are totally ionized at $T_{el} > 20\, eV$. Now the transversal distribution of electron density can, in principle, change without changing of the density of heavy Ar ions. The relatively fast redistribution of protons maintain local neutrality of plasma. Thus, the presence of a light positively charged component opens a channel of the fast immediate exchange of energy between electro - magnetic field and the electron gas. Moreover, in the comprehensive MHD - model, this channel can play role of a trigger for the local fluctuations of plasma parameters and microturbulence phenomena. It is known, that plasma microturbulence, Lengmuir or ion- acoustic, present additional channels of energy transport [9].

The possibility to introduce an arbitrary amount of H_2 molecules gives us one new variable plasma parameter - the relation of electron density to Ar ions density. This can be used, in principle, as additional degree of freedom in searching for the optimal lasing conditions, even in a steady state plasma. We investigated this question using our kinetic code. The numerical experiments show, that the optimal relation is close to the value "8", inherent to the pure Ar - plasma with only Ne- like ions. The considerable amount of extra electrons will quench inversion, thus there must be an optimal amount of H_2, stimulating lasing effect. This was reported in [2] as experimental fact.

WORK in PROGRESS. Definite amount of the highly excited ($n > 3$) Ne- like ions and definite amount of the F- like ions appears when plasma is "hot",that reduces the population of the "working" levels. Our atomic code accounts for the $n > 3$ these states in a certain approximation, but they have been omitted in the kinetic equations. It can be justified to certain degree by the next arguments, that are especially reasonable in the case of low Z- ions.

Under plasma conditions, here treated, the region of phase space related to the states with $n > 3$ is essentially thermalized. We mean local thermadynamical equilibrium. Thus, the population fluxes therefrom are determined mostly by statistical weights of states and do not influence population inversion. The inversion arises due to selective redistribution of these fluxes in the beneath lying energy region.

The role of the highly excited Ne- like ion and of F- like ions is suppressed in the fast running processes because of their creation assumes a certain time - about or more than 10 ns for the cases here considered.

Nevertheless, we are working out now the code accounting for effectively the infinite number of the excited levels of the Ne- like ion and three lowest levels of the F- like ion. Code is based on the generalized Lotz formula [6].

Another direction of our work: evolution of the three component plasma in the magnetic field. Problem includes physical kinetics as well as quantum- mechanical aspects.

[1] Rocca J.J., Cortazar O.D., Szapiro B., Floyd K., a d Tomasel F.G., Phys.Rev. ,47, 1299, 1993.

[2] Rocca J.J. et. al., this issue.

[3] Korobkin V.V., Romanovsky M.Yu., Phys.Rev.,E49, 2316, 1994.

[4] Ivanov L.N., Ivanova E.P., and Knight L.V., Laser-93, Proceedings of the International Conference.

[5] Ivanov L.N., Ivanova E.P., and Knight L.V., Phys.Rev., A48, 4365, 1993.

[6] Ivanov L.N., Ivanova E.P., and Knight L.V., Phys.Rev.E, in print.

[7] Fill K.K., J.Q.S.R.T., 39, 489, 1988.

[8] Shlyaptsev V.N., Nickles P.V., Shlegel T., Kalashnikov M.P., and Osterheld A.L., SPIE, Proceedings, Volume 2012, p. 111, 1993.

[9] Whitney K.G., and Pulsifer P.E., Phys.REV., E47,1968,1993.

Non-Maxwellian Plasma Electrons in the Inversion Population of Ne-Like Ions

E.P. Ivanova

Institute of Spectroscopy of Russian Academy of Science,
Troitsk, Moscow Region, 142092

L.V. Knight and B.G. Peterson

Department of Physics and Astronomy, Brigham Young University, Provo, UT 84602

Abstract: Level populations and line intensities are calculated for a pure Ne-like argon plasma showing their dependence on plasma parameters (electron temperature, density, plasma column diameter, electron energy distribution, etc.). It is shown that intensity ratios for 2-3 resonance transitions in the Ne-like ion can be used for plasma diagnostics. These intensity ratios are sensitive enough to detect the fraction of suprahot electrons in the plasma. Suprahot electrons cause a strong inversion effect for highly excited levels, e.g. 15:2p3p [J=0], 29:2s3s [J=0]. For these levels, collisional excitation strengths are large at almost any energy of impact electron above threshold. The inversion changes in time with a maximum at t ~ 80 psec at an electron density of n_e ~ 8 x 10^{18} cm^{-3}. The inversion increases with electron energy.

INTRODUCTION

Recent results indicate that plasma production based on a two-step process is effective in amplified stimulated emission (ASE) studies at short wavelengths. Such results have been reported for both capillary discharges plasmas and laser produced plasmas [1,2]. The time duration of the secondary discharge (laser pulse), and the time separation between the primary pulse and the secondary pulse are as important as the main discharge (laser pulse) parameters. Such experiments have been reported since 1991. However, there is no comprehensive explanation for experimental results. Typically the secondary discharge (laser pulse) is shorter than the primary discharge. The secondary impulse occurs when Ohmic (laser) heating dominates over electron collisional thermalization. Thus, the electron energy distribution is non-Maxwellian. In [3] the criteria are derived for the strengths of the current density (the laser intensity in a laser produced plasma) that are needed to drive the electrons into a non-Maxwellian state. Suprahot electrons has been studied for over a decade [4,5,6]. However, a description of the space and time evolution of the non-maxwellian electron distribution is an unsolved problem of physical kinetics. The lack of a realistic electron distribution function is one of the reasons that a magneto-hydrodynamic code has not been developed that predicts plasma parameters reliably.

Thus it might be useful to solve the reverse problem—to determine the electron non-Maxwellian distribution by spectroscopic means. The atomic kinetic code is needed to calculate the spectroscopic characteristics for plasma radia-

tive emission. The code can also be used to search for the optimal plasma conditions for ASE. The atomic kinetic code should be based on the following criteria:
i) It must acccount for all important processes connecting the ions of adjacent multiplicity.
ii) The atomic constants for isolated ions and for elementary processes in the plasma should be calculated using the same self-consistent approach.
iii) It must be flexible and noncumbersome (with moderate computer requirements). Thus allowing generalization. Hence, a great deal of the calculation data can be drawn from principal conclusions and laws.

THEORETICAL MODEL DESCRIPTION

Our approach to the atomic kinetic problem is described in [7,8] where the spectra of Ne-like and Na-like ions are studied. Rydberg and autoionizing Rydberg states of Na-like ions were accounted for in [8]. 37 levels of Ne-like ions $2p^5$ 3l;$2s2p^5$ 3l are calculated precisely. The calculations account for all electron collisional and radiative transitions. Level populations for Na-like ions are calculated effectively. The F-like ions are not considered since low electron temperatures are assumed. Ne-like argon spectra at high electron energy are considered for a brief period in the system evolution (t < 1.2 nsec). Reabsorbtion is accounted for in the kinetic equations for all transitions. The key parameters of the model are electron temperature (T_e), density (n_e) and energy distribution. All of the results are for an argon plasma. However, general conclusions can be drawn for any Ne-like plasma. For brevity the key levels are given the following labels: #3—2p3/2 3s1/2 [J=1]; #5—2p1/2 3s1/2 [J=1]; #13—2p1/2 3p3/2 [J=2]; #15—2p3/2 3p3/2 [J=0]; #29—2s1/2 3s1/2 [J=0]; #17—2p3/2 3d3/2 [J=1];—#23 is 2p3/2 3d5/2 [J=1]; #27—2p1/2 3d3/2 [J=1].

RESULTS FOR A MAXWELLIAN ELECTRON DISTRIBUTION

The starting point for our calculations should be based on a well developed theory for plasma diagnostics for a Maxwellian electron energy distribution. We suggest that Ne-like plasma could be diagnosed by measuring the intensity ratios for 2-3 resonant transitions in Ne-like ions. Our model is used to calculate the time evolution of the level populations of Ne-like and Na-like ions. The time needed to reach steady-state conditions for level populations depends on the electron density. (See figure 1.)

Next we calculate the fraction of Ne-like argon as a function of electron temperature. We are looking for the lowest electron temperature at which Ne-like ions dominate the plasma. The results are given in Fig. 2. At $n_e < 10^{19}$ cm^{-3} a low electron temperature (about 20 eV) is enough to produce 80-90% of all ions in the ground Ne-like state. It can be seen by Fig.2. that a significant fraction of Na-like ions can exist at $n_e > 10^{20}$ cm^{-3}. At temperatures above $T_e > 45$ eV Fig. 2

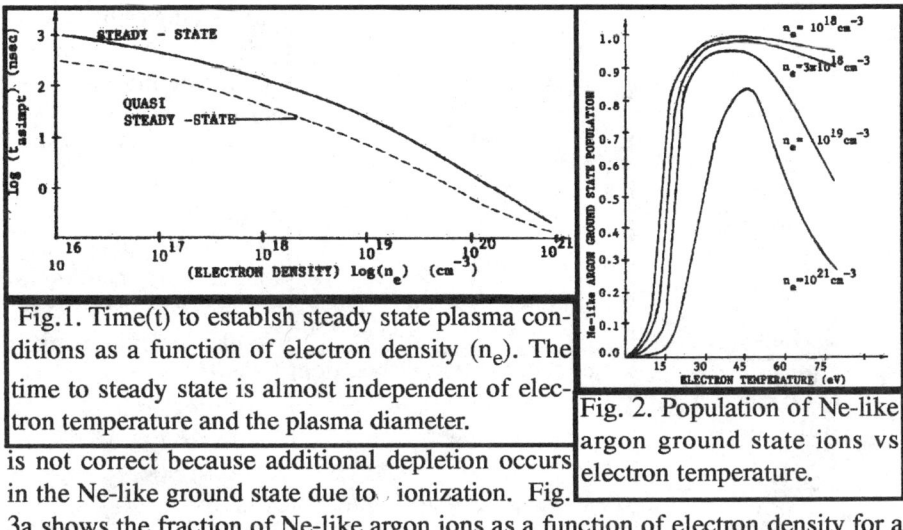

Fig.1. Time(t) to establsh steady state plasma conditions as a function of electron density (n_e). The time to steady state is almost independent of electron temperature and the plasma diameter.

Fig. 2. Population of Ne-like argon ground state ions vs electron temperature.

is not correct because additional depletion occurs in the Ne-like ground state due to ionization. Fig. 3a shows the fraction of Ne-like argon ions as a function of electron density for a steady-state uniform plasma at $T_e=75$ eV. The pre-equilibrium electron density corresponds to the optimal plasma electron density for lasing. The pre-equilibrium electron density increases with the average electron energy in the plasma. The curves for the inversion population (Inv = $P_u - g_u \times P_l / g_l$) as a function of electron density are illuminating. Curves for the inversion population relative to level #5 are shown in Fig. 3b, 4a and 4b for $T_e=$ 75eV, 20eV and 150eV respectively. Note that under certain conditions inversion is possible relative to level #3.

Fig. 3a Level population vs. electron density. $T_e=75$ eV, diameter, $D_e=0.02$ cm.

Fig.3b. Inversion population for levels #13,#15 vs. electron density at Te=75 eV, $D_e=0.02$ cm.

PLASMA DIAGNOSTICS BASED ON 2-3 RESONANCE TRANSITIONS

The seven brightest transitions to the ground state of Ne-like ions have

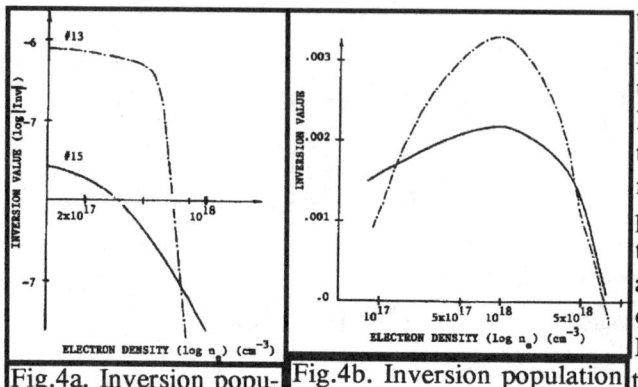

Fig.4a. Inversion population vs. electron density at Te=20 eV, De=0.02 cm.

Fig.4b. Inversion population vs. electron density at T_e=150 eV, D_e=0.02 cm.

been studied both theoretically and experimentally. Line intensities of Ne-like ions are sensitive to the plasma parameters. Thus, the most prominent radiative transitions can be used as a tool for plasma diagnostics. However Ne-like ion line intensities are sensitive to the theoretical model used for their calculation. Hence a spectroscopic measurement of line intensities is also a way to test theory and bootstrap the diagnostic measurements. The first 5 resonance 3-2 [J=1 - J=0] electrical dipole transitions are in the wavelength region λ = 49.185 - 41.471Å for Ne-like argon. Their intensities can be measured simulteniously. Their intensity ratios contain information about electron density, temperature and the plasma diameter. Fig.5,6,7 show the rules for some line intensity ratios at steady-state, uniform conditions.

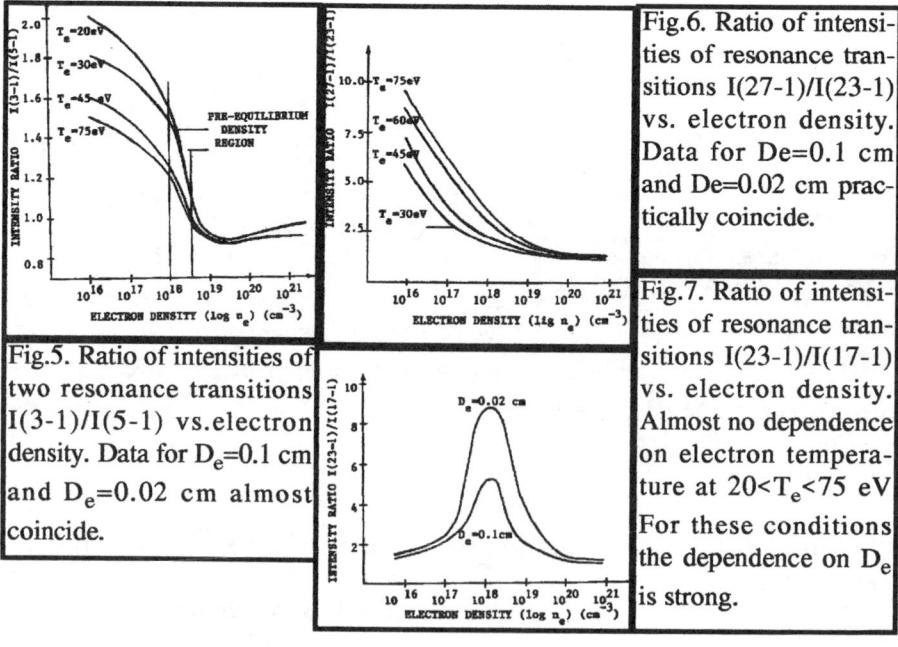

Fig.5. Ratio of intensities of two resonance transitions I(3-1)/I(5-1) vs. electron density. Data for D_e=0.1 cm and D_e=0.02 cm almost coincide.

Fig.6. Ratio of intensities of resonance transitions I(27-1)/I(23-1) vs. electron density. Data for De=0.1 cm and De=0.02 cm practically coincide.

Fig.7. Ratio of intensities of resonance transitions I(23-1)/I(17-1) vs. electron density. Almost no dependence on electron temperature at 20<T_e<75 eV For these conditions the dependence on D_e is strong.

EFFECT OF SUPRAHOT ELECTRONS

Fig. 8 and 9 show the population inversion for levels #13 and #15. These figures plot the same parameters as Fig. 3b and 4a. However, the electron energy distribution in Fig. 8 and 9 each contain a hump due a distribution of suprahot electrons. The inversion population increases dramatically.

Note a high inversion population occurs for level #29 in Fig. 8b. The resonance line intensity ratios (Fig.5-7) also change. Thus, it might be possible to detect suprahot electrons with accurate spectral line intensity measurements. Our theory will be refined by accounting for the F-like ionization stage. However the model is well enough founded to indicate that non-maxwellian high energy electrons can create an inversion population for highly excited levels with J=0. Time for electron thermalization depends on discharge parameters and plasma conditions. However, one can state that at electron velocities of about 10^7 cm/sec and electron densities of 10^{18}-10^{19} cm^{-3} t_{therm} < 2 nsec (t_{therm} is the electron thermalization time.). Thus a short secondary discharge (t_{disch} < 1 nsec) will produce electrons that are high energy and quasimonoenergetic. Fig.10 (a-d) shows the time dependence of inversion population for levels #15 and #29 relative to both low active levels: #3 and #5 for a few values of electron densities.

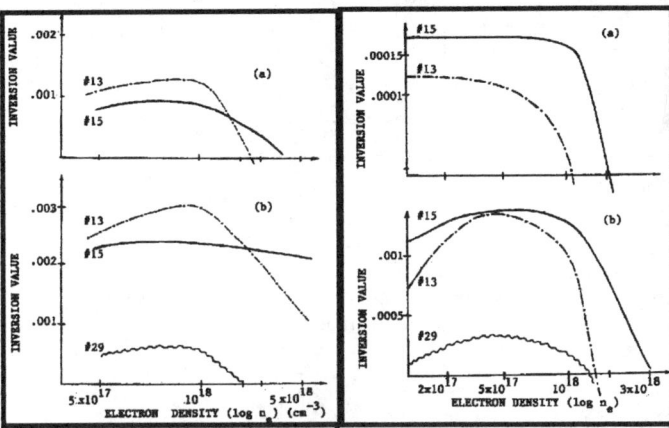

Fig. 8. The inversion population for levels #13 and #15 vs. electron density in the presence of suprahot electrons (a) 0.9 electrons at T_e=75 eV and 0.1 electrons at T_e=450 eV. (b) 0.5 electrons at T_e=75 eV and 0.5 electrons at T_e=450 eV.

Fig.9. The inversion population for levels #13 and #15 vs. electron density in the presence of suprahot electrons (a) 0.9 electrons at T_e=20 eV and 0.1 electrons at T_e=450 eV. (b) 0.5 electrons at T_e=20 eV and 0.5 electrons at T_e=450 eV.

Fig.10 shows that for higher electron density, the shorter time needed to reach maximum inversion.

398 Non-Maxwellian Plasma Electrons

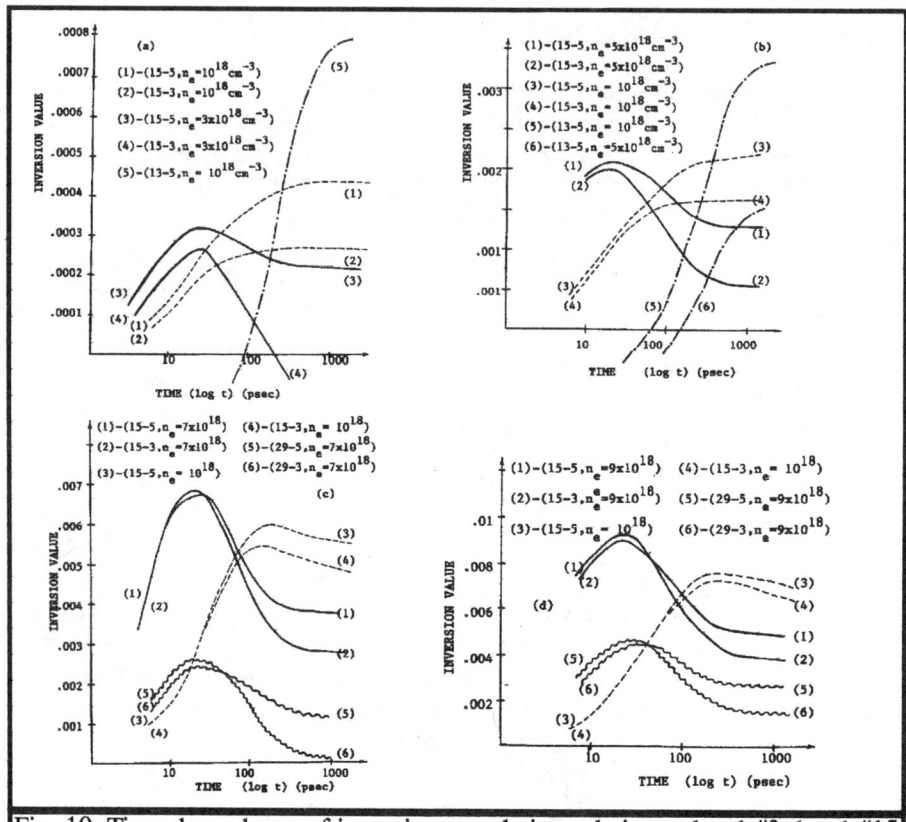

Fig. 10. Time dependence of inversion population relative to level #3, level #15 ($\lambda = 430$ Å) and level #29 ($\lambda = 158$Å) for several values of electron density. (a) $T_e = 75$ eV, (b) $T_e = 150$ eV, (c) $T_e = 450$ eV, and (d) $T_e = 1000$ eV.

REFERENCES.

1. Wan, A.S., Da Silva, L. et. al. "Development of X-ray Laser Architectural Components", presented at *4-th Int.Colloq. on X-ray lasers*, Williamsburg, VA, May 16-20, 1994.
2. Rocca, J.J. et. al., *Phys. Rev.* **E47**, 1299-1311 (1993).
3. Whitney, K.G., Pulsifer, P.E., *Phys. Rev.* **E47**, 1968-1976 (1993).
4. Apruzese, J.P., Davis, J., *Phys. Rev.* **A28**, 3686-3688 (1983).
5. Apruzese, J.P. et. al., *Phys. Rev.* **A39**, 5697-5703 (1989).
6. Davis, J., Clark, R.W., Giuliani, J.L. "Ultrashort Pulse Laser Produced Plasmas", presented at *4-th Int. Colloq. on X-ray Lasers*, Williamsburg, VA, May 16-20, 1994.
7. Ivanov, L.N., Ivanova, E.P., Knight, L.V., *Phys. Rev.* **A48**, 4365-4377 (1993).
8. Ivanov, L.N., Ivanova, E.P., Knight, L.V., *Phys. Rev. E*, (in press).

X-Ray Laser Research Using Z Pinches

J. P. Apruzese

Radiation Hydrodynamics Branch, Plasma Physics Division
Naval Research Laboratory, Washington, D. C. 20375

Abstract. The Z pinch offers both significant advantages and disadvantages when considered as an x-ray laser source. To date, three demonstrations of x-ray fluorescence using resonant photopumping, including one population inversion, have been reported in experiments conducted with Z pinches. Thus far, it has not been possible to demonstrate the robust Ne-like laser on a Z pinch, probably due in part to instabilities which lead to unacceptable inhomogeneity of the plasma. Recent improvement in stability of Kr pinches, demonstrated by R. B. Spielman of Sandia National Laboratories, provides some hope that this problem may be overcome in the future. It is also possible that the intense soft x-ray radiation field in the vicinity of a Z pinch may be able to pump a recombination x-ray laser with gain lasting several ns.

INTRODUCTION

A Z pinch is created by passing a pulsed current, usually hundreds of kA up to 10 MA, through a plasma which collapses on itself due to the $\mathbf{J \times B}$ forces and forms a pinched column, typically 1-4 cm long. The Z pinch offers significant advantages as a source for potentially advancing the state of the x-ray laser art. The pulsed generators which power such pinches are energy-rich, with wall-plug efficiencies exceeding 10%. Their cost per Joule on target is typically at least an order of magnitude lower than that of high-power lasers. A lasing duration of tens of ns, which could well occur in a Z pinch laser (1), would ease difficulties in cavity construction. Unfortunately, there are also serious disadvantages associated with such devices. They are subject to current-driven as well as classical hydrodynamic instabilities. This can result in a highly inhomogeneous medium, compromising or destroying any gain with deleterious refraction. When gas puffs are used as the source of the plasma, it can be difficult to avoid absorption of the laser lines by cold, opaque gas which can get between the spectrometer and the lasing medium. In this article the progress and prospects for Z pinches as an x-ray laser source are reviewed and examined.

NEON-LIKE KRYPTON EXPERIMENTS

The Ne-like ion has proven to be one of the most robust of the lasers scalable to the soft x-ray region. Currently, gain has been demonstrated for atomic numbers ranging from 18 (argon) to 47 (silver), and at least a dozen elements in between. An obvious candidate for a gas-puff Z pinch laser is krypton (2), with an atomic number of 36. In the mid-to-late 1980's the double annulus gas puff had produced the most stable Z pinches. At Physics International Corp. (PI), experiments were conducted (3) on the 3 - 4 MA Double EAGLE device, using a 3 cm long diode. In the resulting implosions, about half the 3 cm column appeared stable for about 5 ns. The laser lines, expected to lie between 17 and 19 nm, could not be identified in time-integrated axial XUV spectra. However, as revealed by keV spectra, the ionization state (mostly F- and Ne-like ions) was very similar to that of experiments at Lawrence Livermore National Laboratory (LLNL) where lasing in Se had occurred. Moreover, it is known from absolute measurements of the keV x-ray yield as well as certain density-sensitive line ratios that the *average* electron density in the pinch is somewhere between 2×10^{20} and 10^{21} cm^{-3}, just what is needed for lasing. Why could lasing not be observed? It is possible that continuum emission time-integrated over tens of ns swamped a laser line if gain only persisted for a few ns. Perhaps the density nonuniformities, seen on pinhole images even in the "stable" regions, produced unacceptable refraction. Another possibility is that cold gas from the initial puff absorbed the XUV laser lines. This would argue for doing an experiment using wires as the load, with zinc (Z=30) as one of several good candidates.

In contrast to the puff-on-puff experments of several years ago, there is now considerable belief in the Z-pinch community that a uniformly filled puff rather than two annuli, is a better bet for a stable implosion and on-axis assembly. A striking experimental support for this thesis has been provided by R. B. Spielman of Sandia National Laboratories (SNL)(4). Time-integrated and time-gated framing keV x-ray images of an imploded Kr puff have been taken on Saturn, a 20 TW, 10 MA accelerator which is the world's most powerful Z pinch driver. The images show excellent linearity and simultaneous assembly, without "zippering".

RESONANT PHOTOPUMPIMG

One method which might surmount some of these difficulties is to use the "unstable" pinch's copious x-rays to pump another plasma. The Na/Ne system, a resonant line coincidence, was proposed in the 1970's (5), and later analyzed by others (6,7,1). The excellent wavelength match between the 1-2 resonance line of He-like Na and the 1-4 line of He-like Ne allows selective pumping of the 4p singlet level of Ne IX, possibly leading to gain in some 4-2, 3-2, and 4-3 singlet transitions. The optimum neon ion density is $\sim 10^{18}$ cm^{-3}, and it is best to keep the neon as cold as possible while preserving the He-like ground state as the dominant species. He-like Ne can be prevalent at 30 eV or lower if there is significant photoionization of Ne VIII by soft x-rays.

Experimental investigation of the Na/Ne system on pulsed-power devices began at the Naval Research Laboratory (NRL) on the 1.2 MA Gamble-II generator (8). Total Na pump line powers of 25 GW were reproducibly achieved, and evidence for fluorescence, in the form of enhancements of the γ to β ratio with the Na pump on, was seen. Later, PI developed a pure Na wire extrusion technology (9) and was able to increase the pump line power to 150 GW at a current of 4 MA. A collaboration among SNL, NRL, and PI resulted in an experiment which employed a static fill Ne gas cell located outside the return current posts on Saturn. An array of 16 Na wires was imploded 2 cm from the Ne cell, typically giving 6 kJ total energy in the pump line and 200 GW or more peak power. Initial diagnostics included time-resolving and time-integrating crystal spectrometers to look at the Ne K shell resonance lines. Later, XUV McPigs and Hettrick time-gated spectrometers were added.

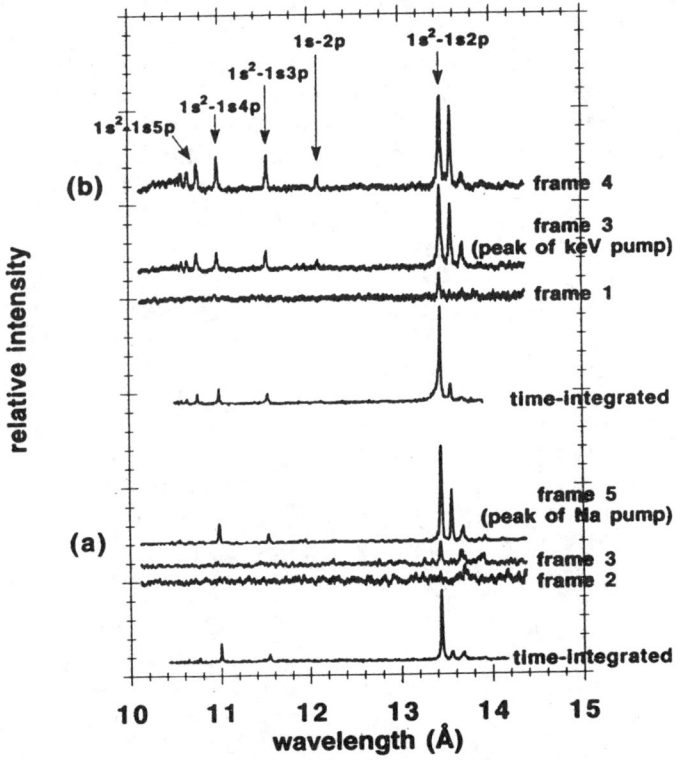

FIGURE 1. From Ref. 10, time-integrated and time-resolved Ne spectra for (a) a typical Na-pinch-Ne-target shot and (b) the Mg-pinch-Ne-target "null" shot. Time increases with increasing frame number.

Figure 1, from Ref. 10, shows time-resolved end-on (frames are 5 ns apart) and time integrated side-on spectra of Ne pumped by both Na and Mg pinches on Saturn. Note that the 1-4 line when pumped by Na is about twice as bright as the 1-3 line. Models developed at SNL, NRL(6), and LLNL(1) agree that these spectra indicate an inversion of the 4p singlet with respect to the 3p singlet level as well as the singlet 4f with respect to 3d level. To the author's knowledge, this is the only demonstration of a population inversion in the soft x-ray region pumped by resonant photoexcitation. It is encouraging also in the sense that only 1% of the sky was filled by Na radiation as seen from the Ne plasma, leaving much room for increasing the pumping rate via improved target design. By contrast, when a Mg pinch is used as the pump for Ne, the He-like Ne resonance line series shows a more normal appearance, with the 1-3 and 1-4 lines about equal, and a strong n=5 line which contrasts with the Na pumped Ne where the 1-5 line is absent. Three-body recombination probably plays a strong role in populating the Mg-pumped Ne, in contrast to the Na-pumped Ne. Three-body rates vary approximately as $n^{5.33}$ for this system, resulting in strong higher Rydberg lines. The Mg lines are of sufficient energy to photoionize He-like Ne and, in corroboration, the Ly α line of Ne X is seen only when Mg is used as the pump. These aspects of the spectra verify the photopumped inversion and lack of recombination when Na pumps Ne, as well as demonstrate the importance of recombination when Mg is the pump. Calculations by the author indicate that the intense soft x-ray radiation within 2 cm of a Saturn pinch may be capable of pumping, via photoionization, Balmer-α recombination lasers in H-like C and B. Further experimental work in the XUV (11) resulted in the detection of Ne IX 4-2 and 3-2 lines whose relative intensities strongly corroborate the inversion first seen in the keV spectra.

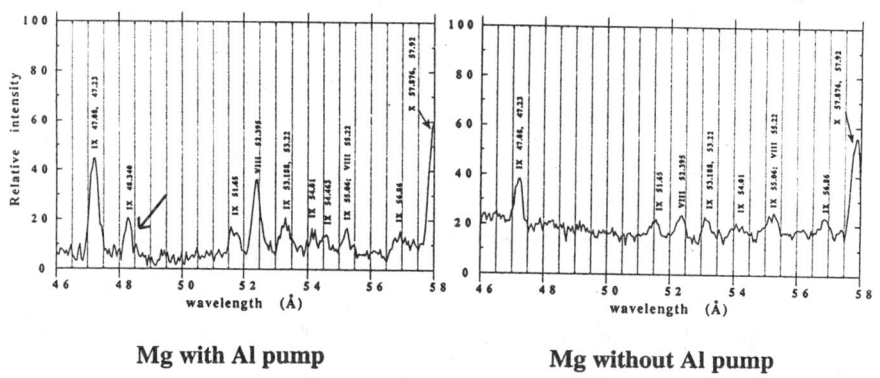

Mg with Al pump   **Mg without Al pump**

FIGURE 2. From Ref. 13, shows that the Mg pumped line at 4.834 nm appears only when the Al pump plasma is present.

In 1984 Krishnan and Trebes pointed out the existence of a class of line coincidences (12) which could pump lasers in Be-like ions ranging from C III to Mg IX. The pumped line is the $2s^2$-2s4p singlet transition. The Al XI-Mg IX system has four times the quantum efficiency of the Na X-Ne IX scheme and uses materials which are more suitable for target fabrication. However, the ion fraction obtainable in the Be-like ground state is a factor of five or so smaller than can be produced in the He-like ground state. The best laser line, as in the Na X-Ne IX system, is the

3d-4f singlet transition, with similar wavelength and oscillator strength.

Recently, Qi, Hammer, Kalantar, and Mittal have employed the 400 kA LION generator at Cornell University to experimentally investigate the Al-Mg system and have succeeded in demonstrating fluorescense (13) of the pumped Mg line using three different initial wire configurations The simplest of these mounts two 50 μm diameter Al wires 0.2 cm apart and a single 75 μm diameter Mg wire 1.0 cm away from them. As shown in Fig. 2, the pumped Mg line at 4.834 nm can only be detected when the Al pump radiation is present. When the Al wires are replaced by Mg (right hand spectrum), the pumped line is not seen. It is evidently too weak to be detected via pure thermal emission in the Mg plasma. Further theoretical work (14) suggests that this laser could be demonstrated at a relatively modest current of 800 kA. Baksht and co-workers (15) have also explored this system using an Al liner at currents of up to 1.5 MA.

ACKNOWLEDGMENT

The x-ray laser research at NRL was supported by BMDO/T/IS and ONR.

REFERENCES

1. Nilsen, J. and Chandler, E., Phys. Rev. A **44**, 4591-4598 (1991).
2. Thornhill, J. W., Apruzese, J. P., Davis, J. and Clark, R. W., J. Appl. Phys. **71**, 4671-4677 (1992).
3. Krishnan, M., Nash, T., LePell, P., and Rodenburg, R., "Krypton on krypton Z pinch x-ray laser experiment," Rep. PIT-87-02 (Physics International Corp., San Leandro, CA, 1987).
4. Spielman, R. B., private communication (1994).
5. Vinogradov, A. V., Sobel'man, I. I., and Yukov, E. A., Sov. J. Quant. Electron. **5**, 59-63 (1975).
6. Apruzese, J. P., Davis, J., and Whitney, K. G., J. Appl. Phys. **53**, 4020-4027 (1982).
7. Hagelstein, P. L., Plasma Phys. **25**, 1345-1367 (1983).
8. Stephanakis, S. J. et al., IEEE Trans. Plasma Sci. **16**, 472-481 (1988).
9. Deeney, C. et al., Appl. Phys. Lett. **58**, 1021-1023 (1991).
10. Porter, J. L. et al., Phys. Rev. Lett. **68**, 796-799 (1992).
11. Nash, T. J., Spielman, R. B., Vargas, M., and Ruggles, L., "Photopumped X-Ray Laser Research on Saturn", in **Ultrashort Wavelength Lasers II**, S. Suckewer, ed., *Proc. SPIE 2012*, 120-130 (1994).
12. Krishnan, M., and Trebes, J., Appl. Phys. Lett. **45**, 189-191 (1984).
13. Qi, N., Hammer, D. A., Kalantar, D. H., and Mittal, K. C., Phys. Rev. A **47**, 2253-2263 (1993).
14. Qi, N., Hammer, D. A., and Apruzese, J. P., J. Appl. Phys. **74**, 4303-4309 (1993).
15. Baksht, R. B. et al., Sov. J. Plasma Phys. **18**, 356-360 (1992).

Heavy Ion Beam Pumping of Charge Transfer Lasers

A. Ulrich, R. Gernhäuser, W. Krötz, M. Salvermoser, J. Wieser
Fakultät für Physik E12, Technische Universität München
James Franck Str. 1, D-85747 Garching, Germany

D. E. Murnick
Department of Physics, Rutgers University
101 Warren Street, Newark, NJ 07102, USA

> In situ production of multiply charged ions by high energy heavy ion beams in target gas mixtures is proposed for application in charge transfer pumping schemes of short wavelength lasers. Charge transfer reactions in Ar-Cs mixtures were studied as a model system using time resolved optical spectroscopy.

The excitation of gas targets with high energy heavy ion beams leads to the formation of a nonthermal plasma. Therefore it is of interest to study this excitation method for the development of new laser systems. Population inversion in an ion beam induced plasma can be obtained from direct collisional excitation [Ulr91]. At the same time the plasma has a pronounced "afterglow" character. Collisional excitation and ionization leads to the formation of multiply ionized target species in a cold, dense environment, a situation which is normally only encountered in the cooling phase of a plasma. Heavy ion beam pumped laser systems based on energy transfer from excimer molecules, which are formed under such conditions, are effective and have low threshold pumping powers in the infrared and visible spectral ranges [Ulr83, Ulr93, Ulr94].

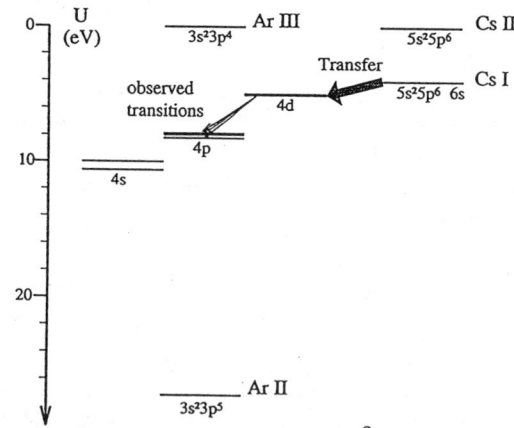

Fig. 1: Schematic level diagram for the Ar^{2+} + Cs pumping scheme.

404   © 1994 American Institute of Physics

We are studying a new approach with promising aspects for extension towards shorter wavelengths. It combines the high multiple ionization cross sections of heavy-ion target-atom collisions with the selectivity and high cross sections of low velocity charge transfer reactions. An intense, well collimated high energy heavy ion beam is sent into a cold, dense gas target producing q- times ionized target species. The target gas is composed of two components. Slow recoil ions in the charge state q of one component (A) are used as acceptor ions for a charge transfer process with neutral species of the other component (B):

$$A^{q+} + B \rightarrow A^{(q-1)+**} + B^+ \qquad (1)$$

The acceptor ion A^{q+} and the donor atom B are selected in such a way that the outer electron of B is transferred into an excited level of an ion $A^{(q-1)+**}$ in an energetically allowed over barrier process with a high cross section. If the $A^{(q-1)+**}$ ion has no direct optically allowed transition to the ground state it can be used as an active laser species for (quasi-) dc laser operation.

Charge transfer has previoulsy been discussed as a laser scheme for short wavelength lasers [Pre71, Vin73, Ols76]. Ion beams sent into neutral gas and ions from a laser plasma expanding into neutral gas were studied for the process. Both techniques were critically reviewed in a paper by Bunkin and coworkers and "in situ" production of the ions by electron collisions was proposed [Bun81]. Beyer and Mann have studied heavy ion beam induced charge transfer processes in gas targets for highly charged recoil ions and measured x-ray emission [Bey84]. They mention potential laser application but no level schemes were selected by criteria necessary for laser action.

Recently we have performed a spectroscopic experiment using an Ar^{2+} + Cs reaction leading to selective population of 4d levels in Ar^+ in an ion beam excited target. A schematic level diagram for the pumping scheme is shown in Fig.1. Population inversion between 4d and 4p levels in ArII could be verified from absolute intensity measurements of 4p-4s transitions. Rate constants to model the laser scheme could be measured or were available from the literature. A pulsed 100MeV ^{32}S beam with 2ns pulse width was used for the excitation of 250hPa Argon gas. The cross section to produce Ar^{2+}-ions in the region of the target which was studied (30MeV projectiles) is $1.7\times10^{-15}cm^2$ [Coc79]. Cesium vapor could be added as an electron donor by heating the target cell connected with a Cesium reservoir. Charge transfer of the outer 6s electrons of Cs into 4d levels of Ar^+ was predicted using a relation given by a transfer model [Bey84, Ols76]:

$$E_{ao} < J_b \, [(1-q) / \{q-(q^{0.5}+1)^2\} + 1]. \qquad (2)$$

Tabulated energy levels E_{ao} were used for the (undisturbed) A^+ ion (here Ar^+). The binding energy of the outer electron of B (here Cs) is J_b. The energies are measured with respect to the ionization limits. The internuclear distance R at which the transfer reaction (1) occurs is given by

$$R = (q-1)/(E_{ao} - J_b) \qquad (3)$$

in this model. A cross section ($\sigma_T = R^2\pi$) of $6\times10^{-14} cm^2$ and a rate constant of $1.3\times10^{-9} cm^3/s$ was predicted for the Ar^{2+} + Cs transfer process using an average thermal velocity of 450m/s to derive the rate constant from the cross section.

The transfer process was observed by time resolved optical spectroscopy and was in excellent agreement with the predictions. A huge intensity increase was found for optical transitions starting at a group of 4d levels of Ar^+ lying within an energy band of 200meV (Fig.2). A rate constant of $(1.4\pm 0.2)\times10^{-9} cm^3/s$ was determined from the decay times of the strong 349.15nm $4d^4D_{7/2}-4p^4P^o_{5/2}$ transition measured for various Cs concentrations. The kinetics of the pumping process can be modeled analytically. It is planned to use a 15MeV/u (3GeV) ^{209}Bi beam from the UNILAC heavy ion linear accelerator at Gesellschaft für Schwerionenforschung (GSI) Darmstadt to demonstrate laser action on the 349.15nm line. A mode lock resonator will be installed in which the round trip time is synchronized with the 27MHz repetition rate of the beam pulses. An optical gain of the order of 5% per pass can be predicted for this experiment with target parameters similar to those used for the spectroscopic experiment described above and 2×10^5 Bi ions per beam pulse. The cross section to form Ar^{2+} ions with these Bi projectiles is of the order of $10^{-14} cm^2$ [Kel85].

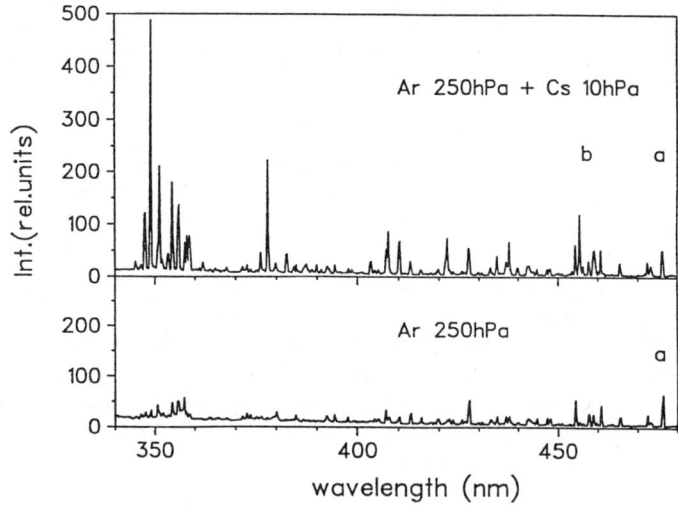

Fig. 2: Wavelength spectra of heavy ion beam excited Argon and Argon Cesium mixture. A group of lines (mainly on the left side) starting at 4d levels show a huge intensity increase when Cs is added. The line (a) is directly excited in projectile-target collisions and shows no change. Cesium resonance lines are indicated by (b).

To proceed towards shorter wavelengths with the pumping scheme described here the processes involved, their limitations, and other mechanisms such as photoabsorption have to be considered. Transitions at shorter wavelengths will require higher charge states q of the acceptors A^{q+}. Multiple ionization cross sections for producing these ions in heavy ion collisions are available up to highest charge states and relativistic energies of the projectiles [Ber93]. Argon, for instance, is singly ionized by 120MeV/u U^{91+} with a cross section of $13 \times 10^{-16} cm^2$, doubly ionized with $8 \times 10^{-16} cm^2$ and Ar^{5+} is produced with a cross section of $1 \times 10^{-16} cm^2$. The range of the projectiles and angular scattering has to be considered in addition to the ionization cross section for a practical laser design.

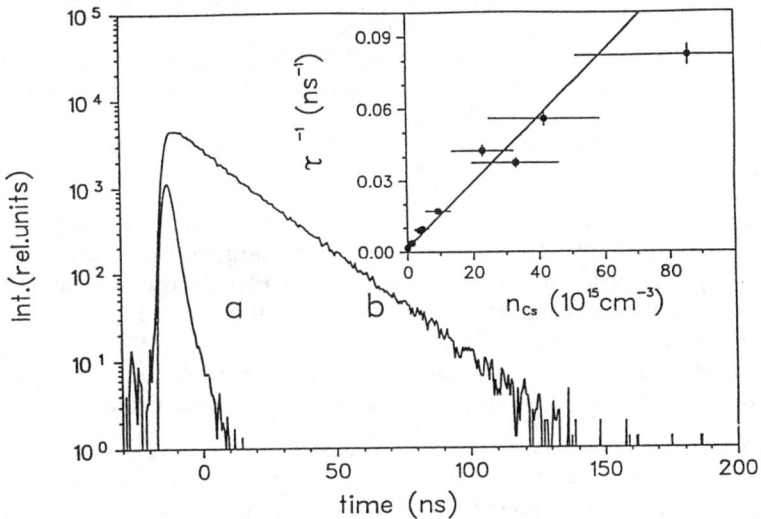

Fig.3: Time spectra measured on the 349.15nm line. Pure Argon is shown as curve a. When Cs is added (b) the decay time is determined by the transfer process. The transfer rate constant can be derived from the inverse decay time constants shown in the inset.

The highest ion beam current densities available must be used to produce high densities of laser species in the target medium. A limitation is given by the single ionization process of the donor atoms which should not lead to ionization densities of more than typically 50% to keep enough neutral atoms in the target for the transfer process. A projectile flux of the order $10^{20} cm^{-2} s^{-1}$ can be expected for the high energy heavy ion synchrotron SIS at GSI after an intensity upgrade (100ns pules, 10^{11} projectiles per pulse, 1mm^2 beam diameter assumed). With single ionization cross sections of the order of $10^{-14} cm^2$ [Ber93] only 10% of neutral target atoms will be ionized. Contribution to the ionization due to the flux of secondary electrons produced in the projectile target collisions must also be considered. The overall energy deposition is of the order of 1.6 eV/atom so that thermal ionization may not be negligible.

Charge transfer processes with higher charge states of the acceptor ions lead to situations where the level scheme of the acceptor is such that more strongly bound electrons in donor atoms can be used. This relaxes the technical problems which are encountered with loosely bound chemically active electrons as in the case of alkali metal vapors. Xenon with $J_B=12.13eV$ is a good candidate as a donor which may be used together with other multiply charged rare gas ions as acceptor species. When well matched acceptor donor system can be found (equality in (2)) the cross sections for charge transfer vary with q and J_B as $\pi(4q-4q^{0.5}+1)/J_B^2$.

Radiative lifetimes and collisional quenching rate constants of the levels involved are needed to calculate stimulated emission cross sections, population densities and the full time dependent kinetics. Measurements using smaller accelerators such as the Munich Tandem van de Graaff can provide these data [Krö94]. Loss processes for the acceptor ions are recombination with free electrons and the formation of molecular ions. In the transfer scheme using Ar^{2+} ions described above Ar_2^{2+} is formed in three body collisions with a rate constant of $(1.46\pm0.12)\times10^{-30}cm^6/s$ [Krö91]. This loss process increases quadratically with the Argon density.

For predictions of optical gain, the line width of the laser transition has to be determined. The initial velocities involved are low. The kinetic energy of recoil ions is of the order of 0.1eV [Ull88]. Repulsion of the $A^{(q-1)+}$ and the B^+ ions after the transfer can lead to Doppler broadening. The potential energy of $Ar^+ + Cs^+$ at the transfer radius R is for example 1eV leading to Doppler broadening of the order of $\Delta\lambda/\lambda=10^{-6}$ when it is fully converted into recoil velocity. Other line broadening mechanisms such as collisional and Stark broadening will have to be studied for each specific case which is considered.

The neutral atoms in the laser medium needed for the charge transfer reaction make photoabsorption an important issue. Three levels of complications exist on the way towards shorter wavelengths: Both acceptor and donor component are transparent to the laser radiation, the donor component with its loosely bound outer electron is already absorbing the laser light and, finally, both acceptor and donor components have their first absorption edge at a wavelength longer than the laser light. In the first case there is no problem but one has to check if the laser line coincides with a resonace absorption line which tend to get dense towards the photoabsorption edge. Cesium, for example, is transparent for the 349.15nm ArII line (photoabsorption edge: 318.4nm, closest Cs absorption line is 6s-10p at 348.0nm). A way to expand the transparent region towards shorter wavelength is to use donors like Xe which have a photoabsorption edge around 100nm. Below this wavelength region the (nonresonant) photoabsorption σ_p has to be compensated by the (resonant) stimulated emission σ_s: $n\sigma_s>(n_A\sigma_{pA}+n_B\sigma_{pB})$. An optimized selection of the laser wavelength and the wavelength dependence of photoabsorption can help. Xenon, for instance, has a small photoabsorption cross section of $2\times10^{-18}cm^2$ between 20 and 35nm [Hud71].

In summary, a pumping scheme is proposed which makes use of the specific qualities of heavy ion beam interaction with matter and may lead to laser systems in the VUV and

XUV with the high repetition rates and homogeneous excitation provided by heavy ion accelerators.

This work has been funded by the German Ministry of Research and Technology, BMFT under contract No. 06TM353/5, GSI-Darmstadt, the Munich Tandem van de Graaff accelerator laboratory, and NATO.

[Ber93] H.E. Berg, Report: GSI-93-12, GSI, P.O. Box 110552, D-64220 Darmstadt
[Bey84] H.F. Beyer and R. Mann, Prog. Atom. Spectr. C, ed.: H.J. Beyer and H. Kleinpoppen, Plenum Press, New York, p. 397-458, (1984)
[Bun81] F.U. Bunkin, V.I. Derzhiev, and S.I. Yakovlenko, Sov. J. Q. El. 11, 981, (1981)
[Coc79] C.L. Cocke, Phys. Rev. A20, 749, (1979)
[Hud71] R.D. Hudson and L.J. Kieffer, Atomic Data 2, 205, (1971)
[Kel85] S. Kelbch, J. Ullrich, R. Mann, P. Richard, and H. Schmidt-Böcking, J. Phys. B18, 323, (1985)
[Krö91] W. Krötz, A. Ulrich, B. Busch, G. Ribitzki, and J. Wieser, Phys. Rev. A43, 6089, (1991)
[Krö94] W. Krötz, A. Ulrich, G. Ribitzki, J.Wieser, and D. E. Murnick, Hyperfine Interactions, in print
[Ols76] R.E. Olson and A. Salop, Phys.Rev. A14, 579, (1976)
[Pre71] L.P. Presnyakov and V.P. and V.P. Shevel'ko, JETP Lett. 14, 203, (1971)
[Ull88] J. Ullrich, M.Horbatsch, V. Dangendorff, S. Kelbch, H. Schmidt-Böcking, J. Phys.B21, 611, (1988)
[Ulr83] A. Ulrich, H. Bohn, P. Kienle, and G. J. Perlow, Appl. Phys. Lett. 42, 782, (1983)
[Ulr91] A. Ulrich, J. Wieser, R. Pfaffenberger, B. Busch, W. Krötz, H.-J. Körner, G. Ribitzki, and D. E. Murnick, Z. Phys. A 341, 111, (1991)
[Ulr93] A. Ulrich, B. Busch, W. Krötz, G. Ribitzki, J. Wieser, and D.E. Murnick, Lasers and Particle Beams 11, 509, (1993)
[Vin73] A. V. Vinogradov and I.I. Sobelman, Sov. Phys. JETP 36, 1115, (1973)

X-RAY LASERS ACTIVE MEDIA FORMATION BY PONDEROMOTIVE FORCE OF COHERENT OPTICAL PUMPING

V.V.Korobkin and M.Yu.Romanovsky.

General Physics Institute of the Russian Academy of Sciences Moscow, Russia 117942, Vavilov str., 38

Abstract. It is shown that the ponderomotive force of a powerful laser beam is capable to confine the plasma with electron density exceeding the critical density for the given radiation. The theory describing force and heat balances of the plasma together with the propagation of the laser radiation is developed. Possible applications for X-ray laser problems are discussed.

INTRODUCTION

First notes about the possibility of the plasma confinement by the ponderomotive force of the laser beam appeared about forty years ago [1,2]. It is in [3] that this process received its fundamental substantiation taking into account competitive effects; preliminary estimates of its use for the controlled thermonuclear fusion (CTF) were given there too. It was assumed in these investigations that the density of the plasma electrons was less than the critical one for the given wavelength of the laser radiation of the confining beam, i.e. that the radiation penetrated all plasma volume.

It seems that an interesting task would be to analyze, first of all from the point of view of possible applications, the case of the possible confinement of the plasma of supercritical electron density. If one uses the steadily confined plasma, consisting now of multiply charged ions, as the active medium of a plasma laser, the increase in the density of this plasma would lead to the augmentation of the power of coherent radiation.

The study presents the analysis of the hydrostatics of the plasma of supercritical density and its heat balance in the field of the ponderomotive force of the laser beam, together with the analysis of the propagation of the pumping beam. This will unable to give some quantitative results on the scaling of the hydrogen plasma for the plasma of multiply charged ions as the active medium of short-wavelength lasers.

THE CONFINEMENT OF THE PLASMA OF SUPERCRITICAL ELECTRON DENSITY

The equation of the local balance of forces in the confined plasma is of the following form [3]:

$$\frac{\partial p}{\partial x} = - \frac{n}{16\pi n_{cr}} \frac{\partial A^2}{\partial x}. \qquad (1)$$

Here p is the gas-kinetic pressure, A is the amplitude of the electromagnetic field in the laser beam, n is the electron density, $n_{cr} = m\omega^2/4\pi e^2$ is the critical electrons density, e and m are the charge and the mass of the electron respectively, $\omega/2\pi$ is the laser radiation frequency. It is assumed that the linearly polarized laser radiation propagates along z axis, the polarization vector is directed along x axis. (It should be noted that for plasma confinement inside the tube-like beam its polarization must be circular. Since the thickness of the transitional area from the maximum plasma density to the minimum one is usually much less than the radius of the plasma column [3], Eq.(1) describes well the situation in this case too).

Eq.(1) shows the dependence of n on A:
$$n = n_* \exp(-A^2/E_{cr}^2), \qquad (2)$$

$E_{cr}^2 = 32\pi T n_{cr}$, T is the plasma temperature (which is to be maintained; it is expressed hereby in energy units). The value n_* is the density of the plasma in the area that the field practically does not reach. If $n_* = n_o < n_{cr}$, $A \to \infty$, at $x \to +\infty$ A tends towards its nonperturbed value E_{max}, and $n \to n_{min} = n_o \exp(-E_{max}^2/E_{cr}^2)$. Thus, for isothermal (stationary [3]) process of confinement none of the fields expels the plasma from the plasma volume completely. Therefore the physically exact statement of the problem will be only the statement assuming that throughout the space there is a (sufficiently rarefied) plasma having the temperature T and some density n_{min}. There are laser beams (a tube-like beam), unlimited from one of its sides. They form a kind of closed (or semi-closed) configuration, confining inside it the plasma with density $n > n_{min}$, the maximum (unperturbed) amplitude of beams E_{max} being connected with the n_{min} by the relation given above. Our aim is to find what is the maximum possible density of this confined plasma n_o?

For that, besides the local balance of forces (1) the general balance of forces must be fulfilled, i.e. the total pressure of plasma must be balanced by the resulting ponderomotive force. In (1) we'll assume $n = p/2T$ (the isothermicity of plasma is implied in the statement of the problem itself) and obtain the dependence of the pressure on the field amplitude:
$$p_{cr}/p_o = \exp(-A^2/E_{cr}^2) \qquad (3)$$
Here p_o is the plasma pressure at $A = E_{min}$ ($A = 0$ at $n_o > n_{cr}$). The relation (3) shows that at $A \gg E_{cr}$ the total balance of forces is fulfilled at any value of p_o, and n_o can exceed n_{cr} if
$$n_{cr}/n_o \geq \exp(-A_{max}^2/E_{cr}^2), \text{ or } A_{max}^2 \geq E_{cr}^2 \ln(n_o/n_{cr}). \qquad (4)$$
Thus, if the maximum amplitude of the field in the laser beam is several times greater than the critical value, the density of the confining plasma can be many times greater than the critical

density due to the logarithmic factor in (4). In (3) only the case $n_o < n_{cr}$ was considered. In the model used in the case $n_o \geqslant n_{cr}$ it is always assumed that there is a sharp boundary between the unperturbed plasma of density n_o and the plasma of density $< n_{cr}$, where the field exists; such a model seems to be realistic due to a most inconsiderable penetration of the field into the plasma of supercritical electron density (see below).

As it has been already noted, no field expels the isothermal plasma completely. The minimum density of plasma n_{min} is

$$n_{min} = n_o \exp(-E_{max}^2/E_{cr}^2). \qquad (5)$$

As the plasma mass is limited, i.e. it cannot occupy all the space with density n_{min} (in practice it is convenient to work in the vacuum with the targets solid-state and of great densities ones, see (3) and below), one should expect the outflow of plasma of density n_{min} into the vacuum along the axis x. This will be taken into account in the course of scaling.

PROPAGATION OF THE RADIATION IN CASE OF $n_o \geqslant n_{cr}$.

In (3) the authors considered the propagation of the radiation in plasma along the axis z, perpendicular to the plasma density gradient, the maximum of density being less than n_{cr}. The wave character of the radiation is kept throughout all the space (even despite strong absorption), the changes of the field amplitude along the z axis were rather slow and therefore the use of the parabolic nonlinear equation for the propagation of radiation was justified. In the case under consideration the real part of the dielectric constant $\mathcal{E}$ can be negative and its changes along the x axis - rather sharp. Therefore the analysis should start from the Maxwell's equations, where only the following simplifications are allowed:

- the plasma is electrically neutral, therefore $\mathrm{div}\vec{D} = 0$ ($\vec{D}$ is the vector of magnetic induction);
- the plasma is non-magnetic, $\mu = 1$ (μ is the magnetic permeability).

We shall refer only the real part of the polarizability to $\mathcal{E}$, and all the real losses will be taken into account through the real conduction of the plasma σ, then the Maxwell's equation can be reduced to the equation for the value of the vector of the electromagnetic field intensity E, which will be initial for the analysis:

$$\Delta \vec{E} - \frac{1}{c^2}\frac{\partial^2 \mathcal{E}\vec{E}}{\partial t^2} - \nabla(\vec{E}\nabla\ln\mathcal{E}) = \frac{4\pi\sigma}{c}\frac{\partial \vec{E}}{\partial t}. \qquad (6)$$

Here c in the unperturbed velocity of light, the expression cannot be reduced to the wave equation due to the third term in the left-hand side (6). For the plasma under consideration directly from (2) it follows that

$$\frac{\partial \mathcal{E}}{\partial x} = \frac{1 - \mathcal{E}}{16\pi T n_{cr}} \frac{\partial A^2}{\partial x}. \tag{7}$$

We will seek stationary solutions of Eq.(6) of the form

$$\vec{E} = A(x,z)\exp[-ik(x)z + i\omega t], \tag{8}$$

(k is the wave vector) taking into consideration that the changes of $\mathcal{E}$ along the axis z are much slower than along the axis x. Since

$$\vec{E} = iE_x = iE;$$

then $\nabla \mathcal{E} = i\partial \mathcal{E}/\partial x$, and we must choose the dependence of the wave vector k on $\mathcal{E}$ of the form

$$k^2(x) = \omega^2 \mathcal{E}(x)/c^2,$$

as to describe the moment of the transition through the density limit $n = n_{cr}$, the spatially inhomogeneous dumping wave. In effect, when $n \geqslant n_{cr}$, the dumping factor of this wave inside dense plasma will be $|k| = =(n_o/n_{cr})^{1/2}\omega/c \gg 1/\lambda$.

Substituting (7) into (6) and using the form of k, one can obtain the nonlinear equation for A. It is extremely complex by form and it is cannot be directly reduced to the Kuramoto-Tsuzuki equation, as it was done in [3]. The simplifications can be reached in case of $A_{max} \gg E_{cr}$. Indeed, for the fulfillment of the equation of the total balance of forces (4), for example, when using the hydrogen plasma of solid-state density ($n_o = 5 \cdot 10^{22}$ cm^{-3}) as the active medium and the radiation of the Nd-laser as the confining radiation ($\lambda = 1,06$ μm), $A^2_{max} \geqslant E^2_{cr}\ln(n_{cr}/n_o) \simeq 4E^2_{cr}$, and for the same density in case of using the radiation of the CO_2-laser $A^2_{max} \geqslant 8,5 \cdot E^2_{cr}$. Later we will see that the conditions where the self-focusing does not develop still more this requirement ($A^2_{max} \gg E^2_{cr}$) to the amplitude of the confining radiation, and the introduction of restriction on A having the form $A \gg E_{cr}$ will lead only to the loss of the accuracy of the solutions being obtained (6) in the thin layer ($<\lambda$) along the axis z near the dividing line of the plasma.

When calculating in (6) the derivative $\partial^2 E/\partial x^2$, together with usual terms the terms of the type of $z(\partial k/\partial x)(\partial A/\partial x)\exp[ik(x)z - i\omega t]$ form owing to the dependence of k on x. From (7) $(\partial k/\partial x) \sim \exp(-A^2/E^2_{cr})$, and the usual terms, corresponding to the parabolic equation, do not contain this factor. Since in our practice we will have to deal with ordinary Gaussian beams, we cannot take z (z is here the confinement length L_{conf}) exceeding L_d - the length of the diffraction divergence of the "outer" ("non-working") edge of the beam (the "working" edge confined the plasma). Therefore the conditions $z\partial k/\partial x \ll k$ and $z^2 \partial^2 k/\partial x^2 \ll k$ are satisfied at $z \ll L_d$, or $L_{conf} \ll r_0 \exp(A_{max}^2/2E_{cr}^2)$, where r_0 is the minimum size of the laser beam (the thickness of the neck of the cylindrically focused or the the thickness of the "wall" of the tube-like beam [3]). Later we will show that this condition follows from the impossibility to use in the experiment all the energy of the laser beam (and nor even the most part of it) for the heating of the plasma to be confined.

Expanding the term with $\ln\mathcal{E}$, using (7) and denoting $a = A/E_{cr}$, we will obtain:

$$2ik(x)\frac{\partial a}{\partial z} + \frac{\partial^2 a}{\partial x^2} +$$

$$+ik(x)\frac{1-\mathcal{E}}{3\mathcal{E}}\frac{\partial a^3}{\partial x} + \frac{1-\mathcal{E}}{3\mathcal{E}}\frac{\partial a^3}{\partial x^2} - \frac{1-\mathcal{E}}{3\mathcal{E}^2}\frac{\partial a^3}{\partial x}\frac{\partial a^2}{\partial x} = \quad (9a)$$

$$= -\frac{4\pi i \sigma \omega}{c^2} a$$

If $a^2 \gg 1$, it is easy to show that $\partial a/\partial x \approx ik\mathcal{E}/a$, the summed up explicit nonlinear terms will give together $(1-\mathcal{E})k^2 a$, and (9a) will reduce to the form:

$$2ik(x)\frac{\partial a}{\partial z} + \frac{\partial^2 a}{\partial x^2} + \exp(-a^2)k^2 a = -\frac{4\pi i \sigma \omega}{c^2} a. \quad (9b)$$

As at $n_0 > n_{cr}$ $1 - \mathcal{E} = \exp(-a^2)$. Now we can proceed to the Kuramoto-Tsuzuki equation [4,5]. Analyzing it, we will not even specify the form of the term $4\pi i \sigma \omega^2/c^2$ (i.e. the form of the laser radiation absorption coefficient δ, we shall only keep in mind that this coefficient is a function of x). It is only for computer calculations that the standard form of δ will be used [6]. For $n_0 > n_{cr}$ δ is assumed to be equal ∞.

The relation (9b) still isn't the Kuramoto-Tsuzuki equation, since k depends on x. However, with our assumption $a^2 \gg 1$ it can be easily shown that $k \simeq k_0 \simeq \omega/c$, and thus we have again reduced the problem to the well-known one [4], from here on we will omit the index on k. Eq. (9b) differs from that derived in [3], since in that work the expansion into a power series a-1 was performed and, besides, Eq.(9b) is not dimensionless in our case. All the considerations about the physical self-similarity of the problem that were presented in [3] are valid in the case under consideration too. Let us assume that, in conformity with [5],

$$a = R(x) \exp[i\Omega z + ib(x)].$$

We obtain the system of equations
$$R''/R - b'^2 + k^2\exp(-R^2) = 2k\Omega,$$

$$2b'R' + b''R' + 2k\delta R = 0. \tag{10}$$

(the sign " ' " signifies the differentiation with respect to x).

Let us make some conclusions about the value of Ω, keeping in mind that in the case under consideration the sufficient criterion of confinement is the condition $\Omega = 0$ (as in [3]; the necessary criterion is the relation (4)). When $x \to \infty$, $R'' \to 0$ and $b'^2 \to -2k\Omega + k^2\exp(-A_{max}^2)$ (of course, $A_{max} = R_{max}$). Since Ω is not negative [5], $0 \leq \Omega \leq k^2\exp(-A_{max}^2)/2$. The second inequality is much stronger than the condition $L_{conf} \ll r_o \exp(A_{max}^2/2E_{cr}^2)$, already used (we shall recall that $\Omega^{-1} = L_{conf}$). Therefore, in effect for the plasma with electron density exceeding the critical density the confinement condition is fulfilled always when $a_{max} \gg 1$.

Let return to the case $\Omega = 0$. From the second equation of system (10), if multiplying all the terms by R, it follows that the value $k\delta R^2$ is the total derivative (and R is the integrating factor):
$$-k\delta R^2 = (b'R^2)'/2. \tag{11a}$$

The following is valid both for the beam focused by a cylindrical lens and for the tube-like beam with its "wall" thickness much less than the tube radius (i.e. in the one-dimensional geometry which will be actual in the majority of cases):
$$b'R^2 = -2\int_0^x k(x)\delta(x)I(x)dx/I_{cr}. \tag{11b}$$

Here I is the laser radiation intensity, $I_{cr} = 4n_{cr}cT = cE_{cr}^2/8\pi$ is the critical intensity.

At $\Omega = 0$ the absorption of the pumping radiation occurs uniform along z - axis. The integral $\iint_\infty \delta(x)I(x)dxdy$ describes the power of the laser radiation, absorbed per unit length from 0 to x in the x direction and from 0 to y in the y direction. If we introduce the total size of the beam in the y direction L_y (for the mentioned tube-like beam it is the length of the tube circle) and the effective length of absorption L_a, the total power of the radiation, absorbed by the plasma between 0 and x coordinates along x - axis P_a within this length is
$$P_a = L_a L_y \int_0^x \delta(x)I(x)dx.$$

If for the upper limit of integration one takes the coordinate x_m, where the field amplitude reaches its maximum (and further in the x direction it does not change, roughly speaking), then P_a will be the total power absorbed within the effective length throughout all transitional area (from 0 to x_m in the x direction). In the area $x > x_m$ the specific absorption will be already inconsiderable due to small density of the residual plasma still not expelled. For real beams, finite over x, such P_a cannot exceed the beam power P, in actual practice it must be less due to the mentioned impossibility to absorb all the laser energy - a part of it (the most part) will be spent to the confinement.

From the first equation of system (10) it follows that

$$\int_0^{x_m} \delta I dx = P/2L_a L_y = -I_{max} b'/2k^* \approx I_{max} k_0 \exp(-a_{max}^2/2)/2, \quad (12)$$

(k^* is some mean value of k within the interval from 0 to x_m in the x direction, $I_{max} = cE_{max}^2/8\pi$) since $b(x = x_m) = -k_0 \exp(-a_{max}^2/2)$. The sign "minus" is chosen taking into account that in (12) all the values except the unknown b' are positive. In real practice, with all our assumptions, one can state with a good accuracy that $k^* = k_0$. Thus,

$$P_a(x = x_m) \simeq I_{max} L_a L_y \exp(-I_{max}/2I_{cr}). \quad (13)$$

But $P > P_a(x = x_m)$, and, since for the real Gaussian beam of zero mode $P = \sqrt{\pi} L_y r_0 I_{max}$, then

$$L_{conf} < L_a \leq \sqrt{\pi} r_0 \exp(I_{max}/2I_{cr}), \quad (14a)$$

Since the confinement length cannot exceed some part of the inserted length of absorption for the reasons mentioned above, then more laser radiation than there exists in reality will be absorbed. The intermediate inequalities can be omitted, and in this case

$$L_{conf} < \sqrt{\pi} r_0 \exp(-I_{max}/2I_{cr}). \quad (14b)$$

The ratio of the experimentally realized L_{conf} to L_a will give the part of the "productively" spent (i.e. for heating) laser pumping radiation. The relations (14) show that the more is I_{max}, the more is the effective length of absorption, since in this case the plasma is more intensively expelled and the volume occupied by a comparably dense plasma, where the most part of absorption just takes place, reduces.

It should be noted that system (10) at $\delta = 0$ is a particular case of the equation of nonlinear optics and describes the radiation refraction. And, analogously/similarly, the more is I_{max}, the less this refraction turns out to be, since, in this case too, the plasma is expelled from the beam better, and (it is) a comparably smaller part of it (that) refracts in this case.

The numerical solutions of system (10) are presented on Fig.1,2. The dependence of the absorption coefficient δ on n and, consequently, on A were chosen in the form [6] at T = 5 keV, i.e. only the bremsstrahlung was taken into account. For the intensities of the laser radiation up to 10^{14} W/cm^2 it is the strongest channel of absorption [7], for the intensities exceeding 10^{14} W/cm^2 the problem of the prevailing mechanism of absorption is discussed in [6], in any case it is known that for the intensities $10^{17} - 10^{18}$ W/cm^2 this channel of losses remains. And the increase of the laser radiation absorption coefficient leads only to the improvement of the situation with the confinement and heating of plasma (the situation with the plasma confinement too, since in this case the refraction is suppressed still better). In general, in contrast to the case of the confinement of the plasma of sub-critical density, in the case under consideration the refraction is always suppressed : $b'(x = x_m) - - k_o \exp(-a_{max}^2/2)$, which means that the laser beams in the bundle are extended in the direction of the negative semi-space x. On Fig.1 the ratios R(x) at $\Omega = 0$ for various a_{max} are presented: 2;3;3.4;3.6 for the wavelength of the confining laser radiation $\lambda = 1.06$ μm. These results must be interpreted as the one-dimensional self-focusing filamentation of the beam. The unit of measurement along the x axis corresponds to 6.6λ or about 7 μm. The beam with $a_{max} = 2$ is split/divided into plates with thicknesses ~ λ, since the expulsion of plasma is inconsiderable even in the field maximum: $n_{min} \simeq n_{cr} \exp(-4) \simeq 2 \cdot 10^{19}$ cm^{-3}, and the self-focusing develops easily [9]. For $a_{max} = 3$ n_{min} is equal already to $\simeq 1.2 \cdot 10^{17}$ cm^{-3}, and the self-focusing weakens: transversal dimensions of the filaments increase up to ~ 13 μm. For $a_{max} = 3.4$ and $a_{max} = 3.6$ the self-focusing already does not take place.

Fig.2 corresponds to $\lambda = 10.6$ μm and $\Omega = 0$. The value corresponding to the unit of measurement in the x axis is 2.1λ or 23 μm. The main distinction of Fig.2 from Fig.1 is that the self-focusing is already absent at somewhat smaller a_{max} - at $a_{max} = 3$.

It is obvious that all the analytical results presented above are valid in the absence of self-focusing, i.e. at $a_{max} \geq 3.5$ for λ = 0.53 and 1.06 μm and $a_{max} \geq 3$ for $\lambda = 10.6$ μm. As it can be seen, the infinitesimal parameter of the approximations made above is just the inverse value $1/a_{max}^2$ (squared), which confirms the calculations performed.

HEAT BALANCE OF ALL THE PLASMA VOLUME.

The ways to obtain the heat balance of all the plasma volume are described in [3]: keeping the plasma at a fixed temperature is possible if the rate of the heating of all the plasma volume Q^+ coincides with the rate of its convective heat transfer Q^-. From equations (10) the rate of the heating of all the volume is determined exactly. Since

$$Q^+ = \int\int_0^{x_m} \delta I \, dx \, ds = S \int_0^{x_m} \delta I \, dx \, ,$$

S is the area of the surface of the plasma which is generated by the laser beams. We ignore here the heating of the residual plasma for reasons mentioned above. So,

$$Q^+ = S I_{max} \exp(-I_{max}/2I_{cr})/2. \qquad (15)$$

The value Q^+/S is the rate of the energy supply per unit surface and depends only on I_{max}.

The rate of the convective heat transfer of all the plasma volume Q^- is determined primarily by the type of ions - those partly ionized or well those totally ionized. The pure hydrogen plasma for the CTF quickly becomes overheated (the channel of the heat losses is only the weak bremsstrahlung), in [3] the ways to increase the convective heat transfer are suggested. In all the cases the convective heat transfer is a volumetric process with the specific rate q^-, $Q^- = \int_V q^- dV = q^- V$ (V is the plasma volume). For the hydrogen plasma one can attain the rise of its losses by mixing to the plasma the atoms of heavy elements with the concentration $n_1 \ll n$ [8]. The equation of the heat balance will be written as

$$Q^+ = q^- V,$$

or

$$S I_{max} \exp(-I_{max}/2I_{cr})/2 = q^- V. \qquad (16a)$$

For the plasma volume of cubic form with the side L [3] the balance is written as follows:

$$\frac{31_{max}}{L} \exp(-a^2_{max}/2) = 7.02 \cdot 10^{-16} n_1 [cm^{-3}] n_e [cm^{-3}] T^{1/2} [eV]. \qquad (16b)$$

Here the temperature T is measured in eV, n_1 and $n_e (= n)$ - in cm^{-3}, and all the right-hand side - in $erg/s \cdot cm^3$. By the appropriate choice of the value of n_1 the equality (16b) can be easily satisfied.

It should be noted that the calculations presented are free from empirical parameters introduced in [3] and give exact heat balance within the framework of the applicability of equations (10).

The heat balance of the plasma of non totally ionized atoms differs from that considered above. The confinement of such a plasma is of interest for the creation of dense stationary uniform active medium of plasma lasers. Here powerful radiation of rather hard quanta in the lines of plasma ions leads to significant heat losses [8]. If we consider the hot plasma immediately as the active medium of the X-ray laser with collision pumping [9] and roughly assume that in the plasma only the ions of necessary multiplicity exist, then

$$q^- \simeq \Delta n B_{10} \hbar \omega_{10}. \qquad (17a)$$

Here Δn is the density of ions which are not in the principal state, B_{10} is the rate of the radiative decay of the lower working level into the principal state (this is the most quick radiative transition), $\hbar$ is the Planck's constant, ω_{10} is the frequency of this transition. An elongated plasma tetrahedral column of length l and thickness D can exemplify the real active medium, confined by the laser radiation. In this case

$$Q^- = lD^2 q^-, \quad Q^+ = 2lDI_{max} \exp(-I_{max}/2I_{cr}),$$

and from the equality $Q^- = Q^+$ one can derive the value of D, at the heat balance is realized:

$$D = 2I_{max} \exp(-I_{max}/2I_{cr})/\Delta n B_{10} \hbar \omega_{10}. \qquad (17b)$$

Thus, for the plasma of non totally ionized atoms the heat balance can be reached only by the variation of the dimensions of the confined plasma (the transverse dimension of the tetrahedral plasma column in the considered case of the active medium of the X-ray laser) or the variation of I_{max}.

SCALING OF PLASMA OF THE NON-BARE IONS

The confinement of the plasma of the non totally stripped ions is somewhat different from the case of the hydrogen-like one. First, the ponderomotive force is proportional to the electron density, and, for the thermodynamically equilibrium plasma, - to some average (in the sense of the Saha distribution) charge of the ions z. As for the plasma pressure, it is the sum of the partial pressures of ions and electrons, i.e. $p = (Z + 1)nT$. Therefore, the condition of the local balance of forces (1) will give E_{cr} in the form

$$E_{cr}^2 = 16(Z + 1)\pi T n_{cr}/Z, \qquad (18)$$

that for the plasma of multiply charged ions, practically means that $E_{cr}^2 = 16Tn_{cr}$, and $I_{cr} = 2cTn_{cr}$ which is twice less as in the case of the hydrogen plasma.

Second, a powerful channel of the heat losses of such a plasma is the ion irradiation in lines. Therefore, to reach the heat balance, one can do nothing except variations of the dimensions of the plasma being confined; the change of the maximum intensity of the confining radiation will influence in some way too.

Since for practical applications (creation of X-ray laser

active medium, for instance) the geometry of the confined plasma is of interest, which represents an elongate, rather long column, and, therefore, we ought to consider both the case of the confinement by the tube-like beam and the confinement taking place when creating a cavity in the form of an elongate polyhedral column by means of a few beams, focused by cylindrical lenses. In the first case the generated coherent X-ray radiation is coaxial the line with the pumping beam, which is not convenient. For that reason we consider the second case, which, experimentally, is much the same as/ is closely similar to the realized case [11], though the necessary number of the pumping beams is more than two in this case.

It is evident that the main advantage of the suggested scheme of the X-ray laser pumping by an electron impact is that the plasma density is larger than in the case with the free expanded plasma. But this density is top limited according to the requirement that the collisions de-excitation of the upper working level should be less than the rate of radiative decay. The equivalence of these rates gives the maximal electron density n_e^{max}. Choose the optimal density n_e^{opt} according to [9]: $n_e^{opt} = n_e^{max}/2$. The scaling in non-bare ions plasma is determined as it was noted above mostly due to the heat balance. Let us known the efficiency of the single-pass X-ray collisionally-pumped lasers with confined active medium.

Let us assume the laser on the Se^{25+} ion. n_e should be equal to $n_e^{opt} = 4 \cdot 10^{20}$ cm^{-3}, T = 850 eV. The pumping rate of the upper working level provides saturated regime, because the gain G = 8 and real length of active medium l ~ 1 cm. The transverse size D should be determined from the heat balance. The power of X-ray laser for such regime [10] $P_{xl} = VB_{21}\hbar\omega_{xl}\Delta n$, Δn - population inversion, B_{21} - rate of radiative decay of working transition, $\omega_{xl}/2\pi$ is the X-ray laser frequency, V-plasma volume, $V = lD^2$. The pumping power in the scheme with four pairs of opposite directed plane laser beams $P_l = 8I_{max}Lr_0\sqrt{\pi}$ (transversal Gauss shape is suggested with the r_0 half-width). D is the length of confinement, $D = 2\pi\sqrt{3}r_0^2/\lambda$, and

$$\frac{P_{xl}}{P_l} = \frac{3\pi\sqrt{\pi}B_{21}\hbar\omega r_0^3 \tilde{\Delta n}}{2\lambda^2 I_{max}} \qquad (19)$$

The value $\lambda^2 I_{max}$ does not depend from the wavelength and is determined of T and scaling conditions: $\ln(n_0/n_{cr})$. The ratio P_{xl}/P_l is mostly depended on r_0 - it is large for long wavelengths, and this ratio (efficiency as [9]) is also large, but it occurs under large P.. For CO_2 pumping laser under extremely sharp focusing $2r_0 \sim 10$ μm, $I_{max} = I_{cr} \ln(n_{opt}/n_{cr}) \simeq 3.1 \cdot 10^{14}$ W/cm^2 and $P_{xl}/P_l \simeq 2,5 \cdot 10^{-7}$ - four times less than for experimentally observed [11]. It is clear because the most part of pumping power is spent for the confinement, and $D \simeq 0,2 \cdot L_a$, that is corresponded to the one-four factor. The advantages can be reached, for instance, for the non-optimally experiments with N-like Mo: obtained $P_{xl}/P_l \sim 10^{-9}$ [12], our scheme calculated $\sim 10^{-7}$.

n_e^{opt} is more for plasma of Ni-like ions. For Ni-like W^{46+} $n_e^{opt} = 0,83 \cdot 10^{22}$ cm^{-3} for $\lambda_x = 49,3$ Å. For CO_2 laser pumping T = 1123 eV [13]; $I_{max} \simeq 2,3 \cdot 10^{16}$ W/cm^2, $D \simeq 10 - 100$ μm. The value $\Delta n = 2 \cdot 10^{17}$ cm^{-3} taking into account the presence of another species of ions in plasma, and $G \simeq 65$. If $l \sim 1$ cm, the saturated regime is realized, $P_{xl}/P_l = 2 \cdot 10^{-6}$ for Nd pumping laser with c $r_0 = 1$ μm (P in one beam $\sim$ 4TW).

REFERENCES

1. Arzimovich L.A., Sagdeev R.Z, *Fizika plazmy dlya fizikov (The plasma physics for physicists)*, Moscow, Atomizdat, 1979 [in Russian].
2. Motz H. and Watson C.J., "The Radio-Frequency confinement and acceleration of plasma", *Adv. in Electronics and Electron Physics*, 23, pp.154-302 (1967)
3. Korobkin V.V., Romanovsky M.Yu., *Phys.Rev.E*, 49, pp. 2316 - 2322 (1994).
4. Kuramoto Y., Tsuzuki T., *Progr.Theor.Phys.*, 54, pp. 687 - 705 (1975).
5. Berman V.S., Danilov Yu.A., *Dokl.AN SSSR [Sov.Phys. - Ac.of Sci. reports]* 258, pp. 67 - 71 (1981).
6. Kaw P., Dawson J.M., *Phys.Fluids* 12, pp.2586 - 2599 (1969).
7. Kruer W.L., Kaw P.K., Dawson J.M., and Oberman C., *Phys.Rev. Letters* 24, pp. 987 - 991 (1970).
8. Koshelev K.N., Krauz V.I., Reshetniak N.G.et al., *Fiz.Plasmy* 15, pp. 1068 - 1076 (1989) [Sov.J.Plasma Phys. 15, (1989)].
9. Elton R.C., *"X-ray lasers"*, NY: Academic Press, Inc., 1990.
10. Yariv A., *"Introduction to Optical Electronics"*, NY: Holt, Rinehart and Winston, nc., 1976.
11. Matthews D.L., Hagelstein P.L., Rosen M.D., et.al., *Phys.Rev. Letters* 54, pp. 110 -114 (1985).
12. MacGowan B.J., Rosen M.D., Eckart M.J., et.al., *J.Appl.Phys.*, 61, pp. 5243 - 5254 (1987).
13. Maxon S. et al, *Phys.Rev.Letters*, 63, pp. 236 - 240 (1989).

422 X-Ray Lasers Active Media Formation

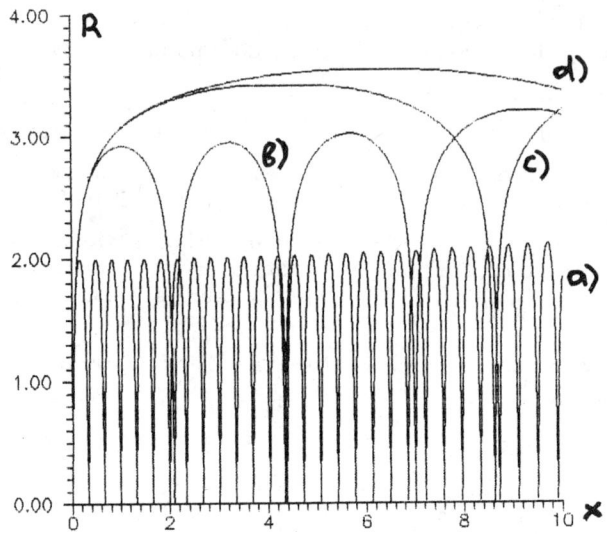

Fig.1. The value $R(x)$ at $\Omega = 0$ for various a_{max}: a) 2; b) 3; c) 3.4; d) 3.6 for the wavelength of the confining laser radiation $\lambda = 1.06$ μm and $\Omega = 0$. $x = 1$ is equal 6.6λ or about 7 μm.

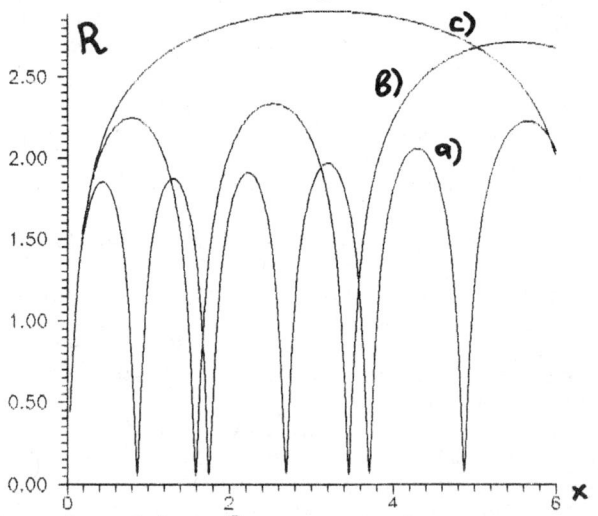

Fig.2. The value $R(x)$ at $\Omega = 0$ for various a_{max}: a) 2; b) 2.6; c) 3 for the wavelength of the confining laser radiation $\lambda = 10.6$ μm and $\Omega = 0$. $x = 1$ is equal 2.1λ or about 23 μm.

Reduction of Driver Energy for X-ray Lasers

M H Key and C G Smith

Central Laser Facility, EPSRC Rutherford Appleton Laboratory, Chilton, Oxon OX11 0QX, England
and
Dept of Atomic and Laser Physics, Clarendon Laboratory, University of Oxford, Parks Road Oxford OX1 3PU

DRIVERS FOR COLLISIONALLY EXCITED LASERS

The widely studied collisionally excited neon-like and nickel-like lasers, driven by typically 0.5 to 1 ns pulses have an efficiency of producing gain, shown in figure 1. Saturation is reached for gain length products, in the range 15-20 and it can be seen from figure 1 that 1 kJ of driver energy is sufficient to reach saturation at wavelengths longer than about 20 nm, but that falling efficiency of producing gain at shorter wavelength leads to a requirement for more than 10 kJ of driver energy for a saturated laser in the water window. This high driver energy requirement has prevented achievement of saturation in the water window so far.

Figure 1 includes two sets of data which differentiate between green-light driven exploding foil targets used principally at the Lawrence Livermore National Laboratory, and infra-red driven slab target used at several other laboratories. There is an improvement in the efficiency of producing gain for the infra-red driven slabs relative to the green light driven foils which, taken together with the loss of energy in conversion to the second harmonic, gives a net advantage to infra-red driven slabs of more than a factor of 5 in driver energy requirement.

The collisional class of X-ray lasers are extremely bright sources of XUV emission and produce megawatt power at saturation. The maximum reported power is 30 MW at 15.5 nm from a neon-like Yttrium laser emitting a 200 psec pulse of over 7 mJ energy[1]. This source gives the brightest X-ray emission

© 1994 American Institute of Physics

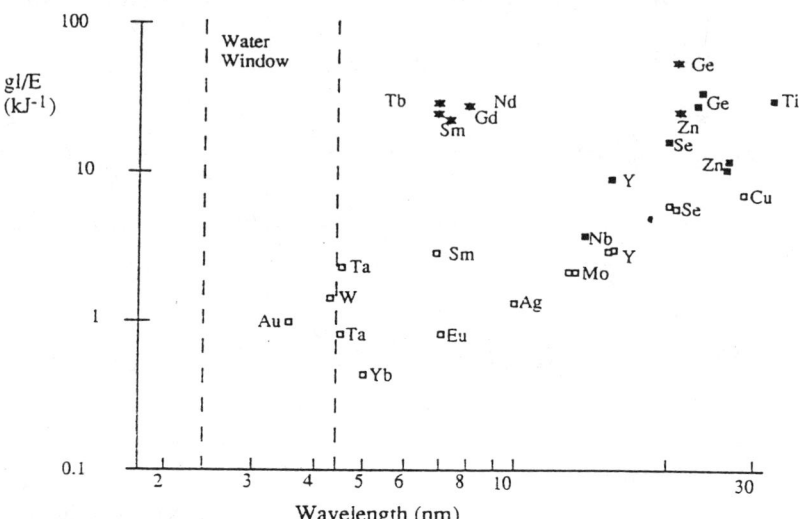

FIGURE 1. Efficiency of gain production gl/E for nanosecond pulse green light driven slabs (open squares), infra red driven slabs (filled squares) and prepulse and multipulse driven slabs (filled stars).

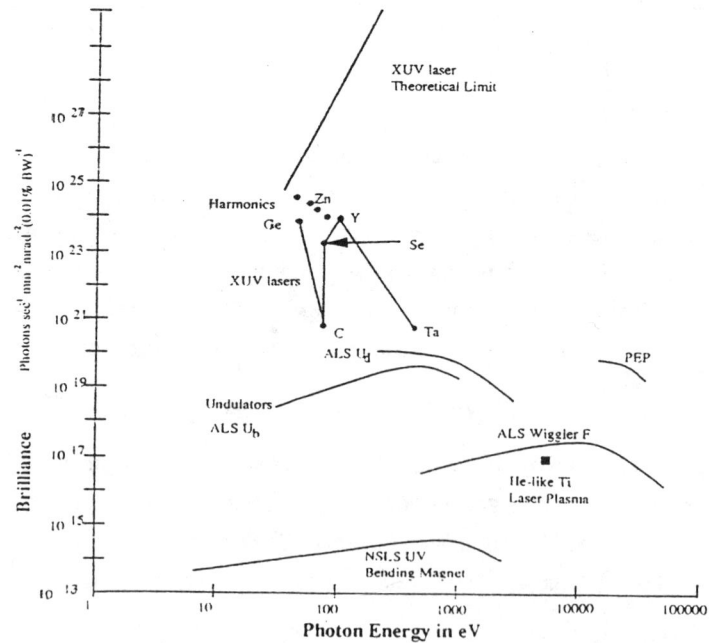

FIGURE 2. Instantaneous brilliance of various XUV sources.

available in the laboratory, as illustrated in figure 2. However, its beam divergence is 10 x 20 mrad from an aperture ~ 120 x 120 µm, which means that in the two perpendicular directions it has 75x and 150x diffraction limited divergence. Thus only a few kilowatts of its output power are fully coherent. There is therefore considerable scope for improving the brilliance of saturated XUV lasers by improving the coherence of the output beam. Several approaches have given progress in improving coherence. Double-passing of the amplifying beam through the plasma using a single mirror has been used to good effect in the Ge XXIII laser (2). Bending of the target to compensate for refraction has led to near diffraction-limited output in the plane perpendicular to the target surface (3). Near diffraction-limited divergence in the plane parallel to the target surface has been achieved by using a separate injector and amplifier plasma (4). Despite this progress it remains a problem to achieve fully coherent output because of the strong refractive disturbance of the laser medium.

REDUCTION OF DRIVER ENERGY USING PREPULSES AND MULTIPLE PULSES

Reduction of driver energy is a major goal because without it XUV lasers will be restricted to a few large installations. The best recent progress with collisional lasers has involved the use of pre-pulses. This work has been of two types.

It has been noted that if an extremely weak pre-pulse is used with an energy content in the range 10^{-4} to 10^{-2} of the main pulse energy, the effect is greatly to enhance the $J = 0\text{-}1$ transition relative to the $J = 2\text{-}1$ transitions of the neon-like laser (5). There is a modest increase in the efficiency of producing gain, as illustrated by the data for slab targets with prepulses in figure 1. The single frequency output of the $J = 0\text{-}1$ transition is also advantageous relative to the 2-frequency output of the $J = 2\text{-}1$. Furthermore, the duration of the $J = 0\text{-}1$ emission is shorter with a faster risetime.

The second type of improvement uses separated shorter pulses of typically 100 psec duration and a train of 2 or 3 such pulses. It is found that the first pulse produces no laser action whereas the second pulse produces very strong $J = 0\text{-}1$ emission with further shortening of the pulse duration. This method has been successfully applied to the Yttrium laser in conjunction with travelling wave excitation to produce an extremely short(20 ps) and powerful pulse. Perhaps its most effective demonstration has been with nickel-like rare earth lasers (7). These recent data from Osaka University are also illustrated in figure 1, and show an exceptional improvement in the efficiency of producing gain and, with 300 J driver energy in two 100 psec pulses of equal energy and 300 psec separation, strong laser action has been observed at wavelengths in the range 6.5 to 9.5 nm.

GENERAL SCALING BEHAVIOUR

A more general view of reduction of driver energy can be taken by considering the power P required to irradiate the target length necessary for saturated output. This can be expressed in terms of the focal length to diameter ratio of the focusing optic, the gL value required for saturation and the efficiency factor g/I relating the gain to the intensity on the target (8)

The necessary power is a function also of the number of beamsand the number of plasma elements. The scaling as $P \approx B^{-1}$ where B is the brightness (W cm^{-2} sterad^{-1}) of the driver is a major consideration. The brightness limits the minimum width of line focus which can be used and therefore the total power required to drive the system. Reduction of the width is constrained by plasma physics consideration as well as by the need to limit coefficient refraction effects. The scaling factors involved have been discussed previously(8) and it has been suggested, for example, that if there is self similarity through simultaneous reduction of pulselength, wavelength and focal line width, that significant increases in the density, gain coefficient, and energy efficiency of producing gain might be obtained.

NEW DRIVER TECHNOLOGY

An interesting technical possibility for such a driver is offered by the KrF Raman laser system (9) which has the characteristic of producing ultra-high brightness (more than 10^{20} W cm^{-2} sterad^{-1}) in relatively short pulses, 10-100 psec at short wavelength to (~ 268 nm).

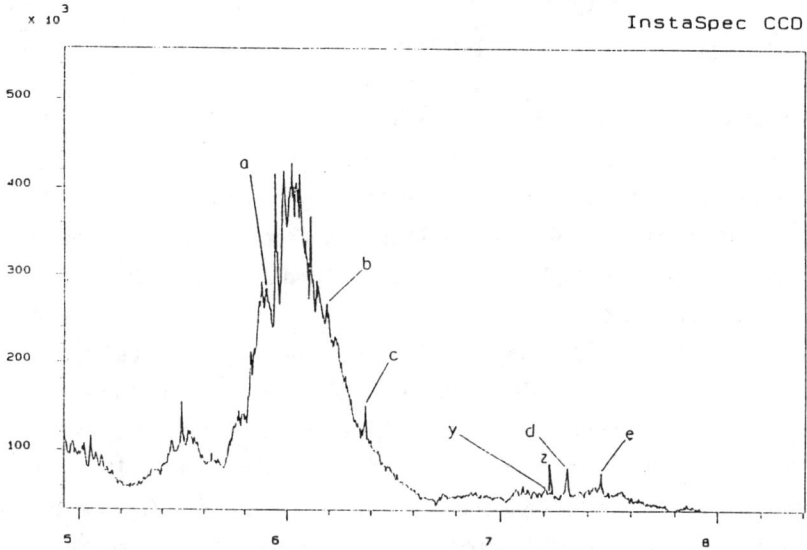

FIGURE 3 Tantalum Spectrum showing Ni-like (a,b,c,d,e) and Co-like (y,z) resonance transitions.

The Titania laser, which will come into operation in two years' time at the Rutherford Appleton Laboratory, will enable full scale tests of the principles involved with 400 J, 100 psec pulses. Trial experiments have already been conducted using a 10 J Raman laser facility, Sprite, to irradiate tantalum targets at 10^{15} W cm^2 with 60 psec pulses at 268 nm in a 15 μm wide line focus. The resulting excitation of the nickel-like ion of tantalum seen in the resonance spectrum recorded in figure 3 indicates production of the plasma temperature necessary for the Ni-like laser.

RECOMBINATION LASERS

Recombination lasers are also of interest as potential systems of reduced driver energy. It is now well established theoretically that reduction of the driver pulse length in adiabatically cooled recombination lasers leads to higher gain coefficient and higher gL/E (11). The recent development of CPA laser systems has opened up the possibility of experiments to test these theoretical predictions and in particular the Vulcan system at the Rutherford Appleton Laboratory had been adapted to produce 35 TW pulses at 40 J in 650 femtoseconds and similar energy in longer picosecond pulses. The laser is equipped with a final compression grating in vacuum and has focusing by an off-axis parabolic mirror to more than 10^{19} W cm^{-2} in a 15 µm diameter focal spot. Imaging of this focal spot by off-axis spherical mirror can be used to produce a 20 µm wide line focus and recent experiments have explored the use of this driver system with 2 psec pulses of 20 J energy at 1.05 µm irradiating 7 µm diameter carbon fibre targets (12). Very high gain has been observed with 5 mm length of plasma as shown in figure 4. The gain coefficient deduced from the change of intensity with length using the Linford formula is more than 12 cm^{-1}. Modelling indicates that the highest gain in space and time is even higher so that saturation of this laser could be expected with less than a factor of 2 increase in its length. The driver energy for a saturated laser at 18.2 nm would then be less than 40 J.

OPTICAL FIELD IONISED RECOMBINATION LASERS

Optical field ionised lasers present a further possibility for still lower energy in the driver at the Joule level. These systems which are designed to give laser action to the ground state of hydrodynamic or Li-like ions can be shown by scaling arguments and modelling (12) to require drivers in the ultraviolet with minimum pulse duration and maximum power. Table-top Titanium Sapphire and LiSaF lasers have been demonstrated to reach 4 TW in 20 femtoseconds (13) and 8 TW in 90 femtoseconds (14) respectively and with third harmonic conversion can operate at 270 nm. Krypton fluoride lasers produce ultraviolet radiation directly at 248 nm and with pulse compression can operate at about 300 femtoseconds. For example, the Sprite facility, with CPA mode of operation and recompression from a stretched pulse of 12 psec duration to a compressed pulse of 300 femtoseconds duration gives 1 TW power beam focusable to 10^{19} W cm^2 with f/4 optics (15). This system is particularly suitable for the investigation of optical field ionised lasers.

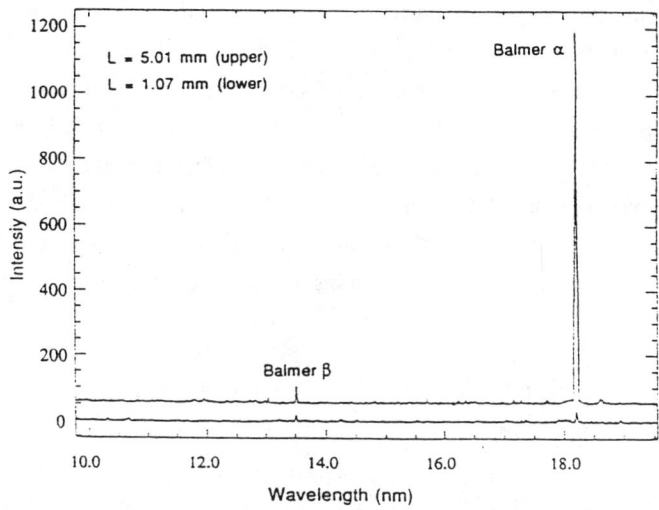

FIGURE 4. CVI recombination laser driven by 2 psec pulse irradiation of a 7 μm fibre target.

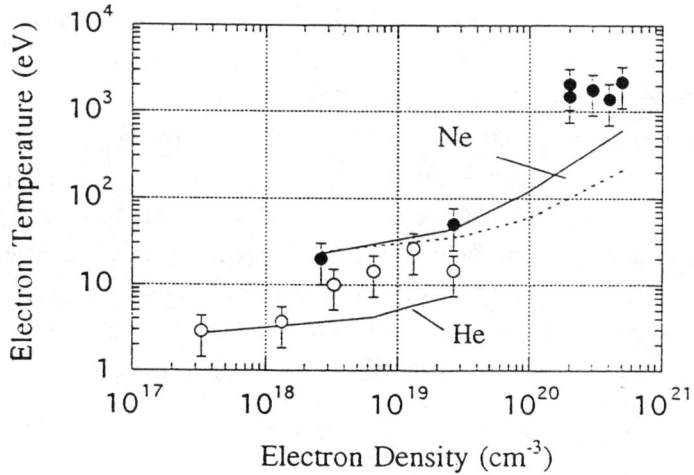

FIGURE 5. Electron temperatures in gaseous targets irradiated by 300 fsec UV pulses at 10^{18} Wcm^{-2} deduced by Thomson scattering.

It seems probable however that constraints imposed by stimulated Raman heating will limit the operation of optical field ionised lasers to wavelengths longer than about 15 nm by the onset of stimulated Raman reflectivity from high intensity interaction at high density in a gas jet. At lower gas pressure, very low electron temperatures necessary for recombination laser action have been observed as illustrated by figure 4. (16) Preliminary results showing gain on the Lyman α transition of hydrogenic lithium with a similar laser system have already given encouraging indications in this direction (17).

HIGH HARMONIC GENERATION

Harmonic generation has been demonstrated hitherto as a source of XUV radiation of very low intensity but recent work has shown that rather efficient conversion to harmonics can be achieved with a shorter wavelength in the fundamental laser. Work by Perry et al (18) comparing conversion at 1.05 µm and 0.53 µm indicates over two orders of magnitude improvement in conversion efficiency with energies in harmonically generated pulses approaching 100 nJ. As the harmonic sources are generated in low density gas, their phase fronts can be rather good so that the brilliance is high and, with quite modest increase in the total conversion efficiency, harmonic sources may reach the same powers as saturated XUV laser with complete coherence. The main difference then between the harmonic source and XUV lasers is that the harmonic source is of short duration. The limited energy may be a problem for some applications but, for applications requiring power, harmonic sources are already competitive with the XUV laser, as shown in figure 2, comparing brilliance of harmonic sources with that of XUV lasers. The main difference is that the XUV lasers are multi-mode whereas the harmonic sources approximate to single mode.

ACKNOWLEDGEMENTS

Many people from other labs have supplied information which has helped in the preparation of this paper, notably from LLNL, D Matthews, L DaSilva, D Eder, J Nielson and M Perry, from ILE Osaka Y Kato and H Daido, and from

Orsay P Jaegle .Users and staff of the Rutherford Appleton lab have also been most helpful and particularly members of my research group at Oxford .

REFERENCES

1) L.B.Silva, B.J.MacGowan, S.Mrowka, J.A.Koch, R.A.London, D.L.Matthews, J.H.Underwood, Opt Lett, 18 pp1174-1176
2) M H Key, A Kidd, P Norreys, R Kodama, H Z Chen, C L S Lewis, D Neely, D O'Neill, J Uhomoibhi, L Dwivedi, J Krishnan, G J Tallents, S A Ramsden, G J Pert, Jie Zhang, A Carillon, P Dhez, P Jaeglé, G Jamelot, A Klisnick, J P Raucourt, Physical Review Letters Vol 68 (no 19) pp 2917-2920 (1992)
3) Y Kato, H Daido, R Kodama, K Murai, G Yuan, M Schulz, M Yamanaka, M Takagi, T Kanabe, S Nakai, D Neely, A MacPhee, C L S Lewis, G Slark, M Niibe, M Tsukamoto, Y Fukuda, H Tsunemi, S Nomoto, I Kodama, T Honda, K Shinohara, H Iwasaki, T Yoshinobu, SPIE Vol 2012 Ultrashort Wavelength Lasers II (1993)
4) G Cairns,D Neely, C L S Lewis, A MacPhee, M Holden, J Krishnan, G Tallents, M H Key, P A Norreys, C G Smith, J Zhang, M T Brown, R E Burge, G Slark, P Holden, G Pert, J Ploues, S A Ramsden Appl Phys B 58 pp51-56.
5) J.Nilsen et al, Private Communication and ibid
6) L.B.DaSilva, R.A.London, B.J.MacGowan, S.Mrowka, D.L.Matthews, G.Frieders, T.L.Wieland, Development of Short Pulse X-ray Lasers for Plasma Probing, To Be Published
7) H.Daido et al, ibid
8) M H Key, IOP Conf Ser No 125, Proceedings of the Colloquium on X-Ray Lasers, Schiersee, 1992, pp171-176
9) M.J.Shaw. G.Bialolenker, G.J.Hirst, C.J.Hooker, M.H.Key, A.K.Kidd, J.M.D.Lister, K.E.Hill, G.H.C.New, D.C.Wilson, Opt Lett 18 pp1320-1322.
10) J.Zhang & M.H.Key, Appl Phys B 58 (1994)
12) D. C. Eder, P. Amendt and S.C.Wilks, Phys Rev A 45 (9) 6761 (1992)
13) C.P.J.Barty, B.E.Lemoff, C.L.Gordon III, IQEC '94, p156
14) M.Richardson, P.Beaud, E.Miesak, B.Chai, IQEC '94, p228
15) I.N.Ross, A.R.Damerell, E.J.Divall, J.Evans, G.J.Hirst, C.J.Hooker, J.R.Houliston, M.H.Key, J.M.K.Lister, K.Osvay, M.J.Shaw, Opt Comm 109 (1994) pp 288-295
16) A.A.Offenberger,W.Blyth, A.E.Dangor, A.Djaoui, M.H.Key, Z.Najmudin, J.S.Wark, Phys Rev Lett, 71 pp 3983-3986
17) Y Nagata, K Midorikawa et al, Phys Rev Lett, 71 (1993) p3774
18) J.K.Crane, M.D.Perry, D.Strickland, S.Herman, R.W.Falcone, IEEE Transactions on Plasma Science, Vol 21, 82 (1993).

On Developing a Table Top Soft X-Ray Laser

A. Morozov, K. Krushelnick, L. Polonsky*, C. H. Skinner, and
S. Suckewer
Princeton University, Princeton, N.J. 08544

C. M. Falco, J. M. Slaughter
University of Arizona, Tucson, AZ

Abstract. We describe progress on the development of a table top soft X-ray laser pumped by a Nd/Glass laser of less than 10J pulse energy. We present recent results on the spatial distribution of soft X-ray radiation from the gain region and near the cooling blades. We also present data on gain measurements for the transition to the ground level in LiIII at 13.5 nm using a powerful subpicosecond laser system.

I. Introduction

Although less than 10 years has passed since the demonstration of soft X-ray lasers (SXLs) we already see tremendous progress towards shorter wavelengths, applications and, very important -the miniaturization of SXLs. In connection with this last subject we have witnessed at this conference very exciting results from Colorado State University[1] demonstrating lasing in a capillary discharge in the XUV spectral region at 46.8 nm in Ne-like Ar. Even more recently (since the conference) another great result was obtained by the Stanford team who demonstrated lasing at a repetition rate of 10Hz in XeIX (Pd-like) also in the XUV region at 41.8 nm using a 70 mJ (40 fsec) pumping laser[2]. This experiment is based on the theoretical development of Corkum and Burnett[3], who suggested using an intense, circularly-polarized, femtosecond laser pulse to tunnel ionize a gaseous target and simultaneously to produce hot electrons for the collisional excitation of the lasing transitions. These two results came after another very promising result from RIKEN[4] indicating the possibilities of obtaining high gain on the transition to the ground state (2-1) at 13.5 nm in LiIII using a subpicosecond (0.5 psec) laser pulse of 50mJ energy. This last result has stimulated work by the Berkeley/Livermore group[5] as well as by our group[6], and our results are presented in the second part of this paper.

The first part of this paper is dedicated to the program on developing a table top SXL primarily at 18.2 nm and 13.5 nm in a rapidly recombining carbon plasma, created by a Nd/Glass laser of less than 10J energy. Although our first SXL at 18.2 nm pumped by a 300J (50 nsec) CO_2 laser in strong magnetic field[7] is not a prohibitively large system (it is housed in a classroom size laboratory), nonetheless it is not what we could call a compact or table top laser. In our approach to a compact SXL we have choosen to continue to work with a recombination system due to its high quantum efficiency. We note however that the MIT table-top experiments pursue both collisional and recombination schemes[8].

II. Experimental arrangement and "one-shot technique" for gain measurements in CVI

The main part of the experimental arrangement with one- and two-target systems and gain measurements for each individual target were presented earlier[9]. Gains of up to ~7 cm^{-1} for 6 mm long carbon targets have been measured for the 3-2 transition in CVI at 18.2 nm. Similarly, high gain (up to 6.5 cm^{-1}) was also observed for the 4-2 transition at 13.5 nm (this surprising result has only been observed recently for multi-fin targets[9]). Gain measurements as a function of distance from the target show a maximum at 0.3 mm - 0.4 mm from the target for the case with iron blades located at 0.6 mm from target. The position of this maximum could be controlled by change of the iron influx from blades: with higher iron influx into the plasma column the gain region was shifted towards the target.

The experimental setup with the addition of a system for the measurement of gain by a "one-shot technique" is shown schematically in Fig. 1. This system consists of a 45° flat multilayer mirror for 18.2 nm (bandwidth ≈ 1 nm), a thinned back illuminated RCA CCD detector, an $A\ell$-filter to block visible and UV light, and data acquisition computer & electronics. The multilayer mirror consists of a repeated Si/B4C bilayer with a spacing of 13.8± 0.2 nm[10]. These mirrors have a first order reflection at 18.2 nm and a 45° angle of incidence. Their 1.1 nm bandpass is much narrower than the typical 2.6 nm bandpass obtained with similar Si/Mo mirrors. In addition the far off-peak rejection is approximately an order-of-magnitude better than for Si/Mo. The S-polarization reflectivity was measured to be 33% at 18.2 nm, somewhat lower than that obtainable from Si/Mo. The narrow bandpass was a significant advantage in isolating the 18.2 gain line from nearby lines. The whole setup is shown in Fig. 2, including a SX reflection microscope and Nd/Glass laser, part of which was used to generate 4-10J for pumping both targets of the SXL (a small size Nd/Glass pumping laser is being constructed, which will fit on the same 4'x10' optical table as the rest of the SXL system). A pinhole camera technique was used to obtain an image on the CCD of radiation from plasma between the blades and target in a relatively narrow spectral region near 18.2 nm (mostly the CVI 18.2 nm line, Fe IX-XI lines, and continuum from the region very close to the surfaces of the target and blades). In Fig. 3a the C-target (6 mm long) is located on the left hand-side and on right hand-side, 0.6 mm away from target surface, are two Fe-blades. The grid pattern is due to the shadow of the mesh supporting the $A\ell$-filter - in the image, the grid period corresponds to 175μ at the target location. One may see very strong soft X-ray emission near the blades cooling region with close to it, the expected gain region. In Fig. 3b is shown the image of radiation emission for a different configuration of blades - the edges of six iron blades are located approximately on semicircular of radius ~0.8 mm centered on the surface of the carbon target, where the pumping Nd/Glass laser beam is focused. Here again we may see very strong SX radiation near the blades close to the expected gain region. We note also that this imaging system allows us to check with high precision the final alignment of two-target assembly.

In order to measure gain with "one-shot technique", the pinhole was retracted from the SX beam, and target II was relocated about 30 cm from target I. This measurement is based on the directionality of stimulated emission in comparison with non-directionality of spontaneous emission[11]. If gain is generated in the "candidate plasma" (from target I) then, the radiation from the "probe plasma"

434 On Developing a Table Top Soft X-Ray Laser

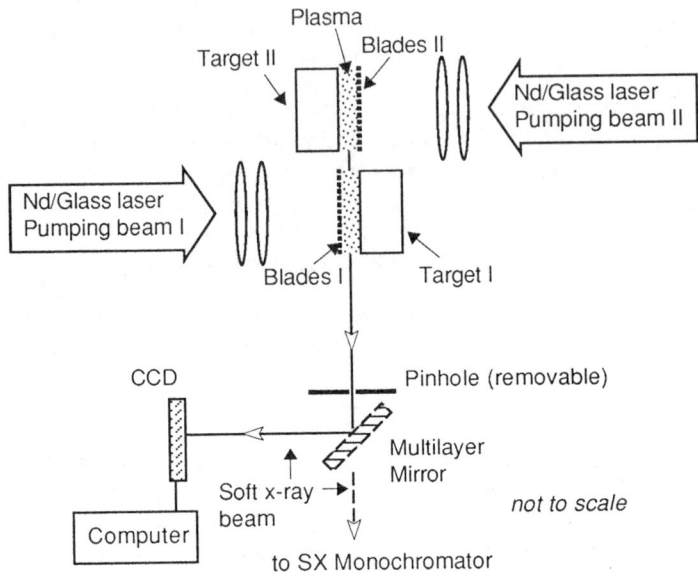

Fig. 1 Experimental setup for two target assembly for imaging (with pinhole) and gain measurement (without pinhole).

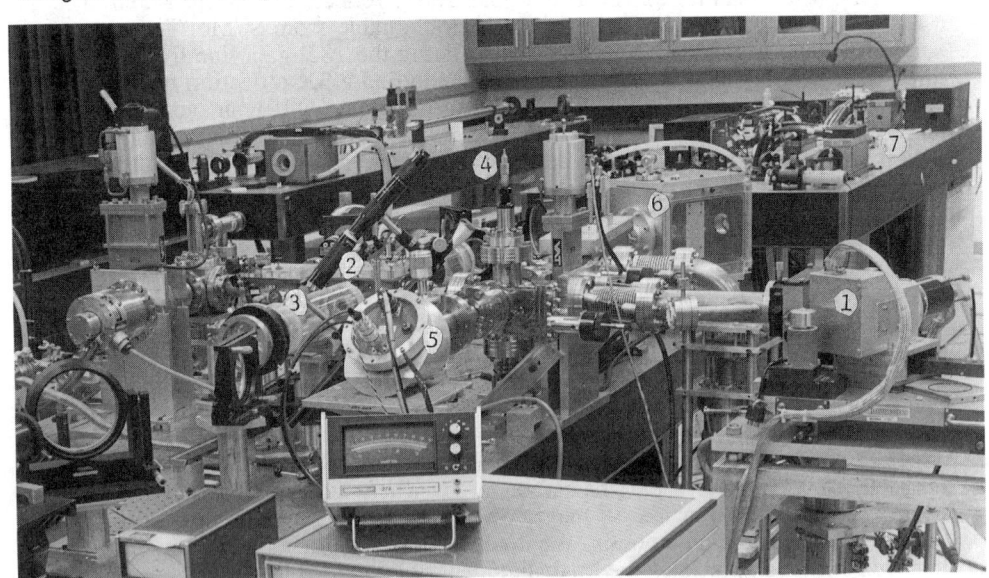

Fig.2 SXL experimental setup: 1-SX spectrometer, 2-targets chamber, 3-pumping beam I, 4-45° multilayer mirror positioner, 5-CCD, 6-SX reflection microscope, 7-Nd/Glass pumping laser.

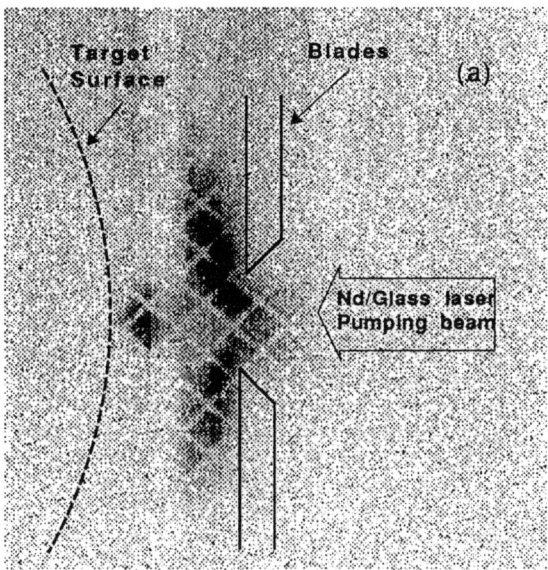

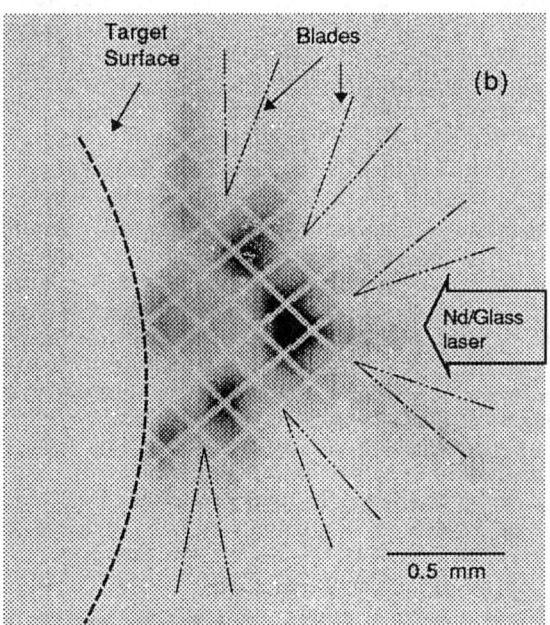

Fig. 3 (a) Image of plasma radiation from target with two vertical blades, (b) target with six radial blades

(from target II) is amplified and propagates in a smaller solid angle than spontaneous emission. On the CCD, an area of enhanced intensity should appear above the otherwise uniform illumination from spontaneous emission. Similarly absorption will appear as a darker region. In the experiment an aperture was used to block the strong emission regions close to the blades in target I. To identify possible variations in sensitivity across the CCD, far field patterns were taken of plasmas I and II separately and together. In these initial experiments, the identification of gain in a single shot was hampered by the poor signal-to-noise ratio in the far field pattern due to strong adsorption in an undesirable surface layer in the CCD. The image in Fig. 4 was obtained by subtraction of the far field illumination patterns from each of the plasmas taken separately, from the far field pattern of both plasmas generated simultaneously. Once again the grid pattern from the $A\ell$-filter support is visible. The higher intensity region corresponds to the region identified in previous experiments as the gain region. Within that region is seen a area corresponding to absorption in the high density plasma very close to the target. Although the clear identification of gain in the far field pattern was not possible, this initial experiment demonstrates the principle and further work is planned with a better CCD.

III. Spectra and gain measurements in OFI scheme

The development of a compact SXL using a powerful subpicosecond laser with a quite low pulse energy appears to be a very promising approach. According to recently developed optical-field-ionization (OFI) theory[12]. it is feasible to create gain for transitions between excited state and ground state (e.g. 2-1 transition) by very fast ionization of the atoms with a high power intensity subpicosecond laser followed by rapid recombination. Following this approach the RIKEN group (4) have demonstrated that optical-field-ionization of LiII ions can produce a rapidly recombining plasma and conditions for gain. They deduced gain $G \approx 20 cm^{-1}$ for the 2-1 transition in H-like LiIII at 13.5 nm for a 2 mm long Li target. For OFI, they used a 50 mJ, 500 fsec KrF laser (power density ~10^{17} W/cm^2), and the initial plasma column of LiII ions was created by a very low power second KrF laser (200 mJ, 20 nsec). However they were not able to obtain gain for plasma lengths longer than 2 mm. In fact, the evaluation of plasma length was quite problematic. Although the initial plasma length was well controlled by the length of line focus of the low power laser, the plasma could expand significantly in all directions during the long delay time (up to 700 nsec) before firing subpicosecond laser, making measurements of gain difficult to interpret.

The goal of our experiment was to confirm the gain observed by RIKEN group at 13.5 nm in LiIII and to increase plasma length by using higher laser power in order to generate higher gain-lengths. Our experimental arrangement, presented schematically in Fig. 5, was only a slight modification of an earlier set up for a fast recombining plasma in a two-laser scheme. The powerful KrF subpicosecond (PSP) laser was operating at a power density ~$1 - 4 \times 10^{17}$ W/cm^2 (max. power density: 2×10^{18} W/cm^2 in 0.3 psec) which was close to optimum for this experiment for a target up to 4 mm long. We used a Nd/Glass laser (operating at 0.8 - 1.5.J in 20 nsec) to create the initial plasma column from a LiF target. The time Δt between these two lasers was adjustable. We found that the optimum conditions for 2-1 gain were created in the plasma at 0.5 mm distance from the

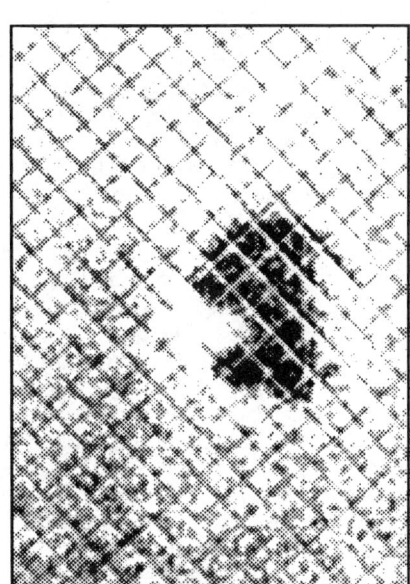

Fig. 4. Demonstration of "one-shot technique" for gain measurements

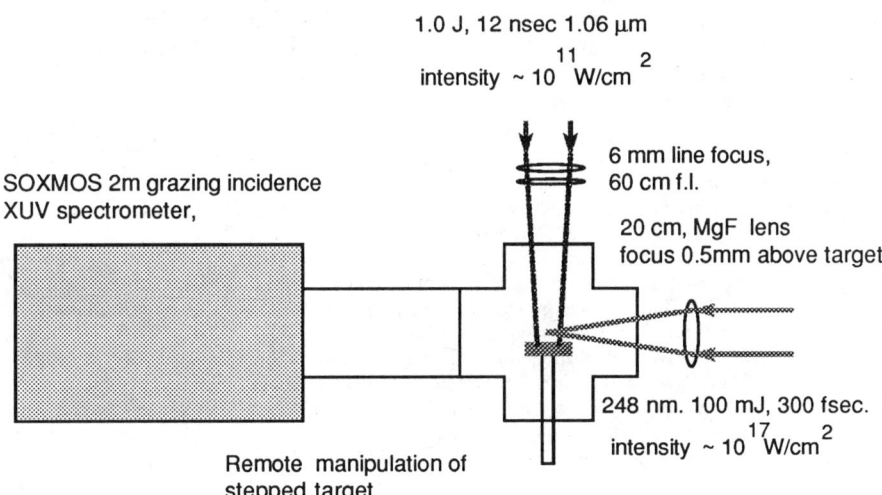

Fig. 5 Experimental setup for optical field ionized recombination scheme using lithium fluoride target

target at a time delay $\Delta t \approx 100 - 150$ nsec by firing the PSP-laser after initiation of the plasma by Nd/Glass laser. This time delay was much shorter than in case of RIKEN experiment, probably due to the use of Nd/Glass laser instead of KrF laser (different plasma expansion time; at a delay of 300 nsec we did not observed any significant intensity of the LiIII 13.5 nm line).

Fig. 6 shows the spectra in the vicinity of the LiIII 11.4 nm and 13. 5 nm lines that were obtained in one shot for each length of the LiF target for a PSP-laser pulse energy of ~100 mJ in 400 fsec and a delay time of $\Delta t \approx 150$ nsec after the Nd/Glass laser pulse (1 J in 20 nsec). As in the RIKEN experiment, we see a rapid increase of intensity of the LiIII 13.5 nm line up to a taret length of 2 mm. Further increase in target length did not provide a significant intensity increase, possibly due to defocussing of the PSP-laser beam in the plasma. Pumping an initially hotter plasma (smaller delay $\Delta t \approx 100$ nsec) with the PSP-laser seemed to generate higher gain at 13. 5 nm (Fig. 7) for the 2 mm plasma. We did not succeed to obtain gain in longer plasmas as can be seen in the absence of a significant increase of LiIII line intensities for targets lengths above 2 mm. However in some cases (like at presented in Fig. 8) we could observe a nonlinear increase of the 13.5 nm/ 11.4 nm intensity ratio in targets up to 4 mm long, although the intensity of the 13.5 nm line did not show such an increase. It seems at present that an extension of the PSP-laser beam propagation to longer distances is crucial for the OFI experiment in order to obtain significantly larger gain-lengths.

Acknowledgments

We would like to thank Bryan Fong for his dedicated analysis of CCD data. We are thankful to T. Lucatorto, R. Watts and C. Tarrio at NIST for providing us with reflectivity measurements of multilayer mirrors. This work was supported by NSF and by the Office of Health and Environmental Research of DOE. L. Polonsky's work was supported by NRL/SDIO.

References

*Also with PXL Inc.
1. Rocca, J. J., Tomasel, F. G., Shlyaptsev,V. A., Cortazar, D. O. and Guidice, G. This Conference; submitted to *Phys. Rev. Lett.*
2. Lemoff, B.E., Lin, G.Y. Gordon III, C.L., Barty, C.P.J. and Harris,S.E., "Demonstration of a 10-Hz, femtosecond-pulse-driven XUV laser at 41.8 nm", submitted to *Phys. Rev. Lett.*
3. Corkum, P.B. and Burnett, N.H. in OSA Proc. on Short Wavelength Coherent Radiation (Falcone, R.W. and Kirz, J. eds.),Washington, D.C, 1988), **2**, p. 225.
4. Nagata,Y.,Midorikawa, K., Kubodera, S., Obara, M.,Tashiro,H. and Toyoda, K., *Phys. Rev. Lett.* **71**, 3774 (1993); also K. Midorikawa et al., This Conference.
5. Falcone, R. ,Private communication (July 1993); T. Donnelly et al., This Conference.
6. Polonsky, L.Y., Park, C.O., Krushelnick, K. and Suckewer, S., Proc. SPIE Conference on Ultrashort Wavelength II (Ed. S.Suckewer), Vol. **2012** (July 1993).
7. Suckewer, S. , Skinner, C.H., Milchberg, H., Keane, C. and Voorhees, D. *Phys. Rev. Lett.* **55**, 1753 (1985); Suckewer, S. et al., *Phys. Rev.Lett.* **57**, 1004 (1986).
8. Hagelstein, P. et al. , This Conference.
9. Park, C.O., Polonsky, L. and Suckewer, S. *Appl. Phys.* **B58**, 19 (1994).
10. Slaughter, J.M., Medower, B. S., Watts, R.N., Tarrio, C.,Lucatorto, T.B. and Falco,C.M., "Si/B4C Narrow-Bandpass Mirrors for the XUV," *Optics Letters* (in press).
11. Skinner, C. H.,*Optics Lett.* **16**, 1266 (1991).
12. Burnett, N. and Corkum, P. B., *J. Opt. Soc. Am.* **B6**, 1195 (1989); Burnett, N. and Enright, G. D., *IEEE J. Quant. Electr.* **26**, 1797 (1990); Amendt, et al.,*Phys.Rev. Lett.* **66**,(1991).

1 mm

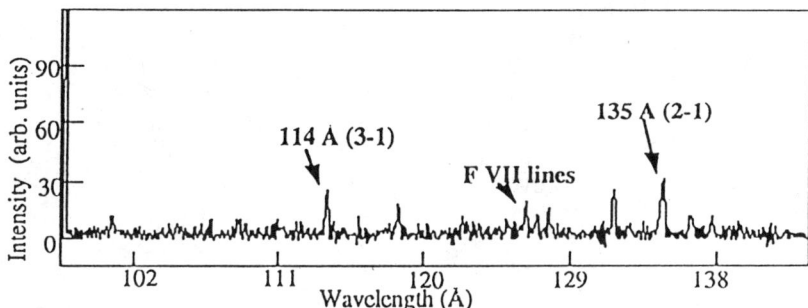

2 mm

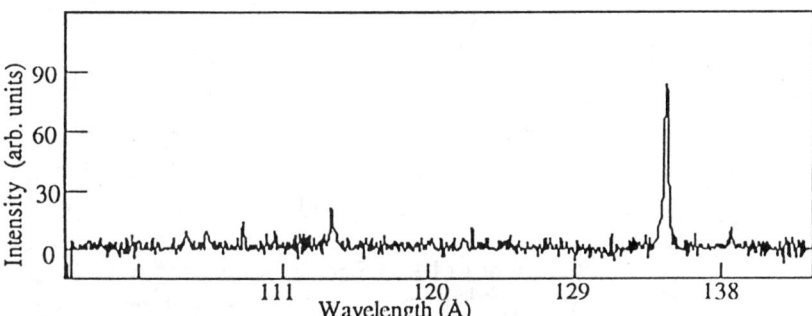

4 mm

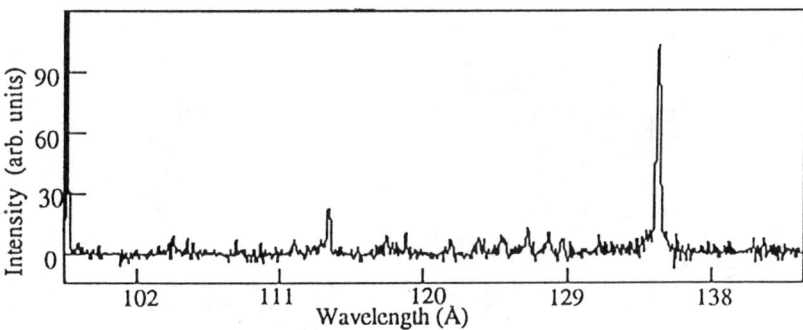

Fig. 6 Spectra from recombining optically ionized plasma for 1mm, 2mm, and 4mm long LiF target. PSP-laser ~100mJ, 400fsec; Nd/Glass laser: 1J, 20nsec; delay Δt=150nsec.

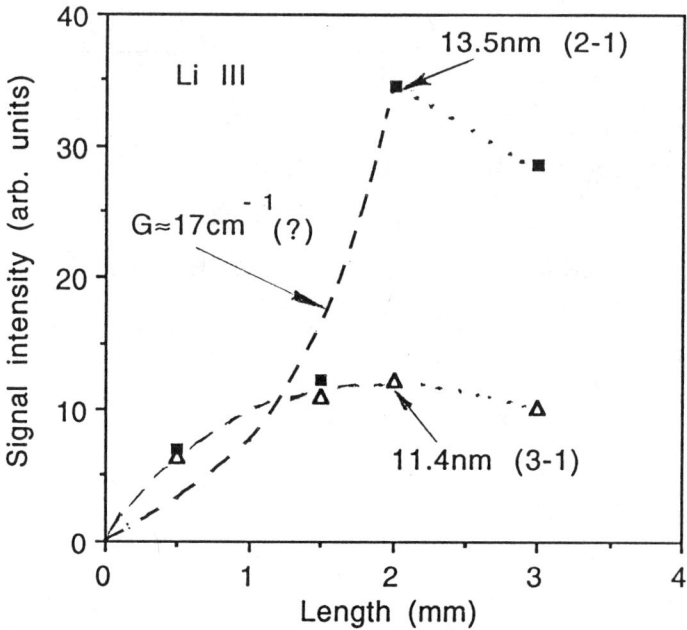

Fig. 7. Intensity of Li III 13.5 and 11.4 nm lines vs LiF target length (PSP-laser: 100 mJ, 400 fsec; Nd/Glass laser: 1 J, 20 nsec; delay time $\Delta t = 100$ nsec)

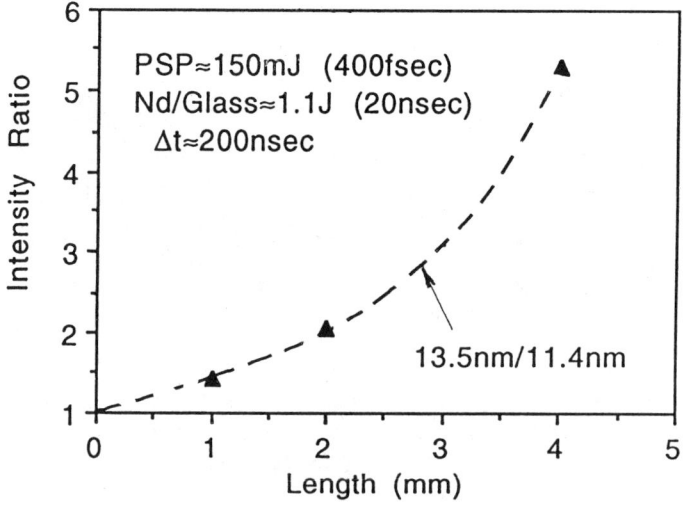

Fig. 8. Ratio of Li III 13.5 nm/11.4 nm line intensities as a function of target length.

CCD Imaging from 20eV to 8keV.

A. G. MacPhee and C. L. S. Lewis

Department of Pure and Applied Physics
The Queens University of Belfast, BT7 1NN, Northern Ireland, UK

Abstract. Pulsed x-ray diagnostics operating in the kilovolt and sub-kilovolt regimes are required in the study of x-ray laser schemes. Sensitivity and dynamic range measurements are presented for position-sensitive detector systems, designed and optimised for these spectral regions. Both systems employ cooled, multi-pinned phase CCD's for image capture. For photon energies from 20eV to 1keV (where direct detection with front illuminated devices is inadequate), a phosphor transducer is used, coupled to the CCD via a fibre optic faceplate with 6µm diameter channels. From 800eV to 8keV, direct detection of electron-hole pairs generated in the depletion region of the CCD is employed. The systems have been tested with single-shot sensitivity using a 10 Hz, 2J/7·5ns injection seeded Nd:YAG laser operated at 2ω, using Bragg crystal and flat field grazing incidence spectrometers to monitor the resonance emission from aluminium and carbon targets irradiated at $\sim 10^{12}$ Wcm^{-2}. At 182Å, the sensitivity and dynamic range are enhanced with respect to that for a standard photographic detector, by factors of 8 and 300 respectively. An absolute calibration of the 700eV-8keV detector system performed at 1.6keV, has been shown to agree with a calculation of sensitivity based on photo-absorption data. For this system, the enhancement in sensitivity and dynamic range over direct exposure X-ray film is 175 and 43 respectively.

INTRODUCTION

The calibration of two X-ray detection systems for sensitivity is described: the first at 33.74Å (368eV) and 182Å (77eV); the second at 1·6keV (7·76Å). The sub-kilovolt system employs a high efficiency, small grain, rare earth phosphor scintillator, to convert uv and soft x-rays to visible photons matched to the sensitivity peak of EEV 15-11 series, ultra-low dark current inverted mode CCDs[1]. The phosphor is coupled to the CCD via 6µm channel diameter, proximity focused fibre bundles. A 150mm×70mm fibre faceplate coated with the phosphor isolates the vacuum system and provide a large area of acceptance for the incident flux. The integral fibre faceplate of the CCD is butt coupled to the region of interest on the phosphor plate for data capture. The kilovolt system[1] employs direct detection of electron-hole pairs generated in the depletion region of the CCD, by photoelectric absorption of 800eV to 8keV photons.

Independent control of the pixel clock phases with respect to those in the readout register, allows the detector to spatially integrate some, or all of the pixel rows in a frame, resulting in a substantial increase in the signal to noise ratio attainable.

DETECTOR CALIBRATION
Sub-kilovolt

The soft x-ray system was calibrated for sensitivity at 33.74Å and 182Å against a pre-calibrated[2] photographic detector (Ilford Q-plate), using hydrogenic carbon resonance emission generated at QUB with a 10Hz, 2J/7·5ns, injection seeded Nd:YAG laser operated at 2ω. The signal was spectrally resolved with a flat-field grazing incidence spectrometer, using a mean 1200 grooves/mm grating with a 200µm entrance slit set 75mm from the source. Films were analysed using a JL Automations MK6 microdensitometer, with matched influx and efflux optics with NA=0.25 and a 5µm × 50µm slit with the long edge parallel to the spectral lines. The optical density was measured at 2µm intervals in the spectral direction, with a separation of 5µm between adjacent scans. The density values were converted to intensity using the fit obtained in (2) to the semi-empirical model of Henke et al[3].

Figure 1 shows a typical single shot emission spectrum snatched from the real time display of the CCD output, where the signal has been spatially integrated on chip between shots and read out at 10Hz.

For the calibration, comparison was made between the signals recorded on each detector, within 1Å of both the third order of the Lyman-α and first order of the Balmer-α emission, integrated along 135µm of each line for 10 shots (table 1).

	33.74Å	182.17Å
$N_{photons}$ (film)	142489	38802
N_{counts} (CCD)	38935	8605

Table 1. Soft X-ray calibration data.

The film calibration employed was based on 33.74Å photons. Hence the above number quoted for the measured 182Å signal on film is the *equivalent* number of 33Å photons necessary to produce the *recorded* density. The film sensitivity is not necessarily independent of photon energy, hence a correction must be made: The phosphor output is known to increase ~linearly with incident photon energy[4]. From table 1, the ratio of the number of *counts* obtained on the CCD for the 33Å and 182Å emission is 4.52, so in terms of the number of incident *photons*, this ratio becomes 0.723. On the film, the same ratio expressed using the quoted *equivalent* number of 33Å photons is 3.67, hence the film sensitivity to 182Å photons is 0.723/3.67≈0.2 that for 33Å, which is reasonable to expect from a film of this type. The magnification of the system from the scintillator to the CCD is ×1, so with a uniform X-ray flux over an equivalent area to one 27μm × 27μm pixels, the sensitivity at 182Å:

$$\frac{38802 \,(33\text{Å photons}) / 729 \mu m^2}{0.2 \,(33\text{Å to }182\text{Å conversion}) \times 8605 \,(\text{CCD counts})}$$

$$= 0.03 \text{ photons}/\mu m^2/\text{count}.$$

The film fog density was measured to have a standard deviation equivalent to an incident 182Å photon intensity of 0.37 photons/μm^2, so that a signal of 1.1 photons/μm^2 would be within 3 standard deviations of the mean. For the CCD system, with standard deviation 1.8 counts (16-bit digitisation), 6 counts corresponds to 0.18 photons per square micron. This calibration is thought to be accurate to within a factor of ~4, based on our knowledge of the sensitivity of the photographic emulsion. The absolute QE of the phosphor alone at 182Å (within 10%)[4], would indicate that for optimum coupling, the sensitivity of the system could be enhanced by a factor of three over our measurement.

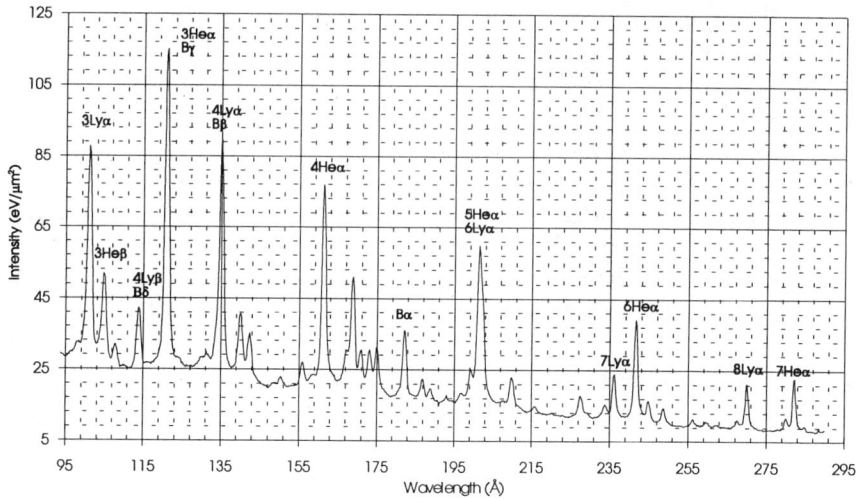

Figure 1. Typical single shot, carbon emission spectrum snatched from the real time display of the CCD output, where the signal has been spatially integrated on chip between shots and read out at 10Hz.

Kilovolt Calibration

The kilovolt system was calibrated against absolutely calibrated (within 15%) Kodak DEF x-ray film[5], using helium like resonant emission from an aluminium plasma. The x-ray film was positioned so as to mask one half of the spectrum falling onto the CCD in a Bragg crystal spectrometer. The emission lines were uniformly continuous across the film / CCD interface, as verified in spectra recorded over the entire device. The number of photons integrated over an area in the vicinity of the interface on the film, was compared to the number of electrons generated in the depletion region of the CCD over an identical area for the same exposure. The standard deviation in the DEF fog density measured over 100 25μm×25μm pixels was equivalent to a 1·6keV photon intensity 0·0233 photons/μm², so that an incident signal 0·07 photons/μm² signal would be within 3 standard deviations of the mean. For the CCD, with standard deviation 1.8 counts, 6 counts was found to correspond to 4.46×10⁻⁴ photons per square micron on film. Therefore at 1.6keV, with a uniform X-ray flux over an area equivalent to one pixel, the CCD system has ~200 times the sensitivity of DEF. To assess the dynamic range of the system, the number of 3 standard deviation signal increments required to bring the DEF film density to the shoulder of the HD curve at roughly OD 5, is 250. The storage capacity of the CCD pixels is 10^6 electrons (increased by a factor of 4 since ref.[1]), hence the corresponding number of 3 standard deviation increments for the 16-bit CCD system is 10800, which is 43 times that of DEF. On this basis, with a system gain of 15 electrons per count, 63% of the incident 1·6keV photon flux generates collectable photo-electrons in the depletion region of the device at 3·65eV / e$^-$, compared to ~68%[5] from theory, which is well within the tolerance of the calibration data.

ACKNOWLEDGEMENTS

The authors wish to thank Andor Technology LTD (10 Malone Rd., Belfast, BT9 5BN), for their provision of the CCD detector systems and many useful discussions.

REFERENCES

[1] MacPhee, A. G., Lewis, C. L. S., 'CCD imaging from 0·7keV to 8keV with Multi-Pinned Phase technology.', SERC Central Laser Facility, Ann. Rept., 135-136, (1993).
[2] Krishnan, J., *et al*, 'Film calibration for soft X-ray wavelengths', SERC Central Laser Facility, Ann.Rept., 22-23, (1992).
[3] Henke, B.L., *et al*, J.Opt.Soc.Am, **1,** 818 (1984)
[4] Husk, D.E., Schnatterly, S.E., 'Quantum efficiency and linearity of 16 phosphors in the soft X-ray regime', J.Opt.Soc.Am.B, **9**(5), (May 1992)
[5] Mather, D., AWE Aldermaston, private communication.
[6] Castelli, C., *et al*, 'Soft X-ray Response of Charge Coupled Devices', NIM **A310**, 240-243 (1991)

X-ray Holography with High Resolution[*]

Jianwen Chen, Peiping Zhu, Tiqiao Xiao, Zhizhan Xu

(Shanghai Institute of Optics and Fine Mechanics, Academia Sinica, P.O.Box 800216, Shanghai, 201800)

Abstract. Some primary factors having effects on the resolution in x-ray holography are discussed. The factors in recording x-ray holograms are the x-ray coherent scattering by the specimen, the recording method and the coherence of the x-ray beam. There are two factors in reconstruction of the hologram. One is that the resolution of the detector may be lower than the spatial frequencies of fringes in x-ray holograms. The other is aberrations. Consequently, some conditions of x-ray holography with high resolution are given.

INTRODUCTION

Holography was invented by Gabor in 1948 to correct the spherical aberration of electron microscope. Only after four years, Baez introduced the idea of holography to x-ray region.

After the development of forty years, x-ray holography has taken a major step forward[1-5], but there are still some barriers which prevent its being put into practice. For example, the brightness and the coherence of the x-ray beam are not high enough, and it is the optical elements that are short of in x-ray region such as x-ray refractive lens. Because of the above reasons, x-ray holography is confined to in-line holography and lensless

---

[*]Supported by the National Nature Science Foundation of China.

Fourier transform holography, and the highest resolution obtained is lower than the theoretical limit by one order of magnitude.

The factors in recording x-ray holograms with high resolution are the x-ray coherent scattering by the specimen, the recording medium, and thecoherence of the x-ray beam. There are two factors in reconstruction of the hologram with high resolution. One is that the resolution of the detector may be lower than the spatial frequency of x-ray holograms. The other is aberrations.

CONDITIONS FOR HIGH RESOLUTION IN HOLOGRAPHY

Holography is a two-step procedure. The first step is recording, and the second is reconstruction. In order to get a holographic image of the specimen with high resolution, the information of high spatial frequency of the specimen must be recorded in the first step, and the information must be read out from the hologram in the second step without loss and distortion. As a result, it is important to make the numerical aperture as big as possible in the hologram recording, and to make the information capacity carried by the reconstruction image wave equal to that recorded in the hologram.

RESOLUTION IN RECORDING

The mechanism for the object wave to carry the information of the specimen is coherent scattering. The electrons around the molecules and atoms will radiate the x-rays of the same frequency of that of the incident x-rays. Because the biological molecules in the specimen distribute in various spatial periods, and the coherent scattering x-rays carry the information of these spatial periods.

In x-ray in-line hologram, the numerical aperture is limited by the cut-off frequency of the medium, which results in that the resolution cannot be higher than the cut-off frequency of recording medium.

In x-ray lensless Fourier transform hologram, the aperture cannot be bigger than the diffraction disk of the point reference beam on the recording medium, so the limit of resolution cannot be smaller than the radius of it.

In order to ensure that the numerical aperture is not limited by the x-ray illumination, the x-rays must be spatially and temporally coherent within the numerical aperture on the hologram.

The spatial coherence length must be longer than the radius of x-ray in-line hologram, and the diameter of the zone plate used in x-ray lensless Fourier transform holography.

The requirement for temporal coherence depends on the maximum path difference between object and reference waves.

Suppose that f and r_z are the focal length and the radius of the Fresnel-zone plate respectively, the temporal coherence length required in x-ray lensless Fourier transform holography with resolution of δ_f is[6]

$$\frac{\lambda^2}{\Delta\lambda} \geq 0.61\frac{\lambda d_1}{\delta_f} + f(\cos^{-1}\theta - 1). \qquad (1)$$

The requirement for the temporal coherence length in x-ray in-line holography with resolution of δ_i is[6]

$$\frac{\lambda^2}{\Delta\lambda} \geq 0.186\frac{\lambda^2 z}{\delta_i^2}. \qquad (2)$$

RESOLUTION IN RECONSTRUCTION

The spatial frequencies, of interference fringes in the outer zones of x-ray in-line hologram, is so high that the hologram must be at first enlarged, and then can be either reconstructed by visible light or read out by a detector and then numerically reconstructed by a computer.

All aberrations can be eliminated when the x-ray hologram is illuminated with the visible light beam, if the magnification of the hologram equals the wavelength ratio. The fact has been neglected in many papers published[7,8], in which paraboloid of revolution instead of sphere is regarded as ideal wavefront according to Gaussian optics.

The spatial frequency of interference fringes of x-ray lensless Fourier transform hologram can be reduced by recording in a long distance. In this way, the hologram can be directly either reconstructed by visible light or recorded by a CCD camera and then numerically reconstructed by a computer.

The fringe spacing in x-ray lensless Fourier transform hologram can be changed in terms of adjusting the distance between the object plane and the hologram. As soon as the fringe spacing is larger than the reconstruction wavelength, x-ray lensless Fourier transform hologram can directly be reconstructed with visible light. According to the theory of third-order wavefront aberration in holography[9], we obtained the result[10] that not only are there no spherical aberration in lensless Fourier transform holography, but also coma can be eliminated if the point beam of visible light is set on the original place of the reference in reconstruction. The remainder aberrations can be expressed as

$$|W| \leq \frac{\pi}{2} \left| 2 \left(\frac{\rho}{z_0}\right)^2 \mu(\mu+1) \left(\frac{x_0}{z_0}\right) \frac{x_0}{\lambda_2} \right. \quad \text{astigmatism}$$

$$+ \left(\frac{\rho}{z_0}\right)^2 \mu(\mu+1) \left(\frac{x_0}{z_0}\right) \frac{x_0}{\lambda_2} \quad \text{field curvature}$$

$$\left. + 2\left(\frac{\rho}{z_0}\right) \mu(\mu^2-1) \left(\frac{x_0}{z_0}\right)^2 \frac{x_0}{\lambda_2} \right| \quad \text{distortion,} \quad (3)$$

where W represents third-order aberrations, ρ the radius of the hologram, x_0 the distance between an object point and the z axis, z_o the distances from object to the hologram, $\mu=\lambda_2/\lambda_1$. Obviously, the resolution of lensless Fourier transform x-ray hologram depends on ρ/z_o, x_o/z_o determines the fringe spacing.

From Eq.(3), we see that astigmatism and field curvature are proportional to the second power of the resolution, and so is distortion the first power. The shorter distance between an object point and the reference point beam is, the smaller is aberrations.

If some parameters are suitably selected, the resolution of tens of nanometer can be achieved according to Rayleigh's criterion of $|W|\leq\lambda/4$.

REFERENCES

1. S.Aoki and S.Kikuta, "X-ray holography". *Jpn.J.Appl.Phys.* **13**(9), 1385-1392 (1974).
2. M.Howells et al., "X-ray holograms at improved resolution: a study of zymogen granules". *Science* **238**, 514-517 (1987).
3. C.Jacobsen et al., "X-ray holographic microscopy using photoresists". *J.Opt.Soc.Am.* A7(10), 1847-1861 (1990).
4. I.McNulty et al., "High-resolution imaging by Fourier transform x-ray holography". *Science* **256**(15), 1009-1012 (1992).
5. J.Solem, G.Chapline, "X-ray biomicroholography". *Opt.Eng.* **23**(2), 193-203 (1984).
6. Peiping Zhu, Jianwen Chen, Zhizhan Xu, Tiqiao Xiao, Leigang Kou, Zhijiang Wang, "Effect of coherence on resolution in x-ray holography". *Acta Optica Sinica*(in Chinese), in press.
7. R.J.Collier, C.B.Burckhardt, L.H.Lin, *Optical holography*, New York and London: Academic press, 1971, pp.68-78.
8. H.J.Caulfield, *Handbook of optical holography*, New York: Academic press, 1979, pp.40-120.
9. R.W.Meier, "Magnification and third-order aberrations in holography". *J.Opt,Soc.Am.* **55**(8), 987-992 (1965).
10. Peiping Zhu, Jianwen Chen, Zhizhan Xu, Tiqiao Xiao, Leigang Kou, Zhijiang Wang, "The study of feasibility of reconstructing x-ray lensless Fourier transform hologram with visible light". *Acta Optica Sinica*(in Chinese), in press.

Space-Resolved Electron Density Measurements Using the Stark-Broadened Line Wings of Hydrogenic Ions in Line-Shaped Laser Plasmas

Ling-qing Zhang, Shen-sheng Han, Zhi-zhan Xu,
Zheng-quan Zhang Pin-zhong Fan and Lan Sun

*Shanghai Institute of Optics and Fine Mechanics,
P.O.Box 800-211, Shnaghai,201800, P.R.China*

Abstract. A program was developed using the Marquardt algorithm method to fit the Stark-broadened line wings of hydrogenic ions,which can be conveniently used to measure electron density in laser-produced plasma. Applications of the method are illustrated using space-resolved 2p-1s resonance lines in line-shaped laser plasmas of Mg slab targets irradiated with different laser intensity. Space-resolved electron densities higher than the critical density are obtained by fitting the Stark-broadened line wings with the program.

I. INTRODUCTION

The possibility of electron density measurements using the Stark-broadened line wings of hydrogenic ions in laser-produced plasmas was first discussed in the work of Smith and Peacock(1978)[1]. This portion of line is necessarily optically thin and densities may be deduced directly since the broadening depends only on the ion-produced electric microfield. Theoretical variations of wing intensity with changes in density were investigated in order to demonstrate the sensitivity of the method. A program was developed through which electron density can be obtained by fitting the experimental line wing using a nonlinear least square method. The program has the advantage of avoiding the normalization of the observed profiles by reference to the intensity of the free-bound continuum at the series limit. Space-resolved electron densities higher than the critical density were obtained using the program when applied it to Ly-α line wings radiated by line-shaped laser plasmas of Mg slab targets irradiated with different laser intensity.

II. THEORY

Taking the effects of Dedye screening and radiator-perturber correlations, the probability distribution of the ion microfield strength can be expressed as[1]

$$\omega(\beta) = 3x^4 exp\left[-\frac{(Z-1)a'^2}{3x}exp[-(1+Z_p)^{1/2}a'x]\right]$$
$$\times \left[a'[2+\frac{2}{a'x}+a'x]exp(-a'x)\right]^{-1} \quad (1)$$

and

$$a' = \frac{r_p}{\lambda_D} \quad (2)$$

$$r_p = \left[\frac{1}{(4/3)\pi N_p}\right]^{1/3} \qquad \lambda_D = \left[\frac{kT_e}{4\pi N_e e^2}\right] \quad (3)$$

where a' is a screening parameter, $x = r/r_p$, λ_D is the Debye screening length, r_p is the averaged separation of ions(perturbers) in the plasma. Z_p is the charge and N_p the number density of the ions of the plasma.

Defining $\beta = E/E_0$, where

$$E_0 = \frac{Z_p e}{r_p} \quad (4)$$

and E is a screened field then

$$E = \frac{Z_p e}{r^2}[1+\frac{r}{\lambda_D}]exp(-r/\lambda_D) \quad (5)$$

$$\beta = \frac{1}{x^2}[1+a'x]exp(-a'x) \quad (6)$$

The important feature to notice when cast in Eq.(1) is the very weak dependence of Z_p; indeed for strong fields, as x tend to zero, the Z_p dependence disappears altogether.

At small values of a' changes of Z_p cause insignificant changes in the distribution so that the question of the accuracy of the correction does not arise. At large $a'(\sim 0.8)$ corrections of the order of 10% are to be expected for $\beta < 10$ and for moderate values of Z_p (10-20). To notice from Eq.(2), $a' \propto N_p^{1/6} Z_p^{1/2}$, at a given value of a', as Z_p increases, the effect of radiator-perturber correlations is reduced at higher Z_p because of the increase of separation of perturbers. When β is at small values, it is necessary to restrict the values of a' and Z to ensure that only small correction may be brought about by Z_p and a'. Within the parameter regime

$$(Z-2)a' \leq 0.3 \quad (7)$$

i.e. with T_e measured in electron volts,

$$\frac{(Z-2)Z_p^{2/3}N_e^{1/3}}{T_e} \leq 5 \times 10^6$$

anywhere. For MgXII, one can get $T_e \geq 100eV$ with $N_e = 10^{21}cm^{-3}$, the accurate upper electron density limit in determination of electron density in H-like magnesium plasma is about $5 \times 10^{21}cm^{-3}$ (At higher densities the method does not become impossible but the accuracy is bound to get worse).

The intensity of emission at any frequency $\Delta\lambda = CE$ in the line wing is proportional to the probability of a plasma microfield; the line-shape function is given by

$$L(\nu) = \frac{k}{CE_0}\omega(\beta) \quad (8)$$

values of C, with suitable Z scaling, may be derived from 'asymptotic Holtsmark coefficients given by Griem(1974)[2], k is a normalization constant such that $\int L(\nu)d\nu = 1$. The theoretical profiles can be calculated from

$$D(\nu) = N_n A_{n1}\frac{h\nu}{4\pi}L(\nu)d\nu G(\nu) \quad (9)$$

where $G(\nu)$ is the response function of the whole detection system which can be derived by experimental calibration. As for a small range of frequency variation, $G(\nu)$ is approximately a constant, N_n is the number density in upper state, A_{n1} is the radiation rate.

Substituting Eq.(1) and Eq.(8) to Eq.(9), and considering $\nu = CE$ and $\beta = \nu/CE_0$, giving

$$D(\nu) = C'\frac{\nu}{CE_0}\omega\left(\frac{\nu}{CE_0}\right) \quad (10)$$

then E_0 can be obtained through fitting of experimental line wing by means of nonlinear least square algorithm. As

$$E_0 = \left[\frac{4\pi}{3}\right]^{2/3} e\left[\sum Z_p^{3/2}N_p\right]^{2/3} \quad (11)$$

$$N_e = \sum Z_p N_p = Z_{eff}N_i \quad (12)$$

so N_e can be derived when knowing the population of the various ionination stages.

Two main ion species in Mg laser plasma are MgXII(Z_p =11) and MgXI(Z_p =10) in our experimental conditions, therefore determination of N_e using $Z_{eff} = 11$ will not bring about significant error when noticing $E_0 \propto Z_p^{1/3}$ from Eq.(11) and Eq.(12).

Fig.1 shows the theoretical intensity profiles of the Ly-α line wings with differrnt N_e. It is interesting to notice the feature that the method is only sensitive to higher densities of the local plasma.

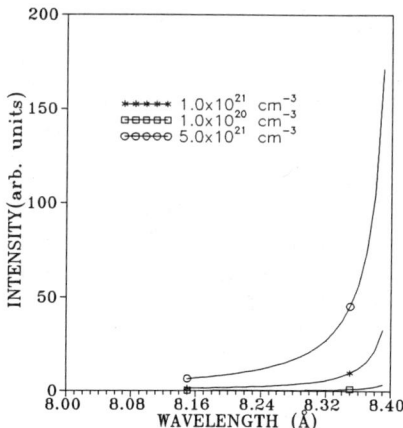

Figure 1: Variations of Ly-α line-wing intensity with electron density

III. EXPERIMENTAL SETUP AND RESULTS

The experiment was carried out at the LF12 Laser Facility of SIOFM. An aspherical lens(f/1.7) is combined with a sylindrical lens array to form a 12mm × 120μm uniform focal line(the non-uniformity of laser illumination along the focal line is within 5%) onto the Mg-slab target surface. In the experimental, 600J and 200J laser beams of 1.05μm with the FWHM of 900ps were used in order to investigate the density profiles as a function of the distance perpendicular to the target surface and the characteristics in the conduction regions of Mg laser plasmas under different laser flux. The corresponding laser intensities on the target surface are $5.0 \times 10^{13} W cm^{-3}$ and $1.7 \times 10^{13} W cm^{-3}$, respectively. The space-resolved spectra(7-10Å) of the line-shaped Mg laser plasmas were taken by a TAP pinhole crystal spectrograph[3] which has spectral and space resolving power of $\lambda/\Delta\lambda \sim 800$ and 50μm,respectively . The spectra were recorded with calibrated Shanghai 5F X-ray film.

For Mg laser plasma the slight change of a' will not bring about obvious effect on the fitting results so far as $a' \leq 0.17$(Eq.(7)) We chose $a'=0.1$ in our data processing. Fitting curves were obtained as shown in Fig.2. The electron density profiles are shown in Fig.3.

It is interesting to notice that each electron density profile has a flat density region near the target surface with widths of($250\pm25)\mu m$ and $(150\pm 25)\mu m$ for laser intensity of 600J and 200J,respectively.These are caused by the motions of the critical surface to the vacuum. This characteristic can be used to study the hydrodynamics in the conduction region of the plasmas.

Three considerations has been taken into account in the data processing: (i)the fitting region should be varying with the distance from the target,(ii)the line wings should be pre-smoothed as the noise is overlapped upon

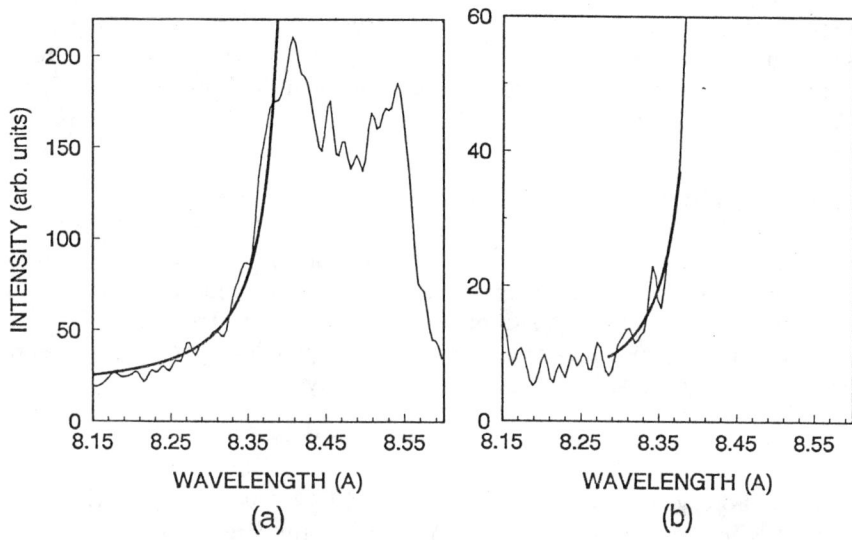

Figure 2: Line wings of Ly-α of MgXII and their fitting curves, (a) 600J, (b) 200J

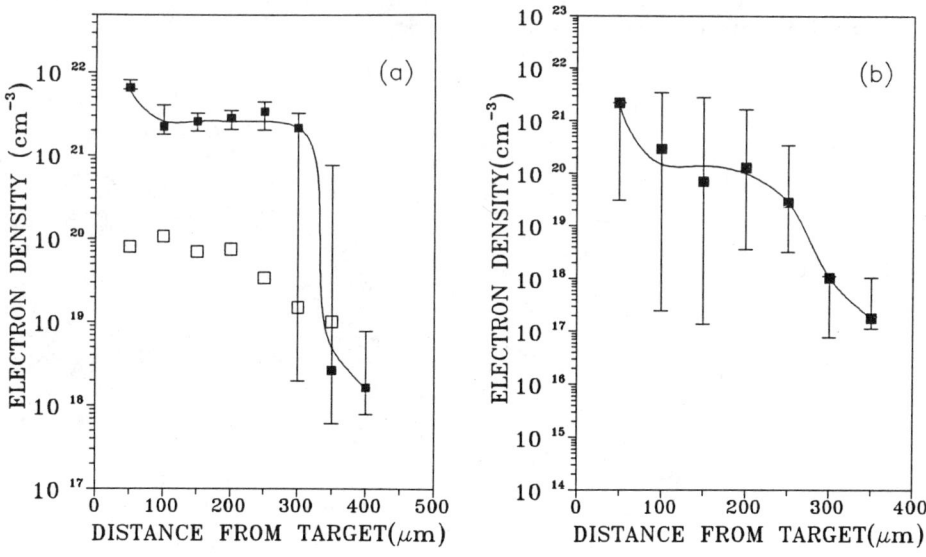

Figure 3: Electron density profiles perpendicular to the target surface, (a) 600J, (b) 200J (where □ was obtained by I_r/I_I of MgXI($1s^2 - 1s2p$))

the spectrum, (iii) taking into account of the uncertainty of the line center position in calibrations, a half FWHM of 2p-1s line was added in the fitting region, so an error bar was obtained for each position from the target surface(as shown in Fig.3).

IV. SUMMARY

The electron density measurement by means of nonlinear least square fitting of the Stark-broadened line wings of Ly-α line in laser plasma is one of the most accuracy method which can be used to measure electron densities above the critical density. Results of electron density profiles of Mg laser plasmas are obtained under 600J and 200J laser intensity.

ACKNOWLEDGMENTS

The authors gratefully acknowledge the LF12 Laser Facility Operation Group for their efforts, and Dr. Wei Yu for helpful discussions.

REFERENCES

1. C.C.Smith and N.J.Peacock, J.Phys. B:Atom.Molec.Phys.,1978,11(15),2749
2. H.R.Griem, Spectral Line broadening by Plasma,1974,313
3. S.S.Han, L.Sun,Z.Z.Xu et al, Inst.Phys.Conf.Ser,1992,No.125:Section 7,383

An Experimental Study on Line-shaped Laser-plasma

Shen-sheng Han, Ling-qing Zhang, Zhi-zhan Xu,
Zheng-quan Zhang, Pin-zhong Fan, and Lan Sun

Shanghai Institute of Optics and Fine Mechanics
P.O.Box 800-211, Shanghai, 201800, P.R.China

Abstract. Spectral diagnostic technique was used to study the properties of line-shaped laser plasmas. Electron temperature and electron density profiles were deduced from the spectra obtained by a pinhole crystal spectrograph. The Stark-broadened line-wings of hydrogenic ions were used to deduce the electron density beyond the critical surface, and the results were compared with the theoretical model of planar laser-driven ablation. Time integrated properties of line-shaped laser plasma were also discussed and compared with those of spot-shaped laser plasma and the laser absorption theoretical estimations.

I. INTRODUCTION

Recently, there has been a considerable interest in distributed absorption of laser light by a plasma[1][2][3], because high-gain laser-fusion pellets and electron collisional X-ray laser targets must be irradiated by long pulse laser light. These pulses produce large scale length plasmas that are cool enough that inverse bremsstrahlung is very strong, so that laser light absorption is not localized at the critical surface, but distributed over densities from well below the critical density to the critical density. However, few experimental data about the conduction region is available in spite of the development of analytical models of distribution absorption. This is mainly due to the difficulties of plasma diagnostics in the conduction region where the electron density is higher than the critical density.

In the present work, a pinhole crystal spectrograph[4] was used to study the properties of a Mg line-shaped laser plasma. The Stark-broadened line-wings of hydrogenic ions were used to deduce the electron density beyond the critical surface, and the results were compared with the theoretical model of distributed absorption of laser light by plasma.

In the next section, we describe the experimental setup and the diagnostic method. Sec.III is devoted to the experimental results and the comparison of experimental data with theoretical results. In section IV, we compared the time integrated properties of line-shaped laser plasma with spot-shaped laser plasma and the theoretical estimations of laser absorption. The summary of the results and a discussion are given in Sec.V.

II. EXPERIMENTAL SET-UP AND THE DIAGNOSTIC METHOD

The experiment was carried out on the LF12 Laser Facility at SIOFM. A spherical lens (f/1.7) was combined with a cylindrical lens array to form a 12mm × 120 µm uniform focal line onto the target surface. A 1.05 µm glass: Nd laser of energy typically 600 J in 900 ps was focused onto a polished magnesium slab with an irradiation of about 5×10^{13} W/cm^{-3}. The length of target was 10mm. A TAP pinhole crystal spectrograph was used to measure the space-resolved X-ray spectra along the normal of target surface. The electron temperature was evaluated from the ratio of intensities of two-electron satellites and the resonance line of H-like ions(5). The electron density was derived from the Stark-broadened line-wings of hydrogenic ions(6)(7) and the ratio of resonance and intercombination lines of He-like ions(8) where the intensity of intercombination line was taken as the difference between the measured line intensity and the calculated intensities of satellites s and t(9). The main advantage of Stark-broadened line-wings diagnostic technique is that: (a) it can reduce the inaccuracy induced by the opacity near the target surface, and (b) because the profile of the line-wing is mainly determined by the highest electron density the spectrograph observed, the diagnostic method has some time-resolving ability in a time-integrated measurement.

III. EXPERIMENTAL RESULTS

The profiles of electron density of the laser plasma irradiated with 600J or 200J pumping energy are given out in Fig.1. A noticeable flat region was observed in both high and low irradiation intensity experiments, which is one of the typical characteristics of conduction region(10). The time needed to establish a steady-state flow structure could be estimated by(1)

$$t_0 \simeq 0.18 nsec \left[\frac{A}{2Z}\right]^{7/6} \left[\frac{ZL}{100\mu m}\right]^{2/3} \left[\frac{I_a}{10^{13}W/cm^2}\right]^{1/3} \left[\frac{\lambda_L}{1\mu m}\right]^{4/3}$$

which equals to 0.88 ns and 0.90 ns for I = 600 J and I = 200 J respectively. The steady state width Δ_s and electron density of conduction region for magnesium target predicted by the model of planar laser-driven ablation are(1)(2)(3)(10):

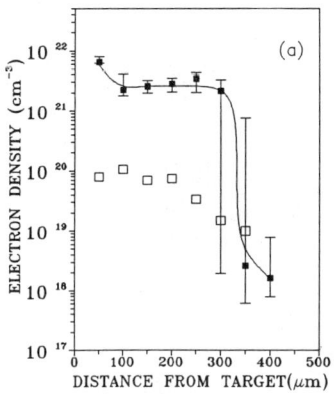 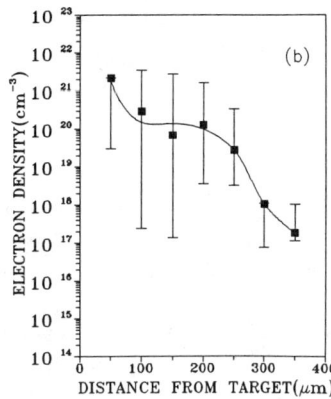

Figure 1: Electron density profiles for magnesium line-shaped laser plasmas. (a) $I_0 = 5 \times 10^{13} W/cm^2$; (b) $I_0 = 1.7 \times 10^{13} W/cm^2$; The error bar was obtained by taking the half of FWHM of 1s-2p line as the wavelength calibration uncertainty in our least-square fitting. (profile □ in (a) was obtained from I_r/I_I of MgXI). adopted from Ref.7

$$\Delta_s = 13.8 \mu m \left[\frac{A}{2Z}\right]^{7/9} \left[\frac{ZL}{100\mu m}\right]^{7/9} \left[\frac{I_a}{10^{13}W/cm^2}\right]^{5/9} \left[\frac{\lambda_L}{1\mu m}\right]^{14/9} \quad (1)$$

$$N_e(x) = 3.64 \times 10^{21} cm^{-3} \left[\frac{A}{2Z}\right]^{1/6} \left[\frac{100\mu m}{ZL}\right]^{1/3} \left[\frac{I_a}{10^{13}W/cm^2}\right]^{1/3}$$

$$\times \left[\frac{1\mu m}{\lambda_L}\right]^{2/3} \left[\frac{x - x_a}{\Delta_s}\right]^{-3/5} \quad (2)$$

The density gradient length in the corona region L is taken here to be constant and equals to the width of focal line.

In Fig.2, the experimental data of the width and the average electron density of conduction region for different irradiation intensities are compared with the theoretical results and their agreement are very well.

Another noticeable character of the electron density profile is the discontinue jump near the sonic point at high irradiation intensity. Here we attribute it to the effect of flux-limited thermal conduction. The upper-limit of flux-limited factor of electron thermal conduction could be estimated from the electron density distribution near the sonic surface by(2):

$$f \leq 0.04 \left[\frac{(Z+1)^{3/2}}{ZA^{1/2}}\right] \left[\frac{N_s}{N_c}\right]^{1/2} \quad (3)$$

where N_s and N_c are electron density at sonic surface and critical surface respectively. From the measured electron density profile, the flux-limited

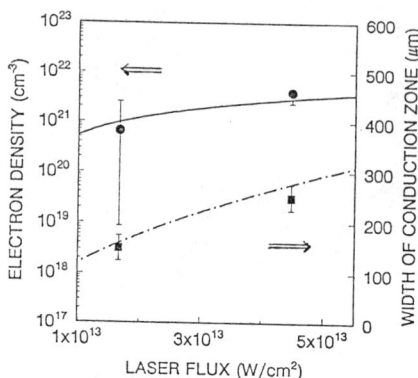

Figure 2: Comparison of experimental and theoretical width and electron density of conduction region as a function of irradiation intensity.($\bar{N}_e = \frac{1}{N}\sum_i^N N_e(x_i)$)

factor was estimated to be less than 0.03 in our high irradiation intensity experiment.

IV. ELECTRON THERMAL PRESSURE COMPARISON OF LINE-SHAPED AND SPOT-SHAPED LASER PLASMA

The time-integrated electron thermal pressure ($P_{th} = N_e T_e$) profiles obtained in different experiments using the same target and diagnostic methods(1)(2) are shown in Fig.3. Electron temperature profile of Mg line-shaped laser plasma is shown in Fig.4. The ratios of T/T_e' and P_{th}/P_{th}' for the same kind of targets and the pumping laser but different irradiation intensity $I(I')$ and pulse length $\tau(\tau')$ can be estimated from(13):

$$\frac{T_e}{T_e'} = \left[\frac{I}{I'}\right]^{1/2}\left[\frac{\tau}{\tau'}\right]^{1/4}, \qquad \frac{P_{th}}{P_{th}'} = \left[\frac{I}{I'}\right]^{3/4}\left[\frac{\tau}{\tau'}\right]^{1/8}$$

here we neglected the difference between the effective charge Z and Z'.

In table 1. we present the theoretical and experimental values of the ratio of electron temperature and thermal electron pressure for the three experiments showed in Fig.3.

TABLE 1.

Ratio	$\langle T_1\rangle/\langle T_2\rangle$	$\langle T_2\rangle/\langle T_3\rangle$	$\langle P_1\rangle/\langle P_2\rangle$	$\langle P_2\rangle/\langle P_3\rangle$
Experimental data	1.53	1.17	9.5	1.3
Theoretical value	1.54	3.76	8.1	5.2

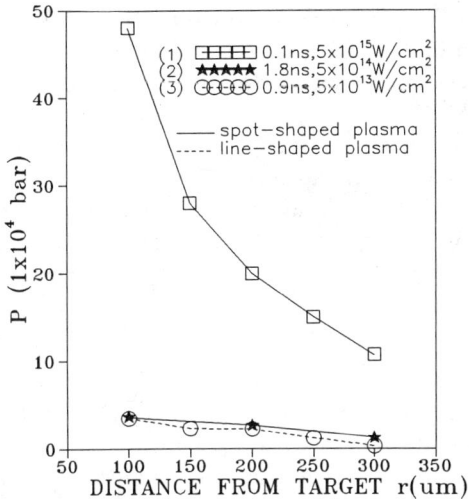

Figure 3: The time-integrated electron thermal pressure profiles obtained in previous and our experiments using the same diagnostic method. (spot diameter (1) 100 μm, (2) 80μm)

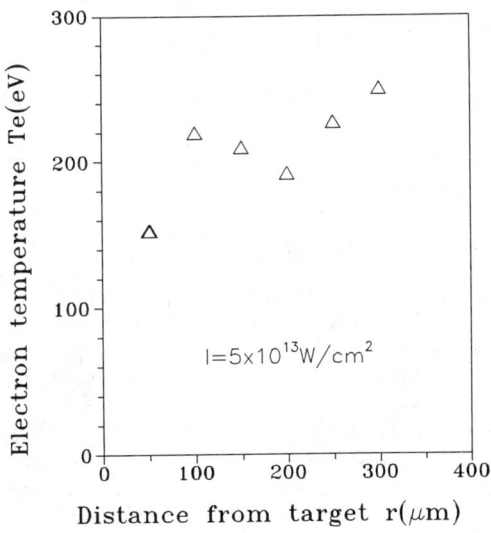

Figure 4: Electron temperature profile obtained by I_r/I_s of MgXII.

With the assumption that the space-averaged $\langle T \rangle$ and $\langle P_{th} \rangle$ being linear proportional to the T_e and P_{th}, the agreement between the experimental data and the theoretical estimated values for two spot-shaped laser plasmas are satisfied, but for $\langle T_2 \rangle \langle P_2 \rangle$ and $\langle T_3 \rangle \langle P_3 \rangle$, which corresponding to a spot-shaped and a line-shaped laser plasma, respectively, the deviation of the experimental data from the theoretical estimation is remarkable. The time-integrated electron temperature and thermal electron pressure in line-shaped laser plasmas are higher than that in the spot-shaped laser plasmas, indicating obviously a different plasma fluid dynamic behavior due to the different expansion dimensions.

IV. SUMMARY

The electron temperature and electron density distribution along the normal of target surface in line-shaped laser plasmas are obtained and compared with the planar laser-driven ablation model of nonlocalized absorption. Also compared are the plasma parameters of line-shaped laser plasma with those of spot-shaped laser plasmas. The experimental data of electron density and the width in the conduction region are in good agreement with the theoretical predictions of planar laser-driven ablation model, and the results also show that the plasma fluid dynamic behavior of line- and spot-shaped laser plasma are quite different.

ACKNOWLEDGMENTS

The author gratefully acknowledge the LF12 Laser Facility Operation Group for their efforts, and also indebted to Dr. Wei Yu and Dr. Hua-Guo Teng for the enlightening discussions. This work was supported by the National High Technology Program of China.

REFERENCES

1. F Dahmani and T Kerdja, Phys. Fluids B, Vol.3, N0.5, 1232 (1991)
2. J.S.De Groot, S.M.Cameren and K.Mizuno, Phys. Fluids B, Vol.3, No.5, 1241 (1991)
3. J.S.De Groot, K.G.Estabrook et al, Phys. Fluids B, Vol.4, No.3, 701 (1992)
4. Shen-sheng Han, Lan Sun, Zhi-Zhan Xu, et al, Inst. Phys. Conf. Ser, 1992, No.125: Section 7, X-ray Lasers 1992, 383
5. E.V.Aglitskii, et al; Sov. J. Quant. Electron., Vol.4, No.3, 322 (1974)
6. C.C.Smith and N.J.Peacock, J. Phys. B: Atom. Molec. Phys., Vol.11, No.15, 2749 (1978)
7. Ling-qing Zhang, et al; "Space-resolved electron density measurements using the Stark-broadened line wings of hydrogenic ions in line-shaped laser plasmas", This colloquium
8. A.V.Vinogradov, et al; Sov. J. Quant. Electron., Vol.5, No.6, 630 (1975)
9. A.A.Ilynkhin, et al; Sov. J. Quant. Electron., Vol.11, No.1, 34 (1981)
10. R.Fabbro, C.Max and E.Fabre, Phys. Fluids, Vol.28, No.5, 1463 (1985)
11. V.I.Bayanov, et al; Sov. J. Quant. Electron., Vol.6, No.10, 1226 (1976)
12. V.A.Boiko, S.A.Pikuz, A.Ya.Faenov; J. Phys. B; Vol.12, 1889 (1979)
13. Patrick Mora; Phys. Fluids, Vol.25, No.6, 1051 (1982)

Observation of resonant photo-excited fluorescence and application to plasma diagnostics

H.Yashiro, T.Tomie, Y.Matsumoto, I.Matsushima and Y.Owadano

Electrotechnical Laboratory, AIST, MITI, Umezono 1-1-4, Tsukuba, Ibaraki 305, Japan

Abstract. Resonantly photo-excited Al He α (7.76 Å) fluorescence is observed in the configuration where the photo-excited Al plasma expanded parallel to the blow off direction of the other Al plasma as a flash lamp. Both plasmas are produced by the irradiation of 10 ns KrF laser pulses on an Al slab target with a step-height of around 500 μm. Doppler de-coupling between two plasmas caused by the steep velocity gradient in expanding plasmas is reduced in this configuration. From the spatial distribution of the fluorescence, the scale length of velocity gradient is evaluated.

INTRODUCTION

Resonant photo-excitation has been studied as one of the pumping scheme of x-ray lasers(1-9). The fluorescence intensity resulting from photo-pumping is proportional to the number of pumped ions, and the excitation efficiency depends on the spectral matching between the pumping line and pumped transition. Hence, resonant photo-excitation can be a very good diagnostic method of laser-produced plasmas. Spatial and temporal variation of ions of different charge number will give an information of the three-body recombination rate which is decided by the electron temperature and density. Spectral mismatching will be mainly caused by the motional Doppler effect caused by the plasma expansion. Therefore, excitation efficiency will give the information on the plasma expansion velocity gradient. In this paper, we report an experiment to demonstrate that resonant photo-excitation is a very good diagnostic of laser-produced plasmas. From the observed Al He α fluorescence (7.76 Å), spatial dependence of the ground state population of He-like Al ion and the scale length of the plasma expansion velocity gradient are evaluated.

EXPERIMENTAL ARRANGEMENT

The schematic diagrams of the experimental setup are illustrated in Fig. 1. Two plasmas blowing off parallel to each other are produced on a step-shaped Al target. This configuration reduces the Doppler de-coupling effect, as explained in b), and enhances resonant-pumping efficiency. Two plasmas were generated using 6 beams of 10 ns KrF laser pulse from ASHURA(10). One laser beam was focused on the lower shelf of the step-shaped target to produce a pumped plasma. The rest 5 beams were focused on the upper shelf of the target to produce a lamp

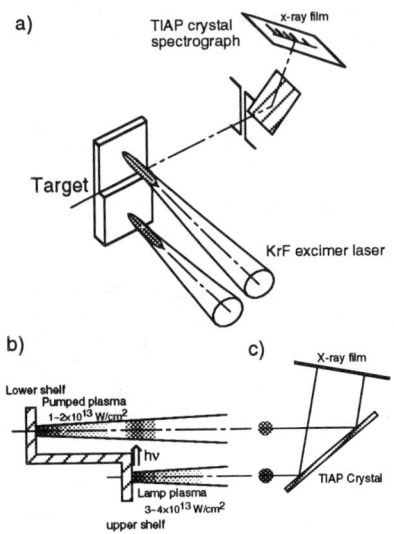

Figure 1 Schematic diagrams of experimental arrangement.

plasma. The separation of two plasmas was 500 μm and the step-height of the target was varied. The x-ray spectra of the Al plasmas were observed using a TlAP (001) crystal spectrograph as shown in Fig. 1a) and recorded on a Kodak DEF film. Space resolved observation was performed by the use of a 25 μm width slit. In this configuration, spatially resolved spectra of two plasmas can be recorded separately. The principle of separating two spectra is explained in Fig. 1c). Because two plasmas are separated along the spectral dispersion direction, spectral lines of the same wavelength from two plasmas are slightly shifted on the film. If the spectral widths are narrow enough, we can distinguish spectral lines emitted from two plasmas. This was possible with the combination of 500 μm plasma separation and the use of TlAP crystal. The irradiance of the laser beams for the production of the pumped and the lamp plasmas were $1\sim2\times10^{13}$ and $3\sim4\times10^{13}$ W/cm^2, respectively. The diameter of each plasma was measured by a x-ray pinhole camera to be about 100 and 130 μm for the pumped and lamp plasmas, respectively.

EXPERIMENTAL RESULTS

The microdensitometer traces of the soft x-ray spectra are shown in Fig. 2 for the case of 540 μm step-height target. The spectrum on the surface of the lower shelf is shown in Fig. 2a). The x-ray spectra at 540 μm from the lower shelf of the target is shown in Fig. 2b). The spectrum seen in Fig. 2a) was emitted from the plasma produced on the lower shelf, and the spectrum in Fig. 2b) was emitted mainly from the plasma on the upper shelf. The reader will notice both spectra are shifted slightly, which was caused by 500 μm separation of two plasmas along the

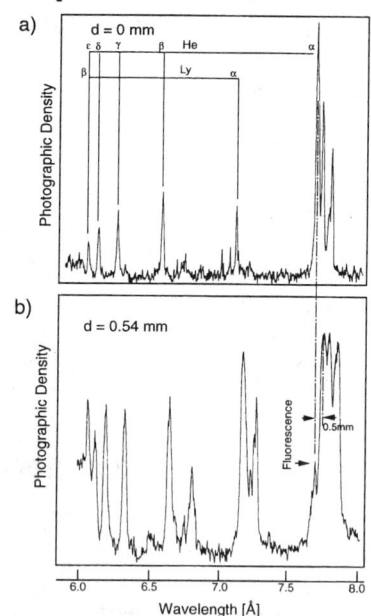

Figure 2. The microdensitometer traces of x-ray spectra. a) On the surface of the lower shelf of the step-target. b) At 0.54 mm from the surface which is step-height of the target.

dispersion direction. The reader will also notice there is an extra component line for He α in Fig. 2b). The position of this extra component in Fig. 2b) is exactly the same with that of the strongest component line of He α in Fig. 2a). The extra component in Fig. 2b) did not appear when the pumped plasma did not exist. Thus we can say, the extra component in Fig. 2b) is the He α emission from the pumped plasma while all other lines in Fig. 2b) are the emissions from the lamp plasma.

The observed He α line intensity emitted from the pumped plasma is plotted in Fig. 3 as a function of the distance from the target surface. The photographic density of the spectral line is converted to intensity by using the published data (11). The He α intensity decreased as the distance from the target increased. However, at 0.4 mm the line intensity increased, peaked at around 0.6 mm and disappeared around 0.8 mm. The position of the peak intensity was close to the surface of the upper shelf of the step-shaped target. When there was no lamp plasma, the He α intensity decayed exponentially. Hence, He α emission peaked at 0.6 mm is clearly the fluorescence photo-excited by the He α spectral line from the lamp plasma.

The fluorescence intensity is proportional to the product of lamp intensity, ground state population of He-like ions, and the excitation efficiency. Therefore, fluorescence intensity variation with the step-height of the target should reflect the ground state population of He-like ions when the laser irradiation condition is maintained. The observed peak intensities of the He α fluorescence are plotted in Fig. 4 as a function of the height of the step-shaped target. Thus, we know the ground state population density of He-like ions decayed with the scale length of about 180 μm.

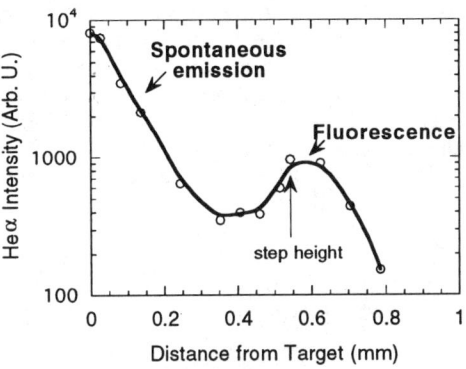

Figure 3. He α intensity from pumped plasma as a function of distance from lower shelf of the step-shaped target. The step-height of the target was 0.54 mm.

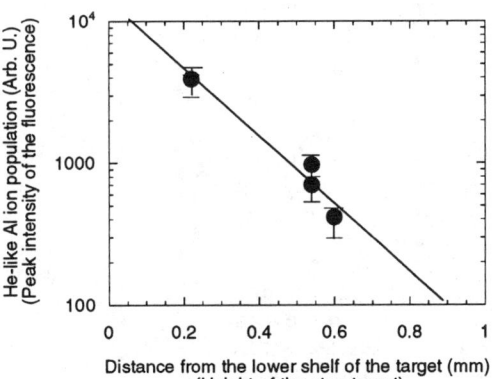

Figure 4. Peak intensity of He α fluorescence as a function of step-height. This result shows the spatial dependence of the He-like Al ion population of the pumped plasma.

DISCUSSION

The photo-excitation efficiency is determined by the spectral matching of the spectral lines from two plasmas. Resonant excitation perpendicular to the blow off direction does not suffer from motional Doppler de-coupling effect. Pumping lines are emitted mainly near the target surface. At the step-height distance from the lower shelf, therefore, the pumped plasma can be resonantly excited effectively by spectral lines from the lamp plasma. However, at further position, the angle between the expanding direction and the pumping direction decreases. Then, the Doppler de-coupling effect becomes significant. Thus, spatial profile of the fluorescence should reflect the velocity distribution. In the following, we try to estimate the velocity gradient.

When spectral profile is assumed Gaussian, the Doppler de-coupling effect is described as follows,

$$\phi(\lambda) = \frac{\Delta\lambda}{2(\ln 2)^{\frac{1}{2}} \pi^{\frac{1}{2}} \Delta\lambda_{\frac{1}{2}D}} \exp[-(\frac{\Delta\lambda}{\Delta\lambda_{\frac{1}{2}D}})^2], \qquad (1)$$

where $\Delta\lambda_{\frac{1}{2}D}$ is the FWHM of the spectral lines. $\Delta\lambda$ is spectral mismatching caused by the difference of expansion velocity components of two plasmas along the pumping direction. In the PET crystal spectrograph observation, the He α spectral line width of the flash lamp was measured to be about 3 mÅ. For the simplicity, we assume that expansion velocity of the plasma is zero on the surface, increases linearly, and reaches the maximum at distance z. For the maximum velocities, we assume 4.1×10^7 and 5.4×10^7 cm/s for the laser irradiance of 1×10^{13} and 4×10^{13} W/cm^2, respectively, which were measured by Faraday cups in our previous experiments.

In Fig. 5, thus calculated photo-pumping efficiencies for several accelerating distances z are shown. The best agreement with the experimentally observed profile was obtained for z = 300 μm. This value is quite reasonable because it is 2 to 3 times of the laser spot diameter.

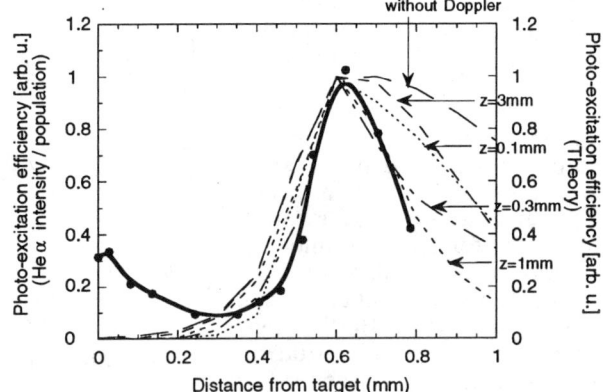

Figure 5. Experimentally observed spatial variation of the photo-excitation efficiency is compared to the calculation of motional Doppler de-coupling. The parameter in the calculation is the scale length of the acceleration of the plasma.

SUMMARY

The fluorescence of the Al XII He α spectral line at 7.76 Å was observed using the step-shaped target by which Doppler de-coupling between the pumped plasma and lamp plasma was greatly reduced. From the fluorescence intensities for various step-heights, spatial distribution of the ground state population of He-like Al ions was obtained. From the fluorescence spatial profile, scale length of the velocity gradient of the plasma was evaluated to be 300 µm.

ACKNOWLEDGMENTS

The authors would like to thank I.Okuda for his efforts in constructing the driver laser ASHURA and technical advice for laser operation and also Dr.E.Miura and Dr.K.Koyama for their helpful discussion concerning this investigation. They also appreciate the production of the experimental devices by the Electrotechnical Laboratory staff.

REFERENCES

1. J.P.Apruzese, J.Davis, and K.G.Whitney, J. Appl. Phys. **53** 4020, (1982).
2. J.P.Apruzese, G.Mehlman, J.Davis, J.E.Rogerson, V.E.Scherrer, S.J.Stepanakis, P.F.Ottinger, and F.C.Young, Phys. Rev. A**35**, 4896, (1987).
3. B.N.Chichkov and E.E.Fill, Phys. Rev. A**42**, 599 (1990).
4. J.Nilsen, Phys. Rev. A**40**, 5440 (1989).
5. J.Nilsen, Opt. Lett. **15**, 798 (1990).
6. J.Nilsen, J.H.Scofield, and E.A.Chandler, Appl. Opt. **31**, 4950, (1992).
7. J.Nilsen, Appl. Opt. **31**, 4957, (1992).
8. P.Beiersdorfer, S.R.Elliott, and J.Nilsen, Phys. Rev. A**49**, 3123 (1994).
9. J.Trebes and M.Krishnan, Phys. Rev. Lett. 50, 679 (1983).
10. Y.Owadano, I.Okuda, Y.Mastumoto, M.Tanimoto, T.Tomie, K.Koyama, and M.Yano, Laser and Particle Beams **7**, 383 (1989).
11. B.L.Henke, J.Y.Uejio, G.F.Stone, C.H.Dittmore and F.G.Fujiwara, J. Opt. Am. B**3**, 1540, (1986).

Higher Order Structure Analysis of Nano-Materials by Spectral Reflectance of Laser-Plasma Soft X-Ray

Hirozumi Azuma, Akihiro Takeichi and Shoji Noda

TOYOTA Central Research and Development Laboratories, Inc., Nagakute, Aichi, 480-11, JAPAN

ABSTRUCT: We have proposed a new experimental arrangement to measure spectral reflectance of nano-materials for analyzing higher order structure with laser-plasma soft X-rays. Structure modification of annealed Mo/Si multilayers and a nylon-6/clay hybrid with poor periodicity was investigated. The measurement of the spectral reflectance of soft X-rays from laser-produced plasma was found to be a useful method for the structure analysis of nano-materials, especially those of rather poor periodicity.

INTRODUCTION

Laser-produced plasma soft X-rays have emerged as a very bright pulse soft X-ray on a laboratory scale since a reliable and compact high power laser has become available. Applications of laser-produced plasma soft X-rays to contact microscopes, lithography and plasma diagonostics have been reported. On the other hand, there has been significant progress in recent years in the development of soft X-ray optics. A Mo/Si periodic multilayer film has been established as a good soft X-ray mirror. The thermal stability of the film is one of the most important factors for ensuring good performance as a mirror. We have proposed a new experimental arrangement for measuring spectral reflectance of multilayers in the soft X-ray region with a single laser shot(1), in which laser-produced plasma is used as a light source. The purpose of this work is to evaluate the spectral reflectance of soft X-rays as a method of analyzing nm-order structure.

In this paper, the results of analysis of nm-structure of Mo/Si multilayers annealed at various temperatures are presented and thermally-induced structure modifications are discussed(2). Also, we applied the spectral reflectance in the soft X-ray region to the structure analysis of a polymer composite(3), a nylon6-clay hybrid (NCH). NCH is prepared by polymerizing ε-caprolactam in the inner spaces between silicate layers of clay (montmorillonite). NCH has multilayer structure composed of nylon-6 and a thin silicate, and its periodicity is about 10 nm but not definite. The properties of NCH depend on the microstructure which depends on the composition and the processing conditions. Thus, it is important to determine the microstructure of NCH produced under a variety of conditions.

EXPERIMENTAL

Spectral reflectance of Mo/Si multilayer mirrors was measured at a given incidence angle in a single laser shot, and polymer composites (NCH) with poor reflectivity at a given incidence angle was measured in 2000 laser shots from a moderate-energy Nd:YAG laser. For soft X-ray generation, an aluminum rod target was irradiated with a pulsed laser of 532 nm wavelength, 8 ns pulse duration and 0.4 J energy on the aluminum target, delivered from a commercial Nd:YAG laser (CONTINUUM, model NY81-10). The laser beam was focused with a spherical lens of 20 cm focal length to spot with a diameter less than 0.007 cm which was estimated from the burn pattern on the target. The irradiation intensity of approximately 1.3×10^{12} W/cm^2 was achieved. The spectral reflectances of multilayer mirrors were measured using a grazing incidence XUV spectrometer that was positioned at an angle of 90° from the laser incidence direction. The optical arrangement of the spectrometer is shown in Fig. 1. The soft X-rays from the laser-produced plasma were collected, with a 2-degree grazing incidence, at an Au-coated toroidal mirror that was located at a distance of 340.8 mm from the source plasma. The toroidal mirror formed a tangential line focus image of the source on the spectrometer slit (located at a distance of 95 mm from the toroidal mirror) and a sagittal line focus image of the source on the detector plane (located at a distance of about 472 mm along the optical path from the slit). The spectrometer had a concave varied-spacing diffraction grating with 1200 grooves/mm (HITACHI, model 001-0437), with which the spectrum was formed on a flat field (the detector plane) perpendicular to the grating surface. A specimen was placed between the grating and the detector plane on a rotatable mount. A photographic plate, which is sensitive in the XUV region (ILFORD, Q-plate), was placed on a flat detector plane parallel to the rotation axis of the specimen. A vertically-dispersed and horizontally-narrow spectrum was recorded on a photographic plate at an angle of 2θ deflected from the optical axis of the spectrometer, where θ is the incidence angle on the specimen projected onto the horizontal plane (see Fig. 1(a)).

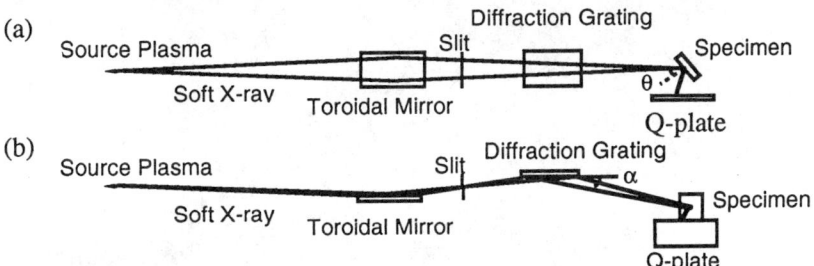

FIGURE 1. Schematic diagram of a grazing-incidence flat field spectrometer equipped with a toroidal mirror used for measuring spectral reflectance of a multilayer mirror and a NCH. (a)top view, (b) side view.

The incidence angle on the mirror, θ_{inc}, is determined by the trigonometric relationship:

$$\theta_{inc} = \tan^{-1}\{(\sin^2\theta + \tan^2\alpha)^{1/2}/\cos\theta\},$$

where α is the wavelength-dependent diffraction angle in the vertical plane which is measured from the grating surface (see Fig. 1(b)). The photographic plate records several reflection spectra, each of which corresponds to the spectrum measured at a specific angle of the specimen. The spectral reflectance was determined by dividing the reflection spectrum by the source spectrum which was determined in a different experiment under the same irradiation conditions. The aluminum target was shifted and realigned for each shot; hence, a fresh surface was irradiated by the laser pulse every time. The spectrum of the XUV emission from the aluminum plasma consists of line emissions from Al-IV, Al-V, Al-VI and Al-VII ions. In the evaluation of the reflection of a specimen, the contribution of the second-order diffraction from the grating was deconvoluted. Reflection spectra of multilayer mirrors were recorded over the incidence angles of $\theta_{inc}=35$-$65°$ with an increment of $\Delta\theta=1°$ and reflection spectra of NCH were recorded over the incidence angles of 40, 50 and 60°. A multilayer mirror containing 30 layer-pairs with Mo and Si layer thicknesses of 4.8 nm and 7.2 nm, respectively, was fabricated on a silicon wafer by RF-sputtering. The multi-layer mirrors were also annealed for 3 h in a vacuum at 200, 300, 500 and 600 °C, and the structure modifications due to the annealing were examined.

Injection-molded NCH (containing 8 wt% montmorillonite) samples were used for the measurement. The cross-sectional TEM image of NCH is shown in Fig. 2. The sheet silicates (dark image) of about 1 nm thickness are observed with about 10nm spacing between adjacent sheet silicates. NCH actually has sea-and-island microstructure: the sea or matrix represents nylon-6 and the island represents a silicate/nylon6 multilayer.

FIGURE 2. Cross-sectional TEM image of NCH. The dark images show silicate layers of about 1 nm thickness.

MULTILAYER MIROR

Figure 3 shows the soft X-ray reflectance at a wavelength of 16.17 nm as a function of the incidence angle for Mo/Si multilayers. Reflectivities of the multilayers before and after annealing were determined by analyzing the reflection spectra using the characteristic response curve of the photographic plate (Q-plate) that was calibrated in a different experiment. The maximum reflectivity is about 10%, and the angular width where the reflectivity becomes half of the peak value is $\Delta\theta=7°$ for the Mo/Si multilayer mirror before annealing at an incidence angle of about 50°. For the multi-layer mirror annealed at 300 °C, the maximum reflectivity decreased at the incidence angle of 49° and another small peak appeared at 45°. This implies that some modification of the nm-structure starts at 300 °C. For the multilayer mirror annealed at 500 °C and 600 °C, the incidence angle for the peak reflectivity is about 43°, and the peak values are 7.4% and about 1% at each annealing temperature. A significant structure modification, shrinkage of the texture, occurred in the Mo/Si multilayer upon annealing at 500 °C. Molybdenum-silicide formation is responsible for the shrinkage. These results were found to be consistent with the theoretical calculation of the reflectivity and the TEM observations. It is clearly shown that the measurement of spectral reflectance of soft X-rays from laser plasma is an effective method for analyzing nm-order multilayer structure modification.

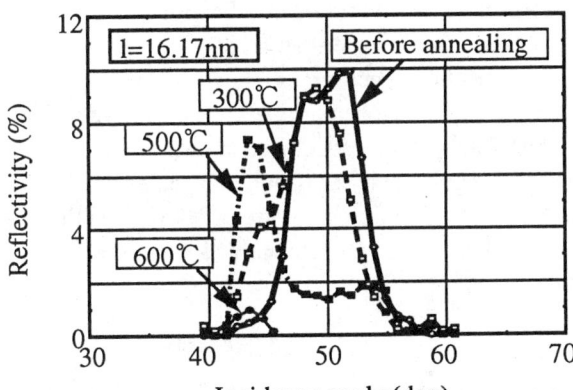

FIGURE 3. The measured reflectivity vs wavelength of 16.17nm for a Mo/Si multilayer mirror before and after annealing.

NCH

Figure 4 shows the soft X-ray reflection spectra of NCH at the incidence angles of 40°, 50° and 60° as a function of the wavelength. The reflectivity of NCH is three orders of magnitude smaller than that of a Mo/Si multilayer film.

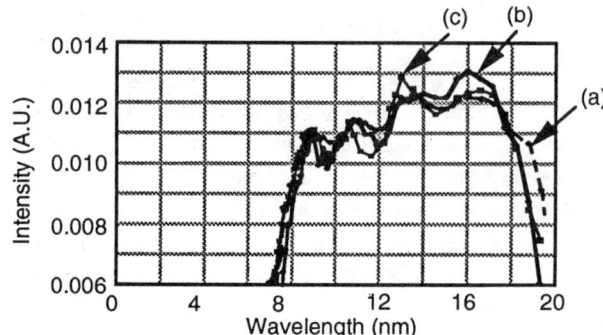

FIGURE 4. Soft X-ray reflection spectra from NCH at incidence angles of 40° (a), 50° (b) and 60° (c) as a function of wavelength.

The reflection spectrum has at least four discernible peaks at wavelengths of around 9 nm, 11 nm, 13 nm and 16 nm. These wavelengths correspond to those of the intense emission line groups. It is noted that spectrum (b) resembles spectrum (a), except for the higher reflectivity at around 16 nm, and that spectrum (c) also resembles spectrum (a), except for the higher reflectivity at around 13 nm. The reflectivity enhancement at those specific wavelengths results from the Bragg reflection of silicate/nylon-6 multilayers in NCH.

The enhancement either at 16nm in spectrum (b) or at 13nm in spectrum (c) indicates that the periodicity of the silicate/nylon-6 multilayer is about 13 nm, according to the Bragg reflection rule. No reflectivity enhancement is observed in spectrum (a) under the present experimental conditions. Spectrum (a) closely corresponds to the reflection spectrum of silicate/nylon-6 mixture without any special microstructure, or to a specular reflectance spectrum. The above results clearly demonstrate that spectral reflectance measurement in the soft X-ray region is useful for evaluating periodicity of nanometer scale in materials.

To deduce the more detailed microstructure of NCH from the experimental spectral reflectance, the spectral reflectance was simulated by assuming a sea-and-island model.

CONCLUSIONS

The measurement of spectral reflectance of soft X-rays and comparison with the calculated reflectance are effective for analysis of nm-order structure modification and for structure analysis of a nm-composite with low reflectivity.

REFERENCES

1. Azuma, H., Watanabe, Y., Kato, Y., Motohiro, H., Noda, S. and Murai, K., *Japnaese Journal of Applied Physics*, 31-2B, L203-L205 (1992)
2. Azuma, H., Takeichi, A and Noda, S., *Japanese Journal of Applied Physics*, 32, 2078 (1993)
3. Azuma, H., Takeichi, A and Noda, S., *Japanese Journal of Applied Physics*, 32, 5558 (1993)

Bright Picosecond X-rays from Intense Sub-Picosecond Laser-Plasma Interactions

A. Maksimchuk, J. Workman, X. Liu, U. Ellenberger,
S. Coe, C.-Y. Chien, and D. Umstadter

Center for Ultrafast Optical Science,
University of Michigan, Ann Arbor, MI 48109-2099
Phone: (313) 764-2284, Fax: (313) 763-4876

Abstract

Short-pulse, high-intensity laser-plasma interactions are investigated experimentally with temporally and spectrally resolved soft x-ray diagnostics. The emitted x-ray spectra from solid targets of various Z are characterized for a range of laser intensities ($I < 5 \times 10^{17}$ W/cm^2) and pulse widths ($\tau \sim 400$ fs). With low contrast (10^5), the x-ray spectrum in the $\lambda = 40$–100 Å spectral region is dominated by line emission, and the x-ray pulse duration is found to be long, $\tau_x > 100$ ps, characteristic of a long-scale-length, low-density plasma. Bright, picosecond, continuum emission, characteristic of a short-scalelength, high-density plasma, is produced only when a high laser contrast (10^{10}) is used. It is demonstrated experimentally that the pulsewidth of laser-produced x-ray radiation may be varied down to the picosecond time-scale by adjusting the incident ultrashort-pulse laser flux. This controls the peak electron temperature relative to the ionization potential, corresponding to the emitted x-ray photon energy of interest. The results are found to be consistent with the predictions of a hydrodynamics code coupled to an average atom model only if non-local thermodynamic equilibrium (NLTE) is assumed.

Short-pulse high-intensity lasers interacting with solid targets make possible the study of a new class of plasmas. They are unique because during the ultrashort laser pulse relatively little expansion occurs and the density scale length remains much less than the laser wavelength. This makes possible the deposition of a significant amount of the laser energy at densities much greater than the critical density. The x-ray pulse will be extremely bright because of the high density, the small dimensions of the laser spot size, and the ultrashort x-ray pulse width. The latter is due to rapid cooling by expansion and diffusion, which results from the steep temperature and density gradients, and the high collision rates at high density. These bright compact sources of ultrashort-pulses of x-rays have applications in time-resolved diffraction, holography, spectroscopy, microscopy or radiography studies of transient biological or physical phenomena.

Picosecond x-ray emission from laser-produced plasmas has been demonstrated in previous investigations. Murnane et al. [1] observed, using a filtered x-ray streak camera, that the laser contrast strongly affects the x-ray pulse width: the higher the contrast, the shorter the pulse width. These results were attributed to the fact that only with high contrast can the laser energy be deposited at high plasma density. Using high-resolution crystal spectroscopy in the keV region, Kieffer et al. [2] and Riley et al. [3] inferred the existence of a solid density plasma from x-ray line widths.

In this paper, we demonstrate the important effects of the laser intensity (I) on the x-ray pulsewidth. Time-resolved x-ray spectra from an intense subpicosecond laser interacting with solid targets of aluminum and gold are characterized with a high temporal and spectral resolution. Bright picosecond continuum emission in the 20–300 Å spectral region, characteristic of a solid density plasma, is measured only when a high laser contrast is used. Otherwise, longer pulse line-radiation, characteristic of a critical density plasma, dominates the emission. It is demonstrated experimentally for the first time that the x-ray pulse duration can be controlled by appropriately adjusting I, and thus the average electron temperature T_e. The plasma will generate x-ray line emission characteristic of the average ionization stage $\bar{Z}$ corresponding to T_0 when $T_e \sim T_0$, where T_0 is an "ionization temperature," which in equilibrium is taken to be approximately one third of the ionization potential for that stage. A given region of the plasma is heated to T_0 either by direct deposition of laser energy or diffusion of heat from neighboring regions. The rise-time of the x-ray pulse is determined by the time it takes to heat the region where the greatest emission occurs. The fall-time is given by the cooling rate, which should decrease exponentially because it is driven by the density and temperature gradients, both of which decrease as the plasma cools either by expansion into the vacuum, or by heat conduction into the colder regions of the solid.

The x-ray pulse should be extremely short when I is low, such that the peak T_e (T_{max}) just barely reaches T_0 and then rapidly decreases below T_0. At higher I such that $T_{max} > T_0$, the plasma emission coressponding to T_0 will continue until T_e drops below T_0 [4]. However, at later times, the gradients relax, and, consequently, so does the slope of the exponential decrease in emission.

The laser used in the experiment is a 400-fs terawatt Nd-glass laser system based on chirped pulse amplification with a contrast ratio of the fundamental 1.06 μm laser light measured to be 5×10^5 and increased to 10^{10} by frequency doubling. An off-axis parabolic mirror is used to focus the laser radiation on a solid target at normal incidence to a minimum spot size of 15 μm, corresponding to a maximum intensity in 2ω of 5×10^{17} W/cm^2. In order to decrease the incident flux on the target the laser spot size was defocused while keeping the total laser energy constant. The soft x-ray emission is dispersed using an imaging flat-field grazing-incidence variable-spaced grating spectrometer. Time-resolved spectra are obtained in a single shot using a streak camera, coupled to the spectrometer, with a 5-ps resolution that is read-out with a CCD camera. An x-ray pinhole camera, filtered with 25 μm of Be and coupled to an intensified microchannel plate detector, was used to monitor the laser spot size. Two absolutely calibrated *pin* diodes filtered with 50 μm and 100 μm of Be were used to monitor keV x-ray emission levels. To receive information about conversion efficiency of the laser radiation into x-rays in the range 1.5-5 keV we used DEF-film with steps of Be-filters of different thicknesses and with known characteristic curves [5].

Figure 1 shows the emission using 1.06 μm laser irradiation on an aluminum target ($Z = 13$), corresponding to low-contrast conditions. The central plot is a single-shot streak picture showing the x-ray amplitude (grayscale in units of CCD counts) versus wavelength (vertical axis in units of Å) and time (horizontal axis in units of picoseconds). To the left is a line-out in wavelength at the time of peak emission, and at the bottom is a line-out in time for $\lambda = 55$ Å ($h\nu = 225$ eV), corresponding to the $2s2p - 2s3d$ transition in AlX. It can be seen that the emission is dominated by lines and that the pulse duration is long, $\tau \sim 70$ ps (FWHM). Figure 1(b) shows aluminum emission using 2ω (0.53 μm) irradiation, corresponding to high-contrast conditions. To the left is a line-out in wavelength at the peak of emission, and at the bottom is a line-out in time for $\lambda = 55$ Å. Note that in the high-contrast case the emission is continuum dominated over the entire wavelength range and that the pulse duration ($\tau \sim 25$ ps) is shorter than in the low-contrast case. In both figures, the sharp cutoff near the C K-edge (44 Å) is due to absorption by the lexan film that supports the potassium bromide photocathode. The falltime is faster for the high-contrast 2ω illumination case, presumably because the plasma density scale length is shorter and also the critical density is higher.

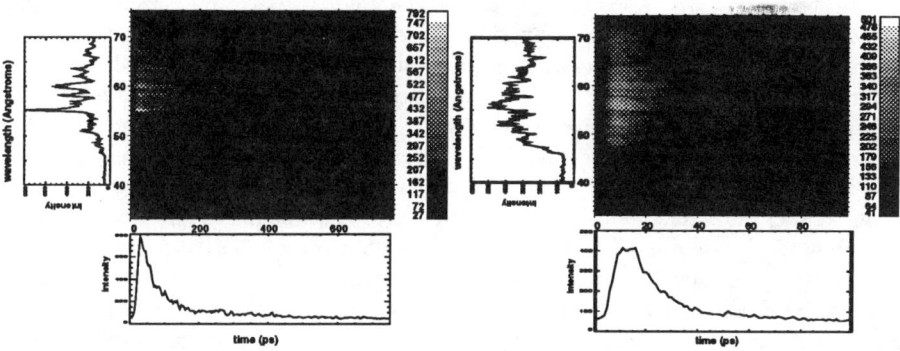

Figure 1: Streaked aluminum x-ray spectra obtained in a single shot with a high-resolution grazing-incidence spectrometer using (a) ω laser irradiation, corresponding to low-contrast conditions, and (b) using 2ω laser irradiation, corresponding to high-contrast conditions.

Figure 2(a) shows a comparison of amplitude-normalized temporal profiles of the AlXI $1s^22s - 1s^23p$ line ($\lambda = 48.3$ Å) obtained for 2ω irradiation for a range of laser flux densities. Note the pronounced decrease of the x-ray pulse width as I is decreased. In the low-I case, the FWHM of the x-ray pulse appears to be streak camera limited. The risetime is short in all cases (the starting positions of the plots relative to the laser pulse and each other are unknown.) The falltime is observed to be shortest in duration in the lowest I case as expected from the previous arguments.

Figure 2(b) shows a similar behavior of amplitude-normalized temporal profiles of the same line ($\lambda = 48.3$ Å) as predicted by a hydrodynamics code coupled to an atomic physics code for a range of laser flux densities. The one-dimensional hydro-

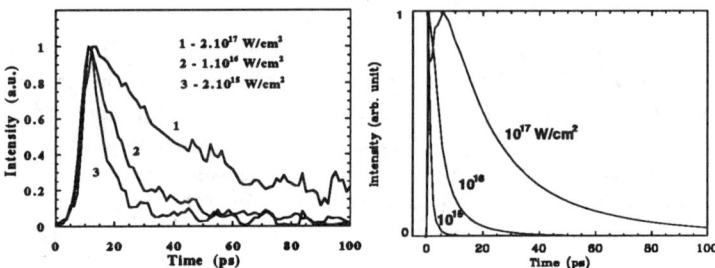

Figure 2: Comparison of amplitude-normalized temporal profiles obtained experimentally (a) and predicted by the numerical code (b).

dynamics code solves the conservation equations and is coupled self-consistently to the Helmholtz equation of the laser field for energy and momentum deposition in the plasma. A non-local thermodynamic equilibrium (NLTE) detailed-configuration time-dependent atomic physics package, FLY, is used to calculate the ionization and spectral emission [6, 7]. Figure 3(a) shows the comparison of measured and simulated x-ray pulse duration as a function of laser flux. The averaged experimental aluminum pulse widths at λ=48.3 Å are shown as crosses. The predicted spatially integrated values for aluminum at this wavelength are shown as pluses and fit with a spline routine. The dashed curve represents the predicted results convolved with the 5-ps response of the x-ray streak camera. Discrepancies between the experimental data and theory may possibly be due in part to opacity effects not included in the model at this time. The model does, however, give us some insight into the physics controlling the pulse duration. The hydro simulation reveals that the electron temperature rises very quickly, as expected, but falls on a time scale that is much faster than that of the x-ray pulse duration. We see that the x-ray pulse duration does not follow the temperature time history profile. The reason for this becomes clear when we look at the time history of $\bar{Z}$, which remains high long after T_e has dropped. The plasma is thus NLTE, and therefore the x-ray emission is no longer dominated by the temperature but by a combination of the electron temperature, electron density and average ionization. X-ray pulse durations at λ=48.3 Å were also measured for a gold target in order to compare the possible differences in brightness from different Z materials. For the same laser intensity the gold x-ray pulse duration was found to be as little as half the aluminum pulse duration. Data on conversion efficiency of laser radiation into x-rays in the range 1.5-5 keV are shown on figure 3(b). Note that conversion efficiency for the gold target is about 4 times higher than for the aluminum target. For the highest laser fluxes on target the conversion efficiency can reach about 1 percent (a few mJ) for the x-ray photons with $h\nu \geq 1$ keV.

To summarize, we have shown that the pulse width of a given range of photon energies of continuum soft x-rays may be minimized by adjusting the laser focal spotsize, and thereby the incident laser intensity, such that $T_{max} \leq T_0$. The results

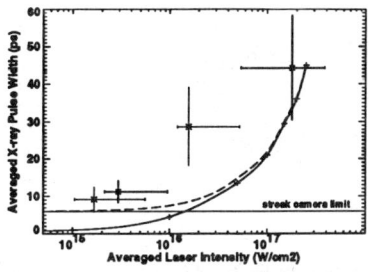 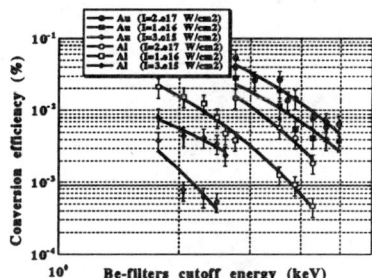

Figure 3: (a) Duration of the $AlXI 1s^22s - 1s^23p$ line ($\lambda= 48.3$ Å) versus laser intensity: experimental (crosses), predicted by the numerical code (solid line w/ pluses), predicted by the numerical code with convolution of streak camera response (dotted line). (b) Measured x-ray conversion efficiency for aluminum and gold solid targets for a range of laser fluxes.

are found to be in close agreement with the predictions of the hydrodynamics code coupled to the atomic physics model only when NLTE effects are included. The conversion efficiency of the high contrast subpicosecond laser pulses into ultrashort soft x-ray pulses can be as high as a few percent. We observed that the emission from gold targets in the x-ray region of 50 Å is an order of magnitude brighter than the emission from the aluminum due both to its shorter pulse duration and higher x-ray yield.

This work was partially funded by the National Science Foundation Center for Ultrafast Optical Science, contract #PHY8920108, and Office of Naval Research, contract #N00014-91-K-2005. The authors would like to thank G.Mourou for useful discussions and his encouragement.

References

[1] M. M. Murnane et al., Phys. Rev. Lett. **62**, 155 (1989).

[2] J. C. Kieffer et al., Bull. Am. Phys. Soc. **37**, 1468 (1992).

[3] D. Riley et al., Phys. Rev. Lett. **69**, 3739 (1992).

[4] D. Umstadter et al., Bull. Amer. Phys. Soc. **34**, 1364 (1989); H. Milchberg et al., Phys. Rev. Lett. **67**, 2654 (1991).

[5] B.L.Henke et al., J. Opt. Soc. Am. B **3**, No.11, 1540 (1986).

[6] S.J.Rose, J.Quant. Spectrosc. Radiat. Transfer **36**, 389 (1986).

[7] R.W.Lee et al., J.Quant. Spectrosc. Radiat. Transfer **32**, 91 (1984).

X-ray + optical nonlinear mixing in plasma

P. L. Shkolnikov and A. E. Kaplan

Electrical and Computer Engineering Department
The Johns Hopkins University, Baltimore, MD 21218

Abstract We have considered feasibility and potential applications of a large variety of nonlinear optical interactions of X-ray laser and optical radiation in plasma.

Recently, feasibility of various X-ray nonlinear optical effects has been theoretically demonstrated (1,2). In particular, it has been shown (2) that difference-frequency mixing of an X-ray laser (XRL) and a longer-wavelength (IR, visible, UV) laser radiation

$$\omega = 2\omega_{XRL} - \omega_{opt} \quad (1)$$

in plasma may generate intense coherent X-rays at almost doubled frequency if energy-level structure of the plasma ions provides for one-, two-, and three-photon resonant enchancement. The other two factors that contribute to predicted high conversion efficiency of the frequency near-doubling are (i) participation of powerful optical lasers, and (ii) feasibility of phase-matching optimization, even for tight focusing, so that high-intensity beams can be employed. In the present paper, we estimate conditions for generating a larger variety of coherent X-ray lines by nonlinear mixing of XRL and longer-wavelength radiation in plasma, not limiting ourselves by one X-ray and one optical line, nor by the third-order processes.

Difference-frequency mixing of one or two X-ray laser lines with the fundamental or/and transformed frequency of the pumping optical laser

The lowest-order nonlinear effect which may transform the frequency of radiation in isotropic media such as plasma is four-wave mixing (FWM) $\omega = \omega_1 \pm \omega_2 \pm \omega_3$. Out of all such processes, we have chosen for X-ray frequency transformations only those which have potential for optimal phase matching in plasma. Of a particular interest are the processes which may efficiently generate coherent X-rays at shorter wavelengths:

$$\omega = \omega_1 + \omega_2 - \omega_{opt} \quad (2)$$

where ω_1 and ω_2 are X-ray frequencies [they are the same if one XRL line participates -- see Eq. (1)], or two lines of a multiline XRL like Se or Ge XRL), and ω_{opt} is the frequency of a longer-wavelength radiation that originates from the pumping laser (usually a Nd laser). For a given combination of the pumping frequencies and an ion, conversion efficiency of the processes Eq. (2), C_{eff}, defined as the ratio of the output power to the incident power of an X-ray line, can be written as:

$$C_{eff} \approx K(N/N_e)^2 I_{XRL} I_{optical} |G|^2 \qquad (3)$$

Here N is the number density of the ions at the initial level of the sequence of resonant transitions, and N_e is the plasma electron density. The phase-matching factor, $|G|^2$, depends only on the geometry of the process (confocal parameters of the beams, their directions, location of the focal points, and the medium length), and of the medium dispersion. The coefficient K is calculated for a given combination of ω_1, ω_2, and ω_{opt}, and the ion species. Generally speaking, this coefficient drops very fast with increasing frequency so that X-ray nonlinear transformations are hardly observable with available now and in the near future XRL output without multiple resonant enhancement. The potential of such enhancement can be seen from the strong dependence of K on the relative detunings δ_j:

$$K \sim (\delta_1 \delta_2 \delta_3)^{-2}, \delta_j = 1 - \omega_{0j}/\omega_j, \qquad (4)$$

where ω_{0j} is a resonant transition frequency in the plasma ion. For instance, triple-resonant enhancement, with $|\delta_j| \sim 10^{-3}$ (which is somewhat larger than typical Doppler width, to prevent resonant absorption and strong ac Stark shift) would increase conversion efficiency by 18 orders of magnitude as compared to nonresonant situation ($|\delta_j| \sim 1$).

The requirement of *multiple* resonant enchancement is a unique feature of X-ray nonlinear optics and is necessitated by the combination of the short wavelength (that is, small transition dipole moments) and relatively low available XRL output. Another unique feature is the necessity to work with plasma. Although some X-ray nonlinear effects are feasible in initially neutral gases and vapors [see e. g. Ref. (3)], and, probably, solids, in general, only plasmas of highly-ionized atoms can provide required resonant conditions, as well as large propagation lengths. Since collective plasma effects are negligible at high frequencies, beam propagation is determined almost totally by free–electron dispersion. As a result, those X-ray nonlinear transformations are more attractive from experimental point of view, whose phase matching is attainable under large positive dispersion, that is, difference-frequency mixings, Eqs. (1), (2) [see Ref. (1)]. Obviously, the frequency ω in Eq. (1) is close to $2\omega_{XRL}$, so that we may call this process "X-ray frequency near-doubling". The same name is applicable to Eq. (2) if ω_1 is close to ω_2 as it is the case for the two most prominent lines of Se or Ge XRL.

Besides "frequency near-doubling", we have considered another kind of X-ray + optical mixing in plasma -- difference-frequency mixing

$$\omega = \omega_{XRL} + \omega_{opt1} - \omega_{opt2} \qquad (5)$$

with $\omega_{opt1} > \omega_{opt2}$. This process may generate a variety of X-ray lines around the pumping X-ray lines, whereby becoming one of the roads to tunable X-ray lasing. Its additional advantages are increased flexibility in search of resonant enhancement and participation of *two* longer-wavelength beams. For this mixing, multiple-resonant nonlinear media can be found not only among Na-like ions, as for X-ray frequency near-doubling, but also among Ne-like and Ar-like ions. For the optical lines, we have considered the fundamental, lowest harmonics, or a Stokes wave of the pumping laser, as well as radiation of a tunable optical laser.

We have found and evaluated plasma media for efficient frequency transformations of available X-ray lasers by large variety of resonant nonlinear difference-frequency mixings (Fig. 1). High conversion efficiencies, up to 1 under idealized conditions, are attainable with available short-wavelength output power and contemporary plasma and X-ray optics technology.

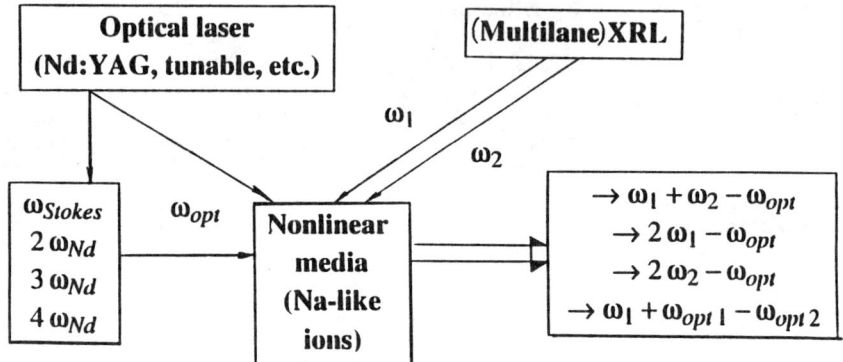

FIGURE 1. Nonlinear difference-frequency mixing of X-ray and optical lasers

Multiphoton processes in X-ray domain

Multiphoton interactions of optical lasers with gases and plasma such as multiphoton ionization, an important new area of atomic physics, or high-order harmonic generation (HHG) (4), a strong manifestation of nonperturbative nonlinear optics and an important new source of short-wavelength coherent radiation, have recently attracted much attention. We believe that parameters of existing X-ray lasers are already close to those required for observing similar multiphoton effects at much shorter wavelength, with similar potential impact on physics of highly-ionized atoms and X-ray nonlinear optics. As always, the easiest to observe are resonantly enhanced processes. For instance, multiphoton absorption $\omega_{C^{5+}XRL} + 4\omega_{4673 Å}$ in Ne-like Mg would be resonantly enhanced at each step so that very strong excitation and ionization of Mg^{2+} to F-like stage would take place even at modest C^{5+} XRL intensity. In the same media, strong multiphoton absorption is expected for Ge XRL + optical pumping.

On the other hand, high-order harmonic generation is a non-resonant process, which, even at longer wavelength, requires high-intensity lasers and, therefore, may seem to be totally out of reach for X-ray lasers. Yet, our estimates based on a two-level model of HHG, Ref. (5), allows one to suggest the feasibility of X-ray HHG at already available XRL output power (but with drastic improvement in the beam quality). Indeed, the most obvious manifestation of HHG is the presence of the "plateau": the intensities of generated harmonics are approximately the same within a large range of harmonics numbers. The model, Ref. (5), approximates a rare gas atom in HHG by a two-level atom and yields for the critical intensity I_{cr}, i. e. the pumping intensity necessary for the plateau formation:

$$I_{cr} \sim |d|^2 \omega_0 \omega \qquad (6)$$

where ω_0 and $|d|$ are the transition frequency and the dipole moment of the model two-level system, respectively. If one assumes that $|d| \sim \omega_0^{-1/2}$, then I_{cr}

scales as ω_3 provided the ratio of the pumping frequency ω to the ionization potential of the medium remains constant. In particular, it follows from the critical intensity being approximately equal to 2×10^{13} W/cm^2 for HHG of an 616 nm laser in neutral argon (6) that the Y^{29+} XRL intensity of about 1.3×10^{18} W/cm^2 might be enough to observe X-ray HHG in Ar-like Kr. Such intensity would be attainable with the *available* Y XRL power of ~40 MW if it becomes possible to focus the beam to e. g. three times diffraction-limited spot of ~3λ. Moreover, a few times larger intensity might generate a fully developed plateau such that the 21st harmonic (7.4 Å) would be as intense as the 5th harmonic.

Feasibility of X-ray laser nonlinear spectroscopy

Already available high brightness of XRL radiation makes X-ray lasers potentially unique tools for nonlinear spectroscopy of plasma. For instance, nonlinear mixings of XRL and longer-wavelength radiation considered above, would yield important spectroscopical information on highly-ionized atoms in relatively dense plasmas (subcritical for optical beams). Moreover, nonlinear frequency transformations may generate tunable, even if only over a narrow frequency range, coherent X-ray radiation to use for both linear and nonlinear spectroscopy of super-dense plasmas such as ones of interest for inertial confinement fusion. In particular, self-induced rotation of the X-ray beam polarization ellipse seems to be promising for XRL nonlinear spectroscopy of plasma: recent experimental and theoretical results on a similar effect in neutral atomic vapors at optical wavelength [see e. g. (7)] allow one to expect that very modest X-ray intensities might be enough to observe large effects near resonances.

Conclusion

In conclusion, we theoretically demonstrated feasibility of a large variety of X-ray + optical nonlinear mixing processes in plasma. We found multiple-resonant media and estimated conversion efficiency of a number of nonlinear transformations of X-ray laser radiation. Experimental realization of such transformations would result in significant increase of the number of available X-ray coherent sources, in generation of coherent radiation at very short wavelengths, and in generation of tunable coherent X-rays. We also suggest nonlinear polarization effects for X-ray laser nonlinear spectroscopy of super-dense plasma.

This work is supported by AFOSR.

References

1. Shkolnikov, P. L. and Kaplan, A. E., Opt. Lett. 16, 1153-1155 (1991); Shkolnikov, P. L. and Kaplan, A. E., Phys. Rev. A 44, 6951-6953 (1991); Shkolnikov, P. L. and Kaplan, A. E., Opt. Lett. 16, 1973-1975 (1991); Hudis, E., Shkolnikov, P. L., and Kaplan, A. E., Appl. Phys. Lett. 64, 818-820 (1994).

2. Shkolnikov, P. L., Kaplan, A. E., Muendel, M. H., and Hagelstein, P. L., Appl. Phys. Lett. 61, 2001-2003 (1992).
3. Hudis, E., Shkolnikov, P. L., and Kaplan, A. E., Appl. Phys. Lett. 64, 818-820 (1994).
4. L'Huillier, A, Lompre, L. A., Mainfray, G., and Manus, C., in *Atoms in Intense Laser Field*, Ed. M. Gavrila (Acad. Press, Inc., Boston, 1992), p.139-206.
5. Kaplan, A. E. and Shkolnikov, P. L., Phys. Rev. A 49, 1275-1280 (1994).
6 Miyazaki, K., Sakai, H., Kim, G., and Takada, H., Phys. Rev 49, 548-557 (1994).
7 Davis, W. V., Gaeta, A. L., and Boyd, R. W., Opt. Lett. 17, 1304-1306 (1992).

Gain measurements and spatial coherence in neon-like x-ray lasers

J. Krishnan,* C. Cairns,† L. Dwivedi,* M. Holden,* M. H. Key,^ C. L. S. Lewis,† A. MacPhee,^ D. Neely,^ P. A. Norreys,^ G. J. Pert,# S. A. Ramsden,# C. G. Smith,^ G. J. Tallents* and J. Zhang^

*Dept. of Physics, University of Essex, Colchester CO4 3SQ, U. K.
^Central Laser Facility, Rutherford Appleton Laboratory, Didcot OX11 0QX, U. K.
†Dept. of Pure & Applied Physics, Queen's University of Belfast, Belfast BT7 1NN, U. K.
#Dept. of Computational Physics, University of York, York, Y01 5DD, U. K.

Abstract. Many of the applications with x-ray lasers require high quality output radiation with properties such as short wavelength and a high degree of coherence (longitudinal and spatial). Ne-like Yttrium (Z=39) is potentially a bright and monochromatic XUV lasing medium. The output at 15.5 nm is monochromatic due to the overlap of the $J = 2-1$ and $J = 0-1$ lines. A gain coefficient of 3 ± 1 was obtained at 15.5 nm by irradiating 100 μm wide yttrium stripes at 6×10^{13} W/cm^2 with 1.06 μm, 650 ps pulses from the Rutherford Appleton Laboratory VULCAN laser.

We have investigated improving x-ray laser spatial coherence utilising a series of amplifiers instead of the standard double target configuration. An 'injector-amplifier' scheme was successfully demonstrated with the Ne-like Ge x-ray laser. A spatially small and coherent part of the 23 nm beam from the standard double target geometry has been relayed using a W/Si multilayer mirror onto a single or double target configuration situated at a distance of ~1.5 m from the mirror and pumped by two 150 mm diameter beams of the VULCAN laser. A beam 'foot-print monitor' was employed with a flat mirror to relay 23 nm output onto a film pack to record the spatial variation of the x-ray laser beam. Analysing the fringes obtained through a cross-wire placed in front of the beam shows that an increase in spatial coherence was achieved by adding amplifiers to the x-ray laser beam line.

Introduction

Many of the previous collisionally excited x-ray laser experiments conducted at the Rutherford Appleton Laboratory (RAL) have concentrated on studying the conditions required to achieve gain and on enhancing the gain production efficiency (1,2,3). A double target configuration has been shown to compensate for x-ray laser beam refraction, thus increasing the achieved gain length product (1). Understanding the operation of double targets has also increased as a result of modelling studies (4).

The direction of collisionally excited x-ray laser research is now moving to investigate and improve the laser properties required for applications. The characterisation and optimisation of the beam profiles and coherence of the neon-like x-ray lasers undertaken here is part of this trend.

© 1994 American Institute of Physics

Gain measurements on neon-like yttrium

To maximise longitudinal coherence, it is necessary that the laser bandwidth is small, i.e. that lasing only occurs on one spectral line. The yttrium laser operates at 15.5 nm and is unique among Ne-like lasers in that the short wavelength $J = 2-1$ line dominates the spectrum probably because of an overlap with a $J = 0-1$ line (4). In other Ne-like systems the relative intensity of the two $J = 2-1$ transitions is within a factor of 10 and usually comparable. This dominance of the short wavelength line at 15.5 nm makes yttrium effectively a monochromatic source of x-rays. The standard double target configuration was employed to compensate for refraction effects (1).

Several experiments in other Ne-like ions have shown that a pre-pulse can lead to higher total x-ray laser emission, to changes in the spectral distribution and to an increase of the pulse duration (5, 6). The initial pulse pre-forms the plasma and then the main pulse produces the optimum conditions for the lasing action. By controlling the delay between the pulses, plasma parameters can be controlled. For example, it has been found that Ge slab targets irradiated with a pre-pulse arriving 5 ns in advance of the main pulse have enhanced brightness of the $J = 0-1$ line (19.6 nm) (5).

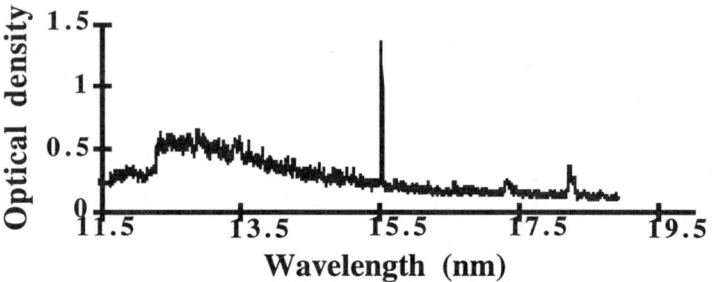

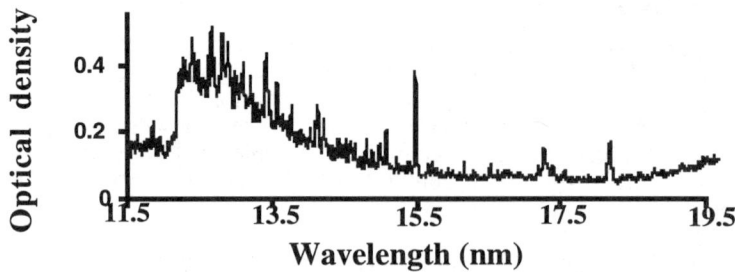

Figure 1 Densitometer traces of the spectra obtained from the flat-field spectrometer showing the Ne-like yttrium lasing line emission at 15.5 nm. For the top spectrum, the pump laser produced two pulses of width ~ 340 ps separated by ~ 275 ps. The bottom spectrum is obtained from a single pump pulse of 650 ps width.

During the yttrium experiment, for one shot, the incident laser mode beated producing two pulses. The duration of pulses was ~ 340±30 ps. The delay between these two pulses were ~ 275 ps and the 15.5 nm line nearly saturated the film through a 0.8 µm thick Al filter. Figure 1 gives the densitometer trace of the mode beated shot and a typical shot. The intensity on film for the double pulse shot corresponded to a gl ~ 8. This result suggests that double pulsing the pump laser energy increases the x-ray laser output, but further investigation of this single observation is, of course, needed.

Injector-amplifier scheme on neon-like germanium

Experiments have previously been performed with Ne-like lasers to increase their efficiency by double passing the gain region (7,8,9) using multilayer mirrors. Saturated output produced from double passed Ne-like media have been demonstrated (7,8). Figure 2 is a schematic of the experimental set-up used for the injector-amplifier scheme reported here.

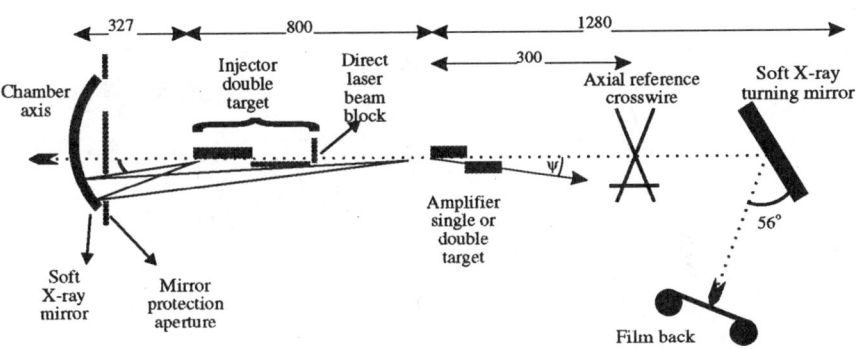

Figure 2 Schematic of the experimental set-up used for coupling between an injector and amplifier plasma using an x-ray multilayer mirror and the footprint monitor employed to measure the x-ray laser beam spatial profile.

A set of reference wires were used in the x-ray mirror alignment. From the analysis of the fringe patterns generated from these wires observed in the x-ray laser beam foot-prints (recorded using a flat multilayer x-ray mirror and film back, see fig. 2), an estimate of the spatial coherence has been made. Figure 3 shows the intensity distribution from one such cross wires. The fringe visibility has been measured for various shots with varying amplifier combination (fig. 4). It is clear from that the visibility and hence spatial coherence increases when additional amplifier targets are inserted in the system. The large error bars are due to the uncertainty in measuring the footprint monitor film background. It is interesting to note that the fringe visibility varied along the vertical cross wire indicating that the coherence is not constant on all parts of the beam. Also, the cross wires which were in the far field of the x-ray laser beam gave better fringe visibility compared to the angled cross wires placed in the near field.

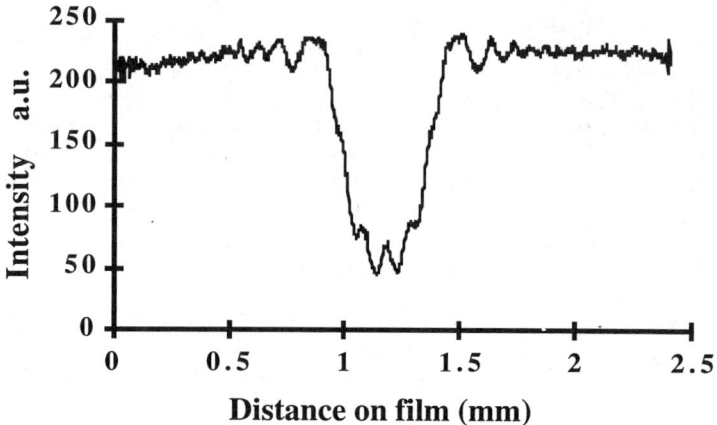

Figure 3 - X-ray laser intensity distribution obtained in the shadow of a 180 μm diameter cross wire.

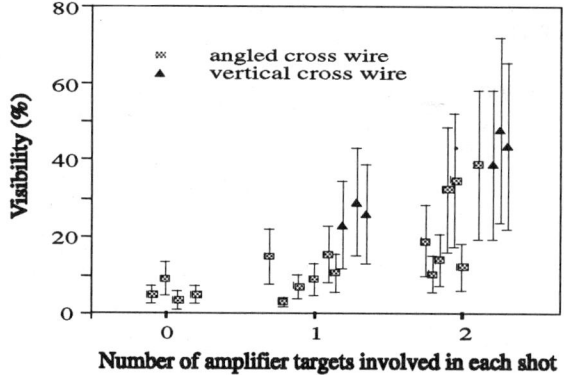

Figure 4 Graph showing the fringe visibility increase when additional x-ray laser amplifiers are introduced.

Fraunhofer diffraction patterns from a wire illuminated by coherent light can be calculated using the formula (10):

$$I(x) = 1 - \frac{2ka^2}{z} \sin\frac{kx^2}{2z} \left[\frac{\sin\left(\frac{kax}{z}\right)}{\frac{kax}{z}}\right] + \frac{k^2 a^4}{z^2} \left[\frac{\sin\left(\frac{kax}{z}\right)}{\frac{kax}{z}}\right]^2 \quad (1)$$

where z is the distance from the wire to the plane of observation, x is the distance on the plane of observation, 2a is the mean diameter of the wire, k =

$2\pi/\lambda$ is the wave number of the x-ray radiation of wavelength λ and $I(x)$ is the intensity at a distance of x on the observation plane. For this experiment, $2a = 180$ μm and $z = 1110$ mm. Figure 5 gives the calculated and the experimentally observed fringe pattern. There is clear agreement of the fringe spacing.

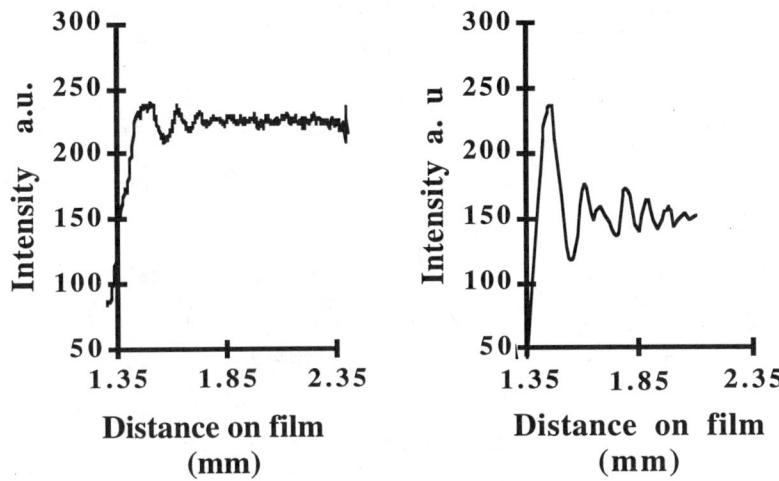

Figure 5 Calculated and experimentally measured fringe patterns produced by diffraction from a 180 μm diameter wire placed in the x-ray laser beam.

Conclusions

Ne-like Y was made to lase at 15.5 nm with a gain coefficient of 3±1 per cm. Double pulsing the pump beam produced higher gain. An injector-amplifier scheme was successfully demonstrated using the Ne-like Ge x-ray laser at 23 nm. An approximate estimate of the increase of coherence by injecting the x-ray laser beam into the amplifiers was obtained by measuring the fringe visibility in the geometrical shadow region of the cross wires. The results indicate that the coherence is increased by the addition of amplifier targets.

References

1. C. L. S. Lewis et al, *Optics Comm.*, **91** p71 (1992).
2. D. M. O'Neill et al., *Opt. Comm.*, **75** p406 1990.
3. D. Neely et al., *Opt. Comm.*, **87** p231 1992.
4. P. B. Holden, G. J. Pert, A. Kingston and E. Robertson, *Appl. Phys.*, **B58** p3209 (1993).
5. J. Nilsen et al., these proceedings.
6. G. Cairns et al., these proceedings.
7. A. Carrillon et al., *Phys. Rev. Lett.*, **8** p618 1991.
8. B. J. MacGowan et al., *Phys. Fluids*, **B4** p2326 1992.
9. N. Ceglio, *J. X-ray Sci. Technol.*, **1** p7 (1989).
10. G. B. Parrent Jr. and B. J. Thompson, Opt. Acta., **11** p183 (1964).

Comparison of the Spectral Response of a Thinned, Backside Illuminated CCD with a CsI Coated MCP System and Kodak 101 Film

Yuelin Li, J. R. Crespo López-Urrutia, G. D. Tsakiris
R. Sigel, R. Volk, and L. Pina[†]

Max-Planck Institut für Quantenoptik, 85748, Garching, Germany
[†]*Czech Technical University, 18040 Prague, Czech Republic*

Abstract. A thinned backside illuminated CCD chip was calibrated by self consistently determining the thickness of its dead layer. Its spectral response and sensitivity were then compared with those of the calibrated Kodak 101 photographic plates and of a CsI coated microchannel plate detection system.

1. INTRODUCTION

Recently, the thinned backside illuminated charge coupled devices(CCDs) have received a great deal of attention because of their potential applications in x-ray microscopy and laser plasma physics(1-3). The spectral response and sensitivity of the sensor is one of the most important factors in those applications. In this paper, we present a method for the calibration of a thinned backside illuminated CCD chip based on the determination of its dead layer thickness. Comparison with spectra recorded on absolutely calibrated Kodak 101 photographic plates shows that this method works quite accurately in the SXR range. We also compared the relative spectral response and sensitivity of the CCD with a CsI coated MCP detection system.

2. QUANTUM EFFICIENCY MODEL FOR A THINNED CCD

A schematic diagram showing the structure of a thinned CCD is presented in Fig. 1. The incidence of the photons on the backside of the device greatly reduces the absorption of the photons before they enter the active region. However, the unavoidable layer of silicon oxide on the back surface traps some positive charges at the interface between silicon and silicon oxide which gives rise to a potential

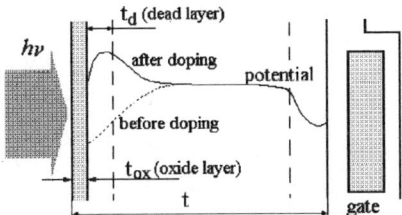

FIGURE 1. Schematic diagram showing the structure of a thinned, backside illuminated CCD.

well. This results some loss of the photoelectrons at the back side. Although some techniques like doping have been developed to change the situation(see Fig. 1: potential distribution before and after doping), there still remains a layer where the photoelectrons diffuse to the backside and are lost. This is the so called *dead layer*, which dominates the spectral response of a thinned CCD in the SXR range.

From the diagram in Fig. 1, one can derive an approximate expression for the quantum efficiency (QE) for a thinned CCD(2). The photons reflected at the backside surface, those going through the chip and those absorbed by the native oxide layer are unable to create photoelectrons, so an interaction efficiency can be written as

$$\eta_{int}(\lambda) = [1-r(\lambda)] \times \{1-\exp[-\alpha(\lambda)t]\} \times \exp[-\alpha_{ox}(\lambda)t_{ox}],$$

where r, t, t_{ox} are, respectively, the reflectivity of the oxide surface, and the thicknesses of the active region and of the oxide layer. α and α_{ox} are the absorption coefficients of silicon and silicon oxide(4). The collection efficiency can be expressed in terms of the transmission of photons through the dead layer, i. e.,

$$\eta_{coll} = \exp[-\alpha(\lambda)t_d],$$

here t_d is the thickness of the dead layer. The loss of the photoelectrons is phenomenologically approximated here by the loss of the photons in the dead layer. The QE is the product of the interaction and collection efficiencies

$$QE = \eta_{coll} \times \eta_{int}.$$

In Fig. 2, the calculated QE as obtained by the model for various thicknesses of the dead layer is given. The oxide layer of SiO_2 was assumed to be 100 Å thick and the CCD active region 10 μm thick. It is apparent that the existence of the dead layer seriously affects the CCD QE in the region between 60-120 Å. Furthermore, one clearly sees the QE jump at the silicon L-edge (~120 Å).

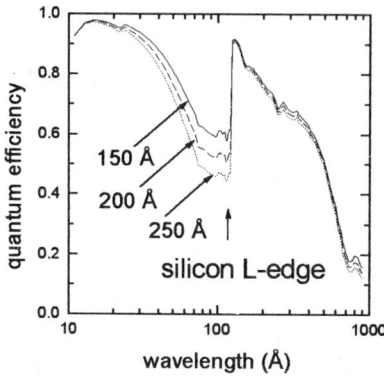
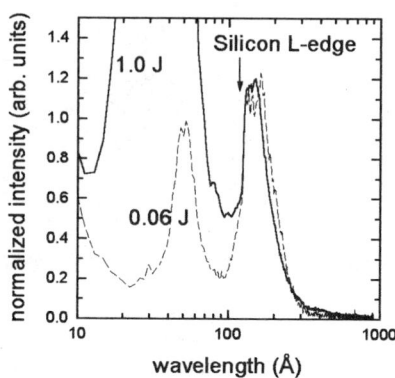

FIGURE 2. Calculated QE of a thinned CCD for various dead layer thicknesses; a 100 Å thick SiO$_2$ layer and a 10 μm thick CCD were assumed. The silicon L-edge at ~120 Å caused by the dead layer is clearly seen.

FIGURE 3. Tungsten spectra measured by CCD for the indicated laser energies (after subtraction of high orders). The spectra show a similar intensity jump of ~40% at the Si L-edge.

3. EXPERIMENT AND RESULTS

The experiment was carried out with a frequency doubled Nd:glass laser system with output energy up to ~ 10 J in 3 ns(FWHM). The laser beam was normally focused on the target surface with a focus size of ~100 μm. Two spectrometers equipped with free standing transmission gratings of 1000 lines/mm were symmetrically positioned at 45° off the target normal. One of them was coupled to a Photometrics AT200 CCD camera system with a thinned backside illuminated chip(TK512B, see Table 1. for the parameters provided by the manufacturer) and the other either to a camera with absolutely calibrated Kodak 101-05 photographic plates or to a CsI coated microchannel plate(MCP) detection system.

Fig. 3 gives typical spectra from a tungsten target recorded by the CCD. We can clearly identify the silicon L-absorption edge at about 120 Å, which is due to the existence of the dead layer in the CCD. It was found that after subtraction of the higher orders, the spectra give an intensity jump of about ~40% at the Si L-

TABLE 1. Parameters of the CCD Camera System

format	512×512 pixel
pixel size	27μm×27μm
dark current	11.5 e$^-$/sec
full well capacity	7.46×10^5 e$^-$
gain	11.54 e$^-$/count(1×)
readout noise	14.5 e$^-$ RMS(1×)
response nonlinearity	0.05%

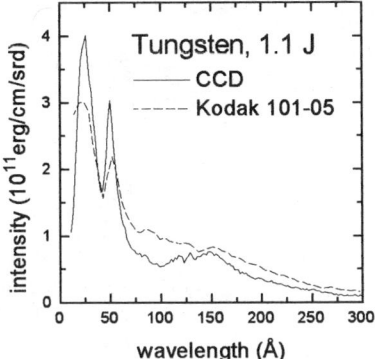 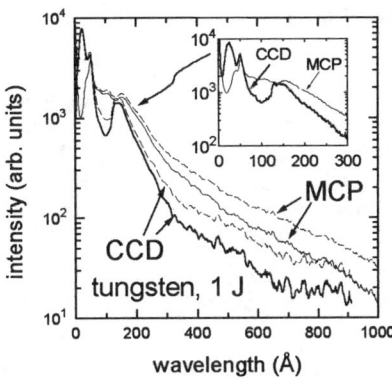

FIGURE 4. Tungsten spectra of one single shot recorded simultaneously by the CCD and the calibrated Kodak 101 photographic plate using similar gratings. Both spectra were unfolded taking into account high orders, detector response, and the geometrical set-up to obtain the absolute intensity.

FIGURE 5. Tungsten spectra taken with the same grating and under similar laser conditions by the CCD and a CsI coated MCP. The intensity scales were normalised such that the intensity is the same at 130 Å. Dashed line: raw spectra; solid line: after subtracting high orders.

edge with deviations of about 10%. Comparing this with the calculated QE curves in Fig. 2, one deduces a dead layer thickness of about 200 Å with a 20% error. This way the chip is semiempirically calibrated.

To test the accuracy of this calibration, we have taken shots simultaneously with the CCD and with a film spectrometer. The typical spectra as recorded by the CCD and by the Kodak 101-05 photographic plate are shown in Fig. 4, which have been converted into absolute intensity(The method for evaluation of the data can be found in Li et al(5)). Calibration data for Kodak 101-01, 07 film from Henke et. al(6) and Kishimoto (7) were used for the film data. The quantitative agreement between the two spectra confirms the accuracy of the method for the calibration of a thinned CCD in the SXR range.

We also compared the spectral response of the CCD chip to that of a the CsI coated MCP system. In Fig. 5 we give the spectra(the intesity is normalised at 130 Å) as recorded by the two detectors, which were taken with the same grating and under similar laser condition. We see that, at wavelength <50 Å the normalised signal from the CCD exceeds that from the MCP. In the range 60-130 Å, the MCP provides a higher signal(see the insert in Fig. 5). Beyond 150 Å, the signal of the CCD drops more rapidly than that from the MCP, with the consequence that the contributions from higher orders become more important. This shows that the CCD is more appropriate for measurements in the region up to 300 Å. In contrast, the MCP exhibits a higher sensitivity at longer wavelengths. The deduced sensitivities of the three detectors at 160 Å are listed in Table 2. The data for Kodak 101 films are from Kishimoto(7). As can be seen, the estimated overall sensitivity for the CCD and the MCP is about two orders of magnitude higher than

TABLE 2. Sensitivity of the CCD, Kodak 101 photographic plate and the CsI coated MCP at 160 Å (~80 eV)

	CCD(gain=1)	Kodak 101	MCP
noise level	3 counts/pixel	0.1-0.5 density	/
reliable signal	10 counts/pixel	0.1 density	/
quantum efficiency	50%	/	40%
detectable electron number	100/pixel	/	1/channel
detectable photon number	10/pixel	500/cm^2	2.5/channel
sensitivity(erg/cm^2)	2×10^{-4}	10^{-2}	5×10^{-5}
dynamic range	$\sim 3\times10^3$	$<10^2$	$\sim 3\times10^3$

that of the Kodak film. Also given are the corresponding dynamic ranges for each detector.

4. SUMMARY

We have calibrated a thinned, backside illuminated CCD chip by determining its dead layer thickness. The accuracy of this calibration was assessed by comparing the spectra recorded by the CCD chip and the absolutely calibrated Kodak 101-05 photographic plates. When compared with a CsI coated MCP system, the CCD seems to have better response at shorter wavelengths, but at long wavelengths (beyond 300 Å) the MCP provides better response. The overall sensitivities of the CCD and CsI coated MCP are nearly two orders of magnitude higher than that of the Kodak 101 film.

ACKNOWLEDGMENTS

The author would like to thank A. Böswald, W. Fölsner and H. Haas for the excellent technical help. One of the author (Li) would also like to thank Dr. S. Witkowski for his hospitality. Li is supported by Alexander von Humboldt Foundation and Crespo López-Urrutia is supported by the European Commission "HCM"-program

REFERENCES

1. Janesick, J. R. et al., *Rev. Sci. Instrum.* **56**, 796-801(1985).
2. Tassin, C. et al., J., *Proc. SPIE* **1140**, 139-145(1989).
3. Salieres, P. et al., in *X-Ray Laser 1992*, Bristol: IOP Publisher, 1992, pp. 367-370.
4. Palik, E. A., *Handbook of Optical Constants of Solids*, London: Academic Press, 1985.
5. Li, Y., et al., *Laser & Part. Beams*, **9**, 787-793(1991).
6. Henke, B. L. et al., *J. Opt. Soc. Am.* **1**, 828-849(1984).
7. Kishimoto, T., *Max-Planck Institut für Quantenoptik Report*, **MPQ 108**(1985).

Study of Multilayer Structures as Soft X-ray Mirrors

D.Kim*, H. W. Lee*, D. Cha*, J. J. Lee†, and J. H. Je†

*Department of Physics and †Department of Material Science and Engineering
Pohang University of Science and Technology, Pohang, Kyungbuk 790-784, KOREA

Abstract. Molybdenum-silicon multilayer as soft x-ray mirrors have been fabricated using a magnetron sputtering system. Their structures have been characterized by x-ray diffraction (XRD) and computer simulation. Reflectivities at normal incidence have been measured by using monochromatized synchrotron radiation in the 18- 24 nm region. A normal incidence reflectivity as high as 40% at 20.8 nm was achieved. Multilayer structural parameters optimized for various soft x-ray laser wavelengths are also given.

INTRODUCTION

Due to their unique properties different from those of bulks and applications to various fields in science and engineering, the interest in multilayer structures has been ever increasing. During the past decades, the rapid advance in the vacuum thin-film technology has allowed the fabrication of multilayer structures with a bi-layer thickness ranging from a few nanometers to a few tens of nanometers.

One important application of multilayer structures is x-ray optics such as normal-incidence x-ray mirrors, x-ray gratings, x-ray diffractometers and spectrometers[1]. A multilayer structure as a high-reflectivity mirror in the soft x-ray region is composed of alternating layers of two elements whose indices of refraction are largely different at a wavelength in interest. Its performance is closely related to the micro-structural parameters of the multilayer structure such as the bi-layer thickness, the number of the bi-layers deposited, the ratio of the thickness of the high-Z element to the bi-layer thickness (multilayer period), the degree of the interface perfection, and the selection of materials [2].

In this paper, we present the normal incidence reflectivity measurement of Mo-Si multilayer structures around 20 nm. The multilayer structures have been characterized by x-ray diffraction using Cu K_a characteristic x-rays and computers simulation. Their x-ray reflectivity around 20 nm region was measured by monochromatized synchrotron radiation.

FABRICATION OF Mo-Si MULTILAYER STRUCTURES

An RF magnetron sputtering system was used in producing Mo-Si multilayer structures. The system has three 3" dia. sputtering guns. The typical base pressure of the chamber was 3 x 10^{-6} torr. During the deposition, the pressure of Ar gas was kept at 5.0 mtorr. The deposition rates of Mo and

Si were measured to be 1.0 ± 0.2 nm / min. and 1.2 ± 0.2 nm / min., respectively, at 20 W input RF power and at the target-to-substrate distance of 6.0 cm. In this measurement, we looked for the difference between the deposition rate of Si on Mo and that of Si on Si substrate. No difference was noticed within our experimental error. The thickness control was done by time. A programmable timer sends a signal at a pre-set time to a motor drive that rotates a substrate platter to the other element.

A series of Mo-Si multilayer structures were fabricated on 10 mm x 15 mm rectangular pieces of single-crystal [100] Si wafers. No attempt was made to cleanse the surface or remove the oxide layer prior to the deposition of multilayers. For the stabilization of the magnetron sputtering system there was a 20 min. pre-conditioning period, during which the shutters covered the Si substrates to block out any deposition. The target-to-substrate distance was fixed at 6.0 cm for all the multilayer depositions in this study. The temperature inside the chamber during the deposition was about 100 °C. No attempt was made to control the substrate temperature throughout the deposition.

CHARACTERIZATION

Characterization of the Mo-Si multilayer structures was performed by the low angle x-ray diffraction using Cu K_α (λ = 0.1542 nm) with a graphite crystal monochromator in the θ - 2θ geometry. In Fig. 1, the small-angle x-ray diffraction data obtained from one of the samples made under this study (the sample number: m930111) is shown. The number of Bragg peaks observed (10 peaks) and their narrow widths indicate that the sample has a relatively well-defined multilayer structure. The angular positions of

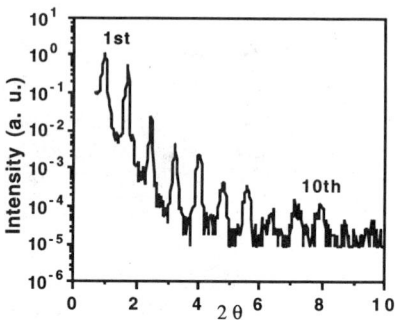

FIGURE 1. X-ray diffraction data obtained from the sample m930111.

Bragg peaks can be used to determine the average period of the multilayer structure. The first-order refraction-corrected Bragg's law(3) is given by

$$\frac{m\lambda}{2\sin\theta_m} = d_m$$

$$= d (1 - \frac{\delta}{\sin^2\theta_m})$$

where m is the order of Bragg peak, d_m the apparent period, d the average real period, θ_m the angular position of the m-th Bragg peak and $1-\delta = n$ the effective real part of the index of refraction of the multilayer structure. The apparent period, d_m, can be found from the angular positions of the Bragg peaks, and from the linear relationship between d_m and $1/\sin^2\theta_m$, the average real period of the multilayer structure, d, can be estimated. From its slope, the decrement of the effective real part of the index of refraction of the multilayer structure can be evaluated, which can then be used to estimate γ, the ratio of the thickness of the Mo layer to the multilayer period: $\gamma = (\delta - \delta_{Si}) / (\delta_{Mo} - \delta_{Si})$. δ_{Mo} and δ_{Si} are the decrements of the effective real part of the index of refraction of Mo and Si, respectively.

TABLE I. Structural parameters of Mo-Si multilayer structures under this study

sample	d(nm)	γ
m930109	11.76	0.28
m930110	10.90	0.32
m930111	11.01	0.21

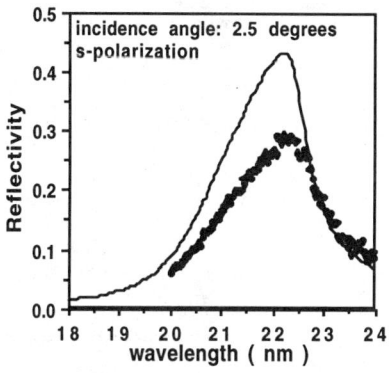

FIGURE 2. The variation of the reflectivity at the incidence angle of 2.5° with the wavelength from 18 to 24 nm for the sample m930109.

The structural parameters measured in this way are listed in Table I.

The reflectivity measurements around 20 nm region at near normal incidence (2.5° off the normal) were carried out with the synchrotron radiation monochromatized by a plane grating monochromator (PGM) at the beam line BL5B at UVSOR, Institute for Molecular Science, Okazaki, Japan(4). The PGM covers a spectral range of 2 - 230 nm with moderate resolving power of 500 ± 200. The output radiation from the PGM was previously measured to be almost linearly polarized(5). The reflectivity measurements were performed for both the s (TE) and p (TM) polarization. Reflected and direct intensities were measured with a gas-flow type proportional counter that has a 0.5 μm polypropylene window. The gas used was the mixture of 90% Ar and 10% CH_4 and was set to flow at the rate of 60 ml / min. The adjustment of the slit width of PGM with an additional 2-μm thick Al filter allowed us to avoid the saturation of the detector and cut out the higher harmonics from the PGM. Pulse height distribution of the signal was monitored and the discriminated signals were counted and recorded as a net intensity. Typical pressure in the calibration chamber was 3×10^{-6} torr due to a gas leakage from the counter.

Figure 2 is the wavelength-scan of the reflectivity for the sample m930109 from 18 to 24 nm at the incidence angle of 2.5°. This angle was the smallest one we could come down to due to the size of the proportional counter. In Fig. 2 is also shown the theoretical calculation (solid line) done for the s-polarized soft x-ray light. Structural parameters of the sample m930109 listed in Table I were used. Ideal interfaces were assumed in these calculations. The calculation uses the Parratt computational recursion method(6). New data for indices of refraction by Henke, et. al. were used(7). Measured and calculated peak reflectivities for the samples under this study are listed in Table II.

For the wavelength-scan, the calculation was done for the s-polarization; however, the angle of incidence (2.5°) is small so that the difference in the calculated reflectivity of a ideal multilayer structure between the s- and the p-polarization becomes negligible. The reflectivity as high as 40% at s-polarized 20.8 nm light for the sample m930111 was observed. This value approaches the theoretical value (46%) for structural parameters given in Table I. Discrepancy between them in the peak reflectivity and the

TABLE II. Measured and calculated peak reflectivities at near normal incidence (2.5°).

sample	Exp.	Calc.
m930109	p pol.: 23% at 22.2nm s pol.: 30% at 22.3nm	43% at 22.25nm
m930110	p pol.: 28% at 20.5nm s pol.: 30% at 20.4nm	46% at 20.60nm
m930111	p pol.: 31% at 20.7nm s pol.: 40% at 20.8nm	46% at 21.00nm

shape may be attributed to the interfacial imperfections such as the interfacial roughness, the thickness error, and the interfacial diffusion not taken into account in the present calculations. Only considering the effect of the interfacial roughness on the reflectivity, the upper bound for the interfacial roughness can be estimated using the equation analogous to the Debye-Waller factor in x-ray diffraction:

$$R_{obs.}(\theta) = R_{cal.}(\theta) \exp\left(-\left(\frac{4\pi\sigma\cos\theta}{\lambda}\right)^2\right)$$

where $R_{cal.}(\theta)$ is the calculated reflectivity at the wavelength of λ for ideal interfaces at the incidence angle of θ, and σ is the rms value of the roughness. Comparing the peak reflectivities for the s-polarization, we obtain $\sigma = 1.1$ nm from for the sample 930111. Similar analyses for other samples were done and their results are listed in Table III.

The computer simulations have been done to optimize the structural parameters of multilayer mirrors for various wavelengths. Due to the limited space, structural parameters for wavelengths, where high GL values have been demonstrated, are listed in Table IV. Note that for wavelengths longer than around 18 nm the Mo/Si combination is not the best.

TABLE III. Estimated average roughness obtained using the Debye-Waller Factor.

sample	σ(nm)
m930109	1.1
m930110	0.8
m930111	0.6

CONCLUSION

Mo-Si multilayer mirrors around 20 nm fabricated by magnetron sputtering were characterized by characteristic x-ray of Cu K_a (0.154 nm) for their structural parameters. It is noticed that the deposition rate for multilayer deposition may be different from that measured by a single-layer deposition. The reflectivity in the range of 18 to 24 nm was measured using monochromatized synchrotron radiations for both the s- and p- polarization. The peak reflectivity at 20.8 nm as high as 40% among the samples studied here was observed.

ACKNOWLEDGMENTS

The authors would like to thank Mr. S. K. Kim for his technical support for the measurement of x-ray diffraction. This work is support by the Pohang Light Source Project, Pohang University of Science and Technology,

TABLE IV. Optimized multilayer structural parameters for soft x-ray laser wavelengths

wavelength (nm)	elements A/B	thickness (nm) $d=d_A+d_B$	Ratio (d_A/d)	Number of layers	R (%)
4.48	C/Li	2.247	0.748	231	78
	C/Co	2.245	0.748	342	57
9.90	Sr/Rh	5.036	0.613	76	73
	Sr/Ag	5.036	0.613	96	71
15.5	Ba/Nb	7.979	0.631	34	62
	Si/Mo	8.107	0.505	42	63
18.2	Al/Mo	9.252	0.685	28	60
	Al/Zr	9.262	0.622	32	65
	Si/Mo	9.457	0.712	32	52
20.6	Al/Y	10.630	0.586	44	56
	Al/Zr	10.641	0.640	25	53
	Si/Mo	10.700	0.766	23	42
20.9	Al/Y	10.804	0.586	35	56
	Al/Zr	10.792	0.658	25	52
	Si/Mo	10.850	0.757	22	40
23.2	Al/Y	12.000	0.658	25	44
	Al/Zr	12.000	0.694	25	44
	Si/Mo	10.700	0.757	11	26
23.6	Al/Y	12.200	0.667	33	42
	Al/Zr	12.200	0.712	32	43
	Si/Mo	12.200	0.766	11	25
26.2	Mg/S	13.240	0.595	56	78
	Mg/La	13.169	0.685	42	69
	Si/Mo	13.499	0.739	20	15

Pohang, Korea and the X-ray Imaging Optics project, Japan.

REFERENCES

1. F. E. Christensen, ed. "X-ray multilayers for Diffractometers, Monochromators, and Spectrometers", Proc. SPIE vol. **984** (1988).
2. A. E. Rossenbluth, Review Phys. Appl. **23**, 1599 (1988).
3. R. W. James, "The Optical Principles of the Diffraction of X-rays" (Ox Bow, Woodbridge, CT, 1982).
4. M. Sakurai, et. al. Rev. Sci. Instrum. **60**, 2089 (1989).
5. M. Sakurai, J. Fujita, K. Yamashita, M. Ohtani, I. Hatsukade, K. Tamura, H. Nagata, Y. Suzuki, and S. Seki, Vacuum **41**, 1234 (1990).
6. L. G. Parrat, Phys. Rev. **95**, 359 (1954).
7. H..L. Henke, J. C. Davis, E. M. Gullikson, and R. C. C. Perera, LBL report, LBL26259 (1988).

Space and Time Resolved Investigation of Recombination X—Ray Lasers

Xiaofang Wang, Zhizhan Xu, Zhengquan Zhang, Pinzhong Fan,
Baifei Shen, Shensheng Han, Lingqing Zhang, Peixiang Lu,
Ruxin Li, Xianping Feng, and Aidi Qian

Shanghai Institute of Optics and Fine Mechanics, Academia Sinica,
P.O. Box 800—211, Shanghai 201800, China

Abstract. Presented are time— and space—resolved investigation of recombination x—ray lasers. An improved method for gain determination was put forward, by which the laser gain of the Li—like Si 4f—3d transition at 13.0 nm was measured, which is in good agreement with theoretical simulation. Plasma parameters of electron density and temperature estimated from spectral Stark broadening and characteristic recombination time, respectively, were in qualitative agreement with the plasma conditions given by the theory for the 4f—3d gain.

INTRODUCTION

Li—like scheme is of relatively higher quantum efficiencies than H—like scheme in recombination mechanism and Ne—like or Ni—like scheme in collisional mechanism. Amplifications have been demonstrated in many laboratories worldwide for the 4f—3d, 5f—3d and even 6f—3d transitions in Li—like ions, which provide a possible approach to small scale x—ray lasers. However, some obstacles hinder the further development of the brightness of this kind of lasers, among which a distinct phenomenon is that gain coefficients decrease with gain medium length increasing [1]. Several reasons such as refraction, self—absorption, and plasma's nonuniformity were considered to explain the phenomenon. Refraction is negligible because lasing usually occurs in the region with electron density $n_e \sim 10^{19}/cm^3$, where the density profile is flat in the case of slab target and long driving pulse (>100 ps). Self—absorption to the lower level of the lasing transition, although not serious with fiber target in H—like scheme [2], should be considered in Li—like scheme in the case of slab target. Plasma's nonuniformity will cause nonuniform plasma conditions, which will cause both gain decreasing and fluctuation of the line intensity. In addition, spontaneous emission from non—lasing region such as the higher density or from non—lasing times is also detrimental to gain measurement. Based on these considerations, we have developed time— and space—resolved spectroscopic diagnostic techniques to investigate both plasma's uniformity and time— and space—discrimnated gain and some interesting results have been obtained [3—5]. We also developed an improved method for gain determination of using line intensity ratio to overcome fluctuations of the line intensity measured from shot to shot. We compare the measured gain values to theoretical simulations for detail understanding of this scheme. In this paper, an improved method for gain determination was put forward. Experiments for Li–like Si 4f–3d lasers were described which included investigation of plasma's uniformity, laser gain, and theoretical simulation.

IMPROVED METHOD FOR GAIN DETERMINATION

Several difficulties arise in determining gain coefficients of recombination lasers. One is spontaneous emission background in nonlasing regime, which is detrimental to accurate measurement of gain. So time and space resolved measurements are needed to discriminate lasing signals from the background. Another is that lasing emission is usually distant from target surface, where the emission is relatively weak compared to that near the target surface. The electron density profile, affected by plasma's nonuniform expansion and laser condition fluctuation, along the normal to the target surface may not be the same for different laser shots or / and plasma lengths, which would cause fluctuations of the line intensities from shot to shot otherwise the influence of plasma length. To overcome this problem, we put forward a new method for gain determination. Recall that the gain of the amplified spontaneous emission (ASE) was usually determined from Linford formula [6]

$$I_{ASE}(x) = \frac{J}{G} \cdot \frac{(e^{GL} - 1)^{\frac{3}{2}}}{(GL \cdot e^{GL})^{\frac{1}{2}}} , \qquad (1)$$

where $J \propto N_u$, N_u is the population density of the upper level of the lasing transition, x means the spatial position away from the target surface. Similarily, for an optically thin emission, its intensity can be expressed as

$$I'_{sp}(x) = J'L , \qquad (2)$$

where $J' \propto N_u'$. From (1) and (2), we have line intensity ratio $I_r(x)$

$$I_r(x) = \frac{I_{ASE}(x)}{I_{sp}(x)} = (\frac{J}{J'}) \cdot \frac{1}{GL} \cdot \frac{(e^{GL} - 1)^{\frac{3}{2}}}{(GL \cdot e^{GL})^{\frac{1}{2}}} . \qquad (3)$$

From the formula above, gain coefficients can be inferred.

For the Li–like Si 4f–3d laser, the intensity ratio of the Li–like 4f–3d to the 4d–3p transitions can be used to determine gain of the 4f–3d laser. From theoretical calculation under quasi–stationary state condition, it is found that the 4d–3p emission can be considered optically thin if the electron density is less than 3×10^{19} / cm^3 and the electron temperature not less than 30 eV. This is a good approximation in the case of long driving pule. We also find from theoretical calculation that in formula (3) the ratio J/J', which is proportional to N_{4f}/N_{4d}, is less sensitive to electron density than the N_{4f} or the N_{4d} themselves. So the intensity ratio $I_{4f}(x)/I_{4d}(x)$ is not sensitive to density fluctuation, and in turn not sensitive to fluctuations from shot to shot.

EXPERIMENTS AND RESULTS

Experiments were carried out at the LF12 Laser Facility at Shanghai Institute of Optics and Fine Mechanics. Driving laser of 1.05 μm and pulse width of 116 ps was used. The uniform line focus was 12 mm $\times$ 100 μm, and the intensity on the target surface was 6.5×10^{12}W/cm^2. Slab targets of Si crystal with polished surface were used.

The primary diagnostics were a stigmatic flat–field grazing incidence flat–field grating spectrograph (SFGS), a pinhole transmission grating spectrograph (PTGS) and a soft x–ray streak camera. There were two modes for recording of the SFGS spectra. Soft x–ray film was used to record space–resolved time–integrated spectra [8]. Soft x–ray streak camera was coupled to the SFGS to record space– and time–resolved spectra [4]. In the latter mode, photocathode slit of 100 μm wide was used and the streak camera collected x–ray emission with zone of observation 50 μm wide spatial region from the plasma along the target normal. The time resolution of the streak camera was 40 ps. The uniformity of the elongated plasma was monitered by the PTGS. The spatial resolution of the PTGS was about 100 μm. There were also two modes for recording of the PTGS spectra [3]. More detail of the experimental arrangement can be found in Ref.[5].

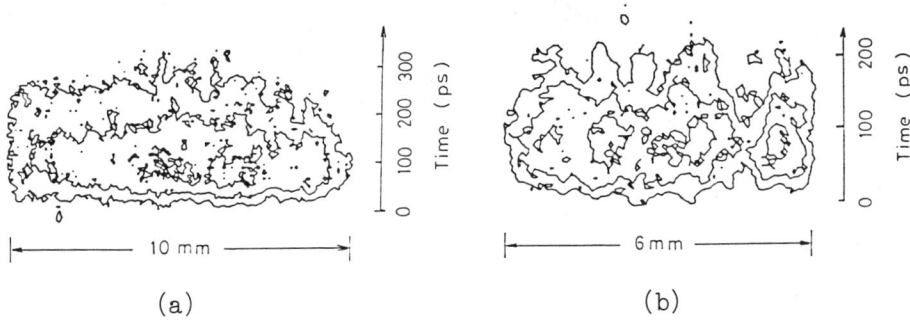

FIGURE 1. Two–dimensional isodensity contours of the time evolution of the spatial uniformities of the zero order of the Si spectra measured by the PTGS.

Investigation of the spatial uniformity of the elongated plasma indicated that the plasma is nearly uniform. But in some cases obvious nonuniformities exist in plasmas, as shown in Fig.1, which was obtained from two laser shots under the same conditions (focusing, target, laser energy output, and pulse shape sampled from whole laser beam) [9]. It is seen that the plasma nonuniformities are different. This may be owed to the driving laser's spatial nonuniformity. As mentioned above, the nonuniformities would cause detrimental effect for gain and its measurement. Thus, not only the uniformity of the line focus but also the laser beam quality should be taken into account to realize a uniform elongated plasma for gain medium.

Figure 2 gives typical space–resolved time–integrated spectra from

SFGS. It shows that candidate lasing lines from the Li–like Si ions occur in nf–3d, nd–3p series and are very evident in the spectra.

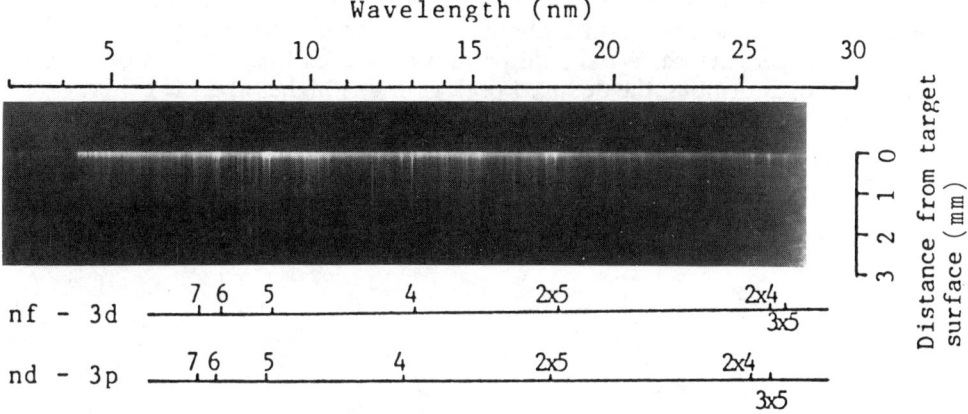

FIGURE 2. Axial spectrum of a 10–mm–long Si plasma.

Figure 3 gives the spatial distribution of some lines with two plasma lengths. Measured here are peak intensities of the spectral lines from the first order of the time–integrated spectra. The film response data for conversion of optical density to intensity were from Ref.[10]. It shows that the peak of the emission of the Li–like Si 3d–2p transition at 4.42 nm moves outwards with plasma length increasing, which may be attributed to intense self–absorption

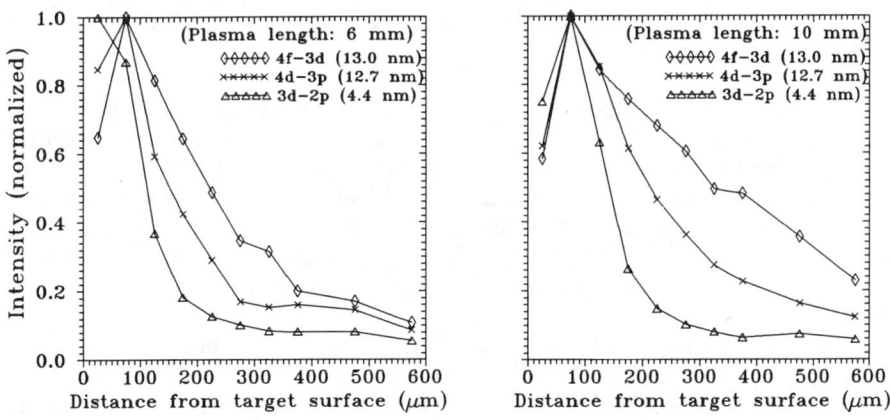

FIGURE 3. Spatial distribution of line emissions from Li–like Si ions.

near the target surface. The intensity profiles of the Li–like Si 4f–3d (13.0 nm) and 4d–3p (12.67nm) emission, are not the same with plasma length increasing. It is seen that after about 300 μm away from the target surface, the 4f–3d emission increases relative to the 4d–3p emission as plasma length increases, which indicates amplification of the 4f–3d emission provided that the 4d–3p line was optically thin in this region. The gain coefficients were determined from the formula (3) with the method described in section 2, and are given in Fig.4. The peak gain is about 2.5 / cm which occurs at about 470 μm from the target surface. We should point out that for the same line but using usual Linford formula, the deduced peak gain was higher and the gain region was much wider than the gain profile here.

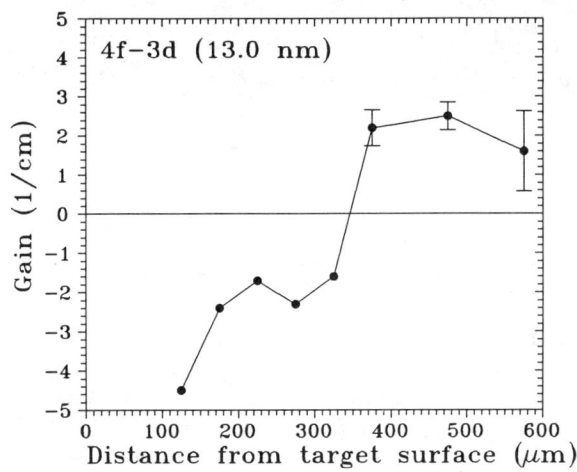

FIGURE 4. Spatial distribution of the gain of the Li–like Si 4f–3d transition. Error bars were estimated by supposing 40% uncertainties of the measured line intensity.

For comparison, we also made theoretical simulation of time- and space-reolved gain for the Li–like Si 4f–3d transition using one-dimensional hydrodynamic code under the same experimental condition (laser intensity, line focus width, target). It is found that the experimentally measured gain value is consistent with the thoretical calculation except that the calculated gain occurs nearer to the target surface. This may be due to the effect of self-absorption in the experiment, which was not considered in the simulation.

Electron densities can be deduced from the Stark broadening of the Li–like Si 5f–3d line [12]. The line width (FWHM) was obtained from the second order of the time-integrated spectra and the electron densities deduced are shown in Fig.5. Limited by the spectral resolving power of the flat-field spectrograph, the electron densities distant from the target surface are somewhat overestimated. In Fig.5 we find that for the 4f–3d transition, the electron densities for the gain are on the order of $\sim 10^{19}$ / cm^3.

Plasma temperature can be estimated from chararcteristic time of recombination emission which was measured by a soft x–ray streak camera coupled to the flat-field spectrograph. The experiemtal result was shown in

Fig.6. The recombination time (FWHM)~ R_3^{-1}, where $R_3 = n_e^2 \cdot \alpha_3$, and α_3 is three—body recombination rate coefficient [13]. With electron density obtained in Fig.5, we estimate that the electron temperature lies between several eV and tens of eV.

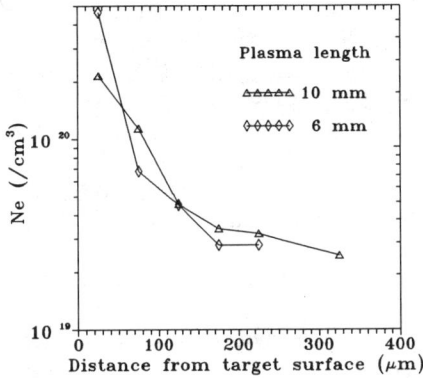

FIGURE 5. Electron density profile of the plasma gain medium deduced from the spectral Stark broadening of the Li—like Si 5f-3d transition.

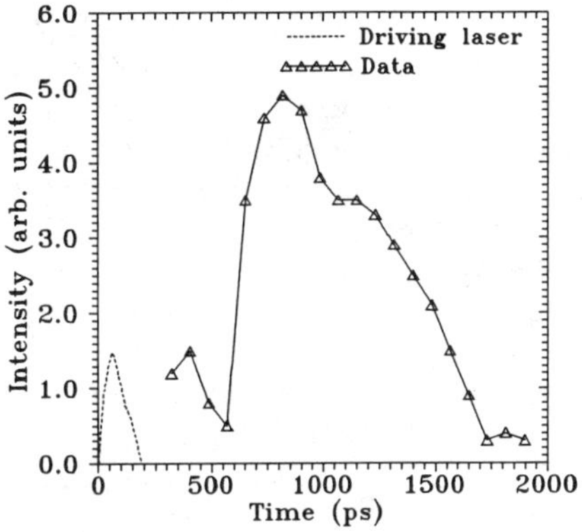

FIGURE 6. Temporal characteristics of the Li—like Si 5f-3d emission from a 6—mm—long plasma. The zone of the observation centered at 150 μm from the target surface.

CONCLUSIONS

Time and space resolved investigation have been made on both recombination Li—like x—ray laser gain and plasma uniformity. It is found that plasma's nonuniformity is an important factor to be considered. To overcome line intensity fluctuations from shot to shot, we developed a new method of using line intensity ratio to determine gain coefficient. Experimental measurement of the Li—like Si 4f-3d laser gain by this method is in good agreement with theoretical simulation. Self—absorption might be considered for better agreement. Plasma parameters estimated from experimental data are also in qualitative coindence with plasma conditions theoretically given for the 4f-3d gain. Further experiment with longer line focus is now in progress and larger gain length product of about 8 might be expected for a 4—cm—long plasma.

ACKNOWLEDGMENTS

The authors wish to express their thanks to the staff of the LF12 Laser Facility for laser operation and technical help. This work was performed under the auspices of China National Advanced Technology Program.

REFERENCES

1. P. Jaegle, A. Carillon, P. Dehz, B. Gauthe, F. Gads, G. Jamelot, and A. Klisnick, Europhys. Lett. 7, 337 (1988); P. Jaegle, A. Carillon, P. Dehz, B. Gauthe, G. Jamelot, A. Klisnick, J.P. Raucourt, in X-Ray Lasers 1990, ed. by G. J. Tallents, Inst. Phys. Conf. Ser. No 116, (IOP, Bristol), p. 43; G. Jamelot, A. Carillon, P. Dehz, B. Gauthe, P. Jaegle, A. Klisnick, J.P. Raucourt, in X-Ray Lasers 1992, ed. by E. Fill, Inst. Phys. Conf. Ser. No. 125, (IOP, Bristol), p.89; Z. Xu, Z. Zhang, P. Fan, X. Wang, R. Li, P. Lu, L. Zhang, A. Qian, C. Jiang, S. Han, and X. Feng, Appl. Phys. B57, 319 (1993).
2. G.J. Pert, in X-Ray Lasers 1990, ed. by G. J. Tallents, Inst. Phys. Conf. Ser. No 116, (IOP, Bristol), p. 143.
3. X.F. Wang, Z. Xu, S. Chen, A. Qian, P. Fan, Z. Zhang, M. Gong, S. Gao, and B. Shan, Appl. Phys. Lett. 58, 2901 (1991); X.Wang, Z. Xu, S. Chen, A. Qian, L. Lin, Y. Li, P. Fan, Z. Zhang, Opt. Eng. 32, 56 (1993).
4. X. Wang, Z. Xu, P. Fan, L. Lin, Y. Li, P. Lu, R. Li, A. Qian, and Z. Zhang, Opt. Commun. 102, 271 (1993).
5. Z.Xu, Z. Zhang, P. Fan, S. Han, R. Li, X. Wang, P. Lu, S. Han, B. Shan, L. Zhang, H. Teng, W. Zhang, X. Feng, and H. Xiang, in this proceedings.
6. G.J. Linford, E.R. Peressini, R. Sooy, and M.L. Spaeth, Appl. Opt. 13, 379 (1974).
7. X.M. Deng, W.Y. Yu, D.Y. Fan, LF12 Research Report, Shanghai Institute of Optics and Fine Mechanics, (1987).
8. P. Fan, Z. Zhang, J. Zhou, Z. Xu, and X. Guo, Acta Optica Sinica, 12, 118 (1992).
9. Z. Xu, Z. Zhang, P. Fan, X. Wang, R. Li, P. Lu, L. Zhang, A. Qian, C. Jiang, S. Han, and X. Feng, Appl. Phys. B57, 319 (1993).
10. P. Lu, R. Li, X. Wang, Y. Li, P. Fan, Z. Zhang, S. Chen, and Z. Xu, in X-Ray Lasers 1992, ed. by E. Fill, Inst. Phys. Conf. Ser. No. 125, (IOP, Bristol), p.379.
11. B. Shen, X. Wang, Z. Xu, and H. Teng, in this proceedings.
12. J.C. Moreno, H. C. Griem, S. Goldsmith, and J. Knauer, Phys. Rev. A 39, 6033 (1989); P. Lu, Z.Zhang, Z. Xu, P. Fan, S. Chen, and B. Shen, Appl. Phys. Lett. 60, 1649 (1992).
13. N.H. Burnett, and G.D. Enright, IEEE J. Quant. Electron. 26, 1797 (1990).

Stark Line Broadening of the n=4 to 3 Transitions in High-Z Heliumlike Ions

P.A. Loboda, V.A.Lykov, and V.V.Popova

Russian Federal Nuclear Center – National Institute of Technical Physics
P.O.Box 245, Chelyabinsk-70, 454070, Russia

Abstract. Stark line broadening of the $n = 4$ to 3 transitions of He-like Ne, Mg, and Al in multicharged ion plasmas is considered. Line profiles calculations involved quasi-static ion broadening, impact electron broadening, natural, and Doppler broadening. Considerable effect of Stark line broadening due to plasma ions for the 4F–3D transitions of He-like Ne is demonstrated at the Ne-plasma parameters yielding a maximum gain in the theoretical modeling of the resonantly photopumped Na-Ne X-ray laser scheme under the conditions of the Saturn experiments. The sensitivity of the calculated line profiles to the intermediate coupling effects and different energy level data is also investigated. Calculated line profiles of the 4F–3D transitions in He-like Mg and Al are compared to the experimental and other theoretical data.

INTRODUCTION

Currently, the experimental and theoretical spectra from the $n = 3 - 5$ to $n' = 2 - 4$ transitions in high-Z H-, He-, and Li-like ions are of particular interest due to extensive investigations of various laboratory X-ray laser schemes utilizing these ions (1–7). Specifically, the resonantly photopumped Na-Ne X-ray laser scheme which would lase on the $n = 4, 3$ to 3,2 transitions of the He-like Ne has been studied experimentally using Z-pinches for the last decade. In recent experiments with powerful Saturn Z-pinch, distinct enhancement of the Ne IX $1s4p - 1s^2$ line relative to the $1s3p - 1s^2$ line has been demonstrated (2,3) thus providing an indication that the 4^1P_1 –states of He-like Ne are photopumped by the radiation of the Na He$_\alpha$ line, $2^1P_1 - 1^1S_0$. Time-dependent modeling (4) of the Ne lasant under the conditions of the Saturn experiments have yielded maximum gain of about 0.2 cm^{-1} on the $4^1F_3 - 3^1D_2$ transition of He-like Ne assuming the Voigt profile of laser line which takes into account the homogeneous and Doppler broadening.

In multicharged ion plasmas, however, the most intense lines of the $n = 5, 4$ to $n' = 4, 3$ transitions in high-Z H-, He-, and Li-like ions may be strongly affected by the ion Stark broadening (7–10). Therefore, detailed Stark-broadening calculations of the He-like Ne $4F - 3D$ lines are required for the evaluation of a potential decreasing of laser gains on the $4^1F_3 - 3^1D_2$ transition relative to the achievable Voigt values.

In present communication, Stark line broadening of the $4F - 3D$ transitions of He-like Ne is considered for the Ne-plasma parameters yielding a maximum gain in the theoretical modeling of the Na-Ne X-ray laser scheme (4). Calculated line profiles of He-like Mg and Al $4F - 3D$ transitions with the nearest Z are compared to the experimental and theoretical data of Ref. (7). The effects due to intermediate coupling and energy level data employed on the line profiles of these transitions are also investigated.

LINE PROFILE CALCULATIONS

Theoretical modeling (4) of the Na-Ne X-ray laser scheme have demonstrated that maximum gain of about 0.2 cm^{-1} on the $4^1F_3 - 3^1D_2$ transition of He-like Ne at $\lambda \cong 231.4$ Å could be achieved at an ion density $N_i = 10^{18}$ cm^{-3} and a temperature of Ne plasma $T = 26.7$ eV. The estimates show that in the Ne-plasma with ion densities $N_i \geq 10^{18}$ cm^{-3} the closely-spaced 4F, 4D, and 4P states should be effectively mixed by plasma ion microfields $F \geq F_0$, since the typical value of interaction matrix element with the Holtzmark field F_0 (11) is comparable to or greater than the energy splittings between the states with the opposite parities. Consequently, at these densities the Stark effect due to plasma ions produces a substantial influence on the profiles of the Ne IX $4F - 3D$ lines. Fig. 1 displays the picture of the Stark effect for the $4F - 3D$ line components $M \rightarrow M'(M, M' = 0, 1)$ at an ion microfields significant for the line broadening calculations. As one can see from the figure, all off the components considered are strongly affected by the ion microfield F and the dependence of the transition energies $E(F)$ goes from quadratic to the linear one. Therefore, no analytic scaling procedure can be applied to estimate the linewidths of the $4F - 3D$ transitions over the plasma parameters domain reasonable for theoretical investigation of the Na-Ne X-ray laser scheme (3,4). For the plasma conditions considered, line broadening parameters evaluations have shown the validity of the quasi-static approximation for plasma ion broadening of the Ne IX $4F - 3D$ line components. Specifically, the characteristic Fourier frequencies for the plasma ion motion $\omega_F \approx v_p / r_p$, where r_p and v_p are the ion-radiator separation and relative velocity, respectively, were found to be smaller than the average Stark shifts

$\Delta\omega_{\alpha\beta}(F_0)$ of the M-components of the $4F - 3D$ lines. The same holds true for the $4F - 3D$ lines of He-like Mg and Al at an electron density $N_e = 10^{20}$ cm^{-3} and plasma temperature $T = 400$ eV. These plasma parameters provided the best fit of the calculated profiles of the $4F - 3D$ Mg XI and Al XII lines to the high-resolution experimental data (7).

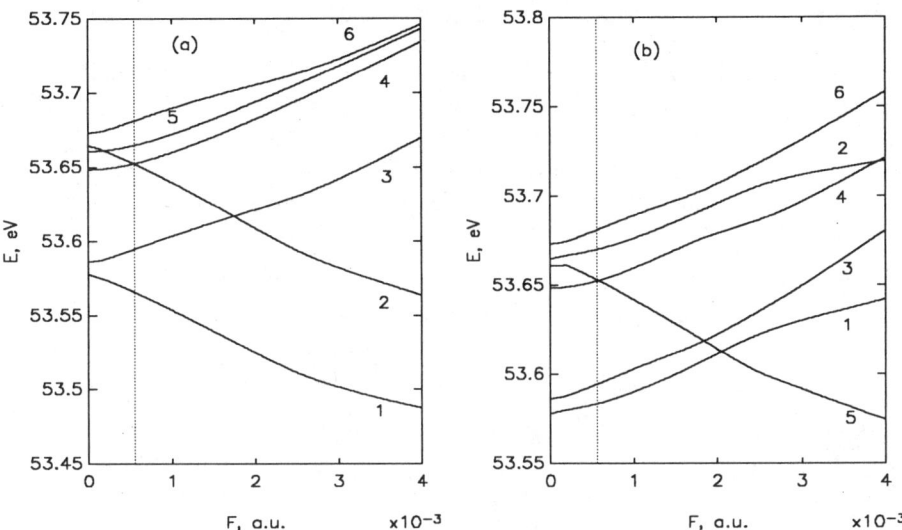

FIGURE 1. Stark effect for the $4F - 3D$ line components $M = 0 \to M' = 0$ (a) and $M = 1 \to M' = 1$ (b). The numbers correspond to the following components: $4^3F_3 - 3^1D_2$ (1); $4^3F_3 - 3^3D_2$ (2); $4^1F_3 - 3^1D_2$ (3); $4^3F_4 - 3^3D_3$ (4); $4^3F_2 - 3^3D_1$ (5); $4^1F_3 - 3^3D_2$ (6). Dotted reference line corresponds to the Holtzmark ion microfield F_0 at an ion density $N_i = 10^{18}$ cm^{-3}.

Line profiles for the $n = 4$ to 3 transitions of He-like Ne, Mg and Al were calculated using HELM code (9) and involved quasi-static ion broadening as well as impact electron broadening, natural, and Doppler broadening. Stark states energies and wavefunctions in the presence of ion microfield were found by solving the eigenvalue problem for the perturbed Hamiltonian energy matrix obtained from the first-order perturbation expansion on the basis of isolated-ion wavefunctions considered within the framework of intermediate coupling scheme. The energies and dipole transition matrix elements for the unperturbed ion states were calculated using multiconfigurational Dirac-Fock code of Grant and co-workers (12,13) and also taken from the Z-expansion data (14,15) The elements of electron-collisional relaxation matrix were calculated with the Stark wavefunctions using the hyperbolic classical paths for nonadiabatic electron broadening.

Line profiles were calculated under the assumption of statistical populations for the upper states with the same $n = 4$. Line profiles $I(\lambda)_{n-n'}$ for the n to n' transitions were obtained by averaging the line profiles, computed for the set of significant ion electric microfield values F over the electric microfields with an appropriate distribution function. For the present calculations, the modified ion microfield distribution of C.F. Hooper (16,17) was employed.

In Fig.2 we show the results of line profile calculations for the $4F - 3D$ transitions of He-like Ne IX at an ion density $N_i = 10^{18}$ cm^{-3} and a temperature of Ne plasma $T = 26.7$ eV at which maximum gain was obtained in the theoretical modeling of the Na-Ne X-ray laser scheme (4). It is clearly seen that the inclusion of Stark broadening due to plasma ions results in the lowering of peak intensity of spectral component related to the lasing transition $4^1F_3 - 3^1D_2$ at $\lambda \cong 231.4$ Å approximately by a factor of 2.6.

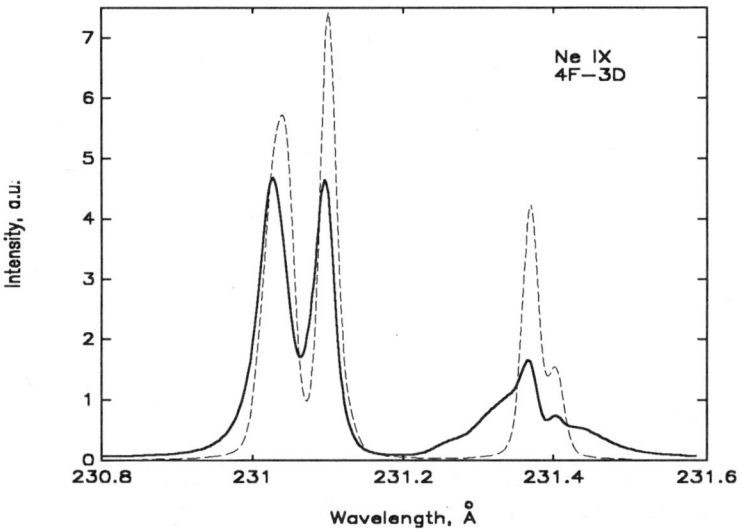

FIGURE 2. Line profiles of the $4F - 3D$ transitions of He-like Ne IX at an ion density $N_i = 10^{18}$ cm^{-3} and a temperature of Ne plasma $T = 26.7$ eV. The solid curve corresponds to the Stark broadened profile, while the dashed curve represents the Voigt profile, allowing for the homogeneous (i.e. electron and natural) and Doppler broadening only.

The comparison of the lineshapes for the Ne IX $4F - 3D$ transitions obtained under the assumptions of intermediate and LS-coupling schemes, (Fig. 3) displays appreciable differences of the line profiles in the range of the spectral feature related to the $4^1F_3 - 3^1D_2$ transition. This is due to the fact that at relatively small temperatures ≤ 30 eV Doppler effect does not provide sufficient smoothing of the detailed structure of the lines considered and so, at these temperatures detailed line-broadening calculations for the He-like $4F - 3D$ transitions require an accurate description of atomic structure even at $Z \geq 10$. The inspection of the line profiles plotted in Fig. 3 also shows that the employment of the different sets of data for the ionic energy levels, resulting only in relative shifts of some components of the Voigt profile, can provide noticeable mismatches of the relevant Stark-broadened profiles at the spectral feature corresponding to the lasing transition.

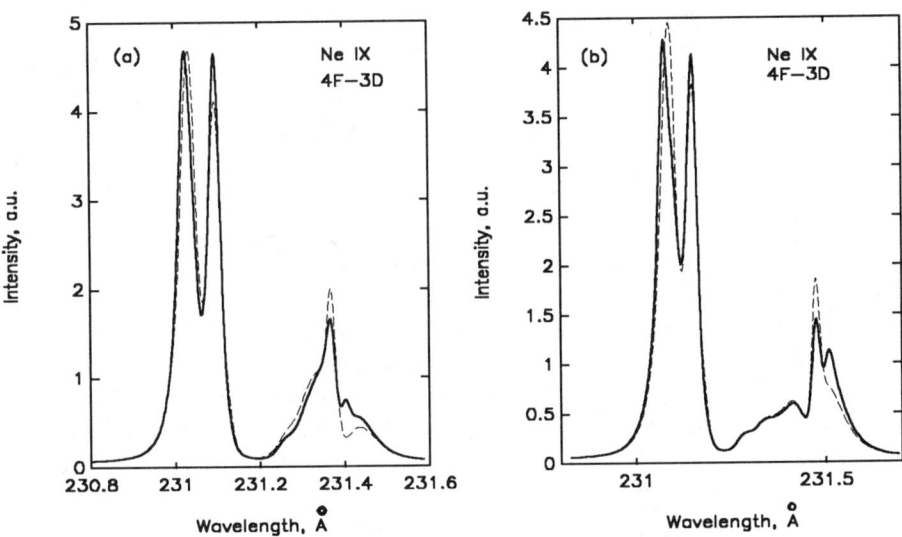

FIGURE 3. Line profiles of the $4F - 3D$ transitions of He-like Ne IX at an ion density $N_i = 10^{18}$ cm^{-3} and a temperature of Ne plasma $T = 26.7$ eV with MZ (13,14) (a) and MCDF (11,12) data (b) for the energies employed. Solid curves represent the calculations with intermediate coupling effects included while dashed ones – the calculations under the assumption of pure LS-coupling.

The comparison of the Mg XI $4F - 3D$ line profiles at an electron density $N_e = 10^{20}$ cm^{-3} and plasma temperature $T = 400$ eV, plotted in Fig. 4(a), shows that our results are in overall good agreement with the experimental and calculated data of Ref. (7). The consistency of the same line profiles for Al XII (Fig. 4(b)) is not as good, however. In our opinion, the reason for the discrepancies in the lineshape details may be attributed mostly to the different atomic data sets employed, since our previous calculations for the $4F - 3D$ line of Li-like Al (10) were found to be in a close agreement with theoretical data obtained with the line-profile code (8) employed to calculate the profiles of the $4F - 3D$ lines of He-like Mg and Al in Ref. (7).

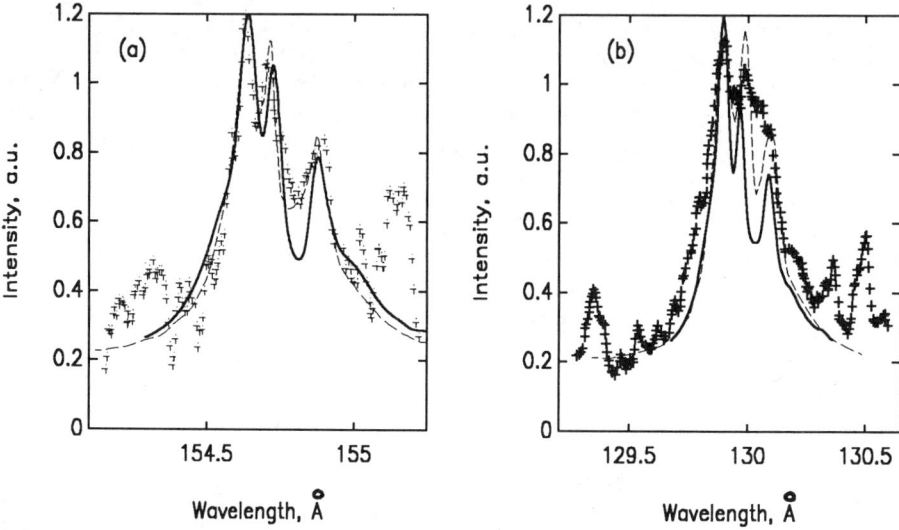

FIGURE 4. Line profiles of the $4F - 3D$ transitions of He-like Mg XI (a) and Al XII (b) at an electron density $N_e = 10^{20}$ cm^{-3} and plasma temperature $T = 400$ eV. Solid curve presents our results, dashed one – theoretical data of Ref. (7). Experimental points of Ref. (7) are displayed as pluses.

CONCLUSION

We have investigated the effect of Stark broadening due to plasma ions on the line profiles for the $4F - 3D$ transitions of He-like Ne for the Ne-plasma parameters yielding a maximum gain in the theoretical modeling (4) of the resonantly photopumped Na-Ne X-ray laser scheme under the conditions of the Saturn experiments (2,3). It is found that the inclusion of Stark broadening due to plasma ions results in the lowering of peak intensity of spectral component related to the lasing transition $4^1F_3 - 3^1D_2$ approximately by a factor of 2.6 and, consequently, to the relevant decreasing of an achievable gains on this transition. The calculations have shown that no analytic scaling procedure can be applied to estimate the linewidths of the $4F - 3D$ transitions over the plasma parameters domain reasonable for theoretical investigation of the Na-Ne X-ray laser scheme (3,4).

Therefore, the ion Stark broadening calculations should be included in the evaluation of the potential laser gains for the further theoretical optimization of the Na-Ne X-ray laser scheme.

Our calculations for the $4F - 3D$ transitions of He-like Ne have also displayed a pronounced sensitivity of the Stark-broadened line profiles to the intermediate coupling effects and different energy level data employed. Calculated Stark-broadened profile of the $4F - 3D$ Mg XI line was found to be in an overall good agreement with the experimental and calculated data of Ref. (7).

ACKNOWLEDGMENTS

This work has been supported in part by the University of California, Lawrence Livermore National Laboratory under Contract No B239707. We are grateful to Prof. L.A. Vainshtein for his kind provision of Z-expansion spectroscopic data on high-Z He-like ions and helpful discussions. We would also like to thank Drs. J. Nilsen, A.L. Osterheld, J.C. Moreno, J. P. Apruzese, and A. Sureau for their interest in this work and useful discussions.

REFERENCES

1. Nilsen J., Scofield J., and Chandler E., *Appl.Opt.*, **31**, 4950–4956, (1992).
2. Apruzese J.P. et al., in *Proc. of the 2nd Int. Colloq. on X-ray lasers*, G.J. Tallents, ed. (IOP, Bristol, England), 1991, p.39–42.
3. Porter J.L. et al., *Phys.Rev.Lett.*, **68**, 796–799, (1992).
4. Nilsen J. and Chandler E., *Phys.Rev.A*, **44**, 4591–4598, (1991).
5. Jaegle P. et al., *J.Opt.Soc.Am.B.*, **4**, 563–574, (1987).
6. Moreno J.C., .Griem H.R, and Goldsmith S., *Phys.Rev.A.*, **39**, 6033–6036, (1989).
7. Moreno J.C., .Griem H.R, Lee R.W., and Seely J.F, *Phys.Rev.A*, **47**, 374–379, (1993).
8. Calisti A., Khelfaoui F., Stamm R., Talin B., and Lee R.W., *Phys.Rev.A.*, **42**, 5433–5440, (1990).
9. Loboda P.A., Lykov V.A., and Popova V.V., in *Proc. of the Int. Symp. on Laser-Plasma Interactions*, Zhijiang Wang, Zhizhan Xu, eds., Proc. SPIE, **1928**, (SPIE, Bellingham), 1992, pp.145–156.
10. Loboda P.A., Lykov V.A., and Popova V.V., in *Proc. of the Ultrashort Wavelength Lasers II*, S. Suckewer, ed., Proc. SPIE, **2012**, (SPIE, Bellingham), 1993, pp.232–239.
11. Griem H.R., *Spectral Line Broadening by Plasmas*, Academic, N.Y., 1974.
12. Grant I.P. et al., *Comput.Phys.Commun.*, **21**, 207–231, (1980).
13. McKenzie et al., *Comput.Phys.Commun.*, **21**, 233–246, (1980).
14. Vainshtein L.A., and Safronova U.I, *Phys.Scr.*, **31**, 519, (1985).
15. Vainshtein L.A., (*Private communication*).
16. Tigne R.J. and Hooper C.F., *Phys.Rev.A*, **15**, 1773–1779, (1977).
17. Akhmedov E.Kh. et al., *Zh.Eksp.Teor.Fiz. (Sov.Phys.: JETP)*, **89**, 470–481, (1985).

# Measurements of Line Overlap for Resonant Spoiling of X-ray Lasing Transitions

P. Beiersdorfer, S. R. Elliott, B. J. MacGowan, and J. Nilsen

Lawrence Livermore National Laboratory, Livermore, CA 94551

Abstract. High-precision measurements are presented of candidate line pairs for resonant spoiling of x-ray lasing transitions in the nickellike W^{46+}, the neonlike Fe^{16+}, and the neonlike La^{47+} x-ray lasers. Our measurements were carried out with high-resolution crystal spectrometers, and a typical precision of 20–50 ppm was achieved. While most resonances appear insufficient for effective photo-spoiling, two resonance pairs are identified that provide a good overlap. These are the $4p_{1/2} \rightarrow 3d_{3/2}$ transition in nickellike W^{46+} with the $2p_{3/2} \rightarrow 1s_{1/2}$ transition in hydrogenic Al^{12+}, and the $3s_{1/2} \rightarrow 2p_{3/2}$ transition in neonlike La^{47+} with the $1^1S_0\text{-}2^1P_1$ line in heliumlike Ti^{20+}.

Introduction

Modification of the x-ray laser kinetics by resonant photo-pumping is of practical and scientific interest. For example, resonant photo-pumping has been proposed as an efficient way to achieve lasing [1]. Conversely, resonant photo-pumping may be used to spoil lasing. This allows tailoring of the laser to yield monochromatic output that suits available x-ray optics. It can also be used to probe the laser kinetics in order to obtain a better understanding of the principles of x-ray lasing.

Photo-pumping requires resonances on the order of a few hundred parts per million (ppm). Candidate resonances must be experimentally verified, as the accuracy achieved by calculations is typically only within a range of a few parts per thousand for the multi-electron ions of interest in x-ray lasing. In the following we present measurements of candidate resonances for spoiling of x-ray laser transitions. The lasers considered are the nickellike W^{46+} laser shown to lase at 43.1 Å [2] and the neonlike Fe^{16+} laser shown to lase at 254.9, 347.6, and 388.9 Å [3]. We also investigated spoiling of a proposed neonlike La^{47+} laser, which might lase at 57, 58 and 128 Å [4]. Lasing in La^{47+} cannot yet be attained with existing facilities, but might be achieved in experiments with the planned National Ignition Facility.

Spoiling of the Nickellike W^{46+} Laser

A diagram showing the levels involved in the resonant spoiling of the nickellike W^{46+} laser is shown in Fig. 1. The 43-Å lasing transition proceeds from level $(3d_{3/2}, 4d_{3/2})_{J=0}$ to level $(3d_{3/2}, 4p_{1/2})_{J=1}$. Raising the population of the lower level will diminish or destroy the population inversion necessary for lasing and thus will spoil the 43-Å laser. This is accomplished by pumping with an appropriate line whose wavelength is nearly coincident with that of the dump transition from level $(3d_{3/2},4p_{1/2})_{J=1}$ to the $^1S_0$ ground level. The energy of the $4p_{1/2} \rightarrow 3d_{3/2}$ dump line was predicted to be 1728 eV, or 7.150 Å [5]. A candidate pump line is provided by the $2p_{3/2} \rightarrow 1s_{1/2}$ Ly-α_1 line in hydrogenic Al^{12+} at 7.1709 Å [6], as shown in Fig. 1. Other candidate pump lines are the $2p_{1/2} \rightarrow 1s_{1/2}$ Ly-α_2 line in Al^{12+} (predicted at 7.1763 Å [6]), the transition from $(2p_{1/2},3d_{3/2})_{J=1}$ to the $^1S_0$ ground level in neonlike Br^{25+} (predicted at 7.1700 Å [7]), and the transition from $(3d_{3/2},4f_{5/2})_{J=1}$ to the $^1S_0$ ground level in nickellike Er^{40+} (predicted at 7.1760 Å [8]).

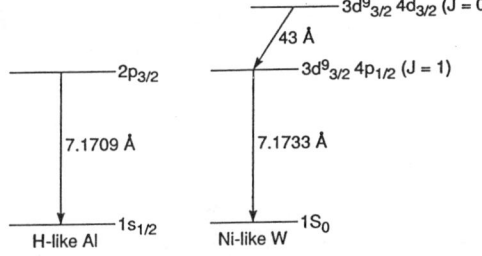

FIGURE 1. Energy level diagram showing the mechanism for spoiling of the 43-Å nickel-like W^{46+} laser by resonant photo-pumping with hydrogen-like Al^{12+}. Increasing the population of the 4p level spoils a possible inversion and thus lasing between the 4p and 4d levels.

The candidate photo-pumping resonances are verified in measurements on the Livermore electron beam ion trap (EBIT). The device uses a monoenergetic electron beam to produce and excite a particular ion species of interest, and a given line can be observed free of blends with collisional or dielectronic satellites [9]. Recording spectra with the evacuated flat-crystal spectrometer described in Ref. [10], and using the Lyman-α lines of Al^{12+} as reference lines, we find 7.1733±0.0003 Å for the tungsten line. Moreover, we measured 7.1837±0.0020 Å for the wavelength of the $4f_{5/2} \rightarrow 3d_{3/2}$ transition in nickellike Er^{40+} and 7.1685±0.0002 Å for that of the $3d_{3/2} \rightarrow 2p_{1/2}$ transition in neonlike Br^{25+}. A summary of the results is given in Table I.

Table I. Comparison of measured and predicted wavelengths of candidate photo resonances in the Ni-like W^{46+} and Ne-like Fe^{16+} and La^{47+} lasers (in Å). Lines used as reference standards are denoted by (*).

Laser	Upper Level	Ion	Theory	Experiment
W^{46+}	$3d_{3/2},4p_{1/2}(J=1)$	W^{46+}	7.150 [Ref.5]	7.1733(3)
	$3d_{5/2},4f_{7/2}(J=1)$	Er^{40+}	7.1760 [Ref.8]	7.1830(20)
	$2p_{1/2},3d_{3/2}(J=1)$	Br^{25+}	7.1700 [Ref.7]	7.1685(2)
	$2p_{3/2}(J=3/2)$	Al^{12+}	7.17091 [Ref.6]	7.17091(*)
	$2p_{1/2}(J=1/2)$	Al^{12+}	7.17632 [Ref.6]	7.17632(*)
Fe^{16+}	$2p_{1/2},3s_{1/2}(J=1)$	Fe^{16+}	16.797 [Ref.7]	16.772(3)
	$1s_{1/2},2p_{3/2}(J=1)$	F^{7+}	16.8064 [Ref.11]	16.8064(*)
La^{47+}	$2p_{3/2},3s_{1/2}(J=1)$	La^{47+}	2.61001 [Ref.13]	2.61015(10)
	$2p_{3/2},3s_{1/2}(J=2)$	La^{47+}	2.61330 [Ref.13]	2.61330(5)
	$1s_{1/2},2p_{3/2}(J=1)$	Ti^{20+}	2.61040 [Ref.11]	2.61040(*)

The measured location of the candidate pump lines relative to the measured position of the $4p_{1/2} \to 3d_{3/2}$ dump transition in tungsten is shown schematically in Fig. 2. The best resonance is found with the $2p_{3/2} \to 1s_{1/2}$ Ly-α_1 line in hydrogenic Al^{12+}. The two lines differ by 2.4±0.3 mÅ or 335 ppm. Figure 2 also gives a schematic overview of the predicted locations of the various lines. A comparison of the predicted and measured locations illustrates the unequivocal need for precise measurements of candidate resonances.

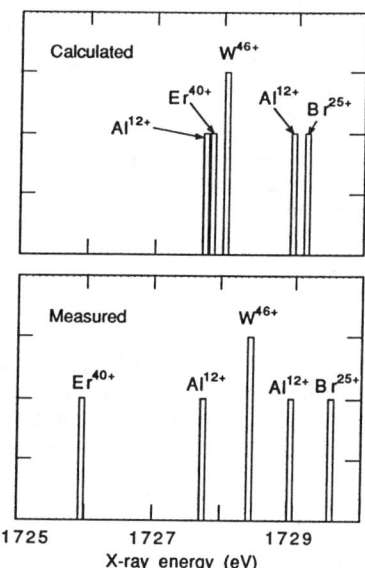

Figure 2. Comparison of predicted and measured location of the 3d-4p transition in nickel-like W^{46+} and of various candidate resonant photo-pumping transitions.

Spoiling of the Neonlike Fe^{16+} Laser

A diagram showing the levels involved in the resonant spoiling of the neonlike Fe^{16+} laser is shown in Fig. 3. Resonant photo-pumping of the $2p_{1/2}–3s_{1/2}$ transition in neonlike iron at 16.772 Å enhances the population of the $(2p_{1/2},3s_{1/2})_{J=1}$ level, which is the lower level for several $3p \to 3s$ lasing transitions, including the 255-Å lasing line observed to dominate lasing in neonlike iron [3]. A candidate pump line is given by the $1^1S_0–2^1P_1$ line in heliumlike F^{7+}.

The candidate photo-pumping resonance is again verified in measurements on the EBIT facility. Using the wavelengths of the $1^1S_0–2^1P_1$ and $1^1S_0–2^3S_1$ lines calculated by Drake [11], (16.8064 and 17.1528 Å, respectively) to calibrate the measurements, we find 16.772±0.003 Å for the $2p_{1/2}–3s_{1/2}$ line in neonlike iron Fe^{16+}. The separation between it and the heliumlike F^{7+} resonance line is 35±3 mÅ, or 2,000 ppm.

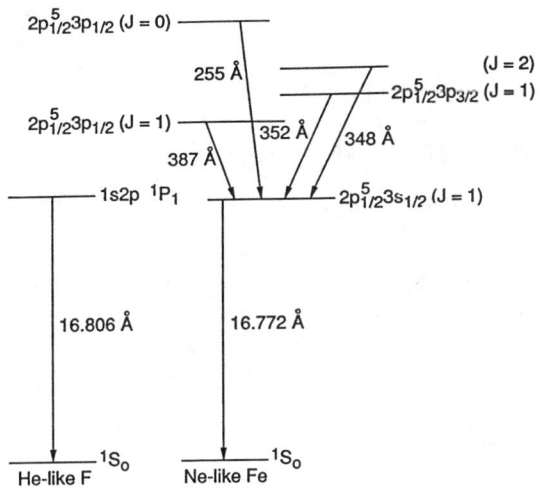

Figure 3. Energy level diagram showing the mechanism for spoiling of lasing in neonlike Fe^{16+} by resonant photo-pumping of the $2p_{1/2}$-$3s_{1/2}$ transition by the $1\,^1S_0$ - $2\,^1P_1$ transition in heliumlike F^{7+}. Increasing the population of the 3s level spoils a possible inversion and thus lasing between the 3s and 3p levels.

Spoiling of the Neonlike La^{47+} Laser

The highest-Z neonlike laser demonstrated so far is neonlike Ag^{37+} with lasing lines at 100 Å [12]. This experiment was carried out on the NOVA laser facility and appears to represent the upper limit of Z that can be made to lase with existing facilities. As a result, a neonlike La^{47+} laser has not yet been demonstrated. Such a laser would, however, be feasible with the planned National Ignition Faciliy. If successful, lasing between the levels $(2p_{1/2},3p_{3/2})_{J=2}$ and $(2p_{1/2},3s_{1/2})_{J=1}$ and between the levels $(2p_{3/2},3p_{1/2})_{J=2}$ and $(2p_{3/2},3s_{1/2})_{J=1}$ would result in lines at 58.3±2 and 128.2±0.8 Å, as inferred from high-resolution measurements of the 2p-3p and 2p-3s transitions [4]. Lasing would also be expected at 57 Å between the levels $(2p_{3/2},3p_{3/2})_{J=2}$ and $(2p_{3/2},3s_{1/2})_{J=1}$ as well as between other levels.

Resonant photo-pumping of the $2p_{3/2}$–$3s_{1/2}$ transition with the $1s2p\ ^1P_1 \rightarrow 1s^2\ ^1S_0$ transition in heliumlike Ti^{20+} could be used to spoil lasing transitions involving the $(2p_{3/2},3s_{1/2})_{J=1}$ dump level. Drake [11] predicted the wavelength of the $1\,^1S_0$-$2\,^1P_1$ line in heliumlike Ti^{20+} to be 2.6104 Å, while Ivanova and Gulov [13] predicted the wavelength of the transition from $(2p_{3/2},3s_{1/2})_{J=1}$ to the $^1S_0$ ground level in neonlike La^{47+} to be 2.6100 Å.

The candidate photo-pumping resonance is verified in measurements on the Princeton Large Torus (PLT) tokamak using a Johann-type Bragg-crystal spectrometer [14]. Using 2.61040 Å for the $1\,^1S_0$-$2\,^1P_1$ line in titanium to calibrate the spectrum, we find 2.61015±0.00010 Å for the wavelength of the transition from $(2p_{3/2},3s_{1/2})_{J=1}$ to the $^1S_0$ ground level in neonlike lathanum La^{47+}. We find 2.61330±0.00005 Å for the neighboring transition from the $(2p_{3/2},3s_{1/2})_{J=2}$ level to ground.

Discussion

Among the candidate resonances investigated only two are found that are sufficiently good to allow photo-pumping. The first is the resonance between the $2p_{3/2} \rightarrow 1s_{1/2}$ Ly-α_1 line in hydrogenic Al^{12+} and the $4p_{1/2} \rightarrow 3d_{3/2}$ transition in nickellike W^{46+}. The two lines differ by 335 ppm, which approximately equals the Doppler broadening of the aluminum line in a 500-eV plasma. Moreover, the tungsten line differs only 420 ppm from the $2p_{1/2} \rightarrow 1s_{1/2}$ Ly-α_2 line in hydrogenic Al^{12+}, and the Ly-α_2 line may also contribute to pumping the nickel-like transition, especially as the two Ly-α components blend into a sinigle line in optically thick plasmas. The fact that the tungsten line lies between the Ly-α lines is especially important in cases where plasma motion shifts the wavelength of the tungsten line (or that of the aluminum lines). Such a shift would improve the resonance, regardless of the relative direction of the motion.

The second, good resonance is found between the $3s_{1/2} \rightarrow 2p_{3/2}$ line in neonlike lathanum La^{47+} and the $2p_{3/2} \rightarrow 1s_{1/2}$ line in heliumlike Ti^{20+}. The two lines differ by only 96 ppm, representing an excellent resonance for photo-pumping provided a neonlike lathanum La^{47+} laser were achieved with the National Ignition Facility or a similar laser facility with the more intensity and energy than currently available.

This work was performed under the auspices of the U.S. Department of Energy by Lawrence Livermore National Laboratory under contract No. W-7405-ENG-48.

References

1. A. V. Vinogradov, I. I. Sobelman, and E. A. Yukov, Sov. J. Quantum Electron. **59**, 5 (1975); B. A. Norton and N. J. Peacock, J. Phys. B **8**, 989 (1975); R. C. Elton, X-ray Lasers, (Academic Press, San Diego, 1990), pp. 99 – 198; C. H. Skinner, Phys. Fluids B **3**, 2420 (1991).
2. B. J. MacGowan, *et al.*, Phys. Rev. Lett. **65**, 420 (1990).
3. J. Nilsen, J. C. Moreno, B. J. MacGowan, and J. A. Koch, Appl. Phys. B **57**, 309 (1993).
4. P. Beiersdorfer, *et al.*, Phys. Rev. A **37**, 4153 (1988).
5. S. Maxon, *et al.*, Rev. Lett. **63**, 236 (1989).
6. W. R. Johnson and G. Soff, At. Data Nucl. Data Tables **33**, 405 (1985).
7. J. A. Cordogan and S. Lunell, Phys. Scr. **33** 406 (1986).
8. J. Nilsen, Phys. Rev. A **40**, 5440 (1989).
9. P. Beiersdorfer, *et al.*, in *UV and X-ray Spectroscopy of Laboratory and Astrophysical Plasmas*, ed. by E. Silver and S. Kahn (Cambridge University Press, Cambridge, 1993), p. 59.
10. P. Beiersdorfer and B. J. Wargelin, Rev. Sci. Instrum. **65**, 13 (1994).
11. G. W. F. Drake, Can. J. Phys. **66**, 586 (1988).
12. D. J. Fields, *et al.*, Phys. Rev. A **46**, 1606 (1992).
13. E. P. Ivanova and A. V. Gulov, At. Data Nucl. Data Tables **49**, 1 (1991).
14. P. Beiersdorfer, Ph. D. thesis, Princeton University (1988).

Generation of Tunable XUV Radiation by High-Order Frequency Mixing

C. Momma, H. Eichmann, and B. Wellegehausen

*Institut für Quantenoptik, Universität Hannover, Welfengarten 1,
D-30167 Hannover, Germany, Tel.: +49-511-762-4406, Fax: -2211*

Abstract. The generation of short-pulse VUV and XUV radiation down to 11 nm by high-harmonic generation and frequency mixing with different pump lasers is described. As short-pulse pump sources excimer lasers, a Ti:Sapphire laser, short-pulse dye lasers, and optical paramtric generators have been used. By mixing the radiation of a fixed-frequency high-power laser with a less powerful but tunable light source, tunable coherent XUV radiation down to 40 nm has been demonstrated so far.

For the generation of powerful coherent radiation in the VUV and XUV spectral range nonlinear optical processes are promising techniques. Light sources with a peak brightness comparable with or larger than synchrotron radiation can be generated using table-top pump laser systems.

The harmonic spectra in the non perturbative regime have the following structure (1): Radiation is generated at the frequencies $n \cdot \omega$ (ω: laser frequency, n odd). The lowest order harmonics show a sharp decrease in intensity with increasing n, followed by harmonics which have nearly constant intensities. This so called plateau ends with a cutoff where the intensity decreases again rapidly. The photon energy at the cutoff can be approximated by

$$E_{cutoff} = I_P + \alpha \cdot U_p \quad , \quad U_P = \frac{e^2 I}{2\epsilon_0 m_e c \omega^2} \tag{1}$$

with I_P being the ionization potential of the medium and U_P the ponderomotive potential. I is the intensity of the laser field which is limited by the saturation intensity due to multi-photon or field-ionization processes at $I_{sat} = 10^{14}\text{-}10^{15}$ W/cm^2, depending on the medium. α is a constant lying in the range of 2 - 3 (Ref. 2).

According to equation 1 highest harmonics can be generated by using nonlinear media with a large ionization potential like helium or neon. On the other hand, the harmonic intensities in media with a low ionization potential, like xenon or

© 1994 American Institute of Physics

krypton, are much more intense due to the higher polarizibility of these media. Shortest wavelengths can be achieved with long wavelength pump lasers due to the frequency scaling of the ponderomotive potential, whereas with short wavelength lasers relatively short harmonic wavelengths can be generated by more efficient low order processes.

These theoretical predictions were confirmed experimentally. For the experiments three pump lasers were used: A commercial KrF excimer laser (3) (30 mJ, 400 fs, 248 nm), a home-made ArF excimer laser (4) (10 mJ, 1 ps, 193 nm), and a commercial Ti:Sapphire laser (B.M.Industries 10 Alpha, 100 mJ, 150 fs, around 800 nm).

With the ArF laser the third harmonic was observed in all noble gases. The 5th harmonic could only be detected in argon as the intensity of the pump laser was not high enough to saturate all media. With the KrF laser up to the 11th harmonic has been detected in helium. In the heavier noble gases the lower harmonics were generated with an intensity of one order of magnitude higher than in helium (more than 10^9 photons per harmonic have been measured using a calibrated semiconductor photodiode, corresponding to peak powers in the kilowatt range). With the Ti:Sapphire laser harmonics up to the 71st order (11 nm) were observed in neon (Fig. 1), limited by the sensitivity of the detection system. With the dependence of the cutoff on the ionization potential and the saturation intensity for the noble gases α was determined to be 2.3 (in good agreement with Ref. 2).

For several applications intense short-pulse radiation at a specific wavelength or in a certain wavelength interval is needed. The excimer lasers have a very limited tunability, so that only harmonics at some fixed frequencies can be produced. Tunable radiation in the XUV can be generated using high-order frequency mixing (5) of high-power fixed frequency pump laser radiation with lower power but tunable radiation. In order to determine the required intensities for the tunable radiation, first mixing experiments were performed with the excimer lasers and a short-pulse dye laser at 496 nm. Fig. 2 shows as an example all 5th order sum-frequency mixing signals according to $n \cdot \omega_{248nm} + m \cdot \omega_{193nm}$ (n+m: odd). Up to eight-wave mixing processes were observed using the KrF and the dye laser

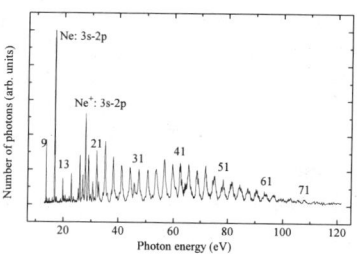
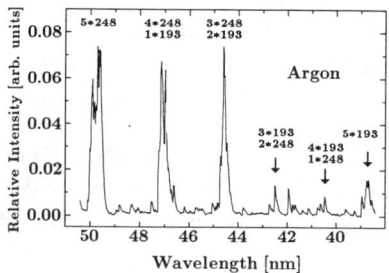

FIGURE 1. Harmonic spectrum obtained by focussing the radiation of a Ti:Sapphire laser into nenon (intensity: 10^{16} W/cm^2).

FIGURE 2. Mixing of KrF (10^{14} W/cm^2) and ArF (10^{14} W/cm^2) short-pulse radiation in argon.

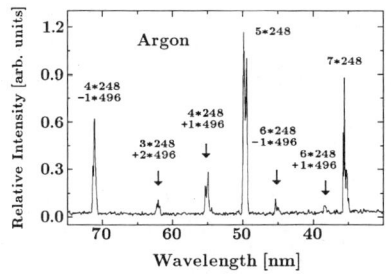

FIGURE 3. Mixing of a dye laser at 496 nm (10^{14} W/cm^2) and the KrF laser ($6 \cdot 10^{14}$ W/cm^2) radiation in Argon.

radiation (Fig. 3). The mixing signals were generated even at dye laser energies below 50 µJ (300 fs), corresponding to an intensity of less than 10^{13} W/cm^2.

Using a Ti:Sapphire laser as the pump source, tunable radiation in the XUV can be generated by tuning the pump laser, because Ti:Sapphire has an amplification bandwidth of at least 300 nm. However, in practice high-power laser systems are optimized at a specific wavelength, and due to the dispersive elements for the chirped pulse amplification a realignment of the whole system is required upon tuning. Therefore, a mixing scheme is much more favorable. For these experiments an optical parametric generator (OPG) was used because of the much larger tuning range compared with dye lasers. Pumped by the second harmonic of the Ti:Sapphire laser (operating at 813 nm), the theoretical tuning range is between 490 nm and 2.5µm. For type I phasematching only one BBO crystal is needed.

The experimental setup is shown in Fig. 4. The second harmonic of the pump laser was split by beamsplitter 2. About 10% of the 2nd harmonic was used to generate parametric fluorescence in a 5 mm BBO crystal. The flourescence was reflected back and amplified with the remaining pump radiation in the same crystal. With this setup radiation with energies up to 350 µJ have been generated

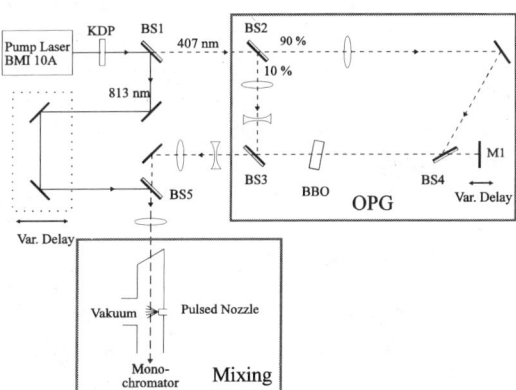

FIGURE 4. Experimental setup for the OPG and the mixing experiment. The radiation of the main pump laser is frequency doubled. The fundamental is focussed into the nonlinear medium together with the output of the OPG, which is pumped by the 2nd harmonic of the pump laser.

so far, tunable between 520 nm and 650 nm and with a pulse width of about 600 fs. This radiation has been mixed with the fundamental of the pump laser in xenon as shown in Fig. 5. Mixing peaks appear between all harmonics of the plateau.

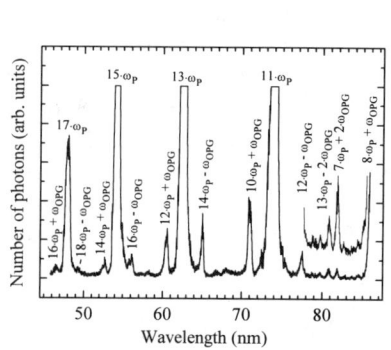

FIGURE 5. Mixing of Ti:Sapphire (813 nm, 10^{15} W/cm^2) and OPG (555 nm) radiation in xenon.

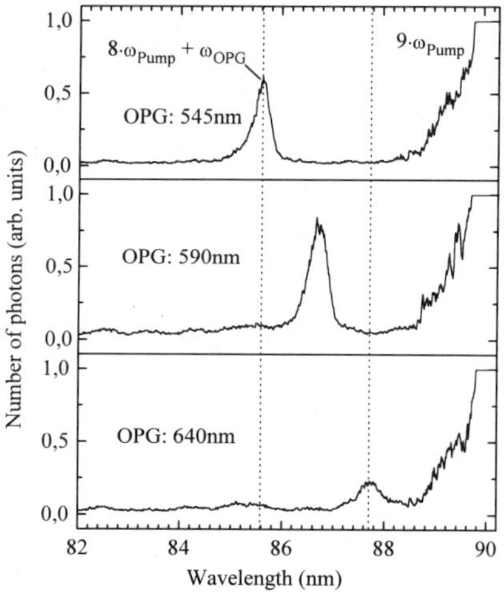

FIGURE 6. Demonstration of the tunability (for further data see Fig. 5).

By tuning the OPG the mixing peaks are tuned simultanously, as demonstrated in Fig. 6. With the present tuning range of the OPG about 30% of the spectral range between two harmonics can be covered by sum- and difference-frequency mixing with one OPG photon and an even number of pump laser photons. The frequency range can extended to more than 60%, if the OPG is operated over the whole theoretical tuning range of about 490 - 810 nm. If the weaker "two OPG photon mixing processes" are also included (see inset of Fig. 5), the whole spectral range between the harmonics can be covered.

The intensity of the generated mixing signals was found to be between 1 and 2 orders lower than the intensity of the harmonics. Increasing the intensity of the mixing signals should be possible by optimizing the OPG (foccussability). With lighter nonlinear media like neon tunable radiation down to 10 nm can be generated.

REFERENCES

1. Atoms in Intense Fields, ed. by M. Gavrila, Advances in Atomic, Molecular, and Optical Physics, Supplement (Academic Press, London, 1992).
2. C.G. Wahlström, J. Larsson, A. Persson, T. Starczewski, S. Svanberg, P. Salières, Ph. Balcou, A. L'Huillier, Phys. Rev. A **48**, 4709 (1993).
3. S. Szatmari, F.P. Schäfer, Opt. Commun. **68**, 196 (1988).
4. C. Momma, H. Eichmann, H. Jacobs, A. Tünnermann, H. Welling, B. Wellegehausen, Opt. Lett. **18**, 516 (1993).
5. M.D. Perry and J.K. Crane, Phys. Rev. A **48**, R4051 (1993).

X-ray nonlinear optics with high-order harmonics

P. L. Shkolnikov and A. E. Kaplan

Electrical and Computer Engineering Department
The Johns Hopkins University, Baltimore, MD 21218

Abstract We demonstrate theoretically the feasibility of nonlinear frequency and polarization transformations of coherent X-rays produced by high-order harmonic generation, and discuss possible applications of these effects to nonlinear spectroscopy of plasma ions.

Feasibility of several X-ray nonlinear optical effects has been demonstrated theoretically for available X-ray laser (XRL) power of order of several MW (1,2). Recently discovered high-order harmonic generation (HHG) of IR, visible, or UV lasers [see e. g. (3)], with its ability to produce simultaneously a large number of X-ray lines (separated by the doubled fundamental frequency) at comparable intensities, and its potential for tunability, could become a very attractive source for X-ray nonlinear optics. Unfortunately, mainly due to the unsolved yet problem of phase-matching optimization, the generated power of high-order harmonics is at least two orders of magnitude lower than that of XRL, so that it may seem unlikely that any nonlinear optical effects can be observed with those harmonics. In the present paper, however, we demonstrate theoretically that several X-ray resonant nonlinear effects in plasma are feasible with already attainable harmonic intensity. In particular, it was shown that XRL frequency could be efficiently near-doubled by difference-frequency mixing (DFM) with an optical laser (most conveniently, with the laser that pumps the XRL) in plasma (2). We evaluate conversion efficiency of a similar process, DFM of two high-order harmonics with the fundamental beam:

$$\omega = \omega_{harm1} + \omega_{harm2} - \omega_{fund} \tag{1}$$

We show that relatively low power of high-order harmonics can be more than compensated for by much larger flexibility in attaining resonant enhancement, mainly due to the harmonic output intensity being of the same order of magnitude for a wide range of harmonic numbers. As a result, for a number of plasma media we expect converted output to be of the same order of magnitude as the harmonic input. Experimental realization of the processes considered would be useful in investigating X-ray nonlinear optical effects with "table-top" equipment, and in generating even shorter coherent X-rays. Our results indicate also that soft-X-ray coherent radiation generated by large-scale frequency transformations of optical lasers might be applied in nonlinear spectroscopy even with already attainable intensities.

Due to narrow bandwidth of existing X-ray mirrors, one should consider DFM of harmonics with close frequencies; the result will be frequency near-doubling of a harmonic. It is reasonable, therefore, to make use of Na-like ions found to be favorable for XRL frequency near-doubling (2). As an example of HHG advantages for applications to nonlinear optics, nonlinear susceptibility for

DFM of the 53rd (153.2 Å) and the 51st (159.2 Å) harmonics of Ti:Sapph laser (8120 Å) in Na-like Ca X ions (78.83 Å output) (see Fig. 1) is 6 orders of magnitude larger than the susceptibility of DFM of Y^{29+} ($\lambda = 154.95$ Å) XRL with Nd:YAG radiation in the same medium, due to much better resonant conditions. The expected conversion efficiency for collinear beams tightly focused in Ca X plasma of ion density 10^{18} cm^{-3} is $C_{eff} \approx 2 \times 10^{-3}$ P_{harm}, where C_{eff} is the ratio of the converted power to harmonic (P_{harm}) power, Ti:Sapph power is taken as 10 GW, optimal phase matching is assumed, and no competing processes except for resonant absorption are accounted for. One could reasonably presume $P_{harm} \sim 1$ kW (4). which, combined with 50% mirror reflectivity, yields $C_{eff} \sim 1$. It is worth noting that both the variety and the output power of X-ray lines produced by large-scale frequency upconversion of optical lasers, may be greatly increased by using, in addition to HHG, high-order sum- and difference-frequency mixing of two optical beams (5).

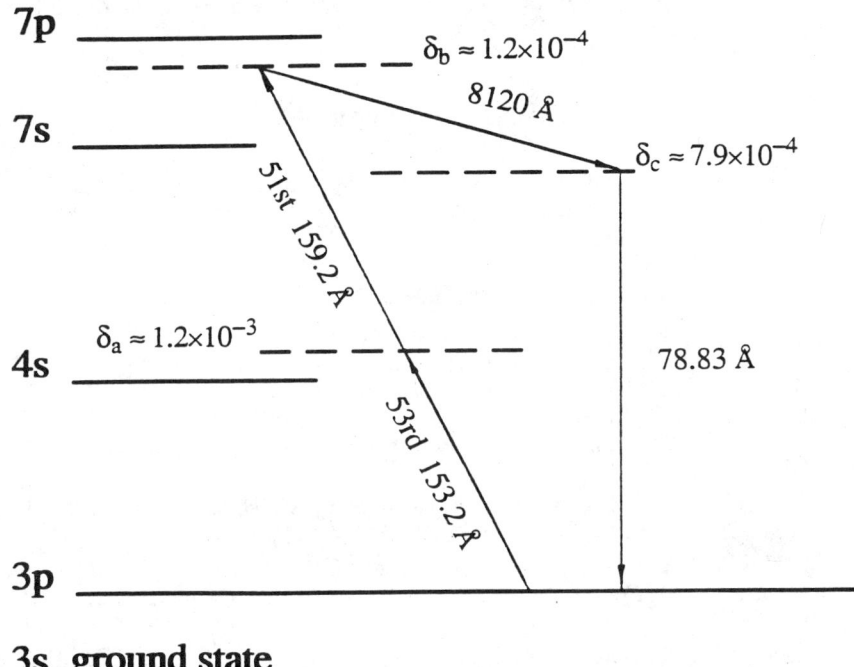

FIGURE 1. Nonlinear difference-frequency mixing of Ti:Sapphire laser with its 51st and 53rd harmonics in Na-like Ar. δ_j are relative detunings.

Another example of X-ray nonlinear optical effects that are feasible at available harmonic intensities is induced optical activity: polarization plane of a weak probe beam may rotate while propagating in a nonlinear medium pumped by strong (control) beam polarized circularly [see e. g. Ref. (6)]. The sign and the angle of the rotation depends on the sign and the size of the detuning from the two-photon resonant transition, and on the control (but not probe) intensity. For the Ti:Sapph laser radiation as the control beam, and its 51st harmonic as the probe, one can estimate, using Ref. (6), the rotation angle, ϕ, per 1 cm propagation length as:

$$\phi(\text{rad/cm}) \sim 2\times 10^{-34} \, I_{opt}(\text{W/cm}^2) \, N_i(\text{cm}^{-3}) / \delta_{2ph}, \qquad (2)$$

where I_{opt} is the intensity of the optical beam, N_i is the ion density, $\delta_{2ph} = 1 - (\omega_{harm} + \omega_{fund})/\omega_0$ is the relative detuning from a two-photon transition ω_0 (transition oscillator strengths are assumed of the order of 0.1). Due to large available I_{opt}, the effect should be readily observable, especially near a 2-photon resonance.

Both considered nonlinear effects strongly depend on resonances with ion transitions in a medium. Together with tunability of the pumping, this make them potentially very useful for nonlinear (in particular, Doppler-free) spectroscopy. Such an application of e. g. induced optical activity is quite common in visible domain [see e. g. Ref. (7)]. In X-ray domain, it may be a unique tool in ionic spectroscopy of very dense plasma (up to the critical density for the optical radiation).

Acknowledgements

Stimulating discussions with R. W. Falcone substantially contributed to this research. This work is supported by AFOSR.

References

1. Shkolnikov, P. L. and Kaplan, A. E., Opt. Lett. 16, 1153-1155 (1991); Shkolnikov, P. L. and Kaplan, A. E., Phys. Rev. A 44, 6951-6953 (1991); Shkolnikov, P. L. and Kaplan, A. E., Opt. Lett. 16, 1973-1975 (1991); Hudis, E., Shkolnikov, P. L., and Kaplan, A. E., Appl. Phys. Lett. 64, 818-820 (1994).
2. Shkolnikov, P. L., Kaplan, A. E., Muendel, M. H., and Hagelstein, P. L., Appl. Phys. Lett. 61, 2001-2003 (1992).
3. L'Huillier, A, Lompre, L. A., Mainfray, G., and Manus, C., in *Atoms in Intense Laser Field*, Ed. M. Gavrila (Acad. Press, Inc., Boston, 1992), p.139-206.
4. Balcou, Ph., Cornaggia, C., Gomes, A. S., Lompre, L. A., and L'Huillier, A., J. Phys. B 25, 4467-4485 (1992).
5. Shkolnikov, P. L., Kaplan, A. E., and Lago, A., Opt. Lett. 18, 1700-1702 (1993); Perry, M. D. and Crane, J. K., Phys. Rev. A 48, R4051 (1993).
6. Liao, P. F. and Bjorklund, G. C., Phys. Rev. A 15, 2009-2018 (1977).
7. Wieman, G. and Hänsch, T., Phys. Rev. Lett. 36, 1170 (1977).

Nonresonant Photopumping Using Heavy Ion Beam Produced Soft X-rays*

W. Krötz, A. Ulrich, M. Salvermoser, J. Wieser
TU-München, Fakultät für Physik E12, D-85748 Garching, Germany
D.E. Murnick
Rutgers University, Department of Physics, Newark, N.J. 07102, USA

It is proposed to use soft x-rays produced by heavy ion beams for nonresonant photopumping of short wavelength lasers. A narrow emission bandwidth is achieved by stopping the ions in an appropriate target material. Spectroscopic experiments have been performed for the well known 108.9nm Xe III laser using a 100MeV ^{32}S beam to produce ~70eV photons in an aluminum target.

In this paper we discuss a novel pumping method which can for instance be used for the 108.9nm XeIII laser line. The laser scheme was proposed and demonstrated by Kapteyn et al. [1]. The inversion mechanism, as shown in fig. 1, starts from Xe I with inner-shell photoionization of a 4d electron, followed by Auger decay of the resulting Xe II, $4d^9 5s^2 5p^6$ $^2D_{3/2,5/2}$ ions into various excited states of Xe III. The branching ratio of the Auger decay to both the upper and lower Xe III laser levels, $5s^0 5p^6$ $^1S_0$ and $5s^1 5p^5$ $^1P_1$, is about 5%. Population inversion results from a higher degeneracy of the lower level [2].

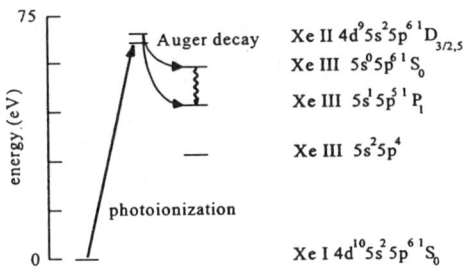

Fig. 1: Partial energy level diagram of xenon showing the pumping scheme for XeIII levels by photoionization followed by Auger decay. The laser transition at 108.9nm is indicated. [1]

This laser scheme has previously been applied in systems, where the soft x-rays were obtained from laser induced plasmas [3,4,5]. Typical laser plasma sources used in these experiments radiate with a spectral distribution corresponding to a blackbody temperature of 30eV. But only radiation with an energy of more than 67 eV can be used for pumping. The intense radiation at lower energy can only photoionize Xe 5s electrons, but not 4d electrons. Filters between the x-ray source and the laser volume are often used to reduce the flux of low energy photons [1,2], because a high density of 5s electrons leads to collisional quenching. Therefore a source of pump radiation with a narrow bandwidth at the right photon energy would be desirable.

We propose an alternative method to produce 70eV radiation with a small bandwidth. Pulsed heavy ion beams stopped in an appropriate target material are used to produce soft x-rays which can photoionize 4d

---

*This work has been funded by the German Ministry of Research under contract Nr. 06TM353/5, the Munich Tandem van de Graaff Accelerator Laboratory, GSI-Darmstadt, and NATO.

electrons in xenon gas.

For a spectroscopic experiment which has been performed at the Munich Tandem van de Graaff accelerator aluminum was selected from x-ray wavelength tables as a target material. A typically 170nA (electr.), 130MeV $^{32}S^{9+}$ ion beam was stopped in this target. The experimental setup consisted of a target cell and a 2.217m grazing incidence monochromator (Mc Pherson modell 247) which measured light emitted perpendicular to the ion beam axis. The emission in a wavelength range between 3 and 125nm was detected with a channeltron operated in photon counting mode. The targets were mounted on a manipulator at an angle of 20^0 with respect to the optical axis and could be moved parallel to the optical and perpendicular to the ion beam axis. The final target position was found by maximizing the count rate at the channeltron detector. In a first experiment the target cell was evacuated ($p \sim 10^{-5}$mbar) and we studied the ion beam induced soft x-ray emission of aluminum. Then the cell was filled with xenon gas at low pressures in order to study the photoionization effect of xenon. During this experiment the target cell was seperated from the beam line by an 1.1mg/cm^2 Ti entrance foil for the ion beam. A LiF window was mounted in front of the entrance slit of the monochromator.

The aluminum targets showed a strong emission between 16.4 and 20.0nm (61.9 to 75.5 eV) and very low intensity below and above this emission band (fig. 2). This emission can be assigned to a V-L_{23} (V=valence band) transition in solid aluminum which is analogous to M-L transitions in gaseous aluminum.

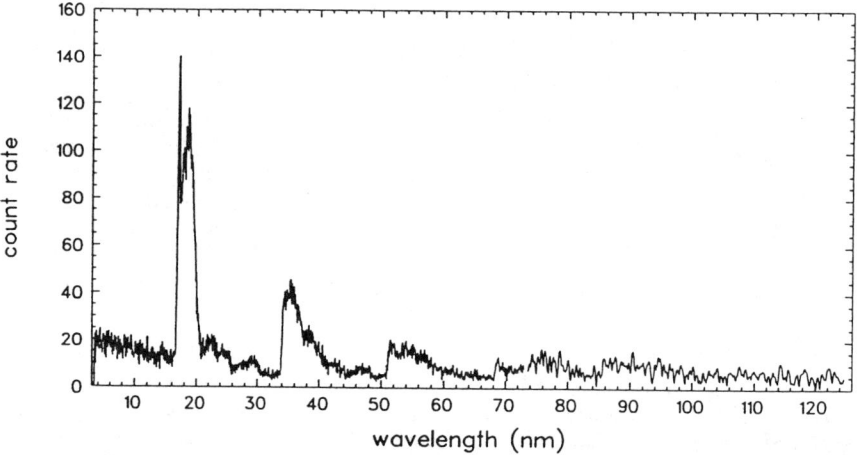

Fig. 2: Emission spectrum of solid aluminum excited by a 130MeV ^{32}S beam. The spectrum is dominated by Al VL_{23} (V=valence band) transitions around 18nm. The features around 36 and 54nm are due to higher orders of the grating monochromator.

An estimate of the total photon flux was deduced from the data in fig. 2., a solid angle of the detection system of 6.6×10^{-7}, a reflectivity of $\sim 25\%$ of the grating, and a detection efficiency of 6% of the channeltron detector [Valvo X914BL]. For an ion beam current of 17nA (particle, $\equiv 1.1 \times 10^{11}$ions/s) 6×10^{12} photons/sec are emitted from the target surface. This is equivalent to a conversion efficiency of the ion beam power into useful soft x-rays of approximately 3×10^{-5}. Due to the target geometry and the fact that the range of the ion beam in aluminum is large compared with the absorption length of the 18nm photons only 2.5% of the photons produced by the ion beam were observed with the present setup. This

may be significantly improved by using a grazing geometry between the ion beam and the target surface. In this case the depth of the excited volume can be matched with the absorption length reducing the reabsorption in the target. A conversion efficiency of approximately 1‰ can be expected for such an optimized setup .
In addition to measurements of the pump radiation the emission from xenon in the wavelength range between 105 and 125 nm was also studied. When 15mb xenon is filled into the target cell the pump photons are reabsorbed within a few centimeters ($\sigma_{abs} \sim$ 6Mbarn) [6].

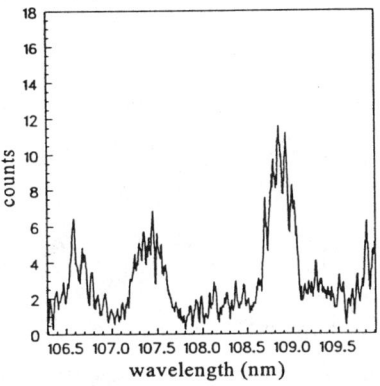

Fig 3a: Xenon, directly excited by the ion beam (Al target removed).

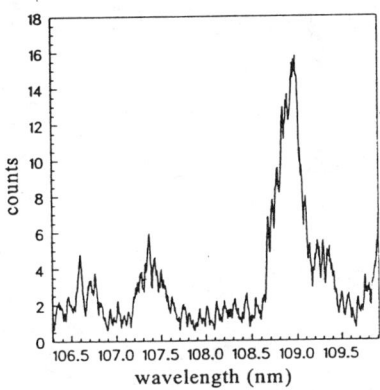

Fig 3b: Same as fig. 3a, but with additional pumping by aluminum x-rays (Al target in place).

The spectra in fig. 3 show the emission of xenon excited by a 130MeV ^{32}S beam with no aluminum target (a) and with the aluminum target at the intersection of the beam and the optical axis (b). In both cases 3 lines are observed in the spectral range between 106 and 110nm. They are obviously populated by direct excitation by the ion beam. The lines at 106.63 and 107.45nm can be assigned to a transition in Xe III ($5p^4\ ^3P - 5p^5\ ^3P^0$) and a ground state transition in Xe II, respectively. The comparison of the spectra in fig. 3 demonstrates the expected photopumping of the 108.9nm laser line by the soft x-ray emission from the aluminum target. In the case of the 108.9nm line the intensity is almost doubled, when the ion beam is stopped in the aluminum target. The intensity of the other lines, however, is reduced, since the solid target blocked part of the ion beam excited xenon volume. This obviously indicates that aluminum x-ray radiation contributes to the pumping of the 108.9nm laser transition.

These experiments show that for photoionization pumping of short wavelength lasers heavy-ion beam produced x-rays provide an interesting alternative to laser plasma x-ray production: The emission bandwidth is very narrow and can be well adjusted to the required wavelength of the pump radiation by choosing an appropriate target material. For the system studied in this paper radiation from silicon targets (91.5eV) might even be better suited for photoionization of Xe 4d levels. It is known from the literature [6,7,8] that the cross section for 4d ionisation increases by a factor of ~ 3.5 if the x-ray energy is shifted from 72.4 to 91.5eV, thus enhancing the population inversion of the laser levels.

Fast heavy ion beams with a high flux necessary for laser design are for example available from the synchrotron SIS at GSI, Darmstadt. These pulsed beams with an energy of typically 300-

1000MeV/nucleon, a pulse length of 70ns, and a beam intensity of 10^{10}-10^{11} ions per pulse can be well collimated to beam diameters of $\sim 200\mu m$ [9]. They are stopped in solid targets within a range of about 10cm which is well matched to the focal length of the beam. Due to the high beam energy and projectile's mass angular straggling is very low in the target and therefore the excited volume is well collimated. This facilitates the design of a setup, where the ion beam is sent into the solid target parallel to its surface and with the maximum particle flux just at the target surface. The distances between this ion beam induced x-ray source and the axis of the photoionization pumped laser can be made very small, typically one millimeter. Due to the range of the high energy ions a 10cm long, homogenously pumped laser volume should be obtained with an expected pump photon flux of 10^{19}-10^{20} photons/cm^{2}s. The geometry is also well suited for the installation of an optical cavity.

References:

[1] H.C. Kapteyn, R.W. Lee, R.W. Falcone, Phys. Rev. Lett. **57** (1986) 2939.
[2] M.H. Sher, J.J. Macklin, J.F. Young, S.E. Harris, Optics Letters **12** (1987) 891.
[3] E.J. Mc Guire, Phys. Rev. Lett. **35** (1975) 844.
[4] M.H. Sher, S.J. Benerofe, J.F. Young, S.E. Harris, J. Opt. Soc. Am. **B8** (1991) 114.
[5] A.J. Mendelsohn, S.E. Harris, Opt. Lett. **10** (1985) 128.
[6] J.B. West, J. Morton, Atomic Data and Nuclear Data Tables **22** (1978) 106.
[7] B.L. Henke, P. Lee, T.J. Tanaka, R.L.Shimabukuro, B.K. Fujikawa, Atomic Data and Nuclear Data Tables **27** (1982) 1.
[8] B.L. Henke, E.M. Gullikson, J.C. Davis, Atomic Data and Nuclear Data Tables **54** (1993) 266.
[9] A. Ulrich, B. Busch, H. Eylers, W. Krötz, R. Miller, R. Pfaffenberger, G. Ribitzki, J. Wieser. D.E. Murnick, Proc. X-ray Lasers 1990, IOP **116** (1990) 267.

Neck type instabilities in axial discharges and population inversion

K. N. Koshelev* and H.-J. Kunze**

*Plasma Spectroscopy Laboratory, Institute of Spectroscopy, Russian Academy of Science, Troitsk, Moscow Reg. 142092, Russia
**Institut für Experimentalphysik V, Ruhr-Universität, 44780 Bochum, Germany

Abstract. "Neck" type instabilities in axial discharges generate strong jets of highly ionized plasma moving with velocities of about 10^7 cm/s. Charge exchange interaction of the fast ions with lower charged ions in the main plasma column can result in population inversion for transitions in the far VUV region.

In 1973 Vinogradov and Sobelman proposed charge exchange (CE) as an effective pumping method for short-wavelength lasers (1). Since then a number of experimental studies were carried out employing laser produced plasmas and an analysis of the prospects of this type of pumping scheme can be found in review publications (2,3). In this letter we discuss the possibility of using strong plasma jets of multiply charged ions which flow out from "neck"-type instabilities of axial discharges. They collide with less ionized atoms in the main plasma column of the discharge and can create population inversion for lasing in the soft X-ray and VUV region through selective recombination (4).

A typical z-pinch is hydromagnetically unstable. The neck of a hydromagnetic sausage instability forms a plasma column which is narrower than the main column. A simplified model of such a geometry is shown in Fig.1. If the height of the neck is h, then the aspect ratio α defined by $h = \alpha \times r$; it is typically in the range $1 \leq \alpha \leq 10$, but mainly it is about 3.

An essential amount of hot plasma flows out from the neck during its development, creating axially directed plasma jets. To estimate the plasma jet parameters, we assume force balance in radial direction (i.e. Bennett equilibrium) which provides a simple relation between the line density of the charged particles (electrons and ions), the plasma temperature and the total current. In numerical form the Bennett relation is:

$$(Z_{eff} + 1) \times N_i \times (kT) = 3.12 \times 10^{21} \times I^2 \qquad (1)$$

where kT is the plasma temperature in eV, N_i is the number of ions per unit length, and I is the pinch current in MA.

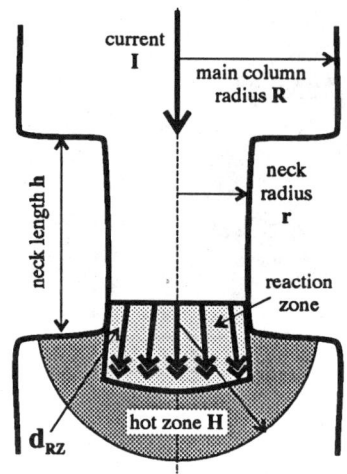

Figure 1. Schematic of a neck type instability

When the temperature reaches a value kT, the total number of ions in the neck is given by $N_{tot} = N_i \times h = N_i\, \alpha\, r$, or with $\alpha \approx 3$

$$N_{tot} = \frac{10^{22} \times I^2 \times r}{kT \times Z} \qquad (2)$$

During the time τ of the subsequent neck evolution almost all these ions will flow out from the neck with a velocity close to the ion sound speed V_s. Numerical simulations and experimental observations (5,6,7) show that in the region away from the neck, the plasma column can have a much larger diameter and a lower temperature. Schematically this is illustrated in Fig.1.

Penetration of the intense fast ion beam into the main plasma leads to heating of the target ions by Coulomb collisions, but the resulting kinetic energy of the beam and the target ions will still be comparatively high. The "stopping" power of the Coulomb collisions starts to play a significant role only when the total number of target ions, which the beam meets on its way, will be comparable with the total number of ions in the beam. The corresponding volume is indicated in Fig.1 as "hot zone" with the radius H.

Electrons moving with the fast ions should have a lower temperature as it is usually the case when a heated plasma expands freely. Therefore heating of the cold target electrons is by collisions with hot ions and occurs at a much longer time scale. The resulting plasma in the target region thus consists of a mixture of hot ions of high and low charge and cold electrons.

If the mean free path of the fast ions for charge exchange recombination in the "target" plasma is smaller than the size of the "hot zone", charge exchange recombination of high Z ions occurs in a smaller than the "hot zone" volume. This is indicated in Fig.1 as "reaction zone".

We now assume that inside the "reaction zone" all beam ions have experienced one selective CE recombination, populating a specific level with principal quantum number u in the resulting ion. If the total number of CE reactions during the time τ in the volume V_{RZ} is equal to N_{tot}, the population density of the excited level u is

$$n_u \approx \frac{N_{tot}}{V_{RZ} \times \tau \times C_u} \qquad (3)$$

where C_u is the decay rate of the level u.

As example we now discuss some numerical estimates of charge-exchange recombination of He-like ions in a Z-pinch discharge with a current of 200 kA. This reaction results in excited Li-like ions: For atoms with nuclear charge $Z \approx 8\text{-}10$ (O, F and Ne) the temperature kT necessary for them to be mainly ionized to the He-like stage is about 120 - 200 eV.

Not many experimental data are available on the size of the "neck" plasma at this temperature. For a current of I = 0.2 MA the radius of the necked region at a temperature of about 150 eV is $r = 5 \times 10^{-2}$ cm (5,8) and its height is $h \approx 0.15$ cm. The line electron density at the Bennett equilibrium is about $N_e \approx 8 \times 10^{17}$ cm^{-1} and the line ion density is about $N_i \approx 10^{17}$ cm^{-1}. The existing densities are $n_e \approx 10^{20}$ cm^{-3} and $n_i \approx 1.3 \times 10^{19}$ cm^{-3}. The total number of ions in the neck is $N_{tot} \approx N_i \times h \approx 1.5 \times 10^{16}$ ions.

At the ion sound velocity of about 10^7 cm/s, the time of the further development of this instability is $\tau \leq 5 \times 10^{-9}$ sec.

For the simplified plasma model presented in Fig.1, we assume that the radius of plasma column is $R \approx 0.5$ cm and the temperature is approximately 5 times lower than in the neck (≈ 30 eV). This yields an electron bulk density in the main column of about 4×10^{18} cm^{-3} and, at an average ion charge $Z \approx 2.5$, the ion density in the main column is about $n_i{}^C \approx 1.5 \times 10^{18}$ cm^{-3}.

The kinetic energy of the fast beam ions moving with a velocity close to 10^7 cm/s is about 600 eV. The cross-section for Coulomb scattering can be as high as 10^{-16} cm^2 and the Coulomb free path ℓ_C of the order of 10 μm. Recent calculations of the charge exchange cross-sections (9) show that for the reaction under discussion respective cross-sections are not lower than 10^{-16} cm^2 at kinetic energies above a few hundred eV. This leads to an upper estimate of the CE free path in the "target" plasma of $\ell_{CE} \leq 100$ μm. This means that after at least a few Coulomb collisions, beam ions will experience charge exchange recombination with an average time constant of about or less than $\ell_{CE}/V_s \approx 10^{-9}$ s. The collisional-recombination pumping rate (see, e.g. (5)) is almost an order of magnitude lower and starts to be comparable with charge exchange recombination at CE cross sections lower than 10^{-17} cm.

This analysis shows that the "reaction zone" is actually much smaller than the "hot zone"; in the simplified geometry (Fig.1) it has a characteristic size of about the "neck" radius r, and the axial extension d_{RZ} is close to the charge-exchange mean free path. The total volume of the "reaction zone" is $V \leq 2\pi r^2 \times \ell_{CE} = 3 \times 10^{-4}$ cm^{-3}.

When discussing the decay of the n=4 levels of Li-like ions in a comparatively dense "target" plasma, we cannot ignore collisional depopulation to higher levels. For a Li-like ion with an effective charge $Z \approx 8$ in a plasma with an electron temperature of about 30 eV, the rate of n=4 to n=5 electron collisional excitation is $X_E \times 1.5 \times 10^{-8}$ cm^3 s^{-1}. For a "target" electron density of 4×10^{18} cm^{-3} we thus have a depopulation rate of $C_u \leq 10^{11}$ s^{-1} in Eq.(3).

For an estimate of the possible inversion we neglect the population of the lower level, i.e. we take the inversion factor to be F=1. Using Eq.(3) we obtain

$n_4 \approx 10^{17}$ cm^{-3} for the population of the n=4 level in Li-like ions with an effective charge $Z_{eff} \approx 8$. The 4f - 3d transition at about 290 Å is the strongest among the $4 \Rightarrow 3$ transitions, and the population density of the upper level 4f is a little less than half of the total population of the n=4 levels assuming thermal equilibrium between the sublevels.

In the case of the Doppler broadening corresponding to the ion velocity of about 10^7 cm/s this gives a gain coefficient $G \approx 250$ cm^{-1}. The size ℓ of the "reaction zone" is $\ell_{CE} \leq \ell < 2r$, i.e. it is in the range 0.02 - 0.1 cm, resulting finally in a "gain-length product" of about $5 \leq GL \leq 25$.

Rather strong assumptions (inclusive the plasma geometry) were made to simplify the present considerations. On the other hand we would like to point out that the chosen parameters of plasma column and neck are far from optimized ones. For example, a "neck" radius $r \approx 0.5$ mm is an upper estimate at the given temperature, while the GL product strongly increases for smaller radii.

There are some experimental results which apparently confirm population inversion by charge exchange in Z-pinch type discharges. Population inversion between n=4 and n=3 levels was derived from line intensities of Li-like ions of N, O and Ne in a gas-puff Z-pinch discharge with a current of about 400 kA (10). Population inversion was detected in the same series of experiments between n=4 and n=3 levels of H-like CVI. We think that quite similar physics is responsible for lasing in a capillary discharge (11). Whether it also applies to lasing reported in the gas-liner discharge (11), has to be established by further investigations.

The authors are thankful to Prof. E. Clothiaux and Dr. A. Schulz for useful discussions. This work was supported by a Collaborative Research Grant of the NATO International Exchange Programs.

REFERENCES

1. Vinogradov, A.V., and Sobelman, I.I., *Sov. Phys. JETP* **36**, 1115 (1973)
2. Bunkin, F.V., Derzhiev, V.I., and Yakovlenko, S.I., *Sov. J. Quant. Electr.* **11**, 981 (1981)
3. Elton, R.C., *X-ray Lasers*, San Diego: Academic Press 1990.
4. Koshelev, K.N., and Kunze, H.-J., "Ion beams from axial discharges and the x-ray laser problem" in *Dense Z-Pinches*, eds. M. Haines and A Knight, AIP Conf. Proc. 299, New York, 1994, pp. 231-235.
5. Koshelev, K.N., and Pereira, N.R., *J. Appl. Phys.* **69**, R21 (1991)
6. Vikhrev, V.V., Ivanov, V.V., Rozanova, G.A., *Nucl. Fusion* **33**, 311 (1993)
7. Pereira, N.R., and Davis, J., *J. Appl. Phys.* **64**, R1 (1988)
8. Kies, W., Decker, G., et al., *J. Appl. Phys.* **70**, 7261 (1991)
9. Uskov, D., *private communication*
10. Koshelev, K.N., Sidelnikov, Yu. V., Churilov, S.S., and Dorokhin, L.A., "Charge exchange of plasma beams from instabilities in axial discharges and population inversion to a highly charged ions" Third Int. Conf. Dense Z-Pinches, London 19-23 April 1993, *Phys. Lett. A* (1994, in press)
11. Kunze, H.-J., Glenzer, S., Steden C., Wieschebrink, H.T., Koshelev, K.N., and Uskov, D., these proceedings

Evaluation of Imaging Properties of Soft-X-Ray Multilayer Mirrors and their Application to Highly Dispersive Spectral Imaging

N.N. Kolachevsky, M.M. Mitropolsky, E.N. Ragozin,
N.N. Salashchenko,* V.A. Slemzin, and I.A. Zhitnik

Optics Division of the P.N. Lebedev Physics Inst., 53 Leninsky ave., 117924 Moscow, Russia
* Institute for Physics of Microstructures, 46 Uljanov Str., 603600 Nizhny Novgorod, Russia.

Abstract. A variety of normal-incidence multilayer mirrors (MMs) intended for studies of astrophysical and laboratory soft-X-ray radiation sources have been synthesized on concave (r = 1.6 - 2.0 m) fused silica substrates. The MMs range in resonance wavelength λ_0 from 4.5 to 31 nm. Their imaging capability has been evaluated from small-source imaging tests employing a laser-plasma broadband XUV radiation source and a high-resolution XUV photographic film. The photographs testify to a subarcsecond angular resolution. For 17.5-nm MMs, a resolution of at least ≈ 0.32 arcseconds has been demonstrated, which is only $2.4\lambda_0/D$ for the MMs involved.

1. INTRODUCTION

Implying investigation of astrophysical and laboratory soft-X-ray radiation sources with a high angular and spectral resolution, a number of focusing normal-incidence multilayer mirrors (MMs) have been fabricated on concave (r = 1.6 - 2.0 m) fused silica substrates. The applications include high-resolution telescopes (1), highly dispersive spectroheliographs (1-4), stigmatic laboratory spectrographs (2, 5), and a multitude of auxiliary configurations used primarily in the characterization of X-ray optical components. In either case, the imaging quality of the MMs involved is crucial to the instrument performance. For instance, in the configuration of a stigmatic high-resolution spectrometer equipped with two focusing normal-incidence MMs and a diffraction grating at grazing incidence (5), the ultimate spectral resolution is $\delta\lambda \sim |d\lambda/d\beta|\delta\beta$, where $d\lambda/d\beta = \sin\beta/mp$ is the reciprocal of the angular dispersion, β the grazing angle of diffraction, m the spectral order, p the groove density, and $\delta\beta$ the angular resolution provided by the MMs. Among the factors which may limit the beam quality, are the small-angle scattering by the interlayer roughness in a multilayer structure and the substrate roughness as well as beam aberrations arising from substrate figure errors. (A comprehensive consideration of these problems was provided by E. Spiller et al. (6)). Specifically, if the angular resolution is proximate to the diffraction limit

($\delta\beta \approx \lambda/D$), the theoretical resolving power attains its maximum value determined by the total number of grooves: $[\lambda/\delta\lambda]_{theor} \approx mpW$ (W is the grating width).

This paper is concerned with imaging quality of the MMs. The method and the results of an evaluation of their spectral properties were presented elsewhere (7, 8). Our technique involves the use of a laser-plasma broadband XUV radiation source.

The evaluation of imaging properties of MMs as well as the realization of a high resolving power inherent in diffraction instruments are conditioned by the capacity to record small-size images in the soft X-ray domain. Specifically, the diffraction-limited dimension of the image of a point source located near the center of curvature of a MM with the radius R = 1 m and the aperture D = 1 cm is about 1 µm. Of all the known XUV radiation detectors, with the exception of a photoresist distinguished by a far lower sensitivity, a photographic film still possesses the best spatial resolution. The domestic UF-4 photographic film, which is usually used in spectroscopy of the XUV range, maintains a resolution of 150 lines/mm. The present paper reports that a new photographic film, UF-n, designed in the NIIKhIMFOTOPROEKT Co. has been used successfully. The UF-n film differs from UF-4 in that its grain size is roughly two times smaller; the new film maintains a resolution of at least 350 lines/mm.

2. EXPERIMENT

A schematic layout of the experiment is shown in Fig. 1. The laser plasma was generated by the second harmonic ($\lambda=0.54$ µm) of a single-mode, passively Q-switched, $Nd:YAlO_3$ crystal laser; the laser energy delivered to the target reached 0.15 J per pulse 5 ns in duration. The beam was focused by an f = 75 mm heavy flintglass lens onto a flat target of stainless steel to a spot with an effective area of $\sim 10^{-5} cm^2$ and a peak flux density of $\sim 2 \cdot 10^{12} W/cm^2$. A step motor rotated the target around the vertical axis. The laser plasma back-illuminated the test-pattern located near the center of curvature of the MM. As a test-pattern, we used either a spectral slit or a fine two-dimensional nickel mesh with a 10-µm period and a 5-µm window superimposed on a rough mesh with a 100-µm period and a 70-µm window. An absorption filter (a free-standing 0.2-µm-thick aluminum film 3 mm in diameter) was placed in the beam path 60 mm away from the test-pattern. The filter served to reject the visible and UV radiation outside the resonance reflection peak of a MM, including the suppression of the non-resonance "Fresnel" reflection in the multilayer structure on the longwave side of the resonance maximum. The filter was void of support structure, which introduces diffraction and distorts the image recorded. The MM imaged the test-pattern with approximately a unit magnification onto the exit plane of the camera, where a photographic film was lo-cated. We used Mo-Si-coated MMs with $\lambda_0 \approx 17.5$ nm deposited over a spherical (R = 1600 mm) or slightly aspherical (toroidal) substrates (R_s = 1620 mm, R_m = 1621 mm) of diameter D = 30 mm. In the former case, the aberration of astigmatism prevails, the astigmatic blur depending on the magnitude of the off-axis angle γ and the MM feed. For $\gamma \sim 0.6°$ and a completely fed MM aperture, the length of the astigmatic blur of the image of a point source is 5 μm. In the latter case, for $\gamma \sim 1.7°$ the resolution can approach the diffraction limit (~1 μm). The spectral width (FWHM) of the resonance reflection maxima is about 0.9 nm for the MMs involved (8).

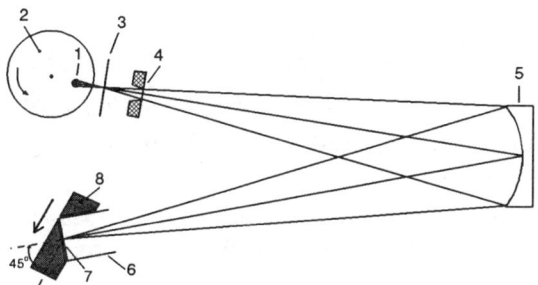

FIGURE 1. Schematic layout of the experiment. Laser plasma (1), rotatable target (2), test-pattern (3), free-standing 0.2-μm thick Al filter (4), focusing MM with $\lambda_0 \approx 17.5$ nm and $\Delta\lambda_{1/2} \approx 0.9$ nm (5), camera (6) with a photographic film (7), table with a translatory motion at $\approx 45°$ to the axis of the incident beam (8).

Since the wavelength of the visible light is more than λ_0 by a factor of 30, a preliminary focusing in the visible light was absolutely insufficient. To ensure a more precise coincidence of the photographic film with the image plane (basically, with an accuracy of $\sim \lambda_0/\Omega^2$, where $\Omega = D/R$), the camera was mounted on a table with its slideway directed at ~45° to the axis of the incident beam. In the experiment, the table was translated in a step-like manner by a distance of about a hundred microns in the intervals between the laser shots. A series of fifty test-pattern images at varying distances from the focus were recorded on the photographic film, the images in the film being shifted in the transverse direction. Two-to-three neighboring images out of this succession possessed the maximum sharpness and were used for quality evaluation of the MMs and the photographic film. All of the components of the setup were mounted on the optical table which measured 3.6 × 0.6 m and was located in a vacuum tank of dimensions 3.8 × 0.9 m equipped with a turbopump. A photograph of the 2-D test-pattern obtained with the aid of a toroidal MM appears in Fig. 2. The 5 × 5-μm squares separated by unexposed 5-μm-wide spacings are clearly visible. The microdensitometer at our disposal did

536 Soft-X-Ray Multilayer Mirrors

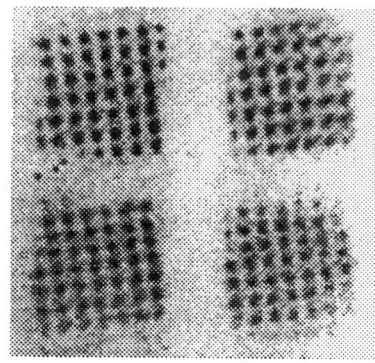

FIGURE 2. Photograph of a two-dimensional test-pattern obtained using a MM with $\lambda_0 \approx 17.5$ nm. The image is formed by the radiation in a band of width ≈ 0.9 nm about λ_0. The five by five micrometer squares separated by unexposed 5-μm-wide spacings are clearly visible. The angular dimension of the squares is 0.6×0.6 arcseconds.

not permit the photographic film density to be analyzed within an area 5 μm wide by 5 μm long. With the microdensitometer slit width corresponding to 1.2 μm in the film, the minimum slit height corresponded to 70 μm, i.e., to seven periods of the fine mesh.

Shown in Fig. 3(a) is a microdensitometer scan of an at-focus image of a 5-μm slit recorded using the new UF-n photographic film. The microdensitometer scan

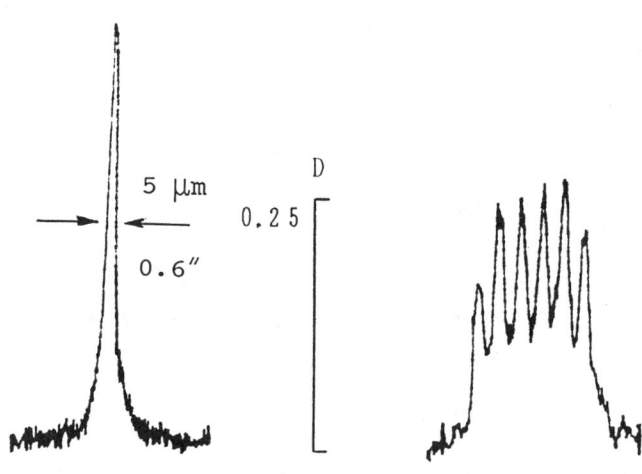

FIGURE 3. a) Microdensitometer scan of an image of a 5-μm slit at focus recorded employing the new UF-n film; b) microdensitometer scan of a 2-D test-pattern image averaged over one dimension. The microdensitometer slit height corresponds to 70 μm, i.e., to seven periods of the fine mesh.

given in Fig. 3(b) represents the density averaged over one dimension in a 2-D image. Assuming the 2-D image to be identical in either direction, the peak density can be restored from the height-averaged density scan. It is safe to say that the peak density is approximately 0.40, the contrast in the 2-D image (the maximum-to- minimum irradiance ratio) therewith being about 3.6. An examination of the images of the slit and the 2-D test-pattern leads to the conclusion that the width (FWHM) of the image of a point source is within 2.5 μm (≈ 0.32 arcseconds, which is only $2.4\lambda_0/D$ in the present case). From the near-to-image areas, we estimate the average grain size in the photo-sensitive layer at 1-to-1.5 µm. Hence it follows that the intrinsic resolution of the MM is likely to be at least 2 µm, or 1/4". The proximity of these figures to the diffraction limit serves as an experimental substantiation for our approach to spectroscopy of the XUV range employing focusing MMs.

3. CONCLUDING REMARKS

The MMs thus tested are currently operating aboard the CORONAS satellite launched on March 2, 1994. The MMs serve in a number of Herschelian telescopes equipped with 2-D X-ray detectors with an electric readout. A highly dispersive spectroheliograph aboard the satellite comprises a plane diffraction grating at grazing incidence with a density of 3600 grooves/mm and a focusing MM with $\lambda_0 \approx 19.5$ nm at near-normal incidence (4). The spectroheliograph forms stigmatic solar images in the radiation of individual spectral lines which are separated spatially owing to dispersion inherent in the configuration.

This work was supported by the Russian Foundation for Basic Research (Project Code: 93-02-15382).

4. REFERENCES

1. I.Sobelman, I.Zhitnik, A.Ignatiev et al., "Diagnostics of the inner corona by XUV-imaging of the Sun," Adv. Space Res., Vol.11, No.1, pp.99-107, 1991.
2. E.N.Ragozin, "Stigmatic spectroscopic instruments for the wavelength range 30 - 300 A with a high angular and spectral resolution using multilayered mirrors," IAU Colloq. 115: **High Resolution X-Ray Spectroscopy of Cosmic Plasmas**, Eds. P.Gorenstein and M.Zombeck, pp.380-383, Cambridge University Press, Cambridge, 1990.
3. Th.Peter, E.N.Ragozin, A.M.Urnov, D.B.Uskov, and D.M.Rust, "Doppler-shifted emission from helium ions accelerated in solar flares," Astrophys. J., Vol.351, No.1, Part 1, p.317, 1990.
4. S.V.Kuzin, A.I.Ignatiev, V.V.Krutov, M.M.Mitropolsky, A.A.Pertsov, E.N.Ragozin, V.A.Slemzin, I.I.Sobelman, I.P.Tindo, I.A.Zhitnik, N.N.Salashchenko, F.A.Sattarov, and R.J.Thomas, "Testing and calibration of the optical elements for XUV spectroheliograph under the CORONAS project," Submitted to **Advances in Multilayer and Grazing Incidence X-Ray/EUV/FUV Optics** (SPIE Conference, 24-29 July 1994, San Diego CA).
5. E.N.Ragozin, N.N.Kolachevsky, M.M.Mitropolsky, A.I.Fedorenko, V.V.Kondratenko, and S.A.Yulin, "Stigmatic high-resolution high-throughput narrow-band diffraction spectrograph employing X-ray multilayer mirrors," Physica Scripta, Vol.47, No.4, pp.495-500, 1993.
6. E.Spiller, D.Stearns, and M.Krumrey, "Multilayer x-ray mirrors: Interfacial roughness, scattering, and image quality," J. Appl. Phys., Vol.74(1), pp.107-118, 1993.
7. N.N.Kolachevsky, V.V.Kondratenko, M.M.Mitropolsky, E.N.Ragozin, A.I.Fedorenko, and S.A.Yulin, "Investigation of imaging Mo-Si multilayer mirrors at $\lambda \approx 135$ A using a laser-plasma XUV radiation source," Kratkie Soobshcheniya po Fizike FIAN, No.7-8, pp.51-56, 1992 (in Russian) (Bulletin of the Lebedev Physics Institute, No.8, 1992, Allerton Press, New York).
8. E.N.Ragozin, N.N.Kolachevskii, M.M.Mitropol'skii, V.A.Slemzin, and N.N. Salashchenko, "Characterization of imaging normal-incidence multilayer mirrors for the 40-300 A range by spectroscopic techniques using a laser-plasma radiation source," Proc. SPIE, Vol.2012, pp.209-218, 1993.

Characterization of a Plasma Produced Using a High Power Laser with a Gas Puff Target for X-ray Laser Experiments

H.Fiedorowicz, A.Bartnik, K.Gac, P.Parys*,
M.Szczurek and J.Tyl

Institute of Optoelectronics, Military University of Technology, 01-489 Warsaw 49, Poland

Institute of Plasma Physics and Laser Microfusion, 01-489 Warsaw 49, Poland

Abstract. A high temperature, high density plasma can be produced by using a nanosecond, high-power laser with a gas puff target. The gas puff target is formed by puffing a small amount of gas from a high-pressure reservoir through a nozzle into a vacuum chamber. In this paper we present the gas puff target specially designed for X-ray laser experiments. The solenoid valve with the nozzle in the form of a slit 0.3-mm wide and up to 40-mm long, allows to form an elongated gas puff suitable for the creation of an X-ray laser active medium by its perpendicular irradiation with the use of a laser beam focused to a line. Preliminary results of the experiments on the laser irradiation of the gas puff targets, produced by the new valve, show that hot plasma suitable for X-ray lasers is created.

INTRODUCTION

Various schemes for producing an X-ray laser active medium require an elongated hot plasma column with a perpendicular scale length of at least 100 μm, having a density on the order of 10^{20} cm^{-3}. A number of methods for producing such plasmas have been investigated. These include laser-irradiated solid slabs, thin foils, fibers or high power electrical discharges.

There were some suggestions of using a gas target (in a form of gas puff or pulsed gas jet) to create an X-ray laser active medium (1,2,3,4), but up to now the demonstration experiment has not been performed.

The possibility to produce a hot plasma by laser irradiation of a gas puff target was demonstrated in the early 1970s (5). In our previous papers (6,7,8) we presented the results of the soft X-ray emission measurements from the laser-irradiated gas puff target in the wavelength range between 5 and 20 Å.

In the experiment the gas puff target was formed by expanding gas from a high pressure reservoir through a nozzle into a vacuum chamber. The flow of the gas was initiated by a fast solenoid valve synchronized with a laser pulse. The gas puff target was irradiated using 1-ns Nd-laser pulses with the intensity in the focus of about 10^{13} Wcm^{-2}. Soft X-ray emission in the softer wavelength range between 25 and 300 Å from the gas puff target irradiated by a nanosecond Nd-laser has been measured by Filbert et al.(9).

In this paper we present the gas puff target predicted for X-ray laser experiments. The specially designed solenoid valve is equipped with a nozzle having a linear exit aperture and forms an elongated gas puff which allows to create an X-ray laser active medium when a laser beam is line-focused by cylindrical optics.

The motivation of this work was manifold. **1.** We have observed in our earlier experiments (8), that the plasma produced by laser-gas puff target interaction has no such large density gradients as plasmas produced by laser irradiation of a solid target. This effect can be important from the point of view of preventing X-rays from refraction out of the gain path. **2.** The created hot-plasma is surrounded by cold gas flowing through the nozzle and some enhancement of the plasma electron density in the outer region of the plasma, has been observed using the laser interferometry(10). This confinement effect can be used to produce the plasma column with negative curvatures of the density profiles to form the waveguide-like active medium. **3.** Up to now, we have no detailed description of the laser-gas puff target interaction process, but there are some experimental evidences that the laser beam interacts with a target, which consists of small droplets created in result of the condensation of the gas flowing through the nozzle (8,10). McPherson et al. described this phenomenon using the cluster beam formation model (11). Adiabatic spherical expansion of the individual droplets can cause additional cooling of the plasma, that enhances recombination. **4.** To increase the cooling rate of the plasma one can add a small amount of high-Z radiators (9), what can be easily performed by a gas puff target produced using a mixture of gases.

DESCRIPTION OF THE GAS PUFF TARGET AND PRELIMINARY EXPERIMENTS

The solenoid gas valve to produce the gas puff target for X-ray laser experiments consists of a nozzle in a form of an elongated slit joined with the valve. Schematic view of the valve is presented in figure 1. The reservoir is capable of operating with pressures up to 5 atm. The gas from the reservoir to the nozzle flows through the 25 mm long channel with rectangular-shape section. The flow of the gas is initiated by the driving coil which opens the diaphragm. The nozzle is made in a replaceable plate mounted at the exit of the channel.

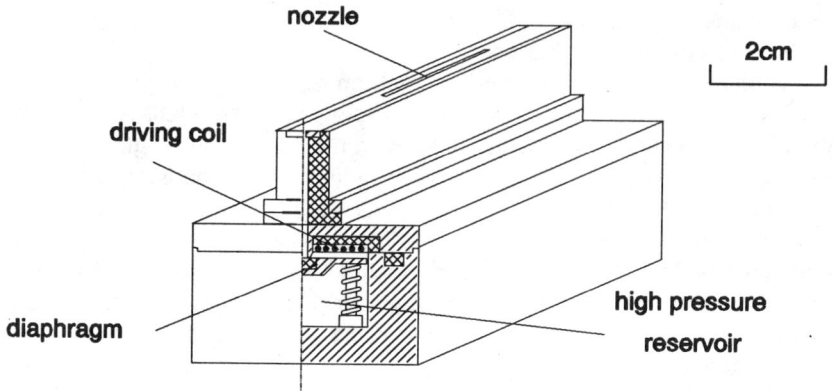

FIGURE 1. Schematic of the solenoid gas valve to produce the gas puff target for X-ray laser experiments.

The valve with the nozzle of 40 mm long and 0.3 mm wide has been tested using a Nd-glass laser produced pulses of 1-ns duration and an energy of nominally 15 J. The circularly-fucused laser beam, with the diameter of the focus about 100 µm, heated only small central part of the puff. In the experiment we have used sulphurhexafluoride (SF_6) to form the gas puff target. To characterize the produced plasma X-ray diagnostic methods, including the spatial, spectral and temporal measurements in the soft X-ray range, have been used.

Pinhole camera X-ray images of the plasma, taken using a pinhole of 30-µm in diameter and a 10-µm-thick beryllium filter, show that a highly uniform hot-plasma was produced (Fig. 2). The size of the plasma is larger than the diameter of the laser focus.

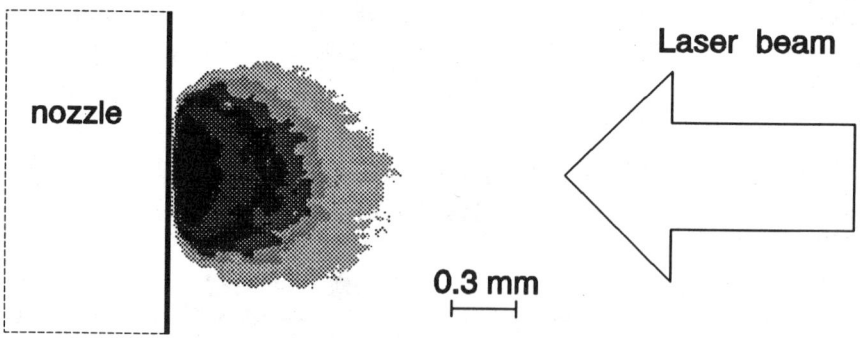

FIGURE 2. Pinhole camera X-ray images of the plasma produced by laser irradiation of SF_6 gas puff target.

X-ray spectra from a SF_6 gas puff target, recorded with the CsAP crystal in the 8 - 20 Å wavelength range, presented the K-line emission of H-like and He-like fluorine ions (Fig. 3). The plasma electron temperature, roughly estimated from the line intensities, is about 150 eV, and the plasma electron density is higher than 5×10^{19} cm^{-3}.

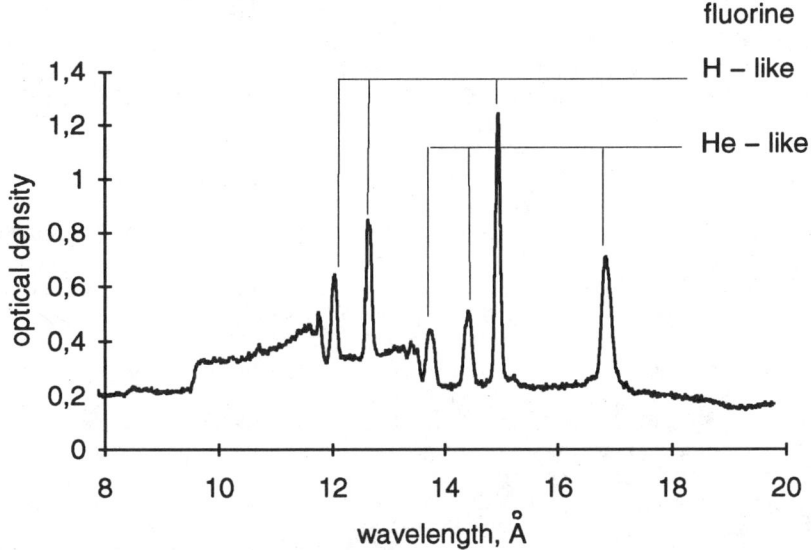

FIGURE 3. X-ray spectrum for SF_6 gas puff target.

In conclusion, we present the solenoid gas valve to produce a gas puff target in a form of an elongated, planar sheet of gas. The preliminary experiments confirmed the vialibility of the developed gas puff valve and its usefulness in producing plasmas suitable for X-ray laser experiments.

ACKNOWLEDGMENTS

The research was sponsored by the State Committee for Scientific Research of Polish Government under grants No. 2 0141 91/p01 and No. 8 8013 91 02. The laser interaction experiments were performed at the Institute of Plasma Physics and Laser Microfusion. The authors would like to thank J. Galik, A. Klimek and J. Makowski for their assistance in conducting the experiment.

REFERENCES

1. Filbert, P. C., and Kohler, D. A., "On the dynamics of a laser-produced Be plasma and its implication for a Be/Ne photopumped X-ray laser scheme", J.Appl.Phys. **68**, 3091-3098 (1990)
2. Eder, D. C., Amendt, P., Bolton, P. R., Guethlein, G., London, R. A., Rosen, M. D., and Wilks, S. C., "Table-top X-ray lasing based on optical-field-induced ionization", in *Proceedings of the 3rd International Colloquium on X-ray lasers 1992*, (IOP Publishing, Bristol, 1992), pp. 177-184
3. Shlyaptsev, V. N., and Gerusov, A. V., "On two methods of table-top X-ray laser design", in *Proceedings of the 3rd International Colloquium on X-ray lasers 1992*, (IOP Publishing, Bristol, 1992), pp. 195-200
4. Falcone, R. W., "Experiments with high-intensity, ultrashort-pulse lasers: interaction with solids and gases", in *Proceedings of the 3rd International Colloquium on X-ray lasers 1992*, (IOP Publishing, Bristol, 1992), pp. 213-218
5. Askarian, G. A., and Tarasova, N. M., "Laser spark in a gas cloud", Pis'ma ZETF **14**, 89-91 (1971) (in Russian)
6. Fiedorowicz, H., Bartnik, A., Parys, P., and Patron, Z., "Laser plasma X-ray source with a gas puff target", in *Proceedings of the 13th International Congress on X-ray Optics and Microanalysis 1992*, (IOP Publishing, Bristol, 1993), pp. 515-518
7. Fiedorowicz, H., Bartnik, A., Patron, Z., and Parys, P., "X-ray emission from laser-irradiated gas puff targets", Appl. Phys. Lett. **62**, 2778-2780 (1993)
8. Fiedorowicz, H., Bartnik, A., Patron, Z., and Parys, P., "Generation of nanosecond soft X-ray pulses in result of interaction of the Nd:glass laser radiation with gas puff targets", Laser and Particle Beams (1994) (in print)
9. Filbert, P. C., Kohler, D. A., and Walton, R. A., "The effect of high-Z atomic radiators on the X-ray emission rate from a laser-produced nitrogen plasma", J.Appl.Phys. **75**, 2332-2338 (1994)
10. Fiedorowicz, H., Bartnik, A., Parys, P., Patron, Z., and Pisarczyk, T., "High intensity laser interaction with gas puff targets", in *Proceedings of the 11th International Conference on Phenomena in Ionized Gases*, (Ruhr University, Bochum, 1993), pp. 132-140
11. McPherson, A., Luk, T. S., Thompson, B. D., Boyer, K., Rhodes, C. K., "Multiphoton-induced X-ray emission and amplification from clusters", Appl. Phys. **B57**, 337-347 (1993)

THREE-DIMENSIONAL X-RAY IMAGING WITH OFF-AXIS X-RAY ZONE PLATE

Heung-Rae Lee, E. Anderson**, L.B. Da Silva, J.E. Trebes

Lawrence Livermore National Laboratory, P.O. Box 808, Livermore, CA 94568
** Center for X-ray Optics, Lawrence Berkeley Laboratory, Berkeley, CA 94720

Fresnel zone plates are now routinely used to produced two dimensional x-ray images of organic and inorganic specimens. We describe one possible approach of achieving three-dimensional x-ray imaging which utilizes non-circular Fresnel zone plates. This approach is ideally suited for applications in the x-ray laser community where specimen and zone plate alignment must be performed off lines.

1] Introduction

Substantial progress has been made in recent years towards the goal of developing x-ray microscopy for biological applications.[1] This work has been primarily motivated by the possibility of producing images with resolutions superior to that obtainable with visible light microscopy while causing less radiation damage to the specimen than conventional electron microscopy.[2] Furthermore, in contrast to electron microscopy, x-ray microscopy has the potential to image organisms hydrated in physiologically normal environments, thus revealing their unperturbed structure.

To date x-ray microscopy has been developed and demonstrated by several groups around the world. Diffraction limited resolution has been reported by Jacobsen et al.[3] and Meyer-Ilse et al.[4] both of which used synchrotron radiation to image biological systems. More recently, x-ray imaging microscopy was demonstrated with an x-ray laser.[5,6] In all this work circular Fresnel zone plate lens were used to obtain two-dimensional images. It has become evident, however, that in order to increase the utility of x-ray microscopy three-dimensional imaging must be developed. A variety of techniques have been discussed for achieving high resolution three-dimensional x-ray imaging. They include multi-view holography[7], x-ray diffraction[8] and multi-view microscopy[9] using conventional zone plates. In this paper we will discuss on approach for performing three-dimensional imaging with multi-view microscopy which is well suited for x-ray laser application.

Biological imaging of specimens in their natural environment remains the main application of three-dimensional imaging microscopy. In order to achieve high resolution it has become clear that x-ray lasers may represent the only promising source. The main advantage of an x-ray laser is that its high brightness (10^{21} photons/(sec-mrad^2-mm^2-0.01% BW)) and short duration allows images to be made with a single ~200 ps exposure.[6] This eliminates image blurring due to both natural motion and chemical decomposition due to ionizing radiation. Unfortunately, this single shot nature precludes the use of multiple exposures to align and focus the zone plate lens. In order to overcome this a focusing technique had to be devised which could be performed prior to the exposure. This is no trivial matter given the small depth of field which for an $f^{\#}$ 10 lens and 44.83 Å x-ray is ~1 μm. The approach we have previously utilized[10] makes use of two white light interferometers to accurately measure the specimen/zone plate separation. Unfortunately this technique is not well suited to the conventional multi-view systems shown in Figure-1a. For this reasons we have investigated the possibility of using off-axis zone plate lens which can all be positioned on one plane as illustrated in Figure-1b. The advantage of this approach is that the alignment technique previously demonstrated for single view images can be equally well applied to the multi-view system.

As a main emphasis of this paper, we determine the accurate shape of each zone by deriving the equation for zone boundaries from optical path length equation. In order to test this solution we wrote a computer program to calculate the Fresnel-Kirchoff integral and determine the intensity distribution at arbitrary positions in space.

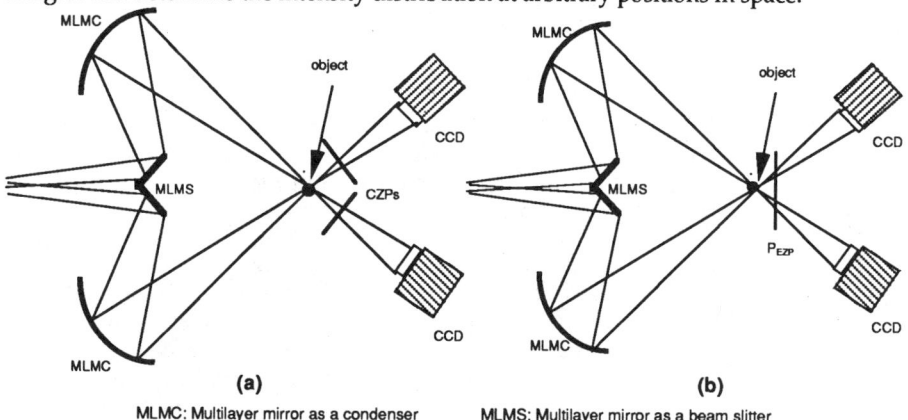

MLMC: Multilayer mirror as a condenser
P_{EZP}: Plane of EZPs
CZPs: Circular zone plates

MLMS: Multilayer mirror as a beam slitter
EZP: Elliptical zone plate on the same plane
CCD: CCD detector

Figure-1a and b. Schematic for Multiview XRM: This diagram represents the simplest system of mutiview imaging. The microbiological object is exposed simultaneously in two different direction with incident angle a and -a; while 1a is the conventional XRM using circular zone plates, 1b is that using off-axis zone plate lens, i.e., elliptical zone plates on the same plane.

2] Formulation

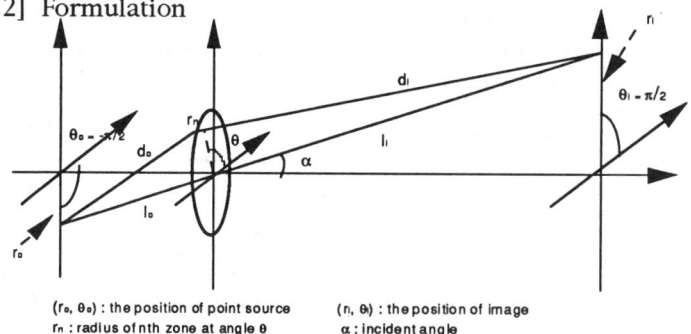

(r_o, θ_o): the position of point source (r_i, θ_i): the position of image
r_n: radius of nth zone at angle θ α: incident angle

Figure-2. Ray-trace Diagram for EZP Formulation: The point source is located at (ro=0.5 mm, qo=-p/2) on the object plane, 1.001 mm from the plane of EZP to make the incident angle a. Its image is positioned at (ri=500 mm, qi=p/2) on the image plane with magnification 1000.

It is clear that the small field of view of conventional zone plates precludes their use for off-axis case. Therefore, what we have investigated is the performance of off-axis zone plates which will be an elliptical zone plates (EZP). In the thin zone approximation which characterizes a zone plate as an amplitude object, zone plates are designed by simply matching optical path lengths. Figure-2 illustrates the formulation of zone plate equation for off-axis incident with the grazing angle α. The requirement for focusing, namely that radiation emitted from an object point P_o is focused by a zone plate to an image point P_i, requires that the path difference is

$$(do + di) - (lo + li) = n\lambda/2 \quad \text{(1)}$$

which, in general, can be rewritten as a quadratic function of zone radius and angle, r_n and θ,

$$(l_o^2 + l_i^2 + c - f_m^2)r_n^2 - (2l_o l_i g + c\, f_p)r_n + (l_o^2 l_i^2 - c^2/4) = 0 \quad \text{(2)}$$

where

$$f_p = r_o \cos(\theta_o - \theta) + r_i \cos(\theta_i - \theta)$$
$$f_m = r_o \cos(\theta_o - \theta) - r_i \cos(\theta_i - \theta)$$
$$g = r_i \cos(\theta_o - \theta) + r_o \cos(\theta_i - \theta)$$
$$c = (n\lambda/2)^2 + 2l_o l_i + n\lambda(l_o + l_i)$$

in which θ_o is $-\pi/2$ and $\theta_i + \pi/2$ for our experiment setup.

In the limit that P_o or P_i is infinite, Equation-2 can be simplified to

$$(1 - \sin^2\alpha\,\sin^2\theta)r^2 \pm (n\lambda\sin\alpha\,\sin\theta)r - \left[(n\lambda/2)^2 + n\lambda f\right] = 0 \quad \text{(3)}$$

where f is the focal length of the zone plate. The plus sign in the first order of zone radius is for the case in which the point source P_o is far from the zone plate, (i.e. low magnification), whereas the minus sign is further case in which the point source is very close to the zone plate, (high magnification). If we convert Equation-3 from polar coordinates to Cartesian,

$$x^2 + \frac{\left(y \pm \dfrac{n\lambda \sin\alpha}{2\cos^2\alpha}\right)^2}{\sec^2\alpha} = (n\lambda/2)^2 \sec^2\alpha + n\lambda f \quad \text{(4)}$$

it becomes clear that each zone is elliptical, with adjacent zones having a shift $\pm \dfrac{n\lambda \sin\alpha}{2\cos^2\alpha}$ and change in eccentricity $\sec\alpha$.

Equation-(5) presents the Fresnel-Kirchoff diffraction integral (FKI) derived from the Green's-theorem, which is used to calculate the complex disturbance at P_i due to a spherical wavefront emitted by P_o..[11] The complex disturbance at P_i relevant to diffraction pattern is

$$\Psi(p_i) = -\left(\frac{iA}{2\lambda}\right) \iint F(r,\theta) \frac{\exp[ik(S_o + S_i)]}{S_o S_i} [\cos(\bar{n}, \bar{s}_i) - \cos(\bar{n}, \bar{s}_o)] da \quad \text{(5)}$$

where A is the amplitude at unit distance from the source P_o, s and s' is the distance of P_o and P_i from the point Q on the diffraction aperture, and $\bar{n}$ is the normal vector to the aperture. $F(r,\theta)$ is the transmission function of the zone plate designed with Equation-(2), which indicates the integral area. We wrote a computer program which includes the subprogram for the design of an EZP and determines the intensity distribution at arbitrary positions.

3] Computation and Results

For the computer simulation with high magnification of 1000, we chose the focal length of 999 μm at 44.83 Å x-ray wavelength and an object distance of 1 mm (i.e. a point source) to the center of EZP. The grazing angle is 20 degrees. Since higher resolutions can be obtained by zone plates with smaller outer zone widths and there is always a low-intensity background caused by the zero diffraction order, the

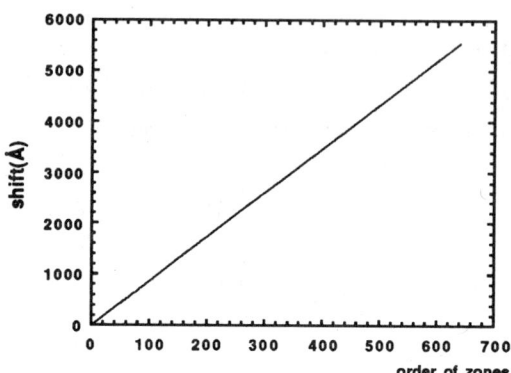

Figure-3: The shift of zone boundaries.

number of zones used in imaging simulation is 200 from 300th and 500th zones; the central obscuration region is from the center to 300th zone. The results of computer calculation for zone plate geometry show the width of the outmost zone on x-axis (minor axis) is 501.2 Å and that on y-axis (major axis) is 572.9 and 547.8 Å. The average size of EZP is ~50 μm in radius. For the close investigation of EZP, we examine Equation-(4) because it is still allowable for the case of high magnification. Figure-3 demonstrates the shift as the function of gazing angle α and that of zone's order n. The shift of the outer zone is ~4300 Å which can significantly affect the resolution at soft x-ray wavelength 44.83 Å.

In Figure-4 we plot the intensity distribution function of the accurate EZP (i.e., Equation-(2)) and that of the approximated (Equation-(3)). Even if the results show very similar point spread functions, the full width half maximum, that is resolution, 400 Å and 600 Å respectively.

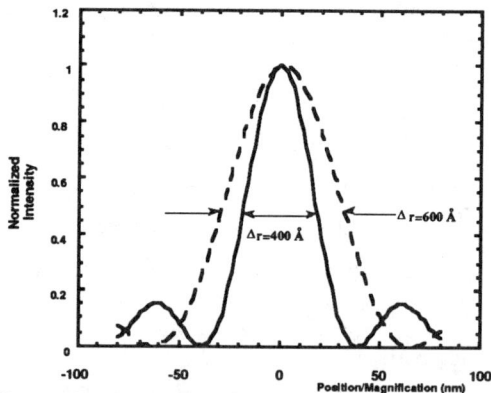

Figure-4. Comparison in resolution between two EZPs: The solid line diffraction pattern is produced by EZP designed with Equation-(2) and the dashed line by EZP designed with Equation-(3); the resolution of the former case is 400 Å and that of the latter is 600 Å.

The latter agrees with Rayleigh's criterion for the limit of resolution in terms of the width δ_{rN} of the outmost zone, $\Delta \geq 1.22 \delta_{rN}$, while the former does not. The reason is that Rayleigh's criterion was derived by far-field approximation which was similarly applied to the formulation of Equation-(3) from Equation-(2). The exact design of a zone plate using Equation-(2) can reduce the resolution limit with two thirds of Rayleigh resolution. If the shift of elliptical zones is ignored, that is, the coefficient of the first order in Equation-(3) is zero, the peak intensity of diffraction pattern distorted by astigmatism will be reduced by 5 order of magnitude.

Another important aspect of an imaging system is its field of view. To examine how effectively EZP can make the image of an object, we estimate the field of view of EZP on the object plane which is defined by two criteria; the half maximum of intensity and the sustenance of Airy-like diffraction pattern at the image plane. For the object simulated in the computer program, we made use of the two-dimensional array of point sources with a 1 μm separation. Each point on the object is imaged by EZP onto the image plane.

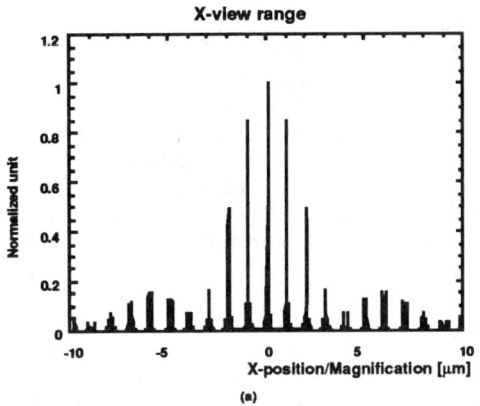

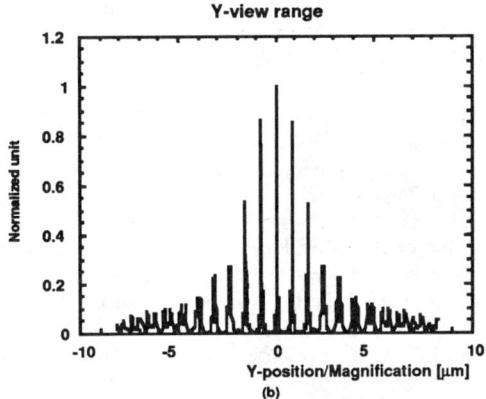

Figure-5a and b. View Range of XRM with Support of Accurate EZP: Two-dimensional object plane on which 20 points are aligned on each axis is used for the computer simulation. The image of each point appears as the diffraction pattern. The view range of XRM on each axis is ~ 6 μm; 5a is for the view range on x-axis and 5b for that on y-axis.

Figure-5a and -5b show, that using our criteria, the field of view is ~6 μm. Such an view range with the attenuation of peak intensities tells us the sensitivity of this imaging system to the grazing angle α. The small variation of few μm along the incident direction, i.e. the variation sustaining α, does not affect too much an image made by EZP due to the characteristics of zone-plate imaging, i.e. the depth of field of a zone plate; the same zone boundaries of EZP can be used to form the image of an off-positioned object at the same distance of a detector. However, if the variation includes the angular one δ_α from the incident direction, the off-positioned object cannot be imaged well because it experiences the different zone boundaries with δ_α. As the grazing angle α increases, the sensitivity of the system to δ_α also increases and reduces the view range.

4] Summary

In this paper we have described a zone plate design which will be useful in the development of three-dimensional x-ray imaging and have examined the performance of EZP. The resolution of this design was found to be comparable to standard zone plates but the field of view is significantly reduced.

What makes this approach practical is that electron beam lithography techniques have evolved to the point where the production of arbitrary shapes is possible. The width of the outer zone on x-axis is 501.2 Å, that on y-axis is 572.9 and 547.8 Å, and the shift of the outmost zone is ~4300 Å. The grazing angle used in multi-view imaging is ultimately constrained by the image reconstruction algorithms and the required three dimensional resolution.

References

1. A.G. Michette, G.R. Morrison, and C.J. Buckley, "X-Ray Microscopy II", Springer Series in Optical Sciences **67** (Springer-Verlag, Berlin, 1992)
2. D. Sayre, J. Kirz, R. Feder, D.M. Kim, and E. Spiller, Science **196**, 1339 (1977)
3. C. Jacobsen, M. Howells, J. Kirz, and S. Rothman, J. Opt. Soc. Am. **A7**, 1847 (1990)
4. W. Meyer-Ilse et al., "X-ray Microscopy Resource Center at the Advanced Light Source" in Soft X-Ray Microscopy, SPIE **1741**, pp. 112-115 (1992)
5. L. B. Da Silva et al., Opt. Lett. **17**, 754 (1992)
6. L. B. Da Silva et al., Science **258**, 269 (1992)
7. I. McNulty and J.E. Trebes et al., "Experimental Demonstration of High Resolution Three-Dimensional X-Ray Holography" in Soft X-Ray Microscopy, SPIE **1741**, pp. 78-84 (1992)
8. J. E. Trebes, Inst. Phys. Conf. Ser. **125**, 265 (1992)
9. W. S. Haddad et al, " Ultra High Resolution Tomography Using a Scanning Transmission X-ray Microscope", to be published
10. L.B. Da Silva et al., "Imaging Microscopy with X-Ray Lasers at LLNL", in Soft X-Ray Microscopy, SPIE 1741, pp. 154-159 (1992)
11. M. Born and E. Wolf, "Principles of Optics", 4th Ed., Pergamon Press, pp. 370-386 (1970)

X-ray laser interferometry experiments at LLNL

P. Celliers, F. Weber, L. B. Da Silva, T. W. Barbee Jr.,
S. Mrowka and J. Trebes

Lawrence Livermore National Laboratory, L-447
Livermore, California U.S.A. 94550

Abstract. We describe initial experiments to produce soft x-ray interferograms using a collisionally-pumped Y soft x-ray laser at 155 Å as the probe source. Successful operation of the interferometer requires tight alignment tolerances: spatial overlap of the beams to within 50 µm, and temporal overlap to within a 150 µm path difference. Beamsplitter vibrations were found to be a strong factor influencing the quality of the interferograms.

Introduction

Soft x-ray laser beams are now being used routinely as a probe for dense plasma experiments[1]. At 155 Å the collisionally pumped Y laser used at LLNL for plasma imaging experiments, the potential exists to diagnose plasmas with electron densities well above 1022 cm-3 exists. Interferometric measurements of electron density using visible and UV wavelengths lasers have become routine for many plasma investigations[2-4]. The maximum electron densities that can be probed with visible and UV lasers are limited to some fraction of the critical electron density corresponding to the probe laser frequency - for visible laser frequencies the limiting density is around 5 x 1021 cm-3. On the other hand, the critical electron density corresponding to the 155 Å wavelength of the Y laser (about 5 x 1024 cm-3) is significantly higher than laser produced ablation plasmas, allowing the beam to penetrate dense plasmas.

Crucial in the design of any interferometric measurement are the coherence properties of the probe beam. The spatial coherence of collisionally pumped x-ray lasers has been previously characterized[5], and has been found to be determined primarily by the transverse dimension of the source region of the laser. A good approximation of the transverse coherence length, L_s of the collimated beam is[6], $L_s = f\lambda / 2\pi r_s$, where r_s is the radius of the source region, f is the focal length of the collimating mirror and λ is the x-ray laser wavelength. For the collimated Y x-ray laser beam at LLNL, $L_s \approx 50-100\,\mu m$; this limited spatial coherence precludes the application of a wavefront shearing configuration. However a carefully aligned Mach-Zehnder configuration will allow good beam overlap, and therefore produce high visibility fringes. The temporal coherence of the beam is also limited, and is most simply estimated from the linewidth, $\Delta\lambda$, of the laser transition[7]; the temporal coherence length is given by the expression[8] $L_t = 0.664\,\lambda^2 / \Delta\lambda$. For the Y laser $L_t \approx 150\,\mu m$. Successful alignment of a Mach-Zehnder configuration must produce path lengths in each arm equal to within L_t and spatial overlap of the scene and reference beams at the object plane with a tolerance less than L_s.

Experiment

A schematic diagram of the interferometer arrangement is displayed in Figure 1. The beams reflect from the mirrors and the beamsplitters at 2 degrees incidence, and are recombined after a path length of 900 mm in each arm. The collapsed parallelogram facilitates placement of the Y x-ray laser and a secondary plasma target close to the center of the Nova two-beam vacuum chamber. A 1 m radius spherical mirror collects and collimates the x-ray laser output beam and injects the collimated beam (about 5 mm diameter) into the interferometer. The combined output beams are collected by a 1 m radius imaging mirror which is focused to project an image of the secondary target onto an x-ray CCD imager located 5 m from the chamber center, at 10x magnification.

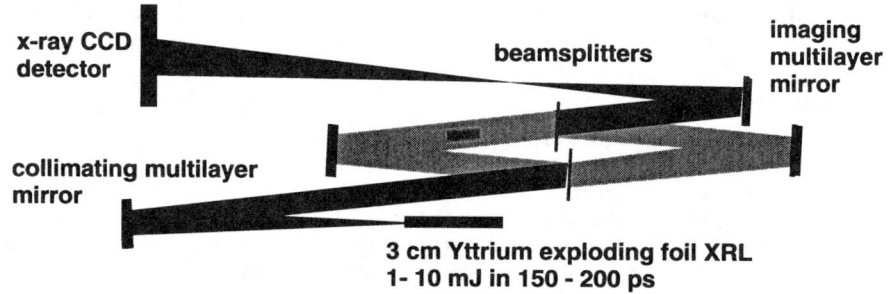

Figure 1. Interferomter arrangement for plasma probing experiments.

Beamsplitters and multilayer mirror components were designed and fabricated to reflect 155 Å laser wavelength at near normal incidence. The beamsplitters are the most critical components in the system. These were fabricated on 1000 Å Si_3N_4 membranes formed within 12 mm (square side or circular diameter) apertures etched out of 250 μm thick Si slabs. Reflectivity at 155 Å was produced by a multilayer coating of 8 Mo/Si layer pairs. Theoretical transmissivity and reflectivity of these components was around 20 - 30%. The beamsplitters are sufficiently transparent to visible light (about ND 1) to allow alignment of the interferometer with visible light beams.

The interferometer components were mounted onto a rigid platform machined out of 25 mm thick aluminum plate. The platform was positioned in the chamber on an adjustable kinematic mount. Alignment of the interferometer was performed off-line prior to placement in the chamber. The optical path lengths of both arms were adjusted to be identical to within ±2 μm as verified by observing fringes produced with an incoherent white light source as the illumination beam. Spatial overlap of the two beams was arranged by projecting an image of a crosshairs onto the plane of the output beamsplitter, and adjusting tilt on the final mirror to produce overlapped beams to within ± 50 μm. This alignment tolerance was similar to the expected spatial coherence of the collimated x-ray beam. Final tilt adjustments were then applied to place the region of fringe localization at the secondary target plane and to produce a finite fringe frequency of around 15 - 20 fringes/mm (at 155 Å). The system was verified to remain aligned while it was moved and placed inside the target chamber.

For plasma imaging experiments a secondary target was mounted in one arm of the interferometer. The plasma is created by irradiating this target with the second beam of the Nova two-beam chamber.

Results

Successful operation of the interferometer is demonstrated in Figure 2. This figure displays a straight fringe pattern recorded during a null test, in which no secondary plasma was placed in the second interferometer path. The interference pattern displays high quality fringes with fringe visibility ranging from 50% to 100%.

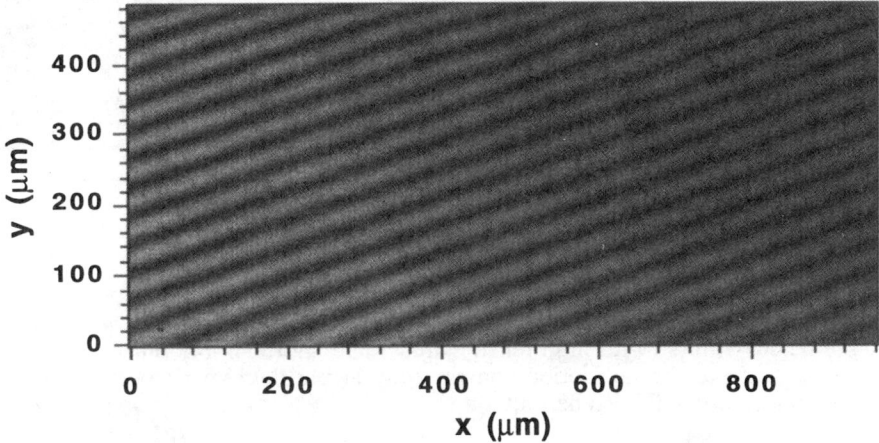

Figure 2. Fringe pattern produced with no secondary plasma target.

During our initial experiments we were unable to produce good quality interferograms reproducibly. Several imaging attempts displayed fringes of poor visibility, fringes with a strong ripple pattern, or no fringes at all. Spatial variations in the transmitted beam intensity were also evident in the image. These were known not to be present in the beam itself, and were also not evident through visible inspection of the beamsplitter prior to the experiment.

We found a correlation between the poor quality interferograms and use of beamsplitters with low membrane tension. The membrane tension produced in the 1000 Å Si_3N_4 substrate is determined by the fabrication process; we found better results with Si_3N_4 membranes fabricated to produce a membrane tension set to about 50% of the Si_3N_4 yield strength. During optical alignment the poorer quality beamsplitters were found to vibrate excessively at acoustic frequencies, with weak damping. We suspect that vibrations coupled into the chamber (from pumps and building vibrations) are sufficient to produce a constant level of motion in the beamsplitter membranes. An example of a poor quality interferogram is shown in Figure 3. This interferogram displays both reduced fringe visibility and a rippled fringe pattern. The ripples have a spatial period of about 200 μm and an amplitude of about 1 fringe shift. This fringe shift is produced by a surface perturbation of approximately 75 Å amplitude. Furthermore, a local shift of the membrane position of about 20 Å *during* the x-ray laser pulse is sufficient to degrade the fringe visibility through motional blurring. For a 200 ps pulse, a localized membrane velocity of only 10 m/s can produce this effect.

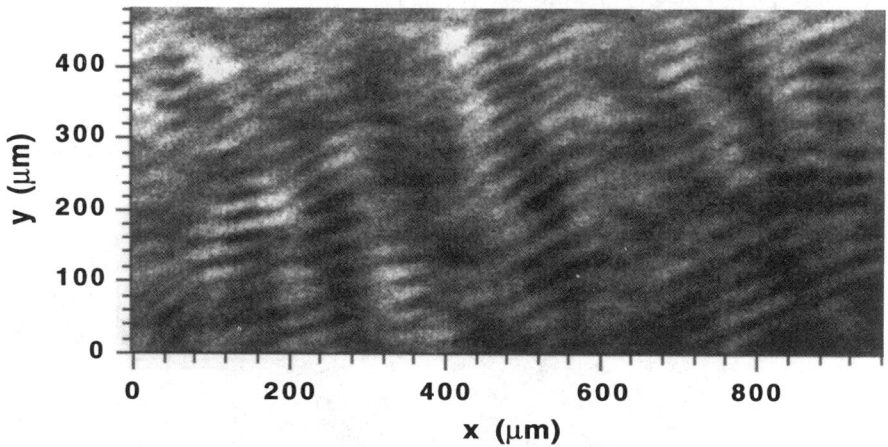

Figure 3. Poor quality fringe pattern produced by beamsplitters with low membrane tension.

In summary, improved results with x-ray laser interferometry will require stable beamsplitters (with high tensile stress) and shorter x-ray pulses to reduce the effects of membrane motion and rippling during the experiment. Improved vibration isolation will also be important.

Acknowledgements

We would like to thank Jim Cox, Sharon Alvarez, Joe Smith and the Nova Experiments Group for their assistance in these experiments. This work was performed under the auspices of the U.S. Department of Energy by the Lawrence Livermore National Laboratory under contract No. W-7405-ENG-48.

References

1. D. Ress, L. B. Da Silva, R. A. London, J. E. Trebes, S. Mrowka, R. J. Procassini, T. W. Barbee Jr., and D. E. Lehr, Science **265**, 514 (1994).

2. H. Houtman, L.E. Legault and J. Meyer, Appl. Opt. **26**, 1106 (1987)

3. R. Fedosejevs, I. V. Tomov, N. H. Burnett, G. D. Enright and M. C. Richardson, Phys. Rev. Lett. **39**, 932 (1977); R. Benattar, C. Popovics and R. Sigel, Rev. Sci. Instrum. **50,** 1583 (1979)

4. D. T. Attwood, D. W. Sweeney, J. M. Auerbach, and P. H. Y. Lee, Phys. Rev. Lett. **40,** 184 (1978); G. E. Busch, C. L. Shepard, L. D. Siebert, and J. A. Tarvin, Rev. Sci. Instrum. **56**, 879 (1985)

5. J. E. Trebes, K. A. Nugent, S. Mrowka, R. A. London, T. W. Barbee, M. R. Carter, J. A. Koch, B. J. MacGowan, D. L. Matthews, L. B. DaSilva, G. F. Stone and M. D. Feit, Phys. Rev. Lett. **68**, 588 (1992)

6. J. W. Goodman, *Statistical Optics*, Wiley, New York (1985), p. 213

7. J. A. Koch, B. J. MacGowan, L. B. DaSilva, D. L. Matthews, J. H. Underwood, P. J. Batson and S. Mrowka, Phys. Rev. Lett. **68,** 3291 (1992)

8. J. W. Goodman, *Statistical Optics*, Wiley, New York (1985), p. 168

X-ray lasers for Imaging and Plasma Diagnostics

L.B. Da Silva, T.W. Barbee, Jr., R. Cauble, P. Celliers, J. Harder,
S. Libby, H.R. Lee, R.A. London, D.L. Matthews, S. Mrowka,
J.C. Moreno, D. Ress, J.E. Trebes, A. Wan, and F. Weber

*Lawrence Livermore National Laboratory, P.O. Box 808, L-059
Livermore, California U.S.A. 94550*

Abstract. Collisionally pumped soft x-ray lasers now operate over a wavelength range extending from 35 - 350 Å. These well characterized sources have high peak brightness (Gev blackbody temperature) and narrow bandwidth making them ideal for x-ray imaging and interferometry. The yttrium neon-like x-ray laser operating at 155 Å has already been used to image directly driven foils with 1-2 μm spatial resolution and probe plasmas at electron densities approaching 10^{22} cm^{-3} with moiré deflectometry. Advances in multilayer mirrors and beam splitters have now also made it possible to develop x-ray laser interferometers. We will describe initial experiments to develop x-ray laser interferometry for probing plasmas relevant to ICF.

The high brightness and comparatively routine operation of existing soft x-ray lasers has opened up the possibility of using these systems for a variety of applications. The use of Ni-like tantalum x-ray laser for biological imaging has already been demonstrated[1] and will ultimately allow us to study biological specimens in a nature environment with resolutions far exceeding that possible with optical techniques. The high brightness of saturated x-ray lasers[2,3,4,5] also makes them an ideal tool for probing high density and large plasmas relevant to astrophysics and ICF. In addition, the wide range of available x-ray laser wavelengths now make it possible to take advantage of absorption edges to enhance contrast.

The Ne-like yttrium x-ray laser is well suited to this application by virtue of its high output energy (~8 mJ)[5] and monochromatic output (i.e. dominated by 155 Å, J=2-1). The wavelength is also well suited to existing multilayer mirror technology which have demonstrated reflectivities of ~60%. At LLNL we have now used this system to image accelerated foils (see Cauble et al. these proc.) and measure electron density profiles[6] in laser irradiated target using moiré deflectometry. More recently we have begun experiments to demonstrate x-ray laser interferometry. In contrast to other techniques interferometry offers the possibility of directly measuring the electron density profile in large and high density plasmas. In figure 1 we show the electron density and plasma length accessible with the yttrium x-ray laser. The parameter space available is ultimately limited by absorption or excessive fringe shifts.

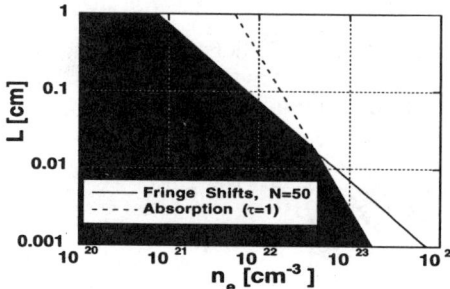

Figure 1 Shaded area is accessible parameter space of electron density and plasma size (L) constrained by absorption (assuming only free-free, T_e=1 keV, $<Z>$=30, λ=155 Å)

The short wavelength of x-ray lasers eliminates the problems associated with critical density layers, which along with high absorption, have made it difficult to probe plasmas with electron densities exceeding ~10^{21} [cm-3][7, 8]. At 155 Å the critical density, $n_{cr} = 1.1 \times 10^{21} \lambda^{-2}$ [cm^{-3}] (λ in µm), is 4.6x10^{24} [cm-3] which is well above most plasmas of interest. Since the index of refraction, n_{ref} in a plasma is given by $n_{ref} = \sqrt{1 - \frac{n_e}{n_{cr}}}$, the adverse effects of refraction[9] can be significantly reduced by using a short wavelength probe. In figure 2 we compare the deflection angle after propagating through a 3 mm plasma for an optical (2650 Å) and xuv (155 Å) probe source. At a fundamental level large deflection angles imply significant spatial blurring and reduced spatial resolution. In addition, since probe rays will propagate through a range of electron densities interpretation of the results will be more difficult. Another important advantage of using xuv probes is that spatial resolutions of better than 1 µm can be easily achieved by using normal incidence spherical multilayer mirrors.

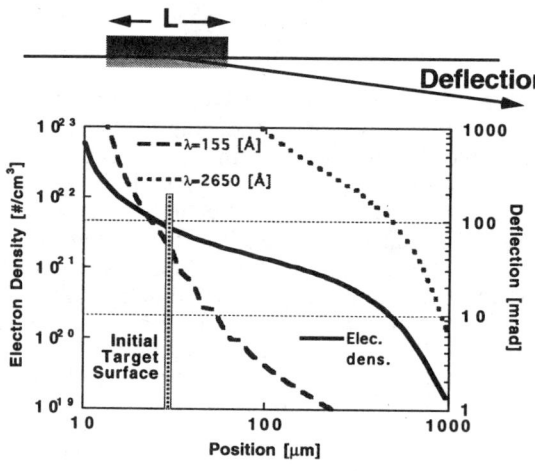

Figure 2 Calculated electron density (t=0.9 ns after start of pulse) for a CH target irradiated at 2.0x10^{13} W/cm^2 and the deflection assuming a 3 mm wide plasma (L) for probe laser wavelength of 2650 Å and 155 Å.

Motivated by these advantages we have over the past couple years started using x-ray lasers for a variety of plasma probing applications. Plasma imaging experiments have been performed using the yttrium x-ray laser to investigate driver uniformity and are described in detail in the article by Cauble et al. in these proceedings. Although direct imaging with x-ray lasers has its advantages it does not require several of the attributes of the x-ray laser, specifically small source size, narrow divergence and narrow bandwidth. The drawback of direct imaging for plasma studies is that in order to determine the electron density of these plasmas it is necessary to have an accurate estimate of the opacity. For high-Z systems this can be a near impossible problem at soft x-ray wavelengths. For this reason we have pushed towards the development of moiré deflectometry (which measures electron density gradients) and interferometry which is a direct measurement of electron density. For a detailed discussion of our progress in moiré deflectometry see the paper by Ress et al.[6].

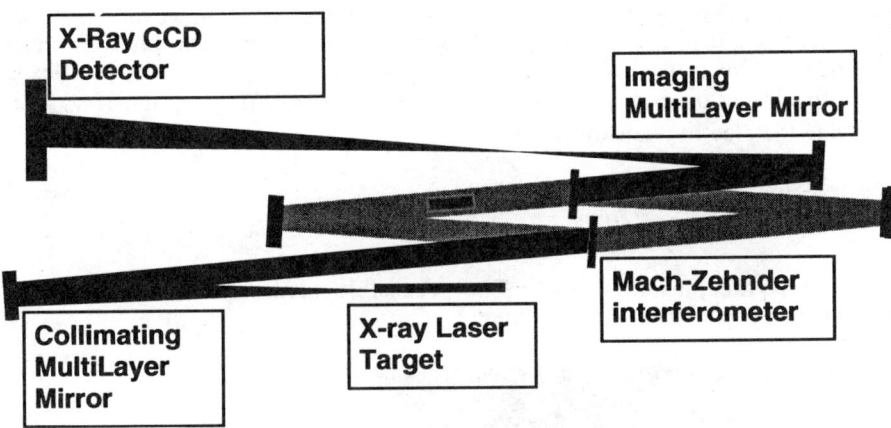

Figure 3. Experimental setup used for xuv interferometry.

The development of xuv interferometry has only recently become possible because of the progress made in the fabrication of high quality beamsplitters. The beamsplitters we utilize have a 1 cm square opening and consist of 1000 Å of silicon Nitride with 8 layer pairs of Mo/Si. These splitters have a reflectivity of 25±5% and transmission of 20±5%. The experimental setup we have used to demonstrate xuv interferometry is shown in figure 3. The x-ray laser beam is collected with a spherical multilayer mirror (50 cm focal length) which collimates the beam and injects the beam into a Mach-Zehnder interferometer. Multilayer mirror damage issues constrained the design of the interferometer to have segment lengths of 50 cm (i.e. total length of each arm is 100 cm). In addition, given the temporal coherence length of ~50 µm for our x-ray laser it was necessary to prealign this interferometer using a white light source(see Cellier et al. in these proceedings). The second plasma is imaged with a spherical mirror onto a backside illuminated CCD detector. The multilayer mirrors consist of 15 layer pairs of Mo/Si and have a measured reflectivity of 60±5 % at normal incidence. In our imaging experiments we have used a variety of imaging mirrors ranging in focal length from 11 cm to 50 cm (f# 20) and magnifications of 3 to 30.

Multilayer mirror damage due to side scattered laser light is a serious problem for mirror to plasma distances less than 25 cm and high optical laser energies (1-2 kJ) on the secondary target. The effective bandpass of the multilayer mirror and filter combination is 8 Å centered at 155 Å. Ideally the bandpass would be comparable to the x-ray laser spectral width of ~7x10^{-3} Å in order to further reduce background emission. The spatial resolution of this imaging system has been measured to be better than 1μm and is limited by the CCD detector pixel size of 24 μm and spherical aberrations. In figure 4 we show an interferogram obtained using this system without a secondary plasma. The results show excellent fringe visibility and prove the viability of this technique. We have observed low fringe visibility on occasions where the beam splitters were under low tension. These splitters were extremely sensitive to vibrations which made optical alignment difficult. The low fringe visibility of these splitters may be due to splitter motion during the x-ray laser pulse. At these wavelengths motions of 20 Å within the 200 ps pulse width correspond to velocities of 10 m/sec and may not be out of the question. By adjusting the manufacturing technique we have, however, been able to produce beamsplitters reliably that avoid this problem. The interferometer developed in this project will now be used to measure electron density profiles in a variety of laser produced plasmas where the number of fringe shifts, N_{Fringe}, is related to the electron density, n_e, by

$$N_{Fringe} = \frac{n_e}{2n_{cr}}\frac{L}{\lambda} = 4.5 \times 10^{-22} n_e L \lambda$$

where L is the plasma length. The accessible parameter space using this technique can be estimated from figure 1.

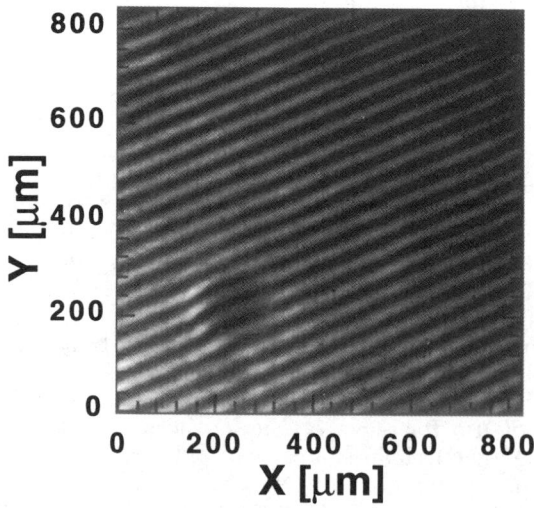

Figure 4. Interferogram obtained using neon-like yttrium x-ray laser λ=155 Å. The pattern at the lower left corner is likely due to a defect on the beam splitter.

Ultimately for two dimensional plasma imaging the spatial resolution is limited by the pulse length of the x-ray laser or alternatively the gate time of the detector.

Gated detectors with large active areas have temporal resolutions at best in the range of 100 to 200 ps. For characteristic expansion velocities of 10^7 cm/s this corresponds to a spatial resolution of 10 - 20 µm which is well below that possible with available imaging optics. In figure 5 we show simulation results showing the electron density and deflection angle at 800 ps and 1000 ps. The results show that significant profile changes occur in the region with electron densities greater than 10^{21} which is the region of interest for xuv probing. Therefore, there exists a real need to produce x-ray lasers with pulse duration shorter than the 200 ps (FWHM) routinely produced with exploding foil targets. In order to produce saturated x-ray lasers with pulse duration shorter than 20 ps (imaging resolutions of ~2 µm) we have begun experiments which irradiate an exploding foil with multiple pulses. The first pulse heats and explodes a thin foil target to produce a plasma with low density gradients. The second and subsequent pulses then ionizes the preformed plasma to produce conditions suitable for high gain. The energy of the first pulse can be significantly lower than the second pulse since high electron temperature are not necessary. This improves the efficiency of this approach by reducing radiation and thermal losses during the hydrodynamic expansion[10]. In addition to using multiple pulse irradiation it was necessary to implement a traveling wave pump to overcome the expected short gain duration. The approach we used for generating a traveling wave pump is described in the article by Moreno et al. in these proceedings. To date using this technique with multiple 100 ps pulses we have generated 45 ps (FWHM) x-ray pulses in neon-like yttrium[11]. Using traveling wave pumping in these experiments led to significant increases in the total energy output and contributes to shortening the output pulse.

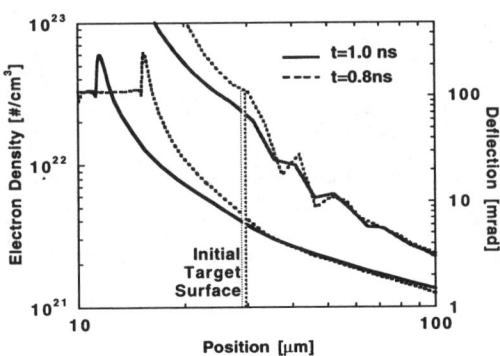

Figure 5 Calculated electron density for a CH target irradiated at 2.0×10^{13} W/cm^2. The calculated deflection assuming a 3 mm wide plasma is also shown. The deflection angle is the quantity measured by moiré deflectometry.

In summary, the high brightness and short wavelength of x-ray lasers make them ideally suited for studying long scalelength and high density plasmas. Using a yttrium x-ray laser we have been able to image high density plasmas with µm resolution and successfully demonstrated xuv interferometry. Using a multi pulse irradiation technique and traveling wave pump we have been able to generate x-ray lasers with pulse widths significantly shorter than conventional exploding foil x-ray lasers. Although our initial results have only achieved ~45 ps output pulses with appropriate pump lasers a double pulse technique should easily achieve the sub 20 ps pulse width desired for high spatial resolution plasma imaging.

ACKNOWLEDGMENTS

We would like to thank Jim Cox, Sharon Alvarez, Joe Smith and the Nova Experiments Group for their assistance in these experiments. This work was performed under the auspices of the U.S. Department of Energy by the Lawrence Livermore National Laboratory under contract No. W-7405-ENG-48.

REFERENCES

1. L. B. Da Silva, J. E. Trebes, R. Balhorn, S. Mrowka, E. Anderson, D. T. Attwood, T. W. Barbee Jr., J. Brase, M. Corzett, J. Gray, J. A. Koch, C. Lee, D. Kern, R. A. London, B. J. MacGowan, D. L. Matthews, and G. Stone, Science **258**, 269-271.

2. A. Carillon, H. Z. Chen, P. Dhez, L. Dwivedi, J. Jacoby, P. Jaegle, G. Jamelot, J. Zhang, M. H. Key, A. Kidd, A. Klisnick, R. Kodama, J. Krishnan, C. L. S. Lewis, D. Neely, P. Norreys, D. O'Neill, G. J. Pert, S. A. Ramsden, J. P. Raucourt, G. J. Tallents, and J. Uhomoibhi, Phys. Rev. Lett. **68**, 2917-2920 (1992).

3. S. Wang, Y. Gu, G. Zhou, Y. Ni, S. Yu, S. Fu, C. Mao, Z. Tao, W. Chen, Z. Lin, D. Fan, G. Zhang, J. Sheng, M. Yang, T. Zhang, Y. Shao, H. Peng, and X. He, Chinese Phys. Lett. **8**, 618 (1991).

4. B. J. MacGowan, L. B. Da Silva, D. J. Fields, C. J. Keane, J. A. Koch, R. A. London, D. L. Matthews, S. Maxon, S. Mrowka, A. L. Osterheld, J. H. Scofield, G. Shimkaveg, J. E. Trebes, and R. S. Walling, Phys. Fluids **4**, 2326-2337 (1992).

5. L. B. Da Silva, B. J. MacGowan, S. Mrowka, J. A. Koch, R. A. London, D. L. Matthews, and J. H. Underwood, Optics Lett. **18**, 1174-1176 (1993).

6. D. Ress, L. B. Da Silva, R. A. London, J. E. Trebes, S. Mrowka, R. J. Procassini, T.W. Barbee Jr., and D. E. Lehr, Science **265**, 514 (1994).

7. G. E. Busch, C. L. Shepard, L. D. Siebert, and J. A. Tarvin, Rev. Sci. Instrum. **56**, 879 (1985).

8. D. T. Attwood, D. W. Sweeney, J. M. Auerbach, and P. H. Y. Lee, Phys. Rev. Lett. **40**, 184 (1978).

9. M. K. Prasad, K. G. Estabrook, J. A. Harte, R. S. Craxton, R. A. Bosch, G. E. Busch, and J. S. Kollin, Phys. Fluids B **4**, 1569 (1992).

10. L. B. Da Silva, B. J. MacGowan, D. L. Matthews, M. D. Rosen, H. A. Baldis, G. D. Enright, B. LaFontaine, and D. M. Villeneuve, in <u>Studies of high Z exploding foils irradiated by combined long (2 ns) and short (10 ps) pulses of 1w light</u>, Proceedings SPIE'S OE/LASE '90 Los Angeles, 1990), vol. 1229, pp. 128. (1992)

11. L. B. Da Silva, R. A. London, B. J. MacGowan, S. Mrowka, D. L. Matthews, and R. S. Craxton, to appear sept. 15 issue of Optics Letters , (1994).

Demonstration of Ultra High Resolution Soft X-Ray Tomography

W. S. Haddad*, I. McNulty**, J. E. Trebes*, E. H. Anderson***, L. Yang**, J. M. Brase*

*University of California, Lawrence Livermore National Laboratory, CA 94551
**Advanced Photon Source, Argonne National Laboratory, Argonne, IL 60439
***Lawrence Berkeley Laboratory, Berkeley, CA 94720

Abstract. Ultra high resolution three dimensional images of a microscopic test object were made with soft x-rays using a scanning transmission x-ray microscope. The test object consisted of two different patterns of gold bars on silicon nitride windows which were separated by ~5μm. Depth resolution comparable to the transverse resolution was achieved by recording nine 2-D images of the object at angles between -50 to +55 degrees with respect to the beam axis. The projections were then combined tomographically to form a 3-D image using an algebraic reconstruction technique (ART) algorithm. We observed a transverse resolution of ~1000 Å. Artifacts in the reconstruction limited the overall depth resolution to ~6000 Å, however some features were clearly reconstructed with a depth resolution of ~1000 Å.

Among the potential applications of x-ray lasers, one of the most exciting is x-ray microscopy which promises both high resolution, and the opportunity to take advantage of natural contrast mechanisms in biological samples. X-ray microscopy has begun to emerge as a viable imaging tool for microbiologists and materials scientists (1-3). Several groups have developed x-ray microscopes of various types using synchrotron sources (4, 5), laser plasma sources (6), and x-ray lasers (7). X-ray lasers are particularly attractive for high resolution biological imaging because of their potential to deliver a large number of coherent photons in a short pulse, circumventing x-ray damage issues. High transverse resolution with near diffraction limited performance has been achieved with scanning transmission x-ray microscopes (STXM) (4, 5). There are however no x-ray optics having sufficiently high numerical aperture (NA) to achieve resolution in depth that is comparable with the transverse resolution. Currently the best x-ray zone plates have a NA < 0.1 for radiation in the water window (8, 9). The ratio of depth resolution to transverse resolution $\partial_l/\partial_t \approx 2/NA > 20$ for present state-of-the-art zone plates.

In order to improve the depth resolution, it is necessary to effectively increase the NA of the imaging system. This can be done by recording several views of the object over a large angular range. If each of the views is taken with low NA optics such that the longitudinal extent of the object is less than the depth resolution of the imaging system, then each view will be a 2-D projection of the object. It is then possible to reconstruct a 3-D image of the object using the principles of tomography with the set of 2-D projections as raw data. We have implemented this approach in conjunction with the STXM on the X1A undulator beamline at the National Synchrotron Light Source in Brookhaven National Laboratory (10). This was done by modifying the STXM to allow the sample to be rotated about an axis perpendicular to the x-ray beam. This modified STXM was then used to make a high resolution 3-D image of a phantom test object.

The phantom consisted of two unique patterns of 650 Å thick gold bars written onto silicon nitride windows which were separated by ~5μm. The two patterns were designed to be simple yet contain a range of features of various sizes including gaps and angles other that 90 degrees. Each pattern was comprised of bar features in the size range between 1000 Å and 5000 Å, and occupied a 2 μm x 2 μm region on the window. Thus the overall dimensions of the phantom were approximately 2 μm x 2 μm x 5 μm. For good 3-D image formation, it was necessary that the object be partially transmissive while providing good contrast. As a result, the thickness of 650 Å was chosen for the gold patterns based on computer simulations.

The STXM forms high transverse resolution 2-D images using coherent x-radiation from the undulator. The x rays are focused to a nearly diffraction limited spot using an x-ray zone plate lens. 2-D images are formed by raster scanning the object through the focus of the zone plate and measuring the transmission. For this experiment the undulator was operated at a wavelength of 36 Å and a spectral width of 1%. Physical limitations of the instrument allowed only a small number of projection images to be taken. In addition, only a limited angular range was available over which to record the projections. A total of nine 2-D images of the object were recorded at angles between -50 to +55 degrees with respect to the beam axis. Since the images were recorded incoherently, diffraction was not a factor and the system was then effectively a first generation tomographic imaging system, i.e.. diffractionless straight-ray tomography (11).

The combination of limited viewing angle and few projections results in a challenging image reconstruction problem. Neither filtered nor unfiltered backprojection methods are satisfactory in this case. Unfiltered backprojection does produce recognizable reconstructions of 2-D slices for simple objects under these conditions, however the reconstructed slices contain so much artifact that a 3-D reconstruction with this method becomes unintelligible. It was therefore necessary to use an alternate method for reconstruction. We used an iterative optimization method known as algebraic reconstruction technique (ART) to form a 3-D image from the nine projections (11, 12). It has been shown that for most limited data tomography problems the algebraic reconstruction technique (ART) yields the best results overall (12).

We observe transverse resolution of ~750 Å in the 2-D projections. Artifacts due to the limited angular range, errors in projection alignment and angle and the ART algorithm limit the depth resolution and produce a loss of transverse resolution in the 3-D reconstruction. The resulting transverse resolution in the 3-D reconstruction is found to be ~1000 Å. The overall depth resolution is ~6000 Å., however some features down to ~1000 Å in depth can clearly be seen. This is approximately an order of magnitude higher resolution than other existing state-of-the-art x-ray tomography systems (13, 14). Our current efforts include improving the methods for projection alignment and measurement of projection angles, and developing better 3-D image reconstruction techniques.

ACKNOWLEDGMENTS

This work was performed under the auspices of the U.S. Department of Energy under contracts W-7405-ENG-48 and W-31-109-ENG-38. The lithography work was supported by the Director, Office of Energy Research, Office of Basic Energy Science, U.S. Department of Energy under contract No. DE-AC03-76SF00098.

REFERENCES

1. A variety of articles in *X-Ray Microscopy*, G. Schmahl and D. Rudolph, Eds. (Springer-Verlag, Berlin, 1984)
2. A variety of articles in *X-Ray Microscopy II*, D. Sayre, M. Howells, J. Kirz, H. M. Rarback, Eds. (Springer-Verlag, Berlin, 1988)
3. A variety of articles in *X-Ray Microscopy III*, A. Michette, G. Morrison, C. Buckley, Eds. (Springer-Verlag, Berlin, 1991)
4. C. Jacobsen et al., *Opt. Commun.* **86**, 351 (1991)
5. W. Meyer-Ilse et al., in *X-Ray Microscopy III*, A. Michette, G. Morrison, C. Buckley, Eds. (Springer-Verlag, Berlin, 1991), pp. 284-289.
6. T. Tomie et al., *Science* **252**, (1991) 691
7. L. B. Da Silva et al., *Science* **258**, (1992) 269
8. J. Thieme et al., in *Proc. 1993 International Conf. on X-Ray Microscopy*, (Chernogolovka, Russia, 1993), in press.
9. C. David et al., *ibid.*
10. H. Rarback et al., *J. X-ray Sci. Tech.* **2**, (1990) 274
11. A. C. Kak and M. Slaney, *Principles of Computerized Tomographic Imaging*, IEEE Press: New York (1988) 275
12. D. Verhoeven, *Appl. Opt.* **32**, (1993) 3736
13. J. H. Kinney et al., *Science* **260**, (1993) 789
14. B. P. Flannery et al., *Science* **237**, (1987) 1439

Micron-scale Resolution Radiography of Laser-accelerated and Laser-exploded Foils Using an Yttrium X-ray Laser

R. Cauble, L. B. Da Silva, T. W. Barbee Jr., P. Celliers,
J. C. Moreno, S. Mrowka, T. S. Perry and A. S. Wan

Lawrence Livermore National Laboratory
University of California
Livermore, CA 94550 USA

Abstract. We have imaged laser-accelerated foils and exploding foils on the few-micron scale using an yttrium x-ray laser (155 Å, 80 eV, ~ 200 ps duration) and a multilayer mirror imaging system. At the maximum magnification of 30, resolution was of order one micron. The images were side-on radiographs of the foils. Accelerated foils showed significant filamentation on the rear-side (away from the driving laser) of the foil, although the laser beam was smoothed. In addition to the narrow rear-side filamentation, some shots revealed larger-scale plume-like structures on the front (driven) side of the Al foil. These plumes seem to be little-affected by beam smoothing and are likely a consequence of Rayleigh-Taylor instability. The experiments were carried out at the Nova two-beam facility.

INTRODUCTION

Production of laboratory x-ray lasers (XRL) is robust enough that the special features of the XRL can now be confidently exploited. One such feature is high brightness, which makes XRLs especially well-suited for imaging other bright sources such as laser plasmas, where the XRL can actually dominate much of the plasma self-emission. We have utilized an imaging system (1), shown in Fig. 1, which operates in the soft x-ray spectral region to obtain high resolution images of laser accelerated and x-ray heated aluminum foils.

The x-ray laser is produced from an exploding yttrium target by one 600 ps long, 0.53 µm wavelength, 2 kilojoule beam of the Nova laser; the other Nova beam is used to create the plasma target. The XRL (2) has a wavelength of 155 Å, a duration of about 200 ps and an energy of about 8 mJ. The collimated XRL in the imaging system has a diameter of approximately 5 mm, which is sufficient to image

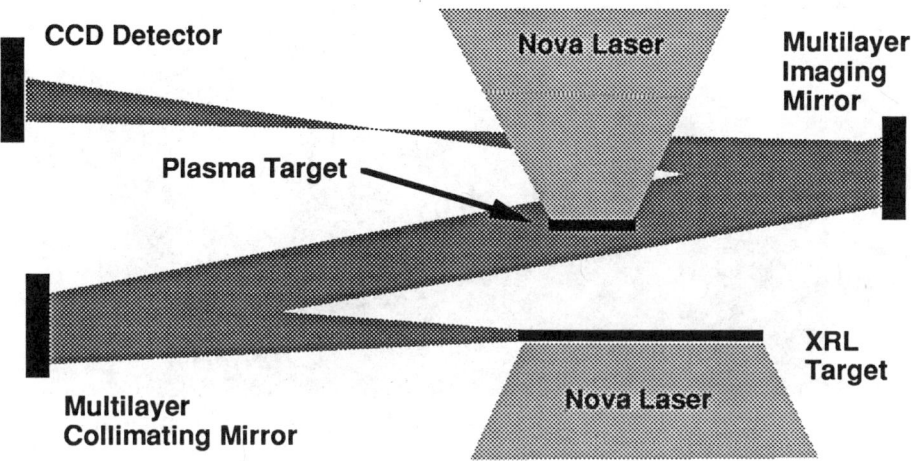

FIGURE 1. Experimental setup for plasma imaging using an x-ray laser on the Nova two-beam facility.

an entire plasma target. The bandpass of the entire system is of about 8 Å. Images were recorded on a 1024x1024 pixel CCD detector.

HIGH RESOLUTION PLASMA IMAGES

Figure 2 shows an image of a laser accelerated foil backlit by the yttrium (Y) XRL halfway through a high energy 1 ns drive. The foil, which was 10 μm of CH on 3 μm Al, was irradiated on the CH side. Horizontal spatial resolution is better than 2 μm. Vertical resolution is limited by the XRL duration to ~ 10^7 cm/s x 200 ps = 20 μm. Although the Nova drive beam was smoothed with steering wedges and a random phase plate, small-scale, 5-6 μm, structure is evident on the rear (Al) side. Larger-scale structure can also be observed. Some of the structure may be an artifact due to refraction of the XRL beam along extreme density gradients in the foil plasma, thus the actual extent of the plasma cannot be quantified from the image. However, the XRL beam is spatially quite smooth so the nonuniformities observed in the image are indeed produced by density variations within the plasma.

Further experiments with these foils have revealed structure on the front (CH) side of an accelerated foil. As before, the plasma is backlit with the Y XRL, but because of higher frontside temperatures, self-emission is very strong. In addition to steering wedges and phase plate, the drive beam was also smoothed by adding 17 Å of frequency dispersion to the beam. In spite of this additional beam smoothing, the front side of the foil shows tens-of-micron size plumes coming off the foil. These are likely a consequence of Rayleigh-Taylor instability.

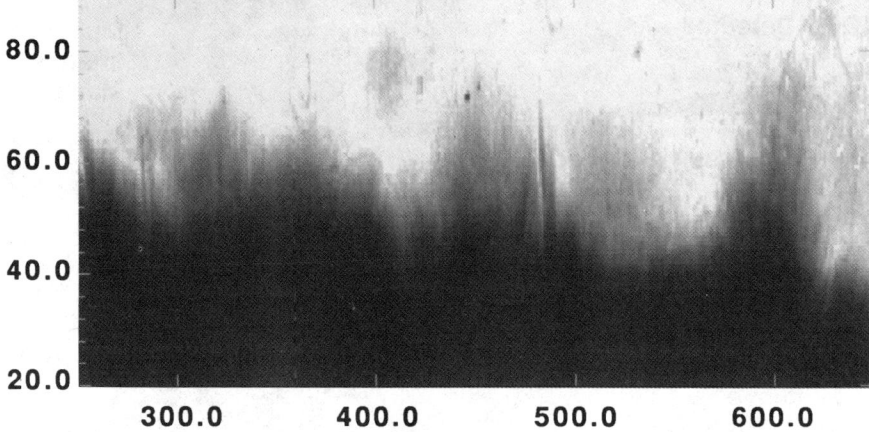

FIGURE 2. X-ray laser backlit image of a foil (10 μm CH on 3 μm Al) accelerated by a 1 ns Nova pulse taken at 0.5 ns. The picture shows an enlarged view of the central region of the foil plasma, which is being driven upward in the picture. The scale is in microns; the 700 μm diameter foil was originally located at zero on the vertical scale. The Nova beam was smoothed, but there are the spatial nonuniformities in the plasma.

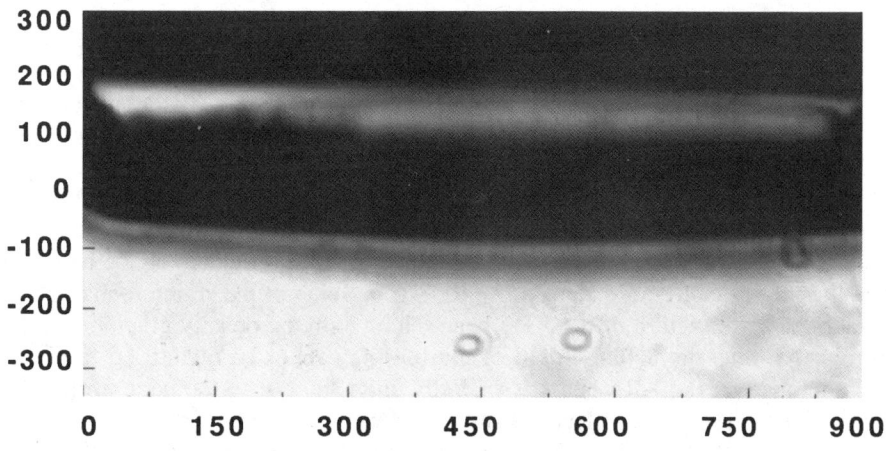

FIGURE 3. X-ray laser backlit image of an exploding foil 1 ns after the end of a 1 ns heating pulse. Zero marks the original foil position. The drive is incident through the support above in the picture. The scale is in microns.

Figure 3 displays an XRL backlit image of an exploding foil. The foil in this case is 3 μm of aluminum. The heating mechanism was indirect absorption of gold M-band radiation (about 2 keV) which was produced by illuminating a 300 μm spot with the remaining Nova beam on one side (away from the exploding foil) of a 1500 Å gold burnthrough disk located 1 mm from the Al foil. X-rays < 2 keV passing through the Au foil were removed by a 50 μm plastic filter. The higher energy x-rays were absorbed volumetrically in the aluminum and the aluminum expanded nearly symmetrically. The Nova drive pulse was 1 ns; the XRL backlight occurred 1 ns after the end of the drive pulse. The image shows the foil expanded to about 200 μm, almost filling the 200 μm gap between the foil and a plastic support above. A single "fringe" can be seen below the plasma. This is believed to be caused by XRL rays, which have been refracted by 1-3 μm in the expanding plasma, re-positioned by the multilayer imaging mirror to a new spot in the image plane. Note that there is sufficient coherence in the XRL to produce circular diffraction patterns in the image plane around small imperfections (dust or oil droplets) on the collimating mirror.

CONCLUSIONS

The high brightness of x-ray lasers makes them ideal for studying plasmas. Some examples of high spatial resolution radiography using an XRL have been discussed here. All are 200 ps snapshots of fast-evolving plasmas. They reveal unexpected structure in the plasma, which may be due to drive beam nonuniformity or initial, undetected imperfections in the targets. More quantitative measurements can be made by combining the XRL's brightness with its property of coherence in deflectometry or interferometry. (1)

ACKNOWLEDGMENTS

We would like to acknowledge highly useful discussions with K. Nugent (University of Melbourne) and J. Trebes (LLNL). This work was performed under the auspices of the U. S. Department of Energy by Lawrence Livermore National Laboratory under contract number W-7405-ENG-48.

REFERENCES

1. L. Da Silva, R. Cauble, G. Frieders, J. Koch, B. MacGowan, D. Matthews, S. Mrowka, D. Ress, J. Trebes, and T. Weiland, in *Ultrashort Wavelength Lasers II*, SPIE Proceedings Vol. **2102**, 158 (1993).
2. L. Da Silva, B. MacGowan, S. Mrowka, J. Koch, R. London, D. Matthews, and J. Underwood, Optics Lett. **18**, 1174 (1993).

Some Potential Applications of X-ray Lasers in Atomic Physics

Bernd Crasemann

Physics Department, University of Oregon, Eugene, Oregon 97403

X-ray lasers can be expected to complement and extend the research potential of synchrotron radiation in atomic and molecular physics. The laser light's extreme brightness, ultra-narrow bandwidth, and coherence will make it possible to conduct very precise measurements of photoelectron angular distributions that reveal departures from electric-dipole patterns, thus pointing to limitations of the central-field approximation, probing screening effects, and testing core relaxation and the transfer of angular momentum to the core during emergence of the photoelectron. Multielectron photoexcitation can be studied, with the possibility of delicate tests of relativistic corrections to the Coulomb interaction. Raman processes, recently demonstrated in atomic inner shells by means of synchrotron radiation, could be studied with x-ray lasers in the much narrower outer levels, provided a slight degree of tunability can be attained. Scattering experiments may lead to accurate measurements of the elastic component due to scattering off virtual electron-positron pairs produced in the field of the nucleus (Delbrück scattering); the monochromaticity of laser radiation may permit experiments that elucidate the fundamental association of elastic and inelastic scattering which arises from the radiative corrections of quantum electrodynamics.

I. INTRODUCTION

History shows that new sources or detectors have invariably led to new physical insights. It is therefore particularly challenging to try to look ahead and anticipate what the advent of x-ray lasers may entail for the understanding of atomic structure and dynamics. Important frontiers of present knowledge in this area include many-body effects that transcend the independent-particle model, and relativistic and quantum electrodynamic effects. Photoexcitation and ionization is an important tool for the investigation of the relevant phenomena, primarily because it is readily tractable in theory. Thus, lasers that produce reasonably hard x radiation (say, of wavelengths down to ~ 1 Å) would complement and extend in important ways the capabilities of synchrotron radiation that have set the field ahead significantly in recent years. Furthermore, important aspects of photon-atom interactions including scattering phenomena would become amenable to exploration with greater precision than can be attained at present. In outlining some of these possibilities, we draw partly upon our previous discussions of the subject (1,2).

II. PHOTOIONIZATION

Basic knowledge of atomic structure and dynamics rests primarily upon experiments in which energy is transferred from photons or charged particles to atoms and photons or electrons are emitted. Excitation by photons rests upon firmer theoretical foundations (3,4) than excitation by particle collisions: it is a relatively weak process; in first approximation, a single photon interacts with a single electron, and perturbation theory is appropriate (5). Near thresholds, however, this apparent simplicity evaporates, as noted below: here the emergence of a slow photoelectron from a many-electron atom tends to be accompanied by complex correlational processes of real and virtual excitations.

It is helpful to distinguish ionization by low-energy photons from high-energy photoionization (2). At low energies, atomic structure is probed at large distances from the nucleus where the potential approaches the r^{-1} form of the residual ion's and the wave functions reflect the interior of the atom through quantum defect; electron-electron correlations are strong, and nonrelativistic matrix elements in the dipole approximation describe the process well. High-energy photons, on the other hand, interact close to the nucleus where the potential approaches the point nuclear Coulomb form and the electron wave functions reflect exterior atomic structure through their normalization and phase. Here the matrix elements display relativistic effects and include higher angular-momentum components. As the linear momentum transfer from photon to atom projected onto the radius vector, $(\mathbf{k} - \mathbf{p}) \cdot \mathbf{r}$, becomes much larger than unity, contributions to the transition matrix element vanish. The innermost regions of the atom saturate in the determination of the total photoionization cross section at a distance of approximately one electron Compton wavelength from the nucleus (2). It is in the high-energy regime that x-ray lasers can contribute significantly to the possibilities of experimental exploration.

A. The "Complete" Experiment

In the low-energy limit, the nonretarded dipole approximation to the electromagnetic field of the photon is adequate:

$$\varepsilon e^{i\mathbf{k}\cdot\mathbf{r}} \approx \varepsilon. \tag{1}$$

We consider the atoms to be randomly oriented. Dipole selection rules imply that photoionization cannot lead to more than 3 continuum states. The process is therefore described by 3 matrix elements. Their angular parts are determined by spherical harmonics; the radial parts contain the physics:

$$R_k \cong |R_k|e^{i\delta_k}, \ k = +, 0, -. \tag{2}$$

One of the phases is arbitrary, hence two phases and three moduli or 5 independent parameters characterize the process completely. To extract this information, it is necessary to measure *5 observables* that contain these quantities in linearly independent combinations. The parameters can, e.g., be extracted from the set of

- the absolute partial cross section σ,
- the angular asymmetry parameter β,
- the 3 spin-polarization parameters ξ, η, ζ of the photoelectron.

Experiments in which such sets of observables are measured have been dubbed "complete" or "perfect" (6,7). Few such measurements have been accomplished (8,9); the potential of x-ray lasers for this purpose is evident.

B. Non-Dipolar Photoionization

The wavelength of visible or ultraviolet light is several orders of magnitude larger than the extent of an atom and the dipole approximation [Eq. (1)] is valid, which implies neglect of retardation ($k \to 0$, $\lambda \to \infty$) and of higher multipoles. For higher photon energies, it is necessary to consider the full retarded multipole expansion of the photon electromagnetic field (10–12). In addition to the 5 parameters of the "perfect" electric-dipole experiment discussed above, a set of quadrupole and other parameters comes into play; such experiments have not yet been performed but may come into reach with lasers.

Most sensitive to higher multipoles is the photoelectron angular distribution, while total cross sections are relatively insensitive to these effects. For unpolarized photons, if the electron spin is not observed, the differential photoionization cross section is

$$\frac{d\sigma}{d\Omega} = \frac{\sigma}{4\pi} \sum_n B_n P_n(\cos \hat{\mathbf{p}} \cdot \hat{\mathbf{k}}). \tag{3}$$

For electric-dipole ($E1$) transitions, only the coefficients $B_0 = 1$ and B_2 of the respective Legendre polynomials P_n do not vanish and the asymmetry parameter is $\beta = -2B_2$. For electric-quadrupole ($E2$) transitions, also B_1 and B_3 are non-zero, breaking the symmetry.

Bechler and Pratt (10) calculated the (nonrelativistic) "first retardation correction" of order kr for $\lim_{k\to 0} E2$. They found the useful result that for small screening the correction can approximately be expressed by a parameter κ in the angular distribution

$$\frac{d\sigma}{d\Omega} \approx A \sin^2 \theta (1 + \kappa \cos \theta). \tag{4}$$

For the $1s$ state, κ is given by

$$\kappa = 2\frac{n_2}{n_1}\left[\left(\frac{Z\alpha}{p}\right)^2 + 4\right]^{\frac{1}{2}} \frac{v}{c}\cos(\delta_2 - \delta_1), \tag{5}$$

where n_ℓ is the ratio of screened to point-Coulomb normalization of the ℓth partial wave which has the phase shift δ_ℓ. Significant deviations from $E1$ distributions are predicted, in terms of κ, even for relatively low atomic numbers and photon energies (10).

In a calculation of leading terms in the full multipole expansion, Cooper (12) points out that magnetic-dipole transitions can make no contribution in the frozen-core model where final and initial wave functions are orthogonal, but can contribute if there is core relaxation; the $E1 \cdot E2$ interference term is shown to give a clear indication of angular-momentum transfer to core electrons.

The approximate results of the foregoing nonrelativistic calculations and Scofield's numerical relativistic calculation (11) of angular and polarization correlations in photoionization illustrate how it will become possible to derive significant atomic-structure information once suitable experiments become feasible.

C. Multiple Photoexcitation

1. Double inner-shell excitation

Several atomic electrons can be promoted upon absorption of one photon due to the combined influences of the photon-one-electron operator and the Coulombic electron-electron interaction. The mechanisms which contribute to the process can be separated within many-body perturbation theory (13). In first order of the combined perturbations by the photon field and electron correlations, the leading contributions come from core rearrangement and initial-state corelations. Core rearrangement effects include *shake processes* (14), in which one electron is excited or emitted through an $E1$ transition while another electron undergoes an electric monople ($E0$) transition to an excited state or into the continuum. (In "conjugate shake," the roles of $E1$ and $E0$ are reversed).

We have recently calculated double inner-shell photoexcitation and ionization cross sections by separately optimizing single-configuration initial and final states and computing the photon-operator matrix elements between these many-electron states (15,16). Results show that for more tightly bound electrons, shakeoff tends to prevail increasingly over shakeup, but that the double photoionization probability exhibits a gradual onset (17) and is exceedingly small. It is for this reason that double inner-shell photoexcitation is very difficult to detect and many reports of multielectron features in absorption spectra have turned out to be erroneous. On the other hand, discrepancies between measured x-ray absorption near thresholds in noble-gas atoms and calculations (16) show that important multielectron processes beyond double excitation play a significant role in these regions. A study of multielectron processes with x-ray lasers could elucidate the complex structure of absorption edges, with considerable relevance to applications in materials science (18). Furthermore, multiple atomic inner-shell vacancies offer some unique possibilities for the study of very fundamental phenomena.

2. Deep double-hole systems

Since it is the Coulomb force that holds atoms, molecules, and solids together, it clearly is of interest to look at manifestations of the lowest-order relativistic correction to the static Coulomb potential, viz., the Breit energy which accounts for the exchange of a single transverse photon and contains the effect of both the current-current interaction and of retardation:

$$H_{Breit}(\omega) = -\frac{1}{r_{ij}}[\alpha_i \cdot \alpha_j \cos\omega r_{ij} + (1 - \cos\omega r_{ij})], \qquad (6)$$

where the α_n are Dirac matrices and r_{ij} is the distance between the two interacting charges; ω is the energy of the virtual photon. In the Pauli limit, the Breit operator corresponds to the orbit-orbit, spin-spin, and spin-other-orbit interactions between two electrons. Ordinarily, the effects of the Breit interaction in atomic structure are rather subtle, as in $\leq 1\%$ shifts of the $1s$ level energies, but they can become pronounced in cases where the static Coulomb interaction cancels out, as in j-splittings of double-hole states (19,20).

As a single example, we mention one effect that arises exclusively from the relativistic Breit-Coulomb Hamiltonian, viz., the splitting of excitations that comprise two vacancies in s states into distinct ^1S and ^3S levels. This splitting contains a dominant contribution from the spin dependence of the relativistic Coulomb interaction and a smaller contribution from the Breit operator. Specifically, we find (16) for the Kr $[1s2s]$ double-vacancy state, $\Delta E(^1S_0 - {}^3S_1)_{Coulomb} = 40.12$ eV and $\Delta E(^1S_0 - {}^3S_1)_{Coulomb+Breit} = 43.00$ eV, i.e., the Breit interaction enhances the splitting by 6.7% in this case. Other splittings are greatly modified by the Breit-Coulomb Hamiltonian and in some cases even their energy ordering is reversed (19). An unusual opportunity for interesting experiments exists here.

III. THRESHOLD RAMAN PROCESSES

The study of many-electron threshold phenomena in atomic deep inner shells, including Raman processes, post-collision interaction, and shake-modified autoionization, was first made possible by synchrotron radiation.

These experiments became feasible because the bandwidth of the probing radiation is narrower than the lifetime width of the initial inner-shell hole state. The yet much higher monochromaticity of short-wavelength lasers, as well as their brightness, lead to the expectation that studies of dynamical processes involving virtual states can be extended outwards from the deepest core levels to longer-lived hole states.

In threshold excitation of atomic inner-shell hole states, ionization and decay cannot be treated as distinct processes. Near the threshold, photoionization and radiative or radiationless deexcitation occur in a single quantum process, the *resonant Raman effect*. Here, the intermediate states are virtual and there is no relaxation phase. The width of the emitted x-ray or Auger line reflects that of the incident radiation and hence can be much narrower than the natural lifetime width of the initial hole state (21,22). As one tunes the energy of the incident photons through a core level and observes a characteristic resonant x-ray or Auger line, the energy of the line displays linear dispersion with incident x-ray energy; the intensity of the line traces out the Lorentzian shape of the core-hole state (17). It has been shown that time-independent resonant scattering theory permits a unified treatment of this process; this subject has recently been reviewed (23).

The applicability of Auger resonant Raman scattering to studies in materials science is foreshadowed by an experiment in which the evolution of K-$L_{2,3}L_{2,3}$ $^1D_2$ radiationless resonant Raman scattering into Auger-electron emission was studied by tuning synchrotron radiation across the K edge of P in InP (24). The spectrum could be interpreted in terms of a two-component model that involves excitation of a photoelectron either into the continuum or to a previously unknown discrete exciton-like state. This finding may prove to be of some relevance in Auger-electron spectrometry of semiconductors and point toward new possibilities of using synchrotron radiation and, once available, x-ray lasers in this type of research.

IV. PHOTON-ATOM SCATTERING

While it is convenient to divide photon-atom interactions into elastic and inelastic processes, this separation is somewhat arbitrary (25): the radiative corrections of quantum electrodynamics and the possibility of emitting very soft photons, as well as target recoil, make all processes in fact inelastic, while experimental comparison of incident- and scattered-photon energies is limited by source bandwidth and detector resolution. It is here that x-ray lasers could serve superbly to advance experimental information (2).

Compton scattering or photon scattering from free electrons at rest is expressed, within lowest-order quantum electrodynamics, by the Klein-Nishina formula. The term is also used for inelastic photon scattering from bound electrons, which approaches the free-electron case when the photon energy greatly exceeds the electron binding energy (25). Compton scattering from bound electrons has extensive applications in the study of electron momentum distributions in atoms and solids (26). An exact second-order S-matrix code for the relativistic numerical calculation of cross sections for Compton scattering of photons by bound electrons has recently been developed within the independent-particle approximation (27). Detailed experimental tests present a challenge.

The elastic photon-atom scattering amplitude, in an approximation that has been called "neither unique nor exact" (28), is the sum of Rayleigh, nuclear Thomson, Delbrück, and nuclear resonance scattering, which are coherent processes (25). Nuclear Thomson scattering is readily calculated. The Rayleigh amplitude for elastic photon scattering from bound atomic electrons, on the other hand, has commonly been estimated only in a simple form-factor approximation. Substantial progress in the understanding of elastic photon-atom scattering was made with the perfection of numerical calculations of Rayleigh amplitudes based on the second-order S matrix of quantum electrodynamics, with relativistic wave functions (28). At energies below the threshold for nuclear resonance scattering it has therefore become possible in principle to measure the Delbrück amplitude, which pertains to photon scattering by virtual electron-positron pairs created in the nuclear Coulomb potential (29). This process belongs to the nonlinear effects of quantum electrodynamics that have no classical analog; although it has been known since 1933, only few experiments have been performed with now superseded techniques. The role of so-called Coulomb corrections in the theory is as yet uncertain (25,30). X-ray lasers would offer a splendid opportunity to study the angular distribution of elastically scattered photons with high precision.

V. CONCLUSION

The few examples presented here, in conjunction with Denise Caldwell's complementary discussion (31), may provide a glimpse of the exciting possibilities that short-wavelength x-ray lasers offer in research on atomic and molecular processes. As suggested by Alexander Vinogradov (32), however, the interaction between atomic physicists and scientists engaged in the invention and development of x-ray lasers is a two-way street; indeed, many of the latter, such as the Chairman of this Conference, arose from the atomic physics community. Continued close interaction between these fields will clearly be of mutual benefit.

The author is indebted to Richard H. Pratt for his collaboration on Ref. (2) and to Teijo Åberg for helpful discussions. This work was supported in part by the National Science Foundation through Grant PHY-9203779.

REFERENCES

1. B. Crasemann, in *Proceedings of the Workshop on Applications of X-ray Lasers,*

San Francisco, California, January 12-14, 1992, edited by R.London and D. Matthews (Lawrence Livermore National Laboratory Report No. CONF-9206170, 1992), p. 67.
2. R. H. Pratt and B. Crasemann, in *Workshop on Scientific Applications of Short Wavelength Coherent Light Sources*, edited by W. Spicer, J. Arthur, and H. Winick (Stanford Linear Accelerator Center Report No. SLAC-414, CONF-92102278, 1992), p. 17.
3. A. F. Starace, in *Corpuscles and Radiation in Matter I*, edited by S. Flügge and W. Mehlhorn, Handbuch der Physik Vol. XXI (Springer, Berlin, 1982), p. 1.
4. M. Ya. Amusia, *Atomic Photoeffect* (Plenum, New York, 1990).
5. J. W. Cooper, in *Atomic Inner-Shell Processes*, edited by B. Crasemann (Academic, New York, 1975), Vol. I, p. 159.
6. J. Kessler, *Polarized Electrons* (Springer, Berlin, 1985), 2nd edition, Sec. 5.2.4.
7. V. Schmidt, Rep. Prog. Phys. **55**, 1483 (1992), Sec. 3.2.1.
8. U. Heinzmann, J. Phys. B **13**, 4353 (1980).
9. F. Schäfers *et al.*, Phys. Rev. A **42**, 2603 (1990).
10. A. Bechler and R. H. Pratt, Phys. Rev. A **39**, 1774 (1989); **42**, 6400 (1990).
11. J. H. Scofield, Phys. Rev. A **40**, 3054 (1989).
12. J. W. Cooper, Phys. Rev. A **42**, 6942 (1990); **45**, 3362 (1992).
13. T. N. Chang, T. Ishihara, and R. T. Poe, Phys. Rev. Let. **27**, 838 (1971).
14. T. Åberg, Phys. Rev. **156**, 35 (1967); Ann Acad. Sci. Fenn. A VI, **308** (1969).
15. D. L. Wark *et al.*, Phys. Rev. Lett. **67**, 2291 (1991).
16. S. J. Schaphorst *et al.*, Phys. Rev. A **47**, 1953 (1993).
17. G. B. Armen *et al.*, Phys. Rev. Lett. **54**, 182 (1985).
18. P. Lagarde, in *X-Ray and Inner-Shell Processes*, edited by T. A. Carlson, M. O. Krause, and S. T. Manson, AIP Conf. Proc. No. 215 (AIP, New York, 199), p. 731.
19. M. H. Chen, B. Crasemann, and H. Mark, Phys. Rev. A **25**, 391 (1982).
20. B. Crasemann, M. H. Chen, and H. Mark, J. Opt. Soc. Am B **1**, 224 (1984).
21. G. S. Brown *et al.*, Phys. Rev. Lett. **45**, 1937 (1980).
22. A. Kivimäki *et al.*, Phys. Rev. Lett. **71**, 4307 (1993).
23. T. Åberg and B. Crasemann, in *X-Ray Anomalous (Resonant) Scattering: Theory and Experiment*, edited by K. Fischer, G. Materlik, and C. Sparks (Elsevier/North Holland, Amsterdam, 1994).
24. H. Wang *et al.*, Phys. Rev. A (in press).
25. L. Kissel and R. H. Pratt, in *Atomic Inner-Shell Physics*, edited by B. Crasemann (Plenum, New York, 1985), Chap. 11.
26. *Compton Scattering*, edited by B. Williams (McGraw-Hill, New York, 1977).
27. T. Surić *et al.*, Phys. Rev. Lett. **67**, 189 (1991).
28. L. Kissel, R. H. Pratt, and S. C. Roy, Phys. Rev. A **22**, 1970(1980).
29. P. P. Kane *et al.*, Phys. Rep. **140**, No. 2, 75 (1986).
30. P. Papatzakos and K. Mork, Phys. Rep. **21**, 81 (1975).
31. C. D. Caldwell, this volume.
32. A. V. Vinogradov, remark following this lecture.

Collisional Redistribution Effects On X-ray Laser Saturation Behavior

J.A. Koch, B.J. MacGowan, L.B. Da Silva, D.L. Matthews, J.H. Underwood[*], P.J. Batson[*], R.W. Lee, R.A. London and S. Mrowka

Lawrence Livermore National Laboratory, P.O. Box 808, L-059 Livermore, California U.S.A. 94550
[*]*Lawrence Berkeley Laboratory, 1 Cyclotron Rd., Berkeley, California U.S.A. 94720*

Abstract. We recently published a detailed summary of our experimental and theoretical research on Ne-like Se x-ray laser line widths, and one of our conclusions was that collisional redistribution rates are likely to have an effect on the saturation behavior of the 206.4 Å Se x-ray laser. In this paper we focus on the effects of collisional redistribution on x-ray laser gain coefficients, and discuss ways of including these effects in existing laser line-transfer models.

Over the past several years, we have obtained high-resolution spectra of several Ne-like x-ray laser transitions, including the (3/2,3/2)J=2 - (3/2,1/2)J=1 lasers in Y[1-2] (155.0 Å) and Se[3-9] (206.4 Å) and the (1/2,1/2)J=0 - (1/2,1/2)J=1 lasers in Se[9] (182.4 Å), Nb[10] (145.9 Å) and Zr[10] (150.4 Å), using a unique high-resolution soft x-ray spectrometer[11,9]. These experiments showed gain-narrowing of the 206.4 Å and 182.4 Å Se lasers with increasing amplifier length, showed no saturation rebroadening of the 206.4 Å Se laser in long, saturated amplifiers, and demonstrated hyperfine line splitting in the soft x-ray regime. It is now possible to measure soft x-ray line profiles with unprecedented resolution, and we have used this technique in other experiments to observe time-dependent Stark broadening in an expanding, rarefying plasma[12] and to measure relative wavelength separations with high precision[2].

We recently published a detailed summary of our experimental and theoretical research on Ne-like Se x-ray laser line widths and their variation with amplifier length[9]. In that research, we measured the line width of the 206.4 Å Se laser from amplifiers of varying lengths, and compared the resulting data to laser line profile and line transfer calculations based on the expected temperature and density of the Se plasma. In particular, we found that the observed lack of inhomogeneous saturation rebroadening of this laser in long, saturated amplifiers could be attributed to: 1) a rapid decrease in saturation rebroadening induced by even relatively small homogeneous lifetime broadening contributions to the intrinsic line profile, and 2) collisional redistribution effects which would be expected to effectively homogenize the Doppler component of the intrinsic line profile. We refer the reader to that paper[9] for the details of the experiments, the calculations and the conclusions, which cannot be reproduced in this short proceedings paper.

Instead, in this paper we discuss in more detail the effects of collisional redistribution on laser gain coefficients, and discuss ways of including these effects in existing amplified spontaneous emission (A.S.E.) laser line transfer models. Again, we refer the reader elsewhere[8,9] for details which we must omit from the present paper due to space constraints.

We begin with a simple three-level collisional excitation model for Ne-like x-ray lasers, where inversion is attained and maintained through collisional excitation of the upper laser level from the ground state, and all other levels are ignored. We will initially ignore collisional redistribution effects, and will derive an expression for the frequency-dependent gain coefficient in their absence.

The velocity-dependent rate equations for a three-level laser can be written as[13,14]

$$\frac{dN_1(v,t)}{dt} = -N_1(v,t)(\Gamma_{12}+\Gamma_{13}) + N_2(v,t)(\Gamma_{21}+A_{21}) + N_3(v,t)\Gamma_{31} \quad (1)$$

$$\frac{dN_2(v,t)}{dt} = N_1(v,t)\Gamma_{12} - N_2(v,t)(\Gamma_{21}+\Gamma_{23}+A_{21}+R_{23}(v))$$
$$+N_3(v,t)(\Gamma_{32}+A_{32}+R_{32}(v)) \quad (2)$$

$$\frac{dN_3(v,t)}{dt} = N_1(v,t)\Gamma_{13} + N_2(v,t)(\Gamma_{23}+R_{23}(v))$$
$$-N_3(v,t)(\Gamma_{31}+\Gamma_{32}+A_{32}+R_{32}(v)) \quad (3)$$

The populations $N_{1,2,3}(v)$ are the population densities of atoms in the ground state, the lower laser level and the upper laser level, respectively, per unit velocity with axial velocity v, and R_{32} and R_{23} are the stimulated emission and absorption rates. The other terms in the equations are the spontaneous emission and electron collisional excitation/de-excitation rates between the levels (A's and Γ's above; the spontaneous 3 - 1 transition is dipole-forbidden in the Se laser). Rates to and from other levels are neglected in order to maintain a closed system, with $N_1+N_2+N_3 = N$ = constant.

The above equations can be solved analytically in steady-state, and the resulting frequency-dependent laser gain coefficient can be written as[9]

$$g(\nu) = \int \sigma_{32}(\nu,v)(N_3(v) - \frac{g_3}{g_2}N_2(v))dv \quad (4)$$

$$= \frac{g_0}{V(\nu_0)} \int_0^\infty \frac{S(u)\phi(\nu,u)du}{1+\frac{1}{I_{sat}}\int_0^\infty I(\nu')\phi(\nu',u)d\nu'} \quad (5)$$

where S(u) is the area-normalized Doppler line profile in inverse-frequency units, $\phi(\nu,u)$ is the area-normalized homogeneous line profile with Doppler-shifted central frequency u, g_0 is the small-signal gain coefficient at line center, I_{sat} is the saturation spectral intensity and $V(\nu_0)$ is the intrinsic line profile function (a Voigt function, the convolution of $\phi(\nu,u)$ and S(u)) evaluated at line center. I(ν) is the laser spectral intensity in W cm^{-2} Hz^{-1}.

The above rate equations are inhomogeneous in the sense that there are no rates transferring population between velocity groups, so that a single velocity is associated with each radiator. This approximation is reasonable when the velocity-changing collision rate (the collisional redistribution rate) is large compared to the inverse of the effective lifetime of the radiator (i.e. the homogeneous line width in frequency units), so that Doppler broadening can be considered to be an inhomogeneous broadening mechanism. However, when this is not the case, essentially different saturation behavior can be expected in A.S.E. lasers, as has been pointed out before[6-9] and was independently noted in a recent paper[15]. The velocity-changing collision rate is the same rate relevant to Dicke line narrowing[16,17], which occurs essentially when this rate Γ is larger than the Doppler width in frequency units. The condition for collisional redistribution effects is less stringent, however, since in the present case the Doppler width of the Se laser (36 mÅ) is several times larger than the homogeneous lifetime-broadening width (14 mÅ), and recent molecular dynamics simulations[18] furthermore suggest that the effective collisional redistribution rate is larger than the lifetime-broadening width but smaller than the Doppler width, so that collisional redistribution effects likely are present but Dicke narrowing effects likely are not.

The essential modification to the above rate equations in order to include collisional redistribution effects is the addition of two terms to each of eqs. (1), (2) and (3) which move population into and out of a particular velocity group[19,20],

$$\frac{dN(v,t)}{dt} \rightarrow \frac{dN(v,t)}{dt} + \int N(v',t)P(v' \rightarrow v)dv' - N(v,t)\int P(v \rightarrow v')dv' \quad (6)$$

where $P(v \rightarrow v')$ is the rate (per unit velocity) at which collisions change a radiator velocity from v to v'. The detailed form of these terms depends on the details of the collisions, and in general gas kinetic-type equations (e.g. Fokker-Plank) should be used[21], but the strong collision limit[19,22] provides a fairly simple means of illustrating the effect. In this limit, the velocity of a radiator after a collision is independent of its velocity prior to a collision, and the above terms are simply related to the collisional re-distribution rate Γ and the velocity distribution function $f(v)$,

$$P(v \rightarrow v') = \Gamma f(v')$$
$$P(v' \rightarrow v) = \Gamma f(v) \quad (7)$$

so that for example eq. (3) becomes, in steady-state,

$$0 = N_1(v)\Gamma_{13} + N_2(v)(\Gamma_{23} + R_{23}(v)) - N_3(v)(\Gamma_{31} + \Gamma_{32} + A_{32} + R_{32}(v) + \Gamma)$$
$$+ \Gamma f(v) \int_0^\infty N_3(v')dv' \quad (8)$$

The modified rate equations can be reduced to integral equations for the level populations, and these can be solved numerically in conjunction with the gain equation (4) to find the gain coefficient $g(v)$ as a function of laser spectral intensity $I(v)$ and collisional frequency Γ. The effects can be illustrated fairly easily; Fig. 1 shows normalized gain coefficient profiles assuming for illustrative purposes a Voigt intrinsic line profile with a 35 mÅ inhomogeneous Gaussian component and a 5 mÅ homogeneous Lorentzian component, for several values of the laser intensity and the collisional redistribution rate. The laser is assumed to have a FWHM of 5 mÅ, and is peaked at line center (206.38 Å). $\Gamma = 0$ in Fig. 1(a), $\Gamma = \Delta v_H$ in Fig.

1(b) and $\Gamma = 5\ \Delta\nu_H$ in Fig. 1(c), where $\Delta\nu_H$ is the homogeneous Lorentzian width in frequency units. The spectral hole-burning evident in Fig. 1(a) is clearly diminished in Fig. 1(b) and becomes almost invisible over the laser intensity range investigated in Fig. 1(c). As Γ increases, spectral hole burning is effectively

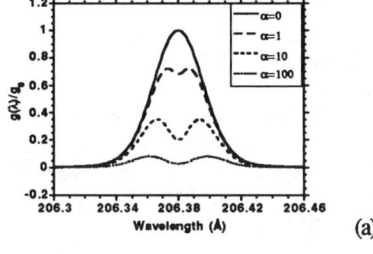

(a)

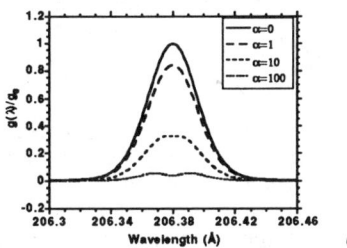

(c)

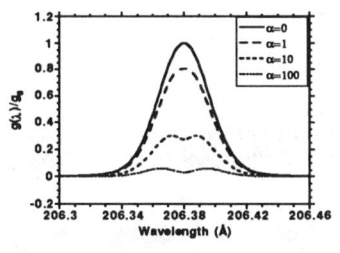

(b)

Figure 1. Normalized gain vs. wavelength for various ratios of peak laser intensity/saturation intensity ($\alpha = I(\nu_0)/I_{sat}$); the velocity changing collision rate is equal to zero in (a), equal to the homogeneous line width in (b), and equal to five times the homogeneous line width in (c).

delayed until farther into saturation; this can be understood to result because the inhomogeneous nature of the Doppler broadening component does not become evident until the stimulated emission/absorption rate induced by the laser field becomes larger than the velocity-changing collision rate (the stimulated rates will essentially set the time-scales with which to compare Γ once they become larger than other transition rates, i.e. once the laser saturates). Since spectral hole burning is the cause of inhomogeneous saturation rebroadening in A.S.E. lasers (whether or not a dip actually occurs in the gain profile), it is clear that a collisional redistribution rate approximately equal to or larger than the homogeneous lifetime width will delay inhomogeneous saturation rebroadening, thus partly homogenizing the Doppler component of the intrinsic line profile.

Inclusion of collisional redistribution effects in A.S.E. laser line-transfer models is difficult because the rate equations for the level populations then become complex and must be solved numerically. The problem is further complicated with the addition of a laser line transfer equation, required to calculate the laser spectral intensity and line profile, and by the necessarily self-consistent solutions to the population rate equations and the line transfer equation. However, the inclusion of collisional redistribution effects in the usual models[13] is in principal straightforward, and presents only numerical difficulties.

We note that our treatment of collisional redistribution is entirely equivalent to that presented in ref. 15. That work is based on an un-closed two-level rate equation model, but the collisional redistribution modifications in eqs. (47) and (48) of that work are equivalent to eq. (8) above when e.g. $N_3(v)$ is written as $N_3 f_3(v)$, so that the integral in eq. (8) is written as N_3. Finally, we note that collisional redistribution effects may be relevant to other A.S.E. laser line width measurements, e.g. those of ref. 23, where experimental measurements showed only slight inhomogeneous saturation re-broadening, in apparent disagreement with calculations[13]. This was also noted in ref. 15, and that author later re-examined[24] the experimental data of ref. 23, finding excellent agreement when collisional redistribution effects are taken into account.

ACKNOWLEDGEMENTS

One of us (JK) wishes to thank Geoff Pert of the University of York for discussions and for the exchange of numerous faxes. This work was performed under the auspices of the U.S. Department of Energy by the Lawrence Livermore National Laboratory under contract No. W-7405-ENG-48 and by the Lawrence Berkeley Laboratory under contract No. DE-AC03-76SF00098.

REFERENCES

1. J.A. Koch et al., *Proc. S.P.I.E.* **1551**, 131 (1991).
2. J. A. Koch et al., *Appl. Phys. B* **58**, 7 (1994).
3. J.A. Koch et al., "High-resolution measurements of x-ray laser line profiles", in *Radiative Properties of Hot Dense Matter, Proceedings of the Fourth International Workshop*, 1991, p. 373.
4. J.A. Koch et al., *Phys. Rev. Lett.* **68**, 3291 (1992).
5. J.A. Koch et al., "Selenium x-ray laser line profile measurements", in *X-ray Lasers 1992*, 1992, p. 67.
6. J.A. Koch et al., "Gain narrowing and saturation behaviour in 206,38 Å neon-like selenium x-ray laser line profiles", in *Spectral Line Shapes Vol. 7*, 1993, p. 205.
7. J.A. Koch et al., "Experimental measurements of selenium x-ray laser spectral line profiles, in *Proceedings of the International Conference on Lasers '92*, 1993, p. 52.
8. J.A. Koch, Ph.D. Thesis, University of California at Davis, 1993, available as Lawrence Livermore National Laboratory Report No. UCRL-LR-116938.
9. J.A. Koch et al., *Phys. Rev. A* (in press).
10. J. Nilsen et al., *Phys. Rev. Lett.* **70**, 3713 (1993).
11. M.C. Hettrick et al., *Appl. Opt.* **27**, 200 (1988).
12. J.C. Moreno et al., manuscript in preparation.
13. L.W. Casperson and A. Yariv, *IEEE J. Quant. Elect.* **QE-8**, 80 (1972); H. Maeda and A. Yariv, *Phys. Lett.* **43A**, 383 (1973).
14. G.B. Rybicki and A.P. Lightman, *Radiative Processes in Astrophysics*, New York: Wiley, 1979.
15. G.J. Pert, *J. Opt. Soc. Am B* (in press).
16. R.H. Dicke, *Phys. Rev.* **89**, 472 (1953).
17. H.R. Griem, *Phys. Rev. A* **33**, 3580 (1986).
18. E.L. Pollack and R.A. London, *Phys. Fluids B* **5**, 4495 (1993).
19. P.W. Smith and T. Hänsch, *Phys. Rev. Lett.* **26**, 740 (1971).
20. A.V. Otieno, *Opt. Commun.* **26**, 207 (1978).
21. see for example E.S.R. Gopal, *Statistical Mechanics and Properties of Matter*, Chichester: Ellis Horwood, 1974.
22. S.G. Rautian, *Sov. Phys. JETP* **24**, 788 (1967).
23. D.H. Schwamb and S.R. Smith, *Phys. Rev. A* **21**, 896 (1980).
24. G.J. Pert, unpublished.

Application of X-Ray Lasers to Current and Future Experiments in Atomic and Molecular Physics

C. Denise Caldwell

Department of Physics, University of Central Florida, Orlando, FL 32816-2385

Abstract. The use of intrinsically narrow-banded, intense x-ray lasers has the potential for a significant impact in atomic and molecular physics. As with any new technology, it is impossible to predict all the new information which may emerge as the technology develops. At least at the beginning it will be important for these lasers to have applicability to existing experimental methods, which can then exploit the new tool for experiments which are currently barely feasible with existing and planned sources of radiation in the high-energy regime. Examples of these are: resonant Auger decay, particularly of dilute species, studied with electron spectrometry; multi-photon processes involving the simultaneous utilization of two laser photons; and fragmentation experiments in which the high-energy photon is one of a pump-probe pair. Results from these experiments will go a long way to suggesting directions for future study.

I. GENERAL CONSIDERATIONS

With any developing technology it is always difficult to predict all the possibilities for experiments which might arise as the unique aspects of the new tool become manifest. There is often a tendency to postulate too many new experiments without careful consideration as to the feasibility and the accompanying problems which may arise. These might be entirely unconnected with the new technology or may be even generated by it. At the same time, unless the new technology leads to new discoveries, there is little point in its development.

In looking over what role x-ray lasers might play in atomic and molecular physics, it is clear that, at the beginning, x-ray lasers should be capable of extending some of the current techniques which are merely on the edge of what is feasible at present sources. Thus, this work will focus on some experiments of a type for which the problems have already been in great part solved but which have certain demands which are unlikely to be met by even the best of the emerging photon sources in this region of the spectrum. Some of these are of the type which are currently underway but which have generated unanswered questions for which improved photon sources will be necessary. Others are examples of experiments for which only proof-of-principle measurements have been made thus far and which remain questionable with existing sources.

Out of all the possibilities which might arise, a number of simplifying assumptions have been made in the selection of topics to be discussed in this work. In all instances the region of photon energy will be restricted to the vacuum ultraviolet (VUV) and soft x-ray regions, that is, energies less than 1000 eV. Bernd Crasemann in the previous contribution (1) has discussed possibilities associated with higher energy photons. As it is also highly likely that an x-ray laser will be a pulsed source, this will be assumed to be the case and also that the pulses may be very fast, even in the femtosecond regime. Finally, the work will proceed from the point of view that the x-ray laser is merely a source. While there are many interesting atomic processes which are associated with development of x-ray lasers in terms of *in situ* excitations and interactions, these will not be touched upon in this work. Rather, the assumption will be made that the photons may be made to impinge onto a well-defined target area. This area may serve as the source region of an electron spectrometer, for example. Thus, focusing will become a critical aspect of the required characteristics of the radiation.

Interaction of an atom or molecule with a photon of moderately high energy in virtually all cases leads to the production of an ion through the photoelectric effect. At the lowest energies the interaction proceeds through direct removal of a valence electron. However, as the energy is increased, the involvement of the inner-shell electrons continues to grow. In the vacuum ultraviolet (VUV) and the soft x-ray regimes the processes focus on the interaction with the inner-shell electrons. Almost immediately one of the more exciting possibilities associated with x-ray lasers becomes apparent. In particular, new phenomena involving excitation of valence electrons of atoms with extremely fast, coherent pulses of visible photons have appeared (2). These have been possible because of the ability to generate these photons with current visible and infra-red laser technology. However, unless and until comparable laser technology is available in the x-ray region of the spectrum, any processes of this type involving inner-shell electrons will remain uninvestigated.

II. RESONANT AUGER DECAY

One of the most interesting and informative aspects of the excitation of inner-shell electrons is the process of autoionization or resonant Auger decay following the excitation of a localized state which lies above one or more ionization continua. In both cases the excited state can decay by electron emission. The two are often distinguished by the fact that autoionization always involves the participation of the excited electron, and it is often accompanied by a strong competing direct ionization which gives rise to the characteristic asymmetric line shape of the feature. Resonant Auger decay may involve the excited electron, but, more often, this electron remains a spectator during the emission process. The result is the production of an excited state of the singly charged ion. Generally the direct production of this state is weak, so the line shapes in resonant Auger decay tend to be symmetric.

Resonant Auger decay can be observed in atoms through electron spectrometry. Although this is already a well-proven technique, the successful implementation of an x-ray laser for studying inner-shell phenomena should allow for the use of electron spectrometry as an analytical tool. It can do this primarily through the possibility which it offers for producing radiation with intrinsically narrow line widths, as good or better than can be achieved through the best modern monochromatization techniques and without the accompanying loss of intensity. In electron spectrometry one common mode of analysis is the recording of the constant ionic state (CIS) spectrum. In this mode the energy of the photon and the kinetic energy of the decay electron are scanned simultaneously so that the observed signal is always that associated with the production of a given ionic state. As long as the electron spectrometer is able to isolate a single ionic state, the resolution of this instrument plays no role in the overall resolution of the experiment; this is determined by the bandpass of the photon source alone. Using this technique it becomes possible to examine resonance excitations through individual decay channels. The advantage of this is that it is generally true that the absorption pattern of the resonance may be dramatically altered as it is examined in the decay into different final channels, particularly for larger atoms having a complex electronic structure. Through examination of a number of final-state channels, it is possible to extract information on the identification of the excited states involved in the absorption.

The elements making up the rare earths, the lanthanides, and the actinides are especially interesting in this regard. They all have partially filled orbitals to which an inner-shell electron can be excited, often by a transition within the same principal shell. Because of the complexity of the structure, the absorption spectrum often contains a broad resonance feature which can be spread over many terms. Isolation of this spectrum into the various decay modes at a resolution adequate for isolation of the individual terms becomes extremely valuable in determining the resulting structure. As an illustration of the value of this analysis we examine the complex $3p \to 3d$ excitation and subsequent decay in manganese (3):

$$Mn(3p^63d^54s^2) + h\nu \to Mn^*(3p^53d^64s^2)$$

followed by:

$$Mn^*(3p^53d^64s^2) \to Mn^+(3p^63d^54s\, ^5S) + Mn^+(3p^63d^54s\, ^5D) + \ldots$$

The absorption spectrum for this excitation consists of a broad (≈ 2 eV) asymmetric feature with one maximum at 50.1 eV separated from a second, weaker maximum at 50.55 eV by a small dip. The electron decay spectrum is extremely complex and contains more than 50 discernable features in the region between 7 and 32 eV. A composite absorption made up by addition of the CIS decay spectra for nine final states of the singly charged ion reproduces the total absorption spectrum relatively well, with the principal contribution arising from decay into the $Mn^+(3p^63d^54s\, ^5D)$ ionic state. The very marked difference which can arise between decays into different final states becomes very apparent from a comparison between the CIS spectra corresponding to the decay into this $Mn^+(3p^63d^54s\, ^5D)$ state and into one of the stronger satellite features, later identified as being a composite of two Mn^+ configurations. Through analysis of the two spectra it was possible to identify one single resonance as appearing as a peak in one decay channel, the satellite channel, and a dip in the other, that leading to the $Mn^+(^5D)$ state. Due to the interaction between this excitation and a second excitation, intensity is removed from the decay into the ^5D state. As production of this state constitutes the principal decay mode, this reduction in intensity appears as a dip near the resonance maximum.

The prevalence of these types of excitation in the elements making up the middle of the periodic table is further evident in the absorption spectra of, for example, the $5p \to 5d$ excitation in ytterbium (4) and the $4d \to 4f$ excitation in europium (5). In these instances the characteristic broad resonances are very similar to that in manganese and lend themselves well to such analyses. The main thrust of the emphasis on the understanding of these complex open-shell atoms is that they form archetypal many-body species. Because the interaction potential among the particles is well-known, the Coulomb interaction, theoretical attention can be focused onto the collective, many-body aspects of the structure. The excitations in these complex resonances are manifestations of strong overlap between various resonances involving a number of electrons so their proper elucidation can have important consequences for the understanding of complex many-body dynamics. With regard to the lanthanides and actinides, there is in addition the fact that these measurements offer access to the f-electrons, which are much less well understood than the more familiar s, p, d electrons.

In order to perform these types of experiment a number of characteristics of the source are necessary. First, it must be possible to select a number of given wavelengths so as to be able to access many different elements. Secondly, it must be possible to tune these wavelengths across the resonance feature. This involves generally tuning over 1-10 eV, depending on the system. Thirdly, the goal of performing these experiments is the resolution which can be achieved with the combination of the CIS scan and the narrow bandpass of the radiation source. Thus, linewidth becomes an important criterion. Finally, as all of these materials are metals, it is necessary to generate the atomic vapor. This is in some cases not trivial, particularly for the actinides, which are only available in small quantities. Having a source which is inherently intense and narrow-banded could be a terrific boon to resonance absorption measurements of very thin targets. Although much is already understood for the lighter elements, a new source would open up areas of the periodic table which have remained untouched. This also holds true for any systems which can only be produced in low concentration. Having access to an x-ray laser could, for example, promote the study of ions, which generally have to be examined at low intensity in a beam, or molecular radicals, which have to be generated by dissociation or various chemical gas phase reactions. It should be noted, however, that these measurements in electron spectrometry generally involve the accumulation of low-level signals over a long period of time so that duty factor and average power become more important than peak power and pulse length.

III. TWO-COLOR EXPERIMENTS

Experiments involving the stepwise or simultaneous interaction with two or more photons have become one of the most valuable sources of analysis of fundamental processes in current use. Such experiments are possible in the visible and near infra-red regions of the spectrum because of the availability of multiple lasers which are easily tuned in energy to the desired spacings and easily synchronized for timing experiments. With this type of experiment it becomes possible to study interactions with selectively excited states, thereby extracting more detailed information by in a sense directing the course of a process. Extending these experiments to include the interaction with inner shells has not been feasible to any great extent. In so doing, one of the photons must at the present time be supplied by a synchrotron. Because of the limited intensity available from the synchrotron and the difficulties associated with synchronizing a laser to the synchrotron, two-color experiments involving even one high energy photon are extremely difficult and currently quite rare. In this area the high intensity and controllable pulse characteristics of the x-ray laser could have a major impact and convert what is presently borderline activity into routine measurements.

An illustration of a two-color experiment which has been successfully performed is provided by the photoionization of sodium in the excited state represented by the process:

$$Na(2p^63s\ ^2S_{1/2}) + h\nu(589nm; P = 1.00) \rightarrow Na(2p^63p\ ^2P_{1/2,3/2}; A_0)$$

followed by:

$$Na(2p^63p\ ^2P_{3/2}) + h\nu(33 - 35eV, P_2) \rightarrow Na^*(2p^53s3p; J)$$

and

$$Na^*(2p^53s3p; J) \rightarrow Na^+(2p^6\ ^1S_0) + e^-(\vec{k})$$

which was carried out at the BESSY storage ring using a cw dye laser to excite the 3p state of the sodium (6). In the above, P is the linear polarization of the laser used for the excitation; A_0 is the alignment of the excited state; P_2 is the linear polarization of the synchrotron radiation; and J is the total angular momentum of the excited state. The goal of the experiment was to carry out a detailed analysis of the dependence of the photoionization process, particularly the angular dependence, on the spatial dependence of the excited state. This was done by measuring the intensity of the decay electrons $I(J, \theta, \eta)$ as a function of A_0 and P_2. θ is the angle between the electron and P_2; η is the angle between P_2 and P. The two photon beams were focused at 180° with respect to each other onto the sodium vapor. By making measurements at the J-level, these investigators were able to extract, among other things, detailed information about the dependence of the angular distribution on the fine structure component. In all cases these results confirmed a concomitant term-dependent Hartree-Fock calculation, which serves to indicate how well theoretical calculations are able to explain the dynamics of alkali photoionization. But what about, for example, photoionization of halogens, in which experiment shows that even LS-coupling breaks down rather readily (7)?

All the combined laser/synchrotron experiments which have been performed (6) have involved the study of alkali metals. Why this is the case can be seen by examining a number of factors for the sodium experiment, although these generally apply to all successful experiments of this type. The first is that one of the photons is generated by a cw laser source operating at what is generally considered to be one of the most easily producible wavelengths, the sodium D-line. The oscillator strength of this transition is also quite strong, and the generation of an adequate concentration of sodium vapor is relatively trivial as the vapor pressure is so high at low temperature. The laser intensity was 500 mW, and the undulator source at BESSY produced 10^{13} photons at 34 eV. If it were desired to extend these measurements to an atom with an order of magnitude reduction in oscillator strength in the desired transition, a very low vapor pressure, and a transition at a wavelength at which only 50 mW of power could be produced, then the required photon flux from the undulator increases correspondingly. Not even the brightest of the new synchrotron sources are predicting 10^{15} photons in any reasonable bandwidth for measurements of this type. However, a pulsed x-ray laser should have the option of producing these types of intensity within the necessary intrinsically narrow band. It could then be synchronized to a second laser which does the excitation, which could also be pulsed. In this fashion it is possible to extend two-color experiments which have been so productive in the valence regime to include the inner shells as well.

IV. FRAGMENTATION EXPERIMENTS

Although the production of an ion in an experiment may be considered a fragmentation, fragmentation experiments in this context will be taken to mean complex processes involving a breakup of the atom or molecule into three or more species. Fragmentation experiments are more and more being studied using coincidence techniques. One area of interest is the double photoionization of atoms. This might be an ionization which proceeds through the simultaneous absorption of one photon to produce two electrons or through a consecutive process consisting of the absorption of a photon followed by the radiationless decay of the ion. Although these processes are generally clearly distinguishable in many instances, they become increasingly smeared when the absorption occurs at the vicinity of an inner-shell threshold. Depending on which of the processes is dominant, the angular distributions of the two electrons when measured in coincidence will be different (8,9). Thus, coincidence experiments become important tools of extapolating the dynamics back to an elucidation of the process by which the fragmentation occurred.

A rich area of application of x-ray photons involves the fragmentation of molecules following upon excitation or ionization of an inner-shell electron. In such an experiment the resulting products may be neutrals or any combinations of singly or doubly charged ions and electrons consistent with charge, energy, and mass conservation. When ions are involved, it is possible to follow the dissociation pathways of the excited molecule through multiple coincidence measurements. Generally the first decay electron gives the signal for starting a time-of-flight detector, and the flight times of one or two of the ionic fragments are measured. In the case of triple coincidence, one electron and two ions, photoelectron-photoion-photoion coincidence (PEPIPICO), the resulting coincidence map of flight times reflects a two-body dissociation, a three-body concerted dissociation, or a three-body stepwise dissociation (10).

These multiple coincidence experiments involve only one exciting photon. However, they do require that ions be produced in the fragmentation. In the case in which fragmentation produces neutrals it becomes necessary to resort to other schemes in order to achieve the detection. In this instance the multicolor experiments merge with the fragmentation experiments in a way that is made-to-order for the application of pulsed x-ray lasers. An illustration of this type of experiment is provided by the measurement of the photodissociation of CF_3Cl with a first laser and resonant enhanced multiphoton excitation to analyze the Cl fragment which is produced (11):

$$CF_3Cl + h\nu(118nm, 125nm) \rightarrow CF_3 + Cl(^2P_{1/2,3/2})$$

followed by:

$$Cl(^2P_{1/2,3/2}) + (2+1)h\nu \to Cl^+ + e^-$$

This group measured the product translational energies and the angular distribution of the Cl fragments as a function of the direction of the polarization of the ionizing photon and the fine-structure component of the Cl. They observed two fragmentation channels, both proceeding through the $4s(a_1)$ and $4p(e)$ Rydberg states of the molecule. The first "slow" channel involved the production of CF_3^* radicals in excited states; the second "fast" channel involved a $C-Cl$ bond rupture followed by a secondary fragmentation of CF_3^* to $CF_2 + F$.

The necessary VUV photons for this experiment were generated by frequency multiplication in a rare gas. This technique is rather straightforward for the production of photons down to about 60 nm. State-selection of the J-value of the Cl atoms was made through the two-step excitation of the Cl resonances as part of the REMPI. All of these type of measurements require at least two lasers which can be synchronized with respect to each other. Currently such experiments are performed in the region in which frequency multiplication works well. While it might be possible to utilize an undulator to generate one of the photons required for this type of experiment, the demands on the resolution (⟨ 0.1 nm in the case of the Cl experiment) and the time synchronization required raise doubts as to whether these types of experiments can become routine at current or planned sources. The extension of these types of measurement to higher energy will require the development of lasers operating at these higher energies, so an x-ray laser could have a notable impact in the area of photochemistry involving inner-shell electrons.

V. SUMMARY

There are clearly a number of areas in which x-ray lasers could have a profound impact in the study of processes in atomic and molecular physics involving the interaction with photons of higher energy. If this is to be the case, then the sources will be required to possess certain basic properties. Some of these exist; others are goals for which to strive. Of those areas which are discussed in this work, in resonance absorption broad tunability and intrinsic bandwidth become important criteria. For experiments exploiting the technique of electron spectrometry, peak power will not play a major role; more important will be the average power and repetition rates. As much work in atomic and molecular physics involves differential analysis in angle, focusing and polarization of the source will be of major concern. The possibilites of exploiting the high intensity to look at dilute species are especially attractive for the x-ray laser as opposed to more conventional sources. Finally, and perhaps most significantly, the application of the x-ray laser to atomic and molecular physics should allow for the study of multi-color and pump-probe processes which are essentially closed to inner-shell electrons. For these the intensity and timing of the lasers involved in the interaction play the major roles. Bandwidth, while it is important, should come as a natural consequence of the fact that the laser has an intrinsically narrow width. The experiments which are listed here are by no means exhaustive. They are meant to serve as a sample of the truly possible, what could be done today if a tunable, narrow-linewidth x-ray laser were suddenly to appear in the laboratory. But awareness should always be made of the fact that any time it becomes possible to access a new domain of time or length, new physics can emerge. Although scales of length and time in the x-ray region have been reached independently, the femtosecond pulsed x-ray laser offers for the first time the possibility of combining the two together. As new analysis techniques appear which can take advantage of this unique mix, new prospects which cannot even be suggested at this stage should become apparent.

ACKNOWLEDGMENTS

This work was supported by the National Science Foundation under grant No. PHY-9207634.

REFERENCES

1. Crasemann, B., this volume.
2. Jones, R.R., Raman, C.S., Schumacher, D.W., and Bucksbaum, P.H., *Phys. Rev. Lett.* **71**, 2575-78 1993).
3. Whitfield, S.B., Krause, M.O., van der Meulen, P., Caldwell, C.D., *Phys. Rev. A* (in press).
4. Tracy, D.H., *Proc. Roy. Soc. Lond. A* **357**, 485-98 (1977).
5. Becker, U., Kerkhoff, H.G., Lindle, D.W., Kobrin, P.H., Ferrett, T.A., Heimann, P.A., Truesdale, C.M., and Shirley, D.A., *Phys. Rev. A* **34**, 2858-64 (1986).
6. Baier, S., Schulze, M., Staiger, H., Zimmermann, P., Lorenz, C., Pahler, M., Rüder, J., Sonntag, B., Costello, J.T., and Kiernan, L., *J. Phys. B* **27** 1341-49 (1994), and references therein.
7. Hansen, J.E., Cowan, R.D., Carter, S.L., and Kelly, H.P., *Phys. Rev. A* **30**, 1540-42 (1984).
8. Kämmerling, B., and Schmidt, V., *Phys. Rev. Lett.* **67**, 1848-51 (1991).
9. Huetz, A., Selles, P., Waymel, D., and Mazeau, J., *J. Phys. B* **24**, 1917-33 (1991).
10. Eland, J.H.D., *Mol. Phys.* **61**, 725-45 (1987).
11. Yen, M-W., Johnson, P.M., and White, M.G., *J. Chem. Phys.* **99** 126-139 (1993).

Applications of Subpicosecond Soft X-ray Sources

R. Haight and P.F. Seidler
IBM T.J. Watson Research Center
PO Box 218 Yorktown Hts. N.Y. 10598

ABSTRACT

We discuss applications of laser based subpicosecond soft X-ray sources presently available in the laboratory to issues in condensed matter and surface physics and surface chemistry. In particular, we describe the production of harmonics of intense subpicosecond lasers and the application of this high energy, narrow bandwidth light to the generation of photoemission spectra of core, valence and transiently occupied conduction states of condensed matter systems.

Introduction

Significant ongoing activity in the study of the interaction of intense laser light with condensed and gas phase media has led to the discovery of physical processes which produce photons of much higher energy than that of the input light. In the case of intense laser pulses interacting with solid matter, high density plasmas radiate broad band continuum and line radiation from recombination occurring within the plasma as it cools. [1] The light radiated is incoherent in nature, but when produced with a femtosecond laser source, can exhibit short time duration over a range of photon energies. This radiation can be collected and refocussed making it extremely useful for applications in a variety of fields including microscopy, lithography, imaging and photoemission.

A second means by which soft X-ray photons can be generated is through a coherent process called harmonic generation. When an intense laser pulse interacts with a rare gas such as Ar, odd multiple harmonics are generated. For modest input power densities at the focus of the input laser, ($< 3 - 5 \times 10^{13}$ watts/cm^2), photons with energies in the 10-15 eV range are produced in sufficient quantities to carry out experiments such as photoemission of electrons from valence and conduction states of solid state systems. [2-5] Unfortunately at these intensities, the significant drop of photon flux with increasing harmonic number renders photoemission experiments at higher photon energies impractical.

More recently, a number of investigators have discovered a plateau in the harmonic generation efficiency for laser power densities exceeding 10^{14} watts/cm^2, a regime where the physics of harmonic generation is distinctly different than at lower intensities. [6-10] In this region, harmonics of high number can be generated. These studies, carried out at relatively low repetition rates (1-10 Hz) have succeeded in producing harmonic photons with energies as high as 150 eV with input laser power densities of 10^{18} watts/cm^2. Theoretical investigations into the nature of the high intensity harmonic generation process have been carried out using a non-perturbative solution of the time-dependent Schrödinger equation for an atom in a time varying electric field. [11, 12] Importantly, these studies have reproduced the plateau behavior observed experimentally. The relative constancy of the generation efficiency over a wide range of harmonics implies that this type of source can be exceptionally useful for spectroscopic applications in condensed matter physics, materials science, photochemistry and in other fields.

For applications of this light in the area of photoemission it is important to emphasize the spectral purity of the harmonic light, the bandwidth of which is determined by the pulsewidth of the input laser light. A transform limited 150 fs hyperbolic secant squared light pulse possesses a bandwidth of 9 meV. Because of the inherent non-linearities of the generation process, the harmonic pulsewidth may be shorter, giving rise to a wider bandwidth than the input pulse. To date, no direct measurement of the harmonic pulsewidths has been carried out in the femtosecond regime and hence this property remains uncertain. In addition to spectral purity, high peak fluxes and spectral brightness are possible since the harmonic photons are concentrated into lines separated by twice the fundamental. In our system, to be described, this separation corresponds to 4.07 eV (for our input of 2.036 eV/photon) resulting in harmonic lines which are easily separated by a normal incidence reflection grating. This contrasts with continuum sources where a narrowing of the bandwidth is achieved by reducing spectrometer slit widths resulting in a loss of photon flux. An added benefit is the relatively constant flux over a wide range of harmonics which provides tunability as an experimental variable. In our work which involves the applications of this light to photoemission investigations of matter, the fundamental properties of this light make it extremely useful.

Experimental Setup

Although a detailed description of the laser source, analysis chamber and detection system has been given elsewhere, [4,13] a short summary is provided here. Figure 1 shows the laser source, grazing incidence grating chamber and ultrahigh vacuum/analysis chamber. The frequency doubled light from a CW-modelocked Nd:YAIG laser is used to synchronously pump a dual jet dye laser with intracavity dispersion compensation, producing 150-200 fs pulses at 610 nm. The 1.06 µ light from a Nd:YAIG regenerative amplifier (seeded by the modelocked Nd:YAIG laser), operating at 540 Hz, is frequency doubled and directed into a 3 stage dye amplifier producing pulses with energies 0.6-0.7 mJ. These pulses are then focussed with a 200 mm lens into the high pressure region of a pulsed valve. We have found that argon is the optimal gas for producing energetic photons up to the 19th harmonic (38.68 eV). To obtain higher photon energies neon is used and harmonics up to 39th (79.5 eV) have been generated. These results are in agreement with other workers. [9] The fundamental and harmonic pulses are separated by gratings. Two types of gratings have been utilized. The first is a 1-meter normal incidence grating blazed at 30.8 nm which is coated with Pt. The normal incidence reflectivity drops precipitously above 38 eV, but for lower energies provides broad tunability for a wide range of experiments. At higher energies, a Au coated grazing incidence toroidal grating is utilized. The enhanced reflectivity of the grazing incidence toroidal grating has resulted in useful photon fluxes to nearly 80 eV. A differentially pumped vacuum beam line is required for generation and transport of the harmonic radiation. A particular harmonic, chosen by angle tuning the grating, is focussed to ~ 1 mm spot size on a sample within an ultrahigh vacuum (UHV) chamber. The harmonic light is incident upon the sample at an angle of 45° with respect to the sample normal. The polarization of the harmonic light can be selected to be either pure s- or mixed s- and p- in this geometry. The UHV chamber houses surface science analytical tools and a multianode time-of-flight (t-o-f) detector which has been described previously. [4] A resolution of 100 meV at 6 eV kinetic energy is achieved with this detector but is observed to degrade at higher kinetic energies.

Surface/Interface States

A number of experiments have been recently carried out using harmonics to investigate surface/interface states and atomic core levels. One example of the use of high harmonics to study interface states is the investigation of the As terminated Ge (111) surface. [14] This surface is produced by the exposure of a clean Ge surface, held at 400°C, to a flux of As_4. The result is an As terminated surface, extremely resistant to subsequent contamination (passivation). The low energy electron diffraction (LEED) pattern observed is a 1x1 with exceptionally low background indicating the formation of a model, reconstructionless surface. By replacing the outer layer of Ge with As the Ge attains a four-fold coordination while the As atoms are three-fold coordinated with a lone pair dangling bond which forms an occupied band 0.4 eV below the Ge valence band maximum. An empty antibonding band is predicted to lie 0.2 eV below the Ge conduction band minimum by quasiparticle calculation (Fig. 2).[15] Electrons which are photoexcited into the bulk conduction band of Ge scatter into the empty surface state and can be observed at a number of photon energies (Fig. 3). Note that a deeper lying feature is observed to disperse with energy; this is the Λ_3 bulk band dispersing from $\Gamma - L$, the bulk symmetry line along the <111> direction. That this feature disperses is a consequence of the three-dimensional nature of the bulk band structure. Conversely neither the occupied As derived surface state at -0.4 eV nor the transiently excited empty state at +0.4 eV disperse with photon energy since they are two-dimensional. Interestingly, contamination with residual gases in the vacuum system (or intentional dosing) in order to determine the surface origin of these features is ineffective due to the passive nature of the surface. Hence photon tunability was critical in this experiment to the identification of this state. Scattering of electrons from the bulk into the surface was observed at early times as a 0.2 eV bulge (Fig.4) in the +0.4 eV signal which was noted to disappear at later times as the electron gas cooled. The 0.2 eV excess energy corresponds to the energy difference between the Ge conduction band minimum and the minimum of the As derived state.

Finally, the time dependence of the As derived signal was measured (Fig. 5). While a rapid rise time was observed, the surface population decreased slowly with a lifetime greater than 200 ps. Since the As derived antibonding state resides within the Ge bulk band gap electrons are prohibited from diffusing into the Ge bulk. Furthermore, the exceptionally high quality of the surface (lack of surface defects) reduces the opportunity for surface recombination through defect states and as a result the surface population decays through e-h optical recombination which is relatively slow. As a result, the surface is not only chemically passive but electronically passive as well.

Atomic Core Levels with High Harmonics

A second example of the use of high harmonics involves the generation of atomic core level spectra. Electron Spectroscopy for Chemical Analysis (ESCA) is a powerful tool used in material science, chemistry, surface science and other fields for determining the presence of particular atomic species and their relative chemical states. Before the advent of harmonic generation, tunability could be achieved only in experiments carried out at a synchrotron facility. A relatively large number of atomic species possessing narrow core level linewidths have binding energies accessible with the harmonic photons we can generate in our apparatus. Core d-levels of many of the group III, IV and V atoms as well as a number of narrow p-levels of the alkali and alkaline earth elements can be spectroscopically observed with our system. Fig. 6 shows a photoemission spectrum of the cleaved (110) surface of GaAs collected with 30 eV light. [16] The spectrum was collected in ~5 minutes and shows the intense spin-orbit split Ga 3d core level at a binding energy of -18.5

eV. The inset shows an expanded view of the core level in which the 3/2-5/2 splitting is clearly resolved. Contributing to the observed linewidth are the lifetime of the d-level, including radiative and Auger recombination, broadening due to surface shifted components, the detector resolution and finally the bandwidth of the harmonic. A photon energy of 30 eV was chosen here in order to produce relatively low kinetic energy core level electrons for which the resolution of the t-o-f detector was satisfactory.

As an example of the importance of the tunability achievable with harmonics, core level photoemission of the Pb 5d spin-orbit split levels are displayed in Fig. 7. These spectra were collected at 30, 34 and 38 eV. The cross-section dependence on photon energy is clearly observed as an increase in emission strength of the 5/2 and 3/2 components (binding energies -18.1 and -20.6 respectively) with increasing photon energy. Tunability with such a source not only provides a means of optimizing cross-section, as is evident in the spectra of Fig. 7, but also allows for varying the electron escape depth in order to enhance surface or bulk features.

Finally we show in Fig. 8 spectra collected from a polycrystalline W foil in which the 4f levels of W are observed. Spectra shown are collected with 67, 71 and 75 eV photons. The high angular momentum of the 4f levels results in a small photoionization cross-section for near threshold energy photons and as a result significantly higher photon energies are required. As a result the kinetic energy of the 4f electrons is quite high and results in relatively poor energy resolution for the 4f's. Improved energy resolution can be achieved with other detector configurations such as electrostatic analyzers or a parabolic mirror t-o-f analyzer.

Summary

In this paper we have described the generation of short wavelength harmonic radiation and applications to the study of electronic states and dynamics in condensed matter. The availability of broadly tunable, short pulsed radiation with the properties described makes possible a wide range of new investigations into the behavior of electrons in matter. In addition, the advantages of generating this light within the laboratory setting means that many experiments previously carried out only at large facilities can now be performed on a much smaller scale. As a result, laboratory based experiments in subpicosecond electron dynamics, tunable core level spectroscopy, photochemistry, lithography and more are achievable in the lab.

References

1. M. Murnane, H. Kapteyn and R. Falcone, IEEE J. Quant. Electr. **25**, 2417 (1989).

2. R. Haight and J. A. Silberman, Phys. Rev. Lett. **62**, 815 (1989).

3. M. Baeumler and R. Haight, Phys. Rev. Lett. **67**, 1153 (1991).

4. R. Haight, J. A. Silberman and M. I. Lilie, Rev. Sci. Instr. **59**, 1941 (1988).

5. J. J.Bokor, R. Storz, R. R. Freeman and P. H. Bucksbaum, Phys. Rev. Lett **57**, 881 (1986).

6. J. K. Crane, S. W. Allendorf, K. S. Budil and M. D. Perry, SPIE **1800**, 146 (1991).

7. J. J. Macklin, J. D. Kmetec and C. L. Gordon III, Phys. Rev. Lett. **70**, 766 (1993).

8. A. L'Huillier and P. Balcou, Phys. Rev. Lett. **70**, 774 (1993).

9. K. Miyazaki and H. Sakai, J. Phys. B, At. Mol. Opt. Phys. **25**, L83 (1992).

10. J. K. Crane, M. D. Perry, S. Herman and R. Falcone, Opt.Lett. **17**, 1256 (1992).

11. K. C. Kulander and B. W. Shore, Phys. Rev. Lett. **62**, 524 (1989).

12. J. L. Krause, K. J. Schafer and K. C. Kulander, Phys. Rev. Lett. **68**, 3535 (1992).

13. R. Haight and D. R. Peale, Rev. Sci. Inst. **65**, 1853 (1994).

14. R. Haight and D. R. Peale, Phys. Rev. Lett. **70**, 3979 (1993).

15. M. S. Hybertsen and S. G. Louie, Phys. Rev. Lett. **58**, 1551 (1987).

[16] R. Haight and P. F. Seidler, Appl. Phys. Lett. **65**, 517 (1994).

Figure Captions

1) Schematic of the laser photoemission laboratory setup including laser source and amplifier, grating chamber, beamlines and UHV/analysis chamber.

2) Top:Surface band structure calculation for the As terminated Ge (111) surface. Hatched regions are the projection of the bulk bands onto the surface. The thick solid curves are the bonding and antibonding As derived surface bands. Bottom: Schematic showing the excitation and relaxation pathways of the photoexcited electrons for the As/Ge (111) system. Note that the excited electrons scatter from the bulk Ge bands into the surface band with an excess energy of 0.2 eV.

3) Photoemission spectra from the As terminated Ge (111) surface for four different photon energies. The peak at 0.4 eV is the transiently excited antibonding surface state while the peak at -0.4 eV is the filled bonding state. The peak at -1.2 eV in the 10 eV spectrum is observed to shift to higher binding energies with increasing photon energy and is the bulk Λ_3 band.

4) Blowup of the 0.4 eV antibonding peak for various time delays. The high energy bulge in the t=0 peak is due to electrons scattering into the surface from the bulk Ge bands 0.2 eV above the surface state minimum.

5) Time dependence of the 0.4 eV peak collected with 18 eV photons.

6) Spectrum of cleaved GaAs showing the valence states and the Ga 3d core level collected with 30 eV light.

7) Spectra of Pb 5d's collected with 30, 34 and 38 eV light.

8) Spectra of W 4f's collected with 67, 71 and 75 eV light.

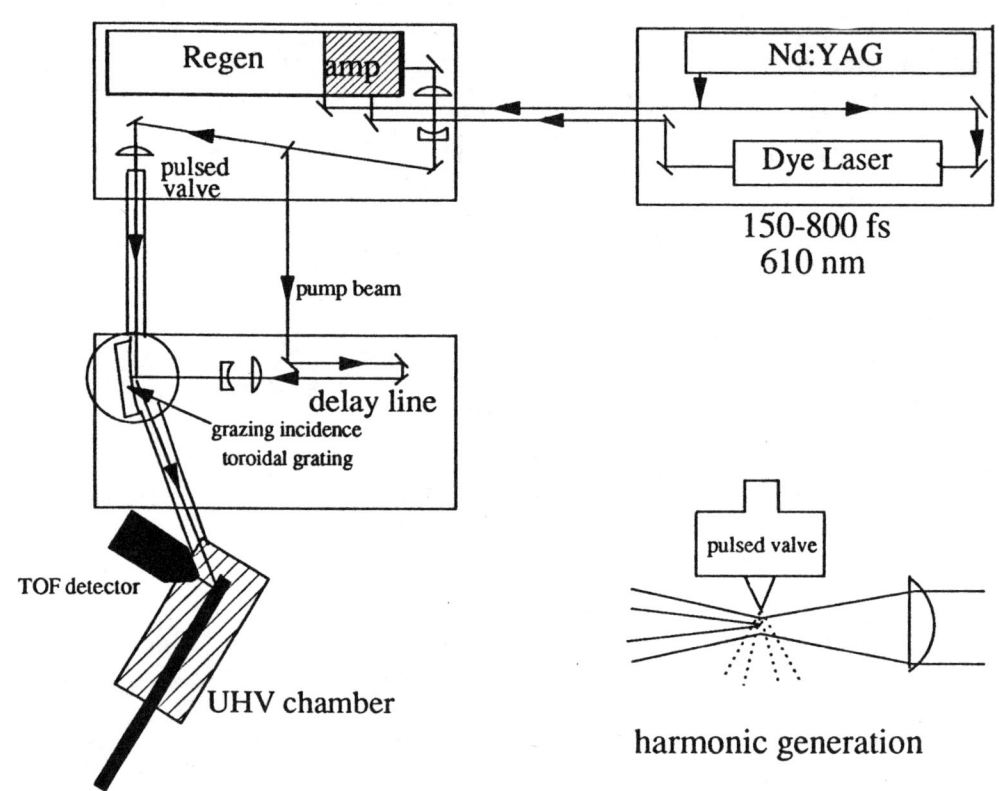

FIGURE 1

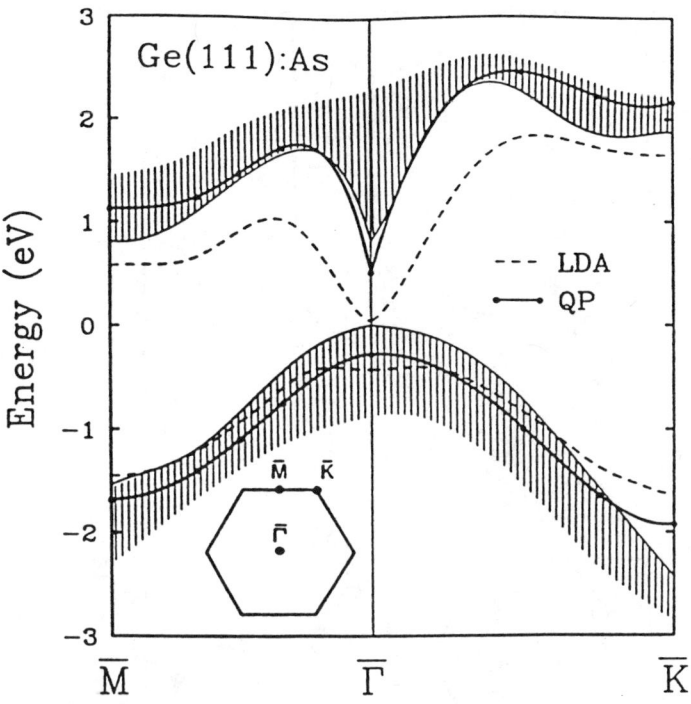

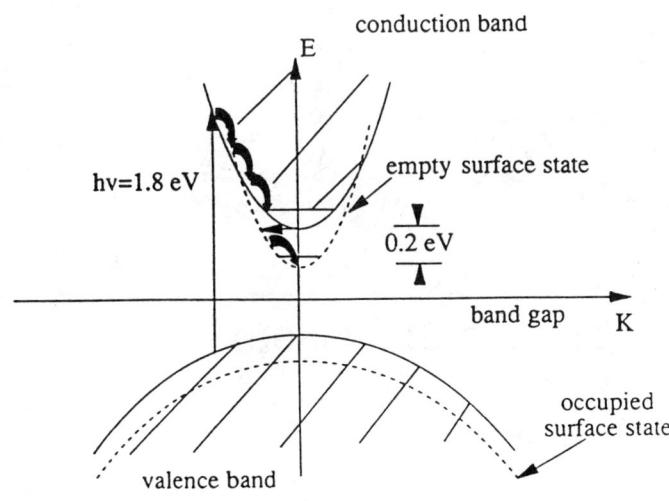

FIGURE 2

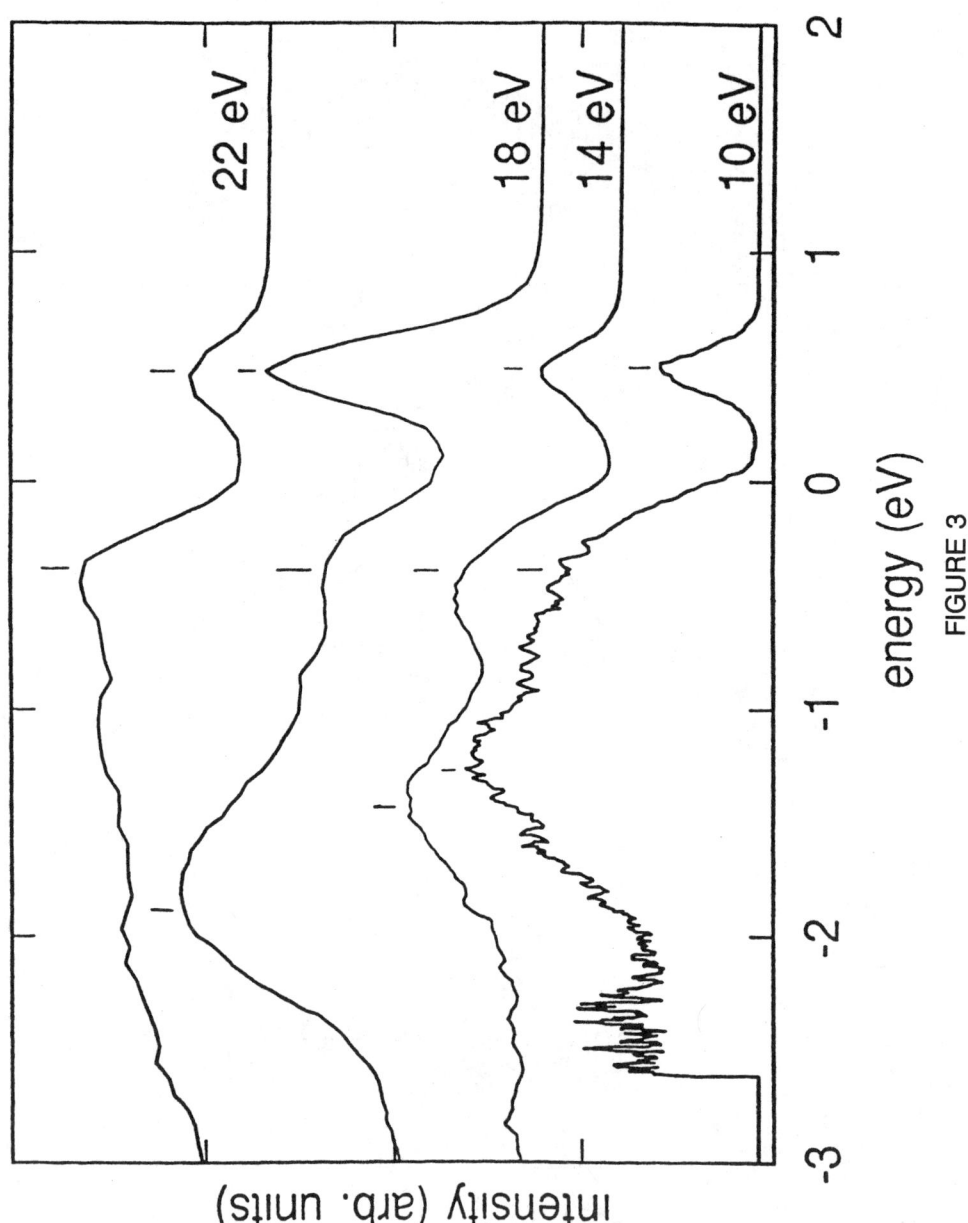

FIGURE 3

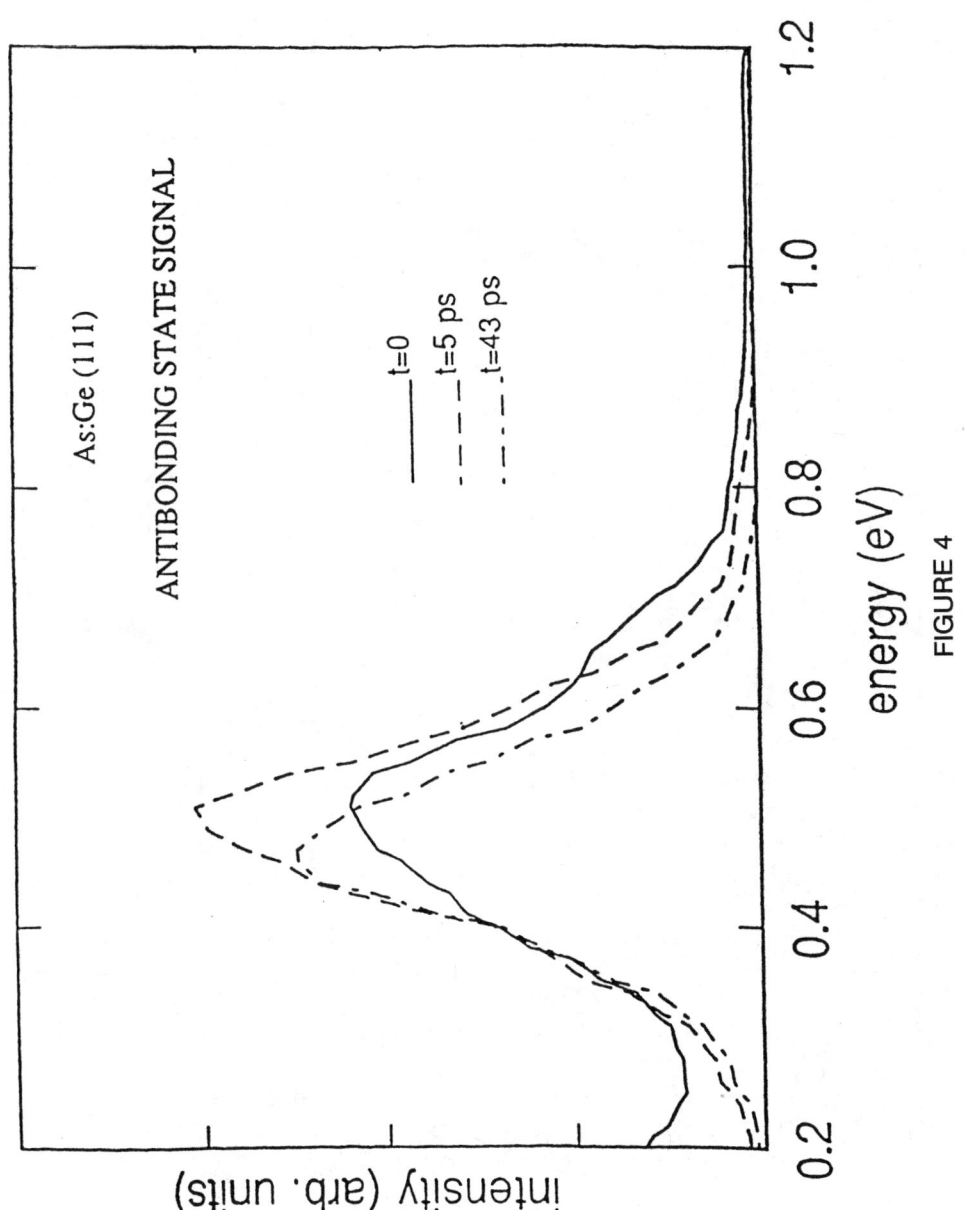

FIGURE 4

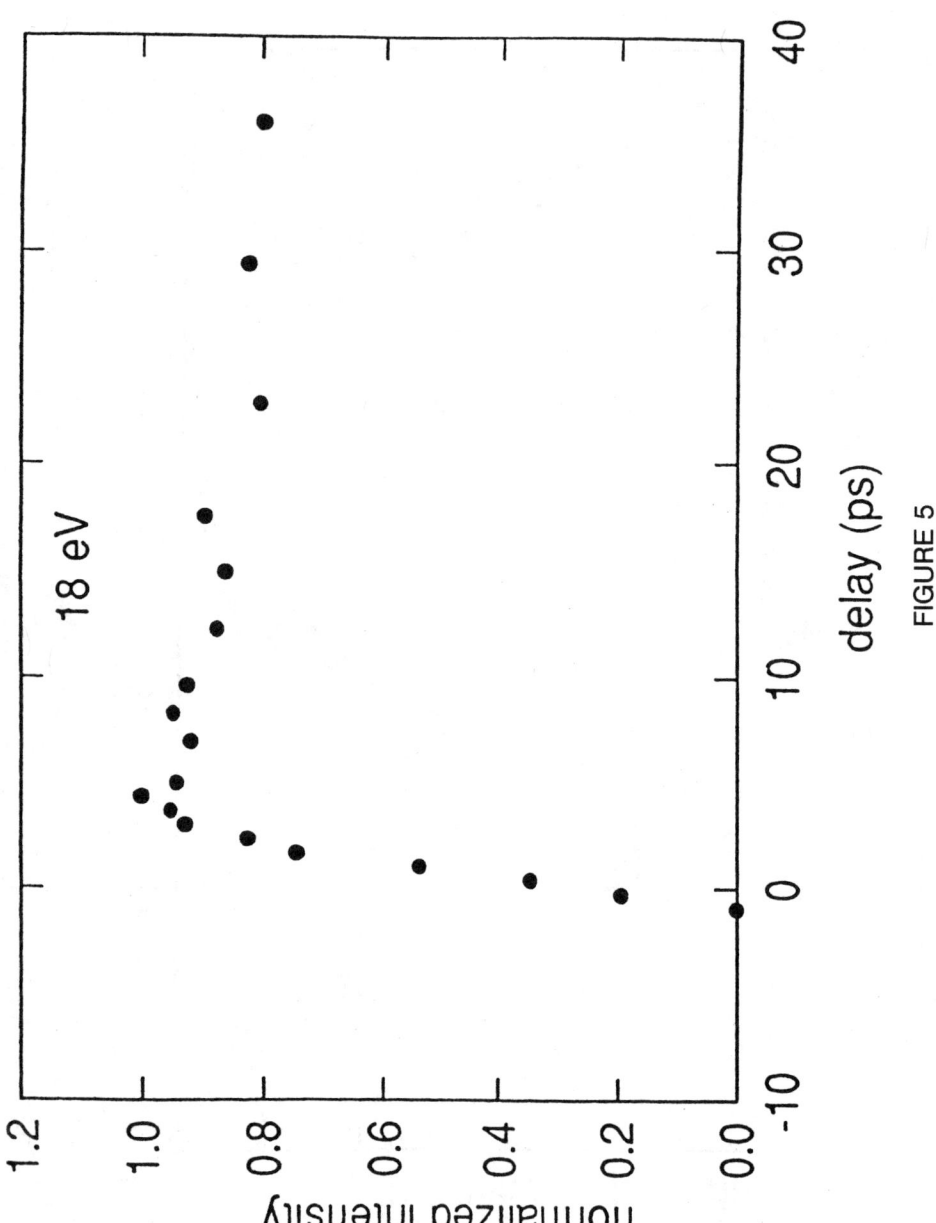

FIGURE 5

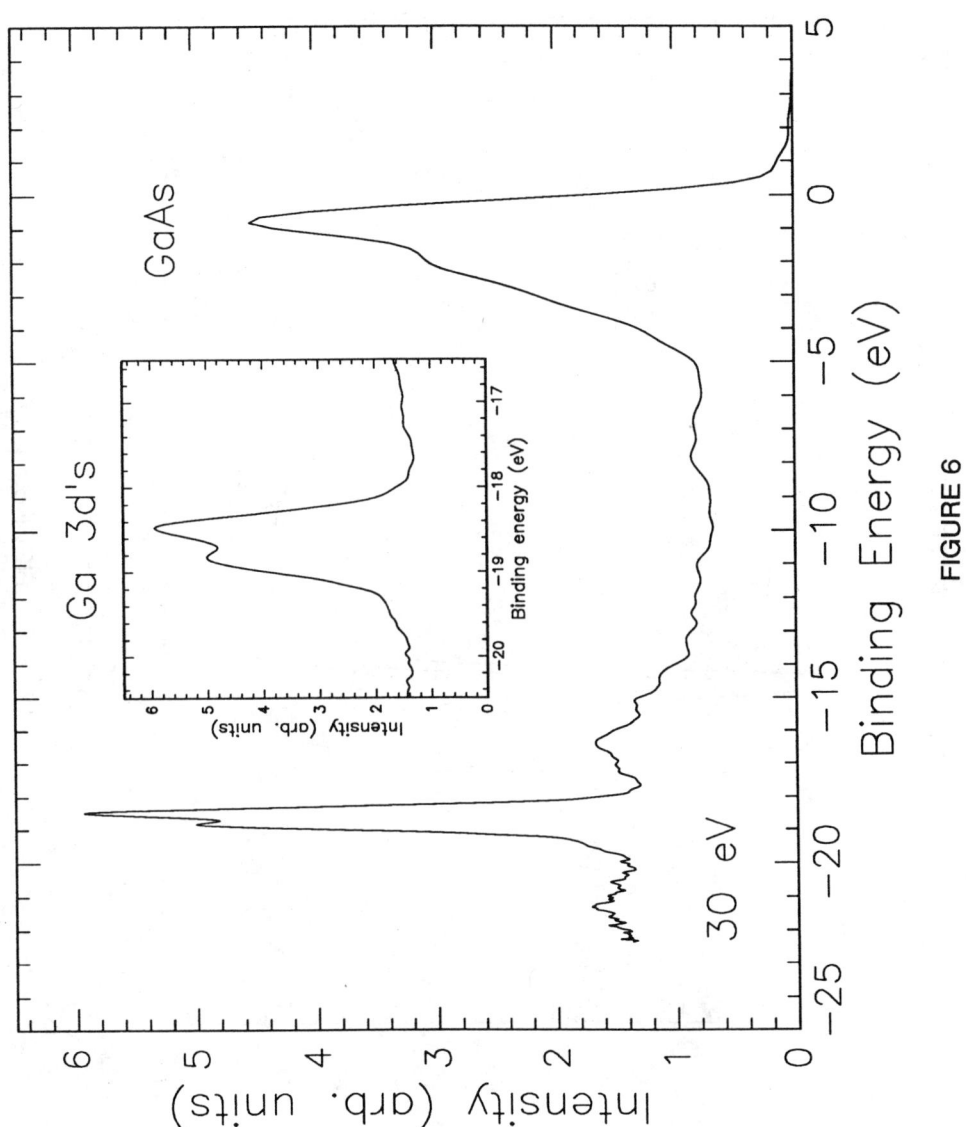

FIGURE 6

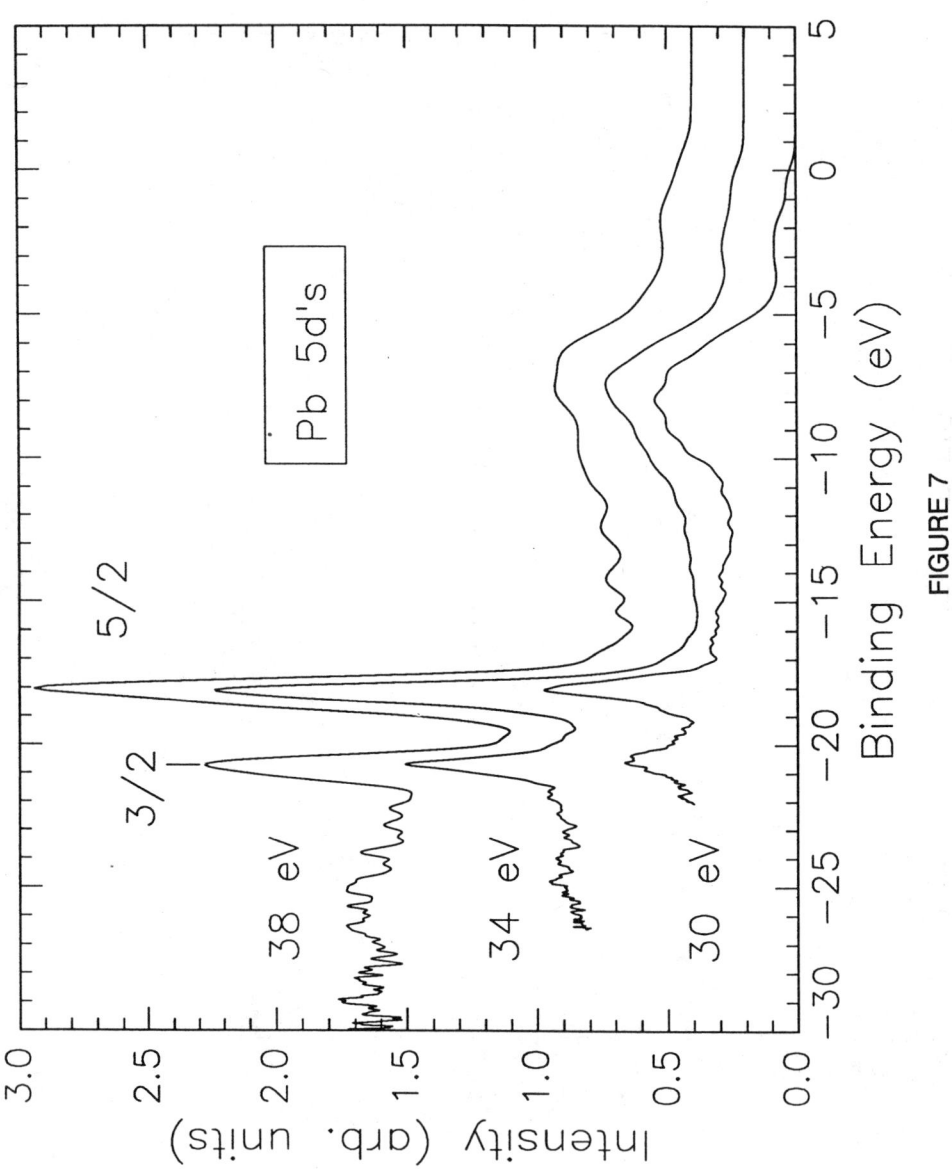

FIGURE 7

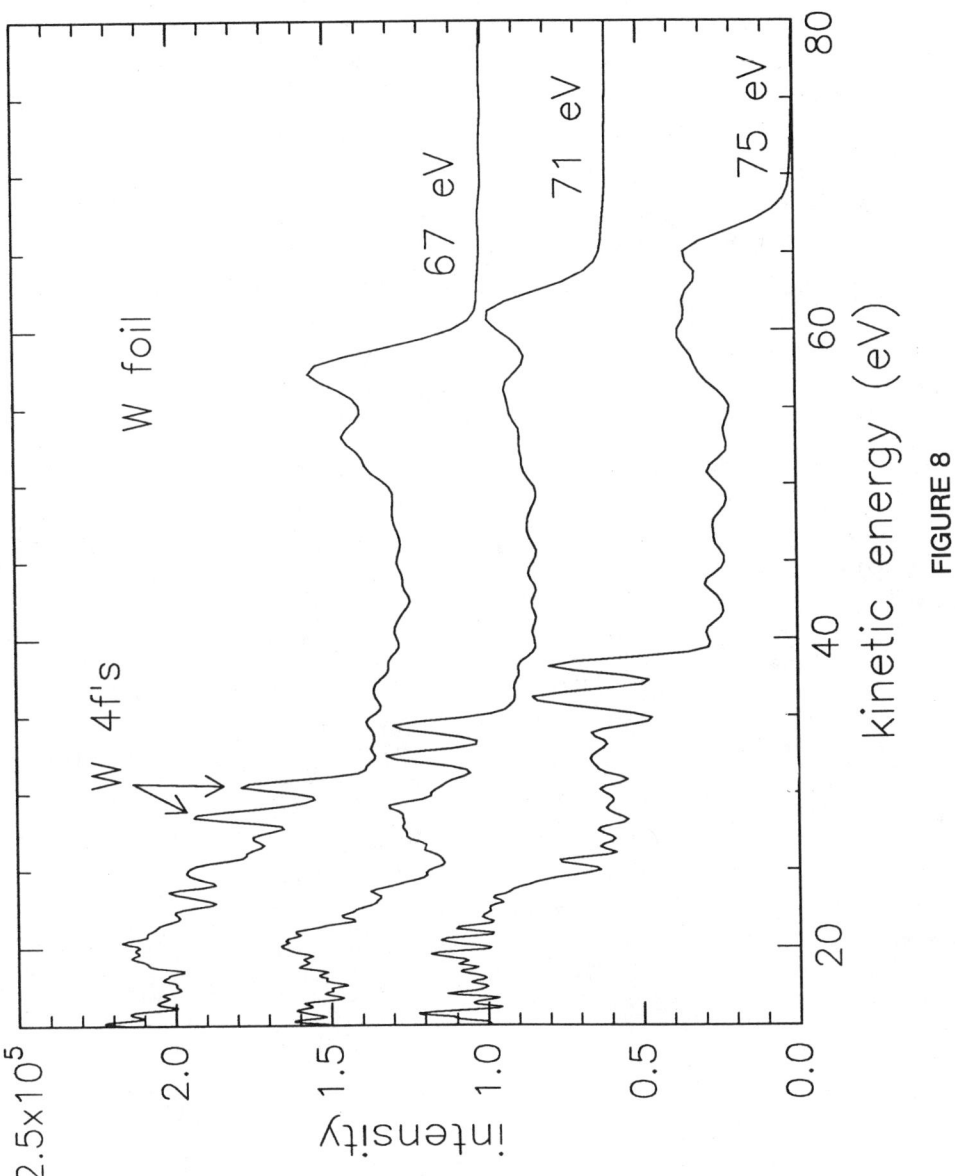

FIGURE 8